T0180585

Springer Complexity

Springer Complexity is an interdisciplinary program publishing the best research and academic-level teaching on both fundamental and applied aspects of complex systems—cutting across all traditional disciplines of the natural and life sciences, engineering, economics, medicine, neuroscience, social and computer science.

Complex Systems are systems that comprise many interacting parts with the ability to generate a new quality of macroscopic collective behavior the manifestations of which are the spontaneous formation of distinctive temporal, spatial or functional structures. Models of such systems can be successfully mapped onto quite diverse "real-life" situations like the climate, the coherent emission of light from lasers, chemical reaction-diffusion systems, biological cellular networks, the dynamics of stock markets and of the Internet, earthquake statistics and prediction, freeway traffic, the human brain, or the formation of opinions in social systems, to name just some of the popular applications.

Although their scope and methodologies overlap somewhat, one can distinguish the following main concepts and tools: self-organization, nonlinear dynamics, synergetics, turbulence, dynamical systems, catastrophes, instabilities, stochastic processes, chaos, graphs and networks, cellular automata, adaptive systems, genetic algorithms and computational intelligence.

The three major book publication platforms of the Springer Complexity program are the monograph series "Understanding Complex Systems" focusing on the various applications of complexity, the "Springer Series in Synergetics", which is devoted to the quantitative theoretical and methodological foundations, and the "Springer Briefs in Complexity" which are concise and topical working reports, case studies, surveys, essays and lecture notes of relevance to the field. In addition to the books in these two core series, the program also incorporates individual titles ranging from textbooks to major reference works.

Understanding Complex Systems

Founding Editor: S. Kelso

Future scientific and technological developments in many fields will necessarily depend upon coming to grips with complex systems. Such systems are complex in both their composition-typically many different kinds of components interacting simultaneously and nonlinearly with each other and their environments on multiple levels-and in the rich diversity of behavior of which they are capable.

The Springer Series in Understanding Complex Systems series (UCS) promotes new strategies and paradigms for understanding and realizing applications of complex systems research in a wide variety of fields and endeavors. UCS is explicitly transdisciplinary. It has three main goals: First, to elaborate the concepts, methods and tools of complex systems at all levels of description and in all scientific fields, especially newly emerging areas within the life, social, behavioral, economic, neuro- and cognitive sciences (and derivatives thereof); second, to encourage novel applications of these ideas in various fields of engineering and computation such as robotics, nano-technology, and informatics; third, to provide a single forum within which commonalities and differences in the workings of complex systems may be discerned, hence leading to deeper insight and understanding.

UCS will publish monographs, lecture notes, and selected edited contributions aimed at communicating new findings to a large multidisciplinary audience.

More information about this series at http://www.springer.com/series/5394

Bruce Edmonds • Ruth Meyer

Editors

Simulating Social Complexity

A Handbook

Second Edition

 Springer

Editors
Bruce Edmonds
Centre for Policy Modelling
Manchester Metropolitan University
Business School
Manchester, UK

Ruth Meyer
Centre for Policy Modelling
Manchester Metropolitan University
Business School
Manchester, UK

ISSN 1860-0832   ISSN 1860-0840   (electronic)
Understanding Complex Systems
ISBN 978-3-319-88352-6   ISBN 978-3-319-66948-9   (eBook)
https://doi.org/10.1007/978-3-319-66948-9

This Springer imprint is published by Springer Nature
The registered company is Springer International Publishing AG
The registered company address is: Gewerbestrasse 11, 6330 Cham, Switzerland

Contents

Part I
Introduction

Chapter 1
Introduction

Bruce Edmonds and Ruth Meyer

Abstract This introduces the themes of the book inherent in its title: *Simulating Social Complexity*. In a deliberate homage to the work of Herbert Simon, it traces the roots of these themes back to his work. It then explains the structure of the handbook with its different parts: introductory, methodological on different kinds of mechanism and applications. It briefly introduces each chapter within this structure.

Why Read This Chapter?
To understand some of the background and motivation for the handbook and how it is structured.

1.1 Simulating Social Complexity

As the title indicates, this book is about *Simulating Social Complexity*. Each of these words is important:

Simulating—the focus here is on individual- or agent-based computational simulation rather than analytic or natural language approaches (although these can be involved). In other words, this book deals with computer simulations where the individual elements of the social system are represented as separate elements of the simulation model. It does not cover models where the whole population of interacting individuals is collapsed into a single set of variables. Also, it does not deal with purely qualitative approaches of discussing and understanding social phenomena, but just those that try to increase their understanding via the construction and testing of simulation models.

Social—the elements under study have to be usefully interpretable as interacting elements of a society. The focus will be on human society but can be extended to include social animals or artificial agents where such work enhances our

B. Edmonds (✉) • R. Meyer
Centre for Policy Modelling, Manchester Metropolitan University Business School, All Saints Campus, Oxford Road, Manchester, M1 6BH, UK
e-mail: bruce@edmonds.name

© Springer International Publishing AG 2017
B. Edmonds, R. Meyer (eds.), *Simulating Social Complexity*,
Understanding Complex Systems, https://doi.org/10.1007/978-3-319-66948-9_1

understanding of human society. Thus, this book does not deal with models of single individuals or where the target system is dealt with as if it *were* a single entity. Rather it is the differing states of the individuals and their interactions that are the focus here.

Complexity—the phenomena of interest result from the interaction of social actors in an essential way and are not reducible to considering single actors or a representative actor and a representative environment. It is this complexity that (typically) makes analytic approaches infeasible and natural language approaches inadequate for relating intricate cause and effect. This complexity is expressed in many different ways, for example, as a macro/micro link, as the social embedding of actors within their society and as emergence. It is with these kinds of complexity that a simulation model (of the kind we are focussing on) helps, since the web of interactions is too intricate and tedious to be reliably followed by the human mind. The simulation allows *emergence* to be captured in a formal model and experimented upon.

Since this area is relatively new, it involves researchers from a wide variety of backgrounds, including computer scientists, sociologists, anthropologists, geographers, engineers, physicists, philosophers, biologists and even economists. The field is starting to mature and this handbook is part of that process. We hope that it will help to introduce and guide newcomers into the field so as to involve more minds and effort in this endeavour, as well as inform those who enter it from one perspective to learn about other sides and techniques.

1.2   The Context: Going Back to Herbert Simon

This handbook is in memory of Herbert Simon, since he initiated several key strands that can be found in the work described here.

He observed how people behave in a social system instead of following some existing framework of assumptions as to how they behave (Simon 1947). That is, he tried to change the emphasis of study from a normative to a descriptive approach—from how academics think people should be behaving to how people are observed to behave. Famously he criticised "armchair" theorising, the attempt to make theories about social phenomena without confronting the theory with observation. There is still a lot of "armchair" theorising in the field of simulating social complexity, with a "Cambrian explosion" of simulation models, which are relatively unconstrained by evidence from social systems. If the development of this work is seen as a sort of evolutionary process, then the forces of variation are there in abundance but the forces of selection are weak or non-existent (Edmonds 2010).

Importantly for the simulation of complex social systems, Simon observed that people act with a procedural rather than substantive rationality—they have a procedure in the form of a sequence of actions that they tend to use to deal with tasks and choices rather than try to find the best or ideal sequence of actions (Simon

1947, 1976; Sent 1997). With the advent of computational simulation, it is now fairly common to represent the cognition of agents in a model with a series of rules or procedures. This is partly because implementing substantive rationality is often infeasible due to the computational expense of doing do, but more importantly it seems to produce results with a greater "surface validity" (i.e. it *looks* right). It turns out that adding some adaptive or learning ability to individuals and allowing the individuals to interact can often lead to effective "solutions" for collective problems (e.g. the entities in Chap. 23). It is not necessary to postulate complex problem-solving and planning by individuals for this to occur.

Herbert Simon observed further that people tend to change their procedure only if it becomes unsatisfactory; they have some criteria of sufficient satisfaction for judging a procedure, and if the results meet this, they do not usually change what they do. Later Simon (1956) and others (e.g. Sargent 1993) focused on the contrast between optimisers and satisficers, since the prevailing idea of decision-making was that many possible actions are considered and compared (using the expected utility of the respective outcomes) and the optimal action was the one that was chosen. Unfortunately it is this later distinction that many remember from Simon, and not the more important distinction between procedural and substantive rationality. Simon's point was that he observed that people use a procedural approach to tasks; the introduction of satisficing was merely a way of modelling this. However, the idea of thresholds, which people only respond to a stimulus when it becomes sufficiently intense, is often credible and is seen in many simulations (for some examples of this, see Chaps. 24 and 27).

Along with Alan Newell, Simon made a contribution of a different kind to the modelling of humans. He produced a computational model of problem-solving in the form of a computer program, which would take complex goals and split them into sub-goals until the sub-goals were achievable (Newell and Simon 1972). The importance of this, from the point of view of this book, is that it was a computational model of an aspect of cognition, rather than one expressed in numerical and analytic form. Not being restricted to models that can be expressed in tractable analytic forms allows a much greater range of possibilities for the representation of human individual and social behaviour. Computational models of aspects of cognition are now often introduced to capture behaviours that are difficult to represent in more traditional analytic models. Computational power is now sufficiently available to enable each represented individual to effectively have its own computational process, allowing a model to be distributed in a similar way to that of the social systems we observe. Thus, the move to a distributed and computational approach to modelling social phenomena can be seen as part of a move away from abstract models divorced from what they model towards a more descriptive type of representation.

This shift towards a more straightforward (even "natural") approach to modelling also allows for more evidence to be applied. In the past, anecdotal evidence, in the form of narrative accounts by those being modelled, was deemed as "unscientific". One of the reasons that such evidence was rejected is that it could not be used to help specify or evaluate formal models; such narrative evidence could only be used within the sphere of rich human understanding and not at the level of a precise model. Computational simulation allows some aspects of individual's narratives to be used to specify or check the behaviour of agents in a model, as well as the results being more readily interpretable by non-experts. This has let such computational simulations to be used in conjunction with stakeholders in a far more direct way than was previously possible. Chapter 12 looks at this approach.

Herbert Simon did not himself firmly connect the two broad strands of his work: the observation of people's procedures in their social context and their algorithmic modelling in computer models. This is not very surprising as the computational power to run distributed AI models (which are essentially what agent-based simulations are) was not available to him. Indeed these two strands of his work are somewhat in opposition to each other, the one attempting to construct a *general* model of an aspect of cognition (e.g. problem-solving) and the other identifying quite specific and limited cognitive procedures. I think it is fair to say that whereas Simon did reject the general economic model of rationality, he did not lose hope of a general model of cognitive processes, which he hoped would be achieved starting from good observation of people. There are still many in the social simulation community who hope for (or assume) the existence of an "off-the-shelf" model of the individuals' cognition which could be plugged into a wider simulation model and get reasonable results. Against any evidence, it is often simply hoped that the details of the individuals' cognitive model will not matter once embedded within a network of interaction. This is an understandable hope, since having to deal with both individual cognitive complexity and social complexity makes the job of modelling social complexity much harder—it is far easier to assume that one or the other does not matter much. Granovetter (1985) addressed precisely this question arguing against both the under-socialised model of behaviour (that it is the individual cognition that matters and the social effects can be ignored) and the over-socialised model (that it is the society that determines behaviour regardless of the individual cognition).

Herbert Simon did not have at his disposal the techniques of individual- and agent-based simulation discussed in this handbook. These allow the formal modelling of socially complex phenomena without requiring the strong assumptions necessary to make an equation-based approach (which is the alternative formal technique) analytically tractable. Without such simulation techniques, modellers are faced with a dilemma: either to "shoehorn" their model into an analytically tractable form, which usually requires them to make some drastic simplifications of what they are representing, or to abandon any direct formal modelling of what they observe. In the latter case, without agent-based techniques, they then would have two further choices: to simply not do any formal modelling at all remaining in the world of natural language or to ignore evidence of the phenomena and instead model their idea concerning the phenomena. In other words, to produce an abstract but strictly *analogical* model—a *way of thinking about the phenomena* expressed as a simulation. This latter kind of simulation does not directly relate to any data derived from observation but to an idea, which, in turn, relates to what is observed in a rich, informal manner. Of course there is nothing wrong with analogical thinking, it is a powerful source of ideas, but such a model is not amenable to scientific testing.

The introduction of accessible agent-based modelling opens up the world of social complexity to formal representation in a more natural and direct manner. Each entity in the target system can be represented by a separate entity (agent or object) in the model, each interaction between entities as a set of messages between the corresponding entities in the model. Each entity in the model can be different, with different behaviours and attributes. The behaviour of the modelled entities can be realised in terms of readily comprehensible rules rather than equations, rules that can be directly compared to accounts and evidence of the observed entities' behaviour. Thus, the mapping between the target system and model is simpler and more obvious than when all the interactions and behaviour are "packaged up" into an analytic or statistical model. Formal modelling is freed from its analytical straight jacket, so that the most appropriate model can be formulated and explored. It is no longer necessary to distort a model with the introduction of overly strong assumptions simply in order to obtain analytic tractability. Also, agent-based modelling does not require high levels of mathematical skill and thus is more accessible to social scientists. The outcomes of such models can be displayed and animated in ways that make them more interpretable by experts and stakeholders (for good and ill).

It is interesting to speculate what Herbert Simon would have done if agent-based modelling was available to him. It is certainly the case that it brings together two of the research strands he played a large part in initiating: algorithmic models of aspects of cognition and complex models that are able to take into account more of the available evidence. We must assume that he would have recognised and felt at home with such kinds of model. It is possible that he would not have narrowed his conception of substantive rationality to that of satisficing if he had other productive ways of formally representing the processes he observed in the way he observed them occurring.

It is certainly true that the battle he fought against "armchair theorising" (working from a neat set of assumptions that are independent of evidence) is still raging. Even in this volume, you will find proponents (let us call them the *optimists*) that still hope that they can find some shortcut that will allow them to usefully capture social complexity within abstract and simple models (theory-like models) and those (the pessimists) that think our models will have to be complex, messy and specific (descriptive models) if they are going to usefully represent anything we observe in the social world. However, there is now the possibility of debate, since we can compare the results and success of the optimistic and pessimistic approaches and indeed they can learn from each other.

It seems that research into social complexity has reached a cusp, between the "revolutionary" and "normal" phases described by Kuhn (1962). A period of exploratory growth, opposed to previous orthodoxies, has occurred over the last 15–20 years, where it was sufficient to demonstrate a new kind of model, where opening up new avenues was more important than establishing or testing ideas about observed systems. Now attention is increasingly turning to the questions such as how to productively and usefully simulate social complexity; how to do it with the greatest possible rigour; how to ensure the strongest possible relation to the evidence; how to compare different simulations; how to check them for unintentional errors; and how to use simulation techniques in conjunction with others (analytic, narrative, statistical, discourse analysis, stakeholder engagement, data collection, etc.). The field—if it is that—is maturing.

This handbook is intended to help in this process of maturation. It brings together summaries of the best thinking and practice in this area, from many of the top researchers. In this way, it aims to help those entering into the field so that they do not have to reinvent the wheel each time. It will help those already in the field by providing accessible summaries of current thought. It aims to be a reference point for best current practice and a standard against which future methodological advances are judged.

1.3   The Structure of the Handbook

The material in this book is divided into four parts: *Introductory*, *Methodology*, *Mechanisms* and *Applications*. We have tried to ensure that each chapter within these parts covers a clearly delineated set of issues. To aid the reader, each chapter starts with a very brief section called "Why read this chapter?" that sums up the reasons you would read it in a couple of sentences. This is followed by an abstract, which summarises the content of the chapter. Each chapter also ends with a section of "Further Reading" briefly describing three to eight things that a newcomer might read next if they are interested. This is separate from the list of references, which contains all the references mentioned in the chapter.

1.3.1   Introductory Part

The introductory part includes four chapters: this chapter, a historical introduction (Chap. 2) that reviews the development of social simulation providing some context for the rest of the book, an overview of the different kinds of simulation (Chap. 3) and an examination of some of the different goals one might have for a simulation model (Chap. 4).

1.3.2 Methodology Part

The next part on methodology consists of 11 chapters that aim to guide the reader through the process of simulating complex social phenomena. It starts with two approaches to designing and building simulation models: formal, i.e. using approaches from computer science (Chap. 6), and informal (Chap. 5). The former is more appropriate where the goals and specification of the proposed simulation are known and fixed, while the latter is more appropriate in the case where possible models are being explored, in other words when the simulation model one wants cannot be specified in advance.

However carefully a modeller designs and constructs such models they are complex entities, which are difficult to understand completely. The next (Chap. 7) guides the reader through the ways in which a simulation model can be checked to ensure that it conforms to the programmer's intentions for it. Chapter 8 looks at the importance of ontological structure for agent-based simulations, contrasting this with approaches that have almost no *a priori* structure. It also takes one through some of the ways of formalising and checking this structure.

Once one has a simulation model one is happy with, then one needs to decide what runs of the model are needed to make one's point. Chapter 11 tackles this subject giving firm guidelines to ensure one has the right "power" that enables the required distinctions to be made, but avoiding showing misleading levels of significance.

Three chapters in this part are concerned with the results of simulations. Chapter 9 concentrates on the validation of simulation models: the many ways in which a model and the possible outputs from simulation runs can be related to data as a check that it is correct for its purpose. Chapter 10 explores ways of analysing and visualising simulation results, which is vital if the programmer or a wider audience are to understand what is happening within complex simulations. Chapter 14 looks at the broader question of the meaning and import of simulations, in other words the philosophy of social simulation including what sort of theorising they imply.

Two other chapters consider separate aspects but ones that will grow in importance over time. Chapter 12 looks at participatory approaches to simulation, that is, ways of involving stakeholders more directly in the model specification and/or development process. This is very different to an approach where the simulation model is built by expert researchers who judge success by the correspondence with data sets and can almost become an intervention within a social process rather than a representation of it. Chapter 13 investigates how analytic approaches can be combined with simulation approaches, both using analytics to approximate and understand a simulation model and using simulation to test the assumptions within an analytic model.

All of the approaches described in these three chapters are aided by good, clear documentation. Chapter 15 describes a way of structuring and performing such documentation that helps to ensure that all necessary information is included without being an overly heavy burden.

1.3.3   Mechanisms Part

The third part considers types of social mechanisms that have been used and explored within simulations. It does not attempt to cover all such approaches, but concentrates upon those with a richer history of use, where knowing about what has been done might be important and possibly useful.

Chapter 16 takes a critical look at mechanisms that may be associated with economics. Although this handbook is not about economic simulation,[1] mechanisms from economics are often used within simulations with a broader intent. Unfortunately, this is often done without thinking so that, for example, an agent might be programmed using a version of economic rationality (i.e. considering options for actions and rating them as to their predicted utility) just because that is what the modellers know or assume. However, since economic phenomena are a subset of social phenomena, this chapter does cover these.

Chapter 17 surveys a very different set of mechanisms, those of laws, conventions and norms. This is where behaviour is constrained from outside the individual in some way (although due to some decision to accept the constraint from the inside to differing degrees). Chapter 18 focuses on trust and reputation mechanisms, how people might come to judge that a particular person is someone they want to deal with.

Chapter 19 looks at a broad class of structures within simulations, those that represent physical space or distribution in some way. This is not a cognitive or social mechanism in the same sense of the other chapters in this part, but has implications for the kinds of interactions that can occur and indeed facilitates some kinds of interaction due to partial isolation of local groups.

The last two chapters in this part examine ways in which groups and individuals might adapt. Learning and evolution are concepts that are not cleanly separable; evolution is a kind of learning by the collection of entities that are evolving and has been used to implement learning within an individual (e.g. regarding the set of competing strategies an individual has) as well as within a society. However, Chap. 20 investigates these concepts primarily from the point of view of algorithms for an individual to learn, while Chap. 21 looks at approaches that explicitly take a population and apply some selective pressures upon it, along with adding some sources of variation.

---

[1]There is an extensive handbook on this (Tesfatsion and Judd 2006).

1.3.4 Applications Part

The last part looks at eight areas where the techniques that have been described are being applied. We chose areas where there has been some history of application and hence some experience of different approaches. Areas of application that are only just emerging are not covered here.

Chapter 22 reviews applications to ecological management. This is one of the oldest and most productive areas where simulation approaches have been applied. Since it is inevitable that the interaction of society and the environment is complex, analytic approaches are usually too simplistic and approaches that are better suited are needed.

Chapter 23 explores how a simulation-based understanding of ICT systems can enable new kinds of distributed systems to be designed and managed, while Chap. 24 looks at how simulation can help us understand animal interaction. Chapter 25 describes agent-based simulations as a useful tool to come to a complex understanding of how markets actually work (in contrast to their economic idealisations). Chapter 26 considers systems where people and/or goods are being moved within space or networks including logistics and supply chains.

The next two chapters look at understanding human societies. Chapter 27 focuses on a descriptive modelling approach to structures of power and authority, with particular reference to Afghanistan, whereas Chap. 28 reviews the different ways in which simulations have been used to understand human societies, briefly describing examples of each.

The final chapter, Chap. 29, looks at some of the pitfalls that can come about when formal models (especially the complex simulation models considered here) can be misused or misunderstood when applied in the policy arena.

1.4 Differences in the Second Edition

This edition of the handbook has a number of new chapters, namely, those on different modelling purposes (Chap. 4), applying computer science to simulation development (Chap. 5), ontological structure (Chap. 8), how many runs one should do (Chap. 11) and the final chapter on pitfalls that can occur when such models are used to inform policy-making or policy delivery (Chap. 29). Furthermore, some of the chapters have been significantly revised, including those on verification and validation (Chap. 9); utility, games and haggling (Chap. 16); social constraint (Chap. 17); reputation (Chap. 17); animal social behaviour (Chap. 24); and human societies (Chap. 28).

References

Edmonds, B. (2010). Bootstrapping knowledge about social phenomena using simulation models. *Journal of Artificial Societies and Social Simulation, 13*(1). http://jasss.soc.surrey.ac.uk/13/1/8.html

Granovetter, M. (1985). Economic action and social structure: The problem of embeddedness. *American Journal of Sociology, 91*(3), 481–510.

Kuhn, T. S. (1962). *The structure of scientific revolutions*. Chicago, IL: University of Chicago Press.

Newell, A., & Simon, H. A. (1972). *Human problem solving*. Englewood Cliffs, NJ: Prentice-Hall.

Sargent, T. J. (1993). *Bounded rationality in macroeconomics: The Arne Ryde Memorial Lectures*. Oxford: Clarendon Press.

Sent, E.-M. (1997). Sargent versus Simon: Bounded rationality unbound. *Cambridge Journal of Economics, 21*, 323–338.

Simon, H. A. (1947). *Administrative behavior: A study of decision-making processes in administrative organizations*. New York: The Free Press.

Simon, H. A. (1956). Rational choice and the structure of the environment. *Psychological Review, 63*(2), 129–138.

Simon, H. A. (1976). *Administrative behavior* (3rd ed.). New York: The Free Press.

Tesfatsion, L., & Judd, K. L. (Eds.). (2006). *Handbook of computational economics, volume 2: Agent-based computational economics, Handbooks in economics* (Vol. 13). Amsterdam: North Holland.

Chapter 2
Historical Introduction

Klaus G. Troitzsch

Abstract This chapter gives an overview of early attempts at modelling social processes in computer simulations. It discusses the early attempts, its successes and its shortcomings and tries to identify some of them as forerunners of modern simulation approaches.

Why Read This Chapter?
To understand the historical context of simulation in the social sciences and thus to better comprehend the developments and achievements in the field.

2.1 Overview

The chapter is organised as follows: the next section will discuss the early attempts at simulating social processes, mostly aiming at prediction and numerical simulation of mathematical models of social processes. Section 3 will then be devoted to the nonnumerical and early agent-based approaches, while Sect. 4 will give a short conclusion followed by some hints at further reading.

2.2 The First Two Decades

Simulation in the social sciences is nearly as old as computer simulation at large. This is partly due to the fact that some of the pioneers of computer science—such as John von Neumann, one of the founders of game theory—were at the same time pioneers in the formalisation of social science. And one must add Herbert A. Simon, one of the pioneers in formalising social science, as another early adopter of computer-assisted methods of building social theories. Thus the first

K.G. Troitzsch (retired)
Universität Koblenz-Landau, Universitätsstraße 1, 56070 Koblenz, Germany
e-mail: klaus.g.troitzsch@bluewin.ch

© Springer International Publishing AG 2017   13
B. Edmonds, R. Meyer (eds.), *Simulating Social Complexity*,
Understanding Complex Systems, https://doi.org/10.1007/978-3-319-66948-9_2

two decades of computational social science saw mathematical models and their inelegant solutions, microsimulation and the first agent-based models before the name of this approach was coined.

Among the first problems tackled with the help of computer simulation, there were already predictions of the future of companies ("industrial dynamics", Forrester 1961), cities ("urban dynamics", Forrester 1969) and the world as a whole ("world dynamics", Forrester 1971) in the early 1960s and 1970s by Jay W. Forrester as well as predictions of the consequences of tax and transfer laws for both the individual household and the national economy in microanalytical simulation, an attempt that started as early as in 1956 (Orcutt 1957). Other early attempts at the prediction of election and referendum campaigns also became known in the 1960s, such as Abelson's and Bernstein's simulation analysis of a fluoridation referendum campaign of the Simulmatics Project directed by de Sola Pool. What all these early simulations have in common is that they were aimed at predicting social and economic processes in a quantitative manner and that computer simulation was seen as a "substitute for mathematical derivations" (Coleman 1964, p. 528), and although Simon and others had already taught computers to deal with nonnumerical problems as early as in 1955 ("the Logic Theorist, the first computer program that solved non-numerical problems by selective search", Simon 1996, pp. 189–190), Coleman still believed in 1964 that "the computer cannot solve problems in algebra; it can only carry out computations when actual numbers are fed in" (Coleman 1964, p. 529).

As system dynamics and microanalytic simulation—simulation approaches that continue to be promoted by learned societies such as the System Dynamics Society, which celebrated its 50th anniversary with an international conference in Boston in July 2007, or the International Microsimulation Association, which also celebrated 50 years of microsimulation with an international conference held in Vienna in August 2007—are not the focus of this handbook, this chapter will only give a short overview of these two approaches and go into the details of some other early models that remained more or less isolated and were even more or less forgotten.

System dynamics was developed by Jay W. Forrester in the mid-1950s as a tool to describe systems which could have modelled with large systems of difference and differential equations containing functions whose mathematical treatment would have been difficult or impossible. The general idea behind system dynamics was and is that a system, without considering its components individually, could be described in terms of its aggregate variables and their changes over time. The best known examples of system dynamic models are Forrester's (1971) and Meadows et al.'s (1974) world models which were inspired by the Club of Rome and won public attention in the 1970s when they tried to forecast the world population, the natural resources, the industrial and agricultural capital and the pollution until the end of the twenty-first century by describing the annual change of these aggregate variables as functions their current states and numerous parameters which had some empirical background.

Microsimulation was first described in papers by Orcutt (1957) who designed a simulation starting with a (sample of a) given population and simulating the individual fate of all the members of this population (sample) with the help of transition probabilities empirically estimated from official statistics. These transitions could be transitions between different jobs and educational levels, or they could represent death or the birth of a child or marriage; these models have mainly been used for predicting demographic changes and the effects of tax and transfer rules. Usually, these models do not take into account that the overall changes of the aggregated variables of the population (or the sample) affect the individual behaviour. Thus in the sense of Coleman (1990, p. 10), these models neglect the "downward causation" (i.e. the influence of the aggregate on the individual) and focus only the "upward causation", namely, the changes on the macro level which are the result of the (stochastically simulated) behaviour of the individuals.

The fluoridation referendum campaign model already mentioned above (Abelson and Bernstein 1963) was one of the first models that can be classified as an early predecessor of today's agent-based models. It consisted of a large number of representatives of people living in a community faced with the option of compulsory fluoridation if drinking water—an issue often discussed in the 1960s—which they would have to vote upon at the end of a longish campaign in which the media and local politicians were publishing arguments in favour of or against this issue. In this model, 500 individuals are exposed to information spread by several communication channels (or sources), and additionally, they also exchange information among themselves. It depends on their simulated communication habits to which extent they actually receive this information and, moreover, to which extent this leads to changes in their attitudes towards the referendum issue. Abelson and Bernstein defined 51 rules of behaviour, 22 of which are concerned with the processing of the information spread over the communication channels, and 27 rules are related to the information exchange among the individuals; another 2 determine the final voting behaviour at the end of the referendum campaign. The rules for processing the information from the public channels and those for processing the information exchanged among the individual citizens are quite similar, one of these rules— A3 and B2, respectively—is, for instance, "Receptivity to [source] s is an inverse function of the extremity of [individual] i's attitude position".

This early model did, of course, not endow the model individuals with an appropriate repertoire of behaviours, but nevertheless it displays a relatively broad range of communication possibilities among the model individuals which was neither aimed at in the classical microanalytical simulation approach nor in the cellular automata approach adopted in the early 1970s in Thomas Schelling's seminal paper on segregation. One of the shortcomings of Abelson's and Bernstein's model in the eyes of its critics was the fact that it "has never been fully tested empirically" (Alker 1974, p. 146), and another is the fact that one never knows "how adequate are the static representations of citizen belief systems defined primarily in terms of assertions held, assertions acceptance predispositions, with associated, more general, conflict levels?" (Alker 1974, p. 146). And, moreover, the assertions are modelled numerically (not a problem with the proponents of a mathematical sociology who would even have used a large system of differential equations to model the citizens' attitude changes) where obviously real citizens' attitudes were never mapped on to the set of integer or real numbers. More reasons for the fact that this approach was given up for decades are given by Nowak et al. (1990, p. 371): "the ad hoc quality of many of the assumptions of the models, perhaps because of dissatisfaction with the plausibility of their outcomes despite their dependence on extensive parameter estimation, or perhaps because they were introduced at a time when computers were still cumbersome and slow and programming time-consuming and expensive."

Simulmatics had mainly the same fate as Abelson's and Bernstein's model: Simulmatics was set up "for the Democratic Party during the 1960 campaign. . . . The immediate goal of the project was to estimate rapidly, during the campaign, the probable impact upon the public, and upon small strategically important groups within the public, of different issues which might arise or which might be used by the candidates" (Ithiel de Pool and Abelson 1961, p. 167). The basic components of this simulation were voter types, 480 of them, not individual voters, with their attitudes towards a number of "issue clusters" (48 of them), "political characteristics on which the voter type would have a distribution". Voter types were mainly defined by region, agglomeration structure, income, race, religion, gender and party affiliation, and from different opinion polls and for different points of time, these voter types were attributed four numbers per "issue cluster": the number of voters in this type and "the percentages pro, anti and undecided or confused on the issue" (168). The simulation then ran in a way that for each voter type, empirical findings about cross-pressure (e.g. anti-Catholic voters who had voted for the Democratic Party in the 1958 congressional elections and were likely to stay at home instead of voting for the Catholic candidate of the Democrats) were used to readjust the preferences of the voters, type by type. It is an open question whether one would call this a simulation in current social simulation communities, but as this approach in some way resembles the classical static microsimulation, where researchers are interested in the immediate consequences of new tax or transfer laws with no immediate feedback, one would classify Simulmatics as a simulation project, though with as little sophistication as static microsimulation has.

Thus the first two decades of computer simulation in the social sciences were mainly characterised by two beliefs: that computer simulations were nothing but the numerical solution of more adequate mathematical models and that they were most useful for predicting the outcome of social processes whose first few phases had already been observed. This was also the core of the discussion that was opened in 1968 by Hayward Alker who analysed, among others, the Abelson-Bernstein community referendum model and came to the conclusion that this "simulation cannot be 'solved': one must project what will be in the media, what elites will be doing, and know what publics already believe before even contingent predictions are made about community decisions. In that sense an open simulation is bad mathematics even if it is a good social system representation" (Alker 1974, p. 153).

In what Federico et al. (1981, p. 515) called "micro-operational computer simulations", they saw the opportunity that "computer modeling [could] contribute to the comprehension of which parameters and variables are most decisive in determining systemic behavior" (Federico et al. 1981, p. 519) and "produc[e] surprising emergent properties" (Federico et al. 1981, p. 518). They predicted that "the future of the social sciences is contingent upon identifying techniques to simultaneously link a multitude of relatively trivial conceptual structures, producing realistic outcomes when no premise alone is powerful enough to determine the state of the system at any moment" (Federico et al. 1981, p. 518). This is certainly a prediction which came true in the decades to come, as agent-based modelling in its various modern approaches is more or less correctly described with Federico's and Figliozzi's words. Nevertheless their "classification of computer simulation studies of psychosocial or sociotechnical systems" (Federico et al. 1981, p. 515) with its double dichotomy of operational and theoretical nature and micro and macro scope is no longer in line with current classifications. Putting, for instance, Abelson's and Bernstein's study (see above for details) in the box of operational (as contrasted to theoretical) macro simulation studies seems strange as this study connects microbehaviour to macrostructures and does not only look at the macro level. The same is true for other studies that fall in this cell of Federico's and Figliozzi's cross table. The reason for this is that what they call "microtheoretical computer simulation studies" is restricted to behaviour in small groups, thus "micro" does no refer to the individual level of social systems (as it usually does today) but to small systems such as Hare's (1961) five person group.

2.3 Computer Simulation in Its Own Right

The Simulmatics Corporation already mentioned in the previous subsection did not only work in the context of election campaigning, but later on also as a consulting agency in other political fields. Crisiscom is another example of an early forerunner of current simulation models of negotiation and decision-making processes. At the same time, it is an early example of a simulation not aimed at prediction but at "our understanding of the process of deterrence by exploring how far the behaviour of political decision makers in crisis can be explained by psychological mechanisms" (Ithiel de Pool and Kessler 1965, p. 31). Crisiscom dealt with messages of the type "actor one is related to actor two", where the set of relations was restricted to just two relations: affect and salience. In some way, Crisiscom could also be used as part of a gaming simulation in which one or more of the actors were represented by human players, whereas the others were represented by the computer programme— thus in a way it can also be classified as a predecessor of participatory simulation (see Chap. 11).

The 1970s and 1980s saw a number of new approaches to simulate abstract social processes, and most of them now were computer simulations in its own right, as—in terms of Thomas Ostrom—they used the "third symbol system" (Ostrom 1988, p. 384) directly without using it as a machine to manipulate symbols of the second symbol system, mathematics, but directly translating their ideas from the first symbol system, natural language, into higher level programming languages. Although this was already true for Herbert Simon's Logic Theorist, the General Problem Solver and other early artificial intelligence programmes, the direct use of the "third symbol system" in social science proper was not introduced before the first multilevel models and cellular automata that integrated at least primitive agents in the sense of software modules with some autonomy.

Cellular automata (Farmer et al. 1984; Ilachinski 2001) are a composition of finite automata which follow the same rule, are ordered in a (mostly) two-dimensional grid and interact with (receive input from) their neighbours. The behavioural rules of the individual cells are quite simple in most cases; they have only a small number of states among which they switch according to relatively simple rules, as in the famous game of life (Gardener 1970), where the cells have only two states, alive and dead, and change their states according to the two simple rules: if the cell is alive, it remains in this state if it has exactly two or three live cells among its eight neighbours—otherwise it dies—and if the cell is dead, it bursts into life if among its eight neighbours there are exactly three live cells. The great variety of outcomes on the level of the cellular automaton as a whole enthused researchers in complexity science and lay the headstone for innumerable cellular automata in one or two dimensions.

One of the first applications of cellular automata to problems of social science is Thomas Schelling's (1971) segregation model, demo versions of which are nowadays part of any distribution of simulation tools used for programming cellular automata and agent-based models—a model that shows impressively that segregation and the formation of ghettos is inevitable even if individuals tolerate a majority of neighbours different from themselves.

Another example is Bibb Latané's dynamic social impact theory with the implementation of the SITSIM model (Nowak and Latané 1994). This model, similar to Schelling's, also ends up in clustering processes and in the emergence of local structures in an initially randomly distributed population, but unlike Schelling's segregation model (where agents move around the grid of a cellular automaton until they find themselves in an agreeable neighbourhood), the clustering in SITSIM comes from the fact that immobile agents change their attitudes according to the attitudes they find in their neighbourhood *and* according to the persuasive strength of their neighbours.

Other cellular automata models dealt with n-person cooperation games and integrated game theory into complex models of interaction between agents and their neighbourhoods, and these models, too, usually end up in emergent local structures (Hegselmann 1996).

And in another computer simulation related to game theory run by Axelrod, it could be shown that the tit-for-tat strategy in the iterated prisoner's dilemma was superior to all other strategies which were represented in a computer tournament (Axelrod 1984). The prisoner's dilemma had served game theorists, economists and social scientists as a prominent model of decision processes under restricted knowledge. The idea stems from the early 1950s, first written down by Albert Tucker, and is about "two men, charged with a joint violation of law, are held separately by the police. Each is told that (1) if one confesses and the other does not, the former will be given a reward ... and the latter will be fined ... (2) if both confess, each will be fined ... At the same time, each has good reason to believe that (3) if neither confesses, both will go clear" (Poundstone 1992, pp. 117–118). In the non-iterated version, the rational solution is that both confess—but if they believe they can trust each other, they can both win, as both will go clear if neither confesses. Axelrod's question was under which conditions a prisoner in this dilemma would "cooperate" (with his accomplice, not with the police) and under which condition they would "defect" (i.e. confess, get a reward and let the accomplice alone in prison). Super-strategies in this tournament had to define which strategy—cooperate or defect—each player would choose, given the history of choices of both players, but not knowing the current decision of the partner. Then every strategy played the iterated game against every other strategy, with identical payoff matrices—and the tit-for-tat strategy proved to be superior to 13 other strategies proposed by economists, game theorists, sociologists, psychologists and mathematicians (and it was the strategy that had the shortest description in terms of lines of code). Although later on several characteristics of several of the strategies proposed could be analysed mathematically, the tournament had at least the advantage of easy understandability of the outcomes—which, by the way is another advantage of the "third symbol system" over the symbol system of mathematics.

Cellular automata later on became the environment of even more complex models of abstract social processes. They serve as a landscape where moving, autonomous, proactive, goal-directed software agents harvest food and trade with it. Sugarscape is such a landscape which serves as a laboratory for a "generative social science" (Epstein and Axtell 1996, p. 19) in which the researcher "grows" the emergent phenomena typical for real-world societies in a way that includes the explanation of these phenomena. In this artificial world, software agents find several types of food which they need for their metabolism, but in different proportions, which gives them an incentive to barter with a kind of food of which they have plenty, for another kind of food which they urgently need. This kind of a laboratory gives an insight under which conditions skewed wealth distributions might occur or be avoided; with some extensions (König et al. 2002), agents can even form teams led be agents who are responsible to spread the information gained by their followers among their group.

2.4 Conclusion and Suggested Further Reading

This short guided tour through early simulation models should have shown the optimism of the early adopters of this method: "If it is possible to reproduce, through computer simulation, much of the complexity of a whole society going through processes of change, and to do so rapidly, then the opportunities to put social science to work are vastly increased" (Ithiel de Pool and Abelson 1961, p. 183). Thirty-five years later, Epstein and Axtell formulate nearly the same optimism when they list a number of problems that social sciences have to face—suppressing real-world agents' heterogeneity, neglecting nonequilibrium dynamics and being preoccupied

Table 2.1 Overview of important approaches to computational social science

Approach	Used since	Characteristics
System dynamics	Mid-1950s	Only one object with a large number of attributes
Microsimulation	Mid-1950s	A large number of objects representing individuals that do not interact, neither with each other nor with their aggregate, with a small number of attributes each, plus one aggregating object
Cellular automata	Mid-1960s	Large number of objects representing individuals that interact with their neighbours, with a very restricted behaviour rule, no aggregating object, thus emergent phenomena have to be visualised
Agent-based models	Early 1990s with some forerunners in the 1960s, afterwards discontinued	Any number of objects ("agents") representing individuals and other entities (groups, different kinds of individuals in different roles) that interact heavily with each other, with an increasingly rich repertoire of changeable behaviour rules (including the ability to learn from other, to change their behavioural rules and to react differently to identical stimuli when the situation in which they are received are different

with static equilibria—and claim "that the methodology developed [in Sugarscape] can help to overcome these problems" (Epstein and Axtell 1996, p. 2).

To complete this overview, Table 2.1 lists the approaches touched in this introductory paper with their main features.

As one easily sees from this table, only the agent-based approach can "cover all the world" (Brassel et al. 1997), as only this one can include the features of all the others, and only this one can meet the needs of social science, as social science cannot content itself with models of individuals which cannot exchange symbolic messages that have to be interpreted by the recipients before they can take effect. If social science deals with large numbers of individuals in comparable situations, then microsimulation, cellular automata, sociophysics models and even systems dynamics can be a good approximation to what happens in human societies. But if we deal with small communities, including the local communities Abelson and Bernstein analysed, then the process of persuasion—which needs at least one persuasive person and one or more persuadable persons—has to be taken into account, and this calls for a richer structure of agents than the early approaches could provide.

Most of the literature suggested for further reading has already been mentioned. Epstein's and Axtells's (1996) work on generating societies gives a broad overview of early applications of agent-based modelling; Epstein (2006) goes even further as he defines this approach as the oncoming paradigm in social science. For the state of the art of agent-based modelling in the social sciences at the onset of this approach, the proceedings of early workshops and conferences on computational social science are still worth reading (Gilbert and Doran 1994; Gilbert and Conte 1995; Conte et al. 1997; Troitzsch et al. 1996). And a very wide overview of topics and approaches can be found in three papers devoted to measuring the "intellectual structures" of two journals which abound in papers on simulation in the social sciences at large (Meyer et al. 2009, 2010; Hauke et al. 2015).

References

Abelson, R. P., & Bernstein, A. (1963). A computer simulation of community referendum controversies. *Public Opinion Quarterly, 27*, 93–122.

Alker Jr., H. R. (1974). Computer simulations: Inelegant mathematics and worse social science. *International Journal of Mathematical Education in Science and Technology, 5*, 139–155.

Axelrod, R. (1984). *The evolution of cooperation*. New York: Basic Books.

Brassel, K. H., Möhring, M., Schumacher, E., & Troitzsch, K. G. (1997). Agents cover all the world? In R. Conte, R. Hegselmann, & P. Terna (Eds.), *Simulating social phenomena, Lecture notes in economics and mathematical systems* (Vol. 456, pp. 55–72). Berlin: Springer.

Coleman, J.S. (1964). Introduction to Mathematical Sociology. New York: The Free Press.

Coleman, J. S. (1990). *The foundations of social theory*. Boston: Harvard University Press.

Conte, R., Hegselmann, R., & Terna, P. (1997). *Simulating social phenomena, Lecture notes in economics and mathematical systems* (Vol. 456). Berlin: Springer.

Epstein, J. M. (2006). *Generative social science. Studies in agent-based computational modeling*. Princeton: Princeton University Press.

Epstein, J. M., & Axtell, R. (1996). *Growing artificial societies. Social science from the bottom up*. Washington, MA/Cambridge, MA: Brookings/MIT Press.

Farmer, D., Toffoli, T., & Wolfram, S. (1984). Cellular automata. In *Proceedings of an interdisciplinary workshop*, Los Alamos, New Mexico, March 7–11, 1983. Amsterdam: North-Holland.

Federico, P., Anthony, P., & Figliozzi, W. (1981). Computer simulation of social systems. *Sociological Methods and Research, 9*(4), 513–533.

Forrester, J. W. (1961). *Industrial dynamics*. Cambridge, MA: MIT/Wright Allen.

Forrester, J. W. (1969). *Urban dynamics*. Cambridge, MA: MIT/Wright Allen.

Forrester, J. W. (1971). *World dynamics*. Cambridge, MA: MIT/Wright Allen.

Gardener, M. (1970). The game of life. *Scientific American, 223*(4), 120–123.

Gilbert, N., & Conte, R. (1995). *Artificial societies: The computer simulation of social life*. London: UCL Press.

Gilbert, N., & Doran, J. E. (1994). *Simulating societies: The computer simulation of social phenomena*. London: UCL Press.

Hare, A. P. (1961). Computer simulation of interaction in small groups. *Behavioral Science, 6*, 261–265.

Hauke, J., Lorscheid, I., & Meyer, M. (2015). The recent development of social simulation as reflected in JASSS from 2008–2014: A citation and co-citation analysis. In *11th conference of the European social simulation association*, Groningen, NL.

Hegselmann, R. (1996). Cellular automata in the social sciences. Perspectives, restrictions, and artefacts. In R. Hegselmann, U. Mueller, & K. G. Troitzsch (Eds.), *Modelling and simulation in the social sciences from the philosophy of science point of view* (pp. 209–234). Dorrecht: Kluwer.

Ilachinski, A. (2001). *Cellular automata. A discrete universe*. Singapore: World Scientific.

Ithiel de Pool, S., & Abelson, R. P. (1961). The simulmatics project. *Public Opinion Quarterly, 25*, 167–183.

Ithiel de Pool, S., & Kessler, A. (1965). The Kaiser, the Czar, and the Computer: Information processing in a crisis. *The American Behavioral Scientist, 8*, 32–38.

König, A., Möhring, M., & Troitzsch, K. G. (2002). Agents, hierarchies and sustainability. In F. Billari & A. Prskawetz-Fürnkranz (Eds.), *Agent based computational demography* (pp. 197–210). Physica: Berlin.

Meadows, D. L., Behrens, W. W., Meadows, D. H., Naill, R. F., Randers, J., & Zahn, E. (1974). *Dynamics of growth in a finite world.* Cambridge, MA: Wright-Allen.

Meyer, M., Lorscheid, I., & Troitzsch, K. G. (2009). The development of social simulation as reflected in the first ten years of JASSS: A citation and co-citation analysis. *Journal of Artificial Societies and Social Simulation 12*(4), 12. http://jasss.soc.surrey.ac.uk/12/4/12.html

Meyer, M., Zaggl, M. A., & Carley, K. M. (2010). Measuring CMOT's intellectual structure and its development. *Computational and Mathematical Organization Theory, 17*, 1–34.

Nowak, A., & Latané, B. (1994). Simulating the emergence of social order from individual behaviour. In N. Gilbert & J. Doran (Eds.), *Simulating societies: The computer simulation of social processes* (pp. 63–84). London: University College of London Press.

Nowak, A., Szamrej, J., & Latané, B. (1990). From private attitude to public opinion: A dynamic theory of social impact. *Psychological Review, 97*, 362–376.

Orcutt, G. (1957). A new type of socio-economic system. *Review of Economics and Statistics, 58*, 773–797.

Ostrom, T. M. (1988). Computer simulation: The third symbol system. *Journal of Experimental Social Psychology, 24*, 381–392.

Poundstone, W. (1992). *Prisoner's dilemma. John von Neumann, game theory, and the puzzle of the bomb.* Oxford: Oxford University Press.

Schelling, T. C. (1971). Dynamic models of segregation. *Journal of Mathematical Sociology, 1*, 143–186.

Simon, H. A. (1996). *Models of my life.* Cambridge, MA: MIT Press.

Troitzsch, K. G., Mueller, U., Gilbert, N., & Doran, J. E. (1996). *Social science microsimulation.* Berlin: Springer.

Chapter 3
Types of Simulation

Paul Davidsson and Harko Verhagen

Abstract This looks at various ways that computer simulations can differ not in terms of their detailed mechanisms but in terms of its broader purpose, structure, ontology (what is represented), and approach to implementation. It starts with some different roles of people that may be concerned with a simulation and goes on to look at some of the different contexts within which a simulation is set (thus implying its use or purpose). It then looks at the kinds of system that might be simulated. Shifting to the modelling process, it looks at the role of the individuals within the simulations, the interactions between individuals, and the environment that they are embedded within. It then discusses the factors to consider in choosing a kind of model and some of the approaches to implementing it.

Why Read This Chapter?
To understand the different ways that computer simulation can differ in terms of (a) purpose, (b) targets for simulation, (c) what is represented, and (d) its implementation and, subsequently, to be more aware of the choices to be made when simulating social complexity.

3.1 Introduction

Simulation concerns the imitation of some aspects of the reality (past, present, or future) for some purpose. We should contrast computer simulation to *physical simulation* in which physical objects are substituted for the real thing. These physical objects are often chosen because they are smaller or cheaper than the actual object or system. When (some of) the objects in a physical simulation are humans, we may refer to this as *human simulation*. However, the focus of this book is on

P. Davidsson (✉)
Malmö University, Malmö, Sweden
e-mail: paul.davidsson@mah.se

H. Verhagen
Stockholm University, Stockholm, Sweden

© Springer International Publishing AG 2017
B. Edmonds, R. Meyer (eds.), *Simulating Social Complexity*,
Understanding Complex Systems, https://doi.org/10.1007/978-3-319-66948-9_3

computer simulation and, in particular, computer simulation of social complexity, which concerns the imitation of the behaviour of one or more groups of social entities and their interaction.

Computer simulation, as any other computer programme, can be seen as a tool, which could be used professionally or used in the user's spare time, e.g. when playing computer games. It is possible to distinguish between different types of professional users, e.g. *scientists* who use simulation in the research process to gain new knowledge, *policy-makers* who use it for making strategic decisions, *managers (of a system)* who use it to make operational decisions, and *engineers* who use it when developing systems. We can also differentiate two user situations, namely, the user as *participant* in the simulation and the user as *observer* of the simulation. Computer games and training settings are examples of the former, where the user is immerged in the simulation. In the case of using simulation as a tool for, say, scientific research or decision support, the user is an outside observer of the simulation. (In other words, we may characterize this difference as that between interactive simulations and batch simulations.)

The main task of computer simulation is the creation and execution of a formal model of the behaviour and interaction (of the entities) of the system being simulated. In scientific research, computer simulation is a research methodology that can be contrasted to empirically driven research.[1] As such, simulation belongs to the same family of research as analytical models. One way of formally modelling a system is to use a mathematical model and then attempt to find analytical solutions enabling the prediction of the system's behaviour from a set of parameters and initial conditions. Computer simulation, on the other hand, is often used when simple closed form analytic solutions are not possible. Although there are many different types of computer simulation, they typically attempt to generate a sample of representative scenarios for a model in which a complete enumeration of all possible states would be prohibitive or impossible.

It is possible to make a general distinction between two ways of modelling the system to be simulated. One is to use mathematical models and is referred to as *equation-based* (or system dynamics or macro-level) simulation. In such models, the set of individuals (the population of the system) is viewed as a structure that can be characterized by a number of variables. In the other way of modelling, which is referred to as *individual-based* (or agent-based or micro-level) simulation, the specific behaviours of specific individuals are explicitly modelled. In contrast to equation-based simulation, the structure is viewed as emergent from the interactions between the individuals, thus exploring the standpoint that complex effects need not have complex causes. We will here, as well as in the remainder of this book, focus on individual-based simulation.

---

[1]This distinction is of course not set in stone. For an example of an evidence-driven approach to computer simulation, see Chap. 27 in this volume (Geller and Moss 2017).

In this chapter, we will describe the main purposes of computer simulation and also give an overview of the main issues that should be regarded when developing computer simulations.

3.2 Purposes of Simulation

We can identify a number of distinct purposes of simulation. In general terms, simulation is almost always used for analysing (some aspects of) a system, typically by predicting future states. More specifically, we may say that in the case when the user is *observing* the simulation, the purpose is often one of the following:

- *Management of a system*, where simulation of (parts of) this system is used to support operational decisions, i.e. which action to take, or strategic decisions, i.e. which policy to use. The chapters on application areas in this book provide some examples of this purpose; e.g. Chap. 22 addresses environmental management (Le Page et al. 2017).
- *Design or* engineering *of a system*, where simulation is used as a tool to support design decisions when developing a system. Chapter 23 illustrates how simulation can help in the design of distributed computer systems (Hales 2017). In fact, many new technical systems are distributed and involve complex interaction between humans and machines, which makes individual-based simulation a suitable approach. The idea is to model the behaviour of the human users, which is useful in situations where it is too expensive, difficult, inconvenient, tiresome, or even impossible for real human users to test out a new technical system. An example of this is the simulation of "intelligent buildings" where software agents model the behaviour of the people in the building (Davidsson 2000).
- Evaluation *and verification*, where simulation is used to evaluate a particular theory, model, hypothesis, or system, or compare two or more of these. Moreover, simulation can be used to verify whether a theory, model, hypothesis, system, or software is correct. An example of this purpose is found in Chap. 4 of this book (Edmonds et al. 2017). More generally, in the context of social theory building, simulations can be seen as an experimental method or as theories in themselves (Sawyer 2003). In the former case, simulations are run, e.g. to test the predictions of theories, whereas in the latter case, the simulations themselves are formal models of theories. Formalizing the ambiguous, natural language-based theories of the social sciences helps to find inconsistencies and other problems and thus contributes to theory building.
- *Understanding*, where simulation is used to gain deeper knowledge of a certain domain. In such explorative studies, there is no specific theory, model, etc. to be verified, but we want to study different phenomena (which may however lead to theory refinement). Chapter 24 in this volume provides a number of examples how simulation has helped in understanding animal social behaviour (Hemelrijk 2017).

The focus of this book is on the user as an observer; the role of the user as participant is just touched upon in Chap. 12 on participatory approaches (Barreteau et al. 2017). However, to give a more complete picture, we have identified the following purposes in the case when the user is participating in the simulation:

- *Education*, where simulation is used to explain or illustrate a phenomenon and deepen the user's theoretical knowledge. An example of this is the recently developed SimPort,[2] a multiplayer serious game where the players have to construct a port area in the vicinity of Rotterdam. One aim of this simulation-based tool is to give its users better insight into any unforeseen, undesirable, and unintentional effects of one or more development strategies and design variations in the medium term (10–30 years) as a result of exogenous uncertainties (economic, market, technological) and due to strategic behaviour of the parties involved. Another example of individual-based simulation for educational purpose is the PSI agent (Künzel and Hämmer 2006) that supports acquiring theoretical insights in the realm of psychological theory. It enables students to explore psychological processes without ethical problems.
- *Training*, where simulation is used to improve a person's practical skills in a certain domain. The main advantage of using simulation for training purposes is to be part of a real-world-like situation without real-world consequences. An early work in this area was a tool to help train police officers to manage large public gatherings, such as crowds and protest marches (Williams 1993). Another example of agent-based simulation for training purposes is Steve, an agent integrated with voice synthesis software and virtual reality software providing a very realistic training environment. For instance, it has been applied to maintenance tasks in nuclear power plants (Méndez et al. 2003).
- *Entertainment*, where simulation is used just to please the user. There are a large number of popular simulation games available. These belong to genres like *construction and management simulations*, where players experience managing a government, a sports team, a business, or a city; *life simulations*, where players manage a life form or ecosystem, such as the well-known "Sims" and its sequels; *vehicle simulations*, where players experience driving a vehicle, such as an airplane or a racing car; and of course different types of *war games*.

3.3 Types of Systems Simulated

It is possible to categorize the systems being simulated:

1. *Human-centred systems*, such as:

- *Human societies*, consisting of a set of persons with individual goals. That is, the goal of different individuals may be conflicting. In Chap. 28 of this book, more information on the simulation of human societies is given (Edmonds et al. 2017).

---

[2]http://www.simport.eu/

– *Organizations*, which we here define as structures of persons related to each other in order to purposefully accomplishing work or some other kind of activity. That is, the persons of an organization share some of their goals. Further details on the modelling and simulation of organizations are provided in (Dignum 2013).
– *Economic systems*, which are organized structures in which actors (individuals, groups, or enterprises) are trading goods or services on a market. Chapter 25 (Rouchier 2017) takes a closer look at markets.

2. *Natural systems*, such as:

– *Animal societies*, which consist of a number of interacting animals, such as an ant colony or a colony of birds. Chapter 24 (Hemelrijk 2017) is devoted to simulation of animal societies.
– *Ecological systems*, in which animals and/or plants are living and evolving in a relationship to each other and in dependence of the environment (even if humans also are part of the ecological system, they are often not part of these simulation models). In Chap. 22 (Le Page et al. 2017) more details on the simulation of ecological systems are discussed.

3. *Socio-technical systems*, which are hybrid systems consisting of both living entities (in most cases humans) and technical artefacts interacting with each other. Examples of this type of system are transportation and traffic systems concerning the movement of people or goods in a transportation infrastructure such as a road network. Chapter 26 (Ramstedt et al. 2017) provides a review of simulation studies in these areas.
4. *Artificial societies*, which consist of a set of software and/or hardware entities, i.e. computer programmes and/or robots, with individual goals. One type of artificial societies, namely, distributed computer systems, is treated in Chap. 23 (Hales 2017).

In addition, there are systems that are interesting to simulate using a micro-level approach but that we do not regard as social systems and are therefore not treated in this book. One class of such systems are *physiological systems*, which consist of functional organs integrated and co-operating in a living organism, e.g. subsystems of the human body. *Physical systems*, which are collections of passive entities following only physical laws, constitute another type of nonsocial systems.

3.4 Modelling

Let us now focus on how to model the system to be simulated. This depends on the type of system and the purpose of the simulation study. An individual- or agent-based model of a system consists of a set of entities and an environment in which the entities are situated. The entities are either *individuals* (agents) that have some decision-making capabilities or *objects* (resources) that have no agency and are purely physical. There are a number of characteristics that can be used to differentiate between different types of models. We will first look at how individuals are being modelled, then on the interaction between the individuals, and finally how the environment is being modelled.

3.4.1   Individuals

A model of an individual can range from being very simple, such a one binary variable (e.g. alive or dead) that is changed using only a single rule, to being very complex. The complexity of the model for a given simulation should be determined by the complexity of the individuals being simulated. Note, however, that very complex collective behaviour could be achieved from very simple individual models, if the number is sufficiently large.

We can distinguish between modelling the *state* of an individual and the *behaviour* of the individual, i.e. the decisions and actions it takes. The state of an individual, in turn, can be divided into the *physical* and the *mental* state. The description of the physical state may include the position of the individual and features such as age, sex, and health status. The physical state is typically modelled as a feature vector, i.e. a list of attribute/value pairs. However, this is not always the case as in some domain the physical state of individual is not modelled at all. An example is the PSI agent mentioned earlier that was used to give students theoretical insights in the area of psychological theory.

Whereas the physical state is often simple to model, representing the mental state is typically much more complex, especially if the individuals modelled are human beings. A common approach is to model the beliefs, desires, and intentions of the individual, for instance, by using the BDI model (Bratman 1987; Georgeff et al. 1998). Such a model may include the social state of the individual, i.e. which norms it adheres to, which coalitions it belongs to, etc. Although the BDI model is not based on any experimental evidence of human cognition, it has proven to be quite useful in many applications. There has also been some work on incorporating emotions in models of the mental state of individuals (cf. Bazzan and Bordini 2001) as well as obligations, like the BOID model (Broersen et al. 2001), which extends the BDI with obligations.

Modelling the behaviours (and decisions) of the individuals can be done in a variety of ways, from simple probabilities to sophisticated reasoning and planning mechanisms. As an example of the former, we should mention *dynamic micro-simulation* (Gilbert and Troitzsch 2005), which was one of the first ways of performing individual-based simulation and is still frequently used. The purpose is to simulate the effect the passing of time has on individuals. Data (feature vectors) from a random sample from the population is used to initially characterize the simulated individuals. A set of *transition probabilities* are then used to describe how these features will change over a time period, e.g. there is a probability that an employed person becomes unemployed during a year. The transition probabilities are applied to the population for each individual in turn and then repeatedly reapplied for a number of simulated time periods. In traditional micro-simulation, the behaviour of each individual is regarded as a "black box". The behaviour is modelled in terms of probabilities, and no attempt is made to justify these in terms of individual preferences, decisions, plans, etc. Thus, better results may be gained if also the cognitive processes of the individuals were simulated.

Opening the black box of individual decision-making can be done in several ways. A basic and common approach is to use decision rules, for instance, in the form of a set of situation-action rules: If an individual and/or the environment is in state X, then the individual will perform action Y. By combining decision rules and the BDI model quite sophisticated behaviour can be modelled. Other models of individual cognition used in agent-based social simulation include the use of Soar, a computer implementation of Allen Newell's unified theory of cognition (Newell 1994), which was used in Steve (discussed above). Another unified theory of individual cognition, for which a computer implementation exists, is ACT-R (Anderson et al. 2004), which is realized as a production system. A less general example is the Consumat model (Janssen and Jager 1999), a meta-model combining several psychological theories on decision-making in a consumer situation. In addition, nonsymbolic approaches such as neural networks have been used to model the agents' decision-making (Massaguer et al. 2006).

As we have seen, the behaviour of individuals could be either *deterministic* or *stochastic*. Also, the *basis* for the behaviour of the individuals may vary. We can identify the following categories:

– *The state of the individual itself*: In most social simulation models, the physical and/or mental state of an individual plays an important role in determining its behaviour.
– *The state of the environment*: The state of the environment surrounding the individual often influences the behaviour of an individual. Thus, an individual may act differently in different contexts although its physical and mental state is the same.
– *The state of other individuals*: One popular type of simulation model, where the behaviour of individuals is (solely) based on the state of other individuals, is those using *cellular automata* (Schiff 2008). Such a simulation model consists of a grid of cells representing individuals, each in one of a finite number of states. Time is discrete and the state of a cell at time t is a function of the states of a finite number of cells (called its neighbourhood) at time $t - 1$. These neighbours are a fixed selection of cells relative to the specified cell. Every cell has the same rule for updating, based on the values in its neighbourhood. Each time the rules are applied to the whole grid, a new generation is created. In this case, information about the state of other individuals can be seen as gained through observations. Another possibility to gain this information is through communication, and in this case, the individuals do not have to be limited to the neighbours.
– *Social states (norms, etc.) as viewed by the agent*: For simulation of social behaviour, the agents need to be equipped with mechanisms for reasoning at the social level (unless the social level is regarded as emergent from individual

behaviour and decision-making). Several models have been based on theories from economy, social psychology, sociology, etc. Guye-Vuillème (2004) provides an example of this with his agent-based model for simulating human interaction in a virtual reality environment. The model is based on sociological concepts such as roles, values, and norms and motivational theories from social psychology to simulate persons with social identities and relationships.

In most simulation studies, the behaviour of the individuals is *static* in the sense that decision rules or reasoning mechanisms do not change during the simulation. However, human beings and most animals do have an ability to adapt and learn. To model *dynamic* behaviour of individuals through learning/adaptation can be done in many ways. For instance, both ACT-R and Soar have learning built in. Other types of learning include the internal modelling of individuals (or the environment) where the models are updated more or less continuously.

Finally, there are some more general aspects to consider when modelling individuals. One such aspect is whether all agents share the same behaviour or whether they behave differently, in other words, representation of behaviour is either *individual* or *uniform*. Another general aspect is the number of individuals modelled, i.e. the *size* of the model, which may vary from a few individuals to billions of individuals. Moreover, the population of individuals could be either *static* or *dynamic*. In dynamic populations, changes in the population are modelled, typically births and deaths.

3.4.2 Interaction Between Individuals

In dynamic micro-simulation, simulated individuals are considered in isolation without regard to their interaction with others. However, in many situations, the interaction between individuals is crucial for the behaviour at system level. In such cases, better results will be achieved if the interaction between individuals was included in the model. Two important aspects of interaction are (a) who is interacting with whom, i.e. the *interaction topology*, and (b) the *form* of this interaction.

A basic form of interaction is *physical interaction* or interaction based on spatial proximity. As we have seen, this is used in simulations based on cellular automata, e.g. in the well-known Game of Life (Gardner 1970). The state of an individual is determined by how many of its neighbours are alive. Inspired by this, work researchers developed more refined models, often modelling the social behaviour of groups of animals or artificial creatures. One example is the BOID model by Reynolds (1987), which simulates coordinated animal motion such as bird flocks and fish schools in order to study emergent phenomena. In these examples, the interaction topology is limited to the individuals immediately surrounding an individual. In other cases, as we will see below, the interaction topology is defined more generally in terms of a *(social) network*. Such a network can be either *static*, i.e. the topology does not change during a simulation, or *dynamic*. In these networks, interaction is typically *language-based*. An example is the work by Verhagen (2001), where agents that are part of a group use direct communication between the group members to form shared group preferences regarding the decisions they make. Communication is steered by the structure of the social network regardless of the physical location of the agents within the simulated world. For a more detailed discussion of the different options to model interaction topologies, see Chap. 19 in this volume (Amblard and Quattrociocchi 2017).

3.4.3 The Environment

The state of the environment is usually represented by a set of (global) parameters, e.g. temperature. In addition, there are a number of important aspects of the environment model, such as:

- *Spatial explicitness:* In some models, there is actually no notion of physical space at all. An example of a scenario where location is of less importance are "innovation networks" (Gilbert et al. 2001). Individual agents are high-tech firms that each have a knowledge base used to develop artefacts to launch on a simulated market. The firms are able to improve their products through research or by exchanging knowledge with other firms. However, in many scenarios, location is very important; thus, each individual (and sometimes objects) is assigned a specific location at each time step of the simulation. In this case, the individuals may be either static (the entity does not change location during the simulation) or mobile. The location could either be specified as an *absolute position* in the environment or in terms of *relative positions* between entities. In some areas, the simulation software is integrated with a Geographical Information System (GIS) in order to achieve closer match to reality (cf. Schüle et al. 2004).
- *Time:* There are in principle two ways to address time, and one is to ignore it. In static simulation, time is not explicitly modelled; there is only a "before" and an "after" state. However, most simulations are dynamic, where time is modelled as a sequence of time steps. Typically, each individual may change state between each time step.
- *Exogenous events:* This is the case when the state of the environment, e.g. the temperature, changes without any influence/action from the individuals. Exogenous events, if they are modelled, may also change the state of entities, e.g. decay of resources, or cause new entities to appear. This is a way to make the environment stochastic rather than deterministic.

3.4.4 Factors to Consider When Choosing a Model

In contrast to some of the more traditional approaches, such as system dynamics, individual-based modelling does not yet have any standard procedures that can support the model development (although some attempts in this direction have been made, e.g. by Grimm et al. (2006), in the area of ecological systems). In addition, it is often the case that the only formal description of the model is the actual programme code. However, it may be useful to use the Unified Modelling Language (UML) to specify the model.

Some of the modelling decisions are determined by the features of the system to be simulated, in particular those regarding the interaction model and the environment model. The hardest design decision is often how the mental state and the behaviour of individuals should be modelled, in particular when representing human beings. For simpler animals or machines, a feature vector combined with a set of transitions rules is often sufficient. Depending on the phenomena being studied, this may also be adequate when modelling human beings. Gilbert (2006) provides some guidelines whether a more sophisticated cognitive model is necessary or not. He states that the most common reason for ignoring other levels is that the properties of these other levels can be assumed constant and exemplifies this by studies of markets in equilibrium where the preferences of individual actors are assumed to remain constant. (Note, however, that this may not always be true.) Another reason for ignoring other levels, according to Gilbert, is when there are many alternative processes at the lower level, which could give rise to the same phenomenon at the macro-level. He illustrates this with the famous study by Schelling (1971) regarding residential segregation. Although Schelling used a very crude model of the mental state and behaviour of the individuals, i.e. ignoring the underlying motivations for household migration, the simulation results were valid (as the underlying motivations were not relevant for the purpose of Schelling's study).

On the other hand, there are many situations where a more sophisticated cognitive model is useful, in particular when the mental state or behaviour of the individual constraints or in other ways influences the behaviour at the system level. However, as Gilbert concludes, the current research is not sufficiently mature in order to give advice on which cognitive model to use (BDI, Soar, ACT-R, or other). Rather, he suggests that more pragmatic considerations should guide the selection.

The model of the environment is mostly dictated by the system to be simulated, with the modeller having to decide on the granularity of the values the environmental attributes can take. The interaction model is often chosen based on the theory or practical situation that lies at the heart of the simulation, but sometimes the limitations of the formal framework used restrict the possibilities. Here, the modeller also has to decide upon the granularity of attribute values.

3.5 Implementation

We will now discuss some issues regarding the implementation (programming and running) of a simulator.

A simulator can be *time-driven*, where the simulated time is advanced in constant time steps, or *event-driven*, where the time is advanced based on the next event. In an event-driven simulation, a simulation engine drives the simulation by continuously taking the first event out of a time-ordered event list and then simulating the effects on the system state caused by this event. Since time segments where no event takes place are not regarded, event-driven simulation is often more efficient than time-driven simulation. On the other hand, since time is incremented at a constant pace during a simulation in time-driven mode, this is typically a better option if the simulation involves user participation.

There are a number of *platforms* or toolkits for agent-based simulation available, such as Swarm, NetLogo, and RePast (see Railsback et al. (2006) for a critical review of these and some other platforms). These are freely available, simplify the programming, and can be of great help, in particular for modellers that are not skilled programmers. However, they all impose some limitations on what can be modelled, which may or may not be crucial for the application at hand. An approach without such limitation is of course to programme the simulator from scratch using ordinary programming languages like Java or C, which is more difficult and time-consuming. In some cases, e.g. if you want to distribute the simulation on a number of computers, it may be appropriate to use an agent platform, such as JADE. In this case, the individuals may be implemented as actual software agents. In particular, when the number of individuals simulated is large and/or the models of individuals are complex, it may be too time-consuming to run the simulation on a single computer. Instead, one may distribute the computational load on several computers in order to get reasonable running times. It should be mentioned that there are some efforts on making agent-based simulation platforms run on large-scale computer networks such as Grids (see, e.g. the work by Chen et al. (2008)).

It is worth noting that the resulting software is an approximation of a simulation model, which in turn is an approximation of the actual system. Thus, there are several steps of verification and validation that need to be addressed in the development of a simulation model, as discussed in Chap. 9 (David et al. 2017).

3.6 Conclusion

As we have seen, there are many different types of individual-based social simulation. In the table below, we provide a summary.

Focus	Aspect	Options
Usage	Users	Scientists
		Policy-makers
		Managers
		Non-professionals
	Purposes	Management of a system
		Design or engineering of a system
		Evaluation and verification
		Understanding
		Education
		Training
		Entertainment
System simulated	Human-centred systems	Human societies
		Organizations
		Economic systems
	Natural systems	Animal societies
		Ecological systems
	Socio-technical systems	
	Artificial systems	
Individual model	Individual physical state	Feature vector
	Individual mental state	Feature vector
		BDI
	Individual behaviour	Transition probabilities
		Decision rules
		Cognitive model (soar, ACT-R, etc.)
	Basis of behaviour	Own state
		State of the environment
		State of other individuals
		Social states
	Uniformity	Uniform/non-uniform
	Population	Static/dynamic
Interaction model	Form of interaction	No interaction
		Physical
		Language-based
	Interaction topology	Static/dynamic
		Neighbourhood/network
Environment model	Spatial explicitness	None
		Relative positions
		Absolute positions
	Time	Static/dynamic
	Exogenous events	Yes/no
Implementation	Simulation engine	Time-driven/event-driven
	Programming	MABS platform (NetLogo, Repast, etc.)
		MAS platform (JADE, etc.)
		From scratch (C, Java, etc.)
	Distributedness	Single computer/distributed

Further Reading

Gilbert and Troitzsch (2005) also have sections that describe the different kinds of simulation available. Railsback and Grimm (2011) present a complementary analysis, coming from ecological modelling. The introductory chapters in (Gilbert and Doran 1994) and (Conte and Gilbert 1995) map out many of the key issues and aspects in which social simulation has developed.

References

Amblard, F., & Quattrociocchi, W. (2017). Social networks and spatial distribution. doi:https://doi.org/10.1007/978-3-319-66948-9_19.

Anderson, J. R., et al. (2004). An integrated theory of the mind. *Psychological Review, 111*(4), 1036–1060.

Barreteau, O., Bots, P., Daniell, K., Etienne, M., Perez, P., Barnaud, C., et al. (2017). Participatory approaches. In B. Edmonds & R. Meyer (Eds.), *Simulating social complexity: A handbook.* Berlin: Springer-Verlag.

Bazzan, A.L.C., & Bordini, R.H. (2001). A framework for the simulation of agents with emotions: Report on experiments with the iterated prisoners dilemma. In *Fifth international conference on autonomous agents*, Montreal, 2001 (pp. 292–299). New York: ACM Press.

Bratman, M. E. (1987). *Intentions, plans, and practical reason.* Cambridge, MA: Harvard University Press.

Broersen, J., Dastani, M., Huang, Z., Hulstijn, J., & Van der Torre, L. (2001). The BOID architecture: Conflicts between beliefs, obligations, intentions and desires. In *Fifth international conference on autonomous agents*, Montreal, 2001 (pp. 9–16). New York: ACM Press.

Chen, D., Theodoropoulos, G. K., Turner, S. J., Cai, W., Minson, R., & Zhang, Y. (2008). Large-scale agent-based simulation on the grid. *Future Generation Computer Systems, 24*(7), 658–671.

Conte, R., & Gilbert, N. (Eds.). (1995). *Artificial societies: The computer simulation of social life.* London: UCL Press.

David, N., Fachada, N., & Rosa, A. C. (2017). Verifying and validating simulations. In B. Edmonds & R. Meyer (Eds.), *Simulating social complexity: A handbook.* Berlin: Springer-Verlag.

Davidsson, P. (2000). Multi agent based simulation: Beyond social simulation. In S. Moss & P. Davidsson (Eds.), *Multi agent based simulation, Lecture notes in computer science* (Vol. 1979, pp. 98–107). Berlin: Springer.

Dignum, V. (2013). Organisational design. In B. Edmonds & R. Meyer (Eds.), *Simulating social complexity – A handbook* (pp. 541–562). Berlin: Springer.

Edmonds, B., Lucas, P., Rouchier, J., & Taylor, R. (2017). Human societies: Understanding observed social phenomena. In B. Edmonds & R. Meyer (Eds.), *Simulating social complexity: A handbook.* Berlin: Springer-Verlag.

Gardner, M. (1970). Mathematical games: The fantastic combinations of John Conway's new solitaire game "Life". *Scientific American, 223*(4), 120–124.

Geller, A., & Moss, S. (2017). Modeling power and authority: An emergentist view from Afghanistan. In B. Edmonds & R. Meyer (Eds.), *Simulating social complexity: A handbook.* Berlin: Springer-Verlag.

Georgeff, M., Pell, B., Pollack, M., Tambe, M., & Wooldridge, M. (1998). The belief-desire-intention model of agency. In J. Muller, M. Singh, & A. Rao (Eds.), *Intelligent agents V, Lecture notes in artificial intelligence* (Vol. 1555, pp. 1–10). Berlin: Springer.

Gilbert, N. (2006). When does social simulation need cognitive models? In R. Sun (Ed.), *Cognition and multi-agent interaction: From cognitive modelling to social simulation* (pp. 428–432). Cambridge: Cambridge University Press.

Gilbert, N., & Doran, J. (Eds.). (1994). *Simulating societies*. London: UCL Press.

Gilbert, N., Pyka, A., & Ahrweiler, P. (2001). Innovation networks: A simulation approach. *Journal of Artificial Societies and Social Simulation, 4*(3). http://jasss.soc.surrey.ac.uk/4/3/8.html

Gilbert, N., & Troitzsch, K. G. (2005). *Simulation for the social scientist* (2nd ed.). Maidenhead: Open University Press & McGraw Hill Education.

Grimm, V., Berger, U., Bastiansen, F., Eliassen, S., Ginot, V., Giske, J., et al. (2006). A standard protocol for describing individual-based and agent-based models. *Ecological Modelling, 198*, 115–126.

Guye-Vuillème, A. (2004). *Simulation of nonverbal social interaction and small groups dynamics in virtual environments*. PhD thesis, École Polytechnique Fédérale de Lausanne, No 2933.

Hales, D. (2017). Distributed computer systems. doi:https://doi.org/10.1007/978-3-319-66948-9_23.

Hemelrijk, C. (2017). Animal social behaviour. doi:https://doi.org/10.1007/978-3-319-66948-9_24.

Janssen M.A., & Jager, W. (1999). An integrated approach to simulating behavioural processes: A case study of the lock-in of consumption patterns. *Journal of Artificial Societies and Social Simulation, 2*(2). http://jasss.soc.surrey.ac.uk/2/2/2.html

Künzel, J., & Hämmer, V. (2006). Simulation in university education: The artificial agent PSI as a teaching tool. *Simulation, 82*(11), 761–768.

Le Page, C., Bazile, D., Becu, N., Bommel, P., Bousquet, F., Etienne, M., et al. (2017). Agent-based modelling and simulation applied to environmental management. doi:https://doi.org/10.1007/978-3-319-66948-9_22.

Massaguer, D., Balasubramanian, V., Mehrotra, S., & Venkatasubramanian, N. (2006, May 8). Multi-agent simulation of disaster response. In N.R. Jennings, M. Tambe, T. Ishida, & S.D. Ramchurn (Eds.), *First international workshop on agent technology for disaster management*, Hakodate, Hokkaido, Japan (pp. 124–130). http://users.ecs.soton.ac.uk/sdr/atdm/ws34atdm.pdf

Méndez, G., Rickel, J., & de Antonio, A. (2003). Steve meets Jack: The integration of an intelligent tutor and a virtual environment with planning capabilities. In *Intelligent virtual agents, Lecture notes on artificial intelligence* (Vol. 2792, pp. 325–332). Berlin: Springer.

Newell, A. (1994). *Unified theories of cognition*. Cambridge, MA: Harvard University Press.

Railsback, S. F., Lytinen, S. L., & Jackson, S. K. (2006). Agent-based simulation platforms: Review and development recommendations. *Simulation, 82*(9), 609–623.

Railsback, S. F., & Grimm, V. (2011). *Agent-based and individual-based modeling: A practical introduction*. Princeton: Princeton University Press.

Ramstedt, L., Törnquist Krasemann, J., & Davidsson, P. (2017). Movement of people and goods. doi:https://doi.org/10.1007/978-3-319-66948-9_26.

Reynolds, C. W. (1987). Flocks, herds, and schools: A distributed behavioural model. *Computer Graphics, 21*(4), 25–34.

Rouchier, J. (2017). Agent-Based simulation as a useful tool for the study of markets. doi:https://doi.org/10.1007/978-3-319-66948-9_25.

Sawyer, R. K. (2003). Artificial societies: Multi-agent systems and the micro-macro link in sociological theory. *Sociological Methods & Research, 31*(3), 325–363.

Schelling, T. C. (1971). Dynamic models of segregation. *Journal of Mathematical Sociology, 1*, 143–186.

Schiff, J. L. (2008). *Cellular automata: A discrete view of the world*. Oxford: Wiley.

Schüle, M., Herrler, R., & Klügl, F. (2004). Coupling GIS and multi-agent simulation: Towards infrastructure for realistic simulation. In G. Lindemann, J. Denzinger, I.J. Timm, & R. Unland (Eds.), *Multiagent system technologies, second German conference*, MATES 2004, *LNCS* (Vol. 3187, pp. 228–242). Berlin: Springer.

Verhagen, H. (2001). Simulation of the learning of norms. *Social Science Computer Review, 19*(3), 296–306.

Williams, R. (1993). An agent based simulation environment for public order management training. In *Western simulation multiconference, object-oriented simulation conference* (pp. 151–156).

Chapter 4
Different Modelling Purposes

Bruce Edmonds

Abstract How one builds, checks, validates and interprets a model depends on its 'purpose'. This is true even if the same model is used for different purposes, which means that a model built for one purpose but now used for another may need to be rechecked, revalidated and maybe even rebuilt in a different way. Here we review some of the different purposes for building a simulation model of complex social phenomena, focussing on five in particular: theoretical exposition, prediction, explanation, description and illustration. The chapter looks at some of the implications in terms of the ways in which the intended purpose might fail. In particular, it looks at the ways that a confusion of modelling purposes can fatally weaken modelling projects, whilst giving a false sense of their quality. This analysis motivates some of the ways in which these 'dangers' might be avoided or mitigated.

Why Read This Chapter?
This chapter will help you understand the importance of clearly identifying one's goal in developing and using a model and the implications of this decision in terms of how the model is developed, checked, validated, interpreted and described. It might thus help you produce models that are more reliable for your intended purpose and increase the reliability of your modelling. It will help you avoid a situation where you partially justify your model with respect to different purposes but succeed at none of them.

B. Edmonds (✉)
Centre for Policy Modelling, Manchester Metropolitan University, All Saints Campus, Oxford Road, Manchester, M1 6BH, UK
e-mail: bruce@edmonds.name

© Springer International Publishing AG 2017
B. Edmonds, R. Meyer (eds.), *Simulating Social Complexity*,
Understanding Complex Systems, https://doi.org/10.1007/978-3-319-66948-9_4

4.1   Introduction

A common view of modelling is that one builds a 'lifelike' reflection of some system, which then can be relied upon to act like that system. This is a correspondence
view of modelling where the details in the model correspond in a one-one manner
with those in the modelling target—as if the model were some kind of 'picture' of
what it models. However, this view can be misleading since models always differ
from what they model, so that they will capture some aspects of the target system but
not others. With complex phenomena, especially social phenomena, it is inevitable
that any model is, at best, a very partial picture of what it represents—in fact I
suggest that this picture analogy is *so* unhelpful that it might be best to abandon it
altogether as more misleading than helpful.[1]

Rather, here I will suggest a more pragmatic approach, where models are viewed
as *tools* designed and useful for specific purposes. Although a model designed for
one purpose may turn out to be OK for another, it is more productive to use a tool
designed for the job in hand. One may be able to use a kitchen knife for shaping
wood, but it is much better to use a chisel. In particular, I argue that even when
a model (or model component) turns out to be useful for more than one purpose,
it needs to be justified and judged with respect to *each* of the claimed purposes
separately (and it will probably require recoding). To extend the previous analogy,
a tool with the blade of a chisel but the handle of a kitchen knife may satisfy some
of the criteria for a tool to carve wood and some of the criteria for a tool to carve
cooked meat but fail at both. If one *did* come up with a new tool that is good at both,
this would be because it could be justified for each purpose separately.

In his paper 'Why Model?', Epstein (2008) lists 17 different reasons[2] for making
a model: from the abstract, 'discover new questions', to the practical 'educate
the general public'. This illustrates both the usefulness of modelling but also the
potential for confusion. As Epstein points out, the power of modelling comes from
making an informal set of ideas formal. That is, they are made precise using
unambiguous code or mathematical symbols. This lack of ambiguity has huge
benefits for the process of science, since it allows researchers to share, critique
and improve models without transmission errors (Edmonds 2010). However, in
many papers on modelling, the purpose that its model was developed for or, more
critically, the purpose under which it is being presented is often left implicit or
confused. Maybe this is due to the prevalence of the 'correspondence picture' of
modelling discussed above, maybe the authors conceive of their creations being
useful in many different ways, or maybe they simply developed the model without a
specific purpose in mind. However, regardless of the reason, the consequence is that
readers do not know how to judge the model when presented. This has the result that
models might avoid proper judgement—demonstrating partial success in different
ways with respect to a number of purposes, but not adequacy against any.

---

[1]With the exception of the purpose of description where a model is *intended* to reflect what is
observed

[2]He discusses 'prediction' and then lists 16 other reasons to model.

Our use of language helps cement this confusion: we talk about a 'predictive model' as if it something in the code that makes it predictive (forgetting all the work in directing and justifying this power)—rather I am suggesting a shift from the code as a thing in itself, to code as a tool for a particular purpose. This marks a shift from programming, where the focus is on the nature and quality of the *code*, to modelling, where the focus is on the *relationship* of the behaviour of some code to what is being modelled. Using terms such as 'explanatory model' is OK, as long as we understand that this is shorthand for 'a model which establishes an explanation' etc.

Producing, checking and documenting code are labour intensive. As a result, we often wish to reuse some code produced for one purpose for another purpose. However, this often causes as much new work as it saves due to the effort required to justify code for a new purpose and—if this is not done—the risk that time and energy of many researchers are wasted due to the confusions and false sense of reliability that can result. In practice, I have seen very little code that does not need to be rewritten when one has a new purpose in mind. Ideas can be transferred and well-honed libraries for very well-defined purposes, but not the core code that makes up a model of complex social phenomena.[3]

In this chapter, I will look at five common modelling purposes: prediction, explanation, theoretical exposition, description and illustration.[4] Each purpose is motivated, defined and illustrated. For each purpose, a 'risk analysis' is presented— some of the ways one might fail to achieve the stated purpose—along with some ways of mitigating these risks. In the penultimate section, some common confusions of purpose are illustrated and discussed, before the chapter concludes with a brief summary and plea to make one's purpose clear.

4.2 Prediction

4.2.1 Motivation

If one can reliably predict anything that is not already known, this is undeniably useful regardless of the nature of the model (e.g. whether its processes are a reflection of what happens in the observed system or not[5]). For instance, the gas laws (stating, e.g. that at a fixed pressure, the increase in volume of gas is proportional to the increase of temperature) were discovered long before the reason why they worked.

---

[3]I am not ruling out the possibility of reusable model components in the future using some clever protocol; it is just that I have not seen any good cases of code reuse and many bad ones.

[4]A later chapter (Chap. 28 (Edmonds et al. 2017)) takes a more fine-grained approach in the context of understanding human societies.

[5]It would not really matter even if the code had a bug in it, if the code reliably predicts (though it might impact upon the knowledge of *when* we can rely upon it or not).

However, there is another reason that prediction is valued: it is considered the gold standard of science—the ability of a model or theory to predict is taken as the most reliable indicator of a model's truth. This is done in two principle ways: (a) model A fits the evidence better than model B, a comparative approach,[6] or (b) model A is falsified (or not) by the evidence, a falsification approach. In either, the idea is that, given a sufficient supply of different models, better models will be gradually selected over time, either because the bad ones are discarded or outcompeted by better models.

Definition

By 'prediction', we mean the ability to reliably anticipate data that is not currently known to a useful degree of accuracy via computations using the model.

Unpacking this definition:

- It has to do it *reliably*—that is, under some known (but not necessarily precise) conditions, the model will work; otherwise one would not know when one could use it.
- The data it anticipates has to be *unknown* to the modeller. 'Predicting' out-of-sample data is not enough, since pressures to redo a model and get a better fit are huge and negative results are difficult to publish.
- The anticipation has to be to a *useful* degree of accuracy. This will depend upon the purpose to which it is being put, e.g. as in weather forecasting.

Unfortunately, there are at least two different uses of the word 'predict'. Almost all scientific models 'predict' in the weak sense of being used to calculate some result given some settings or data, but this is different from correctly anticipating *unknown* data. For this reason, some use the term 'forecast' for anticipating unknown data and use the word 'prediction' for almost any calculation of one aspect from another using a model. However, this causes confusions in other ways, so this does not necessarily make things clearer. *Firstly*, 'forecasting' implies that the unknown data is in the future (which is not always the case), and, *secondly*, large parts of science use the word 'prediction' for the process of anticipating unknown data. For example, if a modeller says their model 'predicts' something when they simply mean that it calculates it, then most of the audience may misunderstand and assume the author is claiming more utility than is intended.

As Watts (2014) points out, useful prediction does not have to be a 'point' prediction of a future event. For example, one might predict that some particular thing will not happen, the existence of something in the past (e.g. the existence of Pluto), something about the shape or direction of trends or distributions or even qualitative facts. The important fact is that what is being predicted is not known beforehand by the modeller and that it can be unambiguously checked when it is known.

An Example Nate Silver aims to predict social phenomena, such as the results of elections and the outcome of sports competitions. This is a data-hungry activity,

---

[6]Where model B may be a random or null model but also might be a rival model

which involves the long-term development of simulations that carefully see what can be inferred from the available data. As well as making predictions, his unit tries to establish the level of uncertainty in those predictions—being honest about the probability of those predictions coming about given the likely levels of error and bias in the data. As described in his book (Silver 2012), this involves a number of properties and activities, including:

- Repeated testing of the models against unknown data
- Keeping the models fairly simple and transparent so one can understand clearly what they are doing (and what they do not cover)
- Encoding into the model aspects of the target phenomena that one is relatively certain about (such as the structure of the US presidential electoral college)
- Being heavily data biased, requiring a lot of data to help eliminate sources of error and bias
- Producing probabilistic predictions, giving a good idea about the level of uncertainty in any prediction
- Being clear about what kinds of factors are not covered in the model, so the predictions are relative to a clear set of declared assumptions and one knows the kind of circumstances in which one might be able to rely upon the predictions

Post hoc analysis of predictions—explaining why it worked or not—is kept distinct from the predictive models themselves; this analysis may inform changes to the predictive model but is not then incorporated into the model. The analysis is thus kept independent of the predictive model, so it can be an effective check. Making a good predictive model requires a lot of time getting it wrong with real, unknown data and trying again before one approaches qualified successful predictions.

4.2.2 Risks

Prediction (as we define it) is *very* hard for any complex social system. For this reason, it is rarely attempted.[7] Many re-evaluations of econometric models against data that has emerged since publication have revealed a high rate of failure (e.g. Meese and Rogoff 1983)—37 out of 40 models failed completely. Clearly, although presented as being predictive models, they did not actually predict unknown data. Many of these used the strategy of first dividing the data into in-sample and out-of-sample data, and then parameterising the model on the former and exhibiting the fit against the latter. Presumably, the apparent fit of the 37 models was not simply a matter of bad luck, but that all of these models had been (explicitly or implicitly) fitted to the out-of-sample data, because the out-of-sample data was known to the modeller before publication. That is, if the model failed to fit the out-of-sample

---

[7]To be precise, some people have claimed to predict various social phenomena, but there are very few cases where the predictions are made public before the data is known and where the number of failed predictions can be checked. Correctly predicting events after they are known is much easier!

data the first time the model was tested, it was then adjusted until it did work, or, alternatively, only those models that fitted the out-of-sample data were published (a publishing bias). Thus, in these cases, the models were not tested against predicting the out-of-sample data even though they were presented as such. Fitting known data is simply not a sufficient test for predictive ability.

There are many reasons why prediction of complex social systems fails, but two of the most prominent are (1) it is unknown what processes are needed to be included in the model and (2) a lack of enough quality data of the right kinds. We will discuss each of these in turn.

1. In the physical sciences, there are often well-validated micro-level models (e.g. fluid dynamics in the case of weather forecasting) that tell us what processes are potentially relevant at a coarser level and which are not. In the social sciences, this is not the case—we do not know what the essential processes are. Here, it is often the case that there are other processes that the authors have not considered that, if included, would completely change the results. This is due to two different causes: (a) we simply do not know much about how and why people behave in different circumstances, and (b) different limitations of intended context will mean that different processes are relevant.
2. Unlike in the physical sciences, there has been a paucity of the kind of data we would need to check the predictive power of models. This paucity can be due to (a) there is not enough data (or data from enough independent instances) to enable the iterative checking and adapting of the models on new sets of unknown data each time we need to, or (b) the data is not of the right kind to do this. What can often happen is that one has partial sets of data that require some strong assumptions in order to compare against the predictions in question (e.g. the data might only be a proxy of what is being predicted, or you need assumptions in order to link sets of data). In the former case, (a), one simply has not enough to check the predictive power in multiple cases, so one has to suspend judgement as to whether the model predicts in general, until the data is available. In the latter case, (b), the success at prediction is relative to the assumptions made to check the prediction.

A more subtle risk is that the conditions under which one can rely upon a model to predict well might not be clear. If this is the case, then it is hard to rely upon the model for prediction in a new situation, since one does not know its conditions of application.

4.2.3   Mitigating Measures

To ensure that a model does indeed predict well, one can seek to ensure the following:

- That the model has been tested on several cases where it has successfully predicted data unknown to the modeller (at the time of prediction)

- That information about the following are included: exactly what aspects it predicts, guidelines on when the model can be used to predict and when not, some guidelines as to the degree or kind of accuracy it predicts with and any other caveats a user of the model should be aware of
- That the model code is distributed so others can explore when and how well it predicts

4.3   Explanation

4.3.1   Motivation

Often, especially with complex social phenomena, one is particularly interested in understanding *why* something occurs—in other words, explaining it. Even if one cannot predict something before it is known, you still might be able to explain it afterwards. This distinction mirrors that in the physical sciences where there are both *phenomenological* and *explanatory* laws (Cartwright 1983)—the former matches the data, whilst the latter explains why that came about. In mature science, predictive and explanatory laws are linked in well-understood ways but with less well-understood phenomena one might have one without the other. For example, the gas laws that link measurements of temperature, pressure and volume were known before the explanation in terms of molecules of gas bouncing randomly around and the formal connection between both accounts only made much later. Understanding is important for managing complex systems as well as understanding when predictive models might work. Whilst generally with complex social phenomena explanation is easier than prediction, sometimes prediction comes first (however, if one can predict then this invites research to explain *why* the prediction works).

If one makes a simulation in which certain mechanisms or processes are built in and the outcomes of the simulation match some (known) data, then this simulation can support an explanation of the data using the built-in mechanisms. The explanation itself is usually of a more general nature, and the traces of the simulation runs are examples of that account. Simulations that involve complicated processes can thus support complex explanations—that are beyond natural language reasoning to follow. The simulations make the explanation explicit, even if we cannot fully comprehend its detail. The formal nature of the simulation makes it possible to test the conditions and cases under which the explanation works and to better its assumptions.

Definition

> By 'explanation' we mean *establishing a possible causal chain from a set-up to its consequences* in terms of the mechanisms in a simulation.

Unpacking some parts of this:

- The *possible causal chain* is a set of inferences or computations made as part of running the simulation—in simulations with random elements, each run will be slightly different. In this case, it is either a possibilistic explanation (A could cause B), in which case one just has to show one run exhibiting the complete chain, or a probabilistic explanation (A probably causes B, or A causes a distribution of outcomes around B) in which case one has to look at an assembly of runs, maybe summarising them using statistics or visual representations.

- For explanatory purposes, the structure of the model is important, because that limits what the explanation consists of. If, for example, the model consisted of mechanisms that are known *not* to occur, any explanation one established would be in terms of these non-existent mechanisms—which is not very helpful. If one has parameterised the simulation on some in-sample data (found the values of the free parameters that made the simulation fit the in-sample data), then the explanation of the outcomes is also in terms of the in-sample data, mediated by these 'magic'-free parameters.[8]

- The *consequences* of the simulations are generally measurements of the outcomes of the simulation. These are compared with the data to see if it 'fits'. It is usual that only some of the aspects of the target data and the data the simulation produces are considered significant—other aspects might not be (e.g. might be artefacts of the randomness in the simulation or other factors extraneous to the explanation). The kind of fit between data and simulation outcomes needs to be assessed in a way that is appropriate to what aspects of the data are significant and which are not. For example, if it is the level of the outcome that is key, then a distance or error measure between this and the target data might be appropriate, but if it is the shape or trend of the outcomes over time that is significant, then other techniques will be more appropriate (e.g. Thorngate and Edmonds 2013).

Example Stephen Lansing spent time in Bali as an anthropologist, researching how the Balinese coordinated their water usage (among other things). He and his collaborator, James Kramer, build a simulation to show how the Balinese system of temples acted to regulate water usage, through an elaborate system of agreements between farmers, enforced through the cultural and religious practices at those temples (Lansing and Kramer 1993). Although their observations could cover many instances of localities using the same system of negotiation over water, they were necessarily limited to all their observations being within the same culture. Their simulation helped establish the nature and robustness of their explanation by exploring a close universe of 'what if' questions, which vividly showed the comparative advantages of the observed system that had developed over a considerable period. The model does not predict that such systems will develop in the same circumstances, but it substantially adds to the understanding of the observed case.

---

[8]I am being a little disparaging here, it may be that these have a definite meaning in terms of relating different scales or some such, but too often, they do not have any clear meaning but just help the model fit stuff.

4.3.2 Risks

Clearly, there are several risks in the project of establishing a complex explanation using a simulation—what counts as a good explanation is not as clear-cut as what is a good prediction.

Firstly, the fit to the target data to be explained might be a very special case. For example, if many other parameters need to have very special values for the fit to occur, then the explanation is, at best, brittle and, at worst, an accident.

Secondly, the process that is unfolded in the simulation might be poorly understood so that the outcomes might depend upon some hidden assumption encapsulated in the code. In this case, the explanation is dependent upon this assumption holding, which is problematic if this assumption is very strong or unlikely.

Thirdly, there may be more than one explanation that fits the target data. So although the simulation establishes one explanation, it does not guarantee that it is the only candidate for this.

4.3.3 Mitigating Measures

To improve the quality and reliability of the explanation being established:

- Ensure that the mechanisms built into the simulation are plausible or at least relate to what is known about the target phenomena in a clear manner.
- Be clear about which aspects of the outcomes are considered significant in terms of comparison to the target data—i.e. exactly which aspects of that target data are being explained.
- Probe the simulation to find out the conditions for the explanation holding using sensitivity analysis, addition of noise, multiple runs, changing processes not essential to the explanation to see if the results still hold and documenting assumptions.
- Do experiments in the classic way, to check that the explanation does, in fact, hold for your simulation code—i.e. check your code and try to refute the explanation using carefully designed experiments with the model.

4.4 Theoretical Exposition

4.4.1 Motivation

If one has a mathematical model, one can do analysis upon its mathematics to understand its general properties. This kind of analysis is both easier and harder with a simulation model—to find out the properties of simulation code, one just has to run the code—but this just gives one possible outcome from one set of initial parameters. Thus, there is the problem that the runs one sees might not be representative of the behaviour in general. With complex systems, it is not easy to understand how the outcomes arise, even when one knows the full and correct specification of their processes, so simply knowing the code is not enough. Thus, with highly complicated processes, where the human mind cannot keep track of the parts unaided, one has the problem of understanding how these processes unfold in general.

Where mathematical analysis is not possible, one has to explore the theoretical properties using simulation—this is the goal of this kind of model. Of course, with many kinds of simulation, one wants to understand how its mechanisms work, but here this is the only goal. Thus, this purpose could be seen as more limited than the others, since some level of understanding the mechanisms is necessary for the other purposes (except maybe black-box predictive models). However, with this focus on just the mechanisms, there is an expectation that a more thorough exploration will be performed—*how* these mechanisms interact and *when* they produce different kinds of outcome.

Thus, the purpose here is to give some more general idea of how a set of mechanisms work, so that modellers can understand them better when used in models for other purposes. If the mechanisms and exploration are limited, this would greatly reduce the usefulness of doing this. General insights are what is wanted here.

In practice, this means a mixture of inspection of data coming from the simulation, experiments and maybe some inference upon or checking of the mechanisms. In scientific terms, one makes a hypothesis about the working of the simulation— why some kinds of outcome occur in a given range of conditions—and then tests that hypothesis using well-directed simulation experiments.

The complete set of simulation outcomes over all possible initialisations (including random seeds) does encode the complete behaviour of simulation, but that is too vast and detailed to be comprehensible. Thus, some general truths covering the important aspects of the outcomes under a given range of conditions are necessary— the complete and certain generality established by mathematical analysis might be infeasible with many complex systems, but we would like something that approximates this using simulation experiments.

Definition

> 'Theoretical exposition' means discovering then *establishing (or refuting) hypotheses* about the *general behaviour* of a set of mechanisms (using a simulation).

Unpacking some key aspects here:

- One may well spend some time illustrating the discovered hypothesis (especially if it is novel or surprising), followed by a sensitivity analysis, but the crucial part is showing these hypotheses are refuted or not by a sequence of simulation experiments.
- The hypotheses need to be (at least somewhat) general to be useful.

- A use of theoretical exposition can be to refute a hypothesis, by exhibiting a concrete counterexample, or to establish a hypothesis.
- Although any simulation has to have some meaning for it to be a model (otherwise it would just be some arbitrary code), this does not involve any other relationship with the observed world in terms of data or evidence.

Example Schelling developed his famous model for a theoretical purpose. He was advising the Chicago district on what might be done about the high levels of segregation there. The assumption was that the sharp segregation observed must be a result of strong racial discrimination by its inhabitants. Schelling's model (Schelling 1969, 1971) showed that segregation could result from just weak preferences of inhabitants for their own kind—that even, a wish for 30% of people of the same trait living in the neighbourhood could result in segregation. This was not obvious without building a model, and Schelling did not rely on the results of his model alone but did extensive mathematical analysis to back up its conclusions.

What the model did not do is say anything about what actually caused the segregation in Chicago—it might well be the result of strong racial prejudice. The model did not predict anything about the level of segregation nor did it explain it. All it did was provide a counterexample to the current theories as to the cause of the segregation, showing that this was not *necessarily* the case.

4.4.2 Risks

In theoretical exposition, one is not relating simulations to the observed world, so it is fundamentally an easier and 'safer' activity.[9] Since a near-complete understanding of the simulation behaviour is desired, this activity is usually concerned with relatively simple models. However, there are still risks—it is still easy to fool oneself with one's own model. Thus, the main risk is that there is a bug in the code, so that what one thinks one is establishing about a set of mechanisms is really about a different set of mechanisms (i.e. those including the bug).

A second area of risk lies in a potential lack of generality or 'brittleness' of what is established. If the hypothesis is true but only holds under very special circumstances, then this reduces the usefulness of the hypothesis in terms of understanding the simulation behaviour.

Lastly, there is the risk of over-interpreting the results in terms of saying anything about the observed world. The model might suggest a hypothesis about the observed world, but it does not provide any level of empirical support for this.

---

[9]In the sense of not being vulnerable to being shown to be wrong later

4.4.3   Mitigating Measures

The measures that should be taken for this purpose are quite general and maybe best understood by the community of simulators.

- One needs to check ones' code thoroughly—see Galán et al. (2017) for a review of techniques.
- One needs to be precise about the code and its documentation—the code should be made publically available.
- Be clear as to the nature and scope of the hypotheses established.
- A very thorough sensitivity check, trying various versions with extra noise added etc.
- It is good practice to illustrate the simulation so that the readers understand its key behaviours but then follow this with a series of attempted refutations of the hypotheses about its behaviour to show its robustness.
- Be very careful about not claiming that this says anything about the observed world.

4.5   Description

4.5.1   Motivation

An important, but currently under-appreciated, activity in science is that of description. Charles Darwin spent a long time sketching and describing the finches he observed on his travels aboard the HMS Beagle. These descriptions and sketches were not measurements or recordings in any direct sense, since he was already selecting from what he perceived and only recording an abstraction of what he thought of as relevant. Later on, these were used to illustrate and establish his theoretical abstraction—his theory of evolution of species by natural selection.

One can describe things using natural language or pictures, but these are inadequate for dynamic and complex phenomena, where the essence of what is being described is how several mechanisms might relate over time. An agent-based simulation framework allows for a direct representation (one agent for one actor) without theoretical restrictions. It allows for dynamic situations as well as complex sets of entities and interactions to be represented (as needed). This can make it an ideal complement to scenario development because it ensures consistency between all the elements and the outcomes. It is also a good base for future generalisations when the author can access a set of such descriptive simulations.

Definition

A description (using a simulation) is an attempt to *partially represent* what is important of *a specific observed case* (or small set of closely related cases).

Unpacking some of this:

- This is not an attempt to produce a one-one representation of what is being observed but only of the features thought to be relevant for the intended kind of study. It will leave out some features; in particular, it may leave out some of the interactions between processes.
- It is not in any sense general, but it seeks to capture a restricted set of cases—it is specific to these, and no kind of generality beyond these can be assumed.
- The simulation has to relate in an explicit and well-documented way to a set of evidence, experiences and data. This is the opposite of theoretical exposition and should have a direct and immediate connection with observation, data or experience.

Example In Moss (1998), Scott Moss describes a model that captures some of the interactions in a water pumping station during crises. This came about through extensive discussions with stakeholders within a UK water company about what happens in particular situations during such crises. The model sought to directly reflect this evidence within the dynamic form of a simulation, including cognitive agents who interact to resolve the crisis. This simulation captured aspects of the physical situation but also tackled some of the cognitive and communicative aspects. To do this, he had represented the problem solving and learning of key actors, so he inevitably had to use some existing theories and structures—namely, Alan Newell and Herbert Simon's 'general problem solving architecture' (Newell and Simon 1972) and Cohen's 'endorsement mechanism' (Cohen 1984a, b). However, this is all made admirably explicit in the paper. The paper is suitably cautious in terms of any conclusions, saying that the simulation 'indicate[s] a clear need for an investigation of appropriate organizational structures and procedures to deal with full-blown crises'.

4.5.2 Risks

Any system for representation will have its own affordances—it will be able to capture some kinds of aspect much more easily than others will. This inevitably biases the representations produced, as those elements that are easy to represent are more likely to be captured than those which are more difficult. Thus, the medium will influence what is captured and what is not.

Since agent-based simulation is not theoretically constrained,[10] there are a large number of ways in which any observed phenomena could be expressed in terms of simulation code. Thus, it is almost inevitable that any modeller will use some

---

[10]To be precise, it does assume there are discrete entities or objects and that there are processes within these that can be represented in terms of computations, but these are not very restrictive assumptions.

structures or mechanisms that they are familiar with in order to write the code. Such a simulation is, in effect, an *abduction* with respect to these underlying structures and mechanisms—the phenomena are seen through these and expressed using them.

Finally, a reader of the simulation may not understand the limitations of the simulation and make false assumptions as to its generality. In particular, the inference within the simulations may not include all the processes that are in what is observed—thus, it cannot be relied upon to either predict outcomes or justify any specific explanation of those outcomes.

4.5.3   Mitigating Measures

As long as the limitations of the description (in terms of its selectivity, inference and biases) are made clear, there are relatively few risks here, since not much is being claimed. If it is going to be useful in the future as part of a (slightly abstracted) evidence base, then its limitations and biases do need to be explicit. The data, evidence or experience it is based upon also need to be made clear. Thus, good documentation is the key here—one does not know how any particular description will be used in the future, so the thoroughness of this is key to its future utility. Here, it does not matter if the evidence is used to specify the simulation or to check it afterwards in terms of the outcomes, all that matters is that the way it relates to evidence is well documented. Standards for documentations (such as the ODD and its various extensions (Grimm et al. 2006, 2010) help ensure that all aspects are covered.

4.6   Illustration

4.6.1   Motivation

Sometimes one wants to make an idea clear, and an illustration is a good way of doing this. It makes a more abstract theory or explanation clear by exhibiting a concrete example that might be more readily comprehended. Complex systems, especially complex social phenomena, can be difficult to describe, including multiple independent and interacting mechanisms and entities. Here a well-crafted simulation can help people see these complex interactions at work and hence appreciate these complexities better. As with description, this purpose does not claim much; it is just a medium for the communication of an idea. If the theory is already instantiated as a simulation (e.g. for theoretical exposition or explanation), then the illustrative simulation might well be a simplified version of this.

Playing about with simulations in a creative but informal manner can be very useful in terms of informing the intuitions of a researcher (Norling et al. 2017).

In a sense, the simulation has illustrated an idea to its creator. One might then exhibit a version of this simulation to help communicate this idea to others. However, this does not mean that the simulation achieves any of the other purposes described above, and it is thus doubtful whether that idea has been established to be of public value (justifying its communication in a publication) until this happens.

This is not to suggest that illustration is not an important process in science. Providing new ways of thinking about complex mechanisms or giving us new examples to consider is a very valuable activity. However, this does not imply its adequacy for any other purpose.

Definition

> An illustration (using a simulation) is to communicate or make clear an idea, theory or explanation.

Unpacking this:

- Here the simulation does not have to fully express what it is illustrating; it is sufficient that it gives a simplified example. So it may not do more than partially capture the idea, theory or explanation that it illustrates, and it cannot be relied upon for the inference of outcomes from any initial conditions or set-up.
- The clarity of the illustration is of overriding importance here, not its veracity or completeness.
- An illustration should not make any claims, even of being a description. If it is going to be claimed that it is useful as a theoretical exposition, explanation or other purposes, then it should be justified using those criteria—that it seems clear to the modeller is not enough.

Example In his book, Axelrod (1984) describes a formalised computational 'game' where different strategies are pitted against each other, playing the iterated prisoner's dilemma. Some different scenarios are described, where it is shown how the 'tit for tat' strategy can survive against many other mixes of strategies (static or evolving). The conclusions are supported by some simple mathematical considerations, but the model and its consequences were not explored in any widespread manner.[11] In the book, the purpose of the model is to illustrate the ideas that the book proposes. The book claims the idea 'explains' many observed phenomena, but in an analogical manner, no precise relationship with any observed measurements is described. There is no validation of the model here or in the more academic paper that described these results (Axelrod and Hamilton 1981). In the academic paper, there are some mathematical arguments which show the plausibility of the model, but the paper, like the book, progresses by showing the idea is coherent with some reported phenomena—but it is the ideas rather than the model that are so related. Thus, in this case, the simulation model is an analogy to support the idea, which is related to evidence in a qualitative manner—the relationship of the model to evidence is indirect (Edmonds 2001). Thus, the role of the simulation model is that

---

[11] Indeed, the work spawned a whole industry of papers doing just such an exploration.

of an illustration of the key ideas and does not qualify for either explaining specific data, predicting anything unknown or exploring a theory.

4.6.2 Risks

The main risk here is that you might deceive people using the illustration into reading more into the simulation than is intended, as these are often quite persuasive in terms of their impact. Such simulations can be used as a kind of analogy—a way of thinking about other phenomena. However, just because you can think about some phenomena in a particular way does not make it true. The human mind is good at creating, 'on the fly', connections between an analogy and what it is considering— so good that it does it almost without us being aware of this process. The danger here is of confusing being able to think of some phenomena using an idea and that idea having any force in terms of a possible explanation or method of prediction. The apparent generality of an analogy tends to dissipate when one tries to precisely specify the relationship of a model to observations, since an analogy has a different set of relationships for each situation it is applied to—it is a supremely flexible way of thinking. This flexibility means that it does not work well to support an explanation or predict well, since both of these necessitate an explicit and fixed relationship with observed data.

There is also a risk of confusion if it is not clear which aspects are important to the illustration and which are not. A simulation for illustration will show the intended behaviour, but (unlike when its theory is being explored) it has been tested only for a restricted range of possibilities; indeed the claimed results might be quite brittle to insignificant changes in assumption.

4.6.3 Mitigating Measures

Be very clear in the documentation that the purpose of the simulation is for illustration only, maybe giving pointers to fuller simulations that might be useful for other purposes. Also be clear in precisely what idea is being communicated and so which aspects of the simulation are relevant for this purpose.

4.7 Some Confusions of Purpose

It should be abundantly clear by now that establishing a simulation for one purpose does not justify it for another and that any assumptions to the contrary risk confusion and unreliable science. However, the field has many examples of such confusions and conflations, so this message is obviously needed. It is true that a simulation model justified for one purpose might be used as *part* of the development of a simulation model for another purpose—this can be how science progresses. However, just because a model for one purpose *suggests* a model for another does not mean it is a good model for the new purpose. If it is being suggested that a model can be used for a new purpose, it has to be justified for this new purpose. To drive home this point further, we look at some common confusions of purpose to underline this danger. Each time some code is mistakenly relied upon for a purpose other than has been established for it.

1. *Theoretical exposition* → *Explanation*. Once one has immersed oneself in a model, there is a danger that the world looks like this model to its author. This is a strong kind of Kuhn's 'theoretical spectacles'[12] and results from the intimate relationship that simulation developers have with their model. Here, the temptation is to jump from a theoretical exposition, which has no empirical basis, to an explanation of something in the world. A simulation can provide a way of looking at some phenomena, but just because one can view some phenomena in a particular way does not make it a good explanation. Of course, one can form a hypothesis from anywhere, including from a theoretical exposition, but it remains only a hypothesis until it is established as a good explanation as discussed above (which would almost certainly involve changing the model).

2. *Description* → *Explanation*. In constructing a simulation for the purpose of describing a small set of observed cases, one has deliberately made many connections between aspects of the simulation and evidence of various kinds. Thus, one can be fairly certain that, at least, some of its aspects are realistic. Some of this fitting to evidence might be in the form of comparing the outcomes of the simulation to data, in which case it is tempting to suggest that the simulation supports an explanation of those outcomes. The trouble with this is twofold: (a) the work to test *which* aspects of that simulation are relevant to the aspects being explained has not been done; and (b) the simulation has not been established against a range of cases—it is not general enough to make a good explanation. An explanation that only explains aspects of a small number of cases using a complex simulation is a bad explanation since there will be many other potentialities in the simulation that are not used for these few cases.

3. *Explanation* → *Prediction*. A simulation that establishes an explanation traces a (complex) set of causal steps from the simulation set-up to outcomes that compare well with observed data. It is thus tempting to suggest that one can use this simulation to predict this observed data. However, the process of using a simulation to establish and understand an explanation inevitably involves iteration between the data being explained and the model specification—that is, the model is fitted to that particular set of data. Model fitting is not a good way to construct a model useful for prediction, since it does not distinguish between

---

[12]Kuhn (1962) pointed out the tendency of scientists to only see the evidence that is coherent with an existing theory—it is as if they have 'theoretical spectacles' that filter out other kinds of evidence.

what is essential for the prediction and the 'noise' (what cannot be predicted). Establishing that a simulation is good for prediction requires its testing against unknown data several times—this goes way beyond what is needed to establish a candidate explanation for some phenomena. This is especially true for social systems, where we often cannot predict events, but we can explain them after they have occurred.

4. *Illustration → Theoretical exposition.* A neat illustration of an idea suggests a mechanism. Thus, the temptation is to use a model designed as an illustration or playful exploration as being sufficient for the purpose of a theoretical exposition. A theoretical exposition involves the extensive testing of code to check the behaviour and the assumptions therein; an illustration, however suggestive, is not that rigorous. For example, it may be that an illustrated process is a very special case and only appears under very particular circumstances, or it may be that the outcomes were due to aspects of the simulation that were thought to be unimportant (such as the nature of a random number generator). The work to rule out these kinds of possibility is what differentiates using a simulation as an illustration from a theoretical exposition.

There is a natural progression in terms of purpose attempted as understanding develops: from illustration to description or theoretical exposition, from description to explanations and from explanations to prediction. However, each stage requires its own justification and probably a complete reworking of the simulation code for this new purpose. It is the lazy assumption that one purpose naturally follows from another that is the danger.

4.8   Conclusion

In Table 4.1, we summarise the most important points of the above discussion. This does not include all the risks of each kind of model but simply picks the most pertinent ones.

As should be clear from the above discussion, being clear about one's purpose in modelling is central to how one goes about developing, checking and presenting the results. Different modelling purposes imply different risks and hence activities to avoid these. If one is intending the simulation to have a public function (in terms of application or publication), then one should not model with unspecified or conflated purposes.[13] Confused, conflated or unclear modelling purpose leads to unreliable models that are hard to check, can create deeply misleading results and is hard for readers to judge—in short, it is a recipe for bad science.

---

[13]This does not include private modelling, whose purpose maybe playful or exploratory; however, in this case one should not present the results or model as if they have achieved anything more than illustration (to oneself). If one finds something of value in the exploration, it should then be redone properly for a particular purpose to be sure it is worth public attention.

Table 4.1 A brief summary of the discussed modelling purposes

Modelling purpose	Essential features	Particular risks (apart from that of lacking the essential features)
Prediction	Anticipates *unknown* data	Conditions of application unclear
Explanation	Uses *plausible* mechanisms to match outcome data in a *well-defined* manner	Model is brittle, so minor changes in the set-up result in bad fit to explained data
Theoretical exposition	Systematically maps out or establishes the consequences of some mechanisms	Bugs in the code; inadequate coverage of possibilities
Description	Relates directly to evidence for a small set of cases	Unclear documentation; over generalisation from cases described
Illustration	Shows an idea clearly	Over interpretation to make theoretical or empirical claims

Acknowledgements Many thanks to all those with whom I have discussed these matters, including Scott Moss, David Hales, Bridget Rosewell and all those who attended the workshop on validation held in Manchester.

Further Reading

Epstein, J. M. (2008). Why model? Journal of Artificial Societies and Social Simulation, 11(4). 12. http://jasss.soc.surrey.ac.uk/11/4/12.html
This gives a brief tour of some of the reasons to simulate other than that of prediction.
Edmonds, B., Lucas, P., Rouchier, J., & Taylor, R. (2017). Understanding human societies. doi:https://doi.org/10.1007/978-3-319-66948-9_28.
In this chapter, some modelling purposes that are specific to human social phenomena are examined in more detail giving examples from the literature.

References

Axelrod, R. (1984). *The evolution of cooperation*. New York, NY: Basic Books.
Axelrod, R., & Hamilton, W. D. (1981). The evolution of cooperation. *Science, 211*, 1390–1396.
Cartwright, N. (1983). *How the laws of physics lie*. Oxford: Oxford University Press.
Cohen, P. R. (1984a). Heuristic reasoning about uncertainty: an artificial intelligence approach. *International Journal of Approximate Reasoning, 1*(2), 243–245.
Cohen, P. R. (1984b). *Heuristic reasoning about uncertainty: an artificial intelligence approach*. Marshfield, MA: Pitman Publishing.
Edmonds, B. (2001). The use of models - making MABS actually work. In S. Moss & P. Davidsson (Eds.), *Multi agent based simulation, Lecture notes in artificial intelligence* (Vol. 1979, pp. 15–32). Berlin: Springer-Verlag.

Edmonds, B. (2010). Bootstrapping knowledge about social phenomena using simulation models. *Journal of Artificial Societies and Social Simulation, 13*(1), 8. http://jasss.soc.surrey.ac.uk/13/1/8.html

Edmonds, B., Lucas, P., Rouchier, J., & Taylor, R. (2017). Understanding human societies. doi:https://doi.org/10.1007/978-3-319-66948-9_28.

Epstein, J. M. (2008). Why model? *Journal of Artificial Societies and Social Simulation, 11*(4), 12. http://jasss.soc.surrey.ac.uk/11/4/12.html

Galán, J. M., Izquierdo, L. R., Izquierdo, S. S., Santos, J. I., del Olmo, R., & López-Paredes, A. (2017a). Checking simulations: Detecting and avoiding errors and artefacts. doi:https://doi.org/10.1007/978-3-319-66948-9_7.

Grimm, V., Berger, U., Bastiansen, F., Eliassen, S., Ginot, V., Giske, J., et al. (2006). A standard protocol for describing individual-based and agent-based models. *Ecological Modelling, 198*, 115–126.

Grimm, V., Berger, U., DeAngelis, D. L., Polhill, J. G., Giske, J., & Railsback, S. F. (2010). The ODD protocol: A review and first update. *Ecological Modelling, 221*, 2760–2768.

Kuhn, T. S. (1962). *The structure of scientific revolutions*. Chicago, IL: University of Chicago Press.

Lansing, J. S., & Kramer, J. N. (1993). Emergent properties of balinese water temple networks: coadaptation on a rugged fitness landscape. *American Anthropologist, 1*, 97–114.

Meese, R. A., & Rogoff, K. (1983). Empirical exchange rate models of the seventies - Do they fit out of sample? *Journal of International Economics, 14*, 3–24.

Moss, S. (1998). Critical incident management: An empirically derived computational model. *Journal of Artificial Societies and Social Simulation, 1*(4), 1. http://jasss.soc.surrey.ac.uk/1/4/1.html

Newell, A., & Simon, H. A. (1972). *Human problem solving*. Englewood Cliffs, NJ: Prentice-Hall.

Norling, E., Meyer, R., & Edmonds, B. (2017). Informal approaches to developing simulations. doi:https://doi.org/10.1007/978-3-319-66948-9_5.

Schelling, T. C. (1969). Models of segregation. *The American Economic Review, 59*(2), 488–493.

Schelling, T. C. (1971). Dynamic models of segregation. *Journal of Mathematical Sociology, 1*(2), 143–186.

Silver, N. (2012). *The signal and the noise: the art and science of prediction*. London: Penguin.

Thorngate, W., & Edmonds, B. (2013). Measuring simulation-observation fit: An introduction to ordinal pattern analysis. *Journal of Artificial Societies and Social Simulation, 16*(2), 14. http://jasss.soc.surrey.ac.uk/16/2/4.html

Watts, D. J. (2014). Common sense and sociological explanations. *American Journal of Sociology, 120*(2), 313–351.

Part II
Methodology

Chapter 5
Informal Approaches to Developing Simulation Models

Emma Norling, Bruce Edmonds, and Ruth Meyer

Abstract This chapter describes an approach commonly taken by most people in the social sciences when developing simulation models instead of following a formal approach of specification, design and implementation. What often seems to happen in practice is that modellers start off in a phase of exploratory modelling, where they don't have a precise conception of the model they want but a series of ideas and/or evidence they want to capture. They then may develop the model in different directions, backtracking and changing their ideas as they go. This phase continues until they think they may have a model or results that are worth telling others about. This then is (or at least should be) followed by a consolidation phase where the model is more rigorously tested and checked so that reliable and clear results can be reported. In a sense what happens in this later phase is that the model is made so that it is *as if* a more formal and planned approach had been taken.

There is a danger of this approach: that the modeller will be tempted by apparently significant results to rush to publication before sufficient consolidation has occurred. There may be times when the exploratory phase may result in useful and influential personal knowledge, but such knowledge is not reliable enough to be up to the more exacting standards expected of publicly presented results. Thus, it is only in combination with a careful consolidation of models that this informal approach to building simulations should be undertaken.

Why Read This Chapter?
To get to know some of the issues, techniques and tools involved in building simulation models using a combination of exploration, checking and consolidation. To understand when a looser, informal style of development might be beneficial and when one needs a more structured approach.

E. Norling (✉)
School of Computing, Mathematics and Digital Technology, Manchester Metropolitan University, Manchester, UK
e-mail: norling@acm.org

B. Edmonds • R. Meyer
Centre for Policy Modelling, Manchester Metropolitan University, All Saints Campus, Oxford Road, Manchester, M1 6BH, UK

© Springer International Publishing AG 2017

B. Edmonds, R. Meyer (eds.), *Simulating Social Complexity*,
Understanding Complex Systems, https://doi.org/10.1007/978-3-319-66948-9_5

5.1 Introduction: Exploration and Consolidation Modelling Phases

Formal approaches to the development of computer programs have emerged through the collective experience of computer scientists (and other programmers) over the past half-century. The experience has shown that complex computer programs are very difficult to understand: once past a certain point, unless they are very careful, programmers lose control over the programs they build. Beyond a certain stage of development, although we may understand each part—each micro-step—completely, we can lose our understanding of the program as a whole; the effects of the interactions between the parts of a program are unpredictable; they are emergent. Thus, computer science puts a big emphasis on techniques that aim to ensure that the program does what it is intended to do as far as possible. However, even with the most careful methodology, it is recognised that a large chunk of time will have to be spent debugging the program—we all *know* that a program cannot be relied on until it has been tested and fixed repeatedly.

However, it is fair to say that most computational modellers do not follow such procedures and methodologies all the time (although since people don't readily admit to how messy their implementation process actually is, we cannot know this, just as one does not know how messy people's homes are when there are no visitors). There are many reasons for this. Obviously, those who are not computer scientists may simply not know these techniques (in which case they should at least read Chap. 6 in this volume). Then there are a large number of modellers who know of these techniques (to some degree) but judge that they are not necessary or not worth the effort. Such a judgement may or may not be correct. Certainly it is the case that people tend to underestimate the complexity of programming and so think they can get away with not bothering with a more careful specification and analysis stage. In some of these cases, the modeller may regret not engaging in more planning, but there may also be other times when there are good reasons not to follow such techniques. Thirdly, a specification and design approach is simply not possible if you don't have a clear idea of your goal. Often, when modelling some complex phenomena (and especially social phenomena), one simply does not know beforehand which parts of the system will turn out to be important to the outcomes and which can be safely omitted. Further, one may not even know what will be possible to model computationally.

One of the big benefits of modelling phenomena computationally is that one learns a lot about what is crucial and possible *in the process of building a simulation model*. This is very unlike the case where one has a functional goal or specification for a program that can be analysed into sub-goals and processes, etc. In (social) simulation, the degree to which formal approaches are useful depends somewhat on the goal of modelling. If the goal is very specific, for example, understanding the effect of the recovery rate on the change in the number of infections in an epidemic, and the basic model structure is known, then what is left is largely an *engineering* challenge. However, if the goal is general understanding of a particular process, then there is no possible way of *systematically* determining what the model should be. Here the modelling is essentially a *creative* process, and the development of the model proceeds in parallel with the development of the understanding of the process; the model is itself a theory under development.

Thus, what often seems to happen in practice is that modellers start off in a phase of exploratory modelling, where they don't have a precise conception of the model they want but a series of ideas and/or evidence they want to capture. They then may develop the model in different directions, backtracking and changing their ideas as they go. This phase continues until they think they may have a model or results that are worth telling others about. This then is (or at least should be) followed by a consolidation phase where the model is more rigorously tested and checked so that reliable and clear results can be reported. In a sense what happens in this later phase is that the model is made so that it is *as if* a more formal and planned approach had been taken.

There is nothing wrong with having an exploratory approach to model development. Unfortunately, it is common to see models and results that are publicly presented without a significant consolidation phase being undertaken. It is very understandable why a researcher might want to skip the consolidation phase: they may have discovered a result or effect that they find exciting and not wish to go through the relatively mundane process of checking their model and results. They may *feel* that they have discovered something that is of more general importance; however, this *personal* knowledge, which may well inform their understanding, is not yet of a standard that makes it worthwhile for their peers to spend time understanding, *until* it has been more rigorously checked.

One of the problems with the activity of modelling is that it *does* influence how the modeller thinks. Paradoxically, this can also be one of the advantages of this approach. After developing and playing with a model over a period of time, it is common to "see" the world (or at least the phenomena of study) in terms of the constructs and processes of that model. This is a strong version of Kuhn's "theoretical spectacles" (Kuhn 1969). Thus, it is common for modellers to be convinced that they have found a *real* effect or principle during the exploration of a model, despite not having subjected their own model and conception to sufficient checking and testing—what can be called *modelling spectacles*. Building a model in a computer is almost always in parallel with the development of one's ideas about the subject being modelled. This is why it is almost inevitable that we think about the subject in terms of our models—this is at once a model's huge advantage but also disadvantage. As long as one is willing to be aware of the modelling spectacles and be critical of them, or try many different sets of modelling spectacles, the disadvantage can be minimised.

Quite apart from anything, presenting papers with no substantial consolidation is unwise. Such papers are usually painfully obvious when presented at workshops and easily criticised by referees and other researchers if submitted to a journal. It is socially acceptable that a workshop paper will not have as much consolidation as might be required of a journal article, since the criticism and evaluation of ideas and models at a workshop are part of its purpose, but presenting a model with an inadequate level of consolidation just wastes the other participants' time.

Fig. 5.1 The exploration and consolidation approach to model development

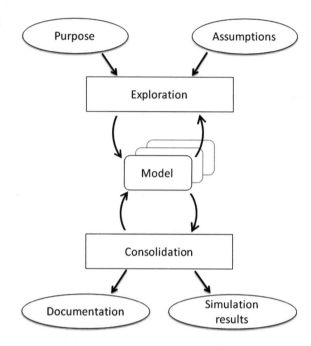

What steps then should modellers who follow such an informal approach take to ensure that their model *is* sufficiently consolidated to present to a wider audience? Firstly, the modeller must have a clear purpose for their model, as described below. Secondly, the modeller must be careful to identify the assumptions that are made during the construction of the model. Thirdly, the modeller must maintain control of the model whilst exploring different possibilities. And fourthly—and this is perhaps the most difficult—the modeller must maintain an understanding of the model. The following sections of this chapter discuss these points in more detail. Then there is the all-important consolidation phase (which may proceed in parallel with the former steps, rather than strictly sequentially), during which the modeller formalises the model in order to ensure that the results are sound and meaningful. Figure 5.1 illustrates this approach to model building.

5.2 Knowing the Purpose of the Model

There are many possible purposes for constructing a model. Although some models might be adapted for different purposes without too much difficulty, at any one time, a model will benefit from having a clear purpose. One of the most common criticisms of modelling papers (after a lack of significant consolidation) is that the author has made a model but is unclear as to its purpose. This can be for several reasons, such as:

- The author may have simply modelled without thinking about why (e.g. having vague ideas about a phenomenon, the modeller decides to construct a model without thinking about the questions one might want to answer about that phenomenon).
- The model might have been developed for one purpose but is being presented as if it had another purpose.
- The model may not achieve any particular purpose and so the author might be forced into claiming a number of different purposes to justify the model.

The purpose of a model will affect how it is judged and hence *should* influence how it is developed and checked.

The classic reason for modelling is to predict some unknown aspect of observed phenomena—usually a future aspect. If you can make a model that does this for unknown data (data not known to the modellers before they published the model), then there can be no argument that such a model is (potentially) useful. Due to the fact that predictive success is a very strong test of a model *for which the purpose is prediction*, this frees one from an *obligation* as to the content or structure of the model.[1] In particular, the assumptions in the model can be quite severe—the model can be extremely abstract as long as it actually does predict.

However, predictive power will not *always* be a measure of a model's success. There are many other purposes for modelling other than prediction. Epstein (2008) lists 16 other purposes for building a model, e.g. explanation, training of practitioners or education of the general public, and it is important to note that the measure of success will vary depending on the purpose.

With an explanatory model, if one has demonstrated that a certain set of assumptions can result in a set of outcomes (e.g. by exhibiting an acceptable fit to some outcome data), this shows that the modelled process is a *possible* explanation for those outcomes. Thus, the model generates an explanation, but only in terms of the assumptions in the setup of the simulation. If these assumptions are severe ones, i.e. the model is very far from the target phenomena, the explanation it suggests in terms of the modelled process will not correspond to a real explanation in terms of observed processes. The chosen assumptions in an explanatory model are crucial to its purpose in contrast to the case of a predictive model—this is an example of how the purpose of a model might greatly influence its construction.

It does sometimes occur that a model made for one purpose can be adapted for another, but the results are often not of the best quality, and it almost always takes more effort than one expects. In particular, using someone else's model is usually not very easy, especially if you are not able to ask the original programmer questions about it and/or the code is not very well documented.

---

[1] Of course a successfully predictive model raises the further question of *why* it is successful, which may motivate the development of further *explanatory* models, since a complete scientific understanding requires both prediction and explanation, but not necessarily from the same models (Cartwright 1983).

Chapter 4 in this volume (Edmonds 2017) goes into five common modelling purposes in more detail, with analyses of the particular risks for each kind of purpose, and the basic steps to mitigate these risks.

5.3 Modelling Assumptions

Whilst the available evidence will directly inform some parts of a model design, other parts will not be so well informed. In fact, it is common that a large part of a complex simulation model is not supported by hard evidence. The second source for design decisions is the conceptions of the modeller, which may have come from ideas or structures that are available in the literature. However, this is *still* not sufficient to get a model working. In order to get a simulation model to run and produce results, it will be necessary to add in all sorts of other details: these might include adding in a random process to "stand in" for an unknown decision or environmental factor, or even be a straight "kludge" because you don't know how else to program something. Even when evidence supports a part of the design, there will necessarily be some interpretation of this evidence. Thus, any model is dependent upon a whole raft of assumptions of different kinds.

If a simulation depends on many assumptions that are not relatable to the object or process it models, it is unlikely to be useful. However, just because a model has *some* assumptions in it, this does not mean it should be disregarded. *Any* modelling is necessarily a simplification of reality, done within some context or other. Hence, there will be the assumption that the details left out are not crucial to the aspect of the results deemed important, as well as those assumptions that are inherent in the specification of the context. This is true for any kind of modelling, not just social simulation. It is not sufficient to complain that a model has assumptions or does simplify, since modelling without this is impossible; one has to argue or show *why* the assumptions included will distort the results. Equally, the author of a model should be able to justify the assumptions that have been made.

However, the use that is made of a simulation will be limited by the strength or weakness of the assumptions taken as a whole. If, for example, the model is going to be used in a policy process that will impact on many people's lives, then a high level of evidential support and validation will be required. If the model is more exploratory—for example, to suggest unconsidered risks or new hypotheses—then more assumptions with weaker evidence might well be acceptable. Chapter 29 in this volume (Edmonds et al. 2017) looks at the dangers when models are used to inform issues of policy importance.

What one can do is to try to make the assumptions as transparent, as clear and as explicit as possible. Thus, future researchers will be better able to judge what the model depends upon and adapt the model if any of the assumptions turns out to be considered bad. The most obvious technique is to try to document and display the assumptions. This not only helps to defend the model against criticism but also helps one to think more clearly about the model.

Particularly in the early stages of constructing a model, it is common to make a number of "assumptions" about various processes that are involved. In a sense, these are not strictly assumptions—they are just necessary simplifications made in order to get *something* running—but nevertheless are included here. The model builder might, for example, include a random term to substitute for an unknown process, or a particular value might be chosen for a constant without knowing if it is a suitable value. The modeller must carefully document such decisions *and be prepared to revisit them and adjust them as necessary.* This is particularly true if the model starts to be used for purposes that go beyond what the model was initially intended for (Chap. 29 in this volume).

The next type of assumption to consider is that which is "forced" by the constraints of the programming system. This might be the simplification of a process due to computational power limitations, restrictions forced upon the modeller due to the data structures and/or algorithms available, or the desire to reuse another (sub-)model. Again, such decisions must be documented. Whilst the modeller may feel that these decisions have been forced, their documentation can serve two purposes. Firstly, other modellers may have insights into the same programming system that will allow them to suggest alternate approaches. Secondly, modellers who wish to replicate the model using an alternate system may be able to better demonstrate the impact of these assumptions.

The third type of assumption to consider is the choice of relevant objects and processes. As mentioned previously, any modelling exercise is necessarily an abstraction, and one must leave out much of the detail of the real world. Of course, it is impractical to document *every* detail that has been omitted, but the modeller should consider carefully which objects and processes may be relevant to the model and document those that have been included and those that have been omitted. This documentation will then prove invaluable in the consolidation phase (see Sect. 6), when the modeller should explicitly test these assumptions.

The most difficult type of assumption to track and document is that which derives from the modeller's own personal biases or "common sense". For example, the modeller may have an innate "understanding" of some social process that is used in the model without question. The modeller may also have been trained within a particular school that embraces a traditional set of assumptions. Such traditional assumptions may be so deeply ingrained that they are treated as fact rather than assumption, making them difficult to identify from within.

This final class of assumption may be difficult for the modeller to identify and document, but all others should be carefully documented. The documentation can then be used in the exploration and consolidation phases (see below), when the modeller checks these assumptions as much as possible, refining the model as necessary. The assumptions should also be clearly stated and justified when reporting the model and results.

5.4   Maintaining Control of the Model Whilst Exploring

The second biggest problem in following the exploration and consolidation approach to model building (after that of giving in to the temptation to promote your results without consolidation) is that one loses control of the model whilst exploring, resulting in a tangle of bugs. Exploration is an essential step, testing the impact of the assumptions that have been made but, if not carefully managed, can result in code that achieves nothing at all. Bugs can creep in, to an extent that fixing one merely reveals another, or the model can become so brittle that any further modifications are impossible to integrate, or the model becomes so flaky that it keeps breaking in unexpected ways. Although interactions between processes might be interesting and the point of exploration, too much unknown interaction can just make the model unusable. Thus, it is important to keep as many aspects as possible under control as you explore, so you are left with something that you *can* consolidate! It is generally helpful to be clear (if this possible) about which aspects one is certain about and which aspects one is exploring. If these are separable, then one can apply the techniques in Chap. 6 of this volume to the parts one knows and constrain the area of uncertainty where the exploration is occurring.

The main technique for maintaining control of a model is doing *some* planning ahead and consolidation *as* you explore. This is a very natural way for a modeller to work—mixing stages of exploration and consolidation as they feel necessary and as matches their ambitions for the model. Each programmer will have a different balance between these styles of work. Some will consolidate immediately after each bit of development or exploration; some will do a lot of exploration, pushing the model to its limits and then reconstruct a large part of the model in a careful and planned way. Some will completely separate the two stages, doing some exploration, and then completely rebuild their ideas in a formal planned way but now having a better idea (if they are correct) of what they are aiming to achieve: what needs to go into the model (*and what not*), what is happening in the model as it runs and which results they need to collect from it. It is a general rule that more checking and consolidation will be required than is generally planned for by modellers.[2]

There is no absolute rule for how careful and planned one should be in developing a model, but roughly the more complex and ambitious, the more careful one should be. Whilst a "quick and dirty" implementation may be sufficient for a simple model, for others it is unlikely to get the desired results: it is too easy to lose understanding and control of the interactions between various parts, and also the model loses the flexibility to be adapted as needed later on. At the other end of the spectrum, one can spend ages planning and checking a model, building in the maximum flexibility and modularity, only to find that the model does not give any useful results. This might be a valuable experience for the programmer but does not produce interesting knowledge about the target phenomenon. This is the fundamental reason why

---

[2]Even when you take this principle into account!

exploration is so important: because one does not know which model to build before it is tried. This is particularly so for models that have emergent effects (like many of the ones discussed in this volume) and also for those where there is no benchmark (either formal or observed) against which to check them.

One important thing about the activity of modelling is that one has to be willing to throw a lot of model versions away. Exploratory modelling is an inherently difficult activity; most of the models built will either be the wrong structure or just not helpful with regard to the phenomena we seek to understand. Further, the modelling is constrained in many ways: in the time available for developing and checking them, in the amount of computational resources they require, in the evidence available to validate the model, in the necessary compromises that must be made when making a model and in the need to understand (at least to some extent) the models we make. Thus, the mark of a good modeller is that he or she throws away a *lot* of models and only publishes the results of a small fraction of those he or she builds. There is a temptation to laziness, to trying to "fix" a model that is basically not right and thus save a lot of time, but in reality this often only wastes time. This relates to the *modelling spectacles* mentioned above: one becomes wedded to the structure one is developing, and it takes a mental effort to start afresh. However, if it is to be effective, a corollary of an exploratory approach is being highly selective about what one accepts—junking a lot of models is an inevitable consequence of this. If one is not following a more formal, planning approach, and one is not throwing a lot of versions away, then you are probably instituting poor modelling decisions into your code.

Whatever balance you choose between exploration and consolidation, it is probably useful to always pause before implementing any *key* algorithm or structure in your model, thinking a little ahead to what might be the best way. This is an ingrained habit for experienced programmers but may take more effort for the beginner. The beginner may not *know* of different ways of approaching a particular bit of programming and so may need to do some research. This is why developing some knowledge of common algorithms and data structures is a good idea. There is a growing body of work on documenting programming "patterns"—which seek to describe programming solutions at a slightly general level—which can be helpful, although none of these pattern catalogues have yet been written specifically with models of social complexity in mind (but see Grimm et al. 2005 for examples from ecology). Increasingly too researchers within this field are making their code, or at least descriptions of the algorithms used, available to wider audiences.

There are dangers of using someone else's code or algorithm though. There is the danger of assuming that one understands an algorithm, relying on someone else's description of it.[3] It is almost inconceivable that there will not be some unforeseen results of applying even a well-known algorithm in some contexts. When it comes to reusing code, the risk is even higher. Just as there are assumptions and

---

[3]Of course, this danger is also there for one's *own* programming: it is more likely, *but far from certain*, that you understand some code you have implemented or played with.

simplifications in one's own code, so there will be in the code of others, and it is important to understand their implications. Parameters may need adjustment, or algorithms tweaking, in order for the code to work in a different context. Thus, one needs to thoroughly understand at the very least the interface to the code and perhaps also its details. In some cases, the cost of doing this may well outweigh the benefits of reuse. One of the advantages of an exploratory process is that it tends to educate the modeller as to the properties of its algorithms in the process.[4]

It is important to note that even though the approach presented here deviates from more *formal* approaches to software development, this does not mean one should ignore the standard "good practices" of computer programming. Indeed, due to the complexity of even the simplest models in this field, it is advisable to do *some* planning and design before coding. In particular, the following principles should be applied:

- Conceptualisation: any model will benefit greatly from developing a clear understanding of the model structure and processes before starting to program. This is often called a *conceptual model* and usually involves some diagramming technique. Whilst computer scientists will tend to use UML for this purpose, any graphical notation that you are familiar with will do to sketch the main entities and their relationships on paper, such as mind maps or flow diagrams. Apart from helping a modeller to better understand what the model is about, this will form a valuable part of the overall model documentation. See Alam et al. (2010; appendix) for an example of using UML class and activity diagrams.
- Modularity: it is not always possible to cleanly separate different functions or parts of a model, but where it is possible, it is hugely advantageous to separate these into different modules, classes or functions. In this way, the interactions with the other parts of your model are limited to what is necessary. It makes it much easier to test the module in isolation, facilitate diagnostics and make the code much simpler and easier to read.
- Clear structures/analogies: it is very difficult to understand what code does and to keep in mind all the relevant details. A clear idea or analogy for each part of the simulation can help you keep track of the details as well as being a guide to programming decisions. Such analogies may already be suggested by the conceptions that the programmer (or others) have of the phenomena under study, but it is equally important not to *assume* that these are always right, even if this was your intention in programming the model.
- Clear benchmarks: if there is a set of reference data, evidence, theory or other models to which the simulation is supposed to adhere, this can help in the development of a model, by allowing one to know when the programming has gone astray or is not working as intended. The clearest benchmark is a set of observed social phenomena, since each set of observations provides a new set of data for benchmarking. Similarly, if a part of the model is supposed to extend

---

[4]Sometimes painfully!

another model, then restricting the new model should produce the same outcomes as the original.[5]

- Self-documentation: if one is continuously programming a simulation that is not very complex, then one *might* be able to recall what each chunk of code does. However, when developing this type of simulation, it is common to spend large chunks of time focusing on one area of a model before returning to another. After such a lapse, one will not necessarily remember the details of the revisited code, but making the code clear and self-documenting will facilitate it. This sort of documentation does not necessarily have to be full documentation but could include using sensible long variable and module names; adding small comments for particularly tricky parts of the code; keeping each module, class, function or method fairly simple with an obvious purpose; and having some system for structuring your code.

- Building in error checking: errors are inevitable in computer code. Even the best programmer can inadvertently introduce errors to his or her code. Some of these will be obvious but some might be subtle, difficult to isolate and time-consuming to eliminate. Detecting such errors as early as possible is thus very helpful and can save a lot of time. Including safeguards within your code that automatically detect as many of these errors as possible might seem an unnecessary overhead, but in the long run can be a huge benefit. Thus, you might add extra code to check that all objects that *should* exist at a certain time do in fact exist or that a message from one object to another is not empty or that a variable that should only take values within a certain range *does* stay within this range. This is especially important in an exploratory development, where one might develop a section of code for a particular purpose, which then comes to be used for another purpose. In other words, the computational context of a method or module has altered.

These matters are covered in Chap. 6 in this volume in greater depth. There are also many techniques that computer scientists may exhort you to use that are not necessarily useful that may be more applicable to the development of software with more clearly defined goals. Thus, do evaluate any such suggested techniques critically and with a large dose of common sense.

5.5 Understanding the Model

Understanding a model is so intertwined with controlling a model that it is difficult to cleanly separate the two. You cannot really control a complex model if you do not at least partially understand it. Conversely, you cannot deeply understand a model

---

[5]What the "same outcomes" here means depends on how close one can expect the restricted new model to adhere to the original, for example, it might be the same but with different pseudorandom number generators.

until you have experimented with it, which necessitates being able to control it to a considerable extent. However, since modelling complex social phenomena requires (at least usually and probably always) complex models, *complete* understanding and/or control is often unrealistic. Nevertheless, understanding your model as much as is practical is key to useful modelling. This is particularly true for exploratory modelling because it is the feedback between trying model variations and building an understanding of what these variations entail that makes this approach useful.

Understanding one's model is a struggle. The temptation is to try shallow approaches by only doing some quick graphs of a few global measures of output, hoping that this is sufficient to give a good picture of what is happening in a complex social simulation. Although information about the relationship of the setup of a simulation and its global outcomes can be useful, this falls short of a full scientific understanding, which must explain *how* these are connected. If you have an idea of what the important features of your simulation model are, you might be able to design a measure that might be suitable for illustrating the nature of the processes in your model. However, a single number is a very thin indication of what is happening—this is OK if you *know* the measure is a good reflection of what is crucial in the simulation—but can tend to obscure the complexity if you are trying to *understand* what is happening.

To gain a deeper understanding, one *has* to look at the details of the interactions between the parts of the simulation as well as the broader picture. There are two main ways of doing this: case studies using detailed traces/records and complex visualisations.

A case study involves choosing a particular aspect of the simulation, say a particular individual, object or interaction, and then following and understanding it, step by step, using a detailed trace of all the relevant events. Many programming environments provide tracing tools as an inbuilt feature, but not all social simulation toolkits have such a feature. In this latter case, the modeller needs to embed the tracing into the model, with statements that will log the relevant data to a file for later analysis. This "zooming in" into the detail is often very helpful in developing a good understanding of what is happening and is well worthwhile, *even* if you don't think you have any bugs in your code. However, in practice, many people seek to avoid this mundane and slightly time-consuming task.

The second way to gain an understanding is to program a good dynamic visualisation of what is happening in the model. What exactly is "good" in this context depends heavily on the nature of the model: it should provide a meaningful view of the key aspects of the model as the simulation progresses. Many social simulation toolkits provide a range of visualisation tools to assist this programming, but the key is identifying the relevant processes and choosing the appropriate visualisation for them—a task that is not amenable to generic approaches. Thus, you could have a 2D network display where each node is an individual, where the size, shape, colour and direction of each node all indicate different aspects of its state, with connections drawn between nodes to indicate interactions, and so on. A good visualisation can take a while to design and program, but it can crucially enhance the understanding of your simulation and in most cases is usable even when you change the simulation setup. Chapter 10 in this volume (Evans et al. 2017) discusses a range of visualisation techniques aimed at aiding the understanding of a simulation model.

5.6 The Consolidation Phase

The consolidation phase should occur after one has got a clear idea about what simulation one wants to run, a good idea of what one wants to show with it and a hypothesis about what is happening. It is in this stage that one stops exploring and puts the model design and results on a more reliable footing. It is likely that even if one has followed a careful and formal approach to model building, some consolidation will still be needed, but it is particularly crucial if one has developed the simulation model using an informal, exploratory approach. The consolidation phase includes processes of *simplification, checking, output collection* and *documentation*. Although the consolidation phase has been isolated here, it is not unusual to include some of these processes in earlier stages of development, intermingling exploration and consolidation. In such circumstances, it is essential that a final consolidation pass is undertaken, to ensure that the model is truly robust.

Simplification is where one decides which features/aspects of the model you need for the particular paper/demonstration you have in mind. In the most basic case, this may just be a decision as to which features to ignore and keep fixed as the other features are varied. However, this is not very helpful to others because (a) it makes the code and simulation results harder to understand (the essence of the demonstration is cluttered with excess detail) and (b) it means your model is more vulnerable to being shown to be brittle (there may be a hidden reliance on some of the settings for the key results). A better approach is to actually remove the features that have been explored but turned out to be unimportant so that only what is important and necessary is left. This not only results in a simpler model for presentation but is also a stronger test of whether or not the removed features were irrelevant.

The *checking* stage is where one ensures that the code does in fact correspond to the original intention when programming it and that it contains no hidden bug or artefact. This involves checking that the model produces "reasonable" outputs for both "standard" inputs and "extreme" inputs (and of course identifying what "standard" and "extreme" inputs and "reasonable" outputs are). Commonly, this involves a series of parameter sweeps, stepping the value of each parameter in turn to cover as wide a combination as possible (limited usually by resources). When possible, the outputs of these sweeps should be compared against a standard, whether that is real-world data on the target phenomenon or data from a comparable (well-validated) model.

The *output collection* stage is where data from the various runs is collected and summarised in such a way that (a) the desired results are highlighted and (b) sufficient "raw" data is still available to understand how these results have been achieved. It would be impractical to record the details of every variable for every run of the simulation, but presenting results in summary form alone may hide essential details. At the very least, it is essential to record the initial parameter settings (including random seeds, if random numbers are used) so that the summary results may be regenerated. It may also be informative to record at least a small number of detailed traces that are illustrative of the simulation process (once one has determined which parameter configurations produce "interesting" results).

Documentation is the last stage to be mentioned here but is something that should be developed throughout the exploration and consolidation of a model. Firstly, as mentioned above, the code should be reasonably self-documenting (through sensible naming and clear formatting) to facilitate the modeller's own understanding. Secondly, the consolidated model should be more formally documented. This should include any assumptions (with brief justifications), descriptions of the main data structures and algorithms and, if third-party algorithms or code have been used, a note to their source. This may seem like unnecessary effort, particularly if the modeller has no intention of publicly releasing the code, but if questions arise some months or years down the track, such documentation can be invaluable, even for the original author's understanding. Chapter 15 in this volume (Grimm et al. 2017) looks at documentation and how one might approach this.

Finally, the modeller must present the model and its results to a wider audience. This is essential to the process of producing a model, since one can only have some confidence that it has been implemented correctly when it has been replicated, examined and/or compared to other simulations by the community of modellers. The distribution of the model should include a description of the model *with sufficient detail* that a reader could re-implement it if desired. It should present the typical dynamics of the system, with example output and summaries of detailed output. The relevant parameters should be highlighted, contrasting those deemed essential to the results with those with little or no impact. The benchmark measurements should be summarised and presented. To maximise a simulation's use in the community, the simulation should be appropriately licensed to allow others to analyse, replicate and experiment with it (Polhill and Edmonds 2007).

5.7   Tools to Aid Model Development

As indicated previously, there is now a variety of systems for aiding the development of complex simulations. These range from programming language-based tracing and debugging tools to frameworks designed explicitly for social simulation, which include libraries of widely used patterns. Learning to use a particular system or framework is a substantial investment, and because of this, most people do not swap from system to system readily once they have mastered one (even when an alternate system may provide a far more elegant solution to a problem). Ideally, a modeller would evaluate a range of systems when embarking on a new project and decide upon the most appropriate one for that project. In practice, most modellers simply continue to use the same system as they have used on previous projects, without considering alternatives. There is no simple answer as to which system is the "best". The available options are constantly changing as new systems are developed and old ones stop being supported. The type of modelling problem will influence the decision. And indeed it is partly a personal decision, depending on the modeller's own personal style and preferences. However, given that such an investment is involved in learning a new system, it is a good idea to make this investment in one that will have large payoffs and that will be useful for developing a wide range of models.

Systems for developing and testing simulations range from the very specific to those that claim to be fairly generally applicable. At the specific end, there are simulators that are designed with a restricted target in mind—such as a grid-based simulation of land use change (e.g. FEARLUS,[6] Polhill et al. 2001, or SLUDGE, Parker and Meretsky 2004)—where most of the structures, algorithms and outputs are already built in. The user has some latitude to adapt the simulation for their own modelling ends, but the ease with which one can make small changes and quickly get some results may be at the cost of being stuck with inbuilt modelling assumptions, which may not be appropriate for the task at hand. The specificity of the model means that it is not easy to adapt the system beyond a certain point; it is not a *universal* system, capable, in principle, of being adapted to any modelling goal. Thus, such a specific modelling framework allows ease of use at the cost of a lack of flexibility.

At the other end of the spectrum are systems that aim to be general systems to support simulation work that can, at least in principle, allow you to build any simulation that can be conceived. Such systems will usually be close to a computer programming language and usually include a host of libraries and facilities for the modeller to use. The difficulty with this type of system is that it can take considerable effort to learn to use it. The range of features, tools and libraries that they provide take time to learn and understand, as does learning the best ways to combine these features. Furthermore, even if a system in principle makes it possible to implement a modelling goal, different systems have different strengths and weaknesses, making any particular system better for some types of models and less good for others. Thus, modellers will sometimes "fight the system", implementing workarounds so that their model *can* be implemented within the system in which they have invested so much time, when in fact the model could more efficiently be implemented in an alternative system.

Between these two extremes lie a host of intermediate systems. Because they are often open source, and indeed more specific modelling frameworks are commonly built *within* one of these generic systems, it is usually possible (given enough time and skill) to "dig down" to the underlying system and change most aspects of these systems. However, the fundamental trade-offs remain—the more of a simulation that is "given", the more difficult it will be to adapt and the more likely it is that assumptions that are not fully understood will affect results.

---

[6]http://www.macaulay.ac.uk/fearlus/

Thus, it is impossible to simply dictate which system is best to use for developing simulation models of social complexity; indeed, there is no single system that is best under all circumstances. However, the sorts of questions one should consider are clearer. They include:

- *Clear structure*: Is the way the system is structured clear and consistent? Are there clear analogies that help "navigate" your way through various choices you need to make? Is it clear how its structures can be combined to achieve more complex goals?
- *Documentation*: Is there a good description of the system? Is there a tutorial to lead you through learning its features? Are there good reference documents where you can look up individual features? Are there lots of well-documented examples you can learn from?
- *Adaptability*: Can the system be adapted to your needs without undue difficulty? Is the way it is structured helpful to what you want to do? Are the structures easily adaptable once implemented in your model? Does the system facilitate the modularisation of your model so that you can change one aspect without having to change it all?
- *Speed*: How long does it take to run a model? Speed of execution is particularly important when a variety of scenarios or parameters need to be explored or when several runs are necessary per parameter configuration due to random processes in the model.
- *User community*: Do many people in your field use the system? Are there active mailing lists or discussion boards where you can ask for help? If you publish a model in that system, is it likely that it will be accessible to others?
- *Debugging facilities*: Does the system provide inbuilt facilities for debugging and tracing your simulation? If not, are there perhaps generic tools that could be used for the purpose? Or would you have to debug/trace your model by manually inserting statements into your code?
- *Visualisation facilities*: Does the system provide tools and libraries to visualise and organise your results? Are there dynamic visualisation tools (allowing one to view the dynamics of the system as it evolves)? How quickly can you develop a module to visualise the key outputs of a simulation?
- *Batch processing facilities*: Is there a means of running the model a number of times, collecting and perhaps collating the results? Is it possible to automatically explore a range of parameters whilst doing this?
- *Data collection facilities*: Are the results collected and stored systematically so that previous runs can easily be retrieved? Is it possible to store them in formats suitable for input into other packages (e.g. for statistical analysis or network analysis)?
- *Portability*: Is the system restricted to a particular platform or does it require special software to run? Even if all your development will be done on one particular machine, in the interests of reusability, it is desirable to use a system that will run on multiple platforms and that is not dependent on specialised commercial software.

- *Programming paradigm*: Different programming paradigms are more appropriate to different types of modelling problems. If, for example, you think of things in terms of "if-then" statements, a rule-based system might be the most appropriate for your modelling. If instead you visualise things as series of (perhaps branching) steps, a procedural one might be more appropriate. In practice, most systems these days are not *purely* one paradigm or another, but they still have leanings one way or another, and this will influence the way you think about your modelling.
- *Timing*: How will time be handled in the simulation? Will it be continuous or stepped or perhaps event-driven? Will all agents act "at once" (in practice, unless each agent is run on a separate processor they will be executed in some sense sequentially, even if conceptually within the model they are concurrent), or do they strictly take turns? Will it be necessary to run the simulation in real time or (many times) faster than real time?

Once one has considered these questions, and decided on the answers for the particular model in mind, the list of potential systems will be considerably shortened, and one should then be able to make an informed choice over the available options. The temptation, particularly when one is beginning to write models, is to go for the option that will produce the quickest results, but it is important to remember that sometimes a small initial investment can yield long-term benefits.

5.8 Conclusion

It is easy to try and rationalise bad practice. Thus, it is tempting to try and prove that some of the more formal techniques of computer science are not applicable to building social simulations just because one cannot be bothered to learn and master them. It *is* true however that not all the techniques suggested by computer scientists are useful in an exploratory context, where one does not know in advance precisely what one wants a simulation to do. In these circumstances, one has to take a looser and less reliable approach but follow it with consolidation once one has a more precise idea of what one wants of the simulation. The basic technique is to mix bits of a more careful approach in with the experimentation in order to keep sufficient control. This has to be weighed against the time that this may take, given one does not know which final direction the simulation will take. There is a danger of this approach: that the modeller will be tempted by apparently significant results to rush to publication before sufficient consolidation has occurred. There may be times when the exploratory phase may result in useful and influential *personal* knowledge, but such knowledge is not reliable enough to be up to the more exacting standards expected of *publicly* presented results. This is particularly true if the model is to be applied in a critical way that has real impacts upon people or the environment. Thus, it is only with careful consolidation of models that this informal approach to building simulations should be undertaken.

Further Reading

Outside the social sciences, simulation has been an established methodology for decades. Thus, there is a host of literature about model building in general. The biggest simulation conference, the annual "Winter Simulation Conference", always includes introductory tutorials, some of which may be of interest to social scientists. Good examples are Law (2008) and Shannon (1998).

For a comprehensive review of the currently existing general agent-based simulation toolkits, see Nikolai and Madey (2009); other reviews focus on a smaller selection of toolkits (e.g. Railsback et al. 2006; Tobias and Hofmann 2004; Gilbert and Bankes 2002).

The chapters in this volume on checking your simulation model (Chap. 7, Galán et al. 2017), documenting your model (Chap. 15, Grimm et al. 2017) and model validation (Chap. 9, David et al. 2017) should be of particular interest for anyone intending to follow the exploration and consolidation approach to model development. However, if you would rather attempt a more formal approach to building an agent-based simulation model, Chap. 6 (Siebers and Klügl 2017) discusses one such approach in detail. You could also consult textbooks on methodologies for the design of multi-agent systems, such as Luck et al. (2004) and Bergenti et al. (2004) or Henderson-Sellers and Giorgini (2005). After all, any agent-based simulation model can be seen as a special version of a multi-agent system.

References

Alam, S. J., Geller, A., Meyer, R., & Werth, B. (2010). Modelling contextualized reasoning in complex societies with "Endorsements". *Journal of Artificial Societies and Social Simulation, 13*(4), 6. http://jasss.soc.surrey.ac.uk/13/4/6.html

Bergenti, F., Gleizes, M.-P., & Zambonelli, F. (Eds.). (2004). *Methodologies and software engineering for agent systems: The agent-oriented software engineering handbook*. Boston: Kluwer Academic.

Cartwright, N. (1983). *How the laws of physics lie*. Oxford: Clarendon Press.

David, N., Fachada, N., & Rosa, A. C. (2017). Verifying and validating simulations. doi:https://doi.org/10.1007/978-3-319-66948-9_9.

Edmonds, B. (2017). Different modelling purposes. doi:https://doi.org/10.1007/978-3-319-66948-9_4.

Epstein, J. M. (2008). Why model? *Journal of Artificial Societies and Social Simulation, 11*(4), 12. http://jasss.soc.surrey.ac.uk/11/4/12.html

Evans, A., Heppenstall, A., & Birkin, M. (2017). Understanding simulation results. doi:https://doi.org/10.1007/978-3-319-66948-9_10.

Galán, J. M., Izquierdo, L. R., Izquierdo, S. S., Santos, J. I., del Olmo, R., & López-Paredes, A. (2017). Checking simulations: Detecting and avoiding errors and artefacts. doi:https://doi.org/10.1007/978-3-319-66948-9_7.

Gilbert, N., & Bankes, S. (2002). Platforms and methods for agent-based modelling. *PNAS, 99*(Suppl. 3), 7197–7198.

Grimm, V., Revilla, E., Berger, U., Jeltsch, F., Mooij, W. M., Railsback, S. F., et al. (2005). Pattern-oriented modeling of agent-based complex systems: lessons from ecology. *Science, 310*, 987–991.

Grimm, V., Polhill, G., & Touza, J. (2017). Documenting social simulation models: The ODD protocol as a standard. doi:https://doi.org/10.1007/978-3-319-66948-9_15.

Henderson-Sellers, B., & Giorgini, P. (Eds.). (2005). *Agent-oriented methodologies*. Hershey, PA: Idea Group Publishing.

Kuhn, T. (1969). *The structure of scientific revolutions*. Chicago, IL: University of Chicago Press.

Law, A.M. (2008). How to build valid and credible simulation models. In S. J. Mason, R. R. Hill, L. Mönch, O. Rose, T. Jefferson, & J. W. Fowler (Eds.), *Proceedings of the 2008 winter simulation conference*. http://www.informs-sim.org/wsc08papers/007.pdf

Luck, M., Ashri, R., & d'Inverno, M. (2004). *Agent-based software development*. London: Artech House.

Nikolai, C., & Madey, G. (2009). Tools of the trade: A survey of various agent based modelling platforms. *Journal of Artificial Societies and Social Simulation, 12*(2). http://jasss.soc.surrey.ac.uk/12/2/2.html

Parker, D. C., & Meretsky, V. (2004). Measuring pattern outcomes in an agent-based model of edge-effect externalities using spatial metrics. *Agriculture, Ecosystems & Environment, 101*(2–3), 233–250.

Polhill, G., Gotts, N., & Law, A. N. R. (2001). Imitative versus non-imitative strategies in a land-use simulation. *Cybernetics and Systems, 32*(1–2), 285–307.

Polhill, J. G., & Edmonds, B. (2007). Open access for social simulation. *Journal of Artificial Societies and Social Simulation, 10*(3). http://jasss.soc.surrey.ac.uk/10/3/10.html

Railsback, S. F., Lytinen, S. L., & Jackson, S. K. (2006). Agent-based simulation platforms: review and development recommendations. *Simulation, 82*, 609–623.

Shannon, R.E. (1998). Introduction to the art and science of simulation. In D.J. Medeiros, E.F. Watson, J.S. Carson, & M.S. Manivannan (Eds.), *Proceedings of the 1998 winter simulation conference*. http://www.informs-sim.org/wsc98papers/001.PDF

Siebers, P.-O., & Klügl, F. (2017). What software engineering has to offer to agent-based social simulation. doi:https://doi.org/10.1007/978-3-319-66948-9_6.

Tobias, R., & Hofmann, C. (2004). Evaluation of free Java-libraries for social-scientific agent based simulation. *Journal of Artificial Societies and Social Simulation, 7*(1). http://jasss.soc.surrey.ac.uk/7/1/6.html

Chapter 6
What Software Engineering Has to Offer to Agent-Based Social Simulation

Peer-Olaf Siebers and Franziska Klügl

Abstract In simulation projects, it is generally beneficial to have a toolset that allows following a more formal approach to system analysis, model design and model implementation. Such formal methods are developed to support a systematic approach by making different steps explicit as well as providing a precise language to express the results of those steps, documenting not just the final model but also intermediate steps. This chapter consists of two parts: the first gives an overview of which tools developed in software engineering can be and have been adapted to agent-based social simulation; the second part demonstrates with the help of an informative example how some of these tools can be combined into an overall structured approach to model development.

Why Read This Chapter?
To get to know the tools and techniques that software engineering has on offer when it comes to taking a more structured approach to model building. This is particularly useful for larger, collaborative and multidisciplinary projects. Resulting models are easy to maintain and extend and are easy to communicate (and consequently to reproduce), even if the models themselves are highly complex.

6.1 Introduction

In most, if not all simulation projects, it is beneficial to proceed in a systematic way, even more for larger, collaborative and multidisciplinary projects. Agent-based social simulation (ABSS) partially suffers from the fact that despite its increasing

P.-O. Siebers (✉)
School of Computer Science, Nottingham University, Nottingham, NG8 1BB, UK
e-mail: pos@cs.nott.ac.uk

F. Klügl
School of Science and Technology, Örebro University, Örebro, Sweden
e-mail: franziska.klugl@oru.se

© Springer International Publishing AG 2017   81
B. Edmonds, R. Meyer (eds.), *Simulating Social Complexity*,
Understanding Complex Systems, https://doi.org/10.1007/978-3-319-66948-9_6

popularity, there is no standard way of addressing model development, simulation handling, etc. Many modellers are basically self-taught when it comes to processes and tools involved in designing and implementing an ABSS model. Developing an ABSS model is anything but a trivial endeavour given the conceptual depth, often unclear level of detail and complexities involved when handing software that contains more than one thread of control. Computer science — in particular software engineering — has developed a set of tools that enables following the so-called "formal" approach to system analysis, model design and implementation. Such elements of a systematic approach make different steps explicit as well as provide clear and precise languages to capture the concepts, content or assumptions of the model, documenting not just the final result but also intermediate steps. Such an approach is naturally used for model development if elements and processes of the targeted system are more or less accessible and empirically well embedded or assumptions to be taken are clear. In the terminology of Boero and Squazzoni (2005), this refers to a type of model more towards the case-based model side of the spectrum of models.

Models on the other end — theoretical abstractions — are more associated with scientific endeavour of hypothesis building and testing. They consequently need a much more exploratory process. Nevertheless, scientific rigor requires a systematic procedure to ensure reproducible results, as also Norling et al. (2017) argue. Formal languages allow to clearly formulate what shall be contained in a model, which not only supports awareness in the overall development process but also facilitates more unambiguous communication between all involved partners, especially between the ones that implement the model and the rest of the group. Thus, exploratory modelling profits as formal approaches support the thoughtfulness of the modelling process. It helps to avoid model artefacts and supports sharing, reproducing and reusing the model. Eventually, the model is transformed into software, and applying software engineering supports the development of well-structured, understandable software that is easy to maintain and easy to extend.

This chapter advertises formal tools for model conceptualisation, software development and project management as offered by software engineering. In the context of social simulation, these tools can either be used individually either to help with specific modelling activities or to guide the entire modelling process. The chapter consists of two parts: In the first part, we provide an overview of tools and techniques used in software engineering which have been used or suggested in a social simulation context. In the second part of this book chapter, we demonstrate with the help of an informative example how some of these tools can be combined into a structured framework that allows a more formal approach to model development. The informative example is based on a real-world study where we aimed to develop a simulation model with a multidisciplinary team to study different facets of normative comparison.

6.2   Review of Formal Approaches to Model Development

Already in 2006, Richardini et al. identified a number of methodological problems supposed to hinder the wider adoption of agent-based modelling and simulation in the social sciences. In contrast to alternative forms of modelling and simulation, ABSS is assumed not to be suitable to follow shared, standardised conventions of how to proceed, how to describe or how to analyse a simulation model due to its exploratory, bottom-up nature aiming at reproducing emergent processes, etc. Building software in general had similar problems for a long time as the early (and ongoing) discussions on the nature of software development (art, engineering or craftsmanship) show (see, e.g. Pyritz (2003)).

There are no underlying principles of physics or other established basic knowledge that could be used for building software in the same way as, for example, rules of statics for building bridges. Nevertheless, there was the need to systematise software development by developing guidelines, conventions and best practices that make the development process more engineering-like, producing software with intended quality in a predictable way. A number of process models have been invented with the waterfall model as the most well-known traditional approach or extreme programming as a more modern, flexible compilation of best practices. Characteristics of the former are a number of steps that express more and more detailed views onto the resulting software product while moving from a clarification of what needs to be eventually implemented, tested and maintained. In modern forms of software development approaches, fast prototyping and frequent testing are in the centre. Iterative development with code refactoring that improves software design replaces clearly structured, systematic larger process steps by more or less organised smaller advance.

Although ABSS has particularities that preclude to take simulation models as just another kind of software, software engineering offers a large repertoire of languages and tools to support systematic and structured system analysis and development: formal and structured text-based and diagram-based languages allow more precise formulations of model elements than natural language would do. By clarifying what needs to be formulated, those languages guide not only model specification and documentation but all phases in the development. Specific process suggestions organise different views and model description elements into a sequence of steps that correspond to some best practice of how to proceed when designing, implementing and working with a simulation model. Those methodologies exist not only for simulation models in general but also for agent-based simulations in particular. At some stage in those processes, best practices on a more technical level support the translation of concepts into program code — such support is given by (software) pattern formalising good solution to recurring problems.

In this section, we will present different contributions of software engineering to support the development of ABSS models. We identified four areas of such formal instruments in the widest sense and organised the section according to these four elements: Methodologies as the first pillar suggest how to manage the overall development process in a structured and aware way, from formulating the objective behind the ABSS study to validating and deploying the simulation results. We hereby concentrate more on the elements of the overall process that relate to the phases from model conceptualisation to software development as we think of software engineering tools most relevant for those. The second pillar are structured and formal languages for expressing the concepts that are seen as relevant in the system under consideration. These languages can be used in the different steps for describing different views or elements of the model in an as unambiguous as possible way. In phases towards implementation of the model, pillars 3 and 4 become essential. Architectures and pattern form a way to capture best practices in model design and implementation, while tools support the implementation process directly.

6.2.1 First Pillar: Development Processes and Software Engineering Methodologies

Social scientists seem to associate software engineering-based approaches with "formal systems" that enforce to apply a prescribed sequence of steps using formal languages far too rigorous to be appropriate for the mostly exploratory nature of simulation model building. Software engineering is an engineering discipline that is concerned with all aspects of software production.[1] In general, it defines a systematic process with steps that guide the developer from requirement elicitation to implementation, validation and sometimes even maintenance of the software.

Nowadays there are a variety of more or less formal approaches in software engineering together with some understanding which of the methods is suitable for which kind of problem. As coined in Sommerville (2016), for example, games are usually developed by producing a sequence of prototypes while safety-critical software development appears to be highly formal with elaborated and analysed specifications. In the same way as there are very different applications of agent-based simulation, one may expect also very different ways of building agent-based simulation software. A formal process here shall ensure that the results possess particular qualities: The resulting software shall be reliable and trustworthy, produced in an economic way. While the latter may mean that the software is based on reusable components in a well-structured way, the former qualities for simulation software refer to reproducibility of results and validity of the implemented simulation model.

---

[1] Sommerville (2016) relates software engineering also to computer science. The latter is focusing more on theory and fundamentals, while software engineering is more practically oriented towards developing and delivering useful software. He also sees software engineering as a part of systems engineering which aims at systems integrating hardware, software and process engineering.

In general, the produced software is usually not used by the programmer. Sommerville (2016) states that the highest costs in a software project are associated with changing the software after it has gone into use. This can also be stated for simulation in general — not just when decisions have been taken supported by results of a simulation study or publications have been published presenting hypotheses and making statements based on the results of a simulation study. Discovering too late that the model contains artefacts or does actually not answer the question it should do can be embarrassing in the best case, deadly in the worst.

6.2.1.1 Generic Processes

Simulation engineering in general has set up a number of generic processes much on the abstraction level of general software engineering activities. Some process models are independent from the actual model paradigm. Basically every simulation textbook proposes a procedure that basically organises activities such as done in Law (2007), Shannon (1998) and Robinson (2004) or the stages of simulation-based research in Gilbert and Troitzsch (2005). Figure 6.1 shows a generic version of this process. This model development cycle starts with an explicit statement of the objective that is behind the simulation study undertaken, i. e. a formulation of the problem actually addressed. In a second step, the system is analysed; that means its basic components and their relations are determined. Reliable information and data sources are to be found for informing the different steps in the model development process. Based on this analysis, a conceptual model is specified elaborating the structure of the model, as well as the dynamics of all the interacting elements.

The conceptual model is hereby particularly important, as it helps not just to understand the system under consideration but also to guide the subsequent phases by documenting the hypotheses taken. In the second part of this chapter, we will demonstrate an example approach in developing such a conceptual model. It is quite common — not just in our example — to elaborate the conceptual model from highly abstract descriptions of model elements and different points of view onto the model to a more and more concrete specification that can directly inform implementation.

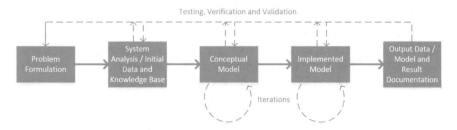

Fig. 6.1 Generic steps in a simulation study

Depending on the tools used for implementation, the production of an executable and thus "simulate-able" representation of the model is achieved as output of the next phase. This phase also usually happens in an iterative way, either by adding more and more details to the model implementation or by fast prototyping and adapting. In the last phase, the model is deployed and experimented with to generate the intended results, which are then documented and used. Each of the model representations produced in the different phases must sufficiently correspond to the original system (validation); this is ensured by testing the models individually and by verifying that one representation is sufficiently related to the other. Figure 6.2 (presented in Sect. 6.3.1) is elaborating this process towards ABSS model development. The focus of this chapter is on these earlier steps of a full study, only indicating how implementation can be achieved and largely omitting running experiments and analysing produced data.

One can also find similar suggestions for structured procedure when developing a simulation study in ABSS. An example is Drogoul et al. (2003). They give more specific detail about types of knowledge and roles of different human experts involved. Activities are more detailed with respect to domain model (real agents), design model (conceptual agents) and operational model (computational agents).

Using the similarity of tools and languages applied to model the conceptual views on software systems consisting of multiple agents and system analysis and model development in ABSS, there are a number of suggestions to extend methodologies developed for agent-oriented software engineering (AOSE). One can see AOSE as an extension of object-oriented software engineering addressing the specific problems that arise when developing multi-agent systems; for an overview of different methodologies, see Bergenti et al. (2004) and Gomez-Sanz and Fuentes-Fernandez (2015). Winikoff and Padgham (2013) give a good introduction into the general principles of AOSE. One of the earliest AOSE methodologies that have been used to develop agent-based simulations was INGENIAS (Gomez-Sanz et al. 2010).

6.2.1.2 Specific Processes for ABSS

Also detailed, formal methodologies that are specific for developing ABSS models have been proposed. Two examples proposing approaches similarly structured to AOSE methodologies are easyABM (Garro and Russo 2010) and MAIA (Ghorbani et al. 2013).

easyABM assumes different phases from system analysis, conceptual system development, simulation design and code generation to simulation setup, execution and results analysis. Particularly elaborated is the conceptual system modelling phase consisting of the development of different partial models that capture relevant views onto the model. An overall metamodel is provided for the different aspects that provides clear high-level language concepts and their relations. The structural system model contains sub-models for each component, determining its abstraction level. The main components are society (composed), agents (active) and artefacts (passive, resource manager). An interaction model describes how intra-

and interrelationships end in interactions. Hereby, a society model describes a society based on its composition, type and rules (safety rules and liveness rules). The central aspect of the agent model is a complex goal model. It also contains a behaviour model composed of activities for achieving these goals, as well as interactions with other agents and artefacts. The latter are specified using behaviour and interactions. UML class and activity diagrams form the main means to express those partial models. Developing the conceptual model further, simulation design is given in a language that resembles elements of the Repast Simphony metamodel and thus enables at least partial code generation. The case study used in Garro and Russo (2010) to exemplify the use of easyABM is a logistics scenario using simulation to test different management policies for vehicles stacking and moving containers. easyABM is already characterised as a model-driven approach and further developed towards MDA4ABMS (see below).

MAIA also focuses on conceptual modelling activities, yet social and society aspects play a specifically central role capturing social phenomena. It guides modelling institutions and social constructs based on a metamodel derived from the Institutional Analysis and Development framework of Ostrom (2005) already used in several agent-based simulation studies. The basic assumption is that social rules and institutions are more easily accessible to modellers than capturing individual behaviour. The MAIA metamodel is organised in five sub-models resembling different aspects of the underlying framework (Ghorbani et al. 2013): (1) collective structure with actors and their attributes; (2) constitutional structure with roles, their dependencies and actions, institutional statements such as norms, shared strategies, etc.; (3) physical structure; (4) operational structure focussing on system dynamics; and (5) evaluation structure containing concepts to evaluate and validate the outcomes of the system. In Ghorbani et al. (2014), these structures were extended by formally grounded operational semantics. This makes the specification given using the MAIA metamodel executable so that a runnable simulation can be directly generated from it.

A purer methodology focussing on interactions is IODA (Kubera et al. 2011). The starting point of this methodology is the identification of interactions that simulated reactive agents exhibit with other agents as well as their simulated environment.

6.2.1.3 Model-Driven Development

The basic idea of model-driven development is that software development may consist of handling models of the intended software starting from a generic level (Stahl et al. 2006). Specifications then can be (semi) automatically transformed into more and more platform-specific representations, eventually generating code. Basically one can see this approach as the currently most formally grounded, controlled evolution of software. Adapting this idea to ABSS means that based on a precise formulation of the conceptual model subsequent, more and more concrete models are elaborated until finally a version that is fully adapted to a particular simulation platform is achieved. The above-described specific methodologies for developing ABSS models, MAIA and easyABM, can be seen as first steps towards model-driven development methodologies. Garro et al. (2013) introduce MDA4ABMS as a complete model-driven approach proposing clearly defined metamodels for each of the major phases of development. There are ABSS-specific metamodels on different levels of abstraction starting from a computation-independent model (CIM) on a conceptual level, platform-independent models (PIM) with more specific architectural and behavioural details to a platform-specific model (PSM) towards realisation for a specific software platform. MDA4ABMS gives also guidelines and rules for the transition between the different phases of development — making even partially automatic transformation possible. The process is exemplified with an extended prisoner dilemma model.

Such methodologies clearly define what elements a system analysis needs to contain — underlying metamodels create a particular awareness behind the conceptualisation. The assumption is that — if the original system is analysed sufficiently thoroughly and the results of this analysis written down in a sufficiently clear way — the simulation model can be communicated and implemented without uncertainties. The critical activity is developing a conceptual model. The formal elements of the methodologies shall sharpen the way the modeller looks onto the system and guide overall model formulation in a reliable way even for models in which the individual agents exhibit complex behaviour. Model-driven development works best in combination with domain-specific languages (DSLs) that provide abstractions specific for a given application domain. Beyond taking an ABSS-specific language as a DSL, there are not many other specific languages yet. The metamodels mentioned above actually provide DSLs for ABSS with a particular perspective in mind. MAIA focuses more on institutions, easyABM more on the complex goal-directed behaviour of individual agents. Franchi (2012) proposed a specific language for agent-based social network modelling. Scherer et al. (2015) describe a model-driven approach for conceptual modelling phases specific for the public policy domain. Their toolset supports a semiautomated transformation of conceptual model representations to formal policy models and then to executable simulations of different scenarios. Their conceptual model is systematically derived from narrative texts. The conceptual model representation at the centre of their approach is specific for public policy development process. This adaptation to the policy domain makes the overall process particularly suitable for involving different stakeholder groups.

6.2.1.4   Agile Approaches

Such structured methodologies seem to resemble more classical waterfall type of software engineering approaches. Knublauch (2002) reports experiences with using extreme programming as a more modern, agile approach to develop agent-based software. Extreme programming (Beck 2004) is more like a collection of best practices and principles such as "on-site customer" resulting in daily contacts with stakeholders to avoid the system developing into something which is actually not intended. Another principle is "simple design"; that means producing software that solves the particular problem and nothing else. With "refactoring", it is ensured that the quality of software design is improved after each iteration in the development cycle. "Short releases" as a principle mean many executable prototypes and software testing is in the centre of the methodology. These principles are as important as the rather more prominent "pair programming" way of implementation, in which two software developers sit in front of the monitor programming together — one coding, the other supporting. Extreme programming as overall approach may fit also to developing model specifications and simulation system formalisations using structured methodologies mentioned above. Short releases and testing would then correspond to running and analysing prototypic simulation runs in an overall iterative approach.

Moyo et al. (2015) organise the development of an agent-based simulation study using SCRUM, an agile approach to manage software development. This article forms a good introduction to agile software development methodologies for simulation in general and gives a case study modelling alcohol consumption dynamics.

6.2.1.5   Formal Methodologies Versus Modelling Principles

None of these more formal methodologies for developing agent-based simulations actually contradicts the principles or informal strategies that are proposed in the social simulation community for model development. Examples for those principles are the KISS principle stating that a model should be as simple as possible. A contrasting principle is the KIDS strategy (Edmonds and Moss 2004) arguing that a model should be preferred that is understandable and descriptive. Simplification should not be exaggerated, especially before fully understanding the system to be modelled. Another strategy is the so-called pattern-oriented modelling (Grimm et al. 2005) that focuses on reproducing all pattern or stylised facts observable in the underlying data. General guidelines from a simulation engineering point of view can be found in Kasaie and Kelton (2015) but also in Richiardi et al. (2006). All these informal strategies can be combined with the more formal methodologies mentioned here. Underlying metamodels are usually very generic and can be used to capture many different societies, agents, etc.

6.2.2   Second Pillar: Structured and Formal Languages

In contrast to natural language, structured and formal languages offer a mean to clearly describe a system. Formal languages form important elements of the methodologies discussed in the last section but have also a value on their own. Syntax and semantics of language elements and their relations are precisely given.

They may be so precise that a model fully described in a formal language may even be automatically processed — execution or analysis may be done without running the description. Often formal languages are distinct from programming languages due to their higher abstraction level enabling more meaningful constructs based on a clearly defined metamodel. Due to this high-level property, descriptions in the formal language can be more compact and focussed on the relevant aspects. Consequently, they are apt for specification and documentation. The clearly defined, underlying metamodel may at first sight be more restrictive than natural language, but the advantage of this restriction is that it may result in a more precise and clearer description.

Some of the languages described below are embedded into frameworks in order to be executable. That means it may be possible to directly run a simulation specified in that language without first translating it into a programming language. If this is not fully possible, there might be a chance to create a code skeleton from the description that can then be complemented for a full implementation. Even without any implementation, specification in some formal languages can be processed directly for deriving properties or for comparing the specified model with likewise formalised high-level system descriptions.

There is a plethora of formal languages that can be used for capturing ABSS models or their elements. Different languages have different foci and are useful for different objectives, or as Edmonds (2004) puts it, "Formal Systems (such as logics) are not the *content* of theory but merely a *tool* for expressing and applying theory in a symbolic way" (p. 1, italics in the original). So they form an instrument for expressing a model or elements of a model. The first group of languages that may come into one's mind when thinking about formal languages are logic based. Many different logical languages exist; each of them focuses on particular elements or uses a different starting point (Fasli 2004).

6.2.2.1 Logic-Based Languages

Languages for logic-based modelling correspond to mathematics as a language for analytical modelling. The language comes with certain constraints limiting the range of particular details that can be formulated. If those details are not relevant when modelling a system, using such a formal language is preferable as it makes tools available for fast or even automated analysis, for fast simulation, etc. An example for a useful tool for ABSS based on logics is the LEADSTO language (Bosse et al. 2005). Its statements as extension of predicate logics can be used for expressing time dependencies between statements. T. Bosse suggests using this logic to describe the overall dynamics of a simulation model so that the output data can be automatically tested for whether the statements hold in the actual simulation runs. In AOSE, logic-based languages play in particular a role in the area of verifiable specification languages (for a review see Mascardi et al. 2004).

6.2.2.2   Algebraic Specification Languages

Besides formal logics, there are many languages that can be used for describing agent-based software as well as ABSS models. d'Inverno and Luck (2001), for example, used the algebraic specification language Z for formally describing different multi-agent systems and their features for clarifying and understanding the core concepts. There are a lot of approaches for formalising particular aspects, such as architectures, organisations, etc. Examples that are relevant as they not just cover agents in isolation are Weyns and Holvoet (2004) and Helleboogh et al. (2007). The language used there are more or less formal algebraic but less structured than, for example, Z. However, the contents are particularly interesting as they show how interactions between agents and between agents and their environment can be captured in a precise and unambiguous way. They also demonstrate on what level of detail a fully clear specification would need to be given.

6.2.2.3   Petri Nets

Another example of a formal language that has been used for specifying multi-agent systems on different levels of aggregation are Petri nets. Köhler et al. (2007) show how social theories can be formalised using this graphical language of "places" and "transitions" with "tokens" traveling through the network. A place may hold tokens, while a transition transports tokens from one place to another on a strictly local basis. In computer science, Petri nets form an established modelling tool for concurrent, interacting processes and their synchronisation. They are amendable for theoretical analysis, but their overall state changes — when becoming too complex for analysis — can also be simulated as places and transitions have a clearly given semantics. For expressing agent interaction and behaviour, complex token structures are needed to actually represent a network on their own.

6.2.2.4   Object-Oriented Simulation and DEVS

In the object-oriented simulation community, formal specification languages have been invented and found wide dissemination. The most prominent example is DEVS (Discrete EVent system Specification) initially introduced by Zeigler (1990), a specification language for object-oriented simulation models that is based on notions from general systems science. Initially restricted to discrete event modelling and simulation, meanwhile it is seen as a more general approach that can also be used for continuous systems. An atomic model consists of a description of the input, state and output variables, a specification of which value combinations of input variables are fed into the running simulation ("input segment"), transition functions for updating state and output variables, as well as a time advance function that characterises how time is updated. Atomic models can be aggregated to composed models. Due to its generality, DEVS was advertised for use in ABSS by Duboz et al. (2006). Hocaoglu et al. (2002) focus on giving more structure to the state of an atomic model in order to allow for more complex agent behaviour. Specifications formulated in DEVS can be executed using specialised environments such as JAMES (Himmelspach et al. 2010).

6.2.2.5   Object-Oriented Software Specification and UML

In AOSE the most prominent language for specifying particular views onto the overall system is UML. UML was developed for supporting software engineering processes (from requirement analysis to implementation and documentation) by providing a language consisting of different specialised diagrams that address different aspects of an object-oriented software system (Fowler 2003). It is actually a semiformal diagram language. That means it allows some extent of vagueness when describing a system. There is an additional formal language — OCL, the object constraint language — that can be used to add information that cannot be expressed in the diagrams directly.

The first edition of UML — used for developing object-oriented software — was defined in the mid-1990s as an integration of different diagram notations from different object-oriented modelling methods. Especially in AOSE, there have been a number of suggestions for extensions, e.g. class diagrams containing information about the particular social organisation, behaviour diagrams containing structures for particular agent architectures, etc. Those extensions were mostly done for specific AOSE methodologies. The best-known extension for software agents was Agent UML (Odell et al. 2000), which mostly pertained to sequence diagrams for enabling the formulation of more flexible and diverse interactions and reactions to messages than simple method calls. Some of these extensions became part of the UML 2 standard, in a graphically different way than originally proposed by Agent UML. Since alternatives and conditional reactions can now be formulated in UML 2, Agent UML has been declared obsolete (Bauer and Odell 2005).

As ABSS in general are often designed and implemented using object-oriented languages and tools, Bommel and Müller (2007) motivate the use of UML diagrams as a suitable tool for communication between different experts involved in a simulation project. A good introduction to UML for ABSS can be found in (Bersini 2012) or (Siebers and Onggo 2014).

UML proposes various types of diagrams to capture different aspects of an overall object-oriented software system. The following diagram types are used mainly in an agent-based simulation context. In the second part of this chapter, we will illustrate their use in more detail.

- *Use case diagrams* show different scenarios of how the user may interact with the system. They could also be applied — on a coarse level — for interactions between an agent and its environment.
- *Class diagrams* show the static structure of the software system by connecting specialised classes to more general ones or showing which classes are composed

of others or how classes are otherwise linked to each other. This type of diagram is not just suitable for depicting an agent's internal setup but also its embedding into an organisation structure.

- *State and activity diagrams* can be used to capture dynamics. They show the states that an (typical) entity can be in, as well as the transitions between them. Activity diagrams focus on behaviour as a flow of activities also in relation to other agents' activities.
- *Sequence diagrams* show how entities interact as a sequence of messages that they exchange.

In addition, other diagram types are proposed to capture details of the package structure, deployment, etc. In the second part of this chapter, we give more details on how to use UML diagrams for model development and embed their use into some form of best-practice process guiding the development of a conceptual model.

6.2.3 Third Pillar: Architectures and Patterns

Methodologies and the use of a formal and precise language to describe different aspects of the conceptual model as well as to capture model specification, etc. are particularly important for people just starting with modelling and simulation as they provide guidelines for managing the development process and support for conceiving a model. A third pillar of software engineering for ABSS is related to best practices in designing models that means they provide advice on how to structure and build the actual software.

In the seminal book on software patterns (Gamma et al. 1994), best practices in (object-oriented) software design have been formalised and captured in such a way that they can easily be communicated and even taught. Over the years software design pattern has been suggested for many problem types, each of them giving a particular abstract "good" solution. Patterns have been also suggested in AOSE (see Juziuk et al. 2014 for a general survey). As North and Macal (2011) state, the standard software patterns are only of limited use for ABSS as the problems addressed by them are of a completely different nature to the ones needing to be solved when developing simulations. For simulation, one may distinguish two different views on design pattern: (1) design pattern that directly relates to particular phenomena to be modelled or (2) design pattern that solves problems on a more technical level. Klügl and Karlsson (2009) give two examples for the first type of pattern, e.g. they describe what agent behaviour can produce exponential agent number growth. North and Macal (2011) give a list of pattern for the second case, e.g. pattern for agent scheduling, how to design spatial environments in an efficient way or the model-view-controller pattern, which is also the most well-known pattern in software engineering, describing how to separate visualisation from application-specific logic.

Agent architectures can be seen as specific pattern for agent-based systems. Depending on whether human decision-making shall be reproduced in ways that resembles how humans think or whether the agents need to exhibit complex and flexible behaviour, different architectures can be used. For the former type, the so-called cognitive architectures such as SOAR (Laird et al. 1987; Wray and Jones 2005) or ACT-R (Anderson et al. 2004; Taatgen et al. 2006) have been suggested (for a short overview, see (Jones 2005)). Those architectures resemble theories from cognitive science supported by results from experiments with humans. Especially SOAR has been used for reproducing human behaviour in military training systems (Wray et al. 2005).

Although often indicated, the so-called BDI architecture is not a cognitive agent architecture but a practical reasoning architecture (Wooldridge 2009). Its underlying motivation consists of a human-inspired means-end analysis separating the decision about which goal ("desire") to pursue from the actual planning towards the goal the agent is committed to achieve ("intention"). The BDI architecture has turned out to be very useful for software agents in general. It also appears to be a reasonable choice for organising the internal decision-making of agents in simulation, especially when more sophisticated agent behaviour needs to be formulated (see, e.g. Joo (2013), Caillou et al. (2015) or Norling (2003)). Even in simulations with rather simple agent behaviour, it is advisable to use an agent architecture to organise the behaviour description, so that the agent program is more transparent, better readable and thus better analysable and maintainable.

Although not introduced as agent architectures, the general setup of rule-based systems, state automata or decision trees can provide important ways to structure agent behaviour descriptions and separate agents' decision-making from the actual processing. A rule-based system contains a set of rules as "if . . . then . . . " constructs and a mechanism that systematically tests the current perception and agent state against the if parts of the constructs. If something is true, the second part, the "then . . . " part, is activated. Using such a setup instead of cascades of if-then-else programming language statements supports clarity of design and extensibility of the decision-making model. Similar are decision trees, which form another way to avoid ugly, inflexible implementations with hard-wired if-then-else cascades. A decision tree is a data structure that organises conditions in nodes and different alternative values for those conditions in the branches out of the node. Another architecture pattern is a state automaton with an explicit representation of the state that the agent is in. The state is associated with particular behaviour. State changes happen based on a trigger relevant in the current state. An older, slightly more complex agent architecture following those ideas is the EMF frame (Drogoul and Ferber 1994). All agent architectures presented in AOSE can also be viewed as local pattern for developing agents. They suggest a structure that supports design and implementation of agents with non-trivial behaviour programs. Clearly, those architectures can be useful for ABSS as well.

In addition to software design patterns and agent architectures, there is another category of (software) pattern relevant for ABSS. These are meta-patterns capturing best practices in working with a model, not directly related to the model design or to a specific methodology. A good example is the ODD protocol (Grimm et al. 2017) for documenting ABSS models. It describes a framework of elements that make up a complete and useful documentation. One can also interpret any description of best practices for model testing, validation, etc. as such a meta-pattern. Rossiter (2015) describes a reference architecture for a simulation system in general, clearly structuring the overall software into different layers of functionality. He also uses this reference architecture to explain the setup of existing platforms and to introduce a new toolkit.

6.2.4   Fourth Pillar: Tools and Development Environments

There are many useful tools available for all phases of developing and using ABSS models. For the purpose of this chapter, we want to single out two particular types: specialised drawing tools and software development platforms.

The diagrams capturing a model in, for example, UML may become quite large and complex. Thus tools that offer specialised shapes and other convenient support such as grid-based layout alignment, automated connections, etc. are highly valuable for making the drawing process more efficient and enable the modeller to concentrate on the important aspects of the description. Especially for UML, there are a number of good tools available, such as Visual Paradigm[2] or Visio.[3] Some platforms for implementing ABSS models, as, for example, Repast (Ozik et al. 2015) or AnyLogic (see below), come with tools for drawing some UML diagram types that are then directly translated into code skeletons.

Professional software development is usually done using an integrated development environment (IDE). This is basically a collection of tools facilitating software development, such as elaborated program editors with built-in syntax checks, code completion, etc. allowing the programmer to concentrate on the semantics of the program rather than its syntax. Prominent IDE examples are Visual Studio[4] or Eclipse.[5] Such development environments also support, for example, code documentation by providing tools that automatically generate UML class diagrams from source code.

Inspired by those general IDEs and in addition to low-level programming support, an IDE for ABSS could contain

- Conceptual views on the implemented model with diagrammatic representations of what happens in the model. Drawing tools can be integrated with automated code generation from diagrams representing agent and organisational structures and agent behaviour and interaction dynamics.

---

[2]www.visual-paradigm.com. A free for non-commercial use community version exists.

[3]products.office.com/en/visio/.

[4]www.visualstudio.com.

[5]eclipse.org.

- Simulation runtime support — tools for handling simulated time and space (maps), animation, inspection tools for individual agents and their interactions.
- Appropriate ways to integrate model documentation, e.g. facilities to add comments or specific elements of an ODD model documentation.
- Automated generation of simulation runs including interfaces for conducting elaborated tests or manipulating model settings during runtime.
- Debugging and validation support.
- Convenient tools for defining experiments and input and output data handling.

Such tools make model handling more convenient and efficient, yet they are built around a particular simulation platform that manages and executes a particular model implementation.

Various specialised platforms for ABSS are available that aim at giving specific support. Over the last decades, hundreds of platforms and tools have been suggested. A Wikipedia page[6] lists 89 tools (in April 2016). Wikipedia also provides an up-to-date list of their attributes. Only a few of them deserve to be called an IDE for ABSS such as Repast Simphony (repast.github.io/repast_simphony.html), AnyLogic (www.anylogic.com/) or SeSAm (www.simsesam.org). In addition to that list, there are a number of partially outdated surveys (Nikolai and Madey 2008; Railsback and Lytinen 2006; Kravari and Bassiliades 2015). The most prominent platforms are NetLogo (ccl.northwestern.edu/netlogo/) and Repast (repast.sourceforge.net/), they are covered in of each of the surveys. Other analysed platforms include AnyLogic, MASON (cs.gmu.edu/~eclab/projects/mason/), Gama (gama-platform.org) or Swarm (http://www.swarm.org).

Which platform to use depends on a variety of factors ranging from the modellers' personal preferences and experience to the properties of the model to be implemented. Also whether the platform is a commercial one or open source often plays a role. Providing general advice about the "best" platform is impossible.

6.3   Illustrative Example: Normative Comparison in an Office Environment

Up to now we have seen that software engineering in general and AOSE in particular offer a lot of support for developing ABSS models. Most of this support can be coined "formal": at the heart are clearly given process models describing the different steps to go through when doing a simulation study. This is particularly important for less experienced modellers as these process models help to solve the problem of translating vague mental representations of models into descriptions that are more and more refined. These methodologies help to know where one should start when doing a simulation study.

---

[6]https://en.wikipedia.org/wiki/Comparison_of_agent-based_modeling_software, accessed 07/05/2016.

In the following we show based on an illustrative example that there is no need to be afraid of formal approaches but that they can indeed be useful to support awareness about the actual model content when developing a model.

6.3.1   Our Structured Approach

When developing ABSS models, one faces the question of how to build them and where to start. This can be challenging not only for novices in the field but also for multidisciplinary teams where it is often difficult to engage everyone in the modelling process. Over the years we have developed a quite sophisticated "plan of attack" in the form of a framework that guides the model development and can be used by either individuals or teams.

When used by individuals, they need to consider the perspective of potential team members (i.e. slip into their roles) during each process step. When used by teams, co-creation is an important aspect. Team members need to be open-minded about the use of new tools and methods and about the collaboration with researchers from other domains and business partners. This is often not easy for researchers trained in more traditional approaches or for business partners who often expect researchers to act like consultants, providing them with a report and a list of recommendations (Mitleton-Kelly 2003).

Our framework, called the "Engineering Agent Based Social Simulation" framework (or EABSS framework for short), supports model reproducibility through rigorous documentation of the conceptual ideas, underlying assumptions and the actual model content. The framework provides a step-by-step guide to conceptualising and designing ABSS models with the support of software engineering tools and techniques. Figure 6.2 provides an overview of the steps that make up the development process.

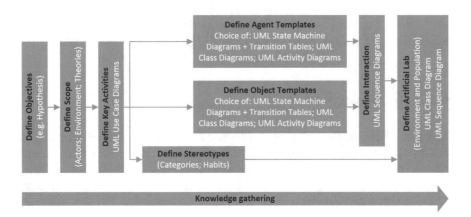

Fig. 6.2 Overview of our EABSS framework

While the framework represents a structured modelling approach, there will always be iterations required by the users to improve definitions from previous tasks. When stepping through the framework, the users may realise that they did not consider important elements/details in a previous step or that they considered too many or that they considered them wrongly. In particular *discussions in focus groups* unearth these kinds of issues and are therefore extremely valuable for the model development process. The framework is a suitable tool for well-organised discussions and to capture the knowledge and ideas coming out of these discussions in a formal way. While there is a given sequence of steps that users should follow, they need to be prepared to go back to a previous task if required and apply changes. Consequently this means that the users do not have to worry too much if in the initial rounds they get things wrong or things feel incomplete. They should simply move on to the next task if they feel that they have some form of contribution. Our experience is that it is necessary to revisit each task four to five times before there is a satisfying result that is acceptable to all stakeholders. In that sense, the approach somewhat resembles "agile" approaches of software engineering with frequent interactions with stakeholders and frequent iterations and not investing a lot of time into specifications that are obsolete after the next discussion.

While this framework will not work perfectly for all possible cases, it provides at least some form of systematic approach. The user should be prepared to adapt it to fit individual needs. In the following we will explain each step (including the necessary tools) and exemplify its application.

In order to demonstrate the use of our structured approach, we use an illustrative example, which is based on work by Zhang et al. (2011), Susanty (2015) and Bedwell et al. (2014). In this example we focus on the simulation model development to support studying the impact of normative comparison amongst colleagues with regards to energy consumption in an office environment. Normative comparison in this context means giving people clear regular personalised insight into their own energy consumption (e.g. "you used x% more energy than usual for this month") and allowing them to compare it to that of their neighbours (e.g. "you used x% more than your efficient neighbours"). A simulation study could compare the impact of "individual apportionment" vs. "group apportionment" of energy consumption information on the actual energy consumption within the office environment.

6.3.2 Gathering Knowledge

The task of knowledge gathering is one that happens throughout the structured modelling approach and in many different ways. The main ones we use in our framework are *literature review*, *focus group discussions*, *observations* and *surveys*. The knowledge gathering is either a prerequisite for tasks (e.g. a literature review) or embedded within the tasks (e.g. focus group discussions). For our study, all focus groups were led by a computer scientist (the initiator of the study), and the participants consisted of a mixture of academics and researchers from the fields of computer science, business management and psychology. In this example study, we did not engage with business partners. The team consisted of five core members who would participate regularly in the focus groups. Over the years we have made the experience that for our purposes smaller focus groups work best. Whenever we describe a task, in the following, we also briefly mentioned when and how the required knowledge was gathered.

6.3.3   Defining the Objectives

The first step within the framework is to define objectives in relation to the aim of the study. In our case this was done through a combination of a *literature review* and *focus group discussions*. After some iteration we came up with the following:

- Our aim is to study normative comparison in an office environment.
- Our objective is to answer the following questions:

  - What are the effects of having the community influence the individual?
  - What is the extent of impact (significant or not)?
  - Can we optimise it using certain interventions?

- Our hypotheses are:

  - Peer pressure leads to greener behaviour.
  - Peer pressure has a positive effect on energy saving.

With the objectives defined, we then need to think about how we can test these objectives. For this we need to consider relevant experimental factors and responses. Experimental factors are the means by which the modelling objectives are to be achieved. Responses are the measures used to identify whether the objectives have been achieved and to identify potential reasons for failure to meet the objectives (Robinson 2004). In other words, experimental factors are simulation inputs that need to be set initially to test different scenarios related to the objectives while responses are simulation outputs that provide insight and show to what level the objectives have been achieved. In our case the hypotheses are very helpful for defining an initial set of experimental factors and responses:

- Experimental factors

  - Initial population composition (categorised by greenness of behaviour)
  - Level of peer pressure ("individual apportionment" vs. "group apportionment")

- Responses

  - Actual population composition (capturing changes in greenness of behaviour)
  - Energy consumption (of individuals and at average)

The experimental factors and responses defined at this stage are still very broad and need to be revisited when more information about the model becomes available.

6.3.4 Defining the Scope

At this stage we are interested in specifying the model scope. This requires some initial knowledge gathering. We did this through a *literature review* and *observation* of the existing system. With the help of the knowledge gathered, we were then able to define the scope of the model. Decisions were made through *focus group discussions*. To guide the discussion and to document the decisions made in a more formal way, we used an adaptation of the conceptual modelling *scope table* proposed by Robinson (2004) specially tailored towards ABSS modelling. The general categories we consider are "Actor", "Physical environment" and "Social/Psychological aspects".

In order to make decisions about including or excluding different elements within these categories, we asked ourselves, amongst others, the following questions:

- What is the appropriate level of abstraction for the objective(s) stated before?

  - This would define the level of abstraction acceptable.

- Do the elements have a relevant impact on the overall dynamics of the system?

  - Then they should be included.

- Do the elements show similar behaviour to other elements?

  - Then they should be grouped.

After some *discussions within the focus group*, we decided that "transparency" would be the key driver for our decision-making and that we want to abstract/simplify as much as possible while still keeping a realistic model (i.e. we aimed to explicitly follow the KISS principle mentioned in Sect. 6.2.1). In order to have easy access to data, we decided to use our own offices (University of Nottingham; School of Computer Science) as the data source. Table 6.1 presents the resulting scope table in which we state for every element whether we want to include or exclude it and why we decided either way.

6.3.5 Defining Key Activities

Interaction can take place between actors and between an actor and the physical environment it is in. Capturing these at a high level can be done with the help of *UML use case diagrams*. In software engineering, *UML use case diagrams* are used to describe a set of actions (use cases) that some system or systems (subject) should or can perform in collaboration with one or more external users of the system (actors). These diagrams do not attempt to represent the order or number of times that the systems actions and subactions should be executed. The relevant components of a use case diagram are depicted and described in Table 6.2.

Table 6.1 Scope table for our illustrative example

Category		Element	Decision	Justification
Actor		Staff	Include as group (User)	Regularly occupy the office building
		Research fellows		
		PhD students		
		UG+MSc students	Exclude	Do not have control over their work environment
		Visitors	Exclude	Insignificant energy use
Physical environment	Appliance	HVAC (Heating + Ventilation + Aircon) system	Exclude	We only need one major energy consumer to test the theory; we decided to go for electricity
		Lighting	Include	Interacts with users on a daily basis; controlled by user
		Computer	Include	Interacts with users on a daily basis; controlled by user
		Monitor	Exclude	Modelled as part of the computer
		Continuously running appliances	Exclude	Constant consumption of electricity; not controllable by individuals
		Personal appliances	Exclude	No way to measure consumption
	Weather	Temperature	Exclude	Not necessary for proof-of-principle
		Natural light level	Exclude	Not necessary for proof-of-principle
	Room	Office	Include	Location where electronic appliances are installed
		Lab	Exclude	Mainly used by UG+MSc
		Kitchen	Include as group (Other room)	Common areas frequently used by "users"
		Toilet		
		Corridor	Include	Commonly used when "users" move around

(continued)

Table 6.1 (continued)

Category	Element	Decision	Justification
Social/Psychological aspect	Comparative feedback	Include	Effective strategy to reduce energy consumption in residential building
	Informative feedback	Include	Effective strategy to remove barriers in performing specific behaviour
	Apportionment level	Include	Potential strategy to reduce energy consumption in office building
	Freeriding	Include	Behaviour that differentiate two apportionment strategy
	Sanction	Include	Factor to encounter freeriding behaviour
	Anonymity	Include	Factor to encounter freeriding behaviour

Table 6.2 Relevant use case diagram components

Component	Symbol	Description
Actors		Entities that interface with the system (this can be people or other systems). Think of actors by considering the roles they play
Use cases		Denotes what the actor wants your system to do for them
System boundary		Indicates the scope of your system: the use cases inside the rectangle represent the functionality that you intend to implement

Relationships	
	`<<Include>>`
	`<<Extend>>`
		There are different types of relationships. In a relationship between use case and actor the associations indicate which actors initiate which use cases. A relationship between two use cases specifies common functionality and simplifies use case flows. We use <<Include>> when multiple use cases share a piece of same functionality which is placed in a separate use case rather than documented in every use case that needs it. We use <<Extend>>when activities might be performed as part of another activity but are not mandatory for a use case to run successfully. We are adding more capability

While in software engineering the actors are outside the system boundaries (they are usually the users of software, and the software represents the system), when using use case diagrams in an ABSS context the actors are inside the system (representing the humans that interact with each other and the environment). The system boundaries are the boundaries of the relevant locations (which in our case would be the building boundaries of the office environment). It is important to understand that the purpose of these diagrams is to promote understanding; as long

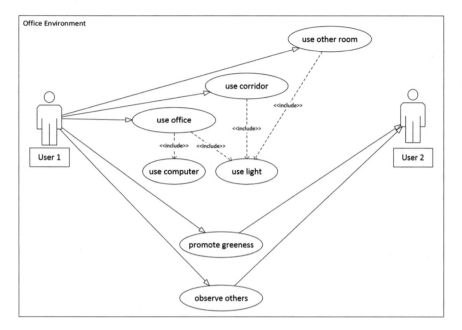

Fig. 6.3 Use case diagram for our illustrative example [drawn with Visio]

as they capture the ideas and help to explain them, they are very useful. The use case diagram which we developed for our illustrative example through *focus group discussions* is depicted in Fig. 6.3.

6.3.6   Defining Stereotypes

In social psychology, a stereotype is a thought (or belief) that can be adopted about specific types of individuals or certain ways of doing things (McGarty et al. 2002). In order to be able to represent a specific population in our simulation models, we define stereotypes that allow us to classify the members of this population. We derived our stereotype templates (categories, habits to be considered and type names) through *focus group discussions* and through considering the *knowledge gathered previously*. Getting the stereotype templates right is more an art than a science. After long debates we decided to have two categories of stereotypes: one related to "work time" and the other related to "energy-saving awareness". Once the categories were identified, we had to come up with the habits that describe these stereotypes:

- Habits for work time category:

  – Arrival time at office
  – Leaving time from office

- Habits for energy-saving awareness category:

  - Energy-saving awareness
  - Likelihood of switching off unused electric appliances
  - Likelihood of promoting greenness

To get the information we needed to fully define the stereotypes, we conducted a *survey* amongst our school's academics, researchers and PhD students, anonymously asking them questions about their habits towards work time and energy-saving awareness. We then *analysed the data* through cluster analysis to come up with the stereotype groups, assigned some speaking name and populated the stereotype tables with the "habit" information. The stereotype definitions we ended up with can be found in Tables 6.3 and 6.4.

6.3.7 Defining Agent and Object Templates

For each of the relevant actor types we have identified in our scope table, we have to develop an agent template containing all information for a prototypical agent. These templates will act as a blueprint when we later create the actor population for each simulation run. When it comes to modelling the environment, we need similar templates for everything relevant we have identified in the scope table that lends itself to be represented as an object (e.g. the appliances). For other things (e.g. the weather), we need to consider other modelling methods. From a technical point of view, there is no big difference between agents and objects. Thus we can use the same types of diagrams to document their design. We will therefore use the term

Table 6.3 User stereotypes defining work time habits

Stereotype	Working days	Arrival time	Leave time
Early bird	Mon–Fri	5 am-9 am	4 pm-7 pm
Time table complier	Mon–Fri	9 am-10 am	5 pm-6 pm
Flexible worker	Mon–Fri	10 am-1 pm	5 pm-11 pm
Hardcore worker	Mon–Fri + Sat	8 am-10 am	5 pm-11 pm

Table 6.4 User stereotypes defining energy-saving habits

Stereotype	Energy saving awareness [0–100]	Probability of switching off unnecessary appliances	Probability of sending emails about energy issues to others
Environmental champion	95–100	0.95	0.9
Energy saver	70–94	0.7	0.6
Regular user	30–69	0.4	0.2
Big user	0–29	0.2	0.05

"entity" when we talk about both. There are three different diagram types that are relevant for defining entity templates: UML class diagrams (to define structure), UML state machine diagrams (to define behaviour) and UML activity diagrams (to define logic). Often only a subset of these is required. When developing the templates, we create the different diagrams in parallel and in an iterative manner as often one informs and inspires the development of the other. As with the stereotypes, getting the entity templates right is not hard science and will therefore require many iterations.

In software engineering *UML class diagrams* are used to define the static structure of the software to be developed by showing classes (which are blueprints to build specific types of objects) and the relationships between classes. These relationships define the logical connections between classes (association, aggregation, composition, generalisation, dependency). UML class diagrams can be very complex, and for our purposes it is often enough to consider individual classes. Therefore we focus on how to define individual classes here. In UML classes are depicted as rectangles with three compartments. The first compartment is reserved for the class name. This is simply the name of the entity as defined in the scope table (e.g. "user" for our user agent template). The second compartment is reserved for attributes (constants and variables). Often we would capture key state variables (e.g. "energy saving awareness"), key parameters and key output variables (e.g. "own energy consumption") here. The third compartment is reserved for operations that the user may perform. For each operation, we define some function names that indicate what kind of additional code we have to produce later (e.g. "moveToNewLocation()"). The brackets indicate that this is a function. Figure 6.4 shows as an example the user class definition we developed in parallel with the other template diagrams in several *focus group discussion* sessions.

In software engineering, *UML state machine diagrams* (sometimes just called "state charts") are used to represent the dependencies between the state of an object and its reaction to messages or other events. State machine diagrams show the states of a single object, the events or the messages that cause a transition from one state to another and the actions that result from a state change. A state machine diagram has exactly one state machine entry pointer which indicates the initial state of the agent. A state in a state machine diagram models a situation during which some invariant condition holds. Usually time is consumed while an object is in a specific state. A simple state is a state that does not have substates, while a composite state is a state that has substates (nested states). The relevant components of a state machine diagram are depicted and described in Table 6.5.

In our case we use state machine diagrams to define the behaviour of our entities. This type of diagram is particularly useful as it can be automatically translated into source code by IDEs who support such features. One can use several diagrams (e.g. one representing physical states and one representing mental states) for the same entity. A state machine diagram is not always meaningful (e.g. if there are no relevant states that need to be represented to capture the behaviour) or necessary (e.g. "energy saving awareness" could be expressed in states "aware" and "not aware" but also as a state variable that represents the level of awareness). There is

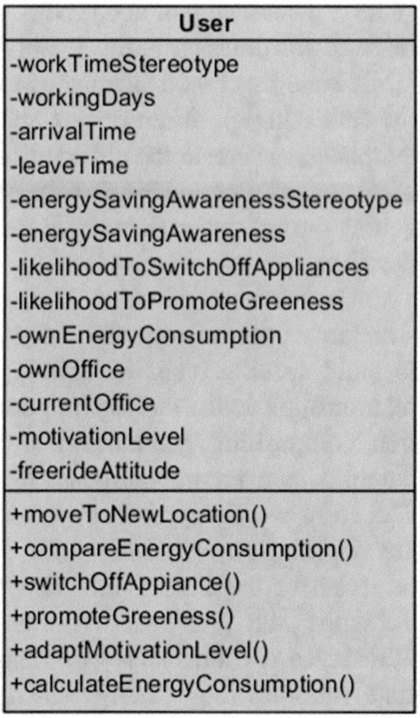

Fig. 6.4 User class definition
[drawn with Visual Paradigm]

User
-workTimeStereotype
-workingDays
-arrivalTime
-leaveTime
-energySavingAwarenessStereotype
-energySavingAwareness
-likelihoodToSwitchOffAppliances
-likelihoodToPromoteGreeness
-ownEnergyConsumption
-ownOffice
-currentOffice
-motivationLevel
-freerideAttitude
+moveToNewLocation()
+compareEnergyConsumption()
+switchOffAppiance()
+promoteGreeness()
+adaptMotivationLevel()
+calculateEnergyConsumption()

Table 6.5 Relevant state machine diagram components

Component	Symbol	Description
Entry pointer	●——	Indicates the initial state after an object is created
State	(□——)	Represents a locus of control with a particular set of reactions to conditions and/or events
Initial states pointer	●——	Points to the initial state within a composite state
Final state	◉	Termination point of a state chart
Transition	——→	Movement between states, triggered by a specific event
Branch	◇	Transition branching and/or connection point
Shallow history	Ⓗ	The state chart remembers the most recent active sub state (but not the lower level sub-states)
Deep history	Ⓗ	The state chart remembers the most recent active sub state (including the lower level sub states)

nothing wrong with having entity templates without state machine diagrams. While for software engineering the descriptions of how transitions are triggered are usually embedded within the diagram (in a rather cryptic language), it might be a good idea to present them in a separate table, to make the diagram easier to understand.

Many people find it difficult to get started with developing the state machine diagrams for agent templates. In order to come up with potential states that an agent can be in, it helps to think in terms of *locations* (e.g. "in office"). The next step would be to think about *key time-consuming activities* within these locations (e.g.

"working with computer"). It is important to consider only key locations and key activities as otherwise the state chart gets too complex. One should only define as much detail as is really necessary for investigating the question studied. The above steps are just suggestions and do not always work. In case they do not work, one has to use intuition and try to draft something that "feels right".

Figure 6.5 shows as an example the "User" state machine diagram we developed in parallel with the other template diagrams in several *focus group discussions*. Here we have defined location states based on the relevant rooms we identified in the scope table and added one location ("outOfOffice") to represent the outside world. The ideas for the activity states stem from our use case diagram (Fig. 6.3). We then added transition arrows to represent the possible transitions between the defined states. Transitions with a question mark symbol are condition triggered while transitions with a clock symbol are time triggered. If there is more than one transition connecting states, we have considered different triggers for state changes.

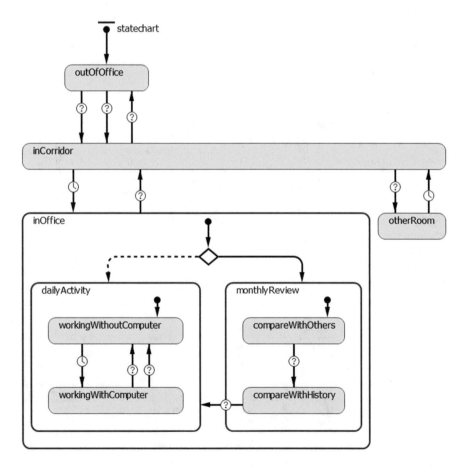

Fig. 6.5 User state machine diagram [drawn with AnyLogic]

This becomes clearer when we look at the transition definitions in Table 6.6. Here we can see that, for example, a state change from "outOfOffice" to "inCorridor" can happen for all user stereotypes during the working week and only for hardcore worker user stereotypes during the weekend.

In software engineering *UML activity diagrams* describe how activities are co-ordinated (the overall flow of control). They represent workflows of stepwise activities (while state machine diagrams show the dynamic behaviour of an object) and actions with support for choice, iteration and concurrency. Often people describe activity diagrams as just being fancy flow charts. The relevant components of an activity diagram are listed in Table 6.7.

Amongst others, we can use these activity diagrams as a formal way to describe a decision-making process (logic flow). In our case we use it to describe the logic flow of the normative comparison process. In order to define the logic flow, we use the information we gathered from our *literature review* on psychological factors in the scoping phase. Figure 6.6 shows as an example the actions happening when the user agent is in the state "compareWithHistory" (which in the model is triggered once per simulated month). It is good practice to provide some evidence from the literature for the rationale behind the decision-making process. This would come from our scoping phase *literature review* but might also require some additional resources. As an example, let's take the case "Less than former month?=no / Group?=yes /

Table 6.6 User state machine transition definitions (excerpt)

From state	To state	Triggered by	When?
outOfOffice	inCorridor	Condition	At typical arrival time during the working week for all
outOfOffice	inCorridor	Condition	At typical arrival time on Saturdays for hard-core workers only
inCorridor	outOfOffice	Condition	At typical leave time
inCorridor	inOffice	Timeout	At average after 5 min
inOffice	inCorridor	Condition	At random while at work or when leaving
inCorridor	otherRoom	Condition	At random while at work
otherRoom	inCorridor	Timeout	At average after 10 min
...	...	...	...

Table 6.7 Relevant activity diagram components

Component	Symbol	Description
Activity		Named box with rounded corners (a state that is left once the activity is finished)
Activity edge	$\longrightarrow$	Arrow (fires when the previous activity completes)
Synchronisation bar		Represent the start (split) or end (join) of concurrent activities
Decision diamond		Used to show decisions
Start marker	◉	Indicate entry point of the diagram
Stop marker	●	Indicate exit point of the diagram

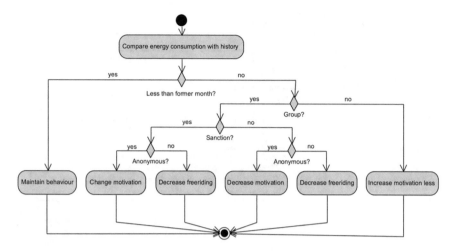

Fig. 6.6 Activity diagram for user agent state "compareWithHistory" [drawn with Visual Paradigm]

Sanction?=yes / Not Anonymous?". In the literature we find that using mechanisms to identify freerides and implement sanctions (social (e.g. gossip) or institutional (e.g. fines)) reduces the likelihood of further freeriding (Fehr et al. 2002). This is our justification for adding the action "decrease freeriding" for this case. In the end we would evaluate our logic flow by *discussing it in the focus group*.

6.3.8   Defining Interactions

As we saw in Sect. 6.3.5, capturing interactions on a high level can be done using *UML use case diagrams*. Capturing interactions in more detail can be done by using *UML sequence diagrams*. These can be used to further specify use cases that involve direct interactions (usually in the form of message passing) between entities (agents and objects).

In software engineering *UML sequence diagrams* are used primarily to show the interactions between objects in the sequential order in which those interactions occur. Often they depict the actors and objects involved in a specific use case realisation and the sequence of messages exchanged between the actors and objects needed to carry out the functionality of the use case realisation. But sometimes they also capture wider scenarios that go beyond a specific use case. The relevant components of a sequence diagram are listed in Table 6.8.

In our case, we discussed the technical way of implementing the "observe others" use case during one of our *focus group discussions*. Figure 6.7 shows the sequence diagram we developed during our discussion for this use case. The entities involved

Table 6.8 Relevant sequence diagram components

Component	Symbol	Description
Lifeline		Named element which represents an individual participant in the interaction
Message		From sender to receiver
Message		Return message
Execution		Represents a period of time in which the participant is active
Message		Self message
Loop		Wrapper for representing loops (has one compartment)
Alternative		Wrapper for representing alternatives (has as many compartments as alternatives exist)

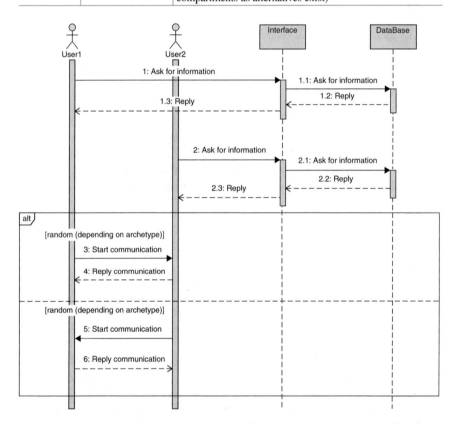

Fig. 6.7 Initial sequence diagram for the use case "observe others" [drawn with Visual Paradigm]

are users and units that provide information. Users interact with information units and with each other. Information units interact with the users and with each other. Creating this diagram sparked a discussion if we should consider a database that stores historic information in our model or not. It is currently not represented in the

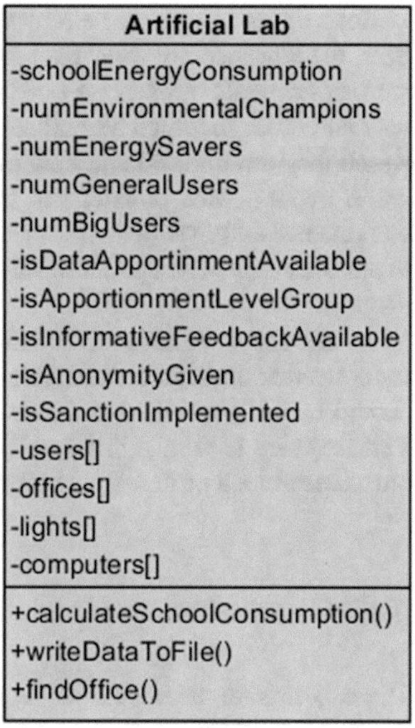

Fig. 6.8 Artificial lab class
definition [drawn with Visual
Paradigm]

Artificial Lab
-schoolEnergyConsumption
-numEnvironmentalChampions
-numEnergySavers
-numGeneralUsers
-numBigUsers
-isDataApportinmentAvailable
-isApportionmentLevelGroup
-isInformativeFeedbackAvailable
-isAnonymityGiven
-isSanctionImplemented
-users[]
-offices[]
-lights[]
-computers[]
+calculateSchoolConsumption()
+writeDataToFile()
+findOffice()

scope table (Table 6.1). In the end we agreed that for our initial model, we will leave it out but keep a record of it in the scope table as it might be something we want to consider in the future. We then removed it from the final version of our sequence diagram.

6.3.9   Defining the Artificial Lab

Finally we need to define an environment in which we can embed all our entities and define some global functionality. We call this environment our "artificial lab". For the development of our artificial lab, we use a class definition as described in Sect. 6.3.7. Within this class definition, we consider things like global variables (e.g. to collect statistics), compound variables (e.g. to store a collection of agents and objects) and global functions (e.g. to read/write to a file). We also need to make sure that we have all variables in place to set the experimental factors and to collect the responses we require for testing our hypotheses. We derive our class content through *focus group discussions*. To inform these discussions, we need to look at our list of objectives (see Sect. 6.3.3) and our scope table (see Sect. 6.3.4). The final class definition should only contain key variables and functions. Figure 6.8 shows the "Artificial Lab" class definition for our illustrative example. Variable names including "[]" represent collection variables.

Sometimes it can be helpful to create a sequence diagram as described in Sect. 6.3.8 to visually show the order of execution describing the actions taken on various elements at each step of the simulation from a high-level approach. The way and order in which all entities are initialised, as well as the way and order in which they are updated and how their interactions are handled, is often not trivial and a major source of artefacts. In such a case, it therefore needs to be clearly documented and specified. Since we do not have any obvious complex dependencies in our illustrative example, it was not necessary to create such a high-level sequence diagram.

At this point, we have all the information for a conceptual model together. Using the collection of diagrams and tables that we produced, the model to be implemented should be fully specified and as well understood as it can be without running it. The next step is to take this specification and either start with the implementation ourselves or let a professional software developer deal with it.

6.4 Conclusion

There seems to be a fear of non-computer scientists with regard to "formal approaches". This might be due to the fact that formal approaches are often presented in a way that makes modelling a very complex and costly task and that seems to take away opportunities for exploratory model development. While this might be true for very large projects, it is usually not the case for smaller ones as tools and techniques do not have to be applied in a dogmatic fashion. They are there to aid the modelling process wherever one thinks it would be appropriate or helpful to use them. Thinking about this as being a *more structured approach* that *adds transparency* to model development rather than a *formal approach* that *makes modelling a complex task* might take away some of the fear. While there will always be a place for informal modelling (in software engineering often coined as "fast prototyping" to quickly try out things), we believe that there is also a place for a more structured approach to modelling.

We have found the framework described in the second part of the chapter very helpful in terms of communicating in multidisciplinary teams during focus group meetings and also for documenting the outcomes of these discussions. It (or parts of it) has been extensively used by the group of PO Siebers (the first author of this chapter) for many different projects, ranging from "Studying People Management Practices in Retail" (Siebers and Aickelin 2011), where we worked with colleagues from Economics and Work Psychology and a leading UK retailer, to "Simulating Peace Building Activities in Africa" (Siebers et al. 2017), where we worked with colleagues from the School of Politics and Psychology. We are currently also applying the framework in several new projects including industrial partners.

So far the feedback from the participating team members has always been very positive. Using these methods has aided "the fun" of collaborative model development. Applying object-oriented principles and tools from software engineering also helped us to develop simulation models that are easy to maintain and easy to extend. Rather than building a model from scratch every time we start a new study, we can reuse previously developed model components with confidence. Using a formal approach to modelling is also a big benefit when it comes to publications as the resulting models are transparent and well documented.

Further Reading

There is a host of literature on the topic of software engineering. A book that provides a comprehensive yet easy to understand entry to most of the software engineering topics discussed in this book chapter is Lethbridge and Laganiere (2005). If you are mainly interested in learning more about UML, then Fowler (2003) is sufficient. A lot of ideas for ABSS stem from the computer science field of artificial intelligence and herein particular multi-agent systems. A good overview on the wide area of topics (including AOSE) is Weiss (2013). Finally, the JASSS special issue "Engineering ABSS" (Siebers and Davidsson 2015) provides lots of information and case studies. The approach contrasts with that described in Chap. 5 in this volume.

References

Anderson, J. R., Bothell, D., Byrne, M. D., Douglass, S., Lebiere, C., & Qin, Y. (2004). An integrated theory of the mind. *Psychological Review, 111*(4), 1036–1060.

Bauer, B., & Odell, J. (2005). UML 2.0 and agents: How to build agent-based systems with the new UML standard. *Journal of Engineering Applications of Artificial Intelligence, 18*(2), 141–157.

Beck, K. (2004). *Extreme programming explained: Embrace change* (2nd ed.). Boston, MA: Addison Wesley.

Bedwell, B., Leygue, C., Goulden, M., McAuley, D., Colley, J., Ferguson, E., et al. (2014). Apportioning energy consumption in the workplace: A review of issues in using metering data to motivate staff to save energy. *Technology Analysis & Strategic Management, 26*(10), 1196–1211.

Bergenti, F., Gleizes, M.-P., & Zambonelli, F. (Eds.). (2004). *Methodologies and software engineering for agent systems: The agent-oriented software engineering handbook.* Boston: Kluwer.

Bersini, H. (2012). UML for ABM. *Journal of Artificial Societies and Social Simulation, 15*(1), 9. http://jasss.soc.surrey.ac.uk/15/1/9.html

Boero, R., & Squazzoni, F. (2005). Does empirical embeddedness matter? Methodological issues on agent-based models for analytical social science. *Journal of Artificial Societies and Social Simulation, 8*(4), 6. http://jasss.soc.surrey.ac.uk/8/4/6.html

Bosse, T., Jonker, C. M., van der Meij, L., & Treur, J. (2005). LEADSTO: A language and environment for analysis of dynamics by simulation. In T. Eymann, F. Klügl, W. Lamersdorf, M. Klusch, & M. N. Huhns (Eds.), *Proc. of the 3rd German Conference on Multi-Agent System Technologies, MATES'05. LNAI 3550* (pp. 165–178). Springer, Berlin, Heidelberg, Germany.

Bommel, P., & Müller, J. P. (2007). An introduction to UML for modelling in the human and social sciences. In D. Phan & F. Amblard (Eds.), *Multi-agent modelling and simulation in the social and human sciences, GEMAS studies in social analysis, Chapter 12.* Bardwell Press, Oxford, United Kingdom.

Caillou, P., Gaudou, B., Grignard, A., Truong, C. Q., & Taillandier, P. (2015, Sep 2015). A simple-to-use BDI architecture for agent-based modeling and simulation. *The Eleventh Conference of the European Social Simulation Association (ESSA 2015)*, Groningen, Netherlands.

d'Inverno, M., & Luck, M. (2001). *Understanding agent systems.* Berlin, Heidelberg, Germany: Springer-Verlag.

Drogoul, A., Vanbergue, A., & Meurisse, T. (2003). *Multi-agent Based Simulation: Where are the agents? Multi-agent Based Simulation II, LNCS 2581* (pp. 1–15). Springer, Berlin, Heidelberg, Germany.

Drogoul, A., & Ferber, J. (1994). Multi-agent simulation as a tool for modelling societies: Application to social differentiation in ant colonies. In C. Chastelfranchi & E. Werner (Eds.), *Artificial social systems -4th European workshop on modelling autonomous agents in a multi-agent world, MAAMAW'92* (pp. 3–23). Heidelberg, Germany: Springer.

Duboz T., Versmisse D., Quesnel G., Muzy A., & Ramat E. (2006, April 2–6). Specification of dynamic structure discrete event multiagent systems. In *Agent-directed simulation (ADS 2006)*, Huntsville, AL, USA.

Edmonds, B. (2004). How formal logic can fail to be useful for modelling or designing MAS. In G. Lindeman et al. (Eds.), *RASTA 2002, LNAI 2934* (pp. 1–15). Berlin, Heidelberg, Germany: Springer-Verlag.

Edmonds, B., & Moss, S. (2004). From KISS to KIDS — an 'anti-simplistic' modelling approach. In P. Davidson et al. (Eds.), *Multi-agent based simulation, LNAI 3415* (pp. 130–144). New York: Springer.

Fasli, M. (2004). Formal systems $\wedge$ agent-based social simulation $= \bot$? *Journal of Artificial Societies and Social. Simulation, 7*(4), 7.

Fehr, E., Fischbacher, U., & Gächter, S. (2002). Strong reciprocity, human cooperation, and the enforcement of social norms. *Human Nature, 13*(1), 1–25.

Fowler, M. (2003). *UML distilled: A brief guide to the standard object modeling language* (3rd ed.). Boston, MA: Pearson Education.

Franchi, E. (2012). A domain specific language approach for agent-based social network modeling. *2012 IEEE/ACM International Conference on Advances in Social Networks Analysis and Mining (ASONAM 2012)*, Istanbul, Turkey.

Gamma, E., Helm, R., Johnson, R., & Vlissides, J. (1994). *Design pattern: Elements of reusable object-oriented software.* Boston, MA: Addison-Wesley.

Garro, A., Parisi, F., & Russo, W. (2013). A process based on the model-driven architecture to enable the definition of platform-independent simulation models. In N. Pina, J. Pacpryzk, & J. Filipe (Eds.), *Simulation and modeling methodologies, technologies and applications SIMULTECH 2011 Noordwijkerhout, The Netherlands, July 2011 revised selected papers* (pp. 113–129). Berlin: Springer.

Garro, A., & Russo, W. (2010). easyABMS: A domain-expert oriented methodology for agent-based modeling and simulation. *Simulation Modelling Practice and Theory, 18*, 1453–1467.

Gilbert, N., & Troitzsch, K. G. (2005). *Simulation for the social scientist* (2nd ed.). Maidenhead, UK: Open University Press.

Ghorbani, A., Bots, P., Dignum, V., & Dijkema, G. (2013). MAIA: a framework for developing agent-based social simulations. *Journal of Artificial Societies and Social Simulation, 16*(2), 9.

Ghorbani, A., Bots, P., Alderwereld, H., Dignum, V., & Dijkema, G. (2014). Model-driven agent-based simulation: procedural semantics of a MAIA model. *Simulation Modelling Practice and Theory, 49*, 27–40.

Gomez-Sanz, J. J., Fernandez, C. R., & Arroyo, J. (2010). Model driven development and simulations with the INGENIAS agent framework. *Simulation Modelling and Practice, 18*(10), 1468–1482.

Gomez-Sanz, J. J., & Fuentes-Fernandez, R. (2015). Understanding agent-oriented software engineering methodologies. *The Knowledge Engineering Review, 30*(4), 375–393.

Grimm, V., Polhill, G., & Touza, J. (2017). Documenting social simulation models: The ODD protocol as standard. doi:https://doi.org/10.1007/978-3-319-66948-9_15.

Grimm, V., Revilla, E., Berger, U., Jeltsch, F., Mooij, W. M., Railsback, S. F., et al. (2005). Pattern-oriented modeling of agent-based complex systems: lessons from ecology. *Science, 310*(5750), 987–991.

Helleboogh, A., Vizzari, G., Uhrmacher, A. M., & Michel, F. (2007). Modeling dynamic environments in multi-agent simulation. *Autonomous Agents and Multi-Agent Systems, 14*(1), 87–116.

Himmelspach, J., Röhl, M., & Uhrmacher, A. M. (2010). Component-based models and simulations for supporting valid multi-agent system simulations. *Applied Artificial Intelligence, 24*(5), 414–442.

Hocaoglu, M. F., Firat, C., & Farjoughian, H. S. (2002). DEVS/RAP: Agent-based simulation. *Proceedings of the 2002 AI, Simulation and Planning in Highly Autonomous Systems conference*, Lisbon, Portugal: IEEE.

Jones, R. M. (2005). An introduction to cognitive architectures for modeling and simulation. *Proceedings of the Interservice/Industry Training/Simulation and Education Conference 2005*, Orlando, FL.

Joo, J. (2013). Perception and BDI reasoning based agent model for human behavior simulation in complex system. In M. Kurosu (Ed.), *Human-computer interaction. Towards intelligent and implicit interaction: 15th Int. Conf., HCI international 2013, Las Vegas, NV, USA, July, 2013, Proc, Part V* (pp. 62–71). Berlin/Heidelberg: Springer.

Juziuk, J., Weyns, D., & Holvoet, T. (2014). Design pattern for multi-agent systems: A systematic literature review. In O. Shehory & A. Sturm (Eds.), *Agent-oriented software engineering: Reflections on architectures, methodologies, languages and frameworks, chapter 5* (pp. 79–99). Berlin, Germany: Springer.

Kasaie, P., & Kelton, W. D. (2015). Guidelines for design and analysis in agent-based simulation studies. In *Proc. of the 2015 Winter Simulation Conference (WSC '15)* (pp. 183–193). Piscataway, NJ: IEEE Press.

Kravari, K., & Bassiliades, N. (2015). A Survey of Agent Platforms. *Journal of Artificial Societies and Social Simulation, 18*(1), 11. http://jasss.soc.surrey.ac.uk/18/1/11.html

Klügl, F., & Karlsson, L. (2009). Towards pattern-oriented design of agent-based simulation models. *Proceedings of the 7th German conference on multiagent system technologies*, Hamburg, Germany.

Knublauch, H. (2002, July 15–19). Extreme programming of multi-agent systems. *Proceedings of AAMAS 2002, Bologna* (pp. 704–711). New York: ACM.

Köhler, M., Langer, R., von Lüde, R., Moldt, D., Rölke, H., & Valk, R. (2007). Socionic multi-agent systems based on reflexive petri nets and theories of social self-organisation. *Journal of Artificial Societies and Social Simulation, 10*(1), 3. http://jasss.soc.surrey.ac.uk/10/1/3.html

Kubera, Y., Mathieu, P., & Picault, S. (2011). IODA: An interaction-oriented approach for multiagent based simulations. *Autonomous Agents and Multi-Agent Systems, 23*(3), 303–343.

Laird, J. E., Newell, A., & Rosenbloom, P. S. (1987). Soar: An architecture for general intelligence. *Artificial Intelligence, 33*, 1–64.

Law, A. M. (2007). *Simulation modeling & analysis* (4th ed.). New York: McGraw-Hill.

Lethbridge, T. C., & Laganiere, R. (2005). *Object-oriented software engineering: Practical software development using UML and Java: Practical software development*. New York: McGraw Hill.

Mascardi, V., Martelli, M., & Sterling, L. (2004). Logic-based specification languages for intelligent software agents. *Theory and Practice of Logic Programming, 4*(4), 429–494.

McGarty, G., Yzerbyt, V. Y., & Spears, R. (2002). Social, cultural and cognitive factors in stereotype formation. In G. McGarty, V. Y. Yzerbyt, & R. Spears (Eds.), *Stereotypes as explanations* (pp. 1–15). Port Chester, NY: Cambridge University Press.

Mitleton-Kelly, E. (2003). Complexity research - approaches and methods: The LSE complexity group integrated methodology. In A. Keskinen, M. Aaltonen, & E. Mitleton-Kelly (Eds.),

Organisational complexity (pp. 56–77). Turku: Tutu Publications. Finland Futures Research Centre, Turku School of Economics and Business Administration.

Moyo, D., Ally, A. K., Brennan, A., Norman, P., Purshouse, R. C., & Strong, M. (2015). Agile development of an attitude-behaviour driven simulation of alcohol consumption dynamics. *Journal of Artificial Societies and Social Simulation, 18*(3), 10. http://jasss.soc.surrey.ac.uk/18/3/10.html

Nikolai, C., & Madey, G. (2008). Tools of the trade: A survey of various agent based modeling platforms. *Journal of Artificial Societies and Social Simulation, 12*(2), 2. http://jasss.soc.surrey.ac.uk/12/2/2.html

Norling, E. (2003). Capturing the quake player: Using a BDI agent to model human behaviour. In J. S. Rosenschein, T. Sandholm, M. Wooldridge, & M. Yokoo, *Proceedings of the 2nd international joint conference on autonomous agents and multiagent systems (AAMAS)*, Melbourne (pp. 1080–1081). New York: ACM.

Norling, E., Edmonds, B., & Meyer, R. (2017). Informal approaches to developing simulation models. doi:https://doi.org/10.1007/978-3-319-66948-9_5.

North, M. J., Macal, C. M. (2011, December 11–14). Product design patterns for agent-based modeling. In S. Jain, R. Creasey, J. Himmelspach, K. P. White, M. C. Fu, *Proc. of the Winter Simulation Conference (WSC '11)* (pp. 3087–3098).

Odell, J., Parunak, H. V. D., & Bauer, B. (2000). Extending UML for agents. In Y. Lesperance, E. Yu, *Proc. of the agent-oriented information systems workshop at the 17th NCAI* (pp. 3–17).

Ostrom, E. (2005). *Understanding institutional diversity*. Princeton, NJ: Princeton University Press.

Ozik, J., Collier, N., Combs, T., Macal, C. M., & North, M. (2015). Repast Simphony Statecharts. *Journal of Artificial Societies and Social Simulation, 18*(3), 11. http://jasss.soc.surrey.ac.uk/18/3/11.html

Pyritz, B. (2003). Craftsmanship versus engineering: Computer programming — An art or a science? *Bell Labs Technical Journal, 8*, 101–104.

Railsback, S. F., & Lytinen, S. L. (2006). Agent-based simulation platforms: review and development recommendations. *SIMULATION, 82*, 609–623.

Richiardi, M., Leombruni, R., Saam, N. J., & Sonnessa, M. (2006). A common protocol for agent-based·social simulation. *Journal of Artificial Societies and Social Simulation, 9*(1), 15. http://jasss.soc.surrey.ac.uk/9/1/15.html

Robinson, S. (2004). *Simulation: The practice of model development and use*. Chichester: Wiley.

Rossiter, S. (2015). Simulation design: Trans-paradigm best-practice from software engineering. *Journal of Artificial Societies and Social Simulation, 18*(3), 9. http://jasss.soc.surrey.ac.uk/18/3/9.html

Scherer, S., Wimmer, M., Lotzmann, U., Moss, S., & Pinotti, D. (2015). Evidence based and conceptual model driven approach for agent-based policy modelling. *Journal of Artificial Societies and Social Simulation, 18*(3), 14. http://jasss.soc.surrey.ac.uk/18/3/14.html

Shannon, R. E. (1998). Introduction to the art and science of simulation. D. J. Medeiros, E. F. Watson, J. S. Carson, M. S. Mannivannan, *Proceedings of the 1998 Winter Simulation Conference* (pp. 7–14).

Siebers, P. O., & Davidsson, P. (2015). Engineering agent-based social simulations: An introduction (Special Issue Editorial). *Journal of Artificial Societies and Social Simulation, 18*(3), 13. http://jasss.soc.surrey.ac.uk/18/3/13.html

Siebers, P. O., & Aickelin, U. (2011). A first approach on modelling staff proactiveness in retail simulation models. *Journal of Artificial Societies and Social Simulation, 14*(2), 2. http://jasss.soc.surrey.ac.uk/14/2/2.html

Siebers, P. O., Onggo, B. S. S. (2014). Graphical representation of agent-based models in operational research and management science using UML. In *Proc. Of the operational research society simulation workshop 2014 (SW14)* (pp. 143–155).

Siebers, P. O., Figueredo, G. P., Hirono, M., & Skatova, A. (2017). Developing agent-based simulation models for social systems engineering studies: A novel framework and its application to modelling peacebuilding activities. In C. Garcia-Diaz & C. Olaya Nieto (Eds.), *Social systems engineering: The design of complexity*. Hoboken, NJ: Wiley.

Sommerville, I. (2016). *Software engineering* (10th ed.). Pearson, Boston, MA.

Stahl, T., Voelter, M., & Czarnecki, K. (2006). *Model-driven software development: Technology, engineering, management*. Hoboken, NJ: Wiley.

Susanty, M. (2015). Adding psychological factors to the model of electricity consumption in office buildings. MSc Dissertation, Nottingham University, School of Computer Science.

Taatgen, N. A., Lebiere, C., & Anderson, J. R. (2006). Modeling paradigms in ACT-R. In R. Sun (Ed.), *Cognition and multi-agent interaction: From cognitive modeling to social simulation* (pp. 29–52). Cambridge: Cambridge University Press.

Weiss, G. (Ed.). (2013). *Multiagent systems* (2nd ed.). Cambridge: MIT Press.

Weyns, D., & Holvoet, T. (2004). A formal model for situated multi-agent systems. *Fundamenta Informaticae, 63*(2–3), 125–158.

Winikoff, M., & Padgham, L. (2013). Agent-oriented software engineering. In G. Weiss (Ed.), *Multiagent systems, Chapter 15* (2nd ed., pp. 695–758). Cambridge: MIT Press.

Wooldridge, M. (2009). *An introduction to multiagent systems*. Hoboken, NJ: Wiley.

Wray, R. E., Laird, J. E., Nuxoll, A., Stokes, D., & Kerfoot, A. (2005). Synthetic adversaries for urban combat training. *AI Magazine, 26*(3), 82–92.

Wray, R. E., & Jones, R. M. (2005). An introduction to Soar as an agent architecture. In R. Sun (Ed.), *Cognition and multi-agent interaction: from cognitive modeling to social simulation* (pp. 53–78). Cambridge: Cambridge University Press.

Zeigler, B. P. (1990). *Object oriented simulation with hierarchical modular models: Intelligent agents and endomorphic systems*. Boston, MA: Academic Press.

Zhang, T., Siebers, P. O., & Aickelin, U. (2011). Modelling electricity consumption in office buildings: An agent based approach. *Energy and Buildings, 43*(10), 2882–2892.

Chapter 7
Checking Simulations: Detecting and Avoiding Errors and Artefacts

José M. Galán, Luis R. Izquierdo, Segismundo S. Izquierdo, José I. Santos, Ricardo del Olmo, and Adolfo López-Paredes

Abstract The aim of this chapter is to simulations. The reader with a set of concepts and a range of suggested activities that will enhance his or her ability to understand agent-based simulations. To do this in a structured way, we review the main concepts of the methodology (e.g. we provide precise definitions for the terms "error" and "artefact") and establish a general framework that summarises the process of designing, implementing, and using agent-based models. Within this framework we identify the various stages where different types of assumptions are usually made and, consequently, where different types of errors and artefacts may appear. We then propose several activities that can be conducted to detect each type of error and artefact.

Why Read This Chapter?

Given the complex and exploratory nature of many agent-based models, checking that the model performs in the manner intended by its designers is a very challenging task. This chapter helps the reader to identify some of the possible types of error and artefact that may appear in the different stages of the modelling process. It will also suggest some activities that can be conducted to detect, and hence avoid, each type.

J.M. Galán (✉) • L.R. Izquierdo • J.I. Santos • R. del Olmo
Department of Civil Engineering, Universidad de Burgos, E-09001, Burgos, Spain
e-mail: jmgalan@ubu.es; luis@izquierdo.name; jisantos@ubu.es; rdelolmo@ubu.es

S.S. Izquierdo • A. López-Paredes
Departamento de Organización de Empresas y C.I.M., Universidad de Valladolid, E-47011, Valladolid, Spain
e-mail: segis@eis.uva.es; adolfo@insisoc.org

© Springer International Publishing AG 2017   119
B. Edmonds, R. Meyer (eds.), *Simulating Social Complexity*,
Understanding Complex Systems, https://doi.org/10.1007/978-3-319-66948-9_7

7.1 Introduction

Agent-based modelling is one of multiple techniques that can be used to conceptualise social systems. What distinguishes this methodology from others is the use of a more direct correspondence between the entities in the system to be modelled and the agents that represent such entities in the model (Edmonds 2001). This approach offers the potential to enhance the transparency, soundness, descriptive accuracy, and rigour of the modelling process, but it can also create difficulties: agent-based models are generally complex and mathematically intractable, so their exploration and analysis often require computer simulation.

The problem with computer simulations is that understanding them in reasonable detail is not as straightforward an exercise as one could think (this also applies to one's own simulations). A computer simulation can be seen as the process of applying a certain function to a set of inputs to obtain some results. This function is usually so complicated and cumbersome that the computer code itself is often not that far from being one of the best descriptions of the function that can be provided. Following this view, understanding a simulation would basically consist in identifying the parts of the mentioned function that are responsible for generating particular (sub)sets of results.

Thus, it becomes apparent that a prerequisite to understand a simulation is to make sure that there is no significant disparity between what we think the computer code is doing and what is actually doing. One could be tempted to think that, given that the code has been programmed by someone, surely there is always at least one person—the programmer—who knows precisely what the code does. Unfortunately, the truth tends to be quite different, as the leading figures in the field report:

> You should assume that, no matter how carefully you have designed and built your simulation, it will contain bugs (code that does something different to what you wanted and expected). (Gilbert 2007)

> An unreplicated simulation is an untrustworthy simulation—do not rely on their results, they are almost certainly wrong. ('Wrong' in the sense that, at least in some detail or other, the implementation differs from what was intended or assumed by the modeller). (Edmonds and Hales 2003)

> Achieving internal validity is harder than it might seem. The problem is knowing whether an unexpected result is a reflection of a mistake in the programming, or a surprising consequence of the model itself. [. . .] As is often the case, confirming that the model was correctly programmed was substantially more work than programming the model in the first place. (Axelrod 1997a)

In the particular context of *agent-based* simulation, the problem tends to be exacerbated. The complex and exploratory nature of most agent-based models implies that, before running a model, there is almost always some uncertainty about what the model will produce. Not knowing a priori what to expect makes it difficult to discern whether an unexpected outcome has been generated as a legitimate result of the assumptions embedded in the model or, on the contrary, it is due to an error or an artefact created in its design, in its implementation, or in the running process (Axtell and Epstein 1994, p. 31; Gilbert and Terna 2000).

Moreover, the challenge of understanding a computer simulation does not end when one is confident that the code is free from errors; the complex issue of identifying what parts of the code are generating a particular set of outputs remains to be solved. Stated differently, this is the challenge of discovering what assumptions in the model are causing the results we consider significant. Thus, a substantial part of this non-trivial task consists in detecting and avoiding artefacts: significant phenomena caused by accessory assumptions in the model that are (mistakenly) deemed irrelevant. We explain this in detail in subsequent sections.

The aim of this chapter is to provide the reader with a set of concepts and a range of suggested activities that will enhance his ability to understand simulations. As mentioned before, simulation models can be seen as functions operating on their inputs to produce the outputs. These functions are created by putting together a range of different assumptions of very diverse nature. Some assumptions are made because they are considered to be an essential feature of the system to be modelled; others are included in a somewhat arbitrary fashion to achieve completeness—i.e. to make the computer model run—and they may not have a clear referent in the target system. There are also assumptions—e.g. the selection of the compiler and the particular pseudorandom number generator to be employed—that are often made, consciously or not, without fully understanding in detail how they work, but *trusting* that they operate in the way we think they do. Finally, there may also be some assumptions in a computer model that not even its own developer is aware of, e.g. the use of floating-point arithmetic, rather than real arithmetic.

Thus, in broad terms, understanding simulations requires identifying what assumptions are being made and assessing their impact on the results. To achieve this, we believe that it is useful to characterise the process by which assumptions accumulate to end up forming a complete model. We do this in a structured way by presenting a general framework that summarises the process of creating and using agent-based models through various stages; then, within this framework, we characterise the different types of assumptions that are made in each of the stages of the modelling process, and we identify the sort of errors and artefacts that may occur; we also propose activities that can be conducted to avoid each type of error or artefact.

The chapter is structured as follows: the following section is devoted to explaining what we understand by modelling, and to argue that computer simulation is a useful tool to explore formal models, rather than a distinctively new symbolic system or a uniquely different reasoning process, as it has been suggested in the literature. In Sect. 7.3 we explain what the essence of agent-based modelling is in our view, and we present the general framework that summarises the process of designing, implementing, and using agent-based models. In Sect. 7.4 we define the concepts of error and artefact, and we discuss their relevance for validation and verification. The framework presented in Sect. 7.3 is then used to identify the various stages of the modelling process where different types of assumptions are made and, consequently, where different types of errors and artefacts may appear. We then propose various activities aimed at avoiding the types of errors and artefacts previously described, and we conclude with a brief summary of the chapter.

7.2   Three Symbolic Systems Used to Model Social Processes

Modelling is the art of building models. In broad terms, a model can be defined as an abstraction of an observed system that enables us to establish some kind of inference process about how the system works or about how certain aspects of the system operate.

Modelling is an activity inherent to every human being: people constantly develop mental models, more or less explicit, about various aspects of their daily life. Within science in particular, models are ubiquitous. Many models in the "hard" sciences are formulated using mathematics (e.g. differential equation models and statistical regressions), and they are therefore formal, but it is also perfectly feasible—and acceptable—to build non-formal models within academia; this is often the case in disciplines like history or sociology, consider, e.g. a model written in natural language that tries to explain the expansion of the Spanish Empire in the sixteenth century or the formation of urban "tribes" in large cities.

We value a model to the extent that it is useful—i.e. in our opinion, what makes a model good is its fitness for purpose. Thus, the assessment of any model can only be conducted relative to a predefined purpose. Having said that, there is a basic set of general features that are widely accepted to be desirable in any model, e.g. accuracy, precision, generality, and simplicity (see Fig. 7.1). Frequently some of these features are inversely related; in such cases the modeller is bound to compromise to find a suitable trade-off, considering the perceived relative importance of each of these desirable features for the purpose of the model (Edmonds 2005).

Some authors (Gilbert 1999; Holland and Miller 1991; Ostrom 1988) classify the range of available techniques for modelling phenomena in which the social dimension is influential according to three symbolic systems.

One possible way of representing and studying social phenomena is through verbal argumentation in natural language. This is the symbolic system traditionally used in historical analyses, which, after a process of abstraction and simplification, describe past events emphasising certain facts, processes, and relations at the expense of others. The main problem with this type of representation is its intrinsic lack of precision (due to the ambiguity of natural language) and the associated difficulty of uncovering the exact implications of the ideas put forward in this way. In particular, using this symbolic system, it is often very difficult to determine the whole range of inferences that can be obtained from the assumptions embedded in the model in reasonable detail; therefore it is often impossible to assess its logical consistency, its scope, and its potential for generalisation in a formal way.

A second symbolic system that is sometimes used in the social sciences, particularly in economics, is the set of formal languages (e.g. leading to models expressed as mathematical equations). The main advantage of this symbolic system derives from the possibility of using formal deductive reasoning to infer new facts from a set of clearly specified assumptions; formal deductive reasoning guarantees that the obtained inferences follow from the axioms with logical consistency. Formal languages also facilitate the process of assessing the generality of a model and its

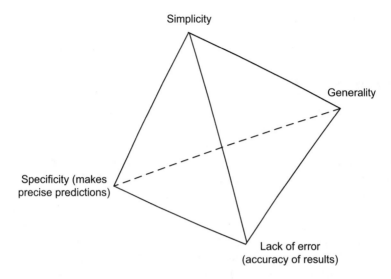

Fig. 7.1 The trade-off between various desirable features depends on the specific case and model. There are not general rules that relate, not even in a qualitative fashion, all these features. The figure shows a particular example from Edmonds (2005) that represents the possible equilibrium relationships between some features in a particular model

sensitivity to assumptions that are allowed to change within the boundaries of the model (i.e. parameter values and nonstructural assumptions).

However, the process of reducing social reality to formal models is not exempt from disadvantages. Social systems can be tremendously complex, so if such systems are to be abstracted using a formal language (e.g. mathematical equations), we run the risk of losing too much in descriptiveness. To make things worse, in those cases where it appears possible to produce a satisfactory formal model of the social system under investigation, the resulting equations may be so complex that the formal model becomes mathematically intractable, thus failing to provide most of the benefits that motivated the process of formalisation in the first place. This is particularly relevant in the domain of the social sciences, where the systems under investigation often include non-linear relations (Axtell 2000). The usual approach then is to keep on adding simplifying hypotheses to the model—thus making it increasingly restrictive and unrealistic—until we obtain a tractable model that can be formally analysed with the available tools. We can find many examples of such assumptions in economics: instrumental rationality, perfect information, representative agents, etc. Most often these concepts are not included because economists think that the real world works in this way, but to make the models tractable (see, for instance, Conlisk 1996; Axelrod 1997a; Hernández 2004; Moss 2001, 2002). It seems that, in many cases, the use of formal symbolic systems tends to increase the danger of letting the pursuit for tractability be the driver of the modelling process.

But then, knowing that many of the hypotheses that researchers are obliged to assume may not hold in the real world, and could therefore lead to deceptive conclusions and theories, does this type of modelling representation preserve its advantages? Quoting G.F. Shove, it could be the case that sometimes "it is better to be vaguely right than precisely wrong".

The third symbolic system, computer modelling, opens up the possibility of building models that somewhat lie in between the descriptive richness of natural language and the analytical power of traditional formal approaches. This third type of representation is characterised by representing a model as a computer program (Gilbert and Troitzsch 1999). Using computer simulation we have the potential to build and study models that to some extent combine the intuitive appeal of verbal theories with the rigour of analytically tractable formal modelling.

In Axelrod's (1997a) opinion, computational simulation is the third way of doing science, which complements induction, the search for patterns in data, and deduction, the proof of theorems from a set of fixed axioms. In his opinion, simulation, like deduction, starts from an explicit set of hypotheses, but, rather than generating theorems, it generates data that can be inductively analysed.

While the division of modelling techniques presented above seems to be reasonably well accepted in the social simulation community—and we certainly find it useful—we do not fully endorse it. In our view, computer simulation does not constitute a distinctively new symbolic system or a uniquely different reasoning process by itself, but rather a (very useful) tool for exploring and analysing formal systems. We see computers as inference engines that are able to conduct algorithmic processes at a speed that the human brain cannot achieve. The inference derived from running a computer model is constructed by example and, in the general case, reads: *the results obtained from running the computer simulation follow (with logical consistency) from applying the algorithmic rules that define the model on the input parameters[1] used.*

In this way, simulations allow us to explore the properties of certain formal models that are intractable using traditional formal analyses (e.g. mathematical analyses), and they can also provide fundamentally new insights even when such analyses are possible. Like Gotts et al. (2003), we also believe that mathematical analysis and simulation studies should not be regarded as alternative and even opposed approaches to the formal study of social systems, but as complementary. They are both extremely useful tools to analyse formal models, and they are complementary in the sense that they can provide fundamentally different insights on one same model.

---

[1] By *input parameters* in this statement, we mean "everything that may affect the output of the model", e.g. the random seed, the pseudorandom number generator employed, and, potentially, information about the microprocessor and operating system on which the simulation was run, if these could make a difference.

To summarise, a computer program is a formal model (which can therefore be expressed in mathematical language, e.g. as a set of stochastic or deterministic equations), and computer simulation is a tool that enables us to study it in ways that go beyond mathematical tractability. Thus, the final result is a potentially more realistic—and still formal—study of a social system.

7.3 Agent-Based Modelling

7.3.1 Concept

As stated before, modelling is the process of building an abstraction of a system for a specific purpose—see Chap. 4 in this volume (Edmonds 2017; Epstein 2008) for a list of potential modelling goals. Thus, in essence, what distinguishes one modelling paradigm from another is precisely the way we construct that abstraction from the observed system.

In our view, agent-based modelling is a modelling paradigm with the defining characteristic that entities within the target system to be modelled—and the interactions between them—are explicitly and individually represented in the model (see Fig. 7.2). This is in contrast to other models where some entities are represented via average properties or via single representative agents. In many other models, entities are not represented at all, and it is only processes that are studied (e.g. a model of temperature variation as a function of pressure), and it is worth noting that such processes may well be already abstractions of the system.[2] The specific process of abstraction employed to build one particular model does not necessarily make it better or worse, only more or less useful for one purpose or another.

The specific way in which the process of abstraction is conducted in agent-based modelling is attractive for various reasons: it leads to (potentially) formal yet more natural and transparent descriptions of the target system, provides the possibility to model heterogeneity almost by definition, facilitates an explicit representation of the environment and the way other entities interact with it, and allows for the study of the bidirectional relations between individuals and groups, and it can also capture emergent behaviour (see Epstein 1999; Axtell 2000; Bonabeau 2002). Unfortunately, as one would expect, all these benefits often come at a price: most of the models built in this way are mathematically intractable. A common approach to study the behaviour of mathematically intractable formal models is to use computer simulation. It is for this reason that we often find the terms "agent-based modelling" and "agent-based simulation" used as synonyms in the scientific literature (Hare and Deadman 2004).

---

[2]The reader can see an interesting comparative analysis between agent-based and equation-based modelling in Parunak et al. (1998).

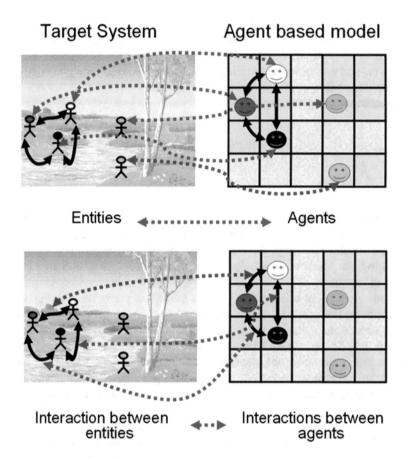

Fig. 7.2 In agent-based modelling, the entities of the system are represented explicitly and individually in the model. The limits of the entities in the target system correspond to the limits of the agents in the model, and the interactions between entities correspond to the interactions of the agents in the model (Edmonds 2001)

Thus, to summarise our thoughts in the context of the classification of modelling approaches in the social sciences, we understand that the essence of agent-based modelling is the individual and explicit representation of the entities and their interactions in the model, whereas computer simulation is a useful tool for studying the implications of formal models. This tool happens to be particularly well suited to explore and analyse agent-based models for the reasons explained above. Running an agent-based model in a computer provides a formal proof that a particular micro-specification is *sufficient* to generate the global behaviour that is observed during the simulation. If a model can be run in a computer, then it is in principle possible to express it in many different formalisms, e.g. as a set of mathematical equations. Such equations may be very complex, difficult to interpret, and impossible to solve, thus making the whole exercise of changing formalism frequently pointless, but what we find indeed useful is *the thought* that such an exercise *could* be undertaken, i.e. an agent-based model that can be run in a computer is not that different from the typical mathematical model. As a matter of fact, it is not difficult to formally characterise most agent-based models in a general way (Leombruni and Richiardi 2005).

7.3.2   Design, Implementation, and Use of an Agent-Based Model

Drogoul et al. (2003) identify three different roles in the design, implementation, and use of a typical agent-based model: the *thematician* (domain expert), the *modeller*, and the *computer scientist*. It is not unusual in the field to observe that one single person undertakes several or even all of these roles. We find that these three roles fit particularly well into the framework put forward by Edmonds (2001) to describe the process of modelling with an intermediate abstraction. Here we marry Drogoul et al.'s and Edmonds' views on modelling by dissecting one of Drogoul et al.'s roles and slightly expanding Edmonds' framework (Fig. 7.3). We then use our extended framework to identify the different types of assumptions that are made in each of the stages of the modelling process, the errors and artefacts that may occur in each of them, and the activities that can be conducted to avoid such errors and artefacts. We start by explaining the three different roles proposed by Drogoul et al. (2003).

The role of the *thematician* is undertaken by experts in the target domain. They are the ones that better understand the target system and, therefore, the ones who carry out the abstraction process that is meant to produce the first conceptualisation of the target system. Their job involves defining the objectives and the purpose of the modelling exercise, identifying the critical components of the system and the linkages between them, and also describing the most prominent causal relations. The output of this first stage of the process is most often a non-formal model expressed in natural language, and it may also include simple conceptual diagrams, e.g. block diagrams. The non-formal model produced may describe the system using potentially ambiguous terms (such as learning or imitation, without fully specifying how these processes actually take place).

The next stage in the modelling process is carried out by the role of the *modeller*. The modeller's task is to transform the non-formal model that the *thematician* aims to explore into the (formal) requirement specifications that the *computer scientist*—the third role—needs to formulate the (formal) executable model. This job involves (at least) three major challenges. The first one consists in acting as a mediator between two domains that are very frequently fundamentally different (e.g. sociology and computer science). The second challenge derives from the fact that in most cases, the *thematician*'s model is not fully specified, i.e. there are many formal models that would conform to it.[3] In other words, the formal model created by the *modeller* is most often just one of many possible particularisations

---

[3]Note that the *thematician* faces a similar problem when building his non-formal model. There are potentially an infinite number of models for one single target system.

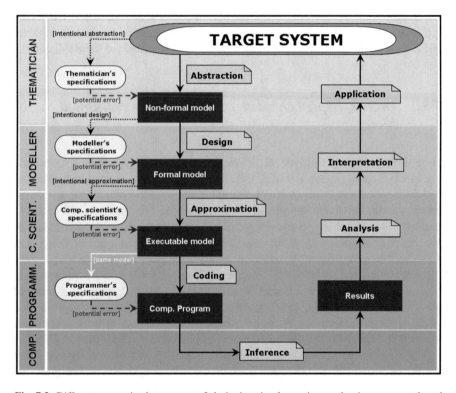

Fig. 7.3 Different stages in the process of designing, implementing, and using an agent-based model

of the *thematician*'s (more general) model. Lastly, the third challenge appears when the *thematician*'s model is not consistent, which may perfectly be the case since his model is often formulated using natural language. Discovering inconsistencies in natural language models is in general a non-trivial task. Several authors (e.g. Christley et al. 2004; Pignotti et al. 2005; and Polhill and Gotts 2006) have identified ontologies to be particularly useful for this purpose, especially in the domain of agent-based social simulation. Polhill and Gotts (2006) write:

> An ontology is defined by Gruber (1993) as 'a formal, explicit specification of a shared conceptualisation'. Fensel (2001) elaborates: ontologies are formal in that they are machine readable; explicit in that all required concepts are described; shared in that they represent an agreement among some community that the definitions contained within the ontology match their own understanding; and conceptualisations in that an ontology is an abstraction of reality. (Polhill and Gotts 2006, p. 51)

Thus, the modeller has the difficult—potentially unfeasible—task of finding a set of (formal and consistent) requirement specifications[4] where each individual require-

---

[4]Each individual member of this set can be understood as a different model or, alternatively, as a different parameterisation of one single—more general—model that would itself define the whole set.

ment specification of that set is a legitimate particular case of the *thematician*'s model and the set as a whole is *representative* of the *thematician*'s specifications (i.e. the set is sufficient to fully characterise the *thematician*'s model to a satisfactory extent).

Drogoul et al.'s third role is the *computer scientist*. Here we distinguish between *computer scientist* and *programmer*. It is often the case that the modeller comes up with a formal model that cannot be implemented in a computer. This could be, for example, because the model uses certain concepts that cannot be operated by present-day computers (e.g. real numbers, as opposed to floating-point numbers) or because running the model would demand computational requirements that are not yet available (e.g. in terms of memory and processing capacity). The job of the *computer scientist* consists in finding a suitable (formal) approximation to the *modeller*'s formal model that can be executed in a computer (or in several computers) given the available technology. To achieve this, the *computer scientist* may have to approximate and simplify certain aspects of the *modeller*'s formal model, and it is his job to make sure that these simplifications are not affecting the results significantly. As an example, Cioffi-Revilla (2002) warns about the potentially significant effects of altering system size in agent-based simulations.

The Navier-Stokes equations of fluid dynamics are a paradigmatic case in point. They are a set of non-linear differential equations that describe the motion of a fluid. Although these equations are considered a very good (formal and fully specified) model, their complexity is such that analytical closed-form solutions are available only for the simplest cases. For more complex situations, solutions of the Navier-Stokes equations must be estimated using approximations and numerical computation (Heywood et al. 1990; Salvi 2002). Deriving such approximations would be the task of the *computer scientist*'s role, as defined here.

One of the main motivations to distinguish between the *modeller*'s role and the *computer scientist*'s role is that, in the domain of agent-based social simulation, it is the description of the *modeller*'s formal model that is usually found in academic papers, even though the *computer scientist*'s model was used by the authors to produce the results in the paper. Most often the *modeller*'s model (i.e. the one described in the paper) simply *cannot* be run in a computer; it is the (potentially faulty) implementation of the *computer scientist*'s approximation to such a model that is really run by the computer. As an example, note that computer models described in scientific papers are most often expressed using equations in real arithmetic, whereas the models that actually run in computers almost invariably use floating-point arithmetic.

Finally, the role of the *programmer* is to implement the *computer scientist*'s executable model. In our framework, by definition of the role *computer scientist*, the model he produces must be executable and fully specified, i.e. it must include all the necessary information so given a certain input the model always produces the same output. Thus, the executable model will have to specify in its definition everything that could make a difference, e.g. the operating system and the specific pseudo-random number generator to be used. This is a subtle but important point, since it implies that the *programmer*'s job does not involve any process of abstraction or simplification; i.e. the executable model and the *programmer*'s specifications are by definition *the same* (see Fig. 7.3). (We consider two models to be *the same* if and only if they produce the same outputs when given the same inputs.) The *programmer*'s job consists "only" in writing the executable model in a programming language.[5] If the *programmer* does not make any mistakes, then the implemented model (e.g. the code) and the executable model will be the same.

Any mismatch between someone's specifications and the actual model he passes to the next stage is considered here an error (see Fig. 7.3). As an example, if the code implemented by the programmer is not the same model as his specifications, then there has been an implementation error. Similarly, if the *computer scientist*'s specifications are not complete (i.e. they do not define a unique model that produces a precise set of outputs for each given set of inputs), we say that he has made an error since the model he is producing is necessarily fully specified (by definition of the role). This opens up the question of how the executable model is defined: the executable model is *the same model* as the code if the *programmer* does not make any mistakes. So, to be clear, the distinction between the role of *computer scientist* and *programmer* is made here to distinguish (a) errors in the implementation of a fully specified model (which are made by the *programmer*) from (b) errors derived from an incomplete understanding of how a computer program works (which are made by the *computer scientist*). An example of the latter would be one where the *computer scientist*'s specifications stipulate the use of real arithmetic, but the executable model uses floating-point arithmetic.

It is worth noting that in an ideal world, the specifications created by each role would be written down. Unfortunately the world is far from ideal, and it is often the case that the mentioned specifications stay in the realm of mental models and never reach materialisation.

The reason for which the last two roles in the process are called "the computer scientist" and the "*programmer*" is because, as mentioned before, most agent-based models are implemented as computer programs and then explored through simulation (for tractability reasons). However, one could also think of, e.g. a mathematician conducting these two roles, especially if the formal model provided by the *modeller* can be solved analytically. For the sake of clarity, and without great loss of generality, we assume here that the model is implemented as a computer program, and its behaviour is explored through computer simulation.

Once the computer model is implemented, it is run, and the generated results are analysed. The analysis of the results of the computer model leads to conclusions on the behaviour of the *computer scientist*'s model, and, to the extent that the *computer scientist*'s model is a valid approximation of the *modeller*'s formal model, these conclusions also apply to the *modeller*'s formal model. Again, to the extent that

---

[5]There are some interesting attempts with INGENIAS (Pavón and Gómez-Sanz 2003) to use modelling and visual languages as programming languages rather than merely as design languages (Sansores and Pavón 2005; Sansores et al. 2006). These efforts are aimed at automatically generating several implementations of one single executable model (in various different simulation platforms).

the formal model is a legitimate particularisation of the non-formal model created by the *thematician*, the conclusions obtained for the *modeller*'s formal model can be interpreted in the terms used by the non-formal model. Furthermore, if the *modeller*'s formal model is representative of the *thematician*'s model, then there is scope for making general statements on the behaviour of the *thematician*'s model. Finally, if the *thematician*'s model is satisfactorily capturing social reality, then the knowledge inferred in the whole process can be meaningfully applied to the target system.

In the following section, we use our extended framework to identify the different errors and artefacts that may occur in each of the stages of the modelling process and the activities that can be conducted to avoid such errors and artefacts.

7.4 Errors and Artefacts

7.4.1 Definition of Error and Artefact and Their Relevance for Validation and Verification

Since the meanings of the terms validation, verification, error, and artefact are not uncontested in the literature, we start by stating the meaning that *we* attribute to each of them. For us, validation is the process of assessing how useful a model is for a certain purpose. A model is valid to the extent that it provides a satisfactory range of accuracy consistent with the intended application of the model (Kleijnen 1995; Sargent 2003).[6] Thus, if the objective is to accurately represent social reality, then validation is about assessing how well the model is capturing the essence of its empirical referent. This could be measured in terms of goodness of fit to the characteristics of the model's referent (Moss et al. 1997).

Verification—sometimes called "internal validation", e.g. by Taylor (1983), Drogoul et al. (2003), Sansores and Pavón (2005), or "internal validity", e.g. by Axelrod (1997a)—is the process of ensuring that the model performs in the manner intended by its designers and implementers (Moss et al. 1997). Let us say that a model is *correct* if and only if it would pass a verification exercise. Using our previous terminology, an expression of a model in a language is correct if and only if it is *the same* model as the developer's specifications. Thus, it could well be the case that a correct model is not valid (for a certain purpose). Conversely, it is also possible that a model that is not correct is actually valid for some purposes. Having said that, one would think that the chances of a model being valid are higher if it performs in the manner intended by its designer. To be sure, according to our definition of validation, what we want is a valid model, and we are interested in its correctness only to the extent that correctness contributes to make the model valid.

---

[6]See a complete epistemic review of the validation problem in Kleindorfer et al. (1998).

We also distinguish between errors and artefacts (Galán et al. 2009). *Errors* appear when a model does not comply with the requirement specifications self-imposed by its own developer. In simple words, an error is a mismatch between what the developer thinks the model is and what it actually is. It is then clear that there is an error in the model if and only if the model is not correct. Thus, verification is the process of looking for errors. An example of an implementation error would be the situation where the *programmer* intends to loop through the whole list of agents in the program, but he mistakenly writes the code so it only runs through a subset of them. A less trivial example of an error would be the situation where it is believed that a program is running according to the rules of real arithmetic, while the program is actually using floating-point arithmetic (Izquierdo and Polhill 2006; Polhill and Izquierdo 2005; Polhill et al. 2005, 2006).

In contrast to errors, *artefacts* relate to situations where there is no mismatch between what the developer thinks a model is and what it actually is. Here the mismatch is between the set of assumptions in the model that the developer thinks are producing a certain phenomenon and the assumptions that are the actual cause of such phenomenon. We explain this in detail. We distinguish between *core* and *accessory* assumptions in a model. *Core* assumptions are those whose presence is believed to be important for the purpose of the model. Ideally these would be the only assumptions present in the model. However, when producing a formal model, it is often the case that the developer is bound to include some additional assumptions for the only purpose of making the model complete. We call these *accessory* assumptions. Accessory assumptions are not considered a crucial part of the model; they are included to *make the model work*. We also distinguish between *significant* and *non-significant* assumptions. A *significant* assumption is an assumption that is the cause of some significant result obtained when running the model. Using this terminology, we define *artefacts* as significant phenomena caused by *accessory* assumptions in the model that are (mistakenly) deemed *non-significant*. In other words, an artefact appears when an accessory assumption that is considered non-significant by the developer is actually significant. An example of an artefact would be the situation where the topology of the grid in a model is accessory; it is believed that some significant result obtained when running the model is independent of the particular topology used (say, e.g. a grid of square cells), but it turns out that if an alternative topology is chosen (say, e.g. hexagonal cells), then the significant result is not observed.

The relation between artefacts and validation is not as straightforward as that between errors and verification. For a start, artefacts are relevant for validation only to the extent that identifying and understanding causal links in the model's referent is part of the purpose of the modelling exercise. We assume that this is the case, as indeed it usually is in the field of agent-based social simulation. A clear example is the Schelling-Sakoda model of segregation, which was designed to investigate the causal link between individual preferences and global patterns of segregation (Sakoda 1971; Schelling 1971, 1978). The presence of artefacts in a model implies that the model is not representative of its referent, since one can change some accessory assumption (thus creating an alternative model which still includes all the core assumptions) and obtain significantly different results. When this occurs, we run the risk of interpreting the results obtained with the (nonrepresentative) model beyond its scope (Edmonds and Hales 2005). Thus, to the extent that identifying causal links in the model's referent is part of the purpose of the modelling exercise, the presence of artefacts decreases the validity of the model. In any case, the presence of artefacts denotes a misunderstanding of what assumptions are generating what results.

7.4.2 Appearance of Errors and Artefacts

The dynamics of agent-based models are generally sufficiently complex that model developers themselves do not understand in exhaustive detail how the obtained results have been produced. As a matter of fact, in most cases if the exact results and the processes that generated them were known and fully understood in advance, there would not be much point in running the model in the first place. Not knowing exactly what to expect makes it impossible to tell whether any unanticipated results derive exclusively from what the researcher believes are the core assumptions in the model or whether they are due to errors or artefacts. The question is of crucial importance since, unfortunately, the truth is that there are many things that can go wrong in modelling.

Errors and artefacts may appear at various stages of the modelling process (Galán and Izquierdo 2005). In this section we use the extended framework explained in the previous section to identify the critical stages of the modelling process where errors and artefacts are most likely to occur.

According to our definition of artefact—i.e. significant phenomena caused by accessory assumptions that are not considered relevant—, artefacts *cannot* appear in the process of abstraction conducted by the *thematician*, since this stage consists precisely in distilling the *core* features of the target system. Thus, there should not be accessory assumptions in the *thematician*'s model. Nevertheless, there could still be issues with validation if, for instance, the *thematician*'s model is not capturing social reality to a satisfactory extent. Errors could appear in this stage because the *thematician*'s specifications are usually expressed in natural language, and rather than being written down, they are often transmitted orally to the modeller. Thus, an error (i.e. a mismatch between the *thematician*'s specifications and the non-formal model received by the *modeller*) could appear here if the *modeller* misunderstands some of the concepts put forward by the *thematician*.

The *modeller* is the role that may introduce the first artefacts in the modelling process. When formalising the *thematician*'s model, the *modeller* will often have to make a number of additional assumptions so the produced formal model is fully specified. By our definition of the two roles, these additional assumptions are not crucial features of the target system. If such accessory assumptions have a significant impact on the behaviour of the model and the *modeller* is not aware of it, then an artefact has been created. This would occur if, for instance, (a) the *thematician* did not specify any particular neighbourhood function, (b) different neighbourhood functions lead to different results, and (c) the *modeller* is using only one of them and believes that they all produce essentially the same results.

Errors could also appear at this stage, although it is not very likely. This is so because the specifications that the *modeller* produces must be formal, and they are therefore most often written down in a formal language. When this is the case, there is little room for misunderstanding between the *modeller* and the computer scientist, i.e. the *modeller*'s specifications and the formal model received by the *computer scientist* would be the same, and thus there would be no error at this stage.

The role of the *computer scientist* could introduce artefacts in the process. This would be the case if, for instance, his specifications require the use of a particular pseudorandom number generator; he believes that this choice will not have any influence in the results obtained, but it turns out that it does. Similar examples could involve the arbitrary selection of an operating system or a specific floating-point arithmetic that had a significant effect on the output of the model.

Errors can quite easily appear in between the role of the *computer scientist* and the role of the programmer. Note that in our framework, any mismatch between the *computer scientist*'s specifications and the executable model received by the *programmer* is considered an error. In particular, if the *computer scientist's* specifications are not executable, then there is an error. This could be, for instance, because the *computer scientist's* specifications stipulate requirements that cannot be executed with present-day computers (e.g. real arithmetic) or because it does not specify all the necessary information to be run in a computer in an unequivocal way (e.g. it does not specify a particular pseudorandom number generator). The error then may affect the validity of the model significantly, or may not.

Note from the previous examples that if the *computer scientist* does not provide a fully executable set of requirement specifications, then he is introducing an error, since in that case, the computer program (which is executable) would be necessarily different from his specifications. On the other hand, if he does provide an executable model but in doing so he makes an arbitrary accessory assumption that turns out to be significant, then he is introducing an artefact.

Finally, the *programmer* cannot introduce artefacts because his specifications are the same as the executable model by definition of the role (i.e. the *programmer* does not have to make any accessory assumptions). However, he may make mistakes when creating the computer program from the executable model.

7.4.3   Activities Aimed at Detecting Errors and Artefacts

In this section we identify various activities that the different roles defined in the previous sections can undertake to detect errors and artefacts. We consider the use of these techniques as a very recommendable and eventually easy to apply practice. In spite of this, we should warn that, very often, these activities may require a considerable human and computational effort.

Modeller's activities:

- Develop and analyse new formal models by implementing alternative accessory assumptions while keeping the core assumptions identified by the *thematician*. This exercise will help to detect artefacts. Only those conclusions which are not falsified by any of these models will be valid for the *thematician*'s model. As an example, see Galán and Izquierdo (2005), who studied different instantiations of one single conceptual model by implementing different evolutionary selection mechanisms. Takadama et al. (2003) conducted a very similar exercise implementing three different learning algorithms for their agents. In a collection of papers, Klemm et al. (2003a, 2003b, 2003c, 2005) investigate the impact of various accessory assumptions in Axelrod's model for the dissemination of culture (Axelrod 1997b). Another example of studying different formal models that address one single problem is provided by Kluver and Stoica (2003).
- Conduct a more exhaustive exploration of the parameter space within the boundaries of the *thematician*'s specifications. If we obtain essentially the same results using the wider parameter range, then we will have broadened the scope of the model, thus making it more representative of the *thematician*'s model. If, on the other hand, results change significantly, then we will have identified artefacts. This type of exercise has been conducted by, e.g. Castellano et al. (2000) and Galán and Izquierdo (2005).
- Create abstractions of the formal model which are mathematically tractable. An example of one possible abstraction would be to study the *expected* motion of a dynamic system (see the studies conducted by Galán and Izquierdo (2005), Edwards et al. (2003), and Castellano et al. (2000) for illustrations of mean-field approximations). Since these mathematical abstractions do not correspond in a one-to-one way with the specifications of the formal model, any results obtained with them will not be conclusive, but they may suggest parts of the model where there may be errors or artefacts.
- Apply the simulation model to relatively well-understood and predictable situations to check that the obtained results are in agreement with the expected behaviour (Gilbert and Terna 2000).

Computer scientist's activities:

- Develop mathematically tractable models of certain aspects, or particular cases, of the modeller's formal model. The analytical results derived with these models should match those obtained by simulation; a disparity would be an indication of the presence of errors.
- Develop new executable models from the modeller's formal model using alternative modelling paradigms (e.g. procedural vs. declarative). This activity will help to identify artefacts. As an example, see Edmonds and Hales' (2003) reimplementation of Riolo et al. (2001) model of cooperation among agents using tags. Edmonds reimplemented the model using SDML (declarative), whereas Hales reprogrammed the model in Java (procedural).

- Rerun the same code in different computers, using different operating systems, with different pseudorandom number generators. These are most often accessory assumptions of the executable model that are considered non-significant, so any detected difference will be a sign of an artefact. If no significant differences are detected, then we can be confident that the code comprises all the assumptions that could significantly influence the results. This is a valuable finding that can be exploited by the programmer (see next activity). As an example, Polhill et al. (2005) explain that using different compilers can result in the application of different floating-point arithmetic systems to the simulation run.

Programmer's activities:

- Reimplement the code in different programming languages. Assuming that the code contains all the assumptions that can influence the results significantly, this activity is equivalent to creating alternative representations of the same executable model. Thus, it can help to detect errors in the implementation. There are several examples of this type of activity in the literature. Bigbee et al. (2007) reimplemented Sugarscape (Epstein and Axtell 1996) using MASON. Xu et al. (2003) implemented one single model in Swarm and Repast. The reimplementation exercise conducted by Edmonds and Hales (2003) applies here too.
- Analyse particular cases of the executable model that are mathematically tractable. Any disparity will be an indication of the presence of errors.
- Apply the simulation model to extreme cases that are perfectly understood (Gilbert and Terna 2000). Examples of this type of activity would be to run simulations without agents or with very few agents, explore the behaviour of the model using extreme parameter values, or model very simple environments. This activity is common practice in the field.

7.5 Summary

The dynamics of agent-based models are usually so complex that their own developers do not *fully* understand how they are generated. This makes it difficult, if not impossible, to discern whether observed significant results are legitimate logical implications of the assumptions that the model developer is interested in or whether they are due to errors or artefacts in the design or implementation of the model.

Errors are mismatches between what the developer believes a model is and what the model actually is. Artefacts are significant phenomena caused by accessory assumptions in the model that are (mistakenly) considered non-significant. Errors and artefacts prevent developers from correctly understanding their simulations. Furthermore, both errors and artefacts can significantly decrease the validity of a model, so they are best avoided.

In this chapter we have outlined a general framework that summarises the process of designing, implementing, and using agent-based models. Using this framework we have identified the different types of errors and artefacts that may occur in each of the stages of the modelling process. Finally, we have proposed several activities that can be conducted to avoid each type of error or artefact. Some of these activities include repetition of experiments in different platforms, reimplementation of the code in different programming languages, reformulation of the conceptual model using different modelling paradigms, and mathematical analyses of simplified versions or particular cases of the model. Conducting these activities will surely increase our understanding of a particular simulation model.

Acknowledgements The authors have benefited from the financial support of the Spanish Ministry of Education and Science (projects CSD2010-00034, DPI2004-06590, DPI2005-05676, and TIN2008-06464-C03-02) and of the Junta de Castilla y León (projects BU034A08 and VA006B09). We are also very grateful to Nick Gotts, Gary Polhill, Bruce Edmonds, and Cesáreo Hernández for many discussions on the philosophy of modelling.

Further Reading

Gilbert (2007) provides an excellent basic introduction to agent-based modelling. Chapter 4 summarises the different stages involved in an agent-based modelling project, including verification and validation. The paper entitled "Some myths and common errors in simulation experiments" (Schmeiser 2001) discusses briefly some of the most common errors found in simulation from a probabilistic and statistical perspective. The approach is not focused specifically on agent-based modelling but on simulation in general. Yilmaz (2006) presents an analysis of the life cycle of a simulation study and proposes a process-centric perspective for the validation and verification of agent-based computational organisation models. An antecedent of this chapter can be found in Galán et al. (2009). Finally, Chap. 9 in this volume (David et al. 2017) discusses validation in detail.

References

Axelrod, R. M. (1997a). Advancing the art of simulation in the social sciences. In R. Conte, R. Hegselmann, & P. Terna (Eds.), *Simulating social phenomena*. (Lecture Notes in Economics and Mathematical Systems, 456) (pp. 21–40). Berlin: Springer.

Axelrod, R. M. (1997b). The dissemination of culture: A model with local convergence and global polarization. *Journal of Conflict Resolution, 41*(2), 203–226.

Axtell, R. L. (2000). Why agents? On the varied motivations for agent computing in the social sciences. In C. M. Macal & D. Sallach (Eds.), *Proceedings of the workshop on agent simulation: applications, models, and tools* (pp. 3–24). Argonne National Laboratory: Argonne, IL.

Axtell, R. L., & Epstein, J. M. (1994). Agent based modeling: Understanding our creations. *The Bulletin of the Santa Fe Institute, 1994*, 28–32.

Bigbee, T., Cioffi-Revilla, C., & Luke, S. (2007). Replication of sugarscape using MASON. In T. Terano, H. Kita, H. Deguchi, & K. Kijima (Eds.), *Agent-based approaches in economic and social complex systems IV: Post-proceedings of the AESCS international workshop 2005* (pp. 183–190). Tokyo: Springer.

Bonabeau, E. (2002). Agent-based modeling: Methods and techniques for simulating human systems. *Proceedings of the National Academy of Sciences of the United States of America, 99*(2), 7280–7287.

Castellano, C., Marsili, M., & Vespignani, A. (2000). Nonequilibrium phase transition in a model for social influence. *Physical Review Letters, 85*(16), 3536–3539.

Christley, S., Xiang, X., & Madey, G. (2004). Ontology for agent-based modeling and simulation. In C. M. Macal, D. Sallach, & M. J. North (Eds.), *Proceedings of the agent 2004 conference on social dynamics: interaction, reflexivity and emergence.* Chicago, IL: Argonne National Laboratory and The University of Chicago. http://www.agent2005.anl.gov/Agent2004.pdf.

Cioffi-Revilla, C. (2002). Invariance and universality in social agent-based simulations. *Proceedings of the National Academy of Sciences of the United States of America, 99*(3), 7314–7316.

Conlisk, J. (1996). Why bounded rationality? *Journal of Economic Literature, 34*(2), 669–700.

David, N., Fachada, N., & Rosa, A. C. (2017). Verifying and validating simulations. doi:https://doi.org/10.1007/978-3-319-66948-9_9.

Drogoul, A., Vanbergue, D., & Meurisse, T. (2003). Multi-agent based simulation: Where are the agents? In J. S. Sichman, F. Bousquet, & P. Davidsson (Eds.), *Proceedings of MABS 2002 multi-agent-based simulation.* (Lecture Notes in Computer Science, 2581) (pp. 1–15). Bologna: Springer.

Edmonds, B. (2001). The use of models: making MABS actually work. In S. Moss & P. Davidsson (Eds.), *Multi-agent-based simulation. (Lecture notes in artificial intelligence, 1979)* (pp. 15–32). Berlin: Springer.

Edmonds, B. (2005). Simulation and complexity: How they can relate. In V. Feldmann & K. Mühlfeld (Eds.), *Virtual worlds of precision: Computer-based simulations in the sciences and social sciences* (pp. 5–32). Lit-Verlag: Münster.

Edmonds, B. (2017). Different modelling purposes. doi:https://doi.org/10.1007/978-3-319-66948-9_4.

Edmonds, B., & Hales, D. (2003). Replication, replication and replication: Some hard lessons from model alignment. *Journal of Artificial Societies and Social Simulation, 6*(4). http://jasss.soc.surrey.ac.uk/6/4/11.html.

Edmonds, B., & Hales, D. (2005). Computational Simulation as Theoretical Experiment. *Journal of Mathematical Sociology, 29*, 1–24.

Edwards, M., Huet, S., Goreaud, F., & Deffuant, G. (2003). Comparing an individual-based model of behaviour diffusion with its mean field aggregate approximation. *Journal of Artificial Societies and Social Simulation, 6*(4). http://jasss.soc.surrey.ac.uk/6/4/9.html.

Epstein, J. M. (1999). Agent-based computational models and generative social science. *Complexity, 4*(5), 41–60.

Epstein, J. M. (2008). Why model?. *Journal of Artificial Societies and Social Simulation, 11*(4), 12. http://jasss.soc.surrey.ac.uk/11/4/12.html.

Epstein, J. M., & Axtell, R. L. (1996). *Growing artificial societies: Social science from the bottom up.* Cambridge, MA: Brookings Institution Press/MIT Press.

Fensel, D. (2001). *Ontologies: A silver bullet for knowledge management and electronic commerce.* Berlin: Springer.

Galán, J. M., et al. (2009). Errors and artefacts in agent-based modelling. *Journal of Artificial Societies and Social Simulation, 12*(1). http://jasss.soc.surrey.ac.uk/12/1/1.html.

Galán, J. M., & Izquierdo, L. R. (2005). Appearances can be deceiving: lessons learned re-implementing Axelrod's 'evolutionary approach to norms'. *Journal of Artificial Societies and Social Simulation, 8*(3). http://jasss.soc.surrey.ac.uk/8/3/2.html

Gilbert, N. (1999). Simulation: A new way of doing social science. *The American Behavioral Scientist, 42*(10), 1485–1487.

Gilbert, N. (2007). *Agent-based models.* London: Sage Publications.

Gilbert, N., & Terna, P. (2000). How to build and use agent-based models in social science. *Mind & Society, 1*(1), 57–72.

Gilbert, N., & Troitzsch, K. G. (1999). *Simulation for the social scientist*. Buckingham: Open University Press.

Gotts, N. M., Polhill, J. G. & Adam, W. J. (2003, 18–21 September). Simulation and analysis in agent-based modelling of land use change. *Online proceedings of the first conference of the European Social Simulation Association*, Groningen, The Netherlands, http://www.uni-koblenz.de/~essa/ESSA2003/gotts_polhill_adam-rev.pdf.

Gruber, T. R. (1993). A translation approach to portable ontology specifications. *Knowledge Acquisition, 5*(2), 199–220.

Hare, M., & Deadman, P. (2004). Further towards a taxonomy of agent-based simulation models in environmental management. *Mathematics and Computers in Simulation, 64*(1), 25–40.

Hernández, C. (2004). Herbert A. Simon, 1916-2001, y el Futuro de la Ciencia Económica. *Revista Europea De Dirección y Economía De La Empresa, 13*(2), 7–23.

Heywood, J. G., Masuda, K., Rautmann, R., & Solonnikov, V. A. (Eds.). (1990). *The Navier-Stokes equations: Theory and numerical methods; Proceedings of a conference held at Oberwolfach, FRG, Sept. 18–24, 1988*. (Lecture Notes in Mathematics, 1431). Berlin: Springer.

Holland, J. H., & Miller, J. H. (1991). Artificial adaptive agents in economic theory. *American Economic Review, 81*(2), 365–370.

Izquierdo, L. R., & Polhill, J. G. (2006). Is your model susceptible to floating point errors? *Journal of Artificial Societies and Social Simulation, 9*(4). http://jasss.soc.surrey.ac.uk/9/4/4.html.

Kleijnen, J. P. C. (1995). Verification and validation of simulation models. *European Journal of Operational Research, 82*(1), 145–162.

Kleindorfer, G. B., O'Neill, L., & Ganeshan, R. (1998). Validation in simulation: Various positions in the philosophy of science. *Management Science, 44*(8), 1087–1099.

Klemm, K., Eguíluz, V., Toral, R., & San Miguel, M. (2003a). Role of dimensionality in Axelrod's model for the dissemination of culture. *Physica A, 327*, 1–5.

Klemm, K., Eguíluz, V., Toral, R., & San Miguel, M. (2003b). Global culture: A noise-induced transition in finite systems. *Physical Review E, 67*(4), 045101.

Klemm, K., Eguíluz, V., Toral, R., & San Miguel, M. (2003c). Nonequilibrium transitions in complex networks: A model of social interaction. *Physical Review E, 67*(2), 026120.

Klemm, K., Eguíluz, V., Toral, R., & San Miguel, M. (2005). Globalization, polarization and cultural drift. *Journal of Economic Dynamics & Control, 29*(1–2), 321–334.

Kluver, J., & Stoica, C. (2003). Simulations of group dynamics with different models. *Journal of Artificial Societies and Social Simulation, 6*(4). http://jasss.soc.surrey.ac.uk/6/4/8.html.

Leombruni, R., & Richiardi, M. (2005). Why are economists sceptical about agent-based simulations? *Physica A, 355*, 103–109.

Moss, S. (2001). Game theory: Limitations and an alternative. *Journal of Artificial Societies and Social Simulation, 4*(2). http://jasss.soc.surrey.ac.uk/4/2/2.html.

Moss, S. (2002). Agent based modelling for integrated assessment. *Integrated Assessment, 3*(1), 63–77.

Moss, S., Edmonds, B., & Wallis, S. (1997). *Validation and verification of computational models with multiple cognitive agents* (Report no. 97–25). Manchester: Centre for Policy Modelling, http://cfpm.org/cpmrep25.html.

Ostrom, T. (1988). Computer simulation: The third symbol system. *Journal of Experimental Social Psychology, 24*(5), 381–392.

Parunak, H. V. D., Savit, R., & Riolo, R. L. (1998). Agent-based modeling vs. equation-based modeling: A case study and users' guide. In J. S. Sichman, R. Conte, & N. Gilbert (Eds.), *Multi-agent systems and agent-based simulation. (Lecture notes in artificial intelligence 1534)* (pp. 10–25). Berlin: Springer.

Pavón, J. & Gómez-Sanz, J. (2003). Agent oriented software engineering with INGENIAS. In V. Marik, J. Müller & M. Pechoucek (Eds.), *Multi-agent systems and applications III, 3rd international central and eastern European conference on multi-agent systems, CEEMAS. (Lecture notes in artificial intelligence, 2691)* (pp. 394–403); Berlin, Heidelberg: Springer.

Pignotti, E., Edwards, P., Preece, A., Polhill, J.G. & Gotts, N.M. (2005). Semantic support for computational land-use modelling. *Proceedings of the 5th international symposium on cluster computing and the grid (CCGRID 2005)* (pp. 840–847). Piscataway, NJ: IEEE Press.

Polhill, J. G. & Gotts, N. M. (2006, August 21–25). A new approach to modelling frameworks. *Proceedings of the first world congress on social simulation.* (Vol. 1, pp. 215–222), Kyoto, Japan.

Polhill, J. G., & Izquierdo, L. R. (2005). Lessons learned from converting the artificial stock market to interval arithmetic. *Journal of Artificial Societies and Social Simulation, 8*(2). http://jasss.soc.surrey.ac.uk/8/2/2.html.

Polhill, J. G., Izquierdo, L. R., & Gotts, N. M. (2005). The ghost in the model (and other effects of floating point arithmetic). *Journal of Artificial Societies and Social Simulation, 8*(1). http://jasss.soc.surrey.ac.uk/8/1/5.html.

Polhill, J. G., Izquierdo, L. R., & Gotts, N. M. (2006). What every agent based modeller should know about floating point arithmetic. *Environmental Modelling & Software, 21*(3), 283–309.

Riolo, R. L., Cohen, M. D., & Axelrod, R. M. (2001). Evolution of cooperation without reciprocity. *Nature, 411*, 441–443.

Sakoda, J. M. (1971). The checkerboard model of social interaction. *Journal of Mathematical Sociology, 1*(1), 119–132.

Salvi, R. (2002). *The Navier-Stokes equation: Theory and numerical methods. (Lecture notes in pure and applied mathematics).* New York: Marcel Dekker.

Sansores, C., & Pavón, J. (2005, November 14–18). Agent-based simulation replication: A model driven architecture approach. In A. F. Gelbukh, A. de Albornoz, & H. Terashima-Marín (Eds.), *Proceedings of MICAI 2005: Advances in artificial intelligence, 4th Mexican international conference on artificial intelligence. (Lecture notes in computer science, 3789)* (pp. 244–253), Monterrey, Mexico. Berlin, Heidelberg: Springer.

Sansores, C., Pavón, J., & Gómez-Sanz, J. (2006, July 25). Visual modeling for complex agent-based simulation systems. In J. S. Sichman & L. Antunes (Eds.), *Multi-agent-based simulation VI, International workshop, MABS 2005, revised and invited papers. (Lecture notes in computer science, 3891)* (pp. 174–189), Utrecht, The Netherlands. Berlin, Heidelberg: Springer.

Sargent, R. G. (2003). Verification and validation of simulation models. In S. Chick, P. J. Sánchez, D. Ferrin, & D. J. Morrice (Eds.), *Proceedings of the 2003 winter simulation conference* (pp. 37–48). Piscataway, NJ: IEEE.

Schelling, T. C. (1971). Dynamic models of segregation. *Journal of Mathematical Sociology, 1*(2), 47–186.

Schelling, T. C. (1978). *Micromotives and macrobehavior.* New York: Norton.

Schmeiser, B. W. (2001, December 09–12). Some myths and common errors in simulation experiments. In B. A. Peters, J. S. Smith, D. J. Medeiros, & M. W. Rohrer (Eds.), *Proceedings of the winter simulation conference* (Vol. 1, pp. 39–46), Arlington, VA.

Takadama, K., Suematsu, Y. L., Sugimoto, N., Nawa, N. E., & Shimohara, K. (2003). Cross-element validation in multiagent-based simulation: Switching learning mechanisms in agents. *Journal of Artificial Societies and Social Simulation, 6*(4). http://jasss.soc.surrey.ac.uk/6/4/6.html.

Taylor, A. J. (1983). The verification of dynamic simulation models. *Journal of the Operational Research Society, 34*(3), 233–242.

Xu, J., Gao, Y. & Madey, G. (2003, April 13–15). A docking experiment: swarm and repast for social network modeling. In *Seventh annual swarm researchers conference (SwarmFest 2003.* Notre Dame, IN.

Yilmaz, L. (2006). Validation and verification of social processes within agent-based computational organization models. *Computational & Mathematical Organization Theory, 12*(4), 283–312.

Chapter 8
The Importance of Ontological Structure: Why Validation by 'Fit-to-Data' Is Insufficient

Gary Polhill and Doug Salt

Abstract This chapter will briefly describe some common methods by which people make quantitative estimates of how well they expect empirical models to make predictions. However, the chapter's main argument is that fit-to-data, the traditional yardstick for establishing confidence in models, is not quite the solid ground on which to build such belief some people think it is, especially for the kind of system agent-based modelling is usually applied to. Further, the chapter will show that the amount of data required to establish confidence in an arbitrary model by fit-to-data is often infeasible, unless there is some appropriate 'big data' available. This arbitrariness can be reduced by constraining the choice of model. In agent-based models, these constraints are introduced by their descriptiveness rather than by removing variables from consideration or making assumptions for the sake of simplicity. By comparing with neural networks, we show that agent-based models have a richer ontological structure. For agent-based models, in particular, this richness means that the ontological structure has a greater significance and yet is all too commonly taken for granted or assumed to be 'common sense'. The chapter therefore also discusses some approaches to validating ontologies.

Why Read This Chapter?

When you have built an agent-based model, you need some way of assessing how 'good' it is. We will tell you how this is done traditionally in empirical contexts, through measures of fit-to-data. You will learn why fitting to data is not enough in the kind of situation where agent-based models are useful and why you also need to assess the model's ontological structure. The chapter will tell you what the ontological structure is, how to assess it and whether and if so how it can be traded off against fit-to-data.

G. Polhill (✉) • D. Salt
The James Hutton Institute, Craigiebuckler, Aberdeen, AB15 8QH, UK
e-mail: gary.polhill@hutton.ac.uk; doug.salt@hutton.ac.uk

© Springer International Publishing AG 2017   141
B. Edmonds, R. Meyer (eds.), *Simulating Social Complexity*,
Understanding Complex Systems, https://doi.org/10.1007/978-3-319-66948-9_8

8.1 Introduction

The chapter argues for the importance of the ontological structure in social simulation – that is, what basic entities exist, their attributes and their relationships with each other. In particular, simply getting a good fit of the outcomes to data is not enough to establish the adequacy of the model. To make this point vivid, it considers the opposite extreme, an example of a machine learning algorithm where the 'model' is simply induced from the data – where there is the minimum predefined ontological structure. The example chosen is that of neural networks, though almost any black-box machine learning approach would have done as well.

Neural networks are universal function approximators (Hornik et al. 1989). This means that given a set of data, they can approximate it to within an arbitrary degree of accuracy simply by adding more parameters. Though it may seem strange to compare neural networks with agent-based models for the purposes of validation and generalization, there are useful lessons from so doing that illustrate where agent-based models add value to traditional modelling approaches and why validation is not so straightforward. The main contrast between neural networks and agent-based models comes down to the 'ontology'. Essentially, apart from the labels assigned to the input and output units of a neural network, neural networks don't have an ontology at all. What they do have is a mathematical structure that allows the number of parameters to be arbitrarily varied and, with that, arbitrary degrees of fit to a set of data to be achieved. By contrast, agent-based models have a rich and highly descriptive ontology but, like neural networks, potentially have a large number of parameters that can be varied (especially if we consider each agent uniquely).

In this chapter, we examine some approaches to validation and generalization in neural networks and consider what they tell us about agent-based modelling. Our arguments are that validation needs to look beyond the relatively trivial question of fit-to-data, especially in non-ergodic complex systems. Rather than being a weakness of agent-based modelling, the challenges of validation and generalization point to its strengths, especially in social systems, where the language used to describe them is influenced by evolving cultural considerations.

The chapter starts with an introduction to neural networks followed by how these are calibrated and validated. It then discusses the issue at the heart of the chapter the importance of predetermined model bias – that is the imposed structure derived from knowledge about what is being modelled. It uses a particular measure (the VC dimension) to show the amount of data needed to infer a good model without imposing such a bias is typically infeasible. It summarizes the various measures one might use for checking fit-to-data. This paves the way for a discussion on validating ontologies discussing a number of approaches and the tools that might be useful for this.

8.1.1   Introduction to Neural Networks

In this section, we will briefly explain what neural networks are, the mathematical formulas that underpin them (in Appendix 1) and the way they are structured. The main points we wish to introduce are that, though neural networks have tremendous potential to approximate data, there is nothing about their structure or the mathematics underpinning their functioning that necessarily reflects any structure or mathematics in whatever system the data were taken from.

Neural networks were originally conceived as simulations of the brains but are essentially networks of nonlinear functions with parameters that are adjusted according to a learning rule. There are several different kinds of neural network mathematically speaking, and for each kind, there can be several different learning rules and minor adaptations and variations thereof. Biologically, a neuron is a cell with axons connecting it to other neurons. In an agent-based simulation of a brain, we would simulate a neuron as an agent and an axon as a link. The behaviour of the neuron is simply to emit an electrical pulse periodically. The more frequent the pulse, the more 'excited' the neuron. Connections between neurons can be excitatory or inhibitory. An excitatory connection means that there is a positive relationship between the excitation of the two connected neurons: all other things being equal, one neuron's excitement increases that of the other. An inhibitory connection means that the relationship is negative – one neuron's excitement decreases that of the other. The connection has a strength – the stronger the connection between one neuron and another, the more significant the relationship is in comparison with other neurons the neuron is connected to.

When simulating neurons, the pulsation is ignored and the frequency of pulsation modelled as a variable. Simulated neurons are typically called *nodes*. The axons form the links in a directed graph connecting the nodes, and the directedness means that nodes have input axons and output axons. Simulated axons are typically called *weights*, largely because it is the value of the weight (representing the strength of the connection) that is of primary interest. The weights of a neural network are its parameters, and the job of the learning algorithm is to determine their values. The qualitative description of the behaviour of neurons is of course given a precise mathematical specification in simulated neural networks; this is provided in Appendix 1 for the benefit of those who are interested.

A further simplification of the structure of the network is to arrange the nodes into distinct layers. (It can be proved that this does not lead to loss of potential functionality.) This simplification means that the choice of network structure is simply a question of determining the number of layers, and for the layers that are not input or output layers (the so-called *hidden* layers), the number of nodes to use in each layer. The number of nodes in the input and output layers is of course determined by the dimensionalities of the domain and range of the function to be approximated. It has been proved (Cybenko 1989; Funahashi 1989; Hornik et al. 1989) that one hidden layer is sufficient to approximate any function. Although having more hidden layers can mean that the contribution of the weights closer to the input units to the difference between the actual and desired output of the network is more diluted, it can also be shown that more efficient network topologies (in terms of number of weights) involving two hidden layers can achieve the same level of accuracy that can be achieved with one hidden layer (Cheng and Titterington 1994; Chester 1990).

The algorithms used to determine the weights such that the network as a whole provides a good fit-to-data are not particularly of interest here. This material is covered in various introductory textbooks on neural networks (e.g. Bishop 1995; Gurney 1997; Hertz et al. 1991). What is of interest is that, having seen the structure of a neural network and what it does, it is immediately clear that there is nothing in that structure that reflects the real world, except for the assignment of input nodes and output nodes to specific variables in the data to be fitted. The numbers of hidden nodes and layers must capture any patterns in how the real-world mechanisms interact, the choice of which essentially reflects how complex the modeller expects the function to fit the data to need to be.

Neural networks have the absolute minimum in the way of ontological structure it is possible to have. Their 'content' comes from the data they are trained to fit. We thus next discuss the principles behind adjusting a model to fit its data, checking a model's fit to available evidence and how this is done in neural networks.

8.1.2   Calibration, Validation and Generalization in Neural Networks

Calibration, validation and generalization are three steps in the development and application of any model. We discuss them here in relation to neural networks, first with a view to clarifying what we mean by those terms and second to discussing some of the ways in which generalization (the application of the model) can go wrong even for a well-validated model.

Since various terms are used in the modelling literature for the three processes intended here by the words 'calibration', 'validation' and 'generalization', it is best to be clear what is meant. The process begins with a set of data, with some explanatory (input) variables and response (output) variables, and a model with a predefined structure that has some parameters that can be adjusted. The data are split into two not necessarily equal parts. The larger part is typically used for *calibration*: adjusting the parameters so that the difference between the model's prediction for the input variables in the data and the corresponding output variables in the data (the *error*) is minimized. In neural networks, this is referred to as *training* and entails adjusting the values of all the weights.

There is a caveat to the use of the term 'minimization'. For reasons such as measurement error in the data, if a function is capable of providing an exact fit to the data, this is potentially undesirable and is seen as *overfitting*. So, when we say we want to minimize the error, it is usually understood that we wish to do so without overfitting.

Bearing this in mind, at the end of the calibration process, you have a parameterized neural network with all the weights specified that you now want to be able to use to make predictions with; except, of course, if you want to have some degree of confidence in those predictions. *Validation* is the process of developing that confidence, and it is achieved by using the data you kept aside and didn't use during calibration to estimate how good your future predictions will be. So, having reached a point where you are happy with the error on the calibration data, you use the validation data to tell you how confident you should be in the model you have fitted: the error rate on the validation set is an estimate of the expected error rate for prediction.

Generalization is the ability of the model to provide output for untrained input. There are two aspects to this. The first is whether the required input can be represented using the formalism provided by the model. In the case of neural networks, the question seems simply to be whether the input can be adequately expressed using the same set of dimensions and any encoding thereof as the data used for calibration and validation. It may seem unfair to expect a model to be able to provide output for cases that cannot be expressed using the 'language' the model was built with. However, sometimes, arguably, that is what happens. Measures of inflation, for example, are based on a 'basket of goods' that changes from time to time as people's buying habits change. This change arguably changes the meaning of inflation. Though something of a straw man, if you have calibrated a model using a measure of inflation that uses one basket of goods and then naively expect it to give meaningful output for a measure of inflation that uses another, then perhaps you are expecting the model to provide output for cases that cannot be expressed using the language the model was built with.[1] Similar problems exist with other social statistics that might be used as variables input to or output from a model, particularly where there are changes in the way the variables are measured from one region to another.

A second problem comes from what you left out of the model when you first built it. Although this too may seem like an unfair criticism, perhaps when you built the original model, a particular variable was not included as an input variable because it was not seen as having any significant relationship with the output. Since the model was calibrated and validated, however, a large change in the ignored variable might have occurred that has affected the relationships between the variables you did include. So, although when you come to compute a prediction for a new input you have all the data you need, and can perform the computation, really, the values for the variables you have as inputs to your model do not adequately reflect the scenario any more. This is known as 'omitted variable bias' in the econometrics literature (see, e.g. Clarke 2005).

---

[1]Less naively, you would use a calculated inflation figure for the old basket of goods as input to the model; however, if people are not buying things in the old basket, the model may still not be providing meaningful output.

A final problem is a consequence of encoding variables that have nominal values. Assuming an appropriate encoding of nominals in the input variables of the model, the calibration and validation data may only have provided a subset of the nominals the variable can have. The generalization may, however, be for a value of the nominal that was not in the data used to construct the model. For neural networks, this is less of an issue than with symbolic AI machine learning algorithms: one of the supposed advantages of neural networks is that they are less 'brittle' with respect to the language of representation of the states of the world, because they do not rely on the language having a specific vocabulary to represent every possible state that might ever be of interest (Aha 1992; Hanson and Burr 1990; Holland 1986).

In essence, calibration is the process of finding the parameters of a neural network (or more generally, any model) that best fit your data. Validation is the process of establishing the confidence you can expect to have in the predictions of the model based on the data you have got. Generalization is the capability of a model to make predictions in new situations. There are various reasons why that capability may be questioned. Apart from the relevance of the data used for calibration and validation in the new context, the reasons relate to how the modeller chose to encode, or represent, the data.

8.1.3   Bias vs. Variance

The representation of the data is not the only choice the modeller makes. This section covers the dilemma a modeller faces when choosing the structure of the model. In the case of neural networks, that structure is the number of layers and hidden units, which collectively determine the number of weights or parameters the model has. The fewer the number of parameters, the easier the model is to calibrate, but there is a risk of oversimplification. Since it is so easy to add more parameters to a neural network, there is a temptation to add more parameters. We introduce some rather advanced mathematics (Vapnik-Chervonenkis theory) to argue that in terms of demand for data, adding more parameters can be exponentially costly.

Not all approaches using mathematical functions are ontology-free in the way neural networks are. If we are modelling oscillatory systems, for example, we might start with trigonometric functions. In general, the set of functions we are willing to consider for modelling a system constitutes our 'bias' – the smaller the set of functions, the greater the bias. Even neural networks have a 'bias' (*not* to be confused with the 'bias' node in the network itself), which is inversely related to the number of parameters (weights) in the network. In the ideal world, we would have a very high bias that constrained the set of functions we would consider so much that calibration, the search for 'the' function we are going to accept as modelling the target system, is trivial. The price to pay for this bias is that the data may not fit very well to the set of functions we are willing to consider; if we were only willing to expand that set of functions more, we would be able to achieve a much better fit to the data. The opposite of this meaning of 'bias' is 'variance'; in neural networks, this variance is directly related to the number of weights in the network. High-variance models can be adjusted using the parameters to realize a wide range of input-output mappings, with the obvious cost of increasing the volume of search space in which to find the optimum such as mapping.

Introducing bias just to make the modelling process feasible is arguably unscientific: you are allowing your chosen modelling technique to drive your analysis of a system, rather than allowing your knowledge of that system to determine the way you describe it in your model. This kind of unscientific bias is one of the practices that has led some in the agent-based modelling community to be critical of making assumptions 'for the sake of simplicity' (e.g. Moss 2002; Edmonds and Moss 2005). Although some of these criticisms are focused on the infeasibility of the analysis itself were a more realistic representation to be used that did not make simplifying assumptions (e.g. the computation is undecidable), the feasibility of an empirical modelling process does depend on the availability of data.

Like neural networks, agent-based models potentially have large numbers of parameters – a multiple of the number of agents and the number of links in the social network. These parameters determine the heterogeneity and interaction dynamics of the model. For more traditional modelling paradigms, having large numbers of parameters is regarded with suspicion. From a practical perspective, there is a good reason for this heuristic: a high-variance model is more challenging to calibrate. Each dimension of parameter space adds exponentially to the scale of the search task and to the requirement for data. Another reason is an interpretation of Ockham's razor in a modelling context: if I have two models with the same behaviour, I prefer the one with fewer parameters. Ockham's razor is often stated as *entia non sunt multiplicanda praeter necessitatum* (literally, entities should not be multiplied more than necessary, or more naturally, explanations should not use unnecessary entities) – were it not for the qualifier, this statement would be the antithesis of agent-based modelling![2]

However, the orthogonality of the parameters in agent-based models may be more questionable than in traditional mathematical models. Essentially, in traditional mathematical modelling, each parameter is contributing to the potential 'wiggliness' (to use a term from the spline literature, e.g. Wood and Augustin 2002) of the function the model realizes. Though it is possible (e.g. Gotts and Polhill 2010), it is not necessarily the case that having another agent in the system will mean that the dynamics of the system as a whole are hugely different; adding another connection in a neural network, by contrast, does increase the 'power' of its function to realize different shapes in the mapping from input to output by adjusting the weights. The suspicion of traditional mathematical modellers towards agent-based models because of the apparently large number of parameters may therefore

---

[2]The case for agent-based modelling being that it is necessary to represent all the agents if you want to understand the emergent system-level dynamics.

not be justified.[3] There may be a way to assess the question of the 'power' a system of interacting agents has to realize different 'shapes' from input to output (however, that is understood in an ABM context) quantitatively. In the early 1970s, Vapnik and Chervonenkis (1971) published a paper that provided a lower bound on the probability that the difference between the actual predictive power of a classifier system and that estimated from calibration is more than a small amount, based on the amount of data it is given and something called the 'Vapnik-Chervonenkis dimension' of the classifier. The inequality is written thus (Vapnik and Chervonenkis 1971, p. 269):

$$P\left(|g - h| > \varepsilon\right) \leq 4m(2n)e^{-\varepsilon^2 n/8} \tag{8.1}$$

where g and h are the actual and estimated generalization ability, respectively (the proportion of instances that are correctly classified), ε is the small amount we want to bound the difference between g and h to, n is the amount of data (as number of instances) and $m(x)$ is a function that tells you the number of different realizations the classifier can make on x datapoints. The function $m()$ is equal to 2^x until $x = d_{VC}$, the Vapnik-Chervonenkis (VC) dimension of the classifier, after which it continues to grow but at a polynomial rate less than 2^x and no more than $x^{d_{VC}} + 1$ (Hertz et al. 1991, p. 154). A rough idea of the shape of the growth function $m()$ can be seen in Fig. 8.1, particularly the red (top) curve when $d_{VC} = 4$. In a log-log plot, $m()$ is convex until a critical point at which it becomes linear; as stated above, this critical point is the VC dimension of the function d_{VC}, but the red curve in Fig. 8.1 is $4 m(2n)$, so in fact the critical point on the red curve should be at $n = 0.5 d_{VC}$. However, since $x^{d_{VC}} + 1 > 2^x$ for lower values of x, the polynomial upper bound on $m()$ isn't informative; the critical point in Fig. 8.1 at which $m()$ becomes linear is therefore higher than would otherwise be expected.

To understand (8.1) a bit better, imagine $\varepsilon = 0.01$. That means you want the difference between the actual and estimated abilities to be less than 0.01 ideally. So, suppose you have a validation ability (h) of 0.95 (5% of the model's predictions on the validation data are wrong); then with $\varepsilon = 0.01$, you are saying you want your future predictions to have an ability (g) in the range [0.94, 0.96]. How certain do you want to be that you have achieved that? Suppose you want to be at least 99.9% certain, so one in a thousand predictions will have an ability outside the above range. Then you want the probability on the left-hand side of (8.1), P, to be 0.001. How can you achieve this? The right-hand side says that the probability can be reduced by using a function with a smaller VC dimension (so $m(2n)$ is smaller), using more data (increasing n) or being less fussy about how close your validation ability is to

---

[3]Part of this is the confusion between 'free parameters', which can be adjusted to make the results fit data, and parameters with values that are, at least in theory, empirically observable, even if currently unknown. Agent-based models have a lot of the latter but relatively few of the former.

Fig. 8.1 Plots showing the two expressions in (8.1). *Coloured* curves are upper bounds 4 m(2x) for d_{VC} in {1 (*blue*), 2, 3, 4 (*red*)}. The *black curves* show 0.001/exp.$(-\varepsilon^2 x/8)$ for ε in {0.05, 0.01, 0.001} (*left to right*, respectively)

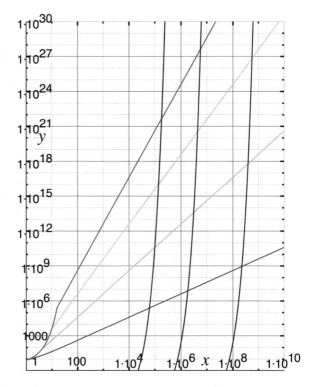

the ability you expect in future predictions (increasing ε). To achieve a probability bound of 0.001, you need exp.$(\varepsilon^2 n/8)$ to be at least a thousand times more than 4 $m(2n)$.

Mapping an ABM context to a classifier one would be somewhat awkward, though we could ask under what conditions (these conditions being the 'input space') the ABM produces a certain outcome – an outcome that either happens or doesn't. However, there is the additional problem that any stochasticity in the model will possibly generate different outcomes given the same conditions. Provided these issues can be addressed, given a thorough exploration of the ABM's parameter space, we may be able to estimate the VC dimension of the model given such an interpretation of its behaviour. We could then see the difference that adding another agent had and compare both with adding a parameter to a neural network, where approaches to estimating the VC dimension or computing it directly have already been investigated (e.g. Abu-Mostafa 1989; Watkin et al. 1993).

One of the rather depressing consequences of using the VC formula is that the value of n needed to get P down to an acceptable level turns out to be rather high, even for models with quite low VC dimension. Figure 8.1 plots expressions in (8.1) on a log-log scale, using the x-axis for n, the amount of data. The coloured curves show upper bounds for 4 $m(2n)$, and the black curves show P/exp.$(-\varepsilon^2 n/8)$ for ε in each of {0.05, 0.01, 0.001} and $P = 0.001$. The intersections of the black and coloured curves show the values of n (on the x-axis) at which P in (8.1) has an upper bound of 0.001. For example, if $d_{VC} = 2$ (cyan curve), and $\varepsilon = 0.05$, then n needs to be roughly 10^5 for P to have an upper bound of 0.001. For quantitative social data, that would be a very simple model for a very expensive questionnaire.

These high estimates are partly a consequence of the fact that the VC formula and growth function $m()$ are both upper bounds. However, the high estimates are also a consequence of the function under scrutiny essentially being an arbitrary choice, without any other information about the system the data have come from or the way the model describes that system. The VC formula is therefore very much a 'worst case', but one that applies to neural networks insofar as relatively little information about the system is encoded in the network's topology. That information is essentially the modeller's assumptions about the appropriate level of 'wiggliness' needed to fit the data – which may be as much about the pragmatics of training the network and the amount of data available as it is a reflection of the system the data have come from.

Using knowledge to constrain the choice of model is one way to reduce the VC estimate. Traditionally, this might be achieved effectively by reducing the VC dimension of the set of models being considered, using the kind of practice criticized above for being 'unscientific'. Introducing bias by removing variables from consideration, reducing the number of parameters on terms using those variables (e.g. by only considering linear models) or making other oversimplifying assumptions is, however, not the only way that we can constrain our choice of model. Though the impact on the VC dimension is less clear, in agent-based models, we can also constrain our choice of model by making it more 'descriptive' (Edmonds and Moss 2005). This essentially amounts to appropriately tuning the model's 'ontology' or 'microworld', but before considering the ontology in more detail, since agent-based models are typically applied to complex systems, we will consider some arguments about validation by fit-to-data in such systems.

8.1.4 Complex Systems and Validation by Fit-to-Data

Since agent-based models are applied to complex systems, this section introduces an important article (Oreskes et al. 1994) posing arguments about the degree to which we should trust fit-to-data as a measure of our confidence in a model's predictions in complex open systems. We move on to criticize Ockham's razor – a heuristic often used by modellers to give preference to simpler models with the same fit-to-data and one that has already been argued against on different grounds by Edmonds (2002).

Naomi Oreskes et al. (1994) have argued eloquently that environmental systems (and hence socio-environmental systems) are 'open', and hence traditional validation expressed as fit-to-data commits a logical fallacy when used as a basis to judge the degree of belief we should have that a model is a 'good' one. Essentially, the fallacious argument affirms the consequent by starting with the observations that

- Good models fit the data ($G \subset F$).
- My model fits the data (F).

and concluding that

- My model is a good model ($\vdash G$).

Oreskes et al. (1994) assert that (prejudices such as Ockham's razor aside) in closed systems, *only* good models fit the data ($G \asymp F$); in open systems, the observed data could have been affected by external influences outside the system. When fitting functions to data from complex open systems (such as social and ecological systems), the ability to exclude or control for external influences is highly constrained. A model of a subsystem that just fits to data will likely also be fitting to external influences on that subsystem.

If a model somehow captures the effect of an external influence that it is not supposed to model, we should be rather suspicious. Further, as Filatova et al. (2016) point out, disturbances to a complex socioecological system need not only arise from exogenous influences but can also grow from endogenous gradual change. If there are multiple 'attractors' and the data have followed one path at a bifurcation but a model follows another, the model will fail to validate. Over multiple runs of the model, of course, it might take the same path as the data did half the time. Given the choice between two models, one of which is simpler, and always follows the path the data did (because it is high bias and doesn't bifurcate), and another of which is more complicated, and only follows the path the data did half the time, Ockham's razor and fit-to-data heuristics tell us to choose the former. However, it is arguably the latter model that has more faithfully captured the underlying dynamics of the system.

The probability of following one trajectory rather than another need not necessarily be 0.5. It could be 1E–6, and it just so happened that this time, the real world followed the one-in-a-million chance trajectory. The model that captures the bifurcation may not be run enough times that the path the data took is observed. The point remains that in complex systems, fit-to-data is not necessarily an indicator that we have a 'good' model. If our model is ontology-free, then it is doubly awful, an oversimplified bendy sheet that hardly reflects the system it is modelling: 'It is a tale told by an idiot, full of sound and fury, signifying nothing'.[4]

To summarize, validation by fit-to-data is not necessarily (on its own) a helpful measure in complex systems. No matter what the outcome, there exists an argument both for and against the model (Table 8.1). Nevertheless, it is still a potentially useful information about a model, and we show in the box various methods for computing validation error on a set of data or otherwise comparing models' expected prediction ability. As is apparent from reading Brewer et al. (2016), there is controversy in some of the modelling literatures about which measure of expected prediction ability is 'best'. This can lead to reviewers complaining that one measure should

---

[4] Macbeth, Act V, Scene V.

Table 8.1 Arguments about validation by fit-to-data and whether the model is 'good' or 'bad'

Validation result	Good model	Bad model
Acceptable	The model has fit the data, and we estimate it will predict accurately in the future	Although the model has fit-to-data, it is oversimplified, relies on unrealistic assumptions, doesn't really explain anything or doesn't allow for the possibility that things could have turned out differently. Its predictions should not be trusted
Not acceptable	The particular course that history took was highly contingent on phenomena that it would not be reasonable to include in any model. There is a 'possible world' in which the model would be right. Alternatively, the model reproduces 'patterns' (as per Grimm et al. 1996) in the data, if not the data itself. It might still be worth considering the model's predictions	The model did not fit the empirical data we have, so it must be rejected and its predictions ignored

have been used rather than another, but since reviewers' statistical fetishes are impossible to predict, we cannot provide guidance as to how to satisfy them. However, we do give a summary of the various measures and their properties in Appendix 2 for reference.

8.1.5 Validating Ontologies

After summarizing the foregoing arguments, this section elaborates more on the structure of the model, which may be referred to as its 'ontology'. After briefly introducing ontologies, we build an argument for why agent-based models have the scope to pay more attention to this side of modelling based on the *expressivity* of a formal language for writing ontologies. We then consider various ways in which ontologies could be 'validated' – in the sense of establishing confidence in them, finding that this is far from being a settled area.

The foregoing pages had two objectives. One was to summarize all the different ways people try to estimate how well their model has fit some empirical data, to give them some kind of (preferably quantitative) idea of how much they should believe in its predictions. (See also Appendix 2.) The other is to argue that there is more to evaluating a model than just looking at its fit-to-data, largely by showing various ways in which fit-to-data may not be as convincing an indicator of a model's suitability as some appear to believe it to be. To summarize the reasons, the first two of which may seem a little 'unfair' but should be anticipated in complex social systems:

- Simplifying assumptions that apply during calibration may not apply at prediction.
- The (formal) language you have used to represent the system during calibration may not be adequate during prediction.
- You may not have enough data to justify a model with a high VC dimension, but using a model with a lower VC dimension would be oversimplifying.
- In complex/non-ergodic systems, at a bifurcation point, the empirical data may have followed a path that had a low probability in comparison with other paths it could have taken.

The various methods for measuring estimated prediction ability say relatively little about the structure of the model itself, except, in the case of metrics like the AIC and BIC, by penalizing models for having too many parameters. In neural networks, this is the number of weights the network has, but assumptions about functional form are embedded in the structure of the network itself – how the nodes are arranged into layers and/or connected to each other. This structure, however, only reflects the flexibility the network will have to achieve certain combinations of outputs on all the inputs it might be given (its 'wiggliness'). This is a rather weak ontological commitment to make to a set of data.

Neural networks are an extreme – one in which there is the minimum representative connection between the empirical world and the nodes and network of connecting weights that determine the behaviour of the model. They are nevertheless useful when there is a large amount of data available for training, the modelled system isn't complex, and one is not particularly concerned about how the input-output mapping is achieved, only that whatever mapping obtained has good prediction ability.

Neural networks are very interesting to contrast with agent-based models, which also feature networks of behaving entities, but where the network of connections and the behaving entities are supposed to have a representative link with the empirical world. In the artificial intelligence community, this representative structure would be referred to as the *microworld* (e.g. Chenoweth 1991) of the simulation. A famous example is Winograd's (1972) blocks world. However, with advances in formal languages for expressing such representative structure, we could also refer to these microworlds as ontologies.

Ontologies in computer science are defined by Gruber (1993) as formal, explicit representations of shared conceptualizations. In general, ontologies cover a broad range of formalized representations, including diagrams, computing code and even the structure of a filesystem, but the development of description logics (Baader and Nutt 2003) means that there are formal languages for ontologies to which automated reasoning can be applied. One of the most popularly used languages for ontologies, which draws on description logics, is the Web Ontology Language (OWL; Cuenca Grau et al. 2008; Horrocks et al. 2003). The application of OWL to agent-based modelling has been discussed by a number of authors (e.g. Gotts and Polhill 2009; Livet et al. 2010), but of particular relevance for our purposes is the application of OWL to representing the structure of agent-based models (Polhill and Gotts 2009).

Mappings between the programming languages used for implementing agent-based models and OWL ontologies are discussed by Polhill (2015) and Troitzsch (2015).

For the purposes of highlighting why the ontology of an agent-based model becomes so much more significant, one of the measures of a description logic is its expressivity. The expressivity of a logic is essentially the various kinds and combinations of axiom it allows you to create whilst still having decidable reasoning. We might compare different modelling approaches according to the ontological expressivity needed to capture descriptions of the states the model can have. Appendix 3 compares the expressivity of the ontologies of various modelling approaches and the corresponding description logics.

The fact that agent-based models have a generally richer expressivity for defining the ontologies over which they operate means that some of the complaints of qualitative social researchers about quantitative social researchers are brought into sharper focus. The ontology of an agent-based model is less constrained by the amount of data available, aesthetic concerns about elegance or the need to reduce the number of variables to enable tractable mathematical evaluation of equilibria.

It is also much clearer that the ontology is by and large a subjective choice. Nevertheless, we wish to have an idea of how 'good' that subjective choice is – something that may be as much about normativity in the community with an interest in the model as (supposedly) objective numerical measures. That said, if we are to move beyond fit-to-data as the sole basis for our belief in the predictions of a model, we still need some ways of assessing the model's ontology as an additional basis for such belief. This is far from being in a position where there are established methods, but four ways in which an ontology can be assessed are:

- Logical consistency
- Populating it with instances
- Stakeholder and/or expert evaluation
- Comparison with existing ontologies

If the ontology can be translated into OWL, the first of these can be achieved using the consistency checking available in reasoning applications such as Pellet (Sirin et al. 2007), FACT++ (Tsarkov and Horrocks 2006), HermiT (Shearer et al. 2008) and Ontop (Bagosi et al. 2014). Though consistency checking ensures we have at least made no logical contradictions in our specification of the ontology, it is rather a low bar to set as it says little about the quality of the representation. Beyond mere logical consistency, there are methodologies such as OntoClean (Guarino and Welty 2009) for validating the ontological adequacy of taxonomies. However, this also says more about the correctness of the operational semantics of a given set of axioms in an ontology, as opposed to addressing the sufficiency of that ontology to represent a given problem domain.

Populating an ontology with instances is another check of the 'validity' of an ontology, as difficulties with so doing, especially with empirical data, can reveal where the ontology is 'awkward' in its specifications. Working ontologies are produced every time a successful IT project is implemented. Any modern enterprise system is usually the result of a problem domain being modelled using some object-oriented analysis and design (Rumbaugh 2003) and as such necessarily involves visual modelling (usually Unified Modelling Language – UML). The resultant conception is then implemented in one of the numerous object-oriented computer languages. Although not formally provable as in any way equivalent, such systems are *prima facie* evidence of the successful construction of working ontologies, albeit normally in UML. Although not equivalent, design practices can be implemented that result in a one-to-one translation between UML and OWL (Object Modelling Group 2014, p. 130). Embedded software systems operating machinery in the real world (e.g. autopilots and control systems) have their ontologies validated every time they send a signal to a servo or relay, which over time constitutes a robust empirical test of their conceptualizations. From an agent-based modelling perspective, where the ontology describes the entities and state variables in the model, pragmatic issues with the ontology could become apparent when trying to populate the model from empirical databases. However, since the schemas of these databases are themselves ontologies, there is the potential to argue that it is those ontologies, or the integration thereof, that is the locus of any problems, rather than with the model itself. Hence, unless the context is embedded software, the ability to initialize a model from empirical data is also a rather weak test of the validity of the model's ontology.

The third idea of stakeholder and/or expert evaluation involves a degree of integration of specific problem-domain knowledge and ontological engineering expertise if we are to be convinced that the evaluators have really understood the implications of the formalization of their knowledge. Sowa (1999, p. 452) points out that knowledge engineering is a specialism requiring skills in logic, language and philosophy that domain experts should not be expected to have. Even if experts agree on a conceptualization of a domain, they will not necessarily be able to construct ontologies of it; this will be done instead by the knowledge engineer. The resulting ontology is the knowledge engineer's conceptualization of the experts' conceptualization and may differ from one knowledge engineer to another. Such problems and in particular their relevance to the veridicality and the actual information content of natural language utterances such as those from domain experts are extensively discussed by Devlin (1991, chaps. 1–2).

There are formal methodologies available for knowledge elicitation, such as On-To-Knowledge (Sure et al. 2004), creating ontologies from existing thesauruses, or taxonomies, as illustrated by Huhn and Schulz (2004) and those listed by Jones et al. (1998). However, such methodologies would normally be associated with model design rather than model validation. Since validation is only really meaningful when using 'out-of-sample' data (i.e. data not used for calibration), we should expect validation of model ontologies to be a process that behaves equivalently, for example, through using different experts during validation than during model design. In the case of peer-reviewed journal articles, this arguably happens automatically assuming that reviewers have had nothing to do with the work. However, validation by peer review detracts from the sense of reporting on a completed piece of work in a journal article and is not something that is typically documented, except in more innovative open access journals such as Earth

System Dynamics,[5] where reviews and authors' responses are also available to read. Whether validating with academic peers or with nonacademic participants or stakeholders in a model, issues with the conceptualization highlighted during validation may reflect controversies and differences in conceptualization in the community rather than issues with the particular conceptualization in the model as such.

Using formal knowledge elicitation methods, such as those listed above, to build new ontologies from the experts involved in model validation rather than those involved in model design may seem excessive. Polhill et al. (2010) document a process by which assumptions in the formalization are converted back to natural language and then 'checked' (they use this somewhat weaker term than 'validation' to describe the process) with domain experts. Since expert validation is, formally or informally, essentially a process of ontology comparison, a rigorous approach to validating ontologies would involve two knowledge elicitation exercises – one during design and one during validation.

Ontology comparison can be seen as matching ontological primitives between at least two differing ontologies. In the world of ontologies, however, such linking of primitives between ontologies is referred to as *interoperability*. Interoperability refers to the conditions under which we can establish a formal correspondence between two ontological primitives. Though interoperability was a motivation for the development of the semantic web (Berners-Lee et al. 2001), interoperability between ontologies has been somewhat intractable historically (Kalfoglou and Schorlemmer 2003) and indeed may have stalled the widespread adoption of ontologies in other application domains.

Pragmatically, interoperability is hampered by issues that come under the heading of *semantic heterogeneity*, in which there are various semantic conflicts (see, e.g. Bellatreche et al. 2006) from the seemingly trivial naming conflicts (the same name for different concepts or different names for the same concepts) to the more significant representation conflicts (concepts are represented in different ways). However, there are also philosophical issues to do with whether ontologies are seen as being 'observed' or 'constructed' (see Klein and Hirschheim 1987). If ontologies are 'observed', then we should expect to find commonality in conceptualizations because we all see the same world and discriminate the same entities in it. If they are 'constructed', such commonality is a function of norms in the way the external world is conceptualized, and any differences are cultural (and hence subject to political connotations if one conceptualization is argued to be 'better' than another). Grubic and Fan (2010, p. 783), reviewing ontologies of supply chains, conclude by noting the need to challenge the perception that building ontologies is simply a problem of terminology – finding the 'right' names for things in the real world.

With all the above caveats in mind, there are a few approaches to ontology interoperability, with some tools listed in Table 8.2:

---

[5]http://www.earth-system-dynamics.net/ <Accessed May 2017>.

Table 8.2 Available ontology interoperability tools, method used for interoperability and licence

Tool	Brief description	Method	Licence
AgreementMakerLight	An automated and efficient ontology matching system derived from AgreementMaker (Faria et al. 2013)	Matching	Apache
COMA++	A schema and ontology matching tool with a comprehensive infrastructure. Its graphical interface supports a variety of interaction (Do and Rahm 2002)	Matching	AGPL
Falcon-AO (finding, aligning and learning ontologies)	This is an automatic ontology matching tool that includes the three elementary matchers of string, virtual documents and graph similarity measures. In addition, it integrates a PBM (partition-based block matching; Hu et al. 2008) algorithm to cope with large-scale ontologies (Hu and Qu 2008)	Matching	Open source
OnAGUI (Ontology Alignment Graphical User Interface)	This is an alignment helper and viewer that also makes automatic discovery of alignment using different kind of algorithms	Manual, graph	GPL
S-Match	Takes any two tree-like structures (such as database schemas, classifications, lightweight ontologies) and returns a set of correspondences between those tree nodes which semantically correspond to one another (Giunchiglia et al. 2012)	Graph	LGPL
YAM++ (Yet Another Matcher)	A self-configuring ontology matching system for discovering semantic correspondences between entities (i.e., classes, object properties and data properties) of ontologies using machine learning (Ngo and Bellahsene 2012)	Machine learning	Open source

The list is based on work by Bergman (2014). The tools are all implemented in Java

- Token matching or token transformation – this makes use of automated token matching via textual analysis or leveraging existing ontologies to provide correspondences between previously unrelated ontological entities. Most of the tools in Table 8.2 use this kind of matching at some level.
- Graph analysis of the ontology – this includes formal concept analysis (FCA), which uses graphs to link informationally related items (Yang and Feng 2012) and other general graph matching or analysis algorithms such as in S-Match (Giunchiglia et al. 2012).
- Machine learning – examples of this include GLUE (Doan et al. 2004) and the more recent YAM++ (Ngo and Bellahsene 2012), both using machine learning to try and create correspondences between ontological elements.
- Information flow (IF) or semantic information content – this has been around quite a while but is still largely theoretical (Barwise and Seligman 1997). Premised by the externalist assumption and the assumption of veridical nature of information, this is treatment of information and its relations using category theory (Kalfoglou and Schorlemmer 2003). There are no useful implementations of this methodology so far.
- Some combination of all the above.

Token matching and graph analysis are most prevalent. An additional review of the available systems and software for ontology interoperability can be found in Shvaiko and Euzenat (2013), and some other, older, but still useful methodologies may also be found in Jean-Mary et al. (2009).

Potentially, therefore, the tools and infrastructure exist to evaluate interoperability between domain and model ontologies. The 'validation data' would comprise a pre-existing domain ontology not used to build the model or a domain ontology obtained through a second knowledge elicitation exercise with experts or stakeholders. The model's ontology could be extracted automatically (e.g. using tools such as Polhill's (2015) NetLogo extension or appropriately designed object-oriented programs enabling exploitation of one-to-one mappings from UML to OWL) or manually, and then applications such as those in Table 8.2 used to assess their interoperability. Such an exercise is rather more effort than fit-to-data validation: the maturity of the area is far from being in a position where it is simply a matter of invoking a function call in the appropriate R library as in the examples in Appendix 2.

There is also the issue that effectively the model is assessed twice, once with respect to its fit-to-data (which is still information, even if arguably not dependable as a sole indicator of how 'good' a model is) and once with respect to its ontology. If we are not to assume that a richer ontology automatically leads to a better fit-to-data, the trade-off between fit-to-data and ontological interoperability is not a trivial choice to make. Even in more established model assessment metrics that use some information about model structure, the differences in penalty of parameters between the AIC and BIC illustrate the scope for potential controversy. Indeed, Brewer et al. (2016) argue that the choice of which of these to use is sensitive to context, suggesting there is no universally applicable trade-off heuristic. In the meantime, in the social simulation community, common practice is either to use stakeholder evaluation in participatory contexts or simply to rely on peer review. Given that ontological expressivity is a major advantage for agent-based modelling, at least as suggested by our categorizations in Appendix 3, the community should set itself the aim of finding ways to quantify that benefit.

We emphasize that our arguments do not mean that fit-to-data can be ignored as a criterion for assessing the confidence we should have but rather that we *also* need to pay attention to the model's ontology. As this section has shown, there are various ways to do this, though the area is far from being sufficiently settled that we can provide 'generally used' quantitative measures of the fit of an ontology to a system. This may reflect the fact that agent-based simulation, a relatively recent development in the world of modelling, has a much greater potential expressivity in its ability to specify ontologies, and the question of model structure has thus far been limited to discussions about numbers of parameters. Further, there is evidence that we should not expect to find a single, general measure that appropriately trades off fit-to-data and ontological fit and provides us with a number that tells us how 'good' a model is.

8.1.6   Conclusion

Summarizing the key arguments in this chapter:

- Methods for validating models have thus far concentrated on fit-to-data.
- There are various ways in which that fit can be assessed.
- However, fit-to-data, though it should not be ignored, cannot be trusted as the sole basis for model validation. Besides questions about comparability of context, the modeller's biases in encoding, or representing, the system need to be questioned.
- In complex open systems, fit-to-data does not resolve whether a model's predictions should be trusted.
- Agent-based models have greater potential ontological expressivity than other modelling approaches, and researchers wishing to validate their models need to pay attention to their ontology as well as their fit-to-data.

The effort involved in building an agent-based model in an empirical context, as opposed to a more traditional aggregate-level mathematical model, is predicated on the empirical world being 'complex'. In such systems, validation by fit-to-data is not, on its own, a sound basis for estimating the ability of a model to make reliable predictions, not least because of issues with path dependency. However, the availability of sufficient data to justify building a model with as many parameters as an agent-based model typically has is a further significant potential issue, at least until methods are developed to assess how flexible agent-based models are in their ability to realize input-output mappings. These points, however, apply just as much to any other modelling approach as they do to agent-based models, noting that models with a number of parameters commensurate with the available data may be oversimplifying.

Unsatisfactory though some will find the idea that a model's ontology might be seen as subjective, the increased expressivity of agent-based models' ontologies over those of other formal modelling approaches places greater onus on the assessment of these ontologies as part of the validation process. Methodologies for assessing ontologies are still not at a sufficiently mature stage that there is a clear 'standard'. We have argued that best practice would involve a separate knowledge elicitation exercise with experts not involved in design and a comparison of the resulting ontology (or an ontology generated from a similar process in by other authors) with that of the model. Given interest in ontologies in other disciplines, there is an opportunity for the agent-based modelling community to contribute to this area, ensuring that tools and techniques can be tailored to meet any specific requirements.

Acknowledgements We acknowledge funding from the Engineering and Physical Sciences Research Council (award no. 91310127), the European Commission Framework Programme 7 'GLAMURS' project (grant agreement no. 613420) and the Scottish Government Rural Affairs, Food and the Environment Strategic Research Programme, Theme 2: Productive and Sustainable Land Management and Rural Economies. We are also grateful to Bruce Edmonds and Mark Brewer for useful comments on earlier drafts of this chapter; any mistakes are of course our own.

Further Reading

Shalizi's (2006) book chapter covers approaches to modelling (and measuring) complex systems in a more formal and comprehensive way, with a focus on more traditional mathematical modelling techniques. However, he also covers issues with validation and penalization of parameters, including discussions of VC theory and Ockham's razor.

Sowa's (1999) book on knowledge representation is a good introduction to various issues in the field and covers various formalisms and underlying philosophical questions that the formal representation of knowledge yields. Baader et al.'s (2003) *Description Logic Handbook* goes in to more details on description logics. Another book, which goes into some depth on controversies in the formal representation of what otherwise seems to be a simple everyday concept, 'if-then', is Evans and Over's (2004) book, and this too is highly recommended.

Since one of the ways of validating ontologies is through engaging with stakeholders, the Companion Modelling school of agent-based modelling, pioneered especially by research teams based in France, is well worth familiarizing yourself with. They have a website[1] and a book (Etienne 2014) as well as several publications

---

[1] https://www.commod.org/en <Accessed May 2017>.

illustrating their work. Since they sometimes use ontologies as part of their methodological approach to modelling with stakeholders, the work of authors such as Jean-Pierre Müller, Nicolas Becu and Pascal Perez and their collaborators are particularly worth investigating. Some example articles include Müller (2010), Becu et al. (2003) and Perez et al. (2009). Companion modellers are not the only ones to apply knowledge elicitation to model design, however – see, for example, Bharwani et al. (2015).

Validation has long been a subject of discussion in agent-based modelling, and this chapter has not dedicated space to reviewing the excellent thinking that has already been done on the topic. The interested reader wanting to access some of this literature is advised to look for keywords such as *validation, calibration* and *verification* in the *Journal of Artificial Societies and Social Simulation*, currently the principal journal for publication of agent-based social simulation work. Notable recent articles include Schulze et al. (2017), Drchal et al. (2016), ten Broeke et al. (2016) and Lovelace et al. (2015). Other older articles worth a read are Elsenbroich (2012), Radax and Rengs (2010) and Rossiter et al. (2010). See also some of the debates such as Thompson and Derr's (2009) critique of Epstein's (2008) article and Troitzsch's (2009) response and Moss's (2008) reflections on Windrum et al.'s (2007) paper. A practical article on one approach to validating agent-based models outwith *JASSS* is Moss and Edmonds (2005).

Appendix 1: Neural Networks

Though there are variants, typically the *excitation*, x_j, of a node j is given by the weighted sum of its inputs (8.2):

$$x_j = \sum_{i \in \text{inputs}} w_{ij} o_i \tag{8.2}$$

where o_i (usually in the range [0, 1], though some formalisms use [−1, 1]) is the output of a node i with a connection that inputs to node j and w_{ij} is the strength (weight) of that connection. Nonlinearity of the behaviour of the node is critical to the power that the neural network has as an information processing system. It is introduced by making the output o_j of a node a nonlinear function of its excitation x_j. There are a number of ways this can be achieved. Since many learning algorithms rely on the differentiability of the output with respect to the weights, the sigmoid function is typically used:

$$o_j = \frac{1}{1 + \exp\left(-x_j\right)} \tag{8.3}$$

So, a neural network essentially consists of a directed graph of nodes, where each of the links has a weight. If the graph is acyclic, the neural network is known as a *feed-forward* network. (If cyclic, the network is *recurrent*.) Nodes with no input connections are *input* nodes; those with no output connections are *output* nodes. Since they have no input connections and hence no excitation, input nodes are often also not given a nonlinear treatment as per (8.3), though this breaks somewhat with the simulation of a neuron. Similarly, nonlinearity may not be applied to output nodes. If there are N input nodes, and M output nodes, then essentially a feed-forward network without nonlinearity on the output nodes is computing a mapping from $\mathbf{R}^N$ to $\mathbf{R}^M$. With nonlinearity, the mapping is from $\mathbf{R}^N$ to $[0, 1]^M$.

Appendix 2: Metrics of and Methods for Validation

Table 8.3 explains various metrics and measures of validation, showing you where to find out more information on them and how to use them with R. For those of you unfamiliar with R, it is a popularly used[1] free (as in open-source and in the financial sense) statistical software package, available for Windows, OS-X and Linux.[2] Each of the examples assumes you are validating against a single variable (unless otherwise stated) for which you have a number of samples from your data and corresponding output from your model. The R variable vdata contains the empirical data to validate against (which must not have been used for calibration – though many of the metrics can of course be applied to the calibration process), whilst the variable model contains the corresponding output from the model. The two variables vdata and model are, in R terms, vectors of equal length. If the model predicted the data perfectly, then for each element i of the two vectors, vdata[i] == model[i]. More information on each of the approaches can be found on Wikipedia,[3] R documentation and in various machine learning and advanced statistical textbooks.

Appendix 3: Expressivity of Various Modelling Approaches

Description logics use a letter-based notation to describe the axioms each logic has (Baader and Nutt 2003; Calvanese and De Giacomo 2003; Baader et al. 2003). Briefly, $\mathcal{AL}$ is a basic description logic, and $^{(\mathcal{D})}$ is for data properties; $\mathcal{C}$ provides more complex class axioms than the basic axioms in $\mathcal{AL}$; r is for complex relationship assertions such as irreflexivity (all NetLogo links are irreflexive, e.g. as you cannot link anything to itself); $\mathcal{O}$ introduces nominals (a bit-like enumerations

---

[1] Its popularity in the social simulation community is reflected by the fact that tools have been built to link it with Wilensky's (1999) Netlogo (Thiele et al. 2012).

[2] http://www.r-project.org/ <Accessed May 2017>.

[3] https://www.wikipedia.org/ <Accessed May 2017>.

Table 8.3 Measures of validation

Metric	Description	R code
vdata[i], model[i] $\in \mathbb{R}$ (cardinal variable)		

L_1 norm	Total absolute distance between the data and model vector elements. Zero implies a perfect fit	`sum(abs(vdata--model))`
		or
		`dm <- rbind(vdata, model)`
		`dist(dm, method = "maximum")`

L_2 norm	The Euclidean distance between the data and model vectors. Zero implies a perfect fit	`sqrt(sum((vdata--model)^2))`
		or
		`dm <- rbind(vdata, model)`
		`dist(dm, method = "euclidean")`

p norm	The pth root of the sum of the pth power of the differences between the data and model vector elements. Zero implies a perfect fit	`dm <- rbind(vdata, model)` `dist(dm, method = "minkowski", p = p)`

L_∞ norm	The maximum difference between any pair of values. Zero implies a perfect fit	`max(abs(vdata--model))`
		or
		`dm <- rbind(vdata, model)`
		`dist(dm, method = "maximum")`

Sum of squared error (SSE)	Sum of squared error is simply the square of the L_2 norm. Zero implies a perfect fit	`sum((vdata--model)^2)`
RMS error	Root-mean-squared error is the L_2 norm corrected for the size of the data. Zero implies a perfect fit	`sqrt(sum((vdata--model)^2)/length(vdata))`

vdata[i], model[i] $\in\{0, 1\}$ (Boolean variable) This case could also be used for nominal variables (i.e. classes, with one number being used for each class), if repeated for each class. Here, for each class j: `vdata[i] <- ifelse(vdata[i] == j, 1, 0)` and `model[i] <- ifelse(model[i] == j, 1, 0)`

(continued)

Table 8.3 (continued)

Metric	Description	R code
Precision or positive predictive value (PPV)	The precision is the cases where the model correctly said a phenomenon occurred as a fraction of all the cases where the model said the phenomenon occurred. Ideally it would be 1	`sum(ifelse(vdata == 1 & model == 1, 1, 0)) / sum(ifelse(model == 1, 1, 0))`
Recall, sensitivity or true positive rate (TPR)	The recall is the cases where the model correctly said a phenomenon occurred as a fraction of all the cases where the data said the phenomenon occurred. Ideally it would be 1	`sum(ifelse(vdata == 1 & model == 1, 1, 0)) / sum(ifelse(vdata == 1, 1, 0))`

F measure or F_1 score	The F measure or F_1 score is the harmonic mean of precision and recall, scaled such that it is 1 in the ideal case and 0 in the worst case	`p <- sum(ifelse(vdata == 1 & model == 1, 1, 0)) / sum(ifelse(model == 1, 1, 0))`
		`r <- sum(ifelse(vdata == 1 & model == 1, 1, 0)) / sum(ifelse(vdata == 1, 1, 0))`
		`2 * (t * p) / (t + p)`

False positive rate (FPR) or fall-out	The false positive rate is the equivalent of Type I error – and is the cases where the model has predicted an occurrence of the phenomenon as a fraction of the number of cases where it doesn't in the data	`sum(ifelse(vdata == 0 & model == 1, 1, 0)) / sum(ifelse(vdata == 0, 1, 0))`
False negative rate (FNR) or miss rate	The false negative rate is the equivalent of Type II error – the cases where the model has predicted the phenomenon does not occur as a fraction of the number of cases where it does in the data	`sum(ifelse(vdata == 1 & model == 0, 1, 0)) / sum(ifelse(vdata == 1, 1, 0))`

Cohen's kappa	Cohen's kappa is a statistic that measures the degree of agreement between the data and the model by comparing the observed agreement with an estimate of the probability of agreement occurring by chance	```require(vcd)```
		```m <- matrix(c(sum(ifelse(vdata == 1 &```
		```model == 1, 1, 0),```
		```sum(ifelse(vdata == 0 & model == 1, 1, 0),```
		```sum(ifelse(vdata == 1 & model == 0, 1, 0),```
		```sum(ifelse(vdata == 0 & model == 0, 1, 0)),```
		```nrow = 2, ncol = 2)```
		```Kappa(m)```

	Various metrics are available for the case where the model is stochastic and has a distribution of outputs for a given parameter setting. In this case, model is now a matrix with the same number of rows as the length of vdata and one column for each replication of the model with the parameter setting being evaluated and inputs corresponding to the row in vdata	

Likelihood	Likelihoods are typically computed using probability distributions, and, in R, are available as computations for specific model-fitting algorithms (such as linear models). Essentially, the likelihood is a function of the parameters of a model given the observed data. For a given set of parameter values, a higher likelihood suggests a better fit to the data. Calibration could search for maximum likelihood parameter values. Unless the distribution of the model's prediction is known (as a function), likelihoods can only be estimated. Log likelihoods are sometimes reported as they can be easier to compute in the case of known distributions	First we build a matrix meq to store the cases where the model has produced the same output as the data:
		```meq <- apply(model, 2, '==', vdata)```
		Strict equality between the model and the data is typically unrealistic; meq would then normally be full of FALSE entries. In mathematical models, the Gaussian 'noise' term usually caters for this issue but requires k in the AIC and BIC to be incremented accordingly. An alternative, allowing an accuracy, epsilon (which similarly counts as another parameter):
		```near.enough <- function(x, y,```
		```epsilon = 0.01) {abs(x--y) < epsilon}```
		```meq <- apply(model, 2, near.enough, vdata)```
		Continuing with the computation of likelihood, ceq is a vector (with one element for each column in model) saying whether *all* the model's outputs are equal (or near enough) to vdata. This can then be used to compute an estimated probability of the model producing acceptable output
		```ceq <- apply(meq, 2, all)est.likelihood <-```
		```sum(ifelse(ceq, 1, 0)) / length(ceq)```

(continued)

Table 8.3 (continued)

Metric	Description	R code
Akaike information criterion (AIC)	The AIC is an information theoretic measure that can be used for comparing various models of the same data. The models must have been calibrated by selecting the maximum likelihood parameterization – a restriction that makes the application to ABM challenging. The absolute value of the AIC is not of any interest outside a model comparison context The AIC is intended to represent the information loss associated with using the model to estimate the data and includes a penalization of the number of parameters the model has (k). For our purposes, k is minimally the number of parameters you have been adjusting in the model during calibration. If your model has samples from probability distributions with hard-coded parameters, k should be incremented for each of those parameters too. (This would be equivalent to incrementing k for mathematical models with a Gaussian 'noise' parameter.) Debate about what might 'count' in your ABM's program code for the purposes of incrementing k could well form the basis of further questioning the applicability of the AIC to ABMs Models with smaller AIC are preferred	Although R provides a function to compute the AIC (the aptly named AIC()), this function uses the likelihood and hence assumes it is being passed a fitted model object, such as a linear model. For ABMs, the AIC has to be computed (estimated) by hand using the estimated likelihood. Here we assume we have two models with output matrices model1 and model2, numbers of parameters k1 and k2 and estimated likelihoods est.lik1 and est.lik2, both of which have been computed using the same vdata:est.aic1 <- 2 * k1-2 * log(est.lik1) est.aic2 <- -2 * k2-2 * log(est.lik2) If est.aic1 < est.aic2, then model2 is exp ((est.aic1--est.aic2) / 2) times as probable as model1 to minimize the information loss with respect to the data. If the two models are 'reasonably close' (e.g. exp ((est.aic1--est.aic2) / 2) > 0.5), and it makes sense to do so, their predictions can be weighted: model1 by 1 and model2 by exp ((est.aic1--est.aic2) / 2); otherwise, the model with higher AIC would be rejected
Bayes information criterion (BIC)	The BIC is based on Bayesian principles for model selection and uses a stronger penalty term for parameters than the AIC: $k \log n$, where k is the number of parameters (as in the AIC and hence with the same issues when applying to ABM), and n is the length of the vdata vector. It is required that $n \gg k$. Just as with the AIC, models with smaller BIC are preferred	est.bic <- k * log(length(vdata))--2 * log(est.likelihood)

Table 8.4 Comparison of expressivity of ontologies of various modelling approaches

Modelling approach	Expressions needed	Description logic	Comments
Neural networks	The concepts of inputs and outputs and data property labels for each node	$\mathcal{AL}^{(\mathcal{D})}$	The only ontologically significant terms are the input and output variables. Rudimentary classes are needed for input and output specifically
ODEs	Data properties for each variable, distinction between exogenous and endogenous variables, causal influence	$\mathcal{ALCOIN}^{(\mathcal{D})}$	Concepts would be needed for each variable so that causal influences can be represented with relationships
System dynamics	As ODEs, but stocks and flows are also relevant concepts	$\mathcal{ALCOIN}^{(\mathcal{D})}$	Stocks and flows as concepts do not add any extra requirements for expressivity
Social network analysis	Individuals and relationships, possibly data properties where attributes of individuals relevant	$\mathcal{ALI}^{(\mathcal{D})}$	Concepts not really needed (other than Top), so $\mathcal{ALI}$ is more expressive than SNA really requires. Data properties optional
Agent-based modelling	Classes, inheritance, individuals, data properties, object properties, lists, arrays, domain and ranges needed	$\mathcal{ALCROIN}\mathcal{F}^{(\mathcal{D})}$	Not all agent-based models will need all the expressivity options. If you have a NetLogo model and want to find out the expressivity of its ontology, you can use Polhill's (2015) automated ontology extraction tool and load the result into Protégé, and the ontology summary tab tells you the description logic needed. For example, Ge and Polhill's (2016) model of commuting has description logic $\mathcal{ALRIF}^{(\mathcal{D})}$

Letters are used to represent terms or groups of terms needed to capture any syntax for the modelling approach's formalism with respect to the real world. See text for an explanation

in Java); $\mathcal{I}$ inverse relationships; $\mathcal{N}$ numerical restrictions on properties; and $\mathcal{F}$ functional properties. Table 8.4 provides an initial indication of the description logic expressivity needed to capture the syntax used to specify the ontologies of various modelling approaches. However, the labels applied in the 'description logic' column do not necessarily mean that the full capabilities of the language are necessarily used.

References

Abu-Mostafa, Y. S. (1989). The Vapnik-Chervonenkis dimension: Information versus complexity in learning. *Neural Computation, 1*(3), 312–317.

Aha, D. W. (1992). Tolerating noisy, irrelevant and novel attributes in instance-based learning algorithms. *International Journal of Man-Machine Studies, 36*(2), 267–287.

Baader, F., & Nutt, W. (2003). Basic description logics. In F. Baader, D. Calvanese, D. L. McGuinness, D. Nardi, & P. F. Patel-Schneider (Eds.), *The description logic handbook* (pp. 43–95). New York, NY: Cambridge University Press.

Baader, F., Küsters, R., & Wolter, F. (2003). Extensions to description logics. In F. Baader, D. Calvanese, D. L. McGuinness, D. Nardi, & P. F. Patel-Schneider (Eds.), *The description logic handbook* (pp. 219–261). New York, NY: Cambridge University Press.

Bagosi, T., Calvanese, D., Hardi, J., Komla-Ebri, S., Lanti, D., Rezk, M., et al. (2014, August 8–12). The ontop framework for ontology based data access. In D. Zhao, J. Du, H. Wang, P. Wang, J. Donghong, & J. Z. Pan (Eds.), *The semantic web and web science. 8th Chinese conference, CSWS, revised selected papers* (pp. 67–77). Berlin: Springer-Verlag, Wuhan, China.

Barwise, J., & Seligman, J. (1997). *Information flow: The logic of distributed systems*. Cambridge: Cambridge University Press.

Bellatreche, L., Xuan Dong, N., Peirra, G., & Hondjack, D. (2006). Contribution of ontology-based data modeling to automatic integration of electronic catalogues within engineering databases. *Computers in Industry, 57*, 711–724.

Becu, N., Bousquet, F., Barreteau, O., Perez, P., & Walker, A. (2003). A methodology for eliciting and modelling stakeholders' representations with agent based modelling. In D. Hales, B. Edmonds, E. Norling, & J. Rouchier (Eds.), *Multi-Agent-Based Simulation III. MABS 2003. Lecture Notes in Computer Science 2927* (pp. 131–148). Berlin, Heidelberg: Springer.

Bergman, M. (2014). *50 ontology mapping and alignment tools*. http://www.mkbergman.com/1769/50-ontology-mapping-and-alignment-tools/. Accessed May 2017.

Berners-Lee, T., Hendler, J., & Lassila, O. (2001). The semantic web: A new form of web content that is meaningful to computers will unleash a revolution of new possibilities. *Scientific American, 284*(5), 28–37.

Bharwani, S., Besa, M. C., Taylor, R., Fischer, M., Devisscher, T., & Kenfack, C. (2015). Identifying salient drivers of livelihood decision-making in the forest communities of Cameroon: Adding value to social simulation models. *Journal of Artificial Societies and Social Simulation, 18*(1), 3. http://jasss.soc.surrey.ac.uk/18/1/3.html. Accessed May 2017.

Bishop, C. M. (1995). *Neural networks for pattern recognition*. Oxford: Oxford University Press.

Brewer, M. J., Butler, A., & Cooksley, S. (2016). The relative performance of AIC, AICC and BIC in the presence of unobserved heterogeneity. *Methods in Ecology and Evolution, 7*, 679–692.

Calvanese, D., & De Giacomo, G. (2003). Expressive description logics. In F. Baader, D. Calvanese, D. L. McGuinness, D. Nardi, & P. F. Patel-Schneider (Eds.), *The description logic handbook* (pp. 178–218). New York, NY: Cambridge University Press.

Cheng, B., & Titterington, D. M. (1994). Neural networks: A review from a statistical perspective. *Statistical Science, 9*(1), 2–30.

Chenoweth, S. V. (1991). On the NP-hardness of blocks world. In *AAAI-91 proceedings* (pp. 623–628).

Chester, D. L. (1990, January 15–19). Why two hidden layers are better than one. In *Proceedings of the international joint conference on neural networks,* (Vol. 1, pp. 265–268), Washington DC.

Clarke, K. A. (2005). The phantom menace: Omitted variable bias in econometric research. *Conflict Management and Peace Science, 22*(4), 341–352.

Cuenca Grau, B., Horrocks, I., Motik, B., Parsia, B., Patel-Schneider, P., & Sattler, U. (2008). OWL 2: The next step for OWL. *Journal of Web Semantics, 6*(4), 309–322.

Cybenko, G. (1989). Approximation by superposition of a sigmoidal function. *Mathematics of Control, Signals, and Systems, 2*(4), 303–314.

Devlin, K. (1991). *Logic and information.* Cambridge, Cambridge University Press.

Do, H.-H., & Rahm, E. (2002, August 20–23) COMA: A system for flexible combination of schema matching approaches. In *VLDB 2002: 28th International Conference on Very Large Data Bases,* Kowloon Shangri-La Hotel, Hong Kong, China. http://www.vldb.org/conf/2002/S17P03.pdf. Accessed May 2017.

Doan, A., Madhavan, J., Domingos, P., & Halevy, A. (2004). Ontology matching: A machine learning approach. In S. Staab & R. Studer (Eds.), *Handbook on ontologies* (pp. 385–403). Berlin: Springer-Verlag.

Drchal, J., Čertický, M., & Jakob, M. (2016). VALFRAM: Validation framework for activity-based models. *Journal of Artificial Societies and Social Simulation, 19*(3), 15. http://jasss.soc.surrey.ac.uk/19/3/15.html. Accessed May 2017.

Edmonds, B. (2002, June 3). Simplicity is not truth-indicative. In *Centre for policy modelling discussion papers CPM-02-99.* http://cfpm.org/discussionpapers/111/simplicity-is-not-truth-indicative. Accessed May 2017.

Edmonds, B., & Moss, S. (2005, July 19). From KISS to KIDS: An 'anti-simplistic' modelling approach. In P. Davidsson, B. Logan, & K. Takadama (Eds.), *Multi-agent and multi-agent-based simulation, joint workshop MABS 2004, Revised selected papers. Lecture notes in artificial intelligence 3415* (pp. 130–114), New York, NY, USA.

Elsenbroich, C. (2012). Explanation in agent-based modelling: Functions, causality or mechanisms? *Journal of Artificial Societies and Social Simulation, 15*(3), 1. http://jasss.soc.surrey.ac.uk/15/3/1.html. Accessed May 2017.

Epstein, J. M. (2008). Why model? *Journal of Artificial Societies and Social Simulation, 11*(4), 12. http://jasss.soc.surrey.ac.uk/11/4/12.html. Accessed May 2017.

Etienne, M. (2014). *Companion modelling: A participatory approach to support sustainable development.* The Netherlands: Springer.

Evans, J. S. B. T., & Over, D. E. (2004). *If.* Oxford: Oxford University Press.

Faria, D., Pesquita, C., Santos, E., Palmonari, M., Cruz, I. F., & Couto, F. M. (2013, September 9–13). The agreementmakerlight ontology matching system. In R. Meersman, H. Panetto, T. Dillon, J. Eder, Z. Bellahsene, N. Ritter, P. De Leenheer, & D. Dou (Eds.), *On the move to meaningful internet systems: OTM 2013 conferences. Confederated international conferences CoopIS, DOA-trusted cloud, and ODBASE 2013, Proceedings. lecture notes in computer science 8185* (pp. 527–541), , Graz, Austria.

Filatova, T., Polhill, J. G., & van Ewijk, S. (2016). Regime shifts in coupled socio-environmental systems: Review of modelling challenges and approaches. *Environmental Modelling and Software, 75,* 333–347.

Funahashi, K. (1989). On the approximate realisation of continuous mappings by neural networks. *Neural Networks, 2*(3), 183–192.

Ge, J., & Polhill, J. G. (2016). Exploring the combined effect of factors influencing commuting patterns and CO_2 emissions in Aberdeen using an agent-based model. *Journal of Artificial Societies and Social Simulation, 19*(3), 11. http://jasss.soc.surrey.ac.uk/19/3/11.html. Accessed May 2017.

Giunchiglia, F., Autayeu, A., & Pane, J. (2012). S-match: An open source framework for matching lightweight ontologies. *Semantic Web, 3*(3), 307–317.

Gotts, N. M., & Polhill, J. G. (2009, October 5–6). Narrative scenarios, mediating formalisms, and the agent-based simulation of land use change. In F. Squazzoni (Ed.), *Epistemological aspects of computer simulation in the social sciences. Second international workshop EPOS, Revised selected and invited papers. Lecture notes in artificial intelligence 5466* (pp. 99–116), Brescia, Italy.

Gotts, N. M., & Polhill, J. G. (2010). Size matters: Large-scape replications of experiments with FEARLUS. *Advances in Complex Systems, 13*(4), 453–467.

Grimm, V., Frank, K., Jeltsch, F., Brandl, R., Uchmański, J., & Wissel, C. (1996). Pattern-oriented modelling in population ecology. *The Science of the Total Environment, 153,* 151–166.

Gruber, T. R. (1993). A translation approach to portable ontology specification. *Knowledge Acquisition, 5*(2), 199–220.

Grubic, T., & Fan, I.-S. (2010). Supply chain ontology: Review, analysis and synthesis. *Computers in Industry, 61*, 776–786.

Guarino, N., & Welty, C. A. (2009). An overview of ontoclean. In S. Staab & R. Studer (Eds.), *Handbook on ontologies* (pp. 201–220). Berlin: Springer Verlag.

Gurney, K. (1997). *An introduction to neural networks.* London: UCL Press.

Hanson, S. J., & Burr, D. J. (1990). What connectionist models learn: Learning and representation in connectionist networks. *The Behavioral and Brain Sciences, 13*, 471–518.

Hertz, J., Krogh, A., & Palmer, R. G. (1991). *Introduction to the theory of neural computation.* Boston, MA: Addison-Wesley.

Holland, J. H. (1986). Escaping brittleness: The possibilities of general-purpose learning algorithms applied to parallel rule-based systems. In R. S. Michalski, J. G. Carbonell, & T. M. Mitchell (Eds.), *Machine learning: An artificial intelligence approach* (Vol. II). Burlington, MA: Morgan Kaufmann.

Hornik, K., Stinchcombe, M., & White, H. (1989). Multilayer feedforward networks are universal approximators. *Neural Networks, 2*(5), 359–366.

Horrocks, I., Patel-Schneider, P. F., & van Harmelen, F. (2003). From SHIQ and RDF to OWL: The making of a web ontology language. *Journal of Web Semantics, 1*(1), 7–26.

Hu, W., & Qu, Y. (2008). Falcon-AO: A practical ontology matching system. *Web Semantics: Science, Services and Agents on the World Wide Web, 6*(3), 237–239.

Hu, W., Qu, Y., & Cheng, G. (2008). Matching large ontologies: A divide-and-conquer approach. *Data & Knowledge Engineering, 67*, 140–160.

Huhn, U., & Schulz, S. (2004). Building a very large ontology from medical thesauri. In S. Staab & R. Studer (Eds.), *Handbook on ontologies* (pp. 133–150). Berlin: Springer-Verlag.

Jean-Mary, Y. R., Shironoshita, E. P., & Kabuka, M. R. (2009). Ontology matching with semantic verification. *Web Semantics: Science, Services and Agents on the World Wide Web, 7*(3), 235–251.

Jones, D. M., Bench-Capon, T. J. M., & Visser, P. R. S. (1998, 31 August–4 September). Methodologies for ontology development. In J. Cuena (Ed.), *IT & knows: Information technologies and knowledge systems. Proceedings of a conference held as part of the XV IFIP world computer congress* (pp. 62–75.), Vienna, Austria and Budapest, Hungary. http://cgi.csc.liv.ac.uk/~tbc/publications/itknows.pdf. Accessed May 2017.

Kalfoglou, Y., & Schorlemmer, M. (2003). Ontology mapping: The state of the art. *The Knowledge Engineering Review, 18*(1), 1–31.

Klein, H. K., & Hirschheim, R. A. (1987). A comparative framework of data modelling paradigms and approaches. *The Computer Journal, 30*(1), 8–15.

Livet, P., Muller, J.-P., Phan, D., & Sanders, L. (2010). Ontology, a mediator for agent-based modeling in social science. *Journal of Artificial Societies and Social Simulation, 13*(1), 3. http://jasss.soc.surrey.ac.uk/13/1/3.html. Accessed May 2017.

Moss, S. (2002). Agent based modelling for integrated assessment. *Integrated Assessment, 3*(1), 63–77.

Moss, S., & Edmonds, B. (2005). Sociology and simulation: Statistical and qualitative cross-validation. *American Journal of Sociology, 110*(4), 1095–1131.

Moss, S. (2008). Alternative approaches to the empirical validation of agent-based models. *Journal of Artificial Societies and Social Simulation, 11*(1), 5. http://jasss.soc.surrey.ac.uk/11/1/5.html. Accessed May 2017.

Müller, J. P. (2010). A framework for integrated modeling using a knowledge-driven approach. In D. A. Swayne, W. Yang, A. A. Voinov, A. Rizzoli, & T. Filatova (Eds.), *Fifth Biennial international congress on environmental modelling and software*, Ottawa, Canada. http:// www.iemss.org/iemss2010/papers/S21/S.21.08.A%20framework%20for%20integrated %20modeling%20using%20a%20knowledgedriven%20approach%20-%20JEAN-PIERRE %20MULLER.pdf. Accessed May 2017.

Ngo, D., & Bellahsene, Z. (2012, October 8–12). YAM++: A multi-strategy based approach for ontology matching task. In A. ten Teije, J. Völker, S. Handschuh, H. Stuckenschmidt, M. d'Acquin, A. Nikolov, N. Aussenac-Gilles, & N. Hernandez (Eds.), *Knowledge engineering and knowledge management. 18th international conference, EKAW. Proceedings. Lecture notes in computer science 7603* (pp. 421–425), Galway City, Ireland.

Object Modelling Group. (2014). Ontology definition metamodel version 1.1. In *OMG Document Number: Formal/2014–09-02*. http://www.omg.org/spec/ODM/1.1/PDF/. Accessed May 2017.

Oreskes, N., Shrader-Frechette, K., & Belitz, K. (1994). Verification, validation, and confirmation of numerical models in the earth sciences. *Science, 263*(5147), 641–646.

Perez, P., Dray, A., Dietze, P., Moore, D., Jenkinson, R., Siokou, C., et al. (2009). An ontology-based simulation model exploring the social contexts of psychostimulant use among young Australians. *International Society for the Study of Drug Policy*. http://ro.uow.edu.au/smartpapers/36. Accessed May 2017.

Polhill, J. G. (2015). Extracting OWL ontologies from agent-based models: A Netlogo extension. *Journal of Artificial Societies and Social Simulation, 18*(2), 15. http://jasss.soc.surrey.ac.uk/18/2/15.html. Accessed May 2017.

Polhill, J. G., & Gotts, N. M. (2009). Ontologies for transparent integrated human-natural systems modelling. *Landscape Ecology, 24*, 1255–1267.

Polhill, J. G., Sutherland, L.-A., & Gotts, N. M. (2010). Using qualitative evidence to enhance an agent-based modelling system for studying land use change. *Journal of Artificial Societies and Social Simulation, 13*(2), 10. http://jasss.soc.surrey.ac.uk/13/2/10.html. Accessed May 2017.

Radax, W., & Rengs, B. (2010). Prospects and pitfalls of statistical testing: Insights from replicating the demographic prisoner's dilemma. *Journal of Artificial Societies and Social Simulation, 13*(4), 1. http://jasss.soc.surrey.ac.uk/13/4/1.html. Accessed May 2017.

Rossiter, S., Noble, J., & Bell, K. R. W. (2010). Social simulations: Improving interdisciplinary understanding of scientific positioning and validity. *Journal of Artificial Societies and Social Simulation, 13*(1), 10. http://jasss.soc.surrey.ac.uk/13/1/10.html. Accessed May 2017.

Rumbaugh, J. (2003). Object-oriented analysis and design (OOAD). In A. Ralston, E. D. Reilly, & D. Hemmendinger (Eds.), *Encyclopedia of computer science* (4th ed., pp. 1275–1279). Chichester: John Wiley and Sons Ltd..

Schulze, J., Müller, B., Groeneveld, J., & Grimm, V. (2017). Agent-based modelling of social-ecological systems: Achievements, challenges, and a way forward. *Journal of Artificial Societies and Social Simulation, 20*(2), 8. http://jasss.soc.surrey.ac.uk/20/2/8.html. Accessed May 2017.

Shalizi, C. R. (2006). Methods and techniques of complex systems science: An overview. In T. S. Deisboeck & J. Y. Kresh (Eds.), *Complex systems science in biomedicine* (pp. 33–114). New York, NY: Springer.

Shearer, R., Motik, B. and Horrocks, I. (2008, 26–27 October). HermiT: A highly-efficient OWL reasoner. In *OWLED 2008. OWL: Experiences and Directions. Fifth International Workshop*, Karlsruhe, Germany. http://webont.org/owled/2008/papers/owled2008eu_submission_12.pdf. Accessed May 2017.

Shvaiko, P., & Euzenat, J. (2013). Ontology matching: State of the art and future challenges. *IEEE Transactions on Knowledge and Data Engineering, 25*(1), 158–176.

Sirin, E., Parsia, B., Cuenca Grau, B., Kalyanpur, A., & Katz, Y. (2007). Pellet: A practical OWL-DL reasoner. *Web Semantics: Science, Services and Agents on the World Wide Web, 5*(2), 51–53.

Sowa, J. (1999). *Knowledge representation: Logical, philosophical, and computational foundations*. Pacific Grove, CA: Brooks/Cole.

Sure, Y., Staab, S., & Studer, R. (2004). On-to-knowledge methodology (OTKM). In S. Staab & R. Studer (Eds.), *Handbook on ontologies* (pp. 117–132). Berlin: Springer-Verlag.

ten Broeke, G., van Voorn, G., & Ligtenberg, A. (2016). Which sensitivity analysis method should I use for my agent-based model? *Journal of Artificial Societies and Social Simulation, 19*(1), 5. http://jasss.soc.surrey.ac.uk/19/1/5.html. Accessed May 2017.

Thiele, J. C., Kurth, W., & Grimm, V. (2012). Agent-based modelling: Tools for linking NetLogo and R. *Journal of Artificial Societies and Social Simulation, 15*(3), 8. http://jasss.soc.surrey.ac.uk/15/3/8.html. Accessed May 2017.

Thompson, N. S., & Derr, P. (2009). Contra Epstein, good explanations predict. *Journal of Artificial Societies and Social Simulation, 12*(1), 9. http://jasss.soc.surrey.ac.uk/12/1/9.html. Accessed May 2017.

Troitzsch, K. G. (2009). Not all explanations predict satisfactorily, and not all good predictions explain. *Journal of Artificial Societies and Social Simulation, 12*(1), 10. http://jasss.soc.surrey.ac.uk/12/1/10.html. Accessed May 2017.

Troitzsch, K. G. (2015). What one can learn from extracting OWL ontologies from a NetLogo model that was not designed for such an exercise. *Journal of Artificial Societies and Social Simulation, 18*(2), 14. http://jasss.soc.surrey.ac.uk/18/2/14.html. Accessed May 2017.

Tsarkov, D., & Horrocks, I. (2006, August 17–20). FaCT++ description logic reasoner: System description. In U. Furbach & N. Shankar (Eds.), *Automated reasoning. Third international joint conference, IJCAR 2006. Proceedings. Lecture notes in computer science 4130* (pp. 292–297), Seattle, WA, USA.

Vapnik, V. N., & Chervonenkis, A. Y. (1971). On the uniform convergence of relative frequencies of events to their probabilities. *Theory of Probability and its Applications, 16*, 264–280.

Watkin, T. L. H., Rau, A., & Biehl, M. (1993). The statistical mechanics of learning a rule. *Reviews of Modern Physics, 65*(2), 499–555.

Windrum, P., Fagiolo, G., & Moneta, A. (2007) Empirical validation of agent-based models: Alternatives and prospects. *Journal of Artificial Societies and Social Simulation 10*(2), 8. http://jasss.soc.surrey.ac.uk/10/2/8.html. Accessed May 2017.

Winograd, T. (1972). *Understanding natural language*. Edinburgh: Edinburgh University Press.

Wilensky, U. (1999). *NetLogo. Center for connected learning and computer-based modeling*. Evanston, IL: Northwestern University. http://ccl.northwestern.edu/netlogo. Accessed May 2017

Wood, S. N., & Augustin, N. H. (2002). GAMs with integrated model selection using penalized regression splines and applications to environmental modelling. *Ecological Modelling, 157(2–3),* 157–177.

Yang, G., & Feng, J. (2012). Database semantic interoperability based on information flow theory and formal concept analysis. *International Journal of Information Technology and Computer Science, 4*(7), 33–42.

Chapter 9
Verifying and Validating Simulations

Nuno David, Nuno Fachada, and Agostinho C. Rosa

Abstract Verification and validation are two important aspects of model building. Verification and validation compare models with observations and descriptions of the problem modelled, which may include other models that have been verified and validated to some level. However, the use of simulation for modelling social complexity is very diverse. Often, verification and validation do not refer to an explicit stage in the simulation development process, but to the modelling process itself, according to good practices and in a way that grants credibility to using the simulation for a specific purpose. One cannot consider verification and validation without considering the purpose of the simulation. This chapter deals with a comprehensive outline of methodological perspectives and practical uses of verification and validation. The problem of evaluating simulations is addressed in four main topics: (1) the meaning of the terms verification and validation in the context of simulating social complexity; (2) types of validation, as well as techniques for validating simulations; (3) model replication and comparison as cornerstones of verification and validation; and (4) the relationship of various validation types and techniques with different modelling strategies.

Why Read This Chapter?

To help you decide how to check your simulation—both against its antecedent conceptual models (verification) and external standards such as data or other simulations (validation)—and in this way help you to establish the credibility of your simulation. In order to do this the chapter will point out the nature of these processes, including the variety of ways in which people seek to achieve them.

The original version of this chapter was revised. An erratum to this chapter can be found at
https://doi.org/10.1007/978-3-319-66948-9_30

N. David (✉)
DINÂMIA'CET - ISCTE-IUL - Centre for Socioeconomic and Territorial Studies, ISCTE-IUL
Instituto Universitário de Lisboa, Av. das Forças Armadas, 1649-026 Lisboa, Portugal
e-mail: nuno.david@iscte.pt

N. Fachada • A.C. Rosa
Institute for Systems and Robotics (ISR/IST), LARSyS, Instituto Superior Técnico, Av. Rovisco
Pais, 1, 1049-001 Lisboa, Portugal
e-mail: nfachada@laseeb.org; acrosa@laseeb.org

© Springer International Publishing AG 2017   173
B. Edmonds, R. Meyer (eds.), *Simulating Social Complexity*,
Understanding Complex Systems, https://doi.org/10.1007/978-3-319-66948-9_9

9.1   Introduction

The terms verification and validation (V&V) are commonly used in science but their meaning may be controversial in the natural and the social sciences. Putting aside the epistemological underpinnings of the terms, in simulation the distinction of meaning has a mere pragmatic nature inherited from computer science and software engineering. Often, *verification* is used in the context of evaluating the computational implementation of a model in terms of the researchers' intentions. In turn, *validation* typically refers to an evaluation of the credibility of the model as a representation of the subject modelled.

In disciplines that make use of computational models, the role of V&V is related to the need of evaluating models along the simulation development process. Basically, the very idea of V&V is comparing models with observations and descriptions of the problem modelled. This may include other models that have been verified and validated to some level, or even the implementation of replications in order to verify and validate models in more depth.

This chapter introduces a methodological perspective on V&V and describes different strategies and techniques to validate models of social complexity. Some aspects of what can be called either verification or validation are also discussed, namely comparison between models and model replication, whereon verification and validation are superimposed or indistinguishable. These are important but frequently neglected methods of promoting V&V, particularly since social simulation models can be very sensitive to implementation details (making them hard to verify), and data from social systems can be difficult or even impossible to collect (making the respective models hard to validate).

The use of simulation for modelling social complexity is very diverse. Often, V&V do not refer to an explicit stage in the simulation development process, but to the modelling process itself according to good practices and in a way that grants credibility to using the simulation for a specific purpose. Normally, the purpose is dependent on different strategies and dimensions, along which simulations can be characterised, with reference to different kinds of claims intended by the modeller, such as theoretical claims, empirical claims or simply subjunctive theoretical claims. The term subjunctive is used when very abstract simulations are used for thinking about scenarios in possible worlds, such as describing "what *would* happen if something were the case." There cannot be V&V without considering the purpose of the simulation.

In the next section of the chapter, we will deal with the meaning of the terms V&V in the context of the simulation development process. In Sect. 9.3, methods and techniques commonly associated with validation are described. The comparison and replication of simulation models as an essential aspect of V&V is discussed in Sect. 9.4. The chapter closes with Sect. 9.5, where the relationship of validation with different modelling strategies is described.

9.2   The Simulation Development Process

Several chains of intermediate models are developed before obtaining a satisfactory verified and validated model. What does it mean to verify and validate a model in social simulation? Is there a fundamental difference between verifying and validating models? The purpose of this section is to define the role of V&V within the scope of the simulation development process.

The most common definitions of V&V are imported from computer science, as well as from technical and numerical simulation,[1] having intended distinct— although epistemologically overlapping—meanings. The reason for distinguishing between the terms derives from the practice of determining the suitability of certain models for representing two distinct subjects of inquiry. This is represented in Fig. 9.1, in which V&V are related to a simplified model development process. Two conceptual models mediate between two subjects of inquiry. The latter are (1) the target theory or phenomenon and (2) the executable computational model. The conceptual model on the right, designated here as the *pre-computational model*, is basically a representation in the minds and writing of the researchers, which presumably represents the target. This model must be implemented as an *executable computational model*, by going through a number of intermediate models such as formal specification or textual programs written in high-level programming languages.

The analysis of the executable model gives rise to one or more conceptual models on the left, here designated as *post-computational* models. They are constructed based on the output of the computational model, often with the aid of statistical

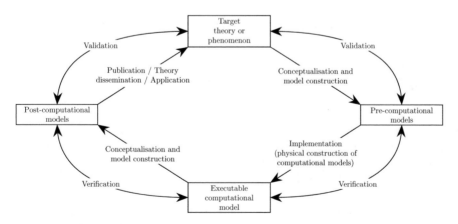

Fig. 9.1 Verification and validation related to the model development process (David 2009)

---

[1]Numerical simulation refers to simulation for finding solutions to mathematical models, normally for cases in which mathematics does not provide analytical solutions. Technical simulation stands for simulation with numerical models in computational sciences and engineering.

packages, graphing and visualisation. The whole construction process results in categories of description that may not have been used for describing the pre-computational model. This is the so-called idea of *emergence*, when interactions among model components specified through pre-computational models at some level of description give rise to different categories of model descriptions identified in the executable model at macro levels of observation, expressed through post-computational models.

As an example consider the culture dissemination model of Axelrod (1997b) which has a goal of analysing the phenomena of social influence. At a micro-level of description, a pre-computational model defines: (a) the concept of *actors* distributed on a grid; (b) the concept of *culture* of each actor, specified as a set of five features; and (c) the *interaction mechanisms* specified with a bit-flipping schema, in which the probability of interaction between two actors is set proportionately to the similarity between two cultures. The executable model is then explored and other categories of descriptions resulting from the interaction of individual cultures may be defined. These are associated with macro properties of interest and conditions in which they form, such as the concepts of *regions* and *zones* on the grid. A great deal of the simulation proposed by Axelrod concerns investigating properties of *regions* and *zones* in the executable model, giving rise to a proposed conceptual, post-computational model, which expresses traits such as the relation between the size of a *region* formed and the number of features per *individual culture*. These concepts are interpreted in relation to the target social phenomena of social influence.

We will now situate the role of V&V in the modelling process of social simulation.

9.2.1 What Does It Mean to Verify a Computational Model?

Computational model *verification* is defined as checking the adequacy among conceptual models and computational models (see also Chap. 7 in this volume, Galán et al. 2017). Consider the lower quadrants of Fig. 9.1. They are concerned with ensuring that the pre-computational model has been implemented *adequately* as an executable computational model, according to the researcher's intentions in the parameter range considered, and also that the post-computational model *adequately* represents the executable model in the parameter range considered.[2] In short, the three models must correspond to each other adequately, relative to the same target they are meant to represent.

At this point you might question the meaning of *adequately*. A minimal definition could be the following: adequateness means that the inputs, outputs and the mechanisms post-computationally modelled from the executable computational model are consistent with the ones specified through the pre-computational models,

---

[2]Verification in the left quadrant of Fig. 9.1 is sometimes known as "internal validation."

in accordance with the researcher's intentions. However, the outcomes of computer programs in social simulation are often unintended or not known a priori and thus the verification process requires more than checking that the executable model does what it was planned to do. The goal of the whole exercise is to assess logical inferences within, as well as between, the pre- and the post-computational models. This requires assessing whether the post-computational model—while expressing emergent concepts that the pre-computational model may not have been intended to express—is consistent with the latter. From a methodological point of view this is a complicated question, but from a practical perspective one might operationally define the verification problem with the following procedures:

(a) For some pre-computational model definable as a set of input/output pairs *in a specified parameter range*, the corresponding executable model is *verified for the range considered* if the corresponding post-computational model expresses the same set of inputs/outputs for the range considered.

(b) For some pre-computational model defined according to the researcher and/or stakeholders' intentions *in a specified parameter range*, the corresponding executable model is *verified for the range considered* if the corresponding post-computational model meets the researchers and/or stakeholders' expectations for the range considered.

Note that both procedures limit the verification problem to a clearly defined parameter range. The first option is appropriate when quantitative data is available from the target with which to test the executable model. This is normally not the case, leaving the second option as the suitable path for the verification process. This is possible since the aim is to assess the appropriateness of the relations that may be established between micro-levels of description specified in the pre-computational model and macro-levels of description expressed through post-computational models, usually amenable to evaluation by researchers and stakeholders.

In any case, the verifiability of a simulation is influenced by the process used to develop that simulation. The tools used to implement the executable computational model are a major factor affecting verification (Sargent 2013). The use of high-level simulation packages has the potential to simplify verification, since the majority of common model building blocks are provided, and these are typically already verified. Arguably, this is even more so in the case of open source toolkits, such as NetLogo (Wilensky 1999) or Repast Simphony (North et al. 2013), where, in addition to the developers themselves, the respective user communities perform verification of the provided simulation blocks and modules. Community members can not only detect bugs, but also correct them due to the open and collaborative nature of these projects. When such modelling toolkits are used, verification mainly consists of guaranteeing that the model has been correctly implemented using the available modules.

However, while the use of modelling toolkits reduces the programming and verification effort, it typically increases simulation times (Fachada et al. 2017a) and limits the modeller's flexibility in implementing non-standard behaviours (Sargent 2013). As such, it is often necessary to directly implement models using general-purpose programming languages. This is not a black or white choice, since several simulation toolkits offer the option of developing models using general-purpose programming languages (e.g. Repast Simphony), and/or provide high-performance and scalable workflows, with Repast HPC (Collier and North 2013) being a case in point.

When the direct use of general-purpose programming languages is involved, the adoption of good programming practices for designing and implementing the model is fundamental. Techniques such as object-oriented design, modularity and encapsulation not only simplify testing and debugging, but also promote incremental model development and the mapping of programming units (e.g. classes or functions) to model concepts, thus making computational models easier to understand, extend and modify. Additionally, defensive programming methodologies, such as assertions and unit tests, are well suited for the exploratory nature of simulation, making models easier to debug and verify.

Two important verification methods, *traces* and *structured walk-throughs*, complement the techniques discussed thus far. The former entails following a specific model variable (e.g. the position of an agent or the value of a simulation output) throughout the execution of the computational model, with the goal of assessing whether the implemented logic is correct and if the necessary precision is obtained. Modelling toolkits and programming language tools typically offer the relevant functionality, making the use of traces relatively simple (Sargent 2013). In turn, structured walk-throughs consist of having more than one person reading and debugging a program. All members of the development team are given a copy of a particular module to be debugged and the module developer goes through the code but does not proceed from one statement to the next until everyone is convinced that a statement is correct (Law 2015).

Nevertheless, and while the techniques described here are an important part of the verification process, a computational model should only be qualified as verified with reasonable confidence if it has been successfully replicated and/or aligned with a valid pre-existing model. We will return to this topic in greater detail in Sect. 9.4.

9.2.2   What Does It Mean to Validate a Model?

Model *validation* is defined as ensuring that both conceptual and computational models are adequate representations of the target. The term "adequate" in this sense may stand for a number of epistemological perspectives. From a practical point of view we could assess whether the outputs of the simulation are close enough to empirical data.

Alternatively, we could assess various aspects of the simulation, such as if the mechanisms specified in the simulation are well accepted by stakeholders involved in a participative-based approach. In Sect. 9.3 we will describe the general idea of validation as the process that assesses whether the pre-computational models—put forward as models of social complexity—can be demonstrated to represent theories or aspects of social behaviour able to give rise to post-computational models that are, at some given level, consistent with the onset theories or similar to real data.

Given the model development process described, is there any fundamental difference between verifying and validating simulations? Rather than being a sharp difference in kind it is a distinction that results from the computational method. Whereas verification is focused on the assessment of micro and macro concepts and inferences in the process of programming, observing and interpreting computational models, validation is focused on the evaluation of such inferences and concepts as representations of the target social phenomenon or theory.

In paraphrasing Axelrod (1997a), at first sight, we could say that the problem is whether an unexpected result is a reflection of the computational model, due to a mistake in the implementation of the pre-computational model, or is a surprising consequence of the pre-computational model itself. Unfortunately, the problem is more complicated than that. In many cases mistakes in the code may not be qualified simply as mistakes, but only as one interpretation among many others possible for implementing a conceptual model. Nevertheless, from a practical viewpoint there may be still good reasons to make the distinction between V&V. A number of established practices exist for the corresponding quadrants of Fig. 9.1. We will address some of these in the following sections.

9.3   Validation Approaches

We offered a conceptual definition of validation in Sect. 9.2.2. Had we given an operational definition, things would have become somewhat problematical. Models of social complexity are diverse and there is no definitive and guaranteed criterion of validity. As Amblard et al. (2007) remarked, "validation suggests a reflection on the intended use of the model in order to be valid, and the interpretation of the results should be done in relation to that specific context."

A specific use may be associated with different methodological perspectives for building the model, with different strategies, types of validity tests, and techniques (Fig. 9.2). Consider the kind of subjunctive, metaphorical models such as Schelling's (1971). In these models there is no salient validation step during the simulation development process. Design and validation walk together. The intended use is not to show that the simulation is plausible against a specific context of social reality but to propose abstract or schematic mechanisms as broad representations of classes of social phenomena. In other cases, the goal may be modelling a specific target domain, full of context, with use of empirical data and significant amounts of rich detail. Whereas in the former a good practice could be modelling with the greatest parsimony possible so as to have a computational model sanctionable by human beings and comparable to other models, parsimony can be in opposition to the goal of descriptive richness and thus inappropriate to the latter case.

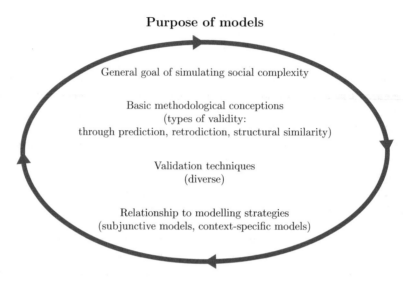

Purpose of models

General goal of simulating social complexity

Basic methodological conceptions
(types of validity:
through prediction, retrodiction, structural similarity)

Validation techniques
(diverse)

Relationship to modelling strategies
(subjunctive models, context-specific models)

Fig. 9.2 Validation implies considering the purpose of the model

There are also different methodological motivations behind the use of a model, such as those conceived to predict or explain and those merely conceived to describe. Regardless of what method is used, the reproduction of characteristics of the object domain is important, but this can be assessed through rather different approaches during the model development process. If it is prediction you are seeking, validation consists of confronting simulated behaviour with the future behaviour of the target system (however, attempting to establish numerical prediction is not a normal goal in simulation). If it is explanation, validation consists of building plausible mechanisms that are able to reproduce simulated behaviour similar to real behaviour. If the goal is the more general aim of descriptiveness, explanation may probably be a goal as well, and a creative integration of ways for assessing the structure and results of the model, from quantitative to qualitative and participatory approaches, will be applied.

In conclusion, one should bear in mind that there is no one special method for validating a model. However, it is important to assess whether the simulation is subjected to good practices during its conception, whether it fits the intended use of the model builder and whether it is able to reproduce characteristics of the object domain. Assessing whether the goals of the modellers are well stated and the models themselves are well described in order to be understood and sanctioned by other model builders are *sine qua non* conditions for good simulation modelling.

In the remainder of this section, we revise the purpose of validating simulations along three dimensions: (1) the general goal of validation in social complexity; (2) basic methodological conceptions of validity types; and (3) typical validation techniques used in social simulation.

9.3.1 The Goal of Validation: Goodness of Description

If one is using a predictive model, then the purpose of the model is to predict either past or future states of the target system. On the other hand, one may strive for a model that is able to describe the target system with satisfactory accuracy in order to become more knowledgeable about the functioning of the system, to exercise future and past scenarios, and to explore alternative designs or inform policies.

The objective in this section is to define the purpose of validation in terms of the purpose of simulating social complexity, which we will define as being of good description. This position entails that there is no single method or technique for validating a simulation. A diversity of methods for validating models is generally applied.

In the rest of this chapter we adopt the agent-based paradigm for modelling. A conceptual understanding of validation, similar but more general than Moss and Edmonds (2005), will be used:

> The purpose of validation is to assess whether the design of micro-level mechanisms, put forward as theories of social complexity validated to arbitrary levels, can be demonstrated to represent aspects of social behaviour and interaction that are able to produce macro-level effects either (i) broadly consistent with the subjacent theories; and/or (ii) qualitatively or quantitatively similar to real data.

By *broad consistency* we mean the plausibility of both micro specification and macro effects accounted as general representations of the target social reality. In its most extreme expression, plausibility may be evaluated on a metaphorical basis. By *qualitative similarity* to real data we mean a comparison with the model in terms of categorical outcomes, accounted as qualitative features, such as the shape of the outcomes, general stylised facts, or dynamical regimes. As for *quantitative similarity* we mean the very unlikely case in which the identification of formal numerical relationships between aggregate variables in the model and in the target—such as enquiring as to whether both series may draw from the same statistical distribution—proves to be possible.

Notice that this definition is general enough to consider both the micro-level mechanisms and macro-level effects assessed on a participatory basis. It is also general enough to consider two methodological practices for building social simulation models, namely the extent to which models should be based on formal theories or on the intuition of the model builders and stakeholders—an issue that we will come back to later. These are omnipresent methodological questions in the social simulation literature and are by no means irrelevant to the purpose of simulation models.

Suppose that on the basis of a very abstract model, such as the Schelling model, you were to evaluate the similarity of its outputs with empirical data. Then you will probably not take issue with the fact that the goal of predicting future states of the target would be out of the scope of simulation research for that kind of modelling. However, despite the belief that other sorts of validation are needed, this does not imply excluding the role of prediction, but simply emphasises the importance of description as the goal of simulating social complexity. In truth, what could be more contentious in assessing the Schelling model is the extreme simplicity used to describe the domain of social segregation. The descriptive power of agent-based models (ABMs) makes them suited to model social complexity. Computational modelling corresponds to a process of abstraction, in that it selects some aspects of a subject being modelled, like entities, relations between entities and change of state, while ignoring those that may be considered less relevant to the questions that are of interest to the model builder. The expressiveness of ABMs allows the researcher to play with intuitive representations of distinct aspects of the target, such as defining societies with different kinds of agents, organisations, networks and environments, which interact with each other and represent social heterogeneity. By selecting certain aspects of social reality into a model, this process of demarcation makes agent-based modelling suited to represent sociality as perceived by researchers and often by the stakeholders themselves.

The descriptive power of simulation is on par with the diversity of ways used for informing the construction and validation of models, from theoretic approaches to the use of empirical data or stakeholder involvement. At any rate, measuring the goodness of fit between the model and real data expressed with data series is neither the unique nor a typical criterion for sanctioning a model. The very idea of using a diversity of formal and informal methods is to assess the credibility of the mechanisms of the model as good descriptions of social behaviour and interaction, which must be shown to be resilient in the face of multiple tests and methods, in order to provide robust knowledge claims and allow the model to be open to scrutiny.

9.3.2   Broad Types of Validity

When we speak about types of validity we mean three general methodological perspectives for assessing whether a model is able to reproduce expected characteristics of an object domain: validation through *prediction*, validation through *retrodiction* and validation through *structural similarity*. Prediction refers to validating a model by comparing the states of a model with future observations of the target system. Retrodiction compares the states of the model with past observations of the target system. Lastly, structural similarity refers to assessing the realism of the structure of the model in terms of empirical and/or theoretical knowledge of the target system (see also Gross and Strand 2000). In practice, all three approaches are interdependent and no single approach is used alone.

9.3.2.1   Validity Through Prediction

Validation through prediction requires matching the model with aspects of the target system before they were observed. The logic of predictive validity is the following: if one is using a *predictive model*—in which the purpose of the model is to predict future states of the target system—and the predictions prove satisfactory in repeated tested events, it may be reasonable to expect the model outcomes to stay reliable under similar conditions (Gross and Strand 2000). The purpose of prediction is somewhat problematic in social simulation:

- Models of social complexity usually show nonlinear effects in which the global behaviour of the model can become path-dependent and self-reinforcing, producing high sensitivity to initial conditions, which limits the use of predictive approaches.
- Many social systems show high volatility with unpredictable events, such as turning points of macroeconomic trade cycles or of financial markets that are in practice (and possibly in principle) impossible to predict; refer to Moss and Edmonds (2005) for a discussion on this.
- Many social systems are not amenable to direct observation, change too slowly, and/or do not provide enough data to be able to compare model outcomes. Most involve human beings and are too valuable to allow repeated intervention, which hinders the acquisition of knowledge about its future behaviour. Policies based on false predictions could have serious consequences, thus making the purpose of prediction unusable (Gross and Strand 2000).

While quantitative prediction of the target system behaviour is rare or simply unattainable, prediction in general is not able to validate per se the *mechanisms* of the model as good representations of the target system. In the words of Troitzsch (2004), "What simulations are useful to predict is only how a target system might behave in the future qualitatively." But a different model using different mechanisms that could lead to the same qualitative prediction may always exist, thus providing a different explanation for the same prediction. More often, the role of predicting future states of the target system becomes the exploration of new patterns of behaviour that were not identified before in the target system, whereby simulation acquires a speculative character useful as a heuristic and learning tool. What we are predicting is really new concepts that we had not realised as being relevant just by looking into the target.

9.3.2.2   Validity Through Retrodiction

The difference from retrodiction to prediction is that in the former the intention is to reproduce *already* observed aspects of the target system. Given the existence of a historical record of facts from the target system, the rationale of *retrodictive* validity for a *predictive* model is the following: If the model is able to reproduce a historical record consistently and correctly, then the model may also be trusted for the future (Gross and Strand 2000). However, as we have mentioned, predictive models of social complexity are uncommon in simulation. Explanation rather than prediction is the usual motive for retrodiction. The logic of retrodictive validity is the following: If a model is able to consistently reproduce a record of past behaviours of the target system, then the mechanisms that constitute the model are *eligible candidates* for explaining the functioning of the target system. Nevertheless, retrodiction alone is not sufficient to assess the validity of the candidate explanations:

- Underdetermination: Given a model able to explain a certain record of behaviours or historical data, there will always be a different model yielding a different explanation for the same record.
- Insufficient quality of data: In many cases it is impossible to obtain long historical series of social facts in the target system. In the social sciences the very notion of social facts or data is controversial, can be subjective, and is not dissociable from effects introduced by the measurement process. Moreover, even when data is available it may not be in a form suitable to be matched to the bulk of data generated by simulation models.

Underdetermination and insufficient data suggest the crucial importance of domain experts for validating the *mechanisms* specified in the model. A model is only valid provided that *both* the generated outcomes and the mechanisms that constitute the model are sanctioned by experts in the relevant domain. The importance of validating the mechanisms themselves leads us to the *structural* validity of the model, which neither predictive nor retrodictive validity is able to assess alone.

9.3.2.3 Validity Through Structural Similarity

In practice, the evaluation of a simulation includes some kind of prediction and retrodiction, based on expertise and experience. Given the implementation of micro-level mechanisms in the simulation, classes of behaviour at the macroscopic scale are identified in the model and compared to classes of behaviour identified in the target. Similarly, known classes of behaviour in the target system are checked for existence in the simulation. The former case is generally what we call the "surprising" character of simulations in which models show something beyond what we expect them to. However, only an assessment of the model from various points of view, including its structure and properties on different grains and levels, will truly determine whether it reflects the way in which the target system operates. For instance, do agents' behaviour, the constituent parts and the structural evolution of the model match the conception we have about the target system with satisfactory accuracy? These are examples of the elements of realism between the model and the system that the researcher strives to find, which requires expertise in the domain on the part of the person who builds and/or validates the model.

9.3.3 Validation Techniques

In this section we describe validation techniques used in social simulation. Some are used as common practices in the literature and most of the terminology has been inhered from simulation in engineering and computer science, particularly from the reviews of validation and verification in engineering by Sargent (2013). All techniques that we describe can be found in the literature, but it would be rare to find a model in which only one technique was used, consistent with the fact that the validation process should be diverse. Also, there are no standard names in the literature and some techniques overlap with others.

9.3.3.1 Face Validity

Face validity is a general kind of test used both before and after the model is put to use. During the model development process, the various intermediate models are presented to persons who are knowledgeable about the problem in order to assess whether they are compatible with the expert's understanding and reasonable for their purpose (Sargent 2013). Face validity may be used for evaluating the conceptual model, the components thereof, and the behaviour of the computational models in terms of categorical outcomes or direct input/output relationships. This can be accomplished via documentation, graphing visualisation models, and animation of the model as it moves through time. Visualisation of model outputs (including a brief look at model animation) is analysed in Chap. 10 of this volume (Evans et al. 2017). Insofar as this is a general kind of test, it is used in several iterations of the model.

9.3.3.2 Turing Tests

People who are knowledgeable about the behaviour of the target system are asked if they can discriminate between system and model outputs (Sargent 2013; Law 2015). The logic of Turing tests is the following: If the outputs of a computational model are qualitatively or quantitatively indistinguishable from the observation of the target system, a substantial level of validation has been achieved.

Note that the behaviour of the target system does not need to be observed directly in the cases where a computational direct representation is available. For example, suppose that videos of car traffic are transformed into three-dimensional scenes, whereby each object in the scene represents a car following the observed trajectory. If an independent investigator is not able to distinguish the computational reproduction from an agent-based simulation of car traffic, then a substantial level of validation has been obtained for the set of behaviours represented in the simulation model.

9.3.3.3 Historical Validity

Historical validity is a kind of retrodiction where the results of the model are compared with the results of previously collected data. If only a portion of the available historical data is used to design the model then a related concept is called *out-of-sample tests* in which the remaining data are used to test the predicative capacity of the model.

9.3.3.4   Event Validity

Event validity compares the occurrence of particular events in the model with the occurrence of events in the source data. This can be assessed at the level of individual trajectories of agents or at any aggregate level. Events are situations that should occur according to pre-specified conditions, although not necessarily predictable. Some events may occur at unpredictable points in time or circumstances. For instance, if the target system data shows arbitrary periods of stable behaviours interwoven with periods of volatility with unpredictable turning points, the simulation should produce similar kinds of unpredictable turning events.

9.3.3.5   Validity of Simulation Output

Since data is hard to collect in social systems, investigating the behaviour of simulation output becomes a crucial model validation technique (Sargent 2013). This can be performed by running the simulation with different parametrisations and checking if the output is reasonable (Law 2015), either based on subjective expert opinion when using "typical" simulations parameters, or by objectively evaluating output behaviour under trivial or extreme parametrisations. For instance, concerning the latter, if interaction among agents is nearly suppressed the modeller should be surprised if such activities as trade or culture dissemination continues in a population.

The concept of internal validity (Sargent 2013) or verification between the executable computational model and post-computational models (lower left quadrant of Fig. 9.1) can also be considered here, since it directly relates to simulation output behaviour. In order to assess the level of stochastic variability in a model, a number of simulation runs are performed using different random number streams. A sizeable level of variability between simulation runs can question the model at different levels. For example, the validity of simulation output for the executable computational model may be disputed, or the stability of a given policy (and the parametrisation that expresses it) in the overall model may be challenged.

For a more in-depth look at issues concerning simulation output behaviour, we refer the reader to the following references. Visualisation-oriented approaches for understanding simulation output are debated in Chap. 10 of this volume (Evans et al. 2017). Visualisation, and statistical and analytical analysis of model outputs are examined and reviewed by Lee et al. (2015). For a pure statistical outlook, Fachada et al. (2015) discuss a generic and systematic approach for evaluating time-series output of simulation models.

9.3.3.6 Solution Space Exploration

The techniques discussed in Sect. 9.3.3.5 are useful for basic output validation under specific parametrisations. However, they do not provide a general understanding of how input parameters influence model behaviour, nor they consider the broader picture of overall model assumptions, which encompass not only input parameters, but also internal model structure, employed submodels and model elements, as well as their inter-relations. In solution space exploration, model assumptions are varied in order to reach a better understanding of how the assumptions of interest affect the model.

The exploration of the solution space can be as simple as testing "what if" scenarios for observing model behaviour under different inputs—similar to what was discussed in the previous subsection—or follow a more systematic approach based on carefully designed experiments (Montgomery 2012). The latter approach aims to get the maximum amount of information from the model with the minimum number of simulation runs (Pereda et al. 2015), and is generally more efficient than hand-guided runs where alternative model configurations are experimented with (Law 2015). Nonetheless basic hand-guided experiments are also valuable for model validation, namely when trying different conceptual- or system-level assumptions. Conceptual-level assumptions include internal mechanisms or submodels that constitute the larger model (e.g. the decision processes of the agents, their learning mechanisms or their interaction topology), while system-level assumptions involve low-level elements of the model (e.g. agent activation regimes). If changing elements at the system-level determines different behaviours of the model that cannot be adequately interpreted, then the validity of the model can be compromised. The case of changing elements at conceptual levels is more subtle and the validity of the results must be assessed by the researcher with reference to the validity of the composing elements of the model. This is basically a kind of cross-model or cross-element validation, as described in Sect. 9.4.

The exploration of the solution space is often undertaken with one or more targeted objectives in mind, especially in the case of formally designed experiments. Typical objectives include optimisation, calibration, uncertainty analysis and sensitivity analysis (Lee et al. 2015). While these objectives may overlap, brief definitions and their potential roles in model validation can be given. In model *optimisation*, the researcher is interested in finding parameters or assumptions that minimise some cost or elicit specific model events or behaviour, which can be directly related with event validity, as discussed in Sect. 9.3.3.4. In turn, *calibration* is concerned with finding the assumptions that maximise the agreement of the model behaviour with the target system behaviour, thus making it a crucial aspect in model validation and in the model development process. *Uncertainty analysis* provides measures related to the reliability of results and how do input uncertainties propagate through to the collected outputs. These measures affect simulation output validity and directly influence the interpretation of data obtained through *sensitivity analysis*. The latter is arguably the most common objective when exploring the solution space of a model. In essence, small perturbations are applied to model assumptions in order to determine which ones have the greatest effect on output behaviour (Evans et al. 2017). This information can be used to improve model accuracy and reduce output variance—issues directly related with model validation—and also to promote model parsimony by fixing inconsequential parameters and simplifying assumptions, reducing dimensionality of the input parameter space and the model's computational cost (Law 2015; Lee et al. 2015). Conversely, sensitivity analysis may also point to underspecified assumptions, which may require additional detail in order to accurately represent some aspect of the target system (Law 2015). If the output remains unpredictable even with controlled changes, the modeller should be concerned about making claims about the model.

A number of techniques for sampling the solution space are described in the modelling and simulation literature. The *one-factor-at-a-time* (OFAT) approach is one of the simplest sampling techniques. The effects of individual assumptions (factors) on model behaviour are analysed in isolation by iterating each one over a set of discretised levels while keeping the other factors unchanged (Lee et al. 2015). Unfortunately, this technique ignores possible interactions between factors (Law 2015). This issue is handled by *factorial*-type designs, for which the different factor levels are combined in specific configurations (e.g. full factorial, fractional factorial or central composite designs) (Pereda et al. 2015). *Space-filling* designs are another type of sampling technique, and aim to cover the solution space more evenly (Pereda et al. 2015). Monte Carlo random sampling is probably the most common space-filling approach, consisting in sampling each parameter range randomly. However, care should be taken with this approach since clustered observations and empty spaces are bound appear by chance. Space-filling alternatives such as quasi-Monte Carlo or Latin Hypercube Sampling (McKay et al. 1979) cover the input space more evenly and are often preferred. In turn, sampling based on *meta-heuristics*, such as genetic algorithms, can search for pre-specified output behaviours. Thus, such techniques are commonly used when the researcher wishes to estimate parameters for calibration and/or optimisation purposes (Miller 1998; Calvez and Hutzler 2005; Stonedahl and Wilensky 2010).

Since the vast majority of the models of interest in social simulation are stochastic, one should also consider the issue of having to perform several runs with different seeds for each sampled assumption set in order to reduce the uncertainty about the expected output value. Consequently, there is a trade-off between assumption space coverage and output accuracy, which can severely limit the exploration of models with long execution times (Pereda et al. 2015). This issue can be minimised with the use of metamodels, which can act as computationally inexpensive proxies of more complex models (Lee et al. 2015). A metamodel, or a model of a model, can be used for predicting the original model's response for non-simulated assumption sets or finding combinations of assumptions that optimise (i.e. minimise or maximise) a response (Law 2015). A metamodel usually takes the form of a regression function relating inputs with an output response, typically a statistic representative of model behaviour. Statistical learning techniques such as regression analysis, Gaussian process modelling (Kriging), neural networks or random forests are commonly used for building the metamodel function (Law 2015; Pereda et al. 2015).

9.3.3.7 Participatory Approaches for Validation

Participatory approaches refer to the involvement of stakeholders both in the design and the validation of a model. Such an approach, also known as Companion Modelling (Barreteau et al. 2001), assumes that model development must be itself considered in the process of social intervention, where dialogue among stakeholders, including both informal and theoretical knowledge, is embedded in the model development process. Rather than just considering the final shape of the model, both the process and the model become instruments for negotiation and decision making. Documentation and visualisation techniques can play a crucial role in bridging the opinions and intentions of all interested parties. Such approaches are particularly suited for policy or strategy development. This topic is discussed in Chap. 12 "Participatory Approaches" (Barreteau et al. 2017).

9.4 Replicating and Comparing Models

Computational models in social science can be very sensitive to implementation details, and the influence that seemingly negligible aspects such as data structures or sequences of events can have on simulation results is striking (Merlone et al. 2008). Furthermore, model implementations can be considerably elaborate, making them prone to programming errors (Will and Hegselmann 2008). This can seriously affect V&V when data from the system being modelled cannot be obtained easily, cheaply or at all—often the case in social simulation. Moreover, even if data were available, the goodness of fit between real and simulated data, albeit reflecting evidence about the validity of the model as a data-generating process, does not provide evidence on how it operates. Model replication—the reimplementation of an existing model and the replication of its results—is a potential but frequently neglected solution to this problem (Will and Hegselmann 2008; Thiele and Grimm 2015). Replicating a model in a different context will sidestep the biases associated with the language or toolkit used to develop the original model, bringing to light inconsistencies between conceptual and computational models (Edmonds and Hales 2003; Wilensky and Rand 2007).

Replication strongly contributes to the V&V of simulation models (Wilensky and Rand 2007; Thiele and Grimm 2015). Verification is improved because if two or more distinct implementations of a conceptual model yield equivalent results, it is more likely that the implemented models correctly describe the conceptual model (Wilensky and Rand 2007). In turn, validation is stimulated since its very idea is comparing models with other descriptions of the problem modelled, and this may include *cross-model validation*, i.e. the comparison with other simulation models that have been validated to some level. Thus, it is reasonable to assume that a computational model cannot be considered fully verified and validated until it has been successfully replicated (Edmonds and Hales 2003). Nonetheless, the most important reason for replicating and comparing models is simply one of good scientific practice, since replication is the gold standard against which scientific claims are evaluated (Peng 2011).

In the remainder of this section we discuss replication and comparison of simulation models under three different perspectives. First, in Sect. 9.4.1, we distinguish the terminology and origins of the different goals related to model replication and comparison. Next, in Sect. 9.4.2, we go over the best practices in developing models so that they may be replicated by other researchers in the future. Finally, in Sect. 9.4.3 we discuss a number of model comparison techniques.

9.4.1 Model Replication, Model Alignment or Submodel Comparison?

A *model replication* study commonly assesses the extent to which building computational models that draw on the same conceptual, usually published, model give results compatible with the ones reported for the latter. If the new results are similar to the published results, then the confidence in the correspondence between the computational and the conceptual models is increased. Replication is represented in Fig. 9.3.

The work of Edmonds and Hales (2003) is particularly informative and worthy of reference. Edmonds and Hales performed two independent replications of a previously published model involving co-operation between self-interested agents. Several shortcomings were found in the original model, leading the authors to conclude that unreplicated simulation models and their results cannot be trusted. The issue was found to be a subtle difference in one of the submodels, which lead to different conclusions about the functioning of the overall model.

The term *model alignment* is frequently used as a synonym for model replication. However, its meaning is somewhat more subtle, as it is more related with the extent to which models can be coupled or docked so that their consequences and results are consistent with each other. In its most general form, this concerns both to V&V. After Axtell et al. (1996), the term became associated with the process of determining whether different published models describing the same class of social phenomena produce the same results. Usually the alignment or docking of

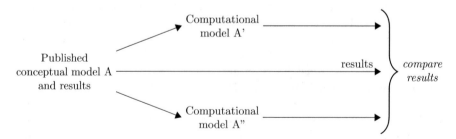

Fig. 9.3 Model replication

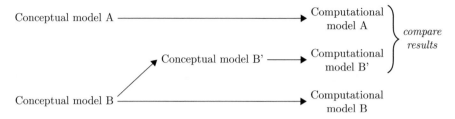

Fig. 9.4 Model alignment, also referred to as docking

two models A and B requires modifying certain aspects of model B—for instance turning off a specific feature—in order to become equivalent to model A. This is represented in Fig. 9.4.

The work of Axtell et al. (1996) is arguably the most-cited attempt to align two distinct but similar models. Rather than re-implementing Axelrod's culture dissemination model, Axtell and colleagues focused on the general case of aligning two models that reflected slightly distinctive mechanisms. For this purpose, Epstein and Axtell's Sugarscape model (1996) was progressively simplified in order to align with the results obtained by Axelrod's culture dissemination model (1997b). They concluded that comparing models developed by different researchers and with different tools (i.e. programming languages and/or modelling environments), can lead to exposing bugs, misinterpretations in model specification, and implicit assumptions in toolkit implementations.

Model alignment has been further investigated in a series of meetings called model-to-model (M2M) workshops (Rouchier et al. 2008). The M2M workshops attract researchers interested in understanding and promoting the transferability of knowledge between model users.

Submodel comparison, often referred to as *cross-element validation*, rather than comparing whole models, compares the results of a model whose architecture of the agents differs only in a few elements. The goal is to assess the extent to which changing elements of the model architecture produces results compatible with the expected results of the (larger) model. It is essentially an exercise in composing different submodels within a larger model, and is related to solution space exploration since submodels are varied, and the consequences of that variation are analysed and compared. In this process, the overall validity of the larger model with reference to the validity of each one of the submodels can also be assessed. For instance, one may study the effects of using a model with agents in a bargaining game employing either evolutionary learning or reinforcement learning strategies, and assess which one of the strategies produces results compatible with theoretical analysis in game theory (Takadama et al. 2003). Submodel comparison is represented in Fig. 9.5.

Submodel comparison can also be used as a model replication or alignment aid. For example, Radax and Rengs (2009) proposed a method for replicating insufficiently described ABMs, consisting in systematically varying ambiguous

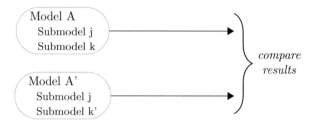

Fig. 9.5 Submodel comparison, also referred to as cross-element validation

model elements in order to align the replicated model with the original one. More generally, if two simulations do not align, trying out different assumptions or submodels is a practical way of finding the source of errors or mismatches. This type of study is greatly facilitated when computational models are implemented in a modular fashion, as discussed in Sect. 9.2.1. If submodels or model elements are implemented as separate modules in the computational model, it becomes much simpler to change or swap them in order to perform submodel comparisons.

9.4.2 Developing Replicable Models

An important aspect when developing a simulation model is to guarantee that it may be replicated by other researchers. Designing and programming for replicability involves a number of aspects that should be considered. Simulations are often a mix of conceptual descriptions and hard technical choices about implementation. The author who reports a model should assume that a replication or alignment may later be tried and thus should be careful about providing detailed information for future use. Some of the best practices include, but are not limited to:

– Effective documentation about the conceptual model should be provided, preferably in the form of a structured natural language description (Müller et al. 2014), such as the ODD protocol, discussed in Chap. 15 (Grimm et al. 2017) of this volume.The ODD protocol (Overview, Design concepts, Details) is one of the most widely used templates for making model descriptions more understandable and complete, providing a comprehensive checklist that covers many of the key features that can define a model. The ODD + D protocol (Müller et al. 2013) extends the ODD protocol for models in which human-decision making is simulated, often the case in social simulation.
– The model's source code should be made available, given that it is the model's definitive implementation, not subject to the vagueness and uncertainty possibly associated with verbal descriptions (Wilensky and Rand 2007; Müller et al. 2014). If possible, an open source simulation platform should be utilised to implement the model, thus fostering software reuse in order to make simulations

reliable and more comparable to each other. Maximum model exposure is achieved if the simulation is runnable on the browser. This is much simpler nowadays, with technologies such as HTML5 and JavaScript dispensing the need for browser plug-ins. ABM toolkits such as AgentScript (Densmore 2016) and AgentBase (Wiersma 2015) use this approach. In any case, making the computational model widely available and easily runnable is crucial for others to be able to experiment with it.

– Besides source code availability, documentation about the computational model should also be provided in the form of (1) detailed source code comments, and (2) a user guide and/or technical report. The former should clearly explain what each code unit (e.g. function or class) does, while the latter should describe the program's architecture, preferably with the aid of visual description standards such as UML diagrams. In either case, the computational model documentation should contain information about technical options where the translation from the conceptual model was neither straightforward nor consensual.

– Detailed information about the results should be made publicly available. This includes statistical methods and/or scripts implementing or using them, raw simulation outputs, distributional information, sensitivity analyses performed or qualitative measures. A number of specialised scientific data repositories exist for this purpose (Assante et al. 2016; Amorim et al. 2015). Furthermore, there is an increasing awareness of how important it is to have published, citable and documented data available in the scholarly record due to its crucial role in reproducible science (Altman et al. 2015; Kratz and Strasser 2014).

The CoMSES Net Computational Model Library (Rollins et al. 2014), an open digital repository for disseminating computational models associated with publications in the social and life sciences, should be highlighted in this regard since it enforces some of the best practices discussed above. Models are organised as searchable entries, by title, author or other relevant metadata. A formatted citation is shown for each entry so that researchers who use the model can easily credit its creators. Model entries have separate sections for code, documentation, generated outputs, solution exploration analyses and other relevant information. The library accepts not only original models, but also explicitly welcomes replications of previous studies. It also offers a certification service that verifies (1) if the model code successfully compiles and runs, and (2) if the model adheres to documentation best practices, with the ODD protocol being the recommended documentation template.

9.4.3 Model Comparison Techniques

Replication is evaluated by comparing the outputs of the original computational model against the output of the replicated implementation (Thiele and Grimm 2015). However, how do we determine whether or not two models produce equivalent output behaviour? Axtell et al. (1996) defined three kinds of equivalence or levels of similarity between model outputs: *numerical identity*, *relational alignment* and *distributional equivalence*. The first, *numerical identity*, implies exact numerical output and is difficult to demonstrate for stochastic models in general and social complexity models in particular. *Relational alignment* between outputs exists if they show qualitatively similar dependencies with input data, which is frequently the only way to compare a model with another which is inaccessible (e.g. implementation has not been made available by the original author), or with a non-controllable "real" social system. Lastly, *distributional equivalence* between implementations is achieved when the distributions of results cannot be statistically distinguished. What this shows is that at conventional confidence probabilities the statistics from different implementations may come from the same distribution, but it does not prove that this is actually the case. In other words, it does not prove that two implementations are algorithmically equivalent. Nonetheless, demonstrating equivalence for a larger number of parametrisations increases the confidence that the implementations are in fact globally equivalent (Edmonds and Hales 2003).

Since numerical identity is difficult to attain, and is not critical for showing that two such models have the same dynamic behaviour, distributional equivalence is more often than not the appropriate standard when comparing two implementations of a stochastic social complexity model. When aiming for distributional equivalence, a set of statistical summaries representative of each output are selected. It is these summaries, and not the complete outputs, that will be compared in order to assess the similarity between the original computational model and the replicated one. As models may produce large amounts of data, the summary measures should be chosen as to be relevant to the actual modelling goal. The summaries of all model outputs constitute the set of focal measures (FMs) of a model (Wilensky and Rand 2007), or more specifically, of a model parametrisation (since different FMs may be selected for distinct parametrisations). However, this process is empirically driven and model-dependent, or even parameter-dependent. Furthermore, it is sometimes unclear as to what output features best describe model behaviour. A possible solution, presented by Arai and Watanabe (2008) in the context of comparing models with different elements, is the automatic extraction of FMs from time-series simulation output using the discrete Fourier transform. Fachada et al. (2017b) proposed a similarly automated method, using principal component analysis to convert simulation output into a set of linearly uncorrelated statistical measures, analysable in a consistent, model-independent fashion. The proposed method was broader in scope—with support for multiple outputs and different types of data—and is available in the form of a software package for the R platform (Fachada et al. 2016; R Core Team 2017).

Once the FMs are extracted from simulation output, there are three major statistical approaches used to compare them: (1) statistical hypothesis tests; (2) confidence intervals; and (3) graphical methods (Balci and Sargent 1984). Statistical hypothesis tests are often used for comparing two or more computational models (Axtell et al. 1996; Wilensky and Rand 2007; Edmonds and Hales 2003; Miodownik et al. 2010; Radax and Rengs 2009; Fachada et al. 2017a,b). More specifically, hypothesis tests check if the statistical summaries obtained from the outputs of two (or more) model implementations are drawn from the same distribution. Confidence intervals are usually preferred for comparing the output of a model with the output of the system being modelled, as they provide an indication of the magnitude by which the statistic of interest differs between the two. Nonetheless, confidence intervals can also be used for model comparison, but in contexts different from replication, such as the evaluation of different models that might represent competing system designs or alternative operating policies (Balci and Sargent 1984; Law 2015). Graphical methods, such as Q–Q plots (e.g. Alberts et al. 2012) or scatter plots (e.g. Arai and Watanabe 2008; Fachada et al. 2017b), can also be employed for comparing output data, though their interpretation is more subjective than the previous methods.

9.5 Modelling Strategies and Its Relationship to Validation

In this section we review the purpose of validation and its relationship to different modelling strategies with respect to the level of descriptive detail embedded in a simulation.

Several taxonomies of modelling strategies have been described in the literature (David et al. 2004; Boero and Squazzoni 2005; Gilbert 2008, pp. 42–44). Normally, the adoption of these strategies does not depend on the class of the target being modelled, but on different ways to address it as the problem domain. For example, if a simulation is intended to model a system for the purpose of designing policies, this implies representing more information and detail than a simulation intended for modelling social mechanisms of the system in a metaphorical way. However, varying levels of model detail imply a trade-off between the effort required for verifying the simulation and the effort required for validating it. As more context and richness are embedded in a model, the more difficult it will be to verify it. Conversely, as one increases the descriptive richness of simulations, more ways will be available to assess its validity. A tension that contrasts the tendency for constraining simulations by formal-theoretical constructs—normally easier to verify—and constraining simulations by theoretical-empirical descriptions—more amenable to validation by empirical and participative-based methods. In the next sections, two contrasting modelling strategies are discussed and the typical cycle of formal and informal approaches for modelling and validation is described.

9.5.1 Subjunctive Agent-Based Models

A popular strategy in social simulation consists of using models as a means for expressing subjunctive moods to talk about possible worlds using what-if scenarios, like "what would happen if something were the case." The goal is building artificial societies for modelling possible worlds that represent classes of social mechanisms, while striving for maximal simplicity and strong generalisation power of the representations used. Reasons for striving for simplicity include the computational tractability of the model and to keep the data analysis as simple as possible.

Simplicity and generalisation power are often seen as elements of elegance in a model. However, making the model simpler in the social sciences does not necessarily make the model more general. More often than not this kind of modelling only makes it metaphorically general, or simply counterfactual (with false assumptions). For example, "What would happen if world geography is regarded as a two-dimensional space arranged on a 10×10 grid, where agents are thought of as independent political units, such as nations, which have specific behaviours of interaction according to simple rules?" To assume that world geography is one-dimensional, as Axelrod (1993) does in his Tribute Model, is clearly a false assumption. Often these models are associated with a design slogan coined by Axelrod (1997a), called the KISS approach—"Keep it Simple Stupid." Despite their simplicity, these kinds of models prove useful for concept formation and theoretical abstraction. The emergence of macro regularities from micro-levels of interaction becomes the fundamental source of concept formation and hypothesis illustration, with the power of suggesting novel theoretical debates.

Given the tendency for simplification and abstraction, mechanisms used in these models are normally described in a formalised or mathematical way. Axelrod's models, such as the culture dissemination model, or Schelling's residential segregation model, are canonical examples. Their simplicity and elegance have been factors for popularity and dissemination that span numerous disciplines and ease replication and verification.

However, whereas simplicity eases verification, the use of metaphorical models also brings disadvantages. Consider a word composed of several attributes representing an agent's culture, such as in Axelrod's culture dissemination model. The attributes do not have any specific meaning and are only distinguishable by their relative position in the word. Thus, they can be interpreted according to a relatively arbitrary number of situations or social contexts. However, such a representation may also be considered too simplified to mean anything relevant for such a complex concept as a cultural attribute. As a consequence, verification is hardly distinguishable from validation, insofar as the model does not represent a specific context of social reality. In such a sense, the researcher is essentially verifying experimentally whether his conceptions are met by an operationalisation that is intentionally and computationally expressed (David et al. 2005). Nevertheless, given their simplicity, subjunctive models can be easily linked and compared to other models, extended with additional mechanisms, as well as modified for model alignment, docking, or replication. Cross-element validation is a widely used technique.

At any rate, the fact that these models are simpler to replicate and compare—but hardly falsifiable by empirically acquired characteristics of social reality—stresses their strong characteristic: when models based on strategies of maximal simplicity become accepted by a scientific community, their influence seems to reach several other disciplines and contexts. Perhaps for this reason, these kinds of models are the most popular in social simulation, and some models are able to reach a considerable impact in many strains of social science.

9.5.2 Context-Specific Agent-Based Models

It would be simplistic to say that models in social simulation can be characterised according to well-defined categories of validation strategies. Even so, the capacity to describe social complexity, whether through simplicity or through rich detail and context, is a determining factor for a catalogue of modelling strategies.

We cannot hope to model general social mechanisms that are valid in all contexts. There are many models that are not designed to be markedly general or metaphorically general, but to stress accurateness, diversity, and richness of description. Instead of using possible worlds representing very arbitrary contexts, models are explicitly bounded to specific contexts. Constraints imposed on these models can vary from models investigating properties of social mechanisms in a large band of situations which share common characteristics, to models with the only ambition of representing a single history, like Dean's retrodiction of the patterns of settlement of the Anasazi in the southwestern United States, household by household (Dean et al. 2000).

Constructing and validating a model of this kind requires the use of empirical knowledge. They are, for this reason, often associated with the idea of "Empirical Validation of Agent-Based Models."

What is the meaning of empirical in this sense? If the goal is to discuss empirical claims, then models should attempt to capture empirically enquired characteristics of the target domain. Specifying the context of descriptions will typically provide more ways for enquiring quantitative and qualitative data in the target, as well as using experimental and participative methods with stakeholders. In this sense, empirical may be understood as a stronger link between the model and a context-specific, well-circumscribed problem domain.

The Anasazi model by Dean et al. (2000) is a well-known and oft-cited example of a highly contextualised model built on the basis of numerous sources, from archaeological data to anthropological, agricultural and ethnographic analyses, in a multidisciplinary context.

Given the higher specificity of the target domain, the higher diversity of ways for enriching the model as well as the increased semantic specificity of the outputs produced by the model, context-specific models may be more susceptible to be compared with empirical results of other methods of social research. On the other hand, comparison with other simulation models is complex and these models are more difficult to replicate and compare.

9.5.3   Modus Operandi: Formal and Informal Approaches

The tension between simplicity and descriptive richness expresses two different ways for approaching the construction and validation of a model. One can start with a rich, complex, realistic description and only simplify it where this turns out to be possible and irrelevant to the target system—known as the KIDS approach (Edmonds and Moss 2005). Or one starts from the outset with the simplest possible description and complexifies it only when it turns out to be necessary to make the model more realistic (Law 2015), nevertheless keeping the model as simple as possible—known as the KISS approach (Axelrod 1997a).

In practice, both trends are used for balancing trades-offs between the model's descriptive accuracy and the practicality of modelling, according to the purpose and the context of the model (Sun et al. 2016). This raises yet another methodological question: the extent to which models ought to be designed on the basis of formal theories, or ought to be constrained by techniques and approaches just on the basis of the intuition of the model builders and stakeholders. As we have seen, strong, subjunctive, ABMs with metaphorical purposes tend to adopt the simplicity motto with extensive use of formal constructs, making the models more elegant from a mathematical point of view, easier to verify, but less liable to validation methods. Game theoretical models, with all their formal and theoretical apparatus, are a canonical example. Results from these models are strongly constrained by the formal theoretical framework used.

A similar problem is found when ABMs make use of cognitive architectures strongly constrained by logic-based formalisms, such as the kind of formalisms used to specify BDI-type architectures. If the cognitive machinery of the agents relies on heuristic approaches that have been claimed valid, many researchers in the literature claim that cognitive ABMs can be validated in the empirical sense of context-specific models. Cited examples of this kind usually point to ABMs based on the Soar cognitive architecture (Laird 2012).

At any rate, context-specific models are normally more eclectic and make use of both formal and informal knowledge, often including informal and stakeholder evidence in order to build and validate the models. Model design tends to be less constrained a priori by formal constructs. In principle, one starts with all aspects of the target domain that are assumed to be relevant and then explores the behaviour of the model in order to find out if there are aspects that do not prove relevant for a particular interval of outcomes. The typical approach the majority of all modelling and validation can be summarised in a cycle with the following iterative and overlapping steps:

(a) *Building and validating pre-computational and computational models*: Several descriptions and specifications are used to build a model, eventually in the form of a computer program, which are micro-validated against a theoretical framework and/or empirical knowledge, usually qualitatively. This may include the individual agents' interaction mechanisms (rules of behaviour for agents or organisations of agents), their internal mechanisms (e.g. their cognitive machinery), the kind of interaction topology or environment, and the passive entities with which the agents interact. The model used should be as general as possible for the context in consideration as well as flexible for testing how parameters vary in particular circumstances. Empirical data—if available— should be used to help configure the parameters. Both the descriptions of the model and the parameters used should be validated for the specific context of the model. For example, suppose empirical data are available for specifying the consumer demand of products. If the demand varies from sector to sector, one may use data to inform the distribution upon which the parameter could be based for each specific sector.

(b) *Specifying expected behaviours of the computational model*: Micro and macro characteristics that the model is designed to reproduce are established from the outset based on theoretical and/or empirical knowledge. Any property, from quantitative to qualitative measures, such as emergent key facts the model should reproduce (stylised facts), the statistical characteristic or shape of time-data series (statistical signatures) and individual agents' behaviour along the simulation (individual trajectories), can be assessed. This may be carried out in innumerable ways, according to different levels of description or grain, and be more or less general depending on the context of the model and the kind of empirical knowledge available. For instance, in some systems it may be enough to predict just a "weak" or "positive" measure on some particular output, such as a positive and weak autocorrelation. Or we might look for the emergence of unpredictable events, such as stable regimes interleaved with periods of strong volatility, and check their statistical properties for various levels of granularity. Or the emergence of different structures or patterns associated with particular kinds of agents, such as groups of political agents with "extremist" or "moderate" individuals.

(c) *Testing the computational model and building and validating post-compu-tational models*: The computational model is executed. Both individual and aggregate characteristics are computed and tested for sensitivity analysis. These are micro-validated and macro-validated against the expected characteristics of the model established in step B according to a variety of validation techniques, as described in the previous sections. A whole process of building post-computational models takes place, possibly leading to the discovery of unexpected characteristics in the behaviour of the computational model which should be assessed with further theoretical or empirical knowledge about the problem domain.

Further Reading

Good introductions to validation and verification of simulation models in general are Sargent (2013) and Troitzsch (2004), the latter with a focus on social simulation. Validation of ABMs in particular is addressed by Amblard et al. (2007).

For readers more interested in single aspects of V&V, with regard to ABMs with applicability in social simulation, the following papers provide highly accessible starting points:

- Edmonds and Hales (2003) demonstrate the importance of model replication (or model alignment) by means of a clear example.
- Boero and Squazzoni (2005) examine the use of empirical data for model calibration and validation and argue that "the characteristics of the empirical target" influence the choice of validation strategies.
- Moss and Edmonds (2005) discuss an approach for cross-validation that combines the involvement of stakeholders to validate the model qualitatively on the micro level with the application of statistical measures to numerical outputs to validate the model quantitatively on the macro level.
- Müller et al. (2014) address the question of whether an ideal standard for describing and documenting models exists, defining different types of model reporting and proposing a minimum description standard for good modelling practice.
- Lee et al. (2015) provide an overview of the state-of-the-art approaches in analysing and reporting ABM outputs, highlighting challenges and issues related to variance stability, sensitivity analysis, spatio-temporal analysis, visualisation, and effective communication of these to non-technical audiences, such as various stakeholders.
- Fachada et al. (2017b) Present a structured approach to designing and performing complete model comparison experiments, using statistical tests to determine if two or more computational models generate distributionally equivalent behaviour.
- Finally, more comprehensive epistemological perspectives on verification and validation are provided in a number of papers published or derived from the Epistemological Perspectives on Simulation (EPOS) workshops, namely Frank and Troitzsch (2005), David (2009), Squazzoni (2009) and David et al. (2010).

Acknowledgements This work was partially funded by the Fundação para a Ciência e a Tecnologia project UID/EEA/50009/2013.

References

Alberts, S., Keenan, M. K., D'Souza, R. M., & An, G. (2012). Data-parallel techniques for simulating a mega-scale agent-based model of systemic inflammatory response syndrome on graphics processing units. *Simulation, 88*(8), 895–907. doi:10.1177/0037549711425180, http://journals.sagepub.com/doi/abs/10.1177/0037549711425180

Altman, M., Borgman, C., Crosas, M., & Matone, M. (2015). An introduction to the joint principles for data citation. *Bulletin of the Association for Information Science and Technology, 41*(3), 43–45. doi:10.1002/bult.2015.1720410313, http://onlinelibrary.wiley.com/doi/10.1002/bult.2015.1720410313/abstract

Amblard, F., Bommel, P., & Rouchier, J. (2007). Assessment and validation of multi-agent models. In *Agent-based modelling and simulation in the social and human sciences* (pp. 93–116). Oxford: Bardwell Press. http://agritrop.cirad.fr/541339/

Amorim, R. C., Castro, J. A., Silva, J. Rd., & Ribeiro, C. (2015). A comparative study of platforms for research data management: Interoperability, metadata capabilities and integration potential. In *New contributions in information systems and technologies* (pp. 101–111). Cham: Springer. doi:10.1007/978-3-319-16486-1_10, https://link.springer.com/chapter/10.1007/978-3-319-16486-1_10

Arai, R., & Watanabe, S. (2008). A quantitative method for comparing multi-agent-based simulations in feature space. In *Multi-agent-based simulation IX* (pp.154–166). Berlin/Heidelberg: Springer. doi:10.1007/978-3-642-01991-3_12, https://link.springer.com/chapter/10.1007/978-3-642-01991-3_12

Assante, M., Candela, L., Castelli, D., & Tani, A. (2016). Are scientific data repositories coping with research data publishing? *Data Science Journal, 15*, 6. doi:10.5334/dsj-2016-006, http://datascience.codata.org/articles/10.5334/dsj-2016-006/

Axelrod, R. (1993). *A Model of the Emergence of New Political Actors*. Working paper 93-11-068, Santa Fe Institute. https://www.santafe.edu/research/results/working-papers/a-model-of-the-emergence-of-new-political-actors

Axelrod, R. (1997a). Advancing the art of simulation in the social sciences. In D. R. Conte, P. D. R. Hegselmann, & P. D. P. Terna (Eds.), *Simulating Social Phenomena*. Lecture notes in economics and mathematical systems (Vol. 456, pp. 21–40). Berlin/Heidelberg: Springer. doi:10.1007/978-3-662-03366-1_2, http://link.springer.com/chapter/10.1007/978-3-662-03366-1_2

Axelrod, R. (1997b). The dissemination of culture: A model with local convergence and global polarization. *Journal of Conflict Resolution, 41*(2), 203–226. doi:10.1177/0022002797041002001, http://dx.doi.org/10.1177/0022002797041002001

Axtell, R., Axelrod, R., Epstein, J. M., & Cohen, M. D. (1996). Aligning simulation models: A case study and results. *Computational & Mathematical Organization Theory, 1*(2), 123–141. doi:10.1007/BF01299065, https://link.springer.com/article/10.1007/BF01299065

Balci, O., & Sargent, R. G. (1984). Validation of simulation models via simultaneous confidence intervals. *American Journal of Mathematical and Management Sciences, 4*(3–4), 375–406. doi:10.1080/01966324.1984.10737151, http://dx.doi.org/10.1080/01966324.1984.10737151

Barreteau, O., Bots, P., Daniell, K., Etienne, M., Perez, P., Barnaud, C., et al. (2017). Participatory approaches. doi:https://doi.org/10.1007/978-3-319-66948-9_12.

Barreteau, O., Bots, P., Daniell, K., Etienne, M., Perez, P., Barnaud, C., et al. (2017). Participatory approaches. In B. Edmonds & R. Meyer (Eds.), *Simulating social complexity. Understanding complex systems* (2nd ed.). Berlin/Heidelberg: Springer. doi:10.1007/978-3-319-66948-9_12

Boero, R., & Squazzoni, F. (2005). Does empirical embeddedness matter? Methodological issues on agent-based models for analytical social science. *Journal of Artificial Societies and Social Simulation, 8*(4), 6. http://jasss.soc.surrey.ac.uk/8/4/6.html

Calvez, B., & Hutzler, G. (2005). Automatic tuning of agent-based models using genetic algorithms. In *Multi-agent-based simulation VI* (pp. 41–57). Berlin/Heidelberg: Springer. doi:10.1007/11734680_4, https://link.springer.com/chapter/10.1007/11734680_4

Collier, N., & North, M. (2013). Parallel agent-based simulation with repast for high performance computing. *Simulation, 89*(10), 1215–1235. doi:10.1177/0037549712462620, http://journals.sagepub.com/doi/abs/10.1177/0037549712462620

David, N. (2009). Validation and verification in social simulation: patterns and clarification of terminology. In *Epistemological aspects of computer simulation in the social sciences* (pp. 117–129). Berlin/Heidelberg: Springer. doi:10.1007/978-3-642-01109-2_9, https://link.springer.com/chapter/10.1007/978-3-642-01109-2_9

David, N., Marietto, M. B., Sichman, J. S., & Coelho, H. (2004). The structure and logic of interdisciplinary research in agent-based social simulation. *Journal of Artificial Societies and Social Simulation, 7*(3), 4. http://jasss.soc.surrey.ac.uk/7/3/4.html

David, N., Sichman, J. S., & Coelho, H. (2005). The logic of the method of agent-based simulation in the social sciences: Empirical and intentional adequacy of computer programs. *Journal of Artificial Societies and Social Simulation, 8*(4), 2. http://jasss.soc.surrey.ac.uk/8/4/2.html

David, N., Caldas, J. C., & Coelho, H. (2010). Epistemological perspectives on simulation III. *Journal of Artificial Societies and Social Simulation, 13*(1). doi:10.18564/jasss.1591, http://jasss.soc.surrey.ac.uk/13/1/14.html

Dean, J. S., Gumerman, G. J., Epstein, J. M., Axtell, R. L., Swedlund, A. C., Parker, M. T., et al. (2000). Understanding Anasazi culture change through agent-based modeling. In T. A. Kohler & G. J. Gumerman (Eds.), *Dynamics in human and primate societies: Agent-based modeling of social and spatial processes. Santa fe institute studies on the sciences of complexity* (pp. 179–205). New York/Oxford: Oxford University Press.

Densmore, O. (2016). AgentScript. http://agentscript.org/

Edmonds, B., & Hales, D. (2003). Replication, replication and replication: Some hard lessons from model alignment. *Journal of Artificial Societies and Social Simulation, 6*(4), 11. http://jasss.soc.surrey.ac.uk/6/4/11.html

Edmonds, B. & Moss, S. (2005). From KISS to KIDS—an 'anti-simplistic' modelling approach. In: P. Davidsson, B. Logan, & K. Takadama (Eds.), *Multi-agent and multi-agent-based simulation* (Vol. 3415, pp. 130–144). Berlin/Heidelberg: Springer. doi:10.1007/978-3-540-32243-6_11. http://link.springer.com/10.1007/978-3-540-32243-6_11

Epstein, J., & Axtell, R. (1996). *Growing artificial societies: Social science from the bottom up.* Washington, DC: Brookings Institution Press; Cambridge, MA: MIT Press.

Evans, A., Heppenstall, A., & Birkin, M. (2017). Understanding simulation results. doi: https://doi.org/10.1007/978-3-319-66948-9_10.

Fachada, N., Lopes, V. V., Martins, R. C., & Rosa, A. C. (2015). Towards a standard model for research in agent-based modeling and simulation. *PeerJ Computer Science, 1*, e36. doi:10.7717/peerj-cs.36, https://peerj.com/articles/cs-36

Fachada, N., Rodrigues, J., Lopes, V. V., Martins, R. C., & Rosa, A. C. (2016). micompr: An R package for multivariate independent comparison of observations. *The R Journal, 8*(2), 405–420. http://journal.r-project.org/archive/2016-2/fachada-rodrigues-lopes-etal.pdf

Fachada, N., Lopes, V. V., Martins, R. C., & Rosa, A. C. (2017a). Parallelization strategies for spatial agent-based models. *International Journal of Parallel Programming, 45*(3), 449–481.

Fachada, N., Lopes, V. V., Martins, R. C., & Rosa, A. C. (2017b). Model-independent comparison of simulation output. *Simulation Modelling Practice and Theory, 72*, 131–149. doi:10.1016/j.simpat.2016.12.013, http://www.sciencedirect.com/science/article/pii/S1569190X16302854

Frank, U., & Troitzsch, K. G. (2005). Epistemological perspectives on simulation. *Journal of Artificial Societies and Social Simulation, 8*(4), 7. http://jasss.soc.surrey.ac.uk/8/4/7.html

Galán, J. M., Izquierdo, L. R., Izquierdo, S. S., Santos, J. I., Olmo, Rd., & López-Paredes, A. (2017). Checking simulations: Detecting and avoiding errors and artefacts. doi: https://doi.org/10.1007/978-3-319-66948-9_9.

Gilbert, N. (2008). *Agent-based models.* Thousand Oaks, CA: SAGE. google-Books-ID: Z3cp0ZBK9UsC.

Grimm, V., Polhill, G., & Touza, J. (2017). Documenting social simulation models: The ODD protocol as a standard. doi: https://doi.org/10.1007/978-3-319-66948-9_10.

Gross, D., & Strand, R. (2000). Can agent-based models assist decisions on large-scale practical problems? A philosophical analysis. *Complexity, 5*(6), 26–33. doi:10.1002/1099-0526(200007/08)5:6<26::AID-CPLX6>3.0.CO;2-G, http://onlinelibrary.wiley.com/doi/10.1002/1099-0526(200007/08)5:6<26::AID-CPLX6>3.0.CO;2-G/abstract

Kratz, J., & Strasser, C. (2014). Data publication consensus and controversies. *F1000Research, 3*, 94. doi:10.12688/f1000research.3979.3, http://f1000research.com/articles/3-94/v3

Laird, J. E. (2012). *The soar cognitive architecture.* Cambridge: MIT Press.

Law, A. M. (2015). Simulation modeling and analysis (5th ed.). New York: McGraw Hill Higher Education.

Lee, J. S., Filatova, T., Ligmann-Zielinska, A., Hassani-Mahmooei, B., Stonedahl, F., Lorscheid, I., et al. (2015). The complexities of agent-based modeling output analysis. *Journal of Artificial Societies and Social Simulation, 18*(4), 4.

McKay, M. D., Beckman, R. J., & Conover, W. J. (1979). Comparison of three methods for selecting values of input variables in the analysis of output from a computer code. *Technometrics, 21*(2), 239–245. doi:10.1080/00401706.1979.10489755, http://dx.doi.org/10.1080/00401706.1979.10489755

Merlone, U., Sonnessa, M., & Terna, P. (2008). Horizontal and vertical multiple implementations in a model of industrial districts. *Journal of Artificial Societies and Social Simulation 11*(2), 5. http://jasss.soc.surrey.ac.uk/11/2/5.html

Miller, J. H. (1998). Active nonlinear tests (ANTs) of complex simulation models. *Management Science, 44*(6), 820–830. doi:10.1287/mnsc.44.6.820, http://pubsonline.informs.org/doi/abs/10.1287/mnsc.44.6.820

Miodownik, D., Cartrite, B., & Bhavnani, R. (2010). Between replication and docking: "adaptive agents, political institutions, and civic traditions" revisited. *Journal of Artificial Societies and Social Simulation, 13*(3), 1.

Müller, B., Bohn, F., Dreßler, G., Groeneveld, J., Klassert, C., Martin, R., et al. (2013), Describing human decisions in agent-based models – ODD + D, an extension of the ODD protocol. *Environmental Modelling & Software, 48*, 37–48. doi:10.1016/j.envsoft.2013.06.003, http://www.sciencedirect.com/science/article/pii/S1364815213001394

Müller, B., Balbi, S., Buchmann, C. M., de Sousa, L., Dressler, G., Groeneveld, J., et al. (2014). Standardised and transparent model descriptions for agent-based models: Current status and prospects. *Environmental Modelling & Software, 55*, 156–163. doi:10.1016/j.envsoft.2014.01.029, http://www.sciencedirect.com/science/article/pii/S1364815214000395

Montgomery, D. C. (2012). *Design and analysis of experiments* (8th ed.). Hoboken: Wiley.

Moss, S., & Edmonds, B. (2005). Sociology and simulation: Statistical and qualitative cross-validation. *American Journal of Sociology, 110*(4), 1095–1131. doi:10.1086/427320, http://www.journals.uchicago.edu/doi/abs/10.1086/427320

North, M. J., Collier, N. T., Ozik, J., Tatara, E. R., Macal, C. M., Bragen, M., et al. (2013). Complex adaptive systems modeling with Repast Simphony. *Complex Adaptive Systems Modeling, 1*(1), 3. doi:10.1186/2194-3206-1-3, http://casmodeling.springeropen.com/articles/10.1186/2194-3206-1-3

Peng, R. D. (2011). Reproducible research in computational science. *Science, 334*(6060), 1226–1227. doi:10.1126/science.1213847, http://science.sciencemag.org/content/334/6060/1226

Pereda, M., Santos, J. I., & Galan, J. M. (2015). *A brief introduction to the use of machine learning techniques in the analysis of agent-based models*. SSRN Scholarly Paper ID 2689676. Rochester, NY: Social Science Research Network. https://papers.ssrn.com/abstract=2689676

R Core Team. (2017). R: A language and environment for statistical computing. https://www.R-project.org/

Radax, W., & Rengs, B. (2009). Prospects and pitfalls of statistical testing: Insights from replicating the demographic prisoner's dilemma. *Journal of Artificial Societies and Social Simulation, 13*(4), 1.

Rollins, N. D., Barton, C. M., Bergin, S., Janssen, M. A., & Lee, A. (2014). A Computational Model Library for publishing model documentation and code. *Environmental Modelling & Software, 61*, 59–64. doi:10.1016/j.envsoft.2014.06.022, http://www.sciencedirect.com/science/article/pii/S1364815214001959

Rouchier, J., Cioffi-Revilla, C., Polhill, J. G., & Takadama, K. (2008). Progress in model-to-model analysis. *Journal of Artificial Societies and Social Simulation, 11*(2), 8. http://jasss.soc.surrey.ac.uk/11/2/8.html

Sargent, R. G. (2013). Verification and validation of simulation models. *Journal of Simulation, 7*(1), 12–24. doi:10.1057/jos.2012.20, https://link.springer.com/article/10.1057/jos.2012.20

Schelling, T. C. (1971). Dynamic models of segregation. *The Journal of Mathematical Sociology,* *1*(2), 143–186. doi:10.1080/0022250X.1971.9989794, http://dx.doi.org/10.1080/0022250X. 1971.9989794

Squazzoni, F. (Ed.). (2009). *Epistemological aspects of computer simulation in the social sciences.* Lecture notes in computer science (Vol. 5466). Berlin/Heidelberg: Springer. doi:10.1007/978-3-642-01109-2, http://link.springer.com/10.1007/978-3-642-01109-2

Stonedahl, F., & Wilensky, U. (2010). Finding forms of flocking: Evolutionary search in ABM parameter-spaces. In *Multi-agent-based simulation XI* (pp. 61–75). Berlin/Heidelberg: Springer. doi:10.1007/978-3-642-18345-4_5, https://link.springer.com/chapter/10.1007/978-3-642-18345-4_5

Sun, Z., Lorscheid, I., Millington, J. D., Lauf, S., Magliocca, N. R., Groeneveld, J., et al. (2016) Simple or complicated agent-based models? A complicated issue. *Environmental Modelling & Software, 86*, 56–67. doi:10.1016/j.envsoft.2016.09.006, http://www.sciencedirect. com/science/article/pii/S1364815216306041

Takadama, K., Suematsu, Y. L., Sugimoto, N., Nawa, N. E., & Shimohara, K. (2003). Cross-element validation in multiagent-based simulation: Switching learning mechanisms in agents. *Journal of Artificial Societies and Social Simulation, 6*(4), 6. http://jasss.soc.surrey.ac.uk/6/4/6.html

Thiele, J. C., & Grimm, V. (2015). Replicating and breaking models: Good for you and good for ecology. *Oikos, 124*(6), 691–696. doi:10.1111/oik.02170, http://onlinelibrary.wiley.com/doi/10.1111/oik.02170/abstract

Troitzsch, K. G. (2004). Validating simulation models. In G. Horton (Ed.), *Proceedings of 18th European Simulation Multiconference, ESM 2004* (pp. 265–270). Magdeburg: SCS Publishing House.

Wiersma, W. (2015). AgentBase: Agent based modelling in the browser. http://wybowiersma.net/pub/papers/Wiersma,Wybo,AgentBase_agent_based_modelling_in_the_browser.pdf

Wilensky, U. (1999). NetLogo. http://ccl.northwestern.edu/netlogo/

Wilensky, U., & Rand, W. (2007). Making models match: Replicating an agent-based model. *Journal of Artificial Societies and Social Simulation, 10*(4), 2. http://jasss.soc.surrey.ac.uk/10/4/2.html

Will, O., & Hegselmann, R. (2008). A replication that failed – On the computational model in 'Michael W. Macy and Yoshimichi Sato: Trust, cooperation and market formation in the U.S. and Japan. Proceedings of the national academy of sciences, May 2002'. *Journal of Artificial Societies and Social Simulation, 11*(3), 3. http://jasss.soc.surrey.ac.uk/11/3/3.html

Chapter 10
Understanding Simulation Results

Andrew Evans, Alison Heppenstall, and Mark Birkin

Abstract Simulation modelling is concerned with the abstract representation of entities within systems and their interrelationships; understanding and visualising these results is often a significant challenge for the researcher. Within this chapter we examine particular issues such as finding "important" patterns and interpreting what they mean in terms of causality. We also discuss some of the problems with using model results to enhance our understanding of the underlying social systems which they represent, and we will assert that this is in large degree a problem of isolating causal mechanisms within the model architecture. In particular, we highlight the issues of equifinality and identifiability—that the same behaviour may be induced within a simulation from a variety of different model representations or parameter sets—and present recommendations for dealing with this problem. The chapter ends with a discussion of avenues of future research.

Why Read This Chapter?
To help you understand the results that a simulation model produces, by suggesting some ways to analyse and visualise them. The chapter concentrates on the internal dynamics of the model rather than its relationship to the outside world.

10.1 Introduction

Simulation models may be constructed for a variety of purposes. Classically these purposes tend to centre on either the capture of a set of knowledge or making predictions. Knowledge capture has its own set of issues that are concerned with structuring and verifying knowledge in the presence of contradiction and uncertainty. The problems of prediction, closely associated with calibration and validation, centre around comparisons with real data, for which the methods covered in Chap. 9 (David et al. 2017) are appropriate. In this chapter, however, we look at

A. Evans (✉) • A. Heppenstall • M. Birkin
School of Geography, University of Leeds, Leeds, UK
e-mail: a.j.evans@leeds.ac.uk

© Springer International Publishing AG 2017   205
B. Edmonds, R. Meyer (eds.), *Simulating Social Complexity*,
Understanding Complex Systems, https://doi.org/10.1007/978-3-319-66948-9_10

what our models tell us through their internal workings and logic, how we might understand/interpret simulation results as results about an attempted simulation of the real world, rather than as results we expect to compare directly with the world. Here then, we tackle the third purpose of modelling: the *exploration* of abstracted systems through simulation. In a sense, this is a purpose predicated only on the limitations of the human mind. By common definition, simulation modelling is concerned with abstract representations of entities within systems and their interrelationships and with the exploration of the ramifications of these abstracted behaviours at different temporal and geographical scales. In a world in which we had larger brains, models would not be required to reveal anything—we would instantly see the ramifications of abstracted behaviours in our heads. To a degree, therefore, models may be seen as replacing the hard joined-up thinking that is required to make statements about the way the world works. This chapter looks at what this simplifying process tells us about the systems we are trying to replicate.

In part, the complications of simulation modelling are a product of the dimensionality of the systems with which we are dealing. Let us imagine that we are tackling a system of some spatio-temporal complexity, for example, the prices in a retail market selling items A, B and C. Neighbouring retailers adjust their prices based on local competition, but the price of raw materials keeps the price surface out of equilibrium. In addition, customers will only buy one of the products at a time, creating a link between the prices of the three items. Here, then, we have three interdependent variables, each of which varies spatio-temporally, with strong auto- and cross-correlations in both time and space. What kinds of techniques can be used to tease apart such complex systems? In Sect. 10.2 of this chapter, we will discuss some of the available methodologies broken down by the dimensionality of the system in question and the demands of the analysis. Since the range of such techniques is extremely sizable, we shall detail a few traditional techniques that we believe might be helpful in simplifying model data that shows the traits of complexity and some of the newer techniques of promise.

Until recently, most social science models represented social systems using mathematical aggregations. We have over 2500 years' worth of techniques to call upon that are founded on the notion that we need to simplify systems as rapidly as we can to the point at which the abstractions can be manipulated within a single human head. As is clear, not least from earlier contributions in this volume, it is becoming increasingly accepted that social scientists might reveal more about systems by representing them in a less aggregate manner. More specifically, the main difference between mathematics and the new modelling paradigm is that we now aspire to work at a scale at which the components under consideration can be represented as having their own discrete histories; mathematics actually works in a very similar fashion to modern models, but at all the other scales. Naturally there are knock-ons from this in terms of the more explicit representation of objects, states and events, but these issues are less important than the additional simulation and analytical power that having a history for each component of a system gives us. Of course, such a "history" may just be the discrete position of an object at a single historical moment, and plainly at this level of complication, the boundary between such models and, for example, Markov models, is somewhat diffuse; however, as the history of components becomes more involved, so the power of modern modelling paradigms comes to the fore. What is lacking, however, are the techniques that are predicated on these new architectures. Whilst models which are specified at the level of individual entities or "agents" may also be analysed using conventional mathematical techniques, in Sect. 10.3 of the chapter, we will discuss some more novel approaches which are moving the direction of understanding the outputs of these new, unaggregated, models on their own terms.

One of the reasons that simulation models are such a powerful methodology for understanding complex systems is their ability to display aggregate behaviour which goes beyond the simple extrapolation of the behaviour of the individual component parts. In mathematical analysis, such as dynamical systems theory, this behaviour tends to be linked to notions of equilibrium, oscillation and catastrophe or bifurcation. Individual- and agent-based modelling approaches have veered more strongly towards the notion of emergence, which can be defined as "an unforeseen occurrence; a state of things unexpectedly arising" (OED 2010). The concept of emergence is essentially a sign of our ignorance of the causal pathways within a system. Nevertheless, emergence is our clearest hope for developing an understanding of systems using models. We hope that emergence will give us a perceptual shortcut to the most significant elements of a system's behaviour. When it comes to applications, however, emergence is a rather double-edged blade: emergence happily allows us to see the consequence of behaviours without us having to follow the logic ourselves; however it is problematic in relying upon us to filter out which of the ramifications are important to us. As emergence is essentially a sign of incomplete understanding, and therefore weakly relative, there is no objective definition of what is "important". one day classification of the kinds of patterns that relate to different types of causal history, but there is no objective manner of recognising a pattern as "important" as such. These two problems, finding "important" patterns (in the absence of any objective way of defining "important") and then interpreting what they mean in terms of causality, are the issues standing between the researcher and perfect knowledge of a modelled system. In the fourth section of this chapter, we will discuss some of the problems with using model results to enhance our understanding of the underlying social systems which they represent, and we will assert that this is in large degree a problem of isolating causal mechanisms within the model architecture. In particular, we highlight the issues of equifinality and identifiability—that the same behaviour may be induced within a simulation from a variety of different model representations or parameter sets—and present recommendations for dealing with this problem. Since recognising emergence and combating the problems of identifiability and equifinality are amongst the most urgent challenges to effective modelling of complex systems, this leads naturally to a discussion of future directions in the final section of the chapter.

10.2   Aggregate Patterns and Conventional Representations of Model Dynamics

Whether a model is based on deductive premises or inferred behaviours, any new understanding of a given modelled system tends to be developed inductively. Modellers examine model outputs, simplify them and then try to work out the cause utilising a combination of hypothesis dismissal, refinement and experimentation. For example, a modeller of a crowd of people might take all the responses of each person over time and generate a single simple mean statistic for each person; these might then be correlated against other model variables. If the correlation represents a real causal connection, then varying the variables should vary the statistic. Proving such causal relationships is not something we often have the ability to do in the real world. During such an analysis, the simplification process is key: it is this that reveals the patterns in our data. The questions are: how do we decide what needs simplifying and, indeed, how simple to make it?

We can classify model results by the dimensionality of the outputs. A general classification for social systems would be:

- Single statistical aggregations (1D)
- Time series of variables (2D)
- The spatial distributions of invariants (2D) or variables (3D)
- Spatio-temporal locations of invariants (3D) or variables (4D)
- Other behaviours in multidimensional variable space (nD)

For simplicity, this assumes that geographical spaces are essentially two-dimensional (while recognising that physical space might also be represented along linear features such as a high street, across networks or within a three-dimensional topographical space for landforms or buildings). It should also be plain that in the time dimension, models do not necessarily produce just a stream of data, but that the data can have complex patternation. By their very nature, individual-level models, predicated as they are on a life cycle, will never stabilise in the way a mathematical model might (Uchmanski and Grimm 1996); instead models may run away or oscillate, either periodically or chaotically.

Methods for aiding pattern recognition in data break down, again, by the dimensionality of the data, but also by the dimensionality of their outputs. It is quite possible to generate a one-number statistic for a 4D spatio-temporal distribution. In some cases, the reduction of dimensionality is the explicit purpose of the technique, and the aim is that patterns in one set of dimensions should be represented as closely as possible in a smaller set of dimensions so they are easier to understand. Table 10.1 below presents a suite of techniques that cross this range (this is not meant to be an exhaustive list; after all, pattern recognition is a research discipline of its own with a whole body of literature including several dedicated journals).

To begin with, let us consider some examples which produce outputs in a single dimension. In other words, techniques for generating global and regional statistics describing the distribution of variables across space, either a physical space or a

Table 10.1 Pattern recognition techniques for different input and output data dimensions

	1D output	2D output	3D output	4D output	ND
1D input					
2D input	Exploratory statistics	Cluster locating Fourier/wavelet transforms			
3D input	Entropy statistics	Phase diagrams Fourier/wavelet transforms			
4D input	Diffusion statistics	Time slices	Recurrence plots		
nD	Network statistics	Eigenvector analysis	Sammon mapping	Animations	Heuristic techniques

variable space. Such statistics generally tend to be single time slice, but can be generated for multiple time slices to gauge overall changes in the system dynamics.

Plainly, standard aggregating statistics used to compare two distributions, such as the variable variance, will lose much of interest, both spatially and temporally. If we wish to capture the distribution of invariants, basic statistics like nearest-neighbour (Clark and Evans 1954) or the more complex patch shape, fragmentation and connectivity indices of modern ecology (for a review and software, see McGarigal 2002) provide a good starting point. Networks can be described using a wide variety of statistics covering everything from shortest paths across a network to the quantity of connections at nodes (for a review of the various statistics and techniques associated with networks, see Boccaletti et al. 2006; Evans 2010). However, we normally wish to assess the distribution of a variable across a surface—for example, a price surface or a surface of predicted retail profitability. One good set of global measures for such distributions are entropy statistics. Suppose we have a situation in which a model is trying to predict the number of individuals that buy product A in one of four regions. The model is driven by a parameter, beta. In two simulations we get the following results: *simulation one* (low beta), 480, 550, 520 and 450 and *simulation two*, (high beta) 300, 700, 500 and 400. Intuitively the first simulation has less dispersal or variability than the second simulation. An appropriate way to measure this variability would be through the use of entropy statistics. The concept of entropy originates in thermodynamics, where gases in a high-entropy state contain dispersed molecules. Thus high entropy equates to high levels of variability. Entropy statistics are closely related to information statistics where a -entropy state corresponds to a high information state. In the example above, *simulation two* is said to contain more "information" than *simulation one*, because if we approximate the outcome using no information, we would have a flat average— 500, 500, 500 and 500—and this is closer to *simulation one* than *simulation two*. Examples of entropy and information statistics include Kolmogorov-Chaitin, mutual information statistics and the Shannon information statistic. Most applications in the literature use customised code for the computation of entropy statistics, although the computation of a limited range of generalised entropy indices is possible within

Stata.[1] Entropy statistics can also be used to describe flows across networks. In this sense they provide a valuable addition to network statistics: most network statistics concentrate on structure rather than the variable values across them. Unless they are looking specifically at the formation of networks over time, or the relationship between some other variable and network structure, modellers are relatively bereft of techniques to look at variation *on* a network.

In the case where variability is caused and constrained by neighbourhood effects, we would expect the variation to be smoother across a region. We generally expect objects in space under neighbourhood effects to obey Tobler's first law of geography (Tobler 1970) that everything is related, but closer things are related more. This leads to spatial auto- or cross-correlation, in which the values of variables at a point reflect those of their neighbours. Statistics for quantifying such spatial auto- or cross-correlation at the global level, or for smaller regions, such as Moran's I and Geary's C, are well established in the geography literature (e.g. Haining 1990); a useful summary can be found in Getis (2007).

Such global statistics can be improved on by giving some notion of the direction of change of the auto- or cross-correlation. Classically this is achieved through semi-variograms, which map out the intensity of correlation in each direction traversed across a surface (for details, see Isaaks and Srivastava 1990). In the case where it is believed that local relationships hold between variables, local linear correlations can be determined, for example, using geographically weighted regression (GWR; for details, see Fotheringham et al. 2002). GWR is a technique which allows the mapping of R^2s calculated within moving windows across a multivariate surface and, indeed, mapping of the regression parameter weights. For example, it would be possible in our retail results to produce a map of the varying relationship between the amount of A purchased by customers and the population density, if we believed these were related. GWR would not just allow a global relationship to be determined, but also how this relationship changed across a country. One important but somewhat overlooked capability of GWR is its ability to assess how the strength of correlations varies with scale by varying the window size. This can be used to calculate the key scales at which there is sufficient overlap between the geography of variables to generate strong relationships (though some care is needed in interpreting such correlations, as correlation strength generally increases with scale: Robinson 1950; Gehlke and Biehl 1934). Plainly, identifying the key scale at which the correlations between variables improve gives us some ability to recognise key distance scales at which causality plays out. In our example, we may be able to see that the scale at which there is a strong relationship between sales of A and the local population density increases as the population density decreases, suggesting rural consumers have to travel further and a concomitant non-linearity in the model components directing competition.

---

[1]Confusingly, "generalised entropy" methods are also widely used in econometrics for the estimation of missing data. Routines which provide this capability, e.g. in SAS, are not helpful in the description of simulation model outputs!

If, on the other hand, we believe the relationships do not vary smoothly across a modelled surface, we instead need to find unusual clusters of activity. The ability to represent spatial clustering is of fundamental importance, for example, within Schelling's well-known model of segregation in the housing market (Schelling 1969). However clustering is often not so easy to demonstrate within both real data and complex simulation outputs. The most recent techniques use, for example, wavelets to represent the regional surfaces, and these can then be interpreted for cluster-like properties. However, for socio-economic work amongst the best software for cluster detection is the geographical analysis machine (GAM), which not only assesses clustering across multiple scales but also allows assessment of clustering in the face of variations in the density of the population at risk. For example, it could tell us where transport network nodes were causing an increase in sales of A, by removing regions with high sales caused by high population density (the population "at risk" of buying A). Clusters can be mapped and their significance assessed (Openshaw et al. 1988).

Often, simulations will be concerned with variations in the behaviour of systems, or their constituent agents, over time. In common with physical systems, social and economic systems are often characterised by periodic behaviour, in which similar states recur, although typically this recurrence is much less regular than in many physical systems. For example, economic markets appear to be characterised by irregular cycles of prosperity and depression. Teasing apart a model can provide nonintuitive insights into such cycles. For example, Heppenstall et al. (2006) considered a regional network of petrol stations and showed within an agent simulation how asymmetric cyclical variations in pricing (fast rises and slow falls), previously thought to be entirely due to a desire across the industry to maintain artificially high profits, could in fact be generated from more competitive profit maximisation in combination with local monitoring of network activity. While it is, of course, not certain these simpler processes cause the pattern in real life, the model exploration does give researchers a new explanation for the cycles and one that can be investigated in real petrol stations.

In trying to detect periodic behaviour, wavelets are rapidly growing in popularity (Graps 2004). In general, one would assume that the state of the simulation can be represented as a single variable which varies over time (let's say the average price of A). A wavelet analysis of either observed or model data would decompose this trend into chunks of time at varying intervals, and in each interval the technique identifies both a long-term trend and a short-term fluctuation. Wavelets are therefore particularly suitable for identifying cycles within data. They are also useful as filters for the removal of noise from data and so may be particularly helpful in trying to compare the results from a stylised simulation model with observed data which would typically be messy, incomplete or subject to random bias. It has been argued that such decompositions are fundamentally helpful in establishing a basis for forecasting (Ramsey 2002).

Wavelets are equally applicable in both two and three dimensions. For example, they may be useful in determining the diffusion of waves across a two-dimensional space and over time and can be used to analyse, for example, the relationship between wave amplitude and propagation distance. Viboud et al. (2006) provide a particularly nice example of such a use, looking at the strength of the propagation of influenza epidemics as influenced by city size and average human travel distances in the USA. Other more traditional statistics, such as the Rayleigh statistic (Fisher et al. 1987; Korie et al. 1998), can also be used to assess the significance of diffusion from point sources.

In addition to global and regional aggregate statistics of single variables or cross-correlations, it may be that there is simply too great a dimensionality to recognise patterns in outputs and relate them to model inputs. At this point it is necessary to engage in multidimensional scaling. If an individual has more than four characteristics, then multidimensional scaling methods can be used to represent the individuals in two or three dimensions. In essence, the problem is to represent the relation between individuals such that those which are most similar in n-dimensions still appear to be closest in a lower-dimensional space which can be visualised more easily. The most popular technique is Sammon mapping. This method relies on the ability to optimise an error function which relates original values in high-dimensional space to the transformed values. This can be achieved using standard optimisation methods within packages such as MATLAB or using a number of bespoke R packages. Multidimensional scaling can be useful in visualising the relative position of different individuals within a search space, for exploring variations in a multi-criteria objective function within a parameter space or for comparing individual search paths within different simulations (Pohlheim 2006).

Eigenvector methods are another form of multidimensional scaling. Any multidimensional representation of data in n-dimensional space can be transformed into an equivalent space governed by n orthogonal eigenvectors. The main significance of this observation is that the principal eigenvector constitutes the most efficient way to represent a multidimensional space within a single value. For example, Moon, Schneider and Carley (Moon et al. 2006) use the concept of "eigenvector centrality" within a social network to compute a univariate measure of relative position based on a number of constituent factors.

Eigenvector analyses, however, can be nonintuitive to those not used to them. Somewhat simpler presentations of multidimensional data can be made using clustering techniques. These collapse multidimensional data so that individual cases are members of a single group or cluster, classified on the basis of a similarity metric. The method may therefore be appropriate if the modeller wishes to understand the distribution of an output variable in relation to the combination of several input variables. Cluster analysis is easy to implement in all the major statistics packages (R, SAS, SPSS). The technique is likely to be most useful in empirical applications with a relatively large number of agent characteristics (i.e. six or more) rather than in idealised simulations with simple agent rules. One advantage of this technique over others is that it is possible to represent statistical variation within the cluster space, for example, by displaying the interquartile variation in the attribute variable within clusters.

10.3   Individual Patterns, Novel Approaches and Visualisation

Plainly aggregate statistics like those above are a useful way of simplifying individual-level data, both in terms of complexity and dimensionality. However, they are the result of over 2500 years of mathematical development in a research environment unsuited to the mass of detail associated with individual-level data. Now, computers place us in the position of being able to cope with populations of individual-level data at a much smaller scale. We still tend to place our own understanding at the end of an analytical trail, constraining the trail to pass through some kind of simplification and higher level of aggregation for the purposes of model analysis. Despite this, it is increasingly true that individual-level data is dealt with at the individual level for the body of the analysis, and this is especially true in the case of individual-level modelling, in which experimentation is almost always enacted at the individual level. Whether it is really necessary to simplify for human understanding at the end of an analysis is not especially clear. It may well be that better techniques might be developed to do this than those built on an assumption of the necessity of aggregation.

At the individual level, we are interested in recognising patterns in space and time, seeing how patterns at different scales affect each other, and then using this to say something about the behaviour of the system/individuals. Patterns are often indicators of the attractors to which individuals are drawn in any given system and present a shortcut to understanding the mass of system interactions. However, it is almost as problematic to go through this process to understand a model as it is, for example, to derive individual-level behaviours from real large-size spatio-temporal datasets of socio-economic attributes. The one advantage we have in understanding a model is that we do have some grip on the foundation rules at the individual scale. Nonetheless, understanding a rule and determining how it plays out in a system of multiple interactions are very different things. Table 10.2 outlines some of the problems.

Despite the above, our chief tool for individual-level understanding without aggregation is, and always has been, the human ability to recognise patterns in

Table 10.2 Issues related to understanding a model at different levels of complexity

Complexity	Issues
Spatial	What is the impact of space (with whom do individuals initiate transactions and to what degree)?
Temporal	How does the system evolve?
Individuals	How do we recognise which individual behaviours are playing out in the morass of interactions?
Relationships	How do we recognise and track relationships?
Scale	How can we reveal the manner in which individual actions affect the large-scale system and vice versa?

masses of data. Visualisation, for all its subjectivity and faults, remains a key element of the research process. The standard process is to present one or more attributes of the individuals in a map in physical or variable space. Such spaces can then be evolved in movies or sliced in either time or space (Table 10.3 shows some examples). In general, we cannot test the significance of a pattern without first recognising it exists, and to that extent significance testing is tainted by the requirement that it tests our competency in recognising the correct pattern as much as that the proposed pattern represents a real feature of the distribution of our data. Visualisation is also a vital tool in communicating results within the scientific community and to the wider public. The former is not just important for the transmission of knowledge, but because it allows others to validate the work. Indeed, the encapsulation of good visualisation techniques within a model framework allows others to gain deeper understanding of one's model, and to experiment at the limits of the model—what Grimm (2002) calls "visual debugging". Good model design starts like the design of any good application, with an outline of what can be done to make it easy to use, trustworthy and simple to understand. Traditionally, user interface design and visualisation have been low on the academic agenda, to the considerable detriment of both the science and the engagement of taxpayers. Fortunately, in the years since the turn of the millennium, there has been an increasing realisation that good design engages the public and that there is a good deal of social science research that can be built on that engagement. Orford et al. (1999) identify computer graphics, multimedia, the World Wide Web and virtual reality as four visualisation technologies that have recently seen a considerable evolution within the social sciences. There is an ever-increasing array of visualisation techniques at our disposal: Table 10.3 presents a classification scheme of commonly used and more novel visualisation methods based on the dimensionality and type of data that is being explored.

Another classification scheme of these techniques that is potentially very useful comes from Andrienko et al. (2003). This classification categorises techniques based on their applicability to different types of data:

- "Universal" techniques that can be applied whatever the data, e.g. querying and animation
- Techniques revealing existential change, e.g. time labels, colouring by age, event lists and space-time cubes
- Techniques about moving objects, e.g. trajectories, space-time cubes and snapshots in time
- Techniques centred on thematic/numeric change, e.g. change maps, time series and aggregations of attribute values

For information on other visualisation schemes, see Cleveland (1983), Hinneburg et al. (1999) and Gahegan (2001).

In each case, the techniques aim to exploit the ease with which humans recognise patterns (Muller & Schumann Müller and Schumann 2003). Pattern recognition is, at its heart, a human attribute, and one which we utilise to understand models, no matter how we process the data. The fact that most model understanding is founded

Table 10.3 Classification of visualisation methods according to dimensionality and type of data

	Method	Pro	Con
Spatial 1D/2D	Map: overlay; animated trajectory representation (e.g. arrows); snapshots	View of whole trajectory of an object	Cannot analyse trajectory of movement. If several objects cross paths, cannot tell whether objects met at crossing point or visited points at different times
	Spatial distribution, e.g. choropleth maps	Gives a snapshot of an area.	Cannot see how a system evolves through time. Aggregate view of area. Only represents one variable; hard to distinguish relationships
Temporal 1D	Time-series graphs/linear and cyclical graphs	Show how the system (or parameters) change over time	No spatial element. Hard to correlate relationships between multivariate variables
	Rank clocks (e.g. Batty 2006)	Good for visualising change over time in ranked order of any set of objects	No spatial element
	Rose diagrams	Good for representation of circular data, e.g. wind speed and direction	No spatial element
	Phase diagram	Excellent for examining system behaviour over time for one or two variables	No spatial element. Gets confusing quickly with more than two variables
Spatio-temporal 3D/4D	Map animation (e.g. Patel and Hudson-Smith 2012)	Can see system evolving spatially and temporally	Hard to quantify or see impacts of individual behaviour, i.e. isolated effects
	Space-time cube (Andrienko et al. 2003)	Can contain space-time paths for individuals	Potentially difficult to interpret
	Recurrence plot	Reveals hidden structures over time and in space	Computationally intensive. Methods difficult to apply. Have to generate multiple snapshots and run as an animation
	Vector plotting/contour slicing (Ross and Vosper 2003)	Ability to visualise 2D or 3D data and multiple dimensional dataset	Hard to quantify individual effects

on a human recognition of a "significant" pattern is somewhat unfortunate, as we will bring our own biases to the process. At worst we only pay attention to those patterns that confirm our current prejudices: what Wyszomirski et al. (1999) call the *WYWIWYG*—What You Want is What You Get—fallacy. At best, we will only recognise those patterns that match the wiring of the human visual system and our cultural experiences. The existence of visualisation techniques generally points up the fact that humans are better at perceiving some patterns than others, and in some media than others—it is easier to see an event as a movie and not a binary representation of the movie file displayed as text. However, in addition to standard physiological and psychological restrictions on pattern recognition consistent to all people, it is also increasing apparent there are cultural differences in perceptions. Whether there is some underlying biological norm for the perception of time and space is still moot (Nisbett and Masuda 2003; Boroditsky 2001), but it is clear that some elements of pattern recognition vary by either culture or genetics (Nisbett and Masuda 2003; Chua et al. 2005). Even when one looks at the representation of patterns and elements like colour, there are clear arguments for a social influence on the interpretation of even very basic stimuli into perceptions (Roberson et al. 2004). Indeed, while there is a clear and early ability of humans to perceive moving objects in a scene as associated in a pattern (e.g. Baird et al. 2002), there are cultural traits associated with the age at which even relatively universal patterns are appreciated (Clement et al. 1970). The more we can objectify the process, therefore, the less our biases will impinge on our understanding. In many respects it is easier to remove human agents from data comparison and knowledge development than pattern hunting, as patterns are not something machines deal with easily. The unsupervised recognition of even static patterns repeated in different contexts is far from computationally solved (Bouvrie and Sinha 2007), though significant advances have been made in recent years (Druzhkov and Kustikova 2016). Most pattern-hunting algorithms try to replicate the process found in humans, and in that sense one suspects we would do better to skip the pattern hunting and concentrate on data consistency and the comparison of full datasets directly. At best we might say that an automated "pattern" hunter that wasn't trying to reproduce the human ability would instead seek to identify attractors within the data.

Figure 10.1 presents several visualisation methods that are commonly found in the literature, ranging from 1D time-series representation (a) to contour plots (d) that could be potentially used for 4D representation.

Visualisations are plainly extremely useful. Here we'll look at a couple of techniques that are of use in deciphering individual-level data: phase maps and recurrence plots. Both techniques focus on the representation of individual-level states and the relationships between stated individuals.

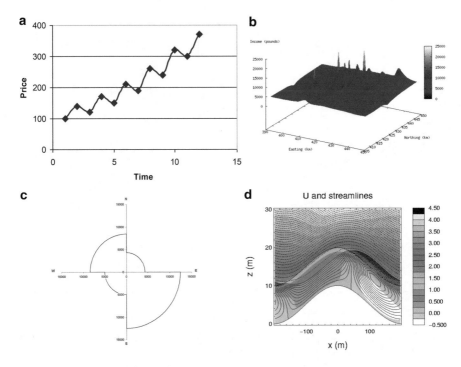

Fig. 10.1 Examples of different visualisation methods. (**a**) 1D Time-series graph (idealised data). (**b**) 3D interpolated map (idealised data). (**c**) Rose diagram. (**d**) Contour plot

10.3.1 Phase Maps

Phase-space maps are commonly used by physicists to study the behaviour of physical systems. In any graphical representation, a phase-space map represents an abstract view of the behaviour of one or more of the system components. These can be particularly useful to us as we can plot the behaviour of our system over time. This allows us to understand how the system is evolving and whether it is chaotic, random, cyclical or stable (Fig. 10.2).

Each of the graphs produced in Fig. 10.2 is a representation of the coincident developments in two real neighbouring city centre petrol stations in Leeds (UK) over a 30-day period (sampled every other day). Figure 10.2a represents a stable system. Here, neither of the stations is changing in price and, thus, a fixed point is produced. However, this behaviour could easily change if one or both of the stations alter it price. This behaviour is seen in Fig. 10.2b. Both stations are changing their prices each day (from 75.1p to 75.2p to 75.1p); this creates a looping effect; the stations are cycling through a pattern of behaviour before returning to their starting point. Note that the graph appears to reveal a causative link between the two stations as they are never simultaneously low. Figure 10.2c, d shows a more varied pattern of behaviour between the stations. In Fig. 10.2c, one point is rising in price, whilst

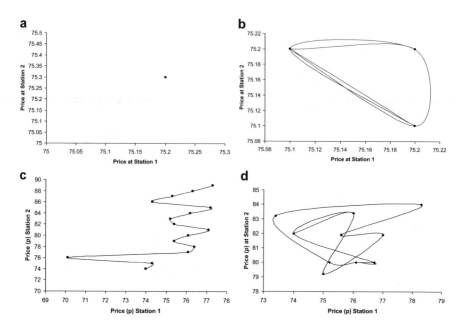

Fig. 10.2 Examples of different types of behaviour found in urban petrol stations (Leeds). (**a**) Stable. (**b**) Looping. (**c**) Two types of behaviour. (**d**) Chaotic

the other is oscillating. In Fig. 10.2(d), there is no apparent pattern in the displayed behaviour. Simply knowing about these relationships is valuable information and allows us a greater understanding of this system, its behaviour and its structure. For example, it may be that the only difference between the graphs is one of distance between the stations, but we would never see this unless the graphs allowed us to compare at a detailed level the behaviours of stations that potentially influence each other.

10.3.2   Recurrence Plots

Recurrence plots (RPs) are a relatively new technique for the analysis of time-series data that allows both visualisation and quantification of structures hidden within data or exploration of the trajectory of a dynamical system in phase space (Eckmann et al. 1987). They are particularly useful for graphically detecting hidden patterns and structural changes in data as well as examining similarities in patterns across a time-series dataset (where there are multiple readings at one point). RPs can be also used to study the nonstationarity of a time series as well as to indicate its degree of aperiodicity (Casdagli 1997; Kantz and Schreiber 1997). These features make RPs a very valuable technique for characterising complex dynamics in the time domain

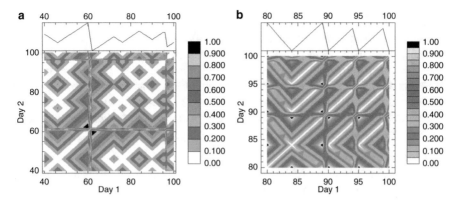

Fig. 10.3 Example of Recurrence Plots. (**a**) RP of the change in price at a retail outlet over 100 days. (**b**) illustrates how oscillations in the change in the price data are represented in the RP

(Vasconcelos et al. 2006), a factor reflected in the variety of applications that RPs can now be found in ranging from climate variation (Marwan and Kruths 2002) and music (Foote and Cooper 2001) to heart rate variability (Marwan et al. 2002).

Essentially a RP is constructed via a matrix where values at a pair of time steps are compared against each other. If the system at the two snapshots is completely different, the result is 1.0 (black), while completely similar periods are attributed the value 0.0 (represented as white). Through this, a picture of the structure of the data is built up. Figure 10.3a shows the RP of the change in price at a retail outlet over 100 days. Above the RP is a time-series graph diagrammatically representing the change in price. Changes in price, either increases, decreases or oscillations, can be clearly seen in the RP. Figure 10.3b illustrates how oscillations in the change in the price data are represented in the RP.

Early work on this area has shown that there is considerable potential in the development and adaptation of this technique. Current research is focused on the development of cross-reference RPs (consideration of the phase-space trajectories of two different systems in the same phase space) and spatial recurrence plots.

10.4   Explanation, Understanding and Causality

Once patterns are recognised, "understanding" our models involves finding explanations highlighting the mechanisms within the models which give rise to these patterns. The process of explanation may be driven with reference to current theory or developing new theory. This is usually achieved through:

1. Correlating patterns visually or statistically with other parts of the model, such as different geographical locations, or with simulations with different starting values.
2. Experimentally adjusting the model inputs to see what happens to the outputs.

3. Tracking the causal processes through the model.

It may seem obvious, and yet it is worth pointing out, that model outputs can only causally relate to model inputs, not additional data in the real world. Plainly insights into the *system* can come from comparison with external data that is correlated or miscorrelated with model outputs, but this is not the same as understanding your model and the way it currently represents the system. One would imagine that this means that understanding of a model cannot be facilitated by comparing it with other, external, data, and yet it can often be worth:

4. Comparing model results with real-world data, because the relationships between real data and both model inputs and model outputs may be clearer than the relationships between these two things within the model.

Let's imagine, for example, a model that predicts the location of burglaries across a day in a city region where police corruption is rife. The model inputs are known offenders' homes, potential target locations and attractiveness, the position of the owners of these targets and the police, who prefer to serve the wealthy. We may be able to recognise a pattern of burglaries that moves, over the course of the day, from the suburbs to the city centre. Although we have built into our model the fact that police respond faster to richer people, we may find, using (1), that our model doesn't show less burglaries in rich areas, because the rich areas are so spatially distributed that the police response times are stretched between them. We can then alter the weighting of the bias away from the wealthy (2) to see if it actually reduces the burglary rate in the rich areas by placing police nearer these neighbourhoods as an ancillary effect of responding to poor people more. We may be able to fully understand this aspect of the model and how it arises (3), but still have a higher than expected burglary rate in wealthy areas. Finally, it may turn out (4) that there is a strong relationship between these burglaries and real data on petrol sales, for no other reason than both are high at transition times in this social system, when the police would be most stretched between regions—suggesting in turn that the change in police locations over time is as important as their positions at any one time.

Let us look at each of these methodologies for developing understanding in turn.

Correlation Most social scientists will be familiar with linear regression as a means for describing data or testing for a relationship between two variables; there is a long scientific tradition of correlating data between models and external variables, and this tradition is equally applicable to intra-model comparisons. Correlating datasets is one of the areas where automation can be applied. As an exploratory tool, regression modelling has its attractions, not least its simplicity in both concept and execution. Simple regressions can be achieved in desktop applications like Microsoft Excel, as well as all the major statistical packages (R, SAS, SPSS, etc.). Standard methodologies are well known for cross-correlation of both continuous normal data and time series. However even for simple analyses with a single input and single output variable, linear regression is not always an appropriate technique. For example, logistic regression models will be more appropriate for binary response data, Poisson models will be superior when values in the dependent tend to be highly clustered, while binomial models may be the most effective when observations are highly dispersed around the mean. An interesting example is Fleming and Sorenson (2001) in which binomial estimates of technological innovation are compared to the complexity of the invention measured by both the number of components and the interdependence between those components. In behavioural space, methodologies such as association rule making (e.g. Hipp et al. 2002) allow the Bayesian association of behavioural attributes. It is worth noting that where models involve a distribution in physical space, this can introduce problems, in particular where the model includes neighbourhood-based behaviours and therefore the potential to develop spatial auto and cross-correlations. These alter the sampling strategies necessary to prove relationships—a full review of the issues and methodologies to deal with them can be found in Wagner and Fortin (2005).

Experimentation In terms of experimentation, we can make the rather artificial distinction between sensitivity testing and "what if?" analyses—the distinction is more one of intent than anything. In sensitivity analysis one perturbs model inputs slightly to determine the stability of the outputs, under the presumption that models should match the real world in being insensitive to minor changes (a presumption not always well founded). In "what if?" analyses, one alters the model inputs to see what would happen under different scenarios. In addition to looking at the output values at a particular time slice, the stability or otherwise of the model, and the conditions under which this varies, also gives information about the system (Grimm 1999).

Tracking Causality Since individual-based models are a relatively recent development, there is far less literature dealing with the tracking of causality through models. It helps a little that the causality we deal with in models, which is essentially a mechanistic one, is far more concrete than the causality perceived by humans, which is largely a matter of the repeated coincidence of events. Nevertheless, backtracking through a model to mark a causality path is extremely hard, primarily for two reasons. The first is what we might call the "find the lady problem"—that the sheer number of interactions involved in social processes tends to be so large we don't have the facilities to do the tracking. The second issue, which we might call the "drop in the ocean problem", is more fundamental as it relates to a flaw in the mathematical representation of objects, that is, that numbers represent aggregated quantities, not individuals. When transacted objects in a system are represented with numbers greater than one, it is instantly impossible to reliably determine the path taken by a specific object through that system. For objects representing concepts, either numerical (e.g. money) or nonnumerical (e.g. a meme), this isn't a problem (one dollar is much like any other; there is only one Gangnam style to know). However, for most objects such aggregations place ambiguous nodes between what would otherwise be discrete causal pathways. Fortunately, we tend to use numbers in agent models as a methodology to cope with our ignorance (e.g. in the case of calibrated parameters) or the lack of the computing power we'd need to deal with individual objects and their transactional histories (e.g. in the case of a variable like "number of children"). As it happens, every day brings improvements to both.

It addition, the last 10 years or so has seen considerable theoretical advances in the determination of the probabilities of causation (e.g. Granger 1980; Pearl and Verma 1991; Greenland and Pearl 2006). For now, however, the tracking of causality is much easier if the models build in appropriate structures from the start. While they are in their infancy, techniques like process calculi (Worboys 2005) and Petri nets show the potential of this area.

The inability to track causality leads to the perennial problem of identifiability, that is, that a single model outcome may have more than one history of model parameters that leads to it. Identifiability is part of a larger set of issues with confirming that the model in the computer accurately reflects the system in the real world—the so-called equifinality issue. These are issues that play out strongly during model construction from real data and when validating a model against real data, and a review of techniques to examine these problems, including using model variation to determine the suitability of variables and parameters, can be found in Evans (2012). At the model stage we are interested in, however, we at least have the advantage that there is only one potential model that may have created the output— the one running. Nevertheless, the identifiability of the parameters in a running model still makes it hard to definitively say when model behaviour is reasonable. For those modelling for prediction, this is of little consequence—as long as the model gives consistently good predictions it may as well be a black box. However, if we wish to tease the model apart and look at how results have emerged, these issues become more problematic.

The mechanisms for dealing with these problems are pragmatic:

1. Examine the stability of the calibration process and/or the state of internal variables that weren't inputs or outputs across multiple runs.
2. Validate internal variables that weren't inputs or outputs used in any calibration against real data.
3. Run the model in a predictive mode with as many different datasets as possible— the more the system can replicate reality at output, the more likely it is to replicate reality internally. If necessary engage in inverse modelling: initialize parameters randomly and then adjust them over multiple runs until they match all known outputs.

Of these, by far the easiest, but the least engaged with, is checking the stability of the model in parameter space (see Evans 2012 for a review). Various AI techniques have been applied to the problem of optimising parameters to fit model output distributions to some predetermined pattern (such as a "real-world" distribution). However, the stability of these parameterizations and the paths AIs take to generate them are rarely used to examine the degree to which the model fluctuates between different states, let alone to reflect on the nature of the system. The assumption of identifiability is that the more parameterized a model, the more likely it is a set of parameter values can be derived which fit the data but don't represent the true values. However, in practice the limits on the range of parameter values within any given model allow us an alternative viewpoint: that the more parameterized rules in a model, the more the system is constrained by the potential range of the elements in its structure and the interaction of these ranges. For example, a simple model $a = b$ has no constraints, but $a = b/c$, *where c = distance between a and b*, adds an additional constraint even though there are more parameters. As such rules build up in complex systems, it is possible that parameter values become highly constrained, even though, taken individually, any given element of the model seems reasonably free. This may mean that if a system is well modelled, exploration of the model's parameter space by an AI might reveal the limits of parameters within the constraints of the real complex system. For example, Heppenstall et al. (2007) use a genetic algorithm to explore the parameterisation of a petrol retail model/market and find that while some GA-derived parameters have a wide range, others consistently fall around specific values that match those derived from expert knowledge of the real system.

The same issues as hold for causality hold for data uncertainty and error. We have little in the way of techniques for coping with the propagation of either through models (see Evans 2012 for a review). It is plain that most real systems can be perturbed slightly and maintain the same outcomes, and this gives us some hope that errors at least can be suppressed; however we still remain very ignorant as to how such homeostatic forces work in real systems and how we might recognise or replicate them in our models. Data and model errors can breed patterns in our model outputs. An important component of understanding a model is understanding when this is the case. If we are to use a model to understand the dynamics of a real system and its emergent properties, then we need to be able to recognise novelty in the system. Patterns that result from errors may appear to be novel (if we are lucky), but as yet there is little in the way of toolkits to separate out such patterns from truly interesting and new patterns produced intrinsically.

Currently our best option for understanding model artefacts is model-to-model comparisons. These can be achieved by varying one of the following contexts while holding the others the same: the model code (the model, libraries and platform), the computer the model runs on or the data it runs with (including internal random number sequences). Varying the model code (for instance, from Java to C++ or from an object-orientated architecture to a procedural one) is a useful step in that it ensures the underlying theory is not erroneously dependent on its representation. Varying the computer indicates the level of errors associated with issues like rounding and number storage mechanisms, while varying the data shows the degree to which model and theory are robust to changes in the input conditions. In each case, a version of the model that can be transferred between users, translated onto other platforms and run on different data warehouses would be useful. Unfortunately, however, there is no universally recognised mechanism for representing models abstracted from programming languages. Mathematics, UML and natural languages can obviously fill this gap to a degree, but not in a manner that allows for complete automatic translation. Even the automatic translation of computer languages is far from satisfactory when there is a requirement that the results be understood by humans so errors in knowledge representation can be checked. In addition, many such translations work by producing the same binary executable. We also need standard ways of comparing the results of models, and these are no more forthcoming. Practitioners are only really at the stage where we can start to talk about model results in the same way (see, e.g. Grimm et al. 2006). Consistency in comparison is still a long way off, in part because statistics for model outputs and validity are still evolving and in part because we still don't have much idea which statistics are best applied and when (for one example bucking this trend, see Knudsen and Fotheringham 1986).

10.5   Future Directions

Recognising patterns in our modelled data allows us to:

1. Compare it with reality for validation.
2. Discover new information about the emergent properties of the system.
3. Make predictions.

Of these, discovering new information about the system is undoubtedly the hardest, as it is much easier to spot patterns you are expecting. Despite the above advances, there are key areas where current techniques do not match our requirements. In particular, these include:

1. Mechanisms to determine when we do not have all the variables we need to model a system and which variables to use.
2. Mechanisms to determine which minor variables may be important in making emergent patterns through non-linearities.
3. The tracking of emergent properties through models.
4. The ability to recognise all but the most basic patterns in space over time.
5. The ability to recognise action across distant spaces over space and time.
6. The tracking of errors, error acceleration and homeostatic forces in models.

While we have components of some of these areas, what we have is but a drop in the ocean of techniques we need. In addition, the vast majority of our techniques are built on the 2500 years of mathematics that resolved to simplify systems that were collections of individuals because we lacked the ability (either processing power or memory) to cope with the individuals *as* individuals. Modern computers have given us this power for the first time, and, as of yet, the ways we describe such systems have not caught up, even if we accept that some reduction in dimensionality and detail is necessary for a human to understand our models. Indeed in the long run, it might be questioned whether the whole process of model understanding and interpretation might be divorced from humans and delegated instead to an artificially intelligent computational agency that can better cope with the complexities directly.

Further Reading

Statistical techniques for spatial data are reviewed by McGarigal (2002) while for network statistics good starting points are Newman (2003) and Boccaletti et al. (2006), with more recent work reviewed by Evans (2010). For information on coping with auto-/cross-correlation in spatial data, see Wagner and Fortin (2005). Patel and Hudson-Smith (2012) provide an overview of the types of simulation tool (virtual worlds and virtual reality) available for visualising the outputs of spatially explicit agent-based models. Evans (2012) provides a review of techniques for analysing error and uncertainty in models, including both environmental/climate models and what they can bring to the agent-based field. He also reviews techniques for identifying the appropriate model form and parameter sets.

References

Andrienko, N., Andrienko, G., & Gatalsky, P. (2003). Exploratory spatio-temporal visualisation: An analytical review. *Journal of Visual Languages and Computing, 14*(6), 503–541.

Baird, A. A., et al. (2002). Frontal lobe activation during object permanence: Data from near-infrared spectroscopy. *NeuroImage, 16*, 1120–1126.

Boccaletti, S., Latora, V., Moreno, Y., Chavez, M., & Hwang, D.-U. (2006). Complex networks: Structure and dynamics. *Physics Reports, 424*(4–5), 175–308.

Boroditsky, L. (2001). Does language shape thought? Mandarin and English speakers' conceptions of time. *Cognitive Psychology, 43*, 1–22.

Batty, M. (2006). Rank clocks. *Nature, 444*, 592–596.

Bouvrie, J. V., & Sinha, P. (2007). Visual object concept discovery: Observations in congenitally blind children, and a computational approach. *Neurocomputing, 70*(13–15), 2218–2233.

Casdagli, M. (1997). Recurrence plots revisited. *Physica D: Nonlinear Phenomena, 108*(1–2), 12–44.

Chua, H. F., Boland, J. E., & Nisbett, R. E. (2005). Cultural variation in eye movements during scene perception. *Proceedings of the National Academy of Sciences of the United States of America, 102*(35), 12629–12633.

Clark, P. J., & Evans, F. C. (1954). Distance to nearest neighbor as a measure of spatial relationships in populations. *Ecology, 35*(4), 445–453.

Clement, D. E., Sistrunk, F., & Guenther, Z. C. (1970). Pattern perception among Brazilians as a function of pattern uncertainty and age. *Journal of Cross-Cultural Psychology, 1*(4), 305–313.

Cleveland, W. S. (1983). *Visualising data*. New Jersey: Hobart Press.

David, N., Fachada, N., & Rosa, A. C. (2017). Verifying and validating simulations. doi:https://doi.org/10.1007/978-3-319-66948-9_9.

Druzhkov, P. N., & Kustikova, V. D. (2016). A survey of deep learning methods and software tools for image classification and object detection. *Pattern Recognition and Image Analysis, 26*(1), 9–15.

Eckmann, J. P., Kamphorst, S. O., & Reulle, D. (1987). Recurrence plots of dynamical systems. *Europhysics Letters, 4*(9), 973–977.

Evans, A. J. (2010). Complex spatial networks in application. *Complexity, 16*(2), 11–19.

Evans, A. J. (2012). Uncertainty and error. In A. J. Heppenstall, A. T. Crooks, L. M. See, & M. Batty (Eds.), *Agent-based models of geographical systems*. Berlin: Springer. Chapter 15.

Fisher, N., Lewis, T., & Embleton, B. (1987). *Statistical analysis of spherical data*. Cambridge: Cambridge University Press.

Fleming, L., & Sorenson, O. (2001). Technology as a complex adaptive system: Evidence from patent data. *Research Policy, 30*, 1019–1039.

Foote, J., & Cooper, M. (2001). Visualising music structure and rhythm via self-similarity. In *Proceedings of the international computer music conference, ICMC'01, Havana, Cuba* (pp. 419–422). San Francisco: ICMA.

Fotheringham, A. S., Brunsdon, C., & Charlton, M. (2002). *Geographically weighted regression: The analysis of spatially varying relationships*. Chichester: Wiley.

Gahegan, M. (2001). Visual exploration in geography: Analysis with light. In H. J. Miller & J. Han (Eds.), *Geographic data mining and knowledge discovery* (pp. 260–287). London: Taylor & Francis.

Gehlke, C. E., & Biehl, H. (1934). Certain effects of grouping upon the size of correlation coefficients in census tract material. *Journal of the American Statistical Association, 29*(Supplement), 169–170.

Getis, A. (2007). Reflections on spatial autocorrelation. *Regional Science and Urban Economics, 37*(4), 491–496.

Granger, C. W. J. (1980). Testing for causality: A personal viewpoint. *Journal of Economic Dynamics and Control, 2*, 329–352.

Graps, A. (2004). *Amara's wavelet page*. http://www.amara.com/current/wavelet.html

Greenland, S., & Pearl, J. (2006). *Causal diagrams* (Technical report, R-332). Los Angeles: UCLA Cognitive Systems Laboratory. http://ftp.cs.ucla.edu/pub/stat_ser/r332.pdf

Grimm, V. (1999). Ten years of individual-based modelling in ecology: What have we learned and what could we learn in the future? *Ecological Modelling, 115*(2), 129–148.

Grimm, V. (2002). Visual debugging: A way of analyzing, understanding, and communicating bottom-up simulation models in ecology. *Natural Resource Modelling, 15*, 23–38.

Grimm, V., et al. (2006). A standard protocol for describing individual-based and agent-based models. *Ecological Modelling, 198*(1–2), 115–126.

Haining, R. (1990). *Spatial data analysis in the social and environmental sciences*. Cambridge: Cambridge University Press.

Heppenstall, A. J., Evans, A. J., & Birkin, M. H. (2006). Using hybrid agent-based systems to model spatially-influenced retail markets. *Journal of Artificial Societies and Social Simulation, 9*(3). http://jasss.soc.surrey.ac.uk/9/3/2.html

Heppenstall, A. J., Evans, A. J., & Birkin, M. H. (2007). Genetic algorithm optimisation of a multi-agent system for simulating a retail market. *Environment and Planning B: Urban Analytics and City Science, 34*(6), 1051–1070.

Hinneburg, A., Keim, D. A., & Wawryniuk, M. (1999). HD-eye: Visual mining of high-dimensional data. *IEEE Computer Graphics and Applications, 19*(5), 22–31.

Hipp, J., Güntzer, U., & Nakhaeizadeh, G. (2002). Data mining of association rules and the process of knowledge discovery in databases. In P. Perner (Ed.), *Advances in data mining. (Lecture Notes in Computer Science, 2394)* (pp. 207–226). Berlin: Springer.

Isaaks, E. H., & Srivastava, R. M. (1990). *Applied geostatistics*. North Carolina: Oxford University Press USA.

Kantz, H., & Schreiber, T. (1997). *Non-linear time series analysis*. Cambridge: Cambridge University Press.

Knudsen, D. C., & Fotheringham, A. S. (1986). Matrix comparison, goodness-of-fit, and spatial interaction modelling. *International Regional Science Review, 10*, 127–147.

Korie, S., et al. (1998). Analysing maps of dispersal around a single focus. *Environmental and Ecological Statistics, 5*(4), 317–344.

Marwan, N., & Kruths, J. (2002). Nonlinear analysis of bivariate data with cross recurrence plots. *Physics Letters A, 302*(5–6), 299–307.

Marwan, N., Wessel, N., Meyerfeldt, U., Schirdewan, A., & Kurths, J. (2002). Recurrence-plot-based measures of complexity and their application to heart-rate-variability data. *Physical Review E, 66*(2), 026702.

McGarigal, K. (2002). Landscape pattern metrics. In A. H. El-Shaarawi & W. W. Piegorsch (Eds.), *Encyclopedia of environmentrics* (Vol. 2, pp. 1135–1142). Chichester: Wiley.

Moon, I.-C., Schneider, M., & Carley, K. (2006). Evolution of player skill in the America's Army game. *SIMULATION, 82*(11), 703–718.

Müller, W, & Schumann, H. S. (2003). Visualisation methods for time-dependent data: An overview. In: S. Chick, P. J. Sánchez, D. Ferrin, & D. J. Morrice (Eds.), *Proceedings of winter simulation 2003*, New Orleans, LA, 7–10 December 2003. http://informs-sim.org/wsc03papers/090.pdf

Newman, M. E. J. (2003). The structure and function of complex networks. *SIAM Review, 45*, 167–256.

Nisbett, R. E., & Masuda, T. (2003). Culture and point of view. *Proceedings of the National Academy of Science USA, 100*(19), 11163–11170.

OED. (2010). *Oxford English dictionary*. http://www.oed.com/

Openshaw, S., Craft, A. W., Charlton, M., & Birch, J. M. (1988). Investigation of leukaemia clusters by use of a geographical analysis machine. *Lancet, 331*(8580), 272–273.

Orford, S., Harris, R., & Dorling, D. (1999). Geography: Information visualisation in the social sciences. *Social Science Computer Review, 17*(3), 289–304.

Patel, A., & Hudson-Smith, A. (2012). Agent tools, techniques and methods for macro and microscopic simulation. In A. J. Heppenstall, A. T. Crooks, L. M. See, & M. Batty (Eds.), *Agent-based models of geographical systems*. Berlin: Springer. chapter 18.

Pearl, J., & Verma, T. S. (1991). A theory of inferred causation. In J. A. Allen, R. Fikes, & E. Sandewall (Eds.), *Proceedings of the 2nd international conference on principles of knowledge representation and reasoning (KR'91), Cambridge, MA, USA, April 22–25, 1991* (pp. 441–452). San Mateo: Morgan Kaufmann.

Pohlheim, H. (2006). Multidimensional scaling for evolutionary algorithms: Visualisation of the path through search space and solution space using Sammon mapping. *Artificial Life, 12*(2), 203–209.

Ramsey, J. B. (2002). Wavelets in economics and finance: Past and future. *Studies in Nonlinear Dynamics & Econometrics, 6*(3), 1–27.

Roberson, D., Davidoff, J., Davies, I. R. L., & Shapiro, L. R. (2004). The development of color categories in two languages: A longitudinal study. *Journal of Experimental Psychology: General, 133*(4), 554–571.

Robinson, W. S. (1950). Ecological correlations and the behaviour of individuals. *American Sociological Review, 15*, 351–357.

Ross, A. N., & Vosper, S. B. (2003). Numerical simulations of stably stratified flow through a mountain pass. *Quarterly Journal of Royal Meteorological Society, 129*, 97–115.

Schelling, T. C. (1969). Models of segregation. *The American Economic Review, 59*(2), 88–493.

Tobler, W. R. (1970). A computer model simulation of urban growth in the Detroit region. *Economic Geography, 46*(2), 234–240.

Uchmanski, J., & Grimm, V. (1996). Individual-based modelling in ecology: What makes the difference? *Trends in Ecology & Evolution, 11*, 437–441.

Vasconcelos, D. B., Lopes, S. R., Kurths, J., & Viana, R. L. (2006). Spatial recurrence plots. *Physical Review E, 73*(5), 056207.

Viboud, C., et al. (2006). Synchrony, waves, and spatial hierarchies in the spread of influenza. *Science, 312*(5772), 447–451.

Wagner, H. H., & Fortin, M.-J. (2005). Spatial analysis of landscapes: Concepts and statistics. *Ecology, 86*, 1975–1987.

Worboys, M. F. (2005). Event-oriented approaches to geographic phenomena. *International Journal of Geographic Information Science, 19*(1), 1–28.

Wyszomirski, T., Wyszomirska, I., & Jarzyna, I. (1999). Simple mechanisms of size distribution dynamics in crowded and uncrowded virtual monocultures. *Ecological Modelling, 115*(2–3), 253–273.

Chapter 11
How Many Times Should One Run a Computational Simulation?

Raffaello Seri and Davide Secchi

Abstract This chapter is an attempt to answer the question "how many runs of a computational simulation should one do," and it gives an answer by means of statistical analysis. After defining the nature of the problem and which types of simulation are mostly affected by it, the chapter introduces statistical power analysis as a way to determine the appropriate number of runs. Two examples are then produced using results from an agent-based model. The reader is then guided through the application of this statistical technique and exposed to its limits and potentials.

Why Read This Chapter?

To understand and reflect on the importance of determining an appropriate number of runs for a simulation of a complex social system, especially agent-based simulation models. Also the chapter guides readers through (a) the issues surrounding this determination, (b) the use of statistical power analysis to identify the number of runs, and (c) two examples to practice the computation.

11.1 Introduction

This chapter explores the issue of how many times a simulation should run. This is an often neglected issue (Ritter et al. 2011) that, sooner or later, all modelers dealing with simulations of complex systems encounter. The literature takes an agnostic stance on how many runs—per configuration of parameters or, as economists put it, *ceteris paribus*—a simulation is to be run. In fact, the focus has mostly been on

R. Seri
University of Insubria, Varese, Italy

D. Secchi (✉)
University of Southern Denmark, Slagelse, Denmark
e-mail: secchi@sdu.dk

© Springer International Publishing AG 2017
B. Edmonds, R. Meyer (eds.), *Simulating Social Complexity*,
Understanding Complex Systems, https://doi.org/10.1007/978-3-319-66948-9_11

defining the "steps," the time, or the interactions within each run through sensitivity and convergence analysis, for example (Mungovan et al. 2011; Robinson 2014; Shimazoe and Burton 2013).

The central assumption of what is proposed in this chapter is that the number of runs in a simulation is often crucial for results to bear some meaning. Of course, this is not true for all simulations and it depends on scope, nature of the simulated phenomenon, purpose, and level of abstraction. We specify these aspects in the following section. For now, it suffices to write that for social simulations with a strong stochastic component, where emergence and complexity cause results to differ even within the same configuration of parameters, knowing how many runs are enough for differences to emerge (or not) becomes an extremely relevant information. This is where this chapter positions itself.

We first try to indicate—very broadly—what type of simulations this approach may apply to. Then, mediating from research on sample size determination for the behavioral sciences (Cohen 1988; Liu 2014), we introduce statistical power analysis and testing theory. The chapter also takes an agent-based model (ABM) with a strong stochastic component and provides two examples that show how crucial the issue is. At the same time, the chapter offers a practical guide on how to conduct the computation. Implications and concluding remarks follow.

11.2   Scope and Nature of Agent-Based Models

In this chapter we identify a particular sub-group of agent-based models that are fit for hypothesis testing. In order to frame the following discussion, we propose a classification of the aims of ABM, with the caveat that the following discussion may not be general or exhaustive. For a more general classification of the types of simulation, one may refer to Chap. 3 of this handbook (Davidsson and Verhagen 2017).

Some agent-based models have the purpose of studying the emergent properties of a system (Anderson 1972; Fioretti 2016). These properties arise when the system as a whole displays a behavior that is not explicit in its single components, in this case, the agents. When this is aimed at establishing whether an outcome is possible, hence the simulation has an exploratory purpose that reflects on theory, then the visual inspection of the trajectories of the simulated system or the computation of some descriptive statistics is sufficient to illustrate the existence of an emergent behavior. For example, Heckbert's (Heckbert 2013) model of the socio-economic system in which the ancient Maya civilization developed and disappeared can be thought of as a simulation of this kind. As a descriptive model, it establishes whether the conditions set in the model offer reasonable explanations of historical facts.

The study of emergent properties is also linked to another objective of some agent-based models, namely hypothesis generation (Bardone 2016; Secchi 2015). A researcher may run a model just to assess whether it is reasonable to suppose that some variables have an impact on a given outcome. The hypotheses obtained in this way may be subject to empirical testing in a future laboratory experiment or through real data. An example of this type of ABM can be the simulation of a team of doctors and nurses working in the emergency room of a hospital, to isolate those socio-cognitive attitudes that may lead to increased performance (Thomsen 2016). Another example comes from political science (de Marchi and Page 2014) and it concerns the model of incumbent advantage in elections proposed by Kollman et al. (1992), that was first studied by simulation and then successfully tested empirically.

The techniques presented in this chapter are not necessarily pertinent to these first two model types described above, because they call for an exploratory approach in which the configurations of parameters and the number of runs are not rigidly chosen in advance, but they may be modified by trial and error while the researcher explores the potential outcomes of the model.[1]

A third objective of ABM is measurement, namely providing a numerical value for a quantity of interest. Since most agent-based models in the social sciences are too simplified a representation of reality to provide accurate estimates of real-world quantities, the models that pursue this objective are generally constrained to specific disciplines in which the rules of behavior of the agents are simple or particularly well known (for an example in biology see Sect. 1.1.1 in Railsback and Grimm (2011); for examples in transportation research, see Maggi and Vallino 2016). In this case, even if statistical tests can still be of interest, the researcher may direct his/her statistical analysis towards different tools. On the one hand, data from an ABM may be compared, through a distance (e.g., Lamperti 2015), with real time series to assess whether the two are similar enough. On the other hand, the researcher may settle on a sample size that guarantees a certain precision in the value computed for the quantity of interest rather than a certain level of power (see Sect. 11.5.1 below).

Finally, a fourth objective of ABMs is to test hypotheses in a controlled environment often emulating, with simplified rules, a real-world situation. The advantage of ABMs in this respect is that they allow the researcher to analyze a realistic situation by removing all the confounding factors arising in the observation of the real-world phenomenon. In this case, agent-based models can be considered the computational equivalent of laboratory experiments (Gilbert and Terna 2000). In the following pages, we explain how the parallel can be established. Note, however, that this is not the only possible setting. It is customary that an ABM has several parameters entering its formulation. The aim of a model can be that of exploring whether these parameters bear any impact on a quantity of interest, obtained as an outcome of the simulations of the model. The usual way is to identify some configurations of parameters that would correspond to different alternative treatments in an experiment, and to run several simulations of the model under each configuration. Each run of the model corresponds to an observation (e.g., a subject) in an experiment: the measured outcome can either be the terminal value of the series or a value computed on (a part of) the trajectory. The presumed independence

---

[1]Note, moreover, that the researcher should not test a hypothesis on the data that have been used to generate it.

of the simulation outcome on the configurations of parameters can then be tested in an ANOVA framework. An example of this type of ABM can be a model of intra-organizational bandwagons (Secchi and Gullekson 2016), in which authors run the simulation multiple times in order to test propositions as guide for future, probably empirical, research.

Another distinction that might be helpful when considering the number of runs can be drawn between models that strive at defining abstract and simple rules of behavior for their agents and those that are more concerned with describing a particular aspect of reality with fine degrees of details. This is the divide between the KISS ("Keep it Simple, Stupid!") and the KIDS ("Keep it Descriptive, Stupid!") principles (Edmonds and Moss 2005). While advocates of the first approach are in line with modeling efforts of the past (Troitzsch 2017; Coen 2009), those who indicate that ABM opens a new way stand with a more descriptive and complex approach to modeling (Edmonds and Moss 2005). If we take that the extreme for a KISS model is a system of deterministic equations and, on the other side, the bound for a KIDS model is the attempt to replicate reality, there is an entire spectrum of models (and ABMs) falling between these two extremes. However, considerations on the number of runs are more likely to become relevant as modelers tend toward increased complexity, without reaching the extreme of full description.

In summary, the determination of the number of runs in a simulation is warranted every time the researcher is seeking to measure—with some degree of confidence—whether different configurations of parameters are more or less likely to affect the outcome.

11.3   Testing Theory: Controlling for Alpha and Beta

In this section, we provide an introduction to testing theory that can be read independently from the rest of the paper. As an example, in this section, the term *parameter* denotes an unknown characteristic of a population, and not a quantity whose value is fixed before data are collected, as customary in ABMs. Therefore, we suppose that the researcher has identified some parameters describing the behavior of the population from which the data have been sampled (as a trivial example, the mean and the variance of the population). We also assume that he/she has formulated a null hypothesis H_0, i.e. an assertion about the value of the parameters.

The original approach to testing, pioneered by K. Pearson and theorized by R.A. Fisher, looks for a statistic T with the following property: when the hypothesis H_0 is verified, the value t that the statistic T assumes in the sample is near to a fixed value, generally identified with 0. Therefore, small values of t appear to bring support to the null hypothesis H_0, while extreme values of t are seen as witnessing a possible violation of H_0. This explains why the most sensible summary of the test, in Fisher's approach, is the p-value, i.e. the probability of observing, under H_0, values of T that

Table 11.1 Table of possible outcomes for a Neyman–Pearson test

	H_0 true	H_1 true
H_0 chosen	True negative	Type-II error or false negative
H_1 chosen	Type-I error or false positive	True positive

are as extreme or more extreme than the one computed on the sample. The *p*-value is sometimes (erroneously) perceived as a measure of strength of support in the null hypothesis. It is however clear that this method does not have the possibility to offer anything more than mild support to H_0, especially because of the absence of an hypothesis that holds true when H_0 does not.

This approach to testing was amended by J. Neyman and E.S. Pearson, who modified it to allow for the possibility of decision and action. The new theory starts with the introduction of the null hypothesis, H_0, and the alternative hypothesis, i.e. H_1, that is supposed to be true when H_0 is not. The hypothesis H_0 is generally, but not always, associated with the absence of an effect (of one variable on another, for example), while H_1 is generally associated with its presence. The researcher is uncertain as to whether H_0 or H_1 holds true. The decision between these two hypotheses is performed, as in a trial, on the basis of the available data (we will see later how).[2] This leads to a table of possible outcomes, see Table 11.1. The use of *positive* and *negative* to denote respectively the choice of H_1 and H_0 comes from the medical use of the same terms, where they indicate the positive or negative result of a medical test. A negative, i.e. a result in which the disease is not detected, can be either true or false, when the unobserved true hypothesis coincides or not with the choice of the procedure; the same holds true for a positive. A *false positive* is also called, with a more statistical term, a Type-I error, while a *false negative* is also called a Type-II error. These two "sources of error" (Neyman and Pearson 1928, p. 177) exist whichever method is used to choose between H_0 and H_1.

The standard procedure to decide between H_0 and H_1 is to consider a statistic T whose distribution is known under H_0 (let us denote the probability as $\mathbb{P}_{H_0}$). The researcher builds an *acceptance region* $\mathcal{A}$ such that, when t belongs to $\mathcal{A}$, then H_0 is chosen as the true hypothesis. The possible values of t that are not contained in $\mathcal{A}$ form a *rejection region* $\mathcal{R}$. Therefore $\mathcal{A}$ and $\mathcal{R}$ make up the entire space in which T varies and are generally chosen in such a way that[3]:

$$\mathbb{P}_{H_0}\{T \in \mathcal{A}\} = 1 - \alpha$$

$$\mathbb{P}_{H_0}\{T \in \mathcal{R}\} = \alpha$$

---

[2] The metaphor of the trial has been introduced in Neyman and Pearson (1933, p. 296) but has been criticized as misleading in Liu and Stone (2007).

[3] Here $\in$ means "belongs to," so that $T \in \mathcal{A}$ means "T belongs to $\mathcal{A}$."

where $\alpha \in [0, 1]$ is called *Type-I error rate*, i.e. the probability of rejecting the null hypothesis when it is true, or *significance level*.[4]

Suppose now that the alternative hypothesis is verified and let $\mathbb{P}_{H_1}$ be the probability distribution under the alternative hypothesis. If the null hypothesis is associated with the absence of an effect (of a variable on the outcome), the alternative hypothesis implies generally that there is an effect. It is generally the case that this effect can be measured through a quantity d called *effect size*,[5] that enters the formulation of $\mathbb{P}_{H_1}$. If we suppose that the alternative hypothesis is true, the probability that T belongs to $\mathcal{A}$ or to $\mathcal{R}$ is:

$$\mathbb{P}_{H_1} \{T \in \mathcal{A}\} = \beta (d)$$

$$\mathbb{P}_{H_1} \{T \in \mathcal{R}\} = 1 - \beta (d)$$

where β, belonging to $[0, 1]$, is the *Type-II error rate*, i.e. the probability of accepting the null hypothesis when it is false. Note that β depends on the effect size d. The quantity $1 - \beta$, especially when seen as a function of the effect size d, is called *(statistical) power* of the test and measures the probability that the test correctly identifies the presence of an effect when there is one.

In a hypothetical simulation, for example, one may want to study how decision making makes employee motivation more effective under conditions of more or less organized corporate structures (Herath et al. 2017). The null hypothesis may be that the average motivation does not vary under alternative corporate structures, and this can be tested using, say, ANOVA. The Type-I error rate or significance level α, that is usually set at 5%, can be used to obtain an acceptance region $\mathcal{A}$. If the null hypothesis is false, the effect size d—i.e. the "strength" of the effect—measures the impact that the different conditions exercise, on average, on the outcome. The probability that the statistic takes a value inside $\mathcal{A}$ under the alternative hypothesis is β and depends on d. If the effect as measured by d is small, the alternative hypothesis is near to the null hypothesis, and the probability β that the statistic T falls inside $\mathcal{A}$ under H_1 is near to the probability $1 - \alpha$ under H_0. If d increases, β decreases.

It is clear that there are several degrees of freedom in the choice of the probability α, of the statistic T for testing H_0, and of the acceptance region $\mathcal{A}$. Now, while the test statistic is often suggested by the problem under scrutiny, the probability α is chosen routinely from a set of possibilities that have been determined by tradition more than by reflection. As to the choice of $\mathcal{A}$ and $\mathcal{R}$, it is often the case that $\mathcal{R}$

---

[4]We note that Neyman (1950, p. 259) used the term "accept" where most modern treatments propose to use "fail to reject" or "do not reject." The original choice of the author is in line with his idea of testing as leading to decision, while the modern use appears to be incorrectly borrowed from Fisher's approach (Fisher 1955, p. 73). However Pearson was more cautious (Pearson 1955, p. 206) and this even suggested to some authors the idea that he had rejected the approach pioneered with Neyman (Mayo 1992).

[5]More generally, the effect size d measures the distance of the true distribution from the distribution under the null hypothesis, and is generally a function of the parameters.

contains the most extreme values of T, i.e. the tails of its distribution. Most users of statistics stop here, and perform a test verifying whether t belongs to $\mathcal{R}$ or to $\mathcal{A}$ and, as a consequence, respectively reject H_0 or fail to reject it, as part of a ritual (Gigerenzer 2004). In this situation, an alternative way of reaching the same result is to compare the p-value, whenever defined, to a fixed threshold α: if the p-value is smaller than α, we reject H_0, otherwise we fail to reject it.

It is interesting to review the relations among the quantities seen until now. We saw before that the effect size d has an impact on β. Since d is a measure of how easy it is to discriminate between H_0 and H_1, it is generally the case that power, $1 - \beta$, increases with d when α is fixed.[6] Another factor affecting α and β is the sample size N. In this case too, $1 - \beta$ generally increases with N, when α is fixed. At last, the formulas $\mathbb{P}_{H_0} \{T \in \mathcal{A}\} = 1 - \alpha$ and $\mathbb{P}_{H_1} \{T \in \mathcal{A}\} = \beta$ show that there is a trade-off between α and β. Indeed, when $\mathcal{A}$ gets larger, α decreases while β increases, and vice versa. This explains why, when N and d are fixed, it is not possible to reduce α without consequences on the Type-II error rate β.[7]

This is the reason why one cannot make α as small as possible, that is because this inflates β. This fact suggests that good results could be achieved by balancing the two error rates. This was indeed proposed by Neyman and Pearson in 1933,[8] and has been revived several times since then. A more recent attempt in this direction is the *compromise power analysis* of Erdfelder (1984). However, the most common approach is to consider the two sources of error differently.

A first approach completely disregards β: a value for α is rigorously fixed (often as $\alpha = 0.05$), and the test checks whether t belongs to $\mathcal{A}$ or not using a sample whose size N has been selected without reference to β. This approach is the one that most closely resembles the original Fisher paradigm, as the alternative hypothesis has practically no role in it. It is based on the fact that, as N increases, β goes to 0, so that a large sample size guarantees that β will be small enough. A second approach supplements this part of the analysis with the computation of power using a value of d estimated on the basis of the data, a procedure called *post hoc power analysis*. Because of the large variability of the estimated effect size, this approach is generally regarded with suspicion by statisticians (Korn 1990; Hoenig and Heisey 2001). In the third approach, the researcher fixes α and β, hypothesizes a value of d, and chooses $\mathcal{A}$ and N so that both $\mathbb{P}_{H_0} \{T \in \mathcal{A}\} = 1 - \alpha$ and $\mathbb{P}_{H_1} \{T \in \mathcal{A}\} = \beta(d)$ hold true. This procedure, called *a priori power analysis*, guarantees that, if d is correctly guessed, the desired values of α and β will be achieved.

---

[6]This also explains why in some cases it is possible to increase the power of a test by designing an experiment in which it is expected that the effect size d, if not null, is large. As an example, in ABM this could be done by setting some of the quantities entering the model to their extreme values.

[7]See also van der Vaart (2000, p. 213) or Choirat and Seri (2012, Proposition 7, p. 285).

[8]The authors say: "The use of these statistical tools in any given case, in determining just how the balance should be struck, must be left to the investigator" (Neyman and Pearson 1933, p. 296).

It is important to recall that the Neyman–Pearson theory of testing is essentially designed to provide the researcher with a decision rule guaranteeing, in the long run, a specified error probability under the null hypothesis. The decision rule equating the rejection of H_0 with the occurrence of a value of t inside $\mathcal{A}$ makes sure that when a large number of tests are performed, the null hypothesis is incorrectly rejected 100α percent of the times, but does not guarantee a good performance in the case of the single test. Otherwise stated, in the Neyman–Pearson approach a controlled long-run performance is obtained if the researcher chooses α and $\mathcal{A}$ and decides on the basis of the fact that t belongs to $\mathcal{A}$ or not (or, equivalently, on the basis of the fact that the p-value is larger than α or not). In general, it is also expected that the researcher sets a value of β and chooses, on the basis of experience or pilot runs, a value of d, and computes N on the basis of these values.

However, this is not the way in which tests are generally performed in practice. Indeed, it is customary that the researcher computes the test statistic t and the p-value and uses the latter as a measure of the support in the null hypothesis. For example, it is quite common that a p-value just under 5% is treated differently than a p-value under 1%, the latter providing a stronger evidential value against the null hypothesis. This is so widespread that some researchers do not report the p-value but only $p < 1\%$ or $p < 5\%$. From the point of view of the Neyman–Pearson theory of testing this is nonsensical. However this has entered common practice and has evolved into an approach of its own, different from the Fisher and the Neyman–Pearson approaches, yet gathering aspects of both, and called *Null-Hypothesis Significance Testing* (NHST). This approach takes from the Fisher approach the emphasis on the p-value and its disregard for power; from the Neyman–Pearson theory, the approach emphasizes the threshold values of α.

In this chapter, we follow more closely the original Neyman–Pearson theory than the NHST. The elements of this approach are the two probabilities of error α and β, a measure of the effect under scrutiny or of the distance between the alternative and the null hypothesis d, and the sample size N. These quantities are linked by some equations. We will see below that determining a value N amounts at choosing some values for the quantities α, β and d, whose interpretation is generally simpler than the one of N.

11.4   The Use of Power in Practice: Two Examples

In order to show how power analysis can help to determine the number of runs in a simulation, we decided to select a model and to proceed with some calculations. The simulation we selected for this computational exercise is an agent-based model that was developed by Fioretti and Lomi (2008, 2010) on the basis of the famous "garbage can" model (Cohen et al. 1972), hereby GCM.

There are several reasons that led to the selection of this ABM. One of the obvious reasons is that it describes a very well-known model that informed the decision making literature and had an extremely significant impact.[9] As a result of this, the basic assumptions of the model should be easy to understand for most scholars. Moreover, the agent-based implementation by Fioretti and Lomi attracted some attention because it does not support all the conclusions of the original model. Another reason—and this is not a secondary reason—is that authors made the code available so that anyone interested could download and run the simulation in NetLogo, an ABM software (Wilensky 1999). Finally, the work of Cohen, March, and Olsen is very much in line with the legacy of Simon (1976, 1978, 1997), thus consistent with the introduction to this handbook (Edmonds and Meyer 2017).

The two examples that follow are both hands-on cases that should inform readers on how to determine the number of runs in an agent-based simulation.[10] In Example 1, the model runs a limited number of times so that insufficient power leads to the risk of not rejecting hypotheses that should be rejected. In Example 2, the model is run a very high number of times to produce over-powered results, reducing to a minimum the likelihood not to make any effect statistically significant.

11.4.1 Short Description of the Model

The "garbage can" is a model of decision making in organizations (Cohen et al. 1972). There are four types of agents: (a) problems, (b) opportunities, (c) solutions, and (d) participants. The overall goal of the model is to determine whether a formal (hierarchic) organizational structure provides the institutional backbone for problem solving that is better than an informal (anarchic) organizational structure or not. In the first case, the four types of agents interact following a specified sequence while in the other they interact at random.

The aim of the model is to match the four elements mentioned above to study the most effective way for an organization to make decisions. Originally, the model was designed to understand whether opportunities become more available to decision makers when organizations relax hierarchical and structural ties. This is what the ABM simulation attempts to study as well. Figure 11.1 shows a screenshot of the model interface; each agent has a different shape and they move on the black environment.

There are two ways in which participants make decisions in the organization. One type of decision is called *by resolution* and it happens when problems are solved once participants match opportunities to the right solutions (Cohen et al. 1972). This happens graphically when the right combination of the four agents are on the same position at the same time (i.e., they overlap, see Fig. 11.1). Another type is

---

[9]The number of citations of the original paper (Cohen et al. 1972) in Google Scholar amounts at 9196 and those from Thomson's Web of Science are 1864.

[10]Even though we use this method for ABM, it may reveal to be useful for any simulation with emergent properties derived from a relevant stochastic component.

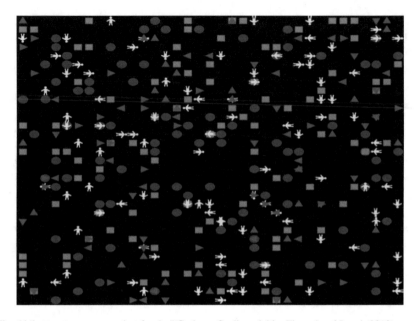

Fig. 11.1 NetLogo screenshot for the "Garbage Can" model by Fioretti and Lomi (2010)

when decisions are made *by oversight* and it is when solutions and opportunities are available to participants but no problems are actually solved (Cohen et al. 1972). Not all problems are solved automatically, just by having opportunities, participants, and solutions available. In fact, all problems have difficulty levels, participants have abilities, and solutions have a certain degree of efficiency. The problem is solved if the match of the participant with an opportunity and a solution is greater than the difficulty of the problem (Fioretti and Lomi 2010).

In the agent-based version of the model, there are three types of structure:

- **Anarchy**. There is no hierarchy so that abilities, efficiencies, and difficulties are randomly distributed among agents.
- **Hierarchy-competence**. The hierarchical structure is such that abilities, efficiencies, and difficulties increase as one moves up the hierarchical ladder.
- **Hierarchy-incompetence**. The hierarchical structure is such that abilities, efficiencies, and difficulties decrease as one moves up the hierarchical ladder.

Finally, the model implements two modes of (not) dealing with problems. One is called *buck passing*, and it happens when one participant has the alternative of passing the decision on a problem to another participant. The other mode is *postpone*, and it refers to problems that are kept on hold by participants and eventually solved at an unspecified future time.

For the purpose of this chapter, we calculate the ratio of decisions made by resolution on those made by oversight in the three cases of anarchy, hierarchy with competence, and hierarchy with incompetence.

11.4.2 Example 1

We performed a simulation for the ABM version of the GCM (Fioretti and Lomi 2010), using the second version of the two models uploaded on the NetLogo community platform. The model has three overall conditions—anarchy, hierarchy-competence, and hierarchy-incompetence—and each of these has four parameter configurations, with *buck passing* [*true, false*], and *postpone* [*true, false*]. We decided to test a simple case, setting both parameters to *false*. This gives a design of 3 configurations of parameters (CoP). Each run had 5000 steps as per the original simulation (Fioretti and Lomi 2010).

Power analysis should be performed before obtaining data from the model to choose how many times a simulation should be run. To do that, there are a few elements to determine. First of all, the researcher should choose a certain number G of configurations of parameters (also called groups). Then, considering the nature of the model or previous simulations one should guess a value of the effect size d that, in the case of ANOVA, is identified by the letter f (Cohen 1988; Liu 2014). At last, one should choose a level for α and a corresponding goal for the level of power— i.e. $1 - \beta$—to be achieved. Although the power threshold of $1 - \beta$ for empirical research is set at 0.80, some (Secchi and Seri 2014, 2017) argue that it can be set at 0.95 for simulations, because the control exerted on variables and parameters is much higher than that usually in place in empirical research. Consistently with this, also the threshold for α can be set at the more stringent level of 0.01 (Secchi and Seri 2017).[11]

As explained above, the dependent variable is the ratio r_{ro} of decisions by resolution in relation to those made by oversight. The differences in its average value across the three CoP can be easily explored by performing a one-way ANOVA with the null hypothesis that the expected value is the same across conditions. We set some notation. If G denotes the number of groups/CoP and n the number of observations per CoP, the sample size N turns out to be $N = n \cdot G$.

---

[11]In an interesting exchange with Bruce Edmonds, we came to realize that this approach might raise some important issues. One of the concerns is that thresholds do not usually adjust because the experiment is so well planned that results come out to be extremely clear; that is to say that good experimental work still accepts or rejects hypotheses at the level $\alpha < 0.05$ with $1 - \beta \approx 0.80$. This implies that adjustments of these levels for simulation work appears to be arbitrary. Our position on this critique is that thresholds actually change as it happens in some medical studies, where $1 - \beta$ raises to 0.90 (Lakatos 2005), or when we listen to the calls *not* to interpret the traditional choices of α levels as absolute from either social scientists (Gigerenzer 2004) or statisticians (Wasserstein and Lazar 2016). While a complete review of the reasons leading to the traditional choices of α and β is in Secchi and Seri (2017), the introduction to testing theory above should have made clear that the fathers of this theory thought of α and β as quantities to be chosen according to the problem at hand. This justifies our proposals as long as we cannot compare artificial computational experiment to real-life experiments because of different variability of observations, observer's control and role, and the usual difficulty of increasing sample size for empirical experiments.

11.4.2.1   Identifying the Appropriate Effect Size

We are left with the problem of guessing a value for the effect size. In the following we propose some reasonings concerning the choice. For this purpose, we need some basic notation. For ANOVA, the effect size f is to be calculated as $f = \frac{\sigma_m}{\sigma}$ (Cohen 1992), where σ_m is the standard deviation of the group means and σ is the within-population (or pooled) standard deviation. The same quantity can be expressed using sums of squares. Let the subscript j, ranging from 1 to G, indicate the j-th group/CoP. Let n be the number of observations in each group/CoP and let i, ranging from 1 to n, denote the i-th observation. Therefore, x_{ij} is the i-th observation in the j-th group. Moreover, we consider the sample means $\bar{x}_j = \frac{1}{n} \sum_{i=1}^{n} x_{ij}$ and $\bar{x} = \frac{1}{nG} \sum_{i=1}^{n} \sum_{j=1}^{G} x_{ij}$, where the latter is the grand mean of all the observations. Now, the variation of the observations between configurations (or groups) is simply the square root of the Sum of Squares Between (SSB), or $n \sum_{j=1}^{G} (\bar{x}_j - \bar{x})^2$, divided by the Sum of Squares Within (SSW) observations, or $\sum_{j=1}^{G} \sum_{i=1}^{n} (x_{ij} - \bar{x}_j)^2$.

A first possibility is to use "canned" effect sizes (Cohen 1988). In this case, verbal descriptions of the strength of the effect (e.g., small, medium and large; see below) are mapped onto numerical values, usually based on effect sizes retrieved from a review of the literature. As per the literature on the GCM, we can expect that many decisions are made by oversight and that the number of difficult problems solved increases under anarchy (Fioretti and Lomi 2008, 2010; Herath et al. 2015). Hence, the "distance" between conditions could be classified as *medium* or *large*, and we can set to an effect size of 0.25 or 0.4 (consistently with Cohen 1992). The computation of the sample size providing the desired level of power can use the formulas in Cohen (1988) as implemented in the R package `pwr` on power analysis (see Champely et al. 2016). For $f = 0.25$, this yields the result $n = 112$ (exactly 111.68). A simpler approximation, proposed in Secchi and Seri (2017), yields $n = 109$ (exactly 109.47).[12] The latter approach makes the relation between components of power more explicit. For $f = 0.40$, R package `pwr` yields $n = 45$ (more precisely 44.58) and our formula yields $n = 43$ (more precisely 43.04).

A second possibility is to guess a value for the effect size by having the simulation run for a pilot study and by calculating the estimated effect size from the results. We will deal below with some problems involved in this approach. For expository purposes, we have decided to run the model for $n = 10$ runs per condition. Taking the definition of f above, the numerator SSB is 0.000813 and the denominator is 0.004341 so that the estimated effect size for these three conditions is 0.43. If we calculate the number of runs reaching a power of 0.95 with such a large effect size as $f = 0.43$, we obtain $n = 38$ (more precisely 38.31) using Cohen's formulas and $n = 37$ (more precisely 36.81) using our approximation.

A third possibility is to use the results of former studies on the same topic. Here, we can use the results of Fioretti and Lomi (2010) in the case without buck

---

[12]This formula can be used in R with an ad hoc function taken from one of our previous publications (Secchi and Seri 2017). See the Appendix for the code for both formulas.

passing and postponement. In that source, the authors state that, based on 100 runs, the average number of decisions by resolution (resp., by oversight) is 43.90 (resp., 779.57) under anarchy (group 1), 24.82 (resp., 461.94) under competent hierarchy (group 2) and 7.71 (resp., 192.77) under incompetent hierarchy (group 3). We can approximate the average value of the ratio r_{ro} through the ratio of the averages, i.e. $\bar{x}_1 \simeq 0.0563$, $\bar{x}_2 \simeq 0.0537$ and $\bar{x}_3 \simeq 0.0400$. Therefore, we expect the difference between the average value of r_{ro} in competent hierarchy with respect to anarchy to be around $\bar{x}_2 - \bar{x}_1 \simeq -0.0026$, and in incompetent hierarchy with respect to anarchy to be around $\bar{x}_3 - \bar{x}_1 \simeq -0.016$. These coefficients are remarkably near to the ones obtained in the tables below. From the Appendix, we can see that:

$$f = \sqrt{\frac{\sum_{j=1}^{G} \left(\bar{x}_j - \bar{x}\right)^2}{\frac{1}{n} \sum_{j=1}^{G} \sum_{i=1}^{n} \left(x_{ij} - \bar{x}_j\right)^2}}.$$

The numbers above allow us to estimate the quantity $\sum_{j=1}^{G} \left(\bar{x}_j - \bar{x}\right)^2$ as 0.000153. Instead, $\frac{1}{n} \sum_{j=1}^{G} \sum_{i=1}^{n} \left(x_{ij} - \bar{x}_j\right)^2$, i.e. SSW divided by n, cannot be estimated from Fioretti and Lomi (2010), but we can use the value from our pilot runs with $n = 10$, i.e. $0.004341/10 = 0.000434$. The final result is $f = 0.594$ that would lead to $n = 21$ (more precisely $n = 21.07$; $n = 19.60$ with our formula). While one should not give too much credit to these numbers, they suggest that the effect size f may be larger than expected.

Another consideration may provide some hints about how to interpret the values provided by the previous three techniques. The standard error associated with estimated effect sizes is generally quite large. Nothing guarantees that the estimated f is indeed equal or even near to the true value. A good idea is therefore to investigate what happens choosing a value f in a neighborhood around the estimate. As an example, if we suppose that f is 0.35 or 0.5, our formula yields respectively n equal to 56 or 28. We will see below that it is generally better to overshoot the correct sample size than to undershoot it. From this point of view, a possibility is to use the estimated effect size to choose a *smallest effect size of interest* (SESOI, see Lakens (2014) for its definition in a different context), i.e. a value of the effect size that is the smallest one for which we want to achieve the desired level of power.[13] This means that for f larger than the SESOI we will experience overpower while for f smaller we will be in underpower. This asymmetry is justified by the fact that values of the effect size under the SESOI are deemed to be improbable or uninteresting. The SESOI is then used in the computation of the sample size. Whether the researcher chooses to use the SESOI or not, the importance of these sensitivity analyses can hardly be exaggerated, as they shed light on the factors that impact the choice of the sample size.

---

[13]A possibility is to choose, as SESOI, the lower bound of a confidence interval on the effect size with a specified confidence probability, e.g., 0.95 or 0.90.

Table 11.2 OLS Regression
Results (DV: decisions by
resolution/decisions by
oversight)

	Model 5	Model 40
(Intercept)	0.052***	0.056***
St. err.	(0.006)	(0.002)
t value	8.721	29.804
Type: HC/AR	−0.005	−0.007**
St. err.	(0.009)	(0.003)
t value	−0.542	−2.692
Type: HI/AR	−0.013	−0.012***
St. err.	(0.009)	(0.003)
t value	−1.481	−4.637
R-squared	0.157	0.156
F-statistic	1.123	10.842
Degrees of freedom	2, 12	2, 117
p-value	0.357	0.000
N	15	120

Note. *HC* hierarchy-competence, *HI* hierarchy-
incompetence, *AR* anarchy
Signif. codes: 0 "***" 0.001 "**" 0.01 " " 1

On the basis of the previous reasonings, taking into account the expository nature of this example, we decided to take $n = 40$, consistently with a value of f around 0.4. In Table 11.2 we reproduce the estimation results for a model with 5 runs (i.e. Model 5), that is clearly under-powered, and for a model with 40 runs (i.e. Model 40), that is correctly powered under an effect size f equal to 0.40. We expect therefore the second model to provide a test of the effect of parameters on the number of decisions by resolution in comparison to those made by oversight, with the desired levels of α and β.

11.4.2.2 The Impact of Under-Power on Outcomes

The previous discussion shows that 5 runs should still be insufficient to provide reliable results. Let us see how. As stated above, we are interested in understanding whether the number of decisions by resolution on those by oversight change (decrease) as we move from anarchy to hierarchy. Hence, we can perform an OLS regression[14] and produce a table with results calculated on 5 and 40 runs, to compare findings from an under-powered to those from an appropriately-powered study. Table 11.2 shows these comparisons and refers to them as Model 5 for the under-powered and Model 40 for the balanced simulation.

---

[14]See the Appendix for details on how the effect size of the ANOVA and OLS regressions map onto each other.

From results in Table 11.2 it is immediately apparent that there are differences between the two models. The under-powered Model 5 is not able to detect some of the effects that are instead captured by the more balanced Model 40. In fact, Model 5 fails to identify the relation between hierarchy with competence (HC) and anarchy (AR) as statistically significant as well as the relation between hierarchy with incompetence (HI) and anarchy (AR). In other words, the null hypothesis was accepted when (probably) false, hence falling into Type-II error. And we know that this is the case because a very similar regression coefficient ($\beta_{HC/AR} = -0.007$, St. err. $= 0.003$) leads instead to the rejection of the null hypothesis—that the corresponding parameter is zero—in Model 40, where it is more reasonable to suppose that power requirements are met. The second coefficient—hierarchy with incompetence on anarchy—is also statistically significant in Model 40 ($\beta_{HI/AR} = -0.012$, St. err. $= 0.003$) as opposed to Model 5 ($\beta_{HI/AR} = -0.012$, St. err. $= 0.008$).

At last, note that in Model 5 the F-statistic for the joint nullity of both effects does not lead to the rejection of the null hypothesis, thus suggesting that there is no effect overall of the structure on problem solving. The conclusion is at odds with the one from Model 40, that leads to the strong rejection of the same hypothesis.

In short, the impact of some of the conditions fails to be acknowledged in the under-powered study with only 5 runs, leaving important and interesting implications out of the study.

11.4.3   Example 2

We also conduct a second example to illustrate the risks and problems of over-powering the simulation. In this example, we over-power the simulation and calculate results on 500 runs, with the same parameter specifications used in the example above.

Results of the two simulations are explored in Table 11.3, where we show the estimation outputs of two OLS regression models. In the table, Model 40 shows results for the correctly-powered simulation while Model 500 refers to the over-powered simulation. The beta coefficients are very close to each other, with a variation that is mostly reflected in the standard errors, that decrease in the case of the over-powered simulation. This leads to a different t value so that the respective probability (the p-value) becomes closer to zero for Model 500 than for Model 40.

From the perspective of accepting or rejecting results in the regression, there is little or no difference. In fact, most values are well below the threshold for statistically significant results. This points at the fact that, if one is interested in accepting or rejecting hypotheses, there is no particular difference between the two.

However, in another article (Secchi and Seri 2017), we warn modelers of the risks of over-power. There we write that over-power hides some dangers because it might be unnecessarily costly (time consuming, for example), it makes small effects as significant as larger ones, and destroys the balance between the two probabilities

Table 11.3 OLS Regression Results (DV: decisions by resolution/decisions by oversight)

	Model 40	Model 500
(Intercept)	0.056***	0.055***
St. err.	(0.002)	(0.001)
t value	29.804	100.12
Type: HC/AR	−0.007**	−0.005***
St. err.	(0.003)	(0.001)
t value	−2.692	−5.99
Type: HI/AR	−0.012***	−0.015***
St. err.	(0.003)	(0.001)
t value	−4.637	−19.18
R-squared	0.156	0.205
F-statistic	10.842	192.497
Degrees of freedom	2, 117	2, 1497
p-value	0.000	0.000
N	120	1500

Note. *HC* hierarchy-competence, *HI* hierarchy-incompetence, *AR* anarchy

Signif. codes: 0 "***" 0.001 "**" 0.01 " " 1

of error α and β,[15] thus decreasing the overall reliability of the model. However, all things considered, Example 2 shows that, in the case of large effect sizes such as this one, overpower does not bear particularly relevant problems besides accuracy. In fact, the two models present results that are close to each other and only differ in the granularity and reliability of details.

One last remark concerns the value of f as estimated from Model 500. In that case, we get $f = 0.51$. This confirms that our initial guess (f between 0.25 and 0.4) was probably an underestimation, and validates with hindsight our choice of focusing on the upper bound of the interval $[0.25, 0.40]$.

11.5 Implications and Conclusions

A few implications can be drawn from the two examples above. The first is that power analysis can guide researchers on establishing the number of times a simulation should run. The most immediate advice to modelers is that using power to compute the number of runs should help avoid under-powered studies. In that

---

[15]Over-power reduces β well below the chosen value of α. This is a problem because Type-I errors are generally perceived as more serious than Type-II errors, and when $\beta \ll \alpha$ we expect exactly a higher incidence of serious errors and a lower incidence of less serious ones. That is the reason why, at least in the intentions of Neyman and Pearson, α and β should have been chosen in a balanced way.

case, Example 1 shows that results are unreliable and one might discard effects that are, in fact, relevant to the study. At the same time, Example 2 shows that—for studies with large effect sizes—overpower does not pose too relevant threats to the overall reliability of a study.

In any case, knowing what makes the ABM more likely to produce reliable results is a relevant information for modelers. It seems more so when modelers perform their simulation a limited number of times per configuration of parameters. But also when too many runs are performed, the absence of power calculations may mislead one's judgement on the effects and actual meaning of the simulation. However, the asymmetry of the effects between under- and over-power suggests that power analysis can be used to provide, if not a guess, at least a lower guess on the number of runs (see the concept of SESOI introduced above). The value that is calculated with the aid of statistical power analysis is a number that—if not taken at face value—should inform the choice on the number of runs, and could at least work as a benchmark.

In a review of models published mostly in *Computational and Mathematical Organization Theory* (CMOT) and in the *Journal of Artificial Societies and Social Simulation* (JASSS) between 2010 and 2013 (Secchi and Seri 2017) it was found that most models are under-powered. If a small effect size $d = 0.1$ is hypothesized, then the average power is $1 - \beta \approx 0.41$, while if a medium effect size $d = 0.3$ is taken, then power becomes $1 - \beta \approx 0.84$ (with $\alpha = 0.01$). In both cases, the review shows that models are under-powered even by the milder standards of $1 - \beta = 0.90$ suggested in Ritter et al. (2011).

11.5.1 Comparing Statistical Power to Other Approaches

Using power is not the only way in which one can determine the number of runs in an experimental study and, in particular, in an ABM.

As an example, another approach sometimes called *accuracy in parameter estimation* (AIPE) (Maxwell et al. 2008) has been proposed. In this approach, first the researcher identifies a quantity of interest (a coefficient in a regression, a correlation, etc.) and chooses the desired width of a confidence interval around this value. Then, the researcher selects the sample size that allows one to reach this objective. The technique is already established, under different names, in medicine (Bland 2009), engineering (Hahn and Meeker 2011, Sect. 8.3), and psychology (Maxwell et al. 2008). A similar approach, putting together AIPE and power analysis, has also been proposed in the context of simulation models in Ritter et al. (2011).

However, we think that, in order to become a feasible option for ABM, this method should overcome some difficulties. First, AIPE may be surely of interest whenever the objective of the analysis is to obtain a precise enough measure of the effect of a treatment (see above for references). However, most ABM studies are not framed in this way (see the distinction between KISS and KIDS above).

The reason is that ABM studies are often simplified representations of reality. Therefore, the effect of a treatment is rarely their desired outcome, as it is clear that the value obtained from an ABM will generally not be the same value observed in reality. Second, even when the outcome of an ABM study is of interest in itself, it is rarely the case that one has a precise idea of what the width of a confidence interval should be. This may be different whenever the outcome variable is measured on a well-known scale, as it is often the case in the disciplines in which AIPE is an established alternative to power analysis. The paper (Schönbrodt and Perugini 2013) (see also Lakens and Evers 2014) provides an interesting example, based on Cohen (1988), of how to determine the width of an interval, but this seems difficult to generalize to other situations.

11.5.2 Concluding Remarks

The message of this article is that statistical power analysis can help modelers to refine their ideas on how many times their ABM simulation should be performed. In this chapter, we first wrote a few notes on the importance of determining the number of runs, and then turned our attention to the type of models that would benefit the most from this approach. The focus is then moved to testing theory so that we could provide an appropriate statistical background for this approach. Finally, some practical examples show the risks and perils of under- or over-estimating the number of runs in a simulation. The implications are then further discussed at the beginning of this section.

As a way to provide a summary of this chapter and, at the same time, help modelers clarify what under- and over-power imply, Table 11.4 shows calculations of power for $\alpha = 0.01$ and $1 - \beta = 0.95$, using the formula that we developed and also appearing in the Appendix.

The left column in Table 11.4 shows the hypothetical number of parameter configurations (or groups G) that a potential ABM could have. Knowing how to determine the appropriate number of configurations is a complex issue that falls beyond the scope of this chapter. However, sensitivity and steady state analyses can provide sound support (Thiele et al. 2015). The table calculates the number of runs that are necessary to reach $1 - \beta = 0.95$ at $\alpha = 0.01$ for five different effect sizes, respectively *ultra-micro* $= 0.01$, *micro* (0.05), *small* (0.1), *medium* (0.2), *large* (0.4), and *huge* (0.8). Results from these calculations confirm with more granularity of details that small simulations, with few configurations of parameters (up to 10) need to be performed many times unless the effect size is large or very large. As the number of configurations grows, the number of runs to perform clearly decreases significantly to the point where one run per configuration is enough when variability is spread to its limits (from 1000 and up) in the presence of large and very large effect sizes.

Table 11.4 A map of statistical power: Number of runs for $\alpha = 0.01$ and $1 - \beta = 0.95$

CoP (G)	Effect sizes f					
	ultra-micro	micro	small	medium	large	huge
2	84,777.89	3,468.39	875.55	221.02	55.79	14.08
3	65,400.97	2,675.65	675.44	170.51	43.04	10.87
4	54,403.07	2,225.71	561.85	141.83	35.80	9.04
5	47,162.95	1,929.51	487.08	122.96	31.04	7.84
10	30,265.08	1,238.19	312.57	78.90	19.92	5.03
20	19,421.49	794.56	200.58	50.63	12.78	3.23
50	10,804.41	442.02	111.58	28.17	7.11	1.79
100	6,933.33	283.65	71.60	18.08	4.56	1.15
200	4,449.21	182.02	45.95	11.60	2.93	0.74
500	2,475.15	101.26	25.56	6.45	1.63	0.41
1000	1,588.33	64.98	16.40	4.14	1.05	0.26
3000	786.29	32.17	8.12	2.05	0.52	0.13
5000	567.02	23.20	5.86	1.48	0.37	0.09
10,000	363.86	14.89	3.76	0.95	0.24	0.06

Note. Effect sizes: *ultra-micro* $= 0.01$, *micro* $= 0.05$, *small* $= 0.1$, *medium* $= 0.2$, *large* $= 0.4$, *huge* $= 0.8$. CoP (G): configuration of parameters (groups)

Clearly, Table 11.4 needs to be taken as an exemplification of how likely it is that a given number of configurations may lead to an under- or over-powered simulation, hence determining the likelihood to make Type-II error or to over-emphasize results. The table can be used as a first indication of how this approach to ABM runs can be applied. More fine grained results may vary depending on the circumstances of each simulation, including the levels of α, β, and the purpose of the model.

Further Reading

Details on several power measures can be found in Cohen (1988) and Liu (2014). Specific information on ABM and power are in Secchi and Seri (2017).

Appendix

Number of Runs Calculations

The following is the R code for a function that calculates the number of runs for the configuration of parameters (G, here G) and effect size (f, here ES), given $1 - \beta = 0.95$, $\alpha = 0.01$:

```
n.runs <- function(G, ES) {
   return(14.091 * G^(-0.640) * ES^(-1.986))
}
```

In the case discussed in Exercise 1 above, the numbers are:

```
n.runs(3, 0.25)
[1]   109.465
```

The same analysis using the exact function of the package `pwr` on power analysis (see Champely et al. 2016) is:

```
pwr.anova.test(f=0.25, k=3, power=0.95,
                  sig.level=0.01)
```

and yields $n = 111.677$.

Effect Size for ANOVA vs OLS Regression

In the text we have used a one-way ANOVA test to estimate the number of runs, taking $1 - \beta = 0.95$, $\alpha = 0.01$ and a given effect size f. However, we then used regression analysis to study the differences between under-, correctly-, and over-powered models.

Since there is transformation between the parameters of ANOVA and OLS regression, it is possible to connect the way effect size is calculated in the first to the second.

As mentioned in the text of the chapter, the effect size for ANOVA is:

$$f = \sqrt{\frac{n \sum_{j=1}^{G} (\bar{x}_j - \bar{x})^2}{\sum_{j=1}^{G} \sum_{i=1}^{n} (x_{ij} - \bar{x}_j)^2}}$$

The quantity under the square root is the SSB divided by the Sum of Squares Within (SSW) or, in Cohen's terms, $f = \frac{\sigma_m}{\sigma}$ (Cohen 1992). The effect size for regression is, according to Cohen (1992), $f^2 = \frac{R^2}{1-R^2}$. It is easy to demonstrate that:

$$f^2 = \frac{R^2}{1 - R^2} = \frac{\text{SSB}}{\text{SSR}}$$

where the SSW in a one-way ANOVA is comparable to the Sum of Squares of Residuals (SSR) in an OLS regression with exactly the same dependent and independent variables.

References

Anderson, P. (1972). More is different. *Science, 177*(4047), 393–396.

Bardone, E. (2016). Intervening via chance-seeking. In D. Secchi & M. Neumann (Eds.), *Agent-based simulation of organizational behavior. New frontiers of social science research* (pp. 203–220). New York: Springer.

Bland, J. M. (2009). The tyranny of power: Is there a better way to calculate sample size? *BMJ, 339*, b3985.

Champely, S., Ekstrom, C., Dalgaard, P., Gill, J., Weibelzahl, S., & Rosario, H. D. (2016). Pwr: Basic functions for power analysis.

Choirat, C., & Seri, R. (2012). Estimation in discrete parameter models. *Statistical Science, 27*(2), 278–293.

Coen, C. (2009). Simple but not simpler. Introduction CMOT special issue–simple or realistic. *Computational and Mathematical Organization Theory, 15*, 1–4.

Cohen, J. (1988). *Statistical power analysis for the behavioral sciences* (2nd ed.). Hillsdale: LEA.

Cohen, J. (1992). A power primer. *Psychological Bulletin, 112*, 155–159.

Cohen, M. D., March, J. G., & Olsen, H. P. (1972). A garbage can model of organizational choice. *Administrative Science Quarterly, 17*(1), 1–25.

Davidsson, P., & Verhagen, H. (2017). Types of simulation. doi: https://doi.org/10.1007/978-3-319-66948-9_3.

de Marchi, S., & Page, S. E. (2014). Agent-based models. *Annual Review of Political Science, 17*(1), 1–20.

Edmonds, B., & Meyer, R. (2017). Introduction to the handbook. doi: https://doi.org/10.1007/978-3-319-66948-9_1.

Edmonds, B., & Moss, S. (2005). From KISS to KIDS — an 'anti-simplistic' modelling approach. In P. Davidson (Ed.), *Multi agent based simulation*. Lecture Notes in Artificial Intelligence (Vol. 3415, pp. 130–144). New York: Springer.

Erdfelder, E. (1984). Zur Bedeutung und Kontrolle des β-Fehlers bei der inferenzstatistischen Prüfung log-linearer Modelle [The significance and control of the β-error during the inference-statistical examination of the log-linear models]. *Zeitschrift für Sozialpsychologie, 15*(1), 18–32.

Fioretti, G. (2016). Emergent organizations. In D. Secchi & M. Neumann (Eds.), *Agent-based simulation of organizational behavior. New frontiers of social science research* (pp. 19–41). New York: Springer.

Fioretti, G., & Lomi, A. (2008). An agent-based representation of the garbage can model of organizational choice. *Journal of Artificial Societies and Social Simulation, 11*(1), 1.

Fioretti, G., & Lomi, A. (2010). Passing the buck in the garbage can model of organizational choice. *Computational and Mathematical Organization Theory, 16*(2), 113–143

Fisher, R. (1955). Statistical methods and scientific induction. *Journal of the Royal Statistical Society. Series B (Methodological), 17*(1), 69–78

Gigerenzer, G. (2004). Mindless statistics. *Journal of Socio-Economics, 33*, 587–606.

Gilbert, N., & Terna, P. (2000). How to build and use agent-based models in social science. *Mind and Society, 1*, 57–72.

Hahn, G. J., & Meeker, W. Q. (2011). *Statistical intervals: A guide for practitioners*. Hoboken: Wiley.

Heckbert, S. (2013). MayaSim: An agent-based model of the ancient Maya social-ecological system. *Journal of Artificial Societies and Social Simulation, 16*(4), 11.

Herath, D., Secchi, D., & Homberg, F. (2015). Simulating the effects of disorganisation on employee goal setting and task performance. In D. Secchi & M. Neumann (Eds.), *Agent-based simulation of organizational behavior. New frontiers of social science research* (pp. 63–84). New York: Springer.

Herath, D., Costello, J., & Homberg, F. (2017). Team problem solving and motivation under disorganization – an agent-based modeling approach. *Team Performance Management, 23*(1/2), 46–65.

Hoenig, J. M., & Heisey, D. M. (2001). The abuse of power. *The American Statistician, 55*(1), 19–24.

Kollman, K., Miller, J. H., & Page, S. E. (1992). Adaptive parties in spatial elections. *The American Political Science Review, 86*(4), 929–937.

Korn, E. L. (1990). Projecting power from a previous study: Maximum likelihood estimation. *The American Statistician, 44*(4), 290–292.

Lakatos, E. (2005). Sample size determination for clinical trials. In *Encyclopedia of biostatistics*. Hoboken: Wiley.

Lakens, D. (2014). Performing high-powered studies efficiently with sequential analyses. *European Journal of Social Psychology, 44*(7), 701–710.

Lakens, D. & Evers, E. R. K. (2014). Sailing from the seas of chaos into the corridor of stability practical recommendations to increase the informational value of studies. *Perspectives on Psychological Science, 9*(3), 278–292.

Lamperti, F. (2015). *An Information Theoretic Criterion for Empirical Validation of Time Series Models*. LEM Papers Series 2015/02, Laboratory of Economics and Management (LEM), Sant'Anna School of Advanced Studies, Pisa, Italy.

Liu, X. S. (2014). *Statistical power analysis for the social and behavioral sciences*. New York: Routledge.

Liu, T., & Stone, C. C. (2007). *Law and statistical disorder: Statistical hypothesis test procedures and the criminal trial analogy*. SSRN Scholarly Paper ID 887964, Social Science Research Network, Rochester, NY.

Maggi, E., & Vallino, E. (2016). Understanding urban mobility and the impact of public policies: The role of the agent-based models. *Research in Transportation Economics, 55*, 50–59.

Maxwell, S. E., Kelley, K., & Rausch, J. R. (2008). Sample size planning for statistical power and accuracy in parameter estimation. *Annual Review of Psychology, 59*(1), 537–563.

Mayo, D. G. (1992). Did pearson reject the neyman-pearson philosophy of statistics? *Synthese, 90*(2), 233–262.

Mungovan, D., Howley, E., & Duggan, J. (2011). The influence of random interactions and decision heuristics on norm evolution in social networks. *Computational and Mathematical Organization Theory, 17*(2), 152–178.

Neyman, J. (1950). *First course in probability and statistics*. New York: Henry Holt and Company.

Neyman, J., & Pearson, E. S. (1928). On the use and interpretation of certain test criteria for purposes of statistical inference: Part I. *Biometrika, 20A*(1/2), 175–240.

Neyman, J., & Pearson, E. S. (1933). On the problem of the most efficient tests of statistical hypotheses. *Philosophical Transactions of the Royal Society of London. Series A, Containing Papers of a Mathematical or Physical Character, 231*, 289–337.

Pearson, E. S. (1955). Statistical concepts in the relation to reality. *Journal of the Royal Statistical Society. Series B (Methodological), 17*(2), 204–207.

Railsback, S. F., & Grimm, V. (2011). *Agent-based and individual-based modeling: A practical introduction* (59468th ed.). Princeton: Princeton University Press.

Ritter, F. E., Schoelles, M. J., Quigley, K. S., & Cousino-Klein, L. (2011). Determining the numbers of simulation runs: Treating simulations as theories by not sampling their behavior. In L. Rothrock & S. Narayanan (Eds.), *Human-in-the-loop simulations: Methods and practice* (pp. 97–116). London: Springer.

Robinson, S. (2014). *Simulation. The practice of model development and use* (2nd ed.). New York: Palgrave.

Schönbrodt, F. D., & Perugini, M. (2013). At what sample size do correlations stabilize? *Journal of Research in Personality, 47*(5), 609–612.

Secchi, D. (2015). A case for agent-based model in organizational behavior and team research. *Team Performance Management, 21*(1/2), 37–50.

Secchi, D., & Gullekson, N. (2016). Individual and organizational conditions for the emergence and evolution of bandwagons. *Computational and Mathematical Organization Theory, 22*(1), 88–133.

Secchi, D., & Seri, R. (2014). 'How many times should my simulation run?' Power analysis for agent-based modeling. In *European Academy of Management Annual Conference, Valencia, Spain*.

Secchi, D., & Seri, R. (2017). Controlling for 'false negatives' in agent-based models: A review of power analysis in organizational research. *Computational and Mathematical Organization Theory, 23*(1), 94–121.

Shimazoe, J., & Burton, R. M. (2013). Justification shift and uncertainty: Why are low-probability near misses underrated against organizational routines? *Computational and Mathematical Organization Theory, 19*(1), 78–100.

Simon, H. A. (1976). How complex are complex systems. In *PSA: Proceedings of the Biennial Meeting of the Philosophy of Science Association* (Vol. 2, pp. 507–522). Baltimore: Philosophy of Science Association.

Simon, H. A. (1978). Rationality as process and a product of thought. *American Economic Review, 68*, 1–14.

Simon, H. A. (1997). *Administrative behavior* (4th ed.). New York: The Free Press.

Thiele, J., Kurth, W., & Grimm, V. (2015). Facilitating parameter estimation and sensitivity analysis of agent-based models: A cookbook using NetLogo and R. *Journal of Artificial Societies and Social Simulation, 17*(3), 11.

Thomsen, S. E. (2016). How docility impacts team efficiency. An agent-based modeling approach. In D. Secchi & M. Neumann (Eds.), *Agent-based simulation of organizational behavior. New frontiers of social science research* (pp. 159–173). New York: Springer.

Troitzsch, K. G. (2017). Historical introduction. doi:https://doi.org/10.1007/978-3-319-66948-9_2.

van der Vaart, A. W. (2000). *Asymptotic statistics*. Cambridge: Cambridge University Press.

Wasserstein, R. L., & Lazar, N. A. (2016). The ASA's statement on p-values: Context, process, and purpose. *American Statistician, 70*(2), 129–133.

Wilensky, U. (1999). Netlogo. Center for Connected Learning and Computer-Based Modeling, Northwestern University, Evanston, IL.

Chapter 12
Participatory Approaches

Olivier Barreteau, Pieter Bots, Katherine Daniell, Michel Etienne,
Pascal Perez, Cécile Barnaud, Didier Bazile, Nicolas Becu,
Jean-Christophe Castella, William's Daré, and Guy Trebuil

Abstract This chapter aims to describe the diversity of participatory approaches in relation to social simulations, with a focus on the interactions between the tools and participants. We consider potential interactions at all stages of the modelling process: conceptual design, implementation, use and simulation outcome analysis. After reviewing and classifying existing approaches and techniques, we describe two case studies with a focus on the integration of various techniques. The first case study deals with fire hazard prevention in Southern France, and the second one with groundwater management on the atoll of Kiribati. The chapter concludes with a discussion of the advantages and limitations of participatory approaches.

O. Barreteau (✉)
IRSTEA, UMR G-EAU, Montpellier, France
e-mail: olivier.barreteau@irstea.fr

P. Bots
Faculty of Technology, Policy and Management, Delft University of Technology,
Delft, The Netherlands

K. Daniell
Research School of Social Sciences, The Australian National University, Canberra, Australia

M. Etienne
Ecodevelopment Unit, National Institute for Agronomic Research, Avignon, France

P. Perez
SMART, University of Wollongong, Wollongong, Australia

C. Barnaud
Dynafor, Centre INRA de Toulouse, Castanet Tolosan, France

D. Bazile • W. Daré • G. Trebuil
Cirad GREEN, TA C-47/F. Campus international de Baillarguet, Montpellier, France

N. Becu
CNRS, Laboratoire de geographie PRODIG 2, Paris, France

J.-C. Castella
IRD (Institute of Research for Development) and CIRAD (UR ÄDA), Vientiane, Laos

© Springer International Publishing AG 2017   253
B. Edmonds, R. Meyer (eds.), *Simulating Social Complexity*,
Understanding Complex Systems, https://doi.org/10.1007/978-3-319-66948-9_12

Why Read This Chapter?
To help you understand how one might involve stakeholders in all stages of the modelling process. This approach allows for including stakeholders' expertise as well as giving them more control over the process.

12.1   Introduction

In this chapter, social simulation is cross-examined with a currently very active trend in policymaking: participation or stakeholder involvement. This cross-examination has two main outputs: the development of tools and methods to improve or facilitate participation and the development of more grounded simulation models through participatory modelling. Technological development provides new devices to facilitate interaction around simulation models: from the phase of conceptual design to that of practical use. In many fields there is a growing requirement from stakeholders and the public to become more actively involved in policymaking and to be aware of probable changing trends due to global policy decisions. New tools and methods related to social simulation have started to be made available for this purpose such as many group decision support systems which use computer simulation, including potentially social items components, to facilitate communication to formulate and solve problems collectively (DeSanctis and Gallupe 1987; Shakun 1996; Whitworth et al. 2000). In addition, simulation of social complexity occurs in models whose validation and suitability depend on their close fit to society, as well as on their acceptability by it. These issues are tackled through the use of participatory modelling, such as group model building (Vennix 1996) or participatory agent-based simulations (Bousquet et al. 1999; Guyot and Honiden 2006; Moss et al. 2000; Pahl-Wostl and Hare 2004; Ramanath and Gilbert 2004). The topic is also related to participatory design as it is a means of involving end users of computer systems in their design, including social simulations focussed ones (Schuler and Namioka 1993).

Group decision support as well as participatory modelling stems from the interactions between simulation models and participants. There is a diversity of ways through which these interactions might take place. They are related to the diversity of approaches to simulate society or to organise participation. It is important to make the choices made for these interactions explicit: for distinction between approaches to be possible, to provide the opportunity for stakeholders to discuss the process and for them to be prepared to be involved in. There is a need to go further than the development of tools as they are liable to create filters that reshape the understanding of social complexity. Description of the mechanisms behind interactions is a way to qualify the potential effects of these interactions.

This chapter aims to describe the diversity of participatory approaches in relation to social simulations, with a focus on the interactions between the tools and participants. This overview is limited to simulation models. Model is considered here as a representation of shared knowledge, which means the gathering of pieces of knowledge and assumptions about a system, written altogether in a model so that they might play or work together. We limit this scope further to simulation model, hence models including the representation of dynamics. We consider here potential interactions among participatory and modelling processes at all stages of the modelling process: conceptual design, implementation, use and simulation outcome analysis.

The first section of this chapter outlines a number of factors which have paved the way for development of the association between social simulation and participation. There is a large body of literature in which authors have developed their own participatory modelling approaches, justified by some specific expectations on participation for modelling or vice versa. This first section makes a synthesis of these expectations and draws out some principles on which various participatory modelling settings should be assessed. The second section describes some existing techniques and approaches. The third section proposes a classification of these participatory approaches according to three dimensions: the level of involvement in the process, the timeliness of involvement and the heterogeneity of population involved. The fourth section describes two case studies with a focus on the integration of various techniques. We discuss the advantages of these approaches but also some limits, according to the expectations and in comparison with more traditional techniques in the fifth section.

12.2 Expectations of Using Participatory Approaches with Simulation of Social Complexity

Joint use of participatory approaches with social simulations is based upon three categories of expectations. They vary according to the target of the expected benefits of the association:

1. Quality of the simulation model per se
2. Suitability of the simulation model for a given use
3. Participation support

These three targets are linked to three different components of a modelling process. Target one is linked to the output, target three to the source system, and target two to the relation between both the output and source system. In this section we further develop these three categories.

12.2.1 Increasing Quality of Simulation Models of Social Complexity

The objective here is to produce a good quality model to simulate social complexity. Participation is then pragmatically assumed to be a means for improving this quality. There is no normative belief which would value participation by itself in this category of expectations.

Quality of the simulation model is understood here rather classically with the following indicators:

- Realism: is the simulation model able to tackle key features of the social complexity it aims to represent?
- Efficiency: is the simulation model representing its target system with a minimum of assumptions and minimal simulation run times?

Quality of the representation according to its use is another classical indicator of a simulation model's quality. It is specifically tackled in the following subsection.

12.2.1.1 Taking Social Diversity and Capacity to Evolve into Account

One of the key features to be taken into account when representing a social system is to deal with its diversity. This diversity is related not only to individual characteristics but also to viewpoints, expectations towards the system and positions in the decision-making processes. Dealing with diversity in simulation of social complexity involves embracing it as well as profiting by its existence.

Classically, dealing with diversity is a process of aggregation or selection. Aggregation consists of the identification of classes of individuals and representatives for them. Selection consists of choosing a few cases with all of their characteristics. This may lead to very simple simulation models with a generic diversity. Aggregation is rather greedy on data and modelling time and is still dependent on the viewpoint of the observers who provide the information leading to the categorisation. Selection is weak to cope with relations among various sources of diversity.

Involvement of stakeholders in the modelling process allows them to bring their own diversity. Concerns over representation are then transferred onto the constitution of the sample of participants. Fischer and colleagues have shown through development of situations to support creativity in various fields, such as art, open source development and urban planning, that diversity, as well as complexity, is important to enhance creativity (Fischer et al. 2005). This creativity is expected to pave the way for surprises in the simulation model.

Involvement of stakeholders in the modelling process is a way to externalise part of this diversity outside the model towards a group of stakeholders. The issue is then to work on the relation between the model and a number of stakeholders to allow a transfer of knowledge and ideas.

Social systems are open and evolving. Their definition depends on the viewpoint of the analyst. As far as simulation is concerned, this means depending on the viewpoint of the model designer(s). This choice means framing: cutting a number of links around the boundaries of the system studied, as well as around the interpretation which might occur based on the simulation outcomes (Dewulf et al. 2006). Firstly, participation provides the opportunity to consider problem boundaries which would be plurally defined, increasing the potential coherence of the model. However, it is still an operation of cutting links out of the real-world situation, even though these chosen cuttings are more grounded and discussed.

Secondly, interactive use of a simulation model is a means to keep some of these links open and active, with participants as driving belts. Stakeholders are embedded in social networks which cross the boundaries into the physical and environmental networks. They make the links come alive, which allows them to function and be updated.

There is thus a need to question the boundaries set in the interactive setting: actors in the neighbourhood, concerns of actors connected to those tackled by the (simulation) model and how these relations are to be mobilised in the interaction.

12.2.1.2 Distribution of Control

A key characteristic of social systems which is to be addressed through social simulation is their complexity. This complexity leads to various consequences, such as the emergence of phenomena, delay effects or discontinuities in some trends, which are present in social systems as in any complex systems. These are usually the effects which one likes to discover or better understand when experimenting with social simulations. From the internal point of view of simulations, Schelling has shown experimentally that reproducing settings with multiple decision centres improves the quality of representation of complexity (Schelling 1961). He could generate complexity through experimental games because of the presence of independent decision centres, the players. This result has also been shown with simulations used for forecasting (Green 2002). Green compared the capacity of forecasting the outcome of past social conflicts with a role-playing game with students, game theorists and a group of experts. He compared the simulated outcomes with those from the real negotiations and found that the role-playing game setting produced the best results. This was the one with the main distribution of decisions among autonomous centres.

The purpose of associating participatory processes and social simulation here is then to increase the complexity through interactive use or implementation of a social model. Unless computational agents are effectively used, which is rare (Drogoul et al. 2003), formal theories of complex systems that are completely embedded in a simulation model do not simulate complex patterns but implement an explanation of a complex pattern. In other words, they should be implemented in a distributed setting with autonomous entities. Participatory approaches provide such settings. There is then an issue of a deep connection between a simulation model and participants in a participatory modelling setting.

12.2.2 Improving Suitability of Simulation Model's Use

Quality of a model is also assessed according to its suitability for its intended use. In this subsection, two cases of use are considered: knowledge increase and policymaking. In both cases, it is expected that involvement of stakeholders at any stage of a modelling process will aid better tuning of the model with its intended use: either through interactions with people represented in the model, or with potential users. Both cases have a major concern with making viewpoints explicit.

12.2.2.1 Case of Increasing Knowledge

The case of use for knowledge increase builds upon the previous subsection. The key element treated here deals with the uncertainty of social systems. The involvement of stakeholders represented in the simulation model is a way to improve its validation or calibration. Participants may bring their knowledge to reduce or better qualify some uncertainties. The simulation model is then expected to give back to the participant's simulation outputs based on the interactions between their pieces of knowledge. On the other hand, this feedback is sometimes difficult to validate (Manson 2002). Its presentation and discussion with stakeholders represented in the simulation model is a way to cope with this issue. This approach has been explored by Barreteau and colleagues to improve the validation of an agent-based model of irrigated systems in Senegal River valley (Barreteau and Bousquet 1999). The format of this feedback, information provided and medium of communication, might make the model really open to discussion.

This joins another expectation which is probably the most common in work that has so far implemented such participatory approaches with a social simulation model: making each participant's assumptions explicit, included the modellers (Fischer et al. 2005; Moss et al. 2000; Pahl-Wostl and Hare 2004). This is a requirement from the simulation modelling community: making stakeholders' beliefs, points of view and tacit knowledge explicit (Barreteau et al. 2001; Cockes and Ive 1996; D'Aquino et al. 2003; McKinnon 2005). Moreover, so that participants might become part of the model, the assumptions behind the model should be made explicit in order to be discussed, as should the outputs of the simulations so that they can also be discussed, transferred and translated in new knowledge. This is to overcome one major pitfall identified with the development of models which is the underuse of decision support models because of their opacity (Loucks et al. 1985; Reitsma et al. 1996). This concern of making explicit assumptions in the modelling process is also at the heart of the participatory approach community. One aim of gathering people together and making them collectively discuss their situation in a participatory setting is to make them aware of others' viewpoints and interests. This process involves and stimulates some explanation of tacit positions.

This means that the interactive setting should allow a bidirectional transfer of knowledge between stakeholders and the simulation model: knowledge elicitation in one direction and validation and explanation of simulation outputs in the other direction.

12.2.2.2 Case of Policymaking

In the case of simulation focusing on policy issues, there is a pragmatic, moral and now sometimes legal need to involve stakeholders, which may lead to open the black box of models of social complexity used in policymaking. Post-normal approaches aim at making the decision process and its tools explicit so that stakeholders can better discuss it and appropriate its outcomes. When this decision process involves the use of decision support tools, which might include social simulation models, this means that the models themselves should be opened to stakeholders (Funtowicz et al. 1999). A simulation model is then expected to be explicit enough so that stakeholders who might be concerned by the implementation of the policy at stake could discuss it. This legitimisation is socially based, while validation, as mentioned with the previous case of use, is scientifically based (Landry et al. 1996). Even though validation is still required in this case of use, because it is the mode of evaluation for some participants, it is rather the legitimisation of the model by the stakeholders which is to be worked out.

Participatory approaches may be a means for opening these models to stakeholders, provided that formats of communication of models' assumptions and structure can be genuinely discussed. Involvement of stakeholders is expected to raise their awareness of the assumptions of the model and potentially able to discuss these and modify them. This includes the evolution of underlying values and choices made in the design of model.

12.2.3 Simulation as a Means to Support Participation

Social simulation might also benefit to participation. While the previous subsection was dedicated to appropriateness between the model and its use as a group decision support tool, we focus here on participation which might be a component of a decision-making process.

Social simulation is seen here as an opportunity to foster participation and cope with some of its pitfalls (Eversole 2003). The use of simulation models may lead to some outcomes such as community building or social learning.

12.2.3.1 Dynamics and Uncertainties

Social systems have to deal with uncertainties just as social simulation models do. This might hamper participatory processes: in wicked problems (Rittel and Webber 1973), encountered in many situations where participatory processes are organised, stakeholders always maintain the opportunity related to these uncertainties to challenge others' viewpoints or observations. As an example, origin, flow and consequences of nonpoint source pollution are uncertain. This leads some farmers to challenge the accusation, made by domestic water companies downstream of their fields, that they are polluting their sources. Sometimes, disparate viewpoints do not conflict. The gathering of these disparate pieces of knowledge is a way to reduce uncertainty and allows the group of stakeholders involved in a participatory process to progress, provided that they can work together.

Another characteristic of any social system which might hamper participation is its dynamicity. Socioecological systems exhibit a range of dynamics, not only social but also natural, which evolve at various paces. In the application developed by Etienne and colleagues in Causse Mejan, pine tree diffusion has a typical time step of 20 years which is long according to the typical time steps of land-use choices and assessment (Étienne et al. 2003). In a participatory process, it might be difficult to put these dynamics on the agenda. Simulation models are known to be good tools to deal with dynamic systems.

Simulation models are therefore a means to gather distributed pieces of knowledge among stakeholders and to cope with scenarios in the face of uncertainties. They can also help make the participants aware of potential changes or regime shifts generated by their interactions (Kinzig et al. 2006).

12.2.3.2 Towards Social Learning

Participation is often linked with the concept of social learning (Webler et al. 1995). However, for social learning to occur, participants should have a good understanding of their interdependencies as well as of the system's complexity. Social simulation can provide these bases, provided that the communication is well developed (Pahl-Wostl and Hare 2004).

This learning comes from exchanges among stakeholders involved in the participatory process but also from new knowledge which emerges in the interaction. Externalisation of tacit knowledge in boundary objects (Star and Griesemer 1989) is useful for both: it facilitates communication in giving a joint framework to make one's knowledge explicit, and it enhances individual, as well as social, creativity (Fischer et al. 2005).

Simulation models are good candidates to become such boundary objects. Agent-based models have long been considered as blackboards upon which various disciplines could cooperate (Hochman et al. 1995). Through simulation outputs, they provide the necessary feedback for reflexivity, be it individual or collective.

The question then remains whether such models constrain the format of knowledge which might be externalised.

12.2.4 Synthesis: A Key Role of the Interaction Pattern Between Model and Stakeholders

These three categories of expectations have led to specific requests for the development of participation in relation to social simulation models. In the following section, we provide an overview of these techniques. On the basis of the previous requests, these techniques and methods have to be analysed according to the following dimensions:

- Set of connections between the participation arena and simulation model: its structure, its content and organisation of its mobilisation
- Control of the process
- Format of information which can travel from one pole to another: openness and suitability to the diversity of stakeholders' competencies.

12.3 A Diversity of Settings

In this section, we describe some examples of participatory techniques and approaches associated with social simulation models. Settings described in this overview stem from various fields and disciplines. Most of these have already produced some reviews on participatory approaches. For the purpose of the discussion in relation with social simulation, a synthesis of these reviews is provided here with a focus on the requests identified in the previous section.

12.3.1 From System Science and Cybernetics

Cybernetics and system sciences have produced a first category of simulation models of social complexity (Gilbert and Troitzsch 1999). These models are based on tools originating from system dynamics, using specific software. They focus on flows of resources and information between stocks which can be controlled.

Two main types of interactions between these models and stakeholders have so far emerged: group model building (Vennix 1996) and management flight simulators or microworlds (Maier and Grössler 2000).

Group model building experiments focus on the interaction with stakeholders in the design stage of a modelling process. It associates techniques of system dynamics modelling with brainstorming tools and other techniques of group work, mainly based on workshops and meetings. This trend consists of integrating future users of the model in the design stage. The participants are supposed to be the clients of the modelling process. Rouwette and colleagues analysed 107 cases of such experiments and proposed a number of guidelines to facilitate consistent reporting on participatory modelling exercises. These guidelines focus on three categories: context, mechanisms and results (Rouwette et al. 2002). The second category focuses predominately on preparation activities and description of meetings, along with factual elements and the modelling process.

This category of participatory modelling deals with the expectations identified in the first section in the following manner:

- The participation arena is constituted of a rather small or medium size well-identified group. The structure of the interaction is rather global: debates tackle the whole model, and participants are supposed to be concerned by the model entity as a whole. The connections may convey information on the tacit knowledge of stakeholders, as well as on their purposes. This is still very diverse among the experiments. The group of stakeholders is mobilised within specific events, workshops, which might be repeated. The aim is to feed the model but also to increase the probability of the use of the models produced.
- The process is predominately controlled by the modellers.
- The format of information is generally not well formalised, even though techniques, such as hexagons brainstorming or causal diagrams (Akkermans 1995), appear to organise the knowledge brought by stakeholders. This low formalisation allows the issues related to stakeholder diversity to be tackled and alleviated in the problem framing phase, but it leaves a large place to the modellers' interpretation.

Management flight simulators or microworlds constitute a complementary technique, which focuses more on the stages of use and simulation outcomes analysis, even though this technique may also be used in a design stage to elicit tacit knowledge. A key characteristic of this type of technique is to encourage learning by doing. Participants, who might be the clients or other concerned people without any formal relation to the modelling team, have to play through a simulation of the model. Martin and colleagues have used this technique to validate a system dynamics model on the hen industry (Martin et al. 2007). Participants were asked to play with some parameters of the model.

When used to elicit knowledge, microworlds attempt to provide events that are similar to those that participants already face or are likely to face in their activities related to the issue at stake in the model. Le Bars and colleagues have thus developed a game setting to lead farmers to understand the dynamics of their territory with regard to water use and changes in EU common agricultural policy (Le Bars et al. 2004). In flight simulator experiments, interaction between stakeholders and the simulation model is structured around future users of the model or people whose stakes are represented in the model, with a slightly deeper connection than with previous group modelling building approaches. Participants are asked to deal with parameters of the model and are framed in the categories used in the model. There is no a priori differentiation among participants. The connections convey information about the object from the model to participants. It also conveys the participants' reactions to this object and some behavioural patterns observed that can provide new information for the modellers. This connection is activated by the participants working through specific events and focus on the use of the tool. Control is still on the side of modellers, who frame the interactions. The format of information is largely formalised from model to stakeholders. It is not formalised from stakeholders to model.

12.3.2 Knowledge Engineering: Between Artificial Intelligence and Social Psychology

Knowledge engineering focuses on a specific time of the interaction between stakeholders and a simulation model in the design stage: the process of translating tacit knowledge into conceptual or sometimes computational models. Many knowledge elicitation techniques are useful in transforming written or oral text into pieces of simulation models. The purpose of these techniques is to separate the contributions made directly to the model from the design of the model itself. Knowledge engineering aims to provide interfaces for this gap.

To deal with this interface, techniques have been developed, grounded in artificial intelligence, (social) psychology and cognitive science. Behavioural patterns in social simulation models are often borrowed in simplified versions from these fields (Moss et al. 2000; Pahl-Wostl and Hare 2004). This cross-pollination of disciplines can be potentially fruitful for model design. As an example, Abel and colleagues have built upon the concept of a mental model. They assume that individuals have representations of their world which may be formalised in causal rules. Working in the Australian bush, they have designed specific individual interview protocols and analysis frameworks to elicit these mental models (Abel et al. 1998). In this case, interaction with the model occurs through the interviewer who in this case was also the modeller. There was no collective interaction. Researchers dealing with the interviews and the corresponding model design clearly guide the process. The format of information is speech (in the form of a transcribed text), which is transformed into a modelling language in this elicitation process.

Building upon Abel's work, Becu has further minimised the involvement of the modeller, still using individual interviews. This has led him to collaborate with an anthropologist and to use ethnographic data as a benchmark. Individual interviews, with the interviewee in the environment suitable to the purpose of the interview, led him to identify objects and relations among these objects. These constitute the initial basis for an exercise, labelled as playable stories: stakeholders, in his case farmers from Northern Thailand, are asked to choose the key elements to describe their world from their own viewpoints (with the possibility of adding new elements), then to draw relations among them and to tell a story with this support (Becu 2006; Becu et al. 2006). In this case, interaction between stakeholders and the simulation model is still on an individual basis. The format of conveyed information is finally less formal, but the work of translation is less important. However, control of the process still remains largely in the hand of the modeller but to a lesser degree than in previous examples. This technique was further associated with semi-automatic ontology building procedures by Dray and colleagues in order to generate collective representations of water management in the atoll of Tarawa (Dray et al. 2006).

With inspiration coming similarly from the domain of ethnography, Bharwani and colleagues have developed the KNeTS method to elicit knowledge. Apart from a first stage with a focus group, this method is also based on individual interviews. As in Becu's work, interaction occurs in two phases: elicitation through questionnaires and involvement in the model design at the validation stage, which is also considered as a learning phase for stakeholders. These authors used an interactive decision tree to check with stakeholders whether the output of simulation would fit their points of view (Bharwani 2006). Control of this process is on the modeller's side. The stakeholders' interaction is marginally deeper in the model than in previous examples, since there is a direct interaction with the model as in management flight simulator. On the other hand, the ontology which is manipulated seems to be poorer, since the categories of choices open in the interaction are rather reduced. The format of information is open in the first phase and very structured in the decision tree in the second phase. The structuration process used in the modelling process occurs outside of the field of interaction with the stakeholders.

On its side, group decision support system design domain is based on a collective interaction with stakeholders as early as the design stage. These systems tend to be used to address higher-level stakeholders. In the method he developed, ACKA, Hamel organised a simulation exercise with the stakeholders of a poultry company. In this exercise, the participants were requested to play their own roles in the company. He constrained the exchanges taken place during the exercise through the use of an electronic communication medium so that he could analyse them and keep track of them later. All of the participants' communication was transformed into graphs and dynamic diagrams (Hamel and Pinson 2005). In this case, the format of information was quite structured.

12.3.3   From Software Engineering

Close to the artificial intelligence trend, working like Hamel and Pinson on the design of agent-based models, there is an emerging trend in computing science based on agent-based participatory simulations (Guyot and Honiden 2006) or participatory agent-based design (Ramanath and Gilbert 2004). This trend focuses on the development of computer tools, multi-agent systems, which originate from software engineering. Guyot proposes the implementation of hybrid agents, with agents in the software controlled by real agents, as avatars (Guyot 2006). These avatars help the players' understanding the system (Guyot and Honiden 2006). They can be thought as learning agents: they learn from choices of their associated player and are progressively designed (Rouchier 2003). The approaches working on hybrid agents implement a deep connection between participants and the social simulation model. Information conveyed in the interaction is relative to the model assumptions, as well as to the model content.

Ramanath and Gilbert have reviewed a number of software engineering techniques which may be coupled to participatory approaches (Ramanath and Gilbert 2004). This union between software design and participatory approaches is based on joint production not only between developers but also with end users. Not only interaction with stakeholders contributes to better software ergonomics—the computer-supported cooperative work (CSCW) workshop series being an example—but their participation tends to improve their acceptance and further appropriation of the mode.

The implementation of interactive techniques may take place at all stages of a software development process. In early stages, joint application design (Wood and Silver 1995) allows issues raised to be dealt with during the software development phase, attributing a champion to each issue. It is also concerned with technical issues. This protocol might involve other developers, as well as potential users. It may also increase the computing literacy of the participants involved in the process. This process is based on the implementation of rather well-framed workshops.

Joint application design is supported by using prototypes. It is here we find a link with a second technique: prototyping. This technique can be used all the way through a software development cycle. It is based around providing rough versions or parts of the targeted product. For example, it allows the pre-product to be criticised, respecified or the interface improved. Quite close to prototyping, in the final stages of the process, user panels can be used to involve end users in assessment of the product. These panels are based on a demonstration or a test of the targeted product.

In these cases, control of the process is dependent on the hiring of a skilful facilitator. Otherwise, control of the process may become rather implicit. The content of the interaction is rather technical, which makes it potentially unbalanced according to participants' literacy in computer science. An assessment of 37 joint application design experiments has shown that the participation of users during the process is actually rather poor, notably due to the technical nature of debates, which is hardly compatible with the time allocated to a joint application design process by users, compared to the time allocated by developers (Davidson 1999). Interaction is rather superficial and needs translation. However, identification of a champion of specific tasks gives a little bit more control to participants, as does involvement in the content of pieces of the tool being developed.

Besides these approaches originating from software engineering, people working in thematic fields such as the environmental sciences propose co-design workshops that focus on the development of simulation models. Such workshops are a type of focus group, organised around the identification of actors, resources, dynamics and interactions, suitable for a set of stakeholders to represent from a socioecological system on which they express their own point of view (Étienne 2006). This approach, which occurs at the design stage of the modelling process, is supposed to lead participants to design the simulation model by themselves, by formalising the conceptual model through a series of diagrams and a set of logical sentences. The final interaction diagram and the attached logical sentences are then translated by the modeller in computer code. It is in this type of process that a deep interaction can occur between participants and the model. This interaction conveys information on the model content, which is attached to the representations and knowledge of each participant.

12.3.4 From Statistical Modelling

Bayesian belief networks have been developed to include in the computation of probabilities, their dependence on the occurrence of any event. They can be useful to represent complex systems and increasingly used in participatory settings because their graphical nature facilitates discussion (Henriksen et al. 2004). A group of participants can be asked individually or collectively to generate relations between events and possibly probabilities as well. Henriksen and his colleagues propose a method in seven stages which alternates between individual and collective assessment and revision of an existent Bayesian belief network diagram.

This approach is reported to still present some difficulties in encouraging strong participant involvement due to the mathematical functions behind the network structure. However, other researchers and practitioners have improved their communication and facilitation of the technique with their own Bayesian belief network processes and are receiving positive stakeholder engagement in the modelling processes (Ticehurst et al. 2005). In the example of Henriksen and colleagues, the process is controlled by the modeller and includes only a rather superficial coupling between participants and the model. The translation of participant-provided information into probabilities is mediated by the modeller and is rather opaque, as in many participatory modelling approaches.

12.3.5 From the Social Sciences

The association of participatory approaches and social simulation modelling also originates from disciplines not focusing on the production of tools but on understanding social systems. Social psychology, economics, management and policy sciences have all developed their own interactive protocols to involve stakeholders in the design and/or use of their models. Sociology is still at the beginning of this process (Nancarrow 2005). These protocols propose a variety of structures of experimental settings, from laboratory to in vivo experiments through interactive platforms (Callon and Muniesa 2006). These three categories vary according to their openness to the influence given to participants. The in vivo category is beyond the scope of this paper since it does not involve modelling: the society in which the experiment is embedded provides its own model (Callon and Muniesa 2006).

Laboratory settings are very controlled experiments, involving human subjects. This is the case for most economic experiments. Participants are encouraged to behave with a given rationality through instructions and payments at the end of the session. In canonical experiments, analysis of the experiments is performed by the scientist. The focus of the analysis is to understand the individual and collective behavioural patterns generated by these settings. The purpose of these experiments is either (i) to test theories and models, (ii) to gain new knowledge on human behavioural patterns in given situations, or (iii) to test new institutional configurations (Friedman and Sunder 1994). These experiments are particularly efficient for situations with strong communication issues or with important interindividual interactivity (Ostrom et al. 1994). The issue of simulating a real situation is not considered but rather the testing of a theoretical model. This field is currently very active and evolves with the emergence of field experiments involving stakeholders concerned by the issues idealised in the model tested, asking them to play in their environment (Cardenas et al. 2000). With this configuration, interactions are rather deep since participants act as parts of the model. The participants convey action choices. However, the experimentalist strongly controls the process.

A platform is an intermediary setting more open to compromise and hybridisation than the laboratory. Heterogeneity of participants is also more welcome, since the setting is designed to enhance sharing interests. Through experimentation, a platform is supposed to bridge the gap between the world of the model and that of the stakeholders (Callon and Muniesa 2006). Policy exercises and role-playing games, as developed in the companion modelling approach, are kinds of these platforms (Richard and Barreteau 2006). Policy exercises embed stakeholders in potential situations they might have to face in the future (Toth 1988). They stem from war games that have been developed since the time of ancient China and are now used in public policy assessment (Duke and Geurts 2004) or environmental foresighting (Mermet 1993). They are actually quite similar to the business games and the system dynamics trend explained previously in Subsect. 12.2.1. However, the underlying social simulation model is rather implicit, though it exists to create the potential situation and to help identify the participants relevant to the exercise. Association with a computer tool tends to be with a simulation model of the environment that does not necessarily involve a social component. The interaction between participants and the social model is rather deep since they are pieces of the model and connect with the model of their environment. Control of the process is rather diffuse. There might be a genuine empowerment of participants since they have the possibility of bringing their own parts of the social model to the process and can adapt it in ways different to what the designers expected. Alike with laboratory settings, platforms provide information to the modeller about behavioural patterns of the participants. Reaction to taboos or innovative behaviours in situations new to the participants, tacit routines and collective behavioural patterns can be elicited using these platforms, while it is difficult with classical interviewing techniques.

Between experimental laboratory settings and policy exercises, the companion modelling approach proposes an association of role-playing games and agent-based simulations (Bousquet et al. 2002). Even though authors in this approach claim not to limit themselves to these two categories of tools, they predominately rest in the trend of participatory agent-based simulations and are thus close to the software design and artificial intelligence trends presented above. This approach makes a full use of similarities in architecture between role-playing games and agent-based simulations (Barreteau 2003). Both implement autonomous agents that interact within a shared dynamic environment. Joint use of both agent-based simulation and role-playing games builds upon these similarities to express the same conceptual model. Authors in this approach use this to reinforce a principle of making all the assumptions underlying a model used or design interactively with stakeholders explicit and understood. At the design stage, this approach aims to incorporate stakeholders' viewpoints in the model. At the model use stage, it aims to improve the appropriation of the tool produced as well as to increase its legitimacy for further operational use. However, this appropriation is still under discussion and might be rather heterogeneous (Barreteau et al. 2005).

12.4 Participation in the Modelling Process: Diversity of Phases and Intensity

While many authors claim to use participatory approaches for the simulation of social complexity, there remains a large diversity of actual involvement of stakeholders and of activities hidden behind this involvement. Associations of participatory methods with social simulation models are rather heterogeneous. It is thus important to qualify the actual involvement of stakeholders in these processes. This level of participation can range from mere information received by concerned parties related to the output of a process to the full involvement of a wide range of stakeholders at all stages of a process. There are also many intermediary situations imaginable. Participation should not be thought of as just talking, and diversity should be made explicit so that criticisms towards participation as a global category (Irvin and Stansbury 2004) can focus on specific implementations. This section explores the potential consequences of this diversity in three dimensions: stage in the modelling process, degree of involvement and heterogeneity of stakeholders involved.

12.4.1 Stages in the Modelling Process

The modelling process can be subdivided into the following stages, with the possibility of iterating along them:

- Preliminary synthesis/diagnosis (through previously available data). This includes making explicit the goal of the modelling process.
- Data collection (specific to the modelling purpose).

- Conceptual model design.
- Implementation.
- Calibration and verification.
- Simulation process (might be running a computer simulation model, playing a game session, etc.).
- Validation.
- Discussion of results.

Involvement of stakeholders in each of the different stages of the modelling process does not generate the same level of empowerment or learning, even if we assume that this involvement is sincere. Preliminary synthesis, conceptual model design, validation and, to some extent, discussion of results are framing stages; stakeholder involvement at these levels gives power to stakeholders to orientate the process. In the preliminary synthesis/diagnosis, stakeholders have the opportunity to play a part in setting the agenda. This is the stage of problem structuring which is identified as a key one in all participatory processes (Daniell et al. 2006). Even if the agenda developed with stakeholder involvement might further evolve, its initialisation generates a strong irreversibility in the process: data collection, participant's selection and (partially) modelling choices (architecture, platform) are related to this agenda and are costly, either directly or through the necessity of reprogramming. The modelling process is a sequential decision process, and as shown in theory of sequential decisions, initial decisions are often at the source of more consequences than envisaged (Henry 1974; Richard and Trometter 2001). Conceptual model design constitutes a landmark in the process. It is the crystallisation of viewpoints that serve as a reference in further stages. Validation is the compulsory stage where stakeholders will have the opportunity to check the effectiveness of the computer model in representing correctly their behaviours and ways of acting. Discussion of results may also constitute a framing phase, according to the purpose of the discussion. If dimensions of discussion are to be defined and model is open to be modified, there is some place for participants to (re-)orientate the modelling process. Otherwise, if the discussion of results aims to choose from a few scenarios, for example, the choice is very narrow and might be completely manipulated. In this regard, it has been shown that for any vote among composite baskets, it is possible to maintain that one item always selected according to the way the baskets are constituted (Marengo and Pasquali 2003). A scenario in this case is a kind of composite basket.

In other stages of a modelling process, the influence of stakeholder involvement on the overall process is less important. When data collection or calibration and verification involve participants, stakeholders tend to take the role of informants. Among the various levels proposed in the classical ladder of participation explained in the following subsection, these stages deal predominately with consultation. Their involvement is framed by the format of information which is expected and on the parts of the model which are to be calibrated or validated. If the process is open to modification in these frames, the level of participation might be higher but still with a limited scope.

Implementation stage is another mean to empower participants. It is often implicitly framing. But empowerment through involving stakeholders in this technical activity is rather to raise their literacy in this part and raise the probability of their appropriation of the model. Simulation stage is basically providing information to stakeholders on what is being done. This is a technical stage (running the simulation) which keeps a part of strategic choices (design of scenarios and indicators to track the simulation progress). Involvement of stakeholders in the technical part, such as through role-playing games, increases their knowledge of the model from inside, provided stakeholders have the literacy for that. Involvement in strategic part is connected to the initial stage which has set the agenda. The further this initialisation has gone in formalising the questions, the less empowering is this involvement.

12.4.2 Level of Involvement

Level of involvement is a more classical dimension. It is inspired by the classical hierarchy of participation levels proposed first by Arnstein (1969). Several reviews and adaptations have been made since then, with the same focus on power issues (Mostert 2006; van Asselt et al. 2001). These works focus on what participation means in decision-making terms (the bases of many political or democratic theories), with democracy cube (Fung 2006) or the work of Pateman (1990) and Rocha (1997). In most of these examples, the emphasis is placed on who (citizens, managers or policymakers) has the balance of power for final decision-making (i.e. the choice phase of a decision process (Simon 1977)), but other issues of process are not specifically mentioned. Such participation classifications, although useful in a very general sense for the question of participation in modelling processes, do not explicitly treat the issue of the place of a modeller or researchers with expert knowledge (Daniell et al. 2006).

On these bases, we consider here the five following levels in which there are at least some interactions between a group of citizens and a group of decision makers:

- Information supply: citizens are provided access to information. This is not genuine participation since it is a one-way interaction.
- Consultation: solicitation of citizens' views.
- Co-thinking: real discussions between both groups.
- Co-design: citizens have an active contribution in policy design.
- Co-decision-making: decisions are taken jointly by members of both groups.

Since a modelling process is a kind of decision process, this hierarchy might apply to modelling process as well. This is a little bit more complicated because two processes are behind the modelling process, and the network of interactions cannot be represented with a group of citizens and a group of decision or policymakers only.

A modelling process with the purpose of simulation has two dimensions along which these scales might be assessed: model content and building on one hand and control over model use on the other. Though these two dimensions are related, it is useful to consider them separately as they provide power and knowledge: either within the process or in the system in which the process takes place. Each of these dimensions is more closely related to specific stages in the modelling process presented in the previous subsection. However, some stages, such as model design or implementation, contribute to both dimensions.

Therefore we consider the following categories:

- Information on a model's content and no control over model use
- Consultation and no control over model use
- Dialogue with modellers and no control over model use
- Dialogue with modellers and control over model use
- Co-building of a model and no control over model use
- Co-building of a model and control over model use

Each category is described in the following subsection by a flow of interactions within an interaction network based on four poles: **A**, **R**, **M** and **P**. **A** stands for all people who are involved in and/or concerned by the social complexity at stake in the modelling process. This includes policymakers and citizens. **R** stands for researchers involved in the modelling process. **M** stands for the model. **P** stands for policymakers. **P** is a subset of **A**, which gathers the actors who might use the model and its output for the design of new regulations or policies concerning the system as a whole. We chose to gather citizens and policymakers in **A**, as in the modelling process they are rather equivalent in their interactions with the researchers about the model. Their distinction is useful for the second dimension: model use and dissemination. We assume that the default situation is an access of **P** members to the output of the modelling process.

12.4.2.1 Information and No Control

Participants are informed about the model's content and the simulation by researchers, who are the only designers. No control over the model's use or dissemination is deputed to participants as such. Whatever the use of the model may be afterwards, citizens become only better aware of the basis on which this model has been built. However, the model exists and can be used by members of **P**. This is the classical situation with simulation demonstration and explanation of a model's assumptions. This explanation might be achieved by more active means, such as a role-playing game. A switch to the following category occurs when this explanation leads to a debate that makes the model open to modifications. Otherwise, it remains mere information.

12.4.2.2 Consultation and No Control

Participants are consulted about the model's content and its simulation that is by the researchers, who are the only designers. They provide information and solicit comments on the model. Mere data collection through a survey does not fall in this category because it assumes active involvement from participants in providing information to the modellers. Some knowledge elicitation techniques, such as BBN design, tend to fall mostly in this category. Translation of the inputs originating from participants into pieces of a model is performed only by researchers. This translation is not necessary transparent. No control over use or dissemination of the model is deputed to participants as such. Compared to previous category, participants have the ability to frame marginally more of what is performed by the model through their inputs to the model's content However, the extent of this ability depends on the participants' skills to identify potential uses of a model. As in any participatory process, when there is an unbalanced power relation between parties, the process is also a way for policymakers to gain information from stakeholders, information that could be used for strategic purposes. This bias can be alleviated if the involvement of **A** includes all members of **A**, including the subset **P**. The constructed model in this case may be used by the members of **P**.

12.4.2.3 Dialogue with Modellers and No Control

In this category, iterative and genuinely interactive processes between stake-holders and modellers start to appear. There is still a translation of inputs from participants into the model through the researchers, but there is feedback about these developments to the stakeholders. This leads to discussion about the model. Convergence of the discussion remains on the researchers' side. Group model building experiments predominately fall into this category. In this case, stakeholders may increase their influence on the framing of the model with better prior assessment of the scope of simulations to be examined. Biases related to strategic information being revealed in the dialogue process are still present if there is unbalanced involvement of members of **A** and notably if members of **P** are less active but still present. However, this category still represents indirect control, and no specification of model use is left open to the stakeholders. At the end of the process, the created model can be used by members of **P** without any control or any road map set by other members of **A**.

12.4.2.4 Dialogue with Modellers and Control

This category is the same as the previous one with translation of stakeholders' inputs and feedback from the researchers about them. However, the output of the discussion, the model, is appropriated by stakeholders. They have control over its use and dissemination of models which may have been produced through the modelling process: who might use them, with which protocol and what is the value of their outputs. They can decide whether the model and simulations are legitimate to be used for the design of policies that may concern them. However, this appropriation raises issues of dialogue between researchers and stakeholders about the suitability of model for various uses. Comparison of several participatory agent-based simulations has shown that there is a need for dialogue about not only a model's content but also on its domain of validity (Barreteau et al. 2005).

12.4.2.5 Co-building of a Model and No Control

A further stage of empowerment of stakeholders through participation in a modelling process is their co-building of the model. The design and/or implementation of such a model are joint activities between the researchers and stakeholders. Co-design workshops or joint application development falls into this category, provided that there is genuinely no translation of stakeholders' inputs by the researchers. Techniques originating from artificial intelligence and knowledge engineering, as presented above, aim to reach this level, either through the implementation of virtual agents extending stakeholders or through constraining the interactions between actors through a computer network. This involvement increases the fidelity of the model to match stakeholders' viewpoints and behavioural patterns. However, at the end of the process, the created model can still be used by members of **P** without any control or any road map set by other members of **A**.

12.4.2.6   Co-building of a Model and Control

This category is the same as the previous one, but actors now have control over use and dissemination of models which may be produced through the process. This leads to possible stakeholder appropriation of the models, raising the same issues as in Sect. 12.4.2.4.

12.4.3   Heterogeneity of Actors

Eversole points out the need for participatory processes to take into account the complexity of the society involved including power relations, institutions and the diversity of viewpoints (Eversole 2003). This is all the more true when applied to the participatory process of social simulation modelling. Most settings presented in Sect. 12.2 have a limited capacity to involve a large numbers of people in interactions with a given version of a model. When interactions convey viewpoints or behavioural patterns, heterogeneity may not appear if no attention is paid to it. Due to limits in terms of number of participants, participatory approaches that deal with social simulation modelling involve usually representatives or spokespeople. The issue of their statistical representativeness is left aside here, as the aim is to comprehend the diversity of possible viewpoints and behavioural patterns. There is still an issue of their representativeness through their legitimacy to speak for the group they represent, as well as their competency to do so. The feedback of these spokespersons to their group should also be questioned. When issues of

Table 12.1 Categories of participation according to level of heterogeneity embraced (from van Daalen and Bots 2006)

Level	Model construction	Model use	
		Computer model	Gaming simulation
1 Individual stakeholders	Knowledge elicitation involving one or more individuals separately, depending on the modelling method this may consist of interviews about (perceptions on) a system or questionnaires related to the aspects being modelled (e.g. Molin 2005)	Model can be executed, and individual stakeholders are informed of the result (e.g. Dudley 2003)	Individual can 'play' an actor in a flight simulator setting (e.g. Maier and Grössler 2000; Sterman 1992)
2 Homogenous group	Same as 1, but group model building includes interaction between stakeholders (e.g. Castella et al. 2005)	The use of a model in a homogenous group means that the model can be run in a workshop setting and model results are discussed (e.g. van Daalen et al. 1998)	Multiplayer gaming simulation can be conducted; the game is followed by a debriefing (e.g. Mayer et al. 2005)
3 Heterogeneous group	Same as 2, but group model building interaction between stakeholders with different perceptions/beliefs (e.g. Van den Belt 2004)	Same as 2, but results discussed with stakeholders with different perceptions/beliefs	Same as 2, but full stakeholder group involved (e.g. Étienne et al. 2003)

empowerment are brought to the fore, the potential for framing or controlling the process is dedicated to the participants. This might induce echoes in power relations within the group, notably due to training that may be induced.

Van Daalen and Bots have proposed a categorisation of participatory modelling according to this dimension with three scales: individual involvement, a group considered as homogeneous and a heterogeneous group (van Daalen and Bots 2006). Table 12.1 provides examples of each level according to the two processes involved that were explained in previous subsection.

These three categories are represented in the diagrams below, as expansions of the relation between A and (M ∪ R) in the previous subsection. The third category corresponds to the deep connection mentioned in the first section (Figs. 12.1, 12.2 and 12.3).

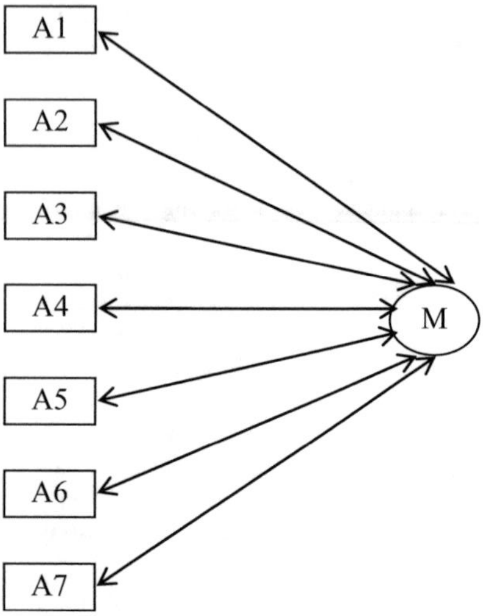

Fig. 12.1 Individual
involvement

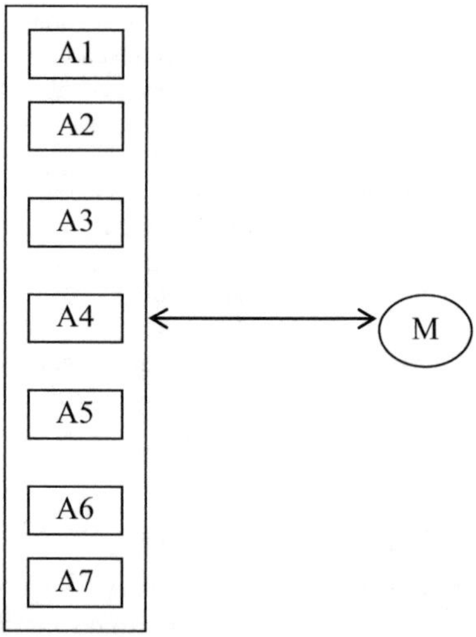

Fig. 12.2 Homogenous
group involvement

Some other ways are currently explored with hybrid agents to technically overcome the difficulty of dealing with representatives: by involving them all in large systems. The internet or mobile phone networks provide the technical substrate for such interaction. A large number of participants have a virtual component in a large system, interacting with other components, possibly with the purpose of building a model (Klopfer et al. 2004). However, in this case it is rather an individual interaction of these participants with the system, than genuine interactions among the participants.

Fig. 12.3 Heterogenous
group involvement

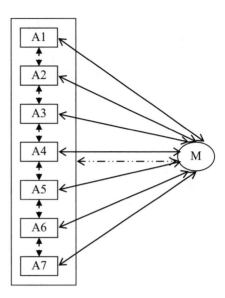

12.4.4 Which Configurations Can Meet the Expectations of the First Section?

In this subsection we revisit the expectations towards the joint use of participatory approaches and social simulation presented in the first section, through the categorisations above. This is a tentative mapping of participatory approach categorisation with model expectations. Table 12.2 below synthesises this mapping.

The two expectations dealing with increasing a model's quality often actually use participants as (sometimes cheap) resources in the simulation modelling process. The most important stage is simulation, because participants are supposed to bring missing information to the simulation, as well as the missing complexity. The minimum level of empowerment is rather low. These processes are hardly participatory in that sense, because participants are not supposed to benefit from the process, except a potential payment. A higher level of empowerment might increase the quality of participants' involvement in the process through a deeper concern in the outcome of the simulation. Finally, the heterogeneous group level is obviously to be respected because it can instil a deep connection between stakeholders and the model still and concurrently profit from their interactions with each other.

To make simulation models match their intended use, the key stage is the design process. Stakeholders are supposed to aid the building of an appropriate model. The main difference between targets of simulation model's use is in the necessity to give control over the process to stakeholders in case of policymaking. New knowledge is of individual benefit to all participants, and the emergence of fruitful interactions can also become an individual benefit. There are few direct consequences of this new knowledge. Therefore, control over the process in this

Table 12.2 Matching expectations on joint use of participatory approaches and social simulation modelling with categories of participation

Expectation	Key stage(s) for participation	Minimum level of empowerment	Level of heterogeneity
Increase model's quality with social diversity and capacity to evolve	Simulation	Information and no control	Heterogeneous group
Increase model's quality through distribution of control	Simulation	Information and no control	Heterogeneous group
Improve suitability of simulation model's use for increasing knowledge	Design	Dialogue and no control	Individual
Improve suitability of simulation model's use for policymaking	Design and discussion of results	Dialogue and control	Homogeneous group
Simulation as a means to support participation to deal with dynamics and uncertainties	Discussion of results	Consultation and no control (depend on participatory process to be supported)	Homogeneous group
Simulation as a means to support participation through social learning	Preliminary diagnosis, design and discussion of results	Co-building and control (to be preferred)	Heterogeneous group

case is useless, and involvement might be individual, as with knowledge elicitation techniques. However, higher level of stakeholder heterogeneity might raise the knowledge acquired in the process.

When simulation is used to support participation, discussion of results is a key stage. Previous stages aid in the problem framing and literacy increase of participants that allow them to reach more solid interpretations. The empowerment level is rather dependent on the participatory process that is being supported. However, consultation in the modelling process should be a minimum requirement so that uncertainties and dynamics tackled by the simulations are relevant to the stakeholders. When focusing on social learning, co-building and control should be preferred because this category increases the potential for exploration and creativity. However, some social learning might take place in lesser levels, provided that group heterogeneity is well encouraged in the process.

12.5 Combining Approaches and Techniques at Work

We present in this section two case studies implementing various methods for joining social simulation modelling and participatory approaches. The first deals with fire hazard prevention in southern France and the second one with groundwater management in the atoll of Kiribati.

12.5.1 The Fire Hazard Case Study

In December 2005, the Forest Service of the Gard Department of Agriculture (DDAF) decided to start a fire prevention campaign focussed on fire hazard at the interface between urban and forest areas. Interested in the participatory approaches developed by INRA researchers on fire prevention and forest management planning (Étienne et al. 2003), they ask for an adaptation of the SylvoPast model to the peri-urban context in order to make local politicians aware of the increasing fire hazard problem. The district of Nîmes city (NM) who was already interested in the use of role-playing games for empowering stakeholders and decision makers asked the Ecodevelopment Unit of the INRA of Avignon to develop a companion modelling approach based on social simulations and a participatory involvement of all the mayors of the district.

The modelling process was subdivided into seven stages:

1. Collection and connection on a GIS of relevant cartographic data on forests, land-use and urbanization and individual interviews with local extensionists on farmers, foresters and property developers practices.
2. Co-construction with DDAF and NM of a virtual but implicit map representing three villages typical from the northern area of Nîmes city and validation of the map (shape, land-use attributes and scale) by a group of experts (EX) covering the main activities of the territory (agriculture and livestock extensionists, forest managers, hunting manager, land tenure regulator, fire brigade captain and town planner).
3. Co-construction with NM, DDAF and experts of a conceptual model accounting for the current functioning of the territory and the probable dynamic trends to occur during the next 15 years. This participatory process followed the ARDI methodology mentioned in Sect. 12.2.3 (Étienne 2006).
4. Implementation of the NimetFeu model on Cormas multi-agent platform by INRA researchers and validation of the model by simulating with the co-construction group, the current situation and its consequences on fire hazard and landscape dynamics for the 15 following years.
5. Co-construction and test of a role-playing game (RPG) using the NimetFeu model as a way to simulate automatically natural processes and some social decisions (vineyard abandonment, horse herding, firefighting). The other social decisions were programmed to be taken directly by the players and used as an input to the model.

6. The use of the RPG during several sessions gathering 6 players (3 mayors, 1 developer, 1 NM representative, 1 DDAF technician) until the 14 villages involved in the project did participate to a session.
7. Adaptation of NimetPasLeFeu to other ecological conditions and decision of the Gard Department to become autonomous in running RPG sessions. A facilitator and a data manager were trained and tested during sessions organised in the framework of an INTERREG project with mayors and fire prevention experts from France, Spain, Italy and Portugal.

The approach is based on a mutual comprehension of the elements of the territory that makes sense with the question asked. This sharing of representations is done by means of a series of collective workshops during which actors, resources, dynamics and interactions (ARDI) which make the stakes of the territory are identified and elicited. To facilitate this sharing, the answers to the questions are formalized into easily comprehensible diagrams, with a minimum of coding making it possible to classify the provided information. The role of the facilitator only consists in calling upon each participant, writing down the proposals in a standard way and asking for reformulating when the proposal is too generic, enounced with a polysemous word or can lend to confusion.

In both models, the environment is divided into three neighbouring villages covering the gradient of urbanization and agricultural land/woodland ratio currently observed around Nîmes city. It is visualised by means of a cellular automaton through a spatial grid representing 18 land-use types that can change according to natural transitions or human activities.

Four categories of social entities are identified: property developers, mayors, farmers and fire prevention managers. The developers propose new urban developments according to social demand and land prices. They have to respect the government regulations (flood hazard, protected areas, urban zoning). Mayors select an urbanization strategy (to densify, to develop on fallow land, olive groves or forests), update their urban zoning according to urban land availability and social demand and make agreements with the developers. When updating the urban zoning, they can create new roads. Farmers crop their fields using or not current practices that impact fire hazard (vineyards weeding, stubble ploughing) or adapt to the economic crisis of certain commodities by uprooting and setting aside lowland vineyards or olive groves near to urban zones. The fire prevention manager establishes a fuel break in a strategic place, selected according to fire hazard ranges in the forest and the possible connections with croplands, as well as available funds and forest cleaning costs.

Four biophysical models issued from previous researches and adjusted to the local conditions are integrated to the MAS to account for fallow development, shrub encroachment, pine overspreading and fire propagation. The model is run at a one-year time step, the state represented on the map corresponding to the land cover at the end of June (beginning of the wildfire period). Each participant was invited to propose a set of key indicators that permit them to monitor key changes on ecological or socio-economic aspects. A common agreement was made on what to measure, on which entity, with which unit and on the way to represent the corresponding qualitative or quantitative value (visualising probes on graphs or viewpoints on maps). They were encouraged to elaborate simple legends, in order to be able to share their point of view with the other participants while running the model.

The first MAS was exclusively used to support the collective thinking on which procedures and agents will be affected to players and which ones would be automatically simulated by the computer. In the RPG model, the playing board was strongly simplified with only four types of land cover. Running the game gives participants the opportunity to play individually or collectively by turns, according to a precisely defined sequence. While the mayors players draw the limits of the urban zone and rank the price of constructible land according to its current land use, the developer player sorts randomly a development demand and elaborates a strategy (village, density, livelihood). Then begin a series of negotiations between the developer and the three mayors in order to decide where to build, at which density and with which type of fire prevention equipment. All the players' decisions are input into the computer, and landscape dynamics are simulated by running the model. Players get different types of output from the simulation run: budget updating, new land-use mapping, popularity scoring. Each round corresponds to a 3-year lapse and is repeated 3–4 times according to players' availability.

A specific effort is made in the RPG design to account for physical remoteness and territory identity among participant: the playing room is set up into three neighbouring but distinct boxes for the three mayors (each box represents one village), one isolated small table for the developer, and another game place with two tables, one small for the DDAF and a huge one for NM. Lastly, in a corner, the computer stuff is placed with an interactive board than can both be used as a screen to project different viewpoints on the map or as an interactive town plan to identify the parcels' number.

At the end of the game, all the participants are gathered in the computer room and discuss collectively, with the support of fast replays of the game played. Different topics are tackled related to ecological processes (effect of fire, main dynamics observed), attitudes (main concerns, changes in practices) and social behaviours (negotiations, alliances, strategies).

Along these various stages, this experiment features a diversity of involvement as well as of structure of interactions. This is synthesised in the Table 12.3 below.

12.5.2   The AtollGame Experiment

This study is carried out in the Republic of Kiribati, on the low-lying atoll of Tarawa. The water resources are predominantly located in freshwater lenses on the largest islands of the atoll. South Tarawa is the capital and main population centre of the Republic. The water supply for the urban area of South Tarawa is pumped from horizontal infiltration galleries in groundwater protection zones. These currently supply about 60% of the needs of South Tarawa's communities. The government's declaration of water reserves over privately owned land has led to conflicts, illegal settlements and vandalism of public assets (Perez et al. 2004).

Table 12.3 Classification of type of participation in various stages of the NimetPasleFeu experiment

	Involvement	Heterogeneity	nb
Preliminary diagnosis	Consultation	Individuals	10
Data collecting	Consultation	Individuals	3
Conceptual model designing	Co-design	Heterogeneous group	14
Implementing	Information	Individuals	2
Calibrating and validating	Co-thinking	Heterogeneous group	14
Role-playing game designing	Co-design	Heterogeneous group	14
RPG playing and debriefing	Co-decision-making	Heterogeneous group	30
Getting self sufficient	Information	Individuals	3

The AtollGame experiment aims at providing the relevant information to the local actors, including institutional and local community representatives, in order to facilitate dialogue and to help devise together sustainable and equitable water management practices. Knowledge elicitation techniques as well as multi-agent-based simulations (MABS) coupled with a role-playing game have been implemented to fulfil this aim. In order to collect, understand and merge viewpoints coming from different stakeholders, the following 5-stage methodology is applied: (1) collecting local and expert knowledge, (2) blending the different viewpoints into a game-based model, (3) playing the game with the different stakeholders, (4) formalising the different scenarios investigated in computer simulations and (5) exploring the simulated outcomes with the different stakeholders (Dray et al. 2006).

Initial knowledge elicitation (stages 1 and 2) relies on three successive methods. First, a Global Targeted Appraisal focuses on social group leaders in order to collect different standpoints and their articulated mental models. These collective models are partly validated through Individual Activities Surveys focusing on behavioural patterns of individual islanders. Then, these individual representations are merged into one collective model using qualitative analysis techniques. This conceptual model is further simplified in order to create a computer-assisted role-playing game (AtollGame). The range of contrasted viewpoints confirms the need for an effective consultation and engagement of the local population in the design of future water management schemes in order to warrant the long-term sustainability of the system. Clear evidence of the inherent duality between land and water use rights on the one hand and between water exploitation and distribution on the other hand provides essential features to frame the computer-assisted role-playing game.

The assistance of a computer is needed as far as interactions between groundwater dynamics and surface water balance involve complex spatial and time-dependent interactions (Perez et al. 2003). The use of agent-based modelling (ABM) enables us to take full advantage of the structure of the conceptual model. We developed the AtollGame simulator with the CORMAS© platform (Bousquet et al. 1998).[1]

---

[1] More details about the AtollGame can be found online at http://cormas.cirad.fr/en/applica/atollGame.htm.

A board game version reproduces the main features of the AtollGame simulator (Dray et al. 2006). Sixteen players—eight on each island—are able to interact according to a set of predefined rules. Their choices and actions are directly incorporated into the simulator at the end of each round of the game. During the game, players can ask for more information from the simulator or discuss the results provided by the simulator (salinity index, global demand). Landowners, traditional or new buyers, are the essential actors in the negotiations with the government. The connection between land tenure issues and water management is essential. It drives the land-use restrictions and land lease discussions. The population increase, mainly through immigration, is perceived as a threat in terms of water consumption, pollution generation and pressure on the land. Financial issues linked with water management usually deal with land leases, equipment investment and, seldom, water pricing. Hence, the model features:

- Agents/players becoming a local landowners
- Land and water allocation conflicting rules and various sources of incomes
- An increasing number of new settlers on agents/players' land

The individual objective of the players is to minimise the number of angry or sick people in their house. People may become *angry* because they didn't have enough water to drink during the round. People may become *sick* if they drank unhealthy (polluted or salty) water during the round. *Pollution* depends on the number of people living on the island and contaminating the freshwater lens. *Salty water* depends on the recharge rate of the fresh water lens and the location of the people on the island.

At first, representatives from the different islands displayed different viewpoints about the water reserves. Hence, the group meetings organised in the villages prior the workshop allowed for a really open debate. On the institutional side, the position of the different officers attending the workshop demonstrated a clear commitment to the project. All the participants showed the same level of motivation either to express their views on the issue or to genuinely try to understand other viewpoints. Participants also accepted to follow the rules proposed by the project team, especially the necessity to look at the problem from a broader perspective. During the first rounds, the players quickly handled the game and entered into interpersonal discussions and comparisons. The atmosphere was good, and the game seemed playful enough to maintain the participants' interest alive. The second day, the introduction of a water management agency and the selection of its (virtual) director created a little tension among the participants. But, after a while, the players accepted the new situation as a gaming scenario and started to interact with the newly created institution. At this stage, players started to mix arguments based on the game with other ones coming directly from the reality. On Island 1, players entered direct negotiations with the (virtual) director of the water management agency. On Island 2, discussions opposed players willing or not to pay the fee.

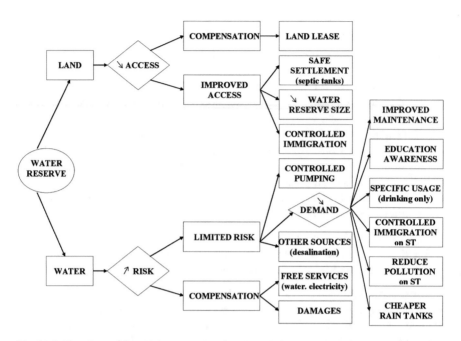

Fig. 12.4 Flowchart of financial, technical and social solutions agreed on by the participants of the AtollGame experiment

Finally, the project team introduced the fact that the water management agency was no longer able to maintain the reticulated system due to a poor recovery of the service fees. It had for immediate consequence a sharp decrease of the water quantity offered on Island 2.

Then, players from both tables were asked to list solutions to improve the situation on their island. When the two lists were completed, the project team and the participants built a flowchart of financial, technical and social solutions, taking into account issues from both islands (Fig. 12.4).

A collective analysis of the flowchart concluded that the actual situation was largely unsustainable either from a financial or social viewpoint. The flowchart above provides a set of interdependent solutions that should be explored in order to gradually address the present situation.

12.6 Discussion: Relations Between Participants and Models

The diverse categories of joint implementation of participatory approaches and social simulation modelling feature a diversity of relations between a set of people, participants and a model.

Classical social simulation models do not feature any participant. People are represented in the model, sometimes from assumed or theoretical behavioural patterns. This entails exploring potential emergent phenomena from interactions among these behavioural patterns. Some participatory approaches involve only an implicit social model. Within this scope, there is a large diversity of relations. This diversity is based on the role undertaken by stakeholders, their actual involvement and issues tackled by the model.

In all the processes allying social simulation models and participation, stakeholders take on various roles: pieces in simulation, interfaces for coupling various sources of knowledge, beneficiaries of the process, key informants... As pointed out by Ryan, managers are overwhelmed by the complexity to be managed. Participation is a way to share this burden (Ryan 2000). Stakeholders provide the missing interactions and add missing pieces of knowledge, such as tacit knowledge (Johannessen et al. 2001). If involvement of stakeholders is useful for principal agents such as managers, we propose it as a rule that they should gain some empowerment in the process.

Stakeholders can be key pieces of the modelling process itself as well. In the simulation they are an alternative to computer code to provide the engine (Hanneman 1995). They provide an answer to issues of coupling several viewpoints (Robinson 1991).

However actual involvement of people in a participatory modelling process might largely differ from formal involvement planned. Leaving aside cases of manipulation and announce effect, people have also to find their place in the participatory process. Suitability of participatory approaches in a specific society has to be taken into account: context (including social) is a key driver for success in stakeholder involvement (Kujala 2003), and practice of interactive policymaking processes depends on local culture (Driessen et al. 2001). Representation mechanisms have already been pointed out as a major factor. It has to be tuned to this local social and cultural context. At a finer grain, facilitator has a key role to lead people towards the level of involvement they are invited (Akkermans and Vennix 1997).

12.7 Conclusion

This chapter provides a review of the diversity of association of participatory approaches and social simulation, for their mutual benefit. This diversity of approaches allows tackling expectations about increasing model's quality, model's suitability to its intended use and improving participation. Their diversity is built upon ingredients coming from various disciplines from social sciences to computer sciences and management. It is expressed according to the implementation of interactions between the participants and the simulation model, the control of the process and the format of information. This leads to expand the classical ladder of participation towards categorization according to the stage in the modelling process when participation takes place and the structure of the interaction to cope with the heterogeneity of stakeholders.

This diversity requires a cautious description of each implementation in situation, so that any evaluation is specific to the implementation of a given association in its context. Generalisation can then be done only on the relation of this practice of participatory simulation and its suitability to its context and purpose. Efficiency to induce changes in practice or knowledge depends on the respect of a triple contingency of collective decision processes: time, people and means (Miettinen and Virkkunen 2005). This means to respect and take into account the own dynamics within the social system at stake, to allow the participation of people with their whole essence (including tacit knowledge, networks, relations to the world) and to be adaptive to means and competences present within the system (Barreteau 2007). Another dimension of evaluation should be democracy, since it is often put to the front. This raises the issue of the existence of a control of the process. Does it rely only on modellers or is it more shared? Finally there is a necessity of being more explicit on the kind of PA which is used because of the potential deconsideration of the whole family if expectations are deceived.

Further Reading

Participatory modelling is increasingly present in special sessions of conferences or special features of scientific journals. A first source of further readings consists in case studies. Among others, Environmental Modelling & Software had a special issue on modelling with stakeholders (Bousquet and Voinov 2010), where readers will find a whole set of well-described case studies using various methods. The biennial international environmental modelling and software conferences have also specific tracks for participatory modelling; proceedings are available online (see http://www.iemss.org/society/ under publications). For specific tools, refer to the papers of a symposium on simulation and gaming in natural resource management, published as a special issue of Simulation & Gaming (volume 38, issues 2 & 3). The introductory paper giving an overview is Barreteau et al. (2007).

Reflexivity is crucial for practitioners of participatory processes, as part of the need for more cautious evaluation of participatory processes as pointed out by Rowe and Frewer (2004). Another direction for reading consists in methods for evaluation and assessment of stakeholder involvement in modelling processes. Etienne edited a whole book aiming at assessing consequences of a specific approach, so-called companion modelling (Étienne 2011).

Readers who are more interested in stakeholder involvement in modelling at a more technical level should go for the review paper of Ramanath and Gilbert (2004) which provides a nice overview of this point of view.

References

Abel, N., Ross, H., & Walker, P. (1998). Mental models in rangeland research, communication and management. *The Rangeland Journal, 20*, 77–91.

Akkermans, H. A. (1995). Developing a logistics strategy through participative business modelling. *International Journal of Operations & Production Management, 15*, 100–112.

Akkermans, H. A., & Vennix, J. A. M. (1997). Clients' opinions on group model building: An exploratory study. *System Dynamics Review, 13*, 3–31.

Arnstein, S. (1969). A ladder of citizen participation. *Journal of the American Planning Association, 35*, 216–224.

Barreteau, O. (2003). The joint use of role-playing games and models regarding negotiation processes: Characterization of associations. *Journal of Artificial Societies and Social Simulations, 6*(2). http://jasss.soc.surrey.ac.uk/6/2/3.html

Barreteau, O. (2007). *Modèles et processus de décision collective: entre compréhension et facilitation de la gestion concertée de la ressource en eau.* HDR thesis, Paris Dauphine University, Paris.

Barreteau, O., & Bousquet, F. (1999). Jeux de rôles et validation de systèmes multi-agents. In M.-P. Gleizes & P. Marcenac (Eds.), *Ingénierie des systèmes multi-agents, actes des 7èmes JFIADSMA* (pp. 67–80). Paris: Hermès.

Barreteau, O., Bousquet, F., & Attonaty, J.-M. (2001). Role-playing games for opening the black box of multi-agent systems: Method and teachings of its application to Senegal River Valley irrigated systems. *Journal of Artificial Societies and Social Simulations, 4*(2). http://jasss.soc.surrey.ac.uk/4/2/5.html

Barreteau, O., Hare, M., Krywkow, J., & Boutet, A. (2005). Model designed through participatory processes: whose model is it? In N. Ferrand, P. Perez, & D. Batten (Eds.), *Joint conference on multiagent modelling for environmental management, CABM-HEMA-SMAGET 2005*, Bourg St Maurice – Les Arcs, France, 21–25 March, 2005.

Barreteau, O., Le Page, C., & Perez, P. (2007). Contribution of simulation and gaming to natural resource management issues: An introduction. *Simulation & Gaming, 38*(2), 185–194.

Becu, N. (2006). *Identification et modélisation des représentations des acteurs locaux pour la gestion des bassins versants.* PhD thesis, Sciences de l'eau, Université Montpellier 2, Montpellier, France.

Becu, N., Barreteau, O., Perez, P., Saising, J., & Sungted, S. (2006). A methodology for identifying and formalizing farmers' representations of watershed management: A case study from Northern Thailand. In F. Bousquet, G. Trebuil, & B. Hardy (Eds.), *Companion modeling and multi-agent systems for integrated natural resource management in Asia* (pp. 41–62). Los Baños: IRRI.

Bharwani, S. (2006). Understanding complex behavior and decision making using ethnographic knowledge elicitation tools (KnETs). *Social Science Computer Review, 24*, 78–105.

Bousquet, F., Bakam, I., Proton, H., & Le Page, C. (1998). Cormas: common-pool resources and multi-agent systems. In A. Pasqual del Pobil, J. Mira, & M. Ali (Eds.), *Tasks and methods in applied artificial intelligence. IEA/AIE 1998, Lecture Notes in Computer Science (Lecture Notes in Artificial Intelligence)* (Vol. 1416, pp. 826–837). Berlin, Heidelberg: Springer.

Bousquet, F., Barreteau, O., d'Aquino, P., Etienne, M., Boissau, S., Aubert, S., et al. (2002). Multi-agent systems and role games: An approach for ecosystem co-management. In M. Janssen (Ed.), *Complexity and ecosystem management: The theory and practice of multi-agent approaches* (pp. 248–285). Northampton, MA: Edward Elgar Publishing.

Bousquet, F., Barreteau, O., Le Page, C., Mullon, C., & Weber, J. (1999). An environmental modelling approach: The use of multi-agent simulations. In F. Blasco & A. Weill (Eds.), *Advances in environmental and ecological modelling* (pp. 113–122). Amsterdam: Elsevier.

Bousquet, F., & Voinov, A. (Eds.). (2010). Thematic issue - Modelling with stakeholders. *Environmental Modelling & Software, 25*(11), 1267–1488.

Callon, M., & Muniesa, F. (2006). Economic experiments and the construction of markets. In D. MacKenzie, F. Muniesa, & L. Siu (Eds.), *Do economists make markets? On the performativity of economics* (pp. 163–189). Princeton, NJ: Princeton University Press.

Cardenas, J.-C., Stranlund, J., & Willis, C. (2000). Local environmental control and institutional crowding-out. *World Development, 28*, 1719–1733.

Castella, J. C., Tran Ngoc, T., & Boissau, S. (2005). Participatory simulation of land-use changes in the northern mountains of Vietnam: The combined use of an agent-based model, a role-playing game, and a geographic information system. *Ecology and Society, 10*(1), 27. http://www.ecologyandsociety.org/vol10/iss1/art27/

Cockes, D., & Ive, J. (1996). Mediation support for forest land allocation: The SIRO-MED system. *Environmental Management, 20*(1), 41–52.

D'Aquino, P., Le Page, C., Bousquet, F., & Bah, A. (2003). Using self-designed role-playing games and a multi-agent system to empower a local decision-making process for land use management: The SelfCormas Experiment in Senegal. *Journal of Artificial Societies and Social Simulations, 6*(3). http://jasss.soc.surrey.ac.uk/6/3/5.html

Daniell, K. A., Ferrand, N., & Tsoukias, A. (2006). Investigating participatory modelling processes for group decision aiding in water planning and management. In S. Seifert & C. Weinhardt (Eds.),*Proceedings of the international conference Group decision and negotiation (GDN) 2006*, Karlsruhe, Germany, June 25–28 2006 (pp. 207–210). Karlsruhe: Universitätsverlag Karlsruhe.

Davidson, E. J. (1999). Joint application design (JAD) in practice. *Journal of Systems and Software, 45*, 215–223.

DeSanctis, G., & Gallupe, R. B. (1987). A foundation for the study of group decision support systems. *Management Science, 33*, 589–609.

Dewulf, A., Bouwen, R., & Tailleu, T. (2006). The multi-actor simulation 'Podocarpus National Park' as a tool for teaching and researching issue framing. In *Proceedings of IACM 2006 Montreal meetings*. http://ssrn.com/abstract=915943

Dray, A., Perez, P., Jones, N., Le Page, C., D'Aquino, P., White, I., et al. (2006). The AtollGame experience: From knowledge engineering to a computer-assisted role playing game. *Journal of Artificial Societies and Social Simulations, 9*(1). http://jasss.soc.surrey.ac.uk/9/1/6.html

Dray, A., Perez, P., Le Page, C., D'Aquino, P., & White, I. (2006). AtollGame: A companion modelling experience in the Pacific. In P. Perez & D. Batten (Eds.), *Complex science for a complex world: Exploring human ecosystems with agents* (pp. 255–280). Canberra: ANU E Press.

Driessen, P. P. J., Glasbergen, P., & Verdaas, C. (2001). Interactive policy making: A model of management for public works. *European Journal of Operational Research, 128*, 322–337.

Drogoul, A., Vanbergue, D., & Meurisse, T. (2003). Multi-agent based simulation: Where are the agents? In J.S. Sichman, F. Bousquet, & P. Davidsson (Eds.), *Multi-agent-based simulation II: third international workshop, MABS 2002*, Bologna, Italy, July 15–16, 2002, revised papers (Lecture notes in computer science, 2581) (pp. 1–15). Berlin: Springer.

Dudley, R. G. (2003). Modeling the effects of a log export ban in Indonesia. *System Dynamics Review, 20*, 99–116.

Duke, R. D., & Geurts, J. L. A. (2004). *Policy games for strategic management*. Amsterdam: Dutch University Press.

Étienne, M. (2006). Companion modelling: A tool for dialogue and concertation in biosphere reserves. In M. Bouamrane (Ed.), *Biodiversity and stakeholders: Concertation itineraries, biosphere reserves - Technical notes 1* (pp. 44–52). Paris: UNESCO.

Étienne, M. (Ed.). (2011). *Companion modeling: A participatory approach to support sustainable development*. Versailles: QUAE.

Étienne, M., Le Page, C., & Cohen, M. (2003). A step by step approach to build up land management scenarios based on multiple viewpoints on multi-agent systems simulations. *Journal of Artificial Societies and Social Simulations, 6*(2). http://jasss.soc.surrey.ac.uk/6/2/2.html

Eversole, R. (2003). Managing the pitfalls of participatory development: Some insight from Australia. *World Development, 31*, 781–795.

Fischer, G., Giaccardi, E., Eden, H., Sugimoti, M., & Ye, Y. (2005). Beyond binary choices: Integrating individual and social creativity. *International Journal of Human-Computer Studies, 63*, 482–512.

Friedman, D., & Sunder, S. (1994). *Experimental methods, a primer for economists*. Cambridge: Cambridge University Press.

Fung, A. (2006). Varieties of participation in complex governance. *Public Administration Review, 66*, 66–75.

Funtowicz, S. O., Martinez-Alier, J., Munda, G., & Ravetz, J. R. (1999). *Information tools for environmental policy under conditions of complexity, Environmental issues series* (Vol. 9). Copenhagen: European Environment Agency.

Gilbert, N., & Troitzsch, K. G. (1999). *Simulation for the social scientist*. Buckingham: Open University Press.

Green, K. C. (2002). Forecasting decisions in conflict situations: A comparison of game theory, role-playing and unaided judgment. *International Journal of Forecasting, 18*, 321–344.

Guyot, P. (2006). *Simulations multi-agents participatives*. PhD thesis, Informatique, Université Paris VI, Paris.

Guyot, P., & Honiden, S. (2006). Agent-based participatory simulations: Merging multi-agent systems and role-playing games. *Journal of Artificial Societies and Social Simulations, 9*(4). http://jasss.soc.surrey.ac.uk/9/4/8.html

Hamel, A., & Pinson, S. (2005). Conception participative de simulations multi-agents basée sur une approche d'analyse multi-acteurs. In A. Drogoul & E. Ramat (Eds.), *Systèmes multi-agents: Vers la conception de systèmes artificiels socio-mimétiques (JFSMA 2005)* (pp. 1–15). Paris: Hermès.

Hanneman, R. A. (1995). Simulation modeling and theoretical analysis in sociology. *Sociological Perspectives, 38*, 457–462.

Henriksen, H.J., Rasmussen, P., Brandt, G., von Bülow, D., & Jensen, F.V. (2004). Engaging stakeholders in construction and validation of Bayesian belief networks for groundwater protection. In *Proceeding of IFAC workshop on modelling and control for participatory planning and managing water systems*, September 29–October 1, 2004, Venice, Italy.

Henry, C. (1974). Investment decisions under uncertainty: The irreversibility effect. *The American Economic Review, 64*, 1006–1012.

Hochman, Z., Hearnshaw, H., Barlow, R., Ayres, J. F., & Pearson, C. J. (1995). X-breed: A multiple domain knowledge based system integrated through a blackboard architecture. *Agricultural Systems, 48*, 243–270.

Irvin, R. A., & Stansbury, J. (2004). Citizen participation in decision making: Is it worth the effort. *Public Administration Review, 64*, 55–65.

Johannessen, J.-A., Olaisen, J., & Olsen, B. (2001). Mismanagement of tacit knowledge: The importance of tacit knowledge, the danger of information technology, and what to do about it. *International Journal of Information Management, 21*, 3–20.

Kinzig, A., Ryan, P., Etienne, M., Allyson, H., Elmqvist, T., & Walker, B. (2006). Resilience and regime shifts: Assessing cascading effects. *Ecology and Society, 11*(1), 20. http://www.ecologyandsociety.org/vol11/iss1/art20

Klopfer, E., Yoon, S., & Rivas, L. (2004). Comparative analysis of Palm and wearable computers for participatory simulations. *Journal of Computer Assisted Learning, 20*, 347–359.

Kujala, S. (2003). User involvement: A review of the benefits and challenges. *Behaviour and Information Technology, 22*, 1–16.

Landry, M., Banville, C., & Oral, M. (1996). Model legitimation in operational research. *European Journal of Operational Research, 92*, 443–457.

Le Bars, M., Le Grusse, P., Allaya, M., Attonaty, J.-M., & Mahjoubi, R. (2004). NECC: Un jeu de simulation pour l'aide à la décision collective; Application à une région méditerranéenne virtuelle. In *Projet INCO-WADEMED, Séminaire Modernisation de l'Agriculture Irriguée*, Rabat, Morocco.

Loucks, D. P., Kindler, J., & Fedra, K. (1985). Interactive water resources modeling and model use: An overview. *Water Resources Research, 21*, 95–102.

Maier, F. H., & Grössler, A. (2000). What are we talking about? A taxonomy of computer simulations to support learning. *System Dynamics Review, 16*, 135–148.

Manson, S. M. (2002). Validation and verification of multi-agent systems. In M. Janssen (Ed.), *Complexity and ecosystem management: The theory and practice of multi-agent approaches* (pp. 63–74). Northampton, MA: Edward Elgar Publishing.

Marengo, L., & Pasquali, C. (2003). How to construct and share a meaning for social interactions? In *Conventions et Institutions: Approfondissements théoriques et Contributions au Débat Politique*, Paris.

Martin, L., Magnuszewski, P., Sendzimir, J., Rydzak, F., Krolikowska, K., Komorowski, H., et al. (2007). Microworld gaming of a local agricultural production chain in Poland. *Simulation and Gaming, 38*(2), 211–232.

Mayer, I. S., van Bueren, E. M., Bots, P. W. G., van der Voort, H. G., & Seijdel, R. R. (2005). Collaborative decision-making for sustainable urban renewal projects: A simulation-gaming approach. *Environment and Planning B – Planning & Design, 32*, 403–423.

McKinnon, J. (2005). Mobile interactive GIS: Bringing indigenous knowledge and scientific information together; a narrative account. In A. Neef (Ed.), *Participatory approaches for sustainable land use in Southeast Asia* (pp. 217–231). White Lotus: Bangkok.

Mermet, L. (1993). Une méthode de prospective: Les exercices de simulation de politiques. *Nature Sciences Sociétés, 1*, 34–46.

Miettinen, R., & Virkkunen, J. (2005). Epistemic objects, artefacts and organizational change. *Organization, 12*, 437–456.

Molin, E. (2005). A causal analysis of hydrogen acceptance. *Transportation Research Records, 1941*, 115–121.

Moss, S., Downing, T., & Rouchier, J. (2000). *Demonstrating the role of stakeholder participation: An agent based social simulation model of water demand policy and response* (CPM Report, 00-76). Manchester: Centre for Policy Modelling, Manchester Metropolitan University. http://cfpm.org/cpmrep76.html

Mostert, E. (2006). Participation for sustainable water management. In C. Giupponi, A. J. Jakeman, D. Karssenberg, & M. P. Hare (Eds.), *Sustainable management of water resources*(pp. 153–176). Northampton, MA: Edward Elgar Publishing.

Nancarrow, B. (2005). When the modeller meets the social scientist or vice-versa. In A. Zerger & R. M. Argent (Eds.), *MODSIM 2005 international congress on modelling and simulation* (pp. 38–44*).* Melbourne, Australia: Modelling and Simulation Society of Australia and New Zealand.

Ostrom, E., Gardner, R., & Walker, J. (1994). *Rules, games and common-pool resources.* Ann Arbor, MI: The University of Michigan Press.

Pahl-Wostl, C., & Hare, M. (2004). Processes of social learning in integrated resources management. *Journal of Community and Applied Social Psychology, 14*, 193–206.

Pateman, C. (1990). *Participation and democratic theory.* Cambridge: Cambridge University Press.

Perez, P., Dray, A., White, I., Le Page, C., & Falkland, T. (2003). AtollScape: Simulating freshwater management in Pacific atolls, spatial processes and time dependence issues. In: D. Post (Ed.), *Proceedings of the international congress on modelling and simulation*, Townsville, Australia, July 14–17 2003 (Vol. 4, pp. 514–518). Townsville: MODSIM.

Perez, P., Dray, A., Le Page, C., D'Aquino, P., & White, I. (2004). Lagoon, agents and kava: A companion modelling experience in the Pacific. In C. van Dijkum, J. Blasius & C. Durand (Eds.), *Recent developments and applications in social research methodology: Proceedings of RC33 sixth international conference on social science methodology*, Amsterdam 2004 (p. 282). Opladen: Barbara Budrich Publishers.

Ramanath, A. M., & Gilbert, N. (2004). The design of participatory agent based simulations. *Journal of Artificial Societies and Social Simulations, 7*(4). http://jasss.soc.surrey.ac.uk/7/4/1.html

Reitsma, R., Zigurs, I., Lewis, C., Wilson, V., & Sloane, A. (1996). Experiment with simulation models in water-resources negotiations. *Journal of Water Resources Planning and Management, 122,* 64–70.

Richard, A., & Barreteau, O. (2006). Concert'eau: un outil de sociologie expérimentale pour l'étude de dispositifs de gestion locale et concertée de l'eau. In *Proceedoing of 2e Congrès de l'Association Française de Sociologie,* Bordeaux, 5–8 Septembre, 2006.

Richard, A., & Trometter, M. (2001). Les caractéristiques d'une décision séquentielle: Effet irréversibilité et endogénéisation de l'environnement. *Revue Economique, 52,* 739–752.

Rittel, H. W. J., & Webber, M. M. (1973). Dilemmas in a general theory of planning. *Policy Sciences, 4,* 155–169.

Robinson, J. B. (1991). Modelling the interactions between human and natural systems. *International Social Sciences Journal, 130,* 629–647.

Rocha, E. M. (1997). A ladder of empowerment. *Journal of Planning Education and Research, 17,* 31–44.

Rouchier, J. (2003). Re-implementation of a multi-agent model aimed at sustaining experimental economic research: The case of simulations with emerging speculation. *Journal of Artificial Societies and Social Simulations, 6*(4). http://jasss.soc.surrey.ac.uk/6/4/7.html

Rouwette, E. A. J. A., Vennix, J. A. M., & van Mullekorn, T. (2002). Group model building effectiveness: A review of assessment studies. *System Dynamics Review, 18,* 5–45.

Rowe, G., & Frewer, L. J. (2004). Evaluating public-participation exercises: A research agenda. *Science, Technology, & Human Values, 29*(4), 512–556.

Ryan, T. (2000). The role of simulation gaming in policy making. *Systems Research and Behavioral Science, 17,* 359–364.

Schelling, T. C. (1961). Experimental games and bargaining theory. *World Politics, 14,* 47–68.

Schuler, D., & Namioka, A. (Eds.). (1993). *Participatory design: Principles and practices.* Hillsdale, NJ: Lawrence Erlbaum Associates.

Shakun, M. E. (1996). Modeling and supporting task-oriented group processes: Purposeful complex adaptive systems and evolutionary systems design. *Group Decision and Negotiation, 5,* 305–317.

Simon, H. (1977). *The new science of management decision.* Englewood Cliffs: Prentice-Hall.

Star, S. L., & Griesemer, J. R. (1989). Institutional ecology, 'Translations' and boundary objects: Amateurs and professionals in Berleley's Museum of Vertebrate Zoology, 1907–39. *Social Studies of Science, 19,* 387–420.

Sterman, J. D. (1992). Teaching takes off – Flight simulators for management education. *OR/MS Today, 35,* 40–44.

Ticehurst, J., Rissik, D., Letcher, R. A., Newham, L. T. H., & Jakeman, A. J. (2005). Development of decision support tools to assess the sustainability of coastal lakes. In A. Zerger & R.M. Argent (Eds.), *MODSIM 2005 international congress on modelling and simulation* (pp. 2414–2420). Melbourne: Modelling and Simulation Society of Australia and New Zealand. http://www.mssanz.org.au/modsim05/papers/ticehurst.pdf

Toth, F. L. (1988). Policy exercises: Objectives and design elements. *Simulation and Games, 19,* 235–255.

Van Asselt, M. B. A., Mellors, J., Rijkens-Klomp, N., Greeuw, S. C. H., Molendijk, K. G. P., Beers, P. J., et al. (2001). *Building blocks for participation in integrated assessment: A review of participatory methods.* Maastricht: ICIS.

van Daalen, C. E., & Bots, P. W. G. (2006). Participatory model construction and model use in natural resource management. In *Proceeding of the workshop on formalised and non-formalised methods in resource management - knowledge and learning in participatory processes,* 21–22 September, 2006, Osnabrück, Germany. http://www.partizipa.uni-osnabrueck.de/wissAbschluss.html

van Daalen, C. E., Thissen, W. A. H., & Berk, M. M. (1998). The Delft process: Experiences with a dialogue between policy makers and global modellers. In J. Alcamo, R. Leemans, & E. Kreileman (Eds.), *Global change scenarios of the 21st century: Results from the IMAGE 2.1 model* (pp. 267–285). London: Elsevier.

Van den Belt, M. (2004). *Mediated modeling: A system dynamics approach to environmental consensus building*. Washington D.C.: Island Press.

Vennix, J. A. M. (1996). *Group model building, facilitating team learning using system dynamics*. Chichester: Wiley.

Webler, T., Kastenholz, H., & Renn, O. (1995). Public participation in impact assessment: A social learning perspective. *Environmental Impact Assessment Review, 15*, 443–463.

Whitworth, B., Gallupe, B., & McQueen, R. (2000). A cognitive three-process model of computer mediated group interaction. *Group Decision and Negotiation, 9*, 431–456.

Wood, J., & Silver, D. (1995). *Joint application development*. New York: Wiley.

Chapter 13
Combining Mathematical and Simulation Approaches to Understand the Dynamics of Computer Models

Luis R. Izquierdo, Segismundo S. Izquierdo, José M. Galán, and José I. Santos

Abstract This chapter shows how computer simulation and mathematical analysis can be used together to understand the dynamics of computer models. For this purpose, we show that it is useful to see the computer model as a particular implementation of a formal model in a certain programming language. This formal model is the abstract entity which is defined by the input–output relation that the computer model executes and can be seen as a function that transforms probability distributions over the set of possible inputs into probability distributions over the set of possible outputs.

It is shown here that both computer simulation and mathematical analysis are extremely useful tools to analyse this formal model, and they are certainly complementary in the sense that they can provide fundamentally different insights on the same model. Even more importantly, this chapter shows that there are plenty of synergies to be exploited by using the two techniques together.

The mathematical analysis approach to analyse formal models consists in examining the rules that define the model directly. Its aim is to deduce the logical implications of these rules for any particular instance to which they can be applied. Our analysis of mathematical techniques to study formal models is focused on the theory of Markov Chains, which is particularly useful to characterise the dynamics of computer models.

In contrast with mathematical analysis, the computer simulation approach does not look at the rules that define the formal model directly but instead tries to infer general properties of these rules by examining the outputs they produce when applied to particular instances of the input space. Thus, conclusions obtained with this approach may not be general. On a more positive note, computer simulation enables us to explore formal models beyond mathematical tractability, and we can

L.R. Izquierdo (✉) • J.M. Galán • J.I. Santos
Departamento de Ingeniería Civil, Universidad de Burgos, E-09001, Burgos, Spain
e-mail: lrizquierdo@ubu.es; jmgalan@ubu.es; jisantos@ubu.es

S.S. Izquierdo
Departamento de Organización de Empresas y C.I.M., Universidad de Valladolid, E-47011, Valladolid, Spain
e-mail: segis@eis.uva.es

© Springer International Publishing AG 2017
B. Edmonds, R. Meyer (eds.), *Simulating Social Complexity*,
Understanding Complex Systems, https://doi.org/10.1007/978-3-319-66948-9_13

achieve any arbitrary level of accuracy in our computational approximations by running the model sufficiently many times.

Bearing in mind the relative strengths and limitations of both approaches, this chapter explains three different ways in which mathematical analysis and computer simulation can be usefully combined to produce a better understanding of the dynamics of computer models. In doing so, it becomes clear that mathematical analysis and computer simulation should not be regarded as alternative—or even opposed—approaches to the formal study of social systems but as complementary. Not only can they provide fundamentally different insights on the same model, but they can also produce hints for solutions for each other. In short, there are plenty of synergies to be exploited by using the two techniques together, so the full potential of each technique cannot be reached unless they are used in conjunction.

Why Read This Chapter?
This chapter is about how to better understand the dynamics of computer models using both simulation and mathematical analysis. The starting point is a computer model which is already implemented and ready to be run; the objective is to gain a thorough understanding of its dynamics. Combining computer simulation with mathematical analysis can help to provide a picture of the model dynamics that could not be drawn by only using one of the two techniques.

13.1   Introduction

This chapter is about how to better understand the dynamics of computer models using both simulation and mathematical analysis. Our starting point is a computer model which is already implemented and ready to be run; our objective is to gain a thorough understanding of its dynamics. Thus, this chapter is *not* about how to design, implement, verify or validate a model; this chapter is about how to better understand its behaviour.

Naturally, we start by clearly defining our object of study: a computer model. The term 'computer model' can be understood in many different ways—i.e. seen from many different perspectives—and not all of them are equally useful for every possible purpose. Thus, we start by interpreting the term 'computer model' in a way that will prove useful for our objective: to characterise and understand its behaviour. Once our object of study has been clearly defined, we then describe two techniques that are particularly useful to understand the dynamics of computer models: mathematical analysis and computer simulation.

In particular, this chapter will show that mathematical analysis and computer simulation should not be regarded as alternative—or even opposed—approaches to the formal study of social systems but as complementary (Gotts et al. 2003, b). They are both extremely useful tools to analyse formal models, and they are certainly complementary in the sense that they can provide fundamentally different insights on the same model. Even more importantly, this chapter will show that there are plenty of synergies to be exploited by using the two techniques together, i.e. the full potential of each technique will not be reached until they are used in conjunction. The remaining of this introduction outlines the structure of the chapter.

Sections 13.2, 13.3 and 13.4 are devoted to explaining in detail what we understand by 'computer model', and they therefore provide the basic framework for the rest of the chapter. In particular, Sect. 13.2 shows that a computer model can be seen as an implementation—i.e. an explicit representation—of a certain deterministic input–output function in a particular programming language. This interpretation is very useful since, in particular, it will allow us to abstract from the details of the modelling platform where the computer model has been programmed and focus on analysing the formal model that the computer model implements. This is clarified in Sect. 13.3, which explains that any computer model can be re-implemented in many different formalisms (in particular, in any sophisticated enough programming language), leading to alternative representations of the same input–output relation.

Most computer models in the social simulation literature make use of pseudorandom number generators. Section 13.4 explains that—for these cases and given our purposes—it is useful to abstract from the details of how pseudorandom numbers are generated and look at the computer model as an implementation of a stochastic process. In a stochastic process, a certain input does not necessarily lead to one certain output only; instead, there are many different paths that the process may take with potentially different probabilities. Thus, in a stochastic process, a certain input will generally lead to a particular probability distribution over the range of possible outputs, rather than to a single output only. Stochastic processes are used to formally describe how a system subjected to random events evolves through time.

Having explained our interpretation of the term 'computer model', Sect. 13.5 introduces and compares the two techniques to analyse formal models that are assessed in this chapter: computer simulation and mathematical analysis. The following two sections sketch possible ways in which each of these two techniques can be used to obtain useful insights about the dynamics of a model. Section 13.8 is then focused on the *joint use* of computer simulation and mathematical analysis. It is shown here that the two techniques can be used together to provide a picture of the dynamics of the model that could not be drawn by using one of the two techniques only. Finally, our conclusions are summarised in Sect. 13.9.

13.2   Computer Models as Input–Output Functions

At the most elementary level, a computer model can be seen as an implementation—i.e. an explicit representation—of a certain deterministic input–output function in a particular programming language. The word 'function' is useful because it correctly conveys the point that any particular input given to the computer model will lead to one and only one output.[1] (Obviously, different inputs may lead to the same output.) Admittedly, however, the word 'function' may also mislead the reader into thinking that a computer model is necessarily simple. The computer model may be as complex and sophisticated as the programmer wants it to be, but ultimately, it is just an entity that associates a specific output to any given input, i.e. a function. In any case, to avoid confusion, we will use the term 'formal model' to denote the function that a certain computer model implements.[2] To be sure, the 'formal model' that a particular computer model implements is the abstract entity which is defined by the input–output relation that the computer model executes.[3]

Thus, running a computer model is just finding out the logical implications of applying a set of unambiguously defined formal rules (which are coded in the program and define the input–output function or formal model) to a set of inputs (Balzer et al. 2001). As an example, one could write the computer program '$y = 4x$' and apply it to the input '$x = 2$' to obtain the output '$y = 8$'. The output ($y = 8$), which is fully and unequivocally determined by the input ($x = 2$) and the set of rules coded in the program ($y = 4x$), can be seen as a theorem obtained by pure deduction ($\{x = 2; y = 4x\} \rightarrow y = 8$). Naturally, there is no reason why the inputs or the outputs should be numbers[4]; they could equally well be, e.g. strings of characters. In the general case, a computer run is a logical theorem that reads: *the output obtained from running the computer simulation follows (with logical necessity) from applying to the input the algorithmic rules that define the model*. Thus, regardless of its inherent complexity, a computer run constitutes a perfectly valid sufficiency theorem (see, e.g. Axtell 2000).

It is useful to realise that we could always apply the same inference rules ourselves to obtain—by logical deduction—the same output from the given input. Whilst useful as a thought, when it comes to actually doing the job, it is much more convenient, efficient and less prone to errors to let computers derive the output for us. Computers are inference engines that are able to conduct many algorithmic processes at a speed that the human brain cannot achieve.

---

[1]Note that simulations of stochastic models are actually using pseudorandom number generators, which are deterministic algorithms that require a seed as an input.

[2]A formal model is a model expressed in a formal system (Cutland 1980). A formal system consists of a formal language and a deductive apparatus (a set of axioms and inference rules). Formal systems are used to derive new expressions by applying the inference rules to the axioms and/or previously derived expressions in the same system.

[3]The mere fact that the model has been implemented and can be run in a computer is a proof that the model is formal (Suber 2002).

[4]As a matter of fact, strictly speaking, inputs and outputs in a computer model are *never* numbers. We may interpret strings of bits as numbers, but we could equally well interpret the same strings of bits as, e.g. letters. More importantly, a bit itself is already an abstraction, an interpretation we make of an electrical pulse that can be above or below a critical voltage threshold.

13.3   Different Ways of Representing the Same Formal Model

A somewhat controversial issue in the social simulation literature refers to the allegedly unique features of some modelling platforms. It is important to realise that any formal model implemented in a computer model can be re-implemented in many different programming languages, leading to exactly the same input–output relation. Different implementations are just different ways of representing one same formal model, much in the same way that one can say 'Spain' or 'España' to express the same concept in different languages: same thing, different representation, that's all.

Thus, when analysing the dynamics of a computer model, it is useful to abstract from the details of the modelling platform that has been used to implement the computer model and focus strictly on the formal model it represents, which could be re-implemented in any *sophisticated enough*[5] modelling platform. To be clear, let us emphasise that *any* computer model implemented in Objective-C (e.g. using Swarm) can be re-implemented in Java (e.g. using RePast or Mason), NetLogo, SDML, Mathematica© or Matlab©. Similarly, *any* computer model can be expressed as a well-defined mathematical function (Epstein 2006; Leombruni and Richiardi 2005; Richiardi et al. 2006).

Naturally, the implementation of a particular formal model may be more straightforward in some programming languages than in others. Programming languages differ in where they position themselves in the well-known trade-offs between ease of programming, functionality and performance; thus, different programming languages lead to more or less natural and more or less efficient implementations of any given formal model. Nonetheless, the important point is this: whilst we may have different implementations of the same formal model, and whilst each of these implementations may have different characteristics (in terms of, e.g. code readability), ultimately they are all just different representations of the same formal model, and they will therefore return the same output when given the same input.

In the same way that using one or another formalism to represent a particular formal model will lead to more or less natural implementations, different formalisms also make more or less apparent certain properties of the formal model they implement. For example, we will see in this chapter that representing a computer

---

[5] A sufficient condition for a programming language to be 'sophisticated enough' is to allow for the implementation of the following three control structures:

- Sequence (i.e. executing one subprogram and then another subprogram),
- Selection (i.e. executing one of two subprograms according to the value of a Boolean variable, e.g. IF[boolean == true]-THEN[subprogram1]-ELSE[subprogram2])
- Iteration (i.e. executing a subprogram until a Boolean variable becomes false, e.g. WHILE[boolean == true]-DO[subprogram])

Any programming language that can combine subprograms in these three ways can implement any computable function; this statement is known as the 'structured program theorem' (Böhm and Jacopini 1966; Harel 1980; Wikipedia 2007).

model as a Markov chain, i.e. looking at the formal model implemented in a computer model through Markov's glasses, can make apparent various features of the computer model that may not be so evident without such glasses. In particular, as we will show later, Markov theory can be used to find out whether the initial conditions of a model determine its asymptotic dynamics or whether they are actually irrelevant in the long term. Also, the theory can reveal whether the model will sooner or later be trapped in an absorbing state.

13.4   'Stochastic' Computer Models as Stochastic Processes

Most computer models in the social simulation literature contain stochastic components. This section argues that, for these cases and given our purposes, it is convenient to revise our interpretation of computer models as *deterministic* input–output relations, abstract from the (deterministic) details of how pseudorandom numbers are generated, and reinterpret the term 'computer model' as an implementation of a stochastic process. This interpretation will prove useful in most cases and, importantly, does not imply any loss of generality: even if the computer model to be analysed does not contain any stochastic components, our interpretation will still be valid.

In the general case, the computer model to be analysed will make use of (what are meant to be) random numbers, i.e. the model will be stochastic. The word 'stochastic' requires some clarification. Strictly speaking, there does not exist a *truly stochastic* computer model, but one can approximate randomness to a very satisfactory extent by using pseudorandom number generators. The pseudorandom number generator is a deterministic algorithm that takes as input a value called the random seed and generates a sequence of numbers that approximates the properties of random numbers. The sequence is not truly random in that it is completely determined by the value used to initialise the algorithm, i.e. the random seed. Therefore, if given the same random seed, the pseudorandom number generator will produce exactly the same sequence of (pseudorandom) numbers. (This fact is what made us define a computer model as an implementation of a certain *deterministic* input–output function in Sect. 13.2.)

Fortunately, the sequences of numbers provided by current off-the-shelf pseudorandom number generators approximate randomness remarkably well. This basically means that, for most intents and purposes in this discipline, it seems safe to assume that the pseudorandom numbers generated in one simulation run will follow the intended probability distributions to a satisfactory degree. The only problem we might encounter appears when running several simulations which we would like to be statistically independent. As mentioned above, if we used the same random seed for every run, we would obtain the same sequence of pseudorandom numbers, i.e. we would obtain exactly the same results. How can we *truly randomly* select a random seed? Fortunately, for most applications in this discipline, the state of the computer system at the time of starting a new run can be considered a truly random

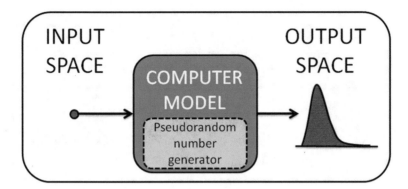

Fig. 13.1 A computer model can be usefully seen as the implementation of a function that transforms any given input into a certain probability distribution over the set of possible outputs

variable, and conveniently, if no seed is explicitly provided to the pseudorandom number generator, most platforms generate a seed from the state of the computer system (e.g. using the time). When this is done, the sequences of numbers obtained with readily available pseudorandom number generators approximate statistical randomness and independence remarkably well.

Given that—for most intents and purposes in this discipline—we can safely assume that pseudorandom numbers are random and independent enough, we dispense with the qualifier 'pseudo' from now on for convenience. Since every random variable in the model follows a specific probability distribution, the computer model will indeed generate a particular probability distribution over the range of possible outputs. Thus, to summarise, a computer model can be usefully seen as the implementation of a stochastic process, i.e. a function that transforms any given input into a certain probability distribution over the set of possible outputs (Fig. 13.1).

Seeing that we can satisfactorily simulate random variables, note that studying the behaviour of a model that has been parameterised stochastically does not introduce any conceptual difficulties. In other words, we can study the behaviour of a model that has been parameterised with probability distributions rather than certain values. An example would be a model where agents start at a random initial location.

To conclude this section, let us emphasise an important corollary of the previous paragraphs: *any statistic that we extract from a parameterised computer model follows a specific probability distribution* (even if the values of the input parameters have been expressed as probability distributions).[6] Thus, a computer model can be seen as the implementation of a function that transforms probability distributions

---

[6]Note that statistics extracted from the model can be of any nature, as long as they are unambiguously defined. For example, they can refer to various time-steps and only to certain agents (e.g. 'average wealth of female agents in odd time-steps from 1 to 99').

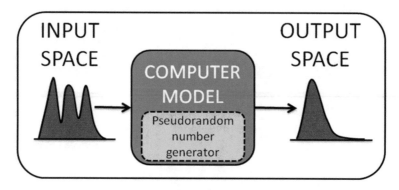

Fig. 13.2 A computer model can be seen as the implementation of a function that transforms probability distributions over the set of possible inputs into probability distributions over the set of possible outputs

over the set of possible inputs into probability distributions over the set of possible outputs (Fig. 13.2). The rest of the chapter is devoted to characterising this function.

13.5   Tools to Understand the Behaviour of Formal Models

Once it is settled that a computer model can be seen as a particular implementation of a (potentially stochastic) function in a certain programming language, let us refer to such a function as the 'formal model' that the computer model implements. As mentioned before, this formal model can be expressed in many different formalisms—in particular, it can always be expressed as a set of well-defined mathematical equations (Leombruni and Richiardi 2005)—and our objective consists in understanding its behaviour. To do that, we count with two very useful tools: mathematical analysis[7] and computer simulation.

The advantages and limitations of these two tools to formally study social systems have been discussed at length in the literature (see, e.g. Axtell 2000; Axtell and Epstein 1994; Edmonds 2005; Gilbert 1999; Gilbert and Troitzsch 1999; Gotts et al. 2003; Holland and Miller 1991; Ostrom 1988). Here we only highlight the most prominent differences between these two techniques (see Fig. 13.3).

In broad terms, when using mathematical analysis, one examines the rules that define the formal model directly and tries to draw general conclusions about these rules. These conclusions are obtained by using logical deduction; hence they follow with logical necessity from the premises of the formal model (and the axioms of

---

[7]We use the term 'mathematical analysis' in its broadest sense, i.e. we do not refer to any particular branch of mathematics, but to the general use of (any type of) mathematical technique to analyse a system.

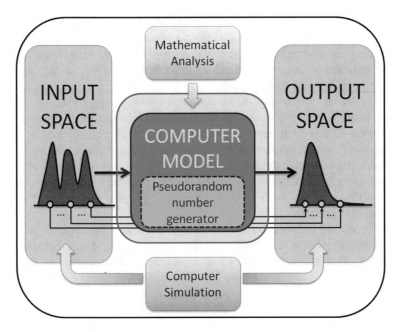

Fig. 13.3 In general terms, mathematical analysis tends to examine the rules that define the formal model directly. In contrast, computer simulation tries to infer general properties of such rules by looking at the outputs they produce when applied to particular instances of the input space

the mathematics employed). The aim when using mathematical analysis is usually to 'solve' the formal system (or, most often, certain aspects of it) by producing general closed-form solutions that can be applied to *any* instance of the whole input set (or, at least, to large portions of the input set). Since the inferences obtained with mathematical analysis pertain to the rules themselves, such inferences can be safely particularised to any specific parameterisation of the model, even if such a parameterisation was never explicitly contemplated when analysing the model mathematically. This greatly facilitates conducting sensitivity analyses and assessing the robustness of the model.

Computer simulation is a rather different approach to the characterisation of the formal model (Epstein 2006; Axelrod 1997a). When using computer simulation, one often treats the formal model as a black box, i.e. a somewhat obscure abstract entity that returns certain outputs when provided with inputs. Thus, the path to understand the behaviour of the model consists in obtaining many input–output pairs and—using generalisation by induction—inferring general patterns about how the rules transform the inputs into the outputs (i.e. how the formal model works).

Importantly, the execution of a simulation run, i.e. the logical process that transforms any (potentially stochastic) given input into its corresponding (potentially stochastic) output, is pure deduction (i.e. strict application of the formal rules that define the model). Thus, running the model in a computer provides a formal proof that a particular input (together with the set of rules that define the model) is sufficient to generate the output that is observed during the simulation. This first part of the computer simulation approach is therefore, in a way, very 'mathematical': outputs obtained follow with logical necessity from applying to the inputs the algorithmic rules that define the model.

In contrast, the second part of the computer simulation approach, i.e. inferring general patterns from particular instances of input–output pairs, can only lead to probable—rather than necessarily true—conclusions.[8] The following section explains how to rigorously assess the confidence we can place on the conclusions obtained using computer simulation, but the simple truth is irrefutable: inferences obtained using generalisation by induction can potentially fail when applied to instances that were not used to infer the general pattern. This is the domain of statistical extrapolation.

So why bother with computer simulation at all? The answer is clear: computer simulation enables us to study formal systems in ways that go beyond mathematical tractability. This role should not be underestimated: most models in the social simulation literature are mathematically intractable, and in such cases computer simulation is our only chance to move things forward. As a matter of fact, the formal models that many computer programs implement are often so complicated and cumbersome that the computer code itself is not that far from being one of the best descriptions of the formal model that can be provided.

Computer simulation can be very useful even when dealing with formal models that are mathematically tractable. Valuable uses of computer simulation in these cases include conducting insightful initial explorations of the model and presenting dynamic illustrations of its results.

And there is yet another important use of computer simulation. Note that understanding a formal model in depth requires identifying the parts of the model (i.e. the subset of rules) that are responsible for generating particular (sub)sets of results or properties of results. Investigating this in detail often involves changing certain subsets of rules in the model, so one can pinpoint which subsets of rules are necessary or sufficient to produce certain results. Importantly, changing subsets of rules can make the original model mathematically intractable, and in such (common) cases, computer simulation is, again, our only hope. In this context, computer simulation can be very useful to produce counterexamples. This approach is very common in the literature of, e.g. evolutionary game theory, where several authors (see, e.g. Hauert and Doebeli 2004; Imhof et al. 2005; Izquierdo and Izquierdo 2006; Lieberman et al. 2009; Nowak and May 1992; Nowak and Sigmund 1992; Nowak and Sigmund 1993; Santos et al. 2006; Traulsen et al. 2006) resort to computer simulations to assess the implications of assumptions made in mathematically tractable models (e.g. the assumptions of 'infinite populations' and 'random encounters').

---

[8]Unless, of course, all possible particular instances are explored.

It is important to note that the fundamental distinction between mathematical analysis and computer simulation as presented here is *not* about whether one uses pen and paper or computers to analyse formal models. We can follow either approach with or without computers, and it is increasingly popular to do mathematical analysis with computers. Recent advancements in symbolic computation have opened up a new world of possibilities to conduct mathematical analyses (using, e.g. Mathematica©). In other words, nowadays it is perfectly possible to use computers to *directly* examine the rules that define a formal model (see Fig. 13.3).

Finally, as so often in life, things are not black or white but involve some shade of grey. Similarly, most models are not tractable *or* intractable in mathematical terms; most often they are *partially* tractable. It is in these cases where an adequate combination of mathematical analysis and computer simulation is particularly useful. We illustrate this fact in Sect. 13.8, but first let us look at each technique separately. The following two sections provide some guidelines on how computer simulation (Sect. 13.6) and mathematical analysis (Sect. 13.7) can be usefully employed to analyse formal models.

13.6   Computer Simulation: Approximating the Exact Probability Distribution by Running the Model

The previous sections have argued that *any* statistic obtained from a (stochastically or deterministically) parameterised model follows a specific probability distribution. The statistic could be anything as long as it is unambiguously defined; in particular, it could refer to one or several time-steps and to one or various subcomponents of the model. Ideally, one would like to calculate the exact probability distribution for the statistic using mathematical analysis, but this will not always be possible. In contrast, using computer simulation we will always be able to approximate this probability distribution to any arbitrary level of accuracy; this section provides basic guidelines on how to do that.

The output probability distribution—which is fully and unequivocally determined by the input distribution—can be approximated to any degree of accuracy by running enough simulation runs. Note that any specific simulation run will be conducted with a particular certain value for every parameter (e.g. a particular initial location for every agent) and will produce one and only one particular certain output (see Fig. 13.3). Thus, in order to infer the probability distribution over the set of outputs that a particular probability distribution over the set of inputs leads to, there will be a need to run the model many times (with different random seeds); this is the so-called Monte Carlo method.

The method is straightforward: obtain as many random samples as possible (i.e. run as many independent simulations as possible), since this will get us closer and closer to the exact distribution (by the law of large numbers). Having conducted a

Fig. 13.4 Snapshot of CoolWorld. Patches are coloured according to their temperature: the higher the temperature, the darker the shade of *red*. Houses are coloured in *orange* and form a *circle* around the central patch. Walkers are coloured in *green*, and represented as a person if standing on a patch without a house, and as a smiling face if standing on a patch with a house. In the latter case, the *white* label indicates the number of walkers in the same house

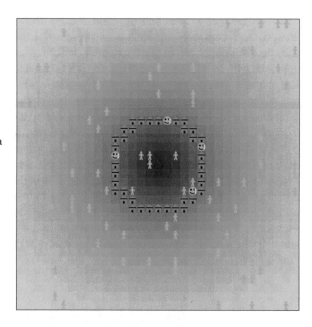

large number of simulation runs, the question that naturally comes to mind is: how close to the exact distribution is the one obtained by simulation?

To illustrate how to assess the quality of the approximation obtained by simulation, we use CoolWorld, a purpose-built agent-based model (Gilbert 2007) implemented in NetLogo 4.0 (Wilensky 1999). A full description of the model, and the source code can be found at the dedicated model webpage https://luis-r-izquierdo.github.io/coolworld/. For our purposes, it suffices to say that in CoolWorld there is a population of agents called walkers, who wander around a two-dimensional grid made of square patches; some of the patches are empty, whilst others contain a house (see Fig. 13.4). Patches are at a certain predefined temperature, and walkers tend to walk towards warmer patches, staying for a while at the houses they encounter in their journey.

Let us assume that we are interested in studying the number of CoolWorld walkers staying in a house in time-step 50. Initial conditions (which involve 100 walkers placed at a random location) are unambiguously defined at the model webpage and can be set in the implementation of CoolWorld provided by clicking on the button 'Special conditions'. Figure 13.4 shows a snapshot of CoolWorld after having clicked on that button.

As argued before, given that the (stochastic) initial conditions are unambiguously defined, the number of CoolWorld walkers in a house after 50 time-steps will follow a specific probability distribution that we are aiming to approximate. For that, let us assume that we run 200 runs and plot the relative frequency of the number of walkers in a patch with a house after 50 time-steps (see Fig. 13.5).

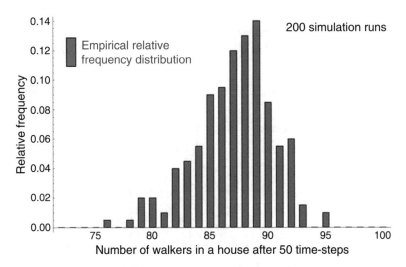

Fig. 13.5 Relative frequency distribution of the number of walkers in a house after 50 time-steps, obtained by running CoolWorld 200 times, with initial conditions set by clicking on 'Special conditions'

Figure 13.5 does not provide all the information that can be extracted from the data gathered. In particular, we can plot error bars showing the standard error for each calculated frequency without hardly any effort.[9] Standard errors give us information about the error we may be incurring when estimating the exact probabilities with the empirical frequencies. Another simple task that can be conducted consists in partitioning the set of runs into two batteries of approximately equal size and comparing the two distributions. If the two distributions are not similar, then there is no point in proceeding: we are not close to the exact distribution, so there is a need to run more simulations.

Figures 13.6 and 13.7 show the data displayed in Fig. 13.5 partitioned in two batteries of 100 simulation runs, including the standard errors. Figure 13.6 and 13.7 also show the exact probability distribution we are trying to approximate, which has been calculated using mathematical methods that are explained later in this chapter.

Figures 13.6 and 13.7 indicate that 100 simulation runs may not be enough to obtain a satisfactory approximation to the exact probability distribution. On the other hand, Figs. 13.8 and 13.9 show that running the model 50,000 times does

---

[9]The frequency of the event 'there are i walkers in a patch with a house' calculated over n simulation runs can be seen as the mean of a sample of n i.i.d. Bernoulli random variables where success denotes that the event occurred and failure denotes that it did not. Thus, the frequency f is the maximum likelihood (unbiased) estimator of the exact probability with which the event occurs. The standard error of the calculated frequency f is the standard deviation of the sample divided by the square root of the sample size. In this particular case, the formula reads:

Std.error$(f, n) = (f(1-f)/(n-1))^{1/2}$

where f is the frequency of the event, n is the number of samples and the standard deviation of the sample has been calculated dividing by $n - 1$.

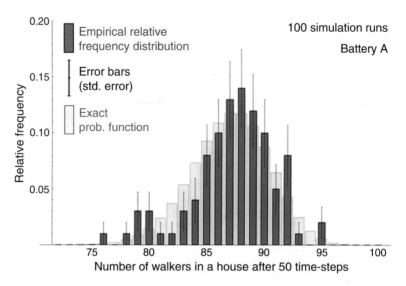

Fig. 13.6 In *blue*: relative frequency distribution of the number of walkers in a house after 50 time-steps, obtained by running CoolWorld 100 times (Battery A), with initial conditions set by clicking on 'Special conditions'. In *grey*: exact probability distribution (calculated using Markov chain analysis)

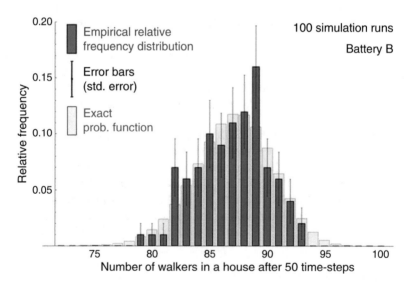

Fig. 13.7 In *blue*: relative frequency distribution of the number of walkers in a house after 50 time-steps, obtained by running CoolWorld 100 times (Battery B), with initial conditions set by clicking on 'Special conditions'. In *grey*: exact probability distribution (calculated using Markov chain analysis)

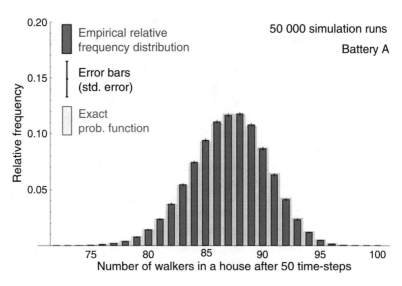

Fig. 13.8 In *blue*: relative frequency distribution of the number of walkers in a house after 50 time-steps, obtained by running CoolWorld 50,000 times (Battery A), with initial conditions set by clicking on 'Special conditions'. In *grey*: exact probability distribution (calculated using Markov chain analysis)

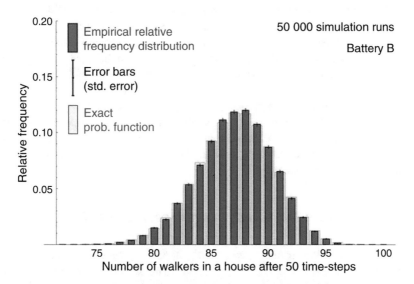

Fig. 13.9 In *blue*: relative frequency distribution of the number of walkers in a house after 50 time-steps, obtained by running CoolWorld 50,000 times (Battery B), with initial conditions set by clicking on 'Special conditions'. In *grey*: exact probability distribution (calculated using Markov chain analysis)

seem to get us close to the exact probability distribution. The standard error, which is inversely proportional to the square root of the sample size (i.e. the number of runs), is naturally much lower in these latter cases.

When, like in this example, the space of all possible outcomes in the distribution under analysis is finite (the number of walkers in a house must be an integer between 0 and 100), one can go further and calculate confidence intervals for the obtained frequencies. This is easily conducted when one realises that the exact probability distribution is a multinomial. Genz and Kwong (2000) show how to calculate these confidence intervals.

To conclude this section, let us emphasise that all that has been written here applies to *any* statistic obtained from *any* computer model. In particular, the statistic may refer to predefined regimes (e.g. 'number of time-steps between 0 and 100 where there are more than 20 walkers in a house') or to various time-steps (e.g. 'total number of walkers in a house in odd time-steps in between time-steps 50 and 200'). These statistics, like any other one, follow a specific probability distribution that can be approximated to any degree of accuracy by running the computer model.

13.7 Mathematical Analysis: Time-Homogenous Markov Chains

The whole range of mathematical techniques that can be used to analyse formal systems is too broad to be reviewed here. Instead, we focus on one specific technique that seems to us particularly useful to analyse social simulation models: Markov chain analysis. Besides, there are multiple synergies to be exploited by using Markov chain analysis and computer simulation together, as we will see in the next section.

Our first objective is to learn how to represent a particular computer model as a time-homogeneous Markov chain. This alternative representation of the model will allow us to use several simple mathematical results that will prove useful to understand the dynamics of the model. We therefore start by describing time-homogeneous Markov chains.

13.7.1 What Is a Time-Homogeneous Markov Chain?

Consider a system that in time-step $n = \{1, 2, 3, \ldots\}$ may be in one of a finite number of possible states $S = \{s_1, s_2, \ldots, s_M\}$. The set S is called the state space; in this chapter, we only consider *finite* state spaces.[10] Let the sequence of random variables $X_n \in S$ represent the state of the system in time-step n. As an example, $X_3 = s_9$ means that at time $n = 3$, the system is in state s_9. The system starts at a

---

[10]The term 'Markov chain' allows for countably infinite state spaces too (Karr 1990).

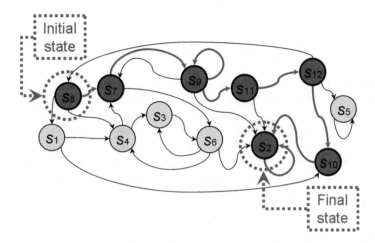

Fig. 13.10 Schematic transition diagram of a Markov chain. *Circles* denote states, and directed *arrows* indicate possible transitions between states. In this figure, *circles* and *arrows* coloured in *red* represent one possible path where the initial state X_0 is s_8 and the final state is s_2

certain initial state X_0 and moves from one state to another. The system is stochastic in that, given the present state, the system may move to one or another state with a certain probability (see Fig. 13.10). The probability that the system moves from state i to state j in one time-step, $P(X_{n+1} = j|X_n = i)$, is denoted by $p_{i,j}$. As an example, in the Markov chain represented in Fig. 13.10, $p_{4,6}$ equals 0 since the system cannot go from state 4 to state 6 in one single time-step. The system may also stay in the same state i, and this occurs with probability $p_{i,i}$. The probabilities $p_{i,j}$ are called transition probabilities, and they are often arranged in a matrix, namely, the transition matrix P. Implicitly, our definition of transition probabilities assumes two important properties about the system:

(a) The system has the **Markov property**. This means that the present state contains all the information about the future evolution of the system that can be obtained from its past, i.e. given the present state of the system, knowing the past history about how the system reached the present state does not provide any additional information about the future evolution of the system. Formally,

$$P\left(X_{n+1} = x_{n+1}|X_n = x_n, X_{n-1} = x_{n-1}, \ldots, X_0 = x_0\right) = P\left(X_{n+1} = x_{n+1}|X_n = x_n\right)$$

(b) In this chapter we focus on **time-homogeneous** Markov chains, i.e. Markov chains with time-homogeneous transition probabilities. This basically means that transition probabilities $p_{i,j}$ are independent of time, i.e. the one-step transition probability $p_{i,j}$ depends on i and j but is the same at all times n. Formally,

$$P(X_{n+1} = j | X_n = i) = P(X_n = j | X_{n-1} = i) = p_{i,j}$$

The crucial step in the process of representing a computer model as a time-homogeneous Markov chain (THMC) consists in identifying an appropriate set of state variables. A particular combination of specific values for these state variables will define one particular state of the system. Thus, the challenge consists in choosing the set of state variables in such a way that the computer model can be represented as a THMC. In other words, the set of state variables must be such that one can see the computer model as a transition matrix that unambiguously determines the probability of going from any state to any other state.

13.7.1.1 Example: A Simple Random Walk

Let us consider a model of a simple one-dimensional random walk and try to see it as a THMC. In this model—which can be run and downloaded at the dedicated model webpage https://luis-r-izquierdo.github.io/random-walk/—there are 17 patches in line, labelled with the integers between 1 and 17. A random walker is initially placed on one of the patches. From then onwards, the random walker will move randomly to one of the spatially contiguous patches in every time-step (staying still is not an option). Space does not wrap around, i.e. patch 1's only neighbour is patch 2 (Fig.

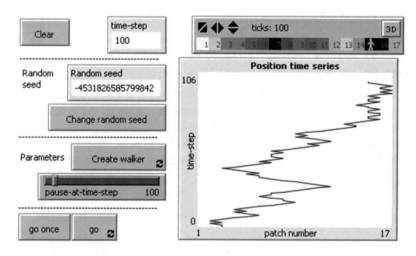

Fig. 13.11 Snapshot of the one-dimensional random walk applet. Patches are arranged in a *horizontal line* on the *top right corner* of the figure; they are labelled with *red* integers and coloured in shades of *blue* according to the number of times that the random walker has visited them: the higher the number of visits, the darker the shade of *blue*. The plot beneath the patches shows the time series of the random walker's position

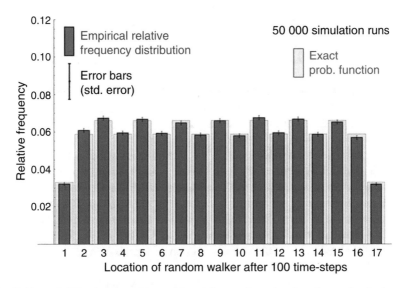

Fig. 13.12 Probability function of the position of the one-dimensional random walker in time-step 100, starting at an initial random location

13.11). This model can be easily represented as a THMC by choosing the agent's position (e.g. the number of the patch she is standing on) as the only state variable. To be sure, note that defining the state of the system in this way, it is true that there is a fixed probability of going from any state to any other state, independent of time. The transition matrix $P = [p_{i,j}]$ corresponding to the model is:

$$P = [P_{i,j}] = \begin{bmatrix} 0 & 1 & 0 & & & & & & \cdots & & 0 \\ 0.5 & 0 & 0.5 & 0 & & & & & & & \vdots \\ 0 & 0.5 & 0 & 0.5 & 0 & & & & & & \\ & 0 & 0.5 & 0 & 0.5 & 0 & & & & & \\ & & \ddots & \ddots & \ddots & \ddots & \ddots & & \ddots & & \\ & & & & 0 & 0.5 & 0 & 0.5 & 0 & & \\ & \vdots & & & & 0 & 0.5 & 0 & 0.5 & \\ & 0 & \cdots & & & & 0 & 1 & 0 & \end{bmatrix} \qquad (13.1)$$

where, as explained above, $p_{i,j}$ is the probability $P(X_{n+1} = j | X_n = i)$ that the system will be in state j in the following time-step, knowing that it is currently in state i.

13.7.2   Transient Distributions of Finite THMCs

The analysis of the dynamics of THMCs is usually divided into two parts: transient dynamics (finite time) and asymptotic dynamics (infinite time). The transient behaviour is characterised by the distribution of the state of the system X_n for a fixed time-step $n \geq 0$. The asymptotic behaviour (see Sects. 13.7.3 and 13.7.4) is characterised by the limit of the distribution of X_n as n goes to infinity, when this limit exists.

This section explains how to calculate the transient distribution of a certain THMC, i.e. the distribution of X_n for a fixed $n \geq 0$. In simple words, we are after a vector $a^{(n)}$ containing the probability of finding the process in each possible state in time-step n. Formally, $a^{(n)} = [a_1^{(n)}, \ldots, a_M^{(n)}]$, where $a_i^{(n)} = P(X_n = i)$ denotes the distribution of X_n for a THMC with M possible states. In particular, $a^{(0)}$ denotes the initial distribution over the state space, i.e. $a_i^{(0)} = P(X_0 = i)$. Note that there is no problem in having uncertain initial conditions, i.e. probability functions over the space of possible inputs to the model.

It can be shown that one can easily calculate the transient distribution in time-step n, simply by multiplying the initial conditions by the n-th power of the transition matrix P.

Proposition 1 $a^{(n)} = a^{(0)} \cdot P^n$ Thus, the elements $p^{(n)}_{i,j}$ of P^n represent the probability that the system is in state j after n time-steps having started in state i, i.e. $p^{(n)}_{i,j} = P(X_n = j | X_0 = i)$. A straightforward corollary of Proposition 1 is that $a^{(n+m)} = a^{(n)} \cdot P^m$.

As an example, let us consider the one-dimensional random walk again. Imagine that the random walker starts at an initial random location, i.e. $a^{(0)} = [1/17, \ldots, 1/17]$. The exact distribution of the walker's position in time-step 100 would then be $a^{(100)} = a^{(0)} \ldots P^{100}$. This distribution is represented in Fig. 13.12, together with an empirical distribution obtained by running the model 50,000 times.

Having obtained the probability function over the states of the system for any fixed n, namely, the probability mass function of X_n, it is then straightforward to calculate the distribution of *any* statistic that can be extracted from the model. As argued in the previous sections, the state of the system fully characterises it, so *any* statistic that we obtain about the computer model in time-step n must be, ultimately, a function of $\{X_0, X_1, \ldots, X_n\}$.

Admittedly, the transition matrix of most computer models cannot be easily derived, or it is unfeasible to operate with it. Nonetheless, this apparent drawback is not as important as one might expect. As we shall see below, it is often possible to infer many properties of a THMC even without knowing the exact values of its transition matrix, and these properties can yield useful insights about the dynamics of the associated process. Knowing the exact values of the transition matrix allows us to calculate the exact transient distributions using Proposition 1; this is desirable but not critical, since we can always approximate these distributions by conducting many simulation runs, as explained in Sect. 13.6.

13.7.3 Important Concepts

This section presents some basic concepts that will prove useful to analyse the dynamics of computer models. The notation used here follows the excellent book on stochastic processes written by Kulkarni (1995).

Definition 1 Accessibility A state j is said to be accessible from state i if starting at state i there is a chance that the system may visit state j at some point in the future. By convention, every state is accessible from itself. Formally, a state j is said to be accessible from state i if for some $n \geq 0$, $p^{(n)}_{i,j} > 0$.

Note that j is accessible from $i \neq j$ if and only if there is a directed path from i to j in the transition diagram. In that case, we write $i \to j$. If $i \to j$ we also say that i leads to j. As an example, in the THMC represented in Fig. 13.10, s_2 is accessible from s_{12} but not from s_5. Note that the definition of accessibility does not depend on the actual magnitude of $p^{(n)}_{i,j}$, only on whether it is exactly zero or strictly positive.

Definition 2 Communication A state i is said to communicate with state j if $i \to j$ and $j \to i$.

If i communicates with j, we also say that i and j communicate and write $i \leftrightarrow j$. As an example, note that in the simple random walk presented in Sect. 13.7.1, every state communicates with every other state. It is worth noting that the relation 'communication' is transitive, i.e.

$$i \leftrightarrow j, j \leftrightarrow k \quad \Rightarrow \quad i \leftrightarrow k.$$

Definition 3 Communicating Class A set of states $C \subset S$ is said to be a communicating class if:

- Any two states in the communicating class communicate with each other. Formally,

$$i \in C, j \in C \quad \Rightarrow \quad i \leftrightarrow j$$

- The set C is maximal, i.e. no strict superset of a communicating class can be a communicating class. Formally,

$$i \in C, i \leftrightarrow j \quad \Rightarrow \quad j \in C$$

As an example, note that in the simple random walk presented in Sect. 13.7.1, there is one single communicating class that contains all the states. In the THMC represented in Fig. 13.10, there are four communicating classes: $\{s_2\}$, $\{s_5\}$, $\{s_{10}\}$, $\{s_1, s_3, s_4, s_6, s_7, s_8, s_9, s_{11}, s_{12}\}$.

Definition 4 Closed Communicating Class (i.e. Absorbing Class): Absorbing State A communicating class C is said to be closed if no state within C leads to any state outside C. Formally, a communicating class C is said to be closed if $i \in C$ and $j \in C$ imply that j is not accessible from i.

Note that once a Markov chain visits a closed communicating class, it cannot leave it. Hence we will sometimes refer to closed communicating classes as 'absorbing classes'. This latter term is not standard in the literature, but we find it useful here for explanatory purposes. Note that if a Markov chain has one single communicating class, it must be closed.

As an example, note that the communicating classes $\{s_{10}\}$ and $\{s_1, s_3, s_4, s_6, s_7, s_8, s_9, s_{11}, s_{12}\}$ in the THMC represented in Fig. 13.10 are not closed, as they can be abandoned. On the other hand, the communicating classes $\{s_2\}$ and $\{s_5\}$ are indeed closed, since they cannot be abandoned. When a closed communicating class consists of one single state, this state is called absorbing. Thus, s_2 and s_5 are absorbing states. Formally, state i is absorbing if and only if $p_{i,i} = 1$ and $p_{i,j} = 0$ for $i \neq j$.

Proposition 2 Decomposition Theorem (Chung 1960) The state space S of any Markov chain can be uniquely partitioned as follows:

$$S = C_1 \cap C_2 \cap \cdots \cap C_k \cap T$$

where $C_1, C_2, \ldots, C_k$ are closed communicating classes, and T is the union of all other communicating classes.

Note that we do not distinguish between non-closed communicating classes: we lump them all together into T. Thus, the unique partition of the THMC represented in Fig. 13.10 is $S = \{s_2\} \cap \{s_5\} \cap \{s_1, s_3, s_4, s_6, s_7, s_8, s_9, s_{10}, s_{11}, s_{12}\}$. The simple random walk model presented in Sect. 13.7.1 has one single (closed) communicating class C_1 containing all the possible states, i.e. $S \equiv C_1$.

Definition 5 Irreducibility A Markov chain is said to be irreducible if all its states belong to a single closed communicating class; otherwise it is called reducible. Thus, the simple random walk presented in Sect. 13.7.1 is irreducible, but the THMC represented in Fig. 13.10 is reducible.

Definition 6 Transient and Recurrent States A state i is said to be transient if, given that we start in state i, there is a non-zero probability that we will never return back to i. Otherwise, the state is called recurrent. A Markov chain starting from a recurrent state will revisit it with probability 1 and hence revisit it infinitely often. On the other hand, a Markov chain starting from a transient state has a strictly positive probability of never coming back to it. Thus, a Markov chain will visit any transient state only finitely many times; eventually, transient states will not be revisited anymore.

Definition 7 Periodic and Aperiodic States: Periodic and Aperiodic Communicating Classes A state i has period d if any return to state i must occur in multiples of d time-steps. If $d = 1$, then the state is said to be aperiodic; otherwise ($d > 1$), the state is said to be periodic with period d. Formally, state i's period d is the greatest common divisor of the set of integers $n > 0$ such that $p^{(n)}_{i,i} > 0$. For our purposes, the concept of periodicity is only relevant for recurrent states. As an example, note that every state in the simple random walk presented in Sect. 13.7.1 is periodic with period 2.

An interesting and useful fact is that if $i \leftrightarrow j$, then i and j must have the same period (see theorem 5.2 in Kulkarni 1995). In particular, note that if $p_{i,i} > 0$ for any i, then the communicating class to which i belongs must be aperiodic. Thus, it makes sense to qualify communicating classes as periodic with period d or aperiodic. A closed communicating class with period d can return to its starting state only at times $d, 2d, 3d, \ldots$.

The concepts presented in this section will allow us to analyse the dynamics of any finite Markov chain. In particular, we will show that, given enough time, any finite Markov chain will necessarily end up in one of its closed communicating classes (i.e. absorbing classes).

13.7.4   Limiting Behaviour of Finite THMCs

This section is devoted to characterising the limiting behaviour of a THMC, i.e. studying the convergence (in distribution) of X_n as n tends to infinity. Specifically, we aim to study the behaviour of $a_i^{(n)} = P(X_n = i)$ as n tends to infinity. From Proposition 1 it is clear that analysing the limiting behaviour of P^n would enable us to characterise $a_i^{(n)}$. There are many introductory books in stochastic processes that offer clear and simple methods to analyse the limiting behaviour of THMCs when the transition matrix P is tractable (see, e.g. Chap. 5 in (Kulkarni 1999), Chaps. 2–4 in (Kulkarni 1995), Chap. 3 in (Janssen and Manca 2006) or the book chapter written by Karr (1990)). Nonetheless, we focus here on the general case, where operating with the transition matrix P may be computationally unfeasible.

13.7.4.1   General Dynamics

The first step in the analysis of any THMC consists in identifying all the closed communicating classes, so we can partition the state space S as indicated by the decomposition theorem (see Proposition 2). The following proposition (Theorems 3.7 and 3.8 in Kulkarni 1995) reveals the significance of this partition:

Proposition 3 General Dynamics of Finite THMCs Consider a finite THMC that has been partitioned as indicated in Proposition 2. Then:

1. All states in T (i.e. not belonging to a closed communicating class) are transient.

2. All states in C_v (i.e. in any closed communicating class) are recurrent; $v \in \{1, 2, \ldots, k\}$.

Proposition 3 states that sooner or later the THMC will enter one of the absorbing classes and stay in it forever. Formally, for all $i \in S$ and all $j \in T$: $\lim_{n \to \infty} p_{i,j}^{(n)} = 0$, i.e. the probability of finding the process in a state belonging to a non-closed communicating class goes to zero as n goes to infinity. Naturally, if the initial state already belongs to an absorbing class C_v, then the chain will never abandon such a class. Formally, for all $i \in C_v$ and all $j \notin C_v$, $p^{(n)}_{i,j} = 0$ for all $n \geq 0$.

As an example of the usefulness of Proposition 3, consider the THMC represented in Fig. 13.10. This THMC has only two absorbing classes: $\{s_2\}$ and $\{s_5\}$. Thus, the partition of the state space is: $S = \{s_2\} \cap \{s_5\} \cap \{s_1, s_3, s_4, s_6, s_7, s_8, s_9, s_{10}, s_{11}, s_{12}\}$. Hence, applying Proposition 3 we can state that the process will eventually end up in one of the two absorbing states, s_2 or s_5. The probability of ending up in one or the other absorbing state depends on the initial conditions $a^{(0)}$ (and on the actual numbers $p_{i,j}$ in the transition matrix, of course). Slightly more formally, the limiting distribution of X_n exists, but it is not unique, i.e. it depends on the initial conditions.

13.7.4.2 Dynamics Within Absorbing Classes

The previous section has explained that any simulation run will necessarily end up in a certain absorbing class; this section characterises the dynamics of a THMC that is already 'trapped' in an absorbing class. This is precisely the analysis of irreducible Markov chains, since irreducible Markov chains are, by definition, Markov chains with one single closed communicating class (see Definition 5). In other words, one can see any THMC as a set of transient states T plus a finite number of irreducible Markov sub-chains.

Irreducible THMCs behave significantly different depending on whether they are periodic or not. The following sections characterise these two cases.

Irreducible and Aperiodic THMCs

Irreducible and aperiodic THMCs are often called ergodic. In these processes the probability function of X_n approaches a limit as n tends to infinity. This limit is called the **limiting distribution** and is denoted here by π. Formally, the following limit exists and is unique (i.e. independent of the initial conditions $a_i^{(0)}$):

$$\lim_{n \to \infty} a_i^{(n)} = \pi_i > 0 \quad i \in S$$

Thus, in ergodic THMCs the probability of finding the system in each of its states in the long run is strictly positive and independent of the initial conditions

(Theorems 3.7 and 3.15 in Kulkarni 1995). As previously mentioned, calculating such probabilities may be unfeasible, but we can estimate them sampling many simulation runs at a sufficiently large time-step.

Importantly, in ergodic THMCs the limiting distribution π coincides with the **occupancy distribution** π^*, which is the long-run fraction of the time that the THMC spends in each state.[11] Naturally, the occupancy distribution π^* is also independent of the initial conditions. Thus, in ergodic THMCs, running just one simulation for long enough (which enables us to estimate π^*) will serve to estimate π just as well.

The question that comes to mind then is: How long is long enough? i.e. when will I know that the empirical distribution obtained by simulation resembles the limiting distribution π? Unfortunately there is no answer for that. The silver lining is that knowing that the limiting and the occupancy distribution coincide, that they must be stable in time and that they are independent of the initial conditions enables us to conduct a wide range of tests that may tell us when it is certainly *not* long enough. For example, we can run a battery of simulations and study the empirical distribution over the states of the system across samples as time goes by. If the distribution is not stable, then we have not run the model for long enough. Similarly, since the occupancy distribution is independent of the initial conditions, one can run several simulations with widely different initial conditions and compare the obtained occupancy distributions. If the empirical occupancy distributions are not similar, then we have not run the model for long enough. Many more checks can be conducted.

Admittedly, when analysing a computer model, one is often interested not so much in the distribution over the possible states of the system but rather in the distribution of a certain statistic. The crucial point is to realise that if the statistic is a function of the state of the system (and all statistics that can be extracted from the model are), then the limiting and the occupancy distributions of the statistic exist, coincide and are independent of the initial conditions.

Irreducible and Periodic THMCs

In contrast with aperiodic THMCs, the probability distribution of X_n in periodic THMCs does not approach a limit as n tends to infinity. Instead, in an irreducible THMC with period d, as n tends to infinity, X_n will in general cycle through d probability functions depending on the initial distribution. As an example, consider the simple random walk again (which is irreducible and periodic, with period 2), and assume that the random walker starts at patch number 1 (i.e. $X_0 = 1$). Given

---

[11]Formally, the occupancy of state i is defined as:

$$\pi_i^* = \lim_{n \to \infty} \frac{E(N_i(n))}{n+1}$$

where $N_i(n)$ denotes the number of times that the THMC visits state i over the time span $\{0, 1, \ldots, n\}$.

these settings, it can be shown that:

$$\lim_{n \to \infty} a_i^{(2n)} = \left[\frac{1}{16}, 0, \frac{1}{8}, 0\frac{1}{8}, 0, \frac{1}{8}, 0, \frac{1}{8}, 0, \frac{1}{8}, 0, \frac{1}{8}, 0, \frac{1}{16} \right]$$

$$\lim_{n \to \infty} a_i^{(2n+1)} = \left[0, \frac{1}{8}, 0, \frac{1}{8}, 0, \frac{1}{8}, 0\frac{1}{8}, 0, \frac{1}{8}, 0, \frac{1}{8}, 0, \frac{1}{8}, 0 \right]$$

In particular, the limits above show that the random walker cannot be at a patch with an even number in any even time-step, and she cannot be at a patch with an odd number in any odd time-step. In contrast, if the random walker started at patch number 2 (i.e. $X_0 = 2$), then the limits above would be interchanged. Fortunately, every irreducible (periodic or aperiodic) THMC does have a unique occupancy distribution $\pi*$, independent of the initial conditions (see Theorem 5.19 in Kulkarni 1999). In our particular example, this is:

$$\pi^* = \left[\frac{1}{32}, \frac{1}{16}, \frac{1}{16}, \frac{1}{16}, \frac{1}{16}, \frac{1}{16}, \frac{1}{16}, \frac{1}{16}, \frac{1}{16}, \frac{1}{16}, \frac{1}{16}, \frac{1}{16}, \frac{1}{16}, \frac{1}{16}, \frac{1}{16}, \frac{1}{32} \right]$$

Thus, the long-run fraction of time that the system spends in each state in any irreducible THMC is unique (i.e. independent of the initial conditions). This is a very useful result, since any statistic which is a function of the state of the system will also have a unique occupancy distribution independent of the initial conditions. As explained before, this occupancy distribution can be approximated with one single simulation run, assuming it runs for long enough.

13.8   Synergies Between Mathematical Analysis and Computer Simulation

In this section, we present various ways in which mathematical analysis and computer simulation can be combined to produce a better understanding of the dynamics of a model. Note that a full understanding of the dynamics of a model involves not only characterising (i.e. describing) them but also finding out *why* such dynamics are being observed, i.e. identifying the subsets of rules that are necessary or sufficient to generate certain aspects of the observed dynamics. To do this, one often has to make changes in the model, i.e. build supporting models that differ only slightly from the original one and may yield useful insights about its dynamics. These supporting models will sometimes be more tractable (e.g. if heterogeneity or stochasticity is averaged out) and sometimes more complex (e.g. if interactions that were assumed to be global in the original model may only take place locally in the supporting model). Thus, for clarity, we distinguish three different cases and deal with them in turn (see Fig. 13.13):

1. Characterisation of the dynamics of a model.

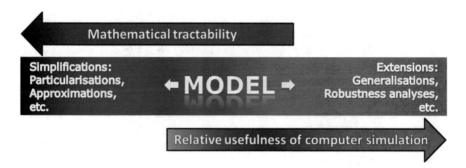

Fig. 13.13 To fully understand the dynamics of a model, one often has to study supporting models that differ only slightly from the original one. Some of these supporting models may be more tractable, whilst others may be more complex

2. Moves towards greater mathematical tractability. This involves creating and studying supporting models that are simpler than the original one.
3. Moves towards greater mathematical complexity. This involves creating and studying supporting models that are less tractable than the original one.

13.8.1 Characterising the Dynamics of the Model

There are many types of mathematical techniques that can be usefully combined with computer simulation to characterise the dynamics of a model (e.g. stochastic approximation theory (Benveniste et al. 1990; Kushner and Yin 1997)), but for limitations of space, we focus here on Markov chain analysis only.

When using Markov chain analysis to characterise the dynamics of a model, it may happen that the transition matrix can be easily computed, and we can operate with it or it may not. In the former case—which is quite rare in social simulation models—one can provide a full characterisation of the dynamics of the model just by operating with the transition matrix (see Proposition 1 and the beginning of Sect. 13.7.4 for references). In general, however, deriving and operating with the transition matrix may be unfeasible, and it is in this common case where there is a lot to gain in using Markov chain analysis and computer simulation together. The overall method goes as follows:

- Use Markov chain analysis to assess the relevance of initial conditions and to identify the different regimes in which the dynamics of the model may end up trapped.
- Use the knowledge acquired in the previous point to design suitable computational experiments aimed at estimating the exact probability distributions for the relevant statistics (which potentially depend on the initial conditions).

The following describes this overall process in greater detail. Naturally, the first step consists in finding an appropriate definition of the state of the system, as explained in Sect. 13.7.1. The next step is to identify all the closed communicating (i.e. absorbing) classes in the model C_v ($v \in \{1, 2, \ldots, k\}$). This allows us to partition the state space of the Markov chain as the union of all the closed communicating classes $C_1, C_2, \ldots, C_k$ in the model plus another class T containing all the states that belong to non-closed communicating classes. Izquierdo et al. (2009) illustrate how to do this in ten well-known models in the social simulation literature.

In most cases, conducting the partition of the state space is not as difficult as it may seem at first. In particular, the following proposition provides some simple sufficient conditions that guarantee that the computer model contains one single aperiodic absorbing class, i.e. the finite THMC that the computer model implements is irreducible and aperiodic (i.e. ergodic).

Proposition 4 Sufficient Conditions for Irreducibility and Aperiodicity

1. If it is possible to go from any state to any other state in one single time-step ($p_{i,j} > 0$ for all $i \neq j$) and there are more than two states, then the THMC is irreducible and aperiodic.
2. If it is possible to go from any state to any other state in a finite number of time-steps ($i \leftrightarrow j$ for all $i \neq j$) and there is at least one state in which the system may stay for two consecutive time-steps ($p_{i,i} > 0$ for some i), then the THMC is irreducible and aperiodic.
3. If there exists a positive integer n such that $p^{(n)}_{i,j} > 0$ for all i and j, then the THMC is irreducible and aperiodic (Janssen and Manca 2006, p. 107).

If one sees the transition diagram of the Markov chain as a (directed) network, the conditions above can be rewritten as:

1. The network contains more than two nodes and there is a directed link from every node to every other node.
2. The network is strongly connected and there is at least one loop.
3. There exists a positive integer n such that there is at least one walk of length n from any node to every node (including itself).

Izquierdo et al. (2009) show that many models in the social simulation literature satisfy one of these sufficient conditions (e.g. Epstein and Axtell's (1996) Sugarscape, Axelrod's (1986) metanorms models, Takahashi's (2000) model of generalised exchange and Miller and Page's (2004) standing ovation model with noise). This is important since, as explained in Sect. 13.7.4.2, in ergodic THMCs, the limiting and the occupancy distributions of any statistic exist, coincide and are independent of the initial conditions (so running just one simulation for long enough, which enables us to estimate the occupancy distribution, will serve to estimate the limiting distribution just as well).

Let us return to the general case. Having partitioned the state space, the analysis of the dynamics of the model is straightforward: all states in T (i.e. in any finite communicating class that is not closed) are transient, whereas all states in C_v (i.e. in any finite closed communicating class) are recurrent. In other words, sooner or later any simulation run will enter one of the absorbing classes C_v and stay in it forever.

Here computer simulation can play a crucial role again, since it allows us to estimate the probability of ending up in each of the absorbing classes for any (stochastic or deterministic) initial condition we may be interested in. A case in point would be a model that has only a few absorbing states or where various absorbing states are put together into only a few groups. Izquierdo et al. (2009) analyse models that follow that pattern: Axelrod's (1997b) model of dissemination of culture, Arthur's (1989) model of competing technologies and Axelrod and Bennett's (1993) model of competing bimodal coalitions. CharityWorld (Polhill et al. 2006; Izquierdo and Polhill 2006) is an example of a model with a unique absorbing state.

The following step consists in characterising the dynamics of the system within each of the absorbing classes. Once the system has entered a certain absorbing class C_v, it will remain in it forever exhibiting a unique conditional[12] occupancy distribution π_v^* over the set of states that compose C_v. Naturally, the same applies to any statistic we may want to study, since all statistics that can be extracted from the model are a function of the state of the system.

The conditional occupancy distribution π_v^* denotes the (strictly positive) long-run fraction of the time that the system spends in each state of C_v given that the system has entered C_v. Importantly, the conditional occupancy distribution π_v^* is the same regardless of the specific state through which the system entered C_v. The role of simulation here is to estimate these conditional occupancy distributions for the relevant statistics by running the model for long enough.

Finally, recall that some absorbing classes are periodic and some are aperiodic. Aperiodic absorbing classes have a unique conditional limiting distribution π_v denoting the long-run (strictly positive) probability of finding the system in each of the states that compose C_v given that the system has entered C_v. This conditional limiting distribution π_v coincides with the conditional occupancy distribution π_v^* and, naturally, is also independent of the specific state through which the system entered C_v. (Again, note that this also applies to the distribution of any statistic, as they are all functions of the state of the system, necessarily.)

In contrast with aperiodic absorbing classes, periodic absorbing classes do not generally have a unique limiting distribution; instead, they cycle through d probability functions depending on the specific state through which the system entered C_v (where d denotes the period of the periodic absorbing class). This is knowledge that one must take into account at the time of estimating the relevant probability distributions using computer simulation.

Thus, it is clear that Markov chain analysis and computer simulation greatly complement each other. Markov chain analysis provides the overall picture of the dynamics of the model by categorising its different dynamic regimes and identifying when and how initial conditions are relevant. Computer simulation uses this information to design appropriate computational experiments that allow us to quantify the probability distributions of the statistics we are interested in. As

---

[12]Given that the system has entered the absorbing class C_v

explained above, these probability distributions can always be approximated with any degree of accuracy by running the computer model several times.

There are several examples of this type of synergetic combination of Markov chain analysis and computer simulation in the literature. Galán and Izquierdo (2005) analysed Axelrod's (1986) agent-based model as a Markov chain, and concluded that the long-run behaviour of that model was independent of the initial conditions, in contrast to the initial conclusions of the original analysis. Galán and Izquierdo (2005) also used computer simulation to estimate various probability distributions. Ehrentreich (2002, 2006) used Markov chain analysis on the artificial stock market (Arthur et al. 1997; LeBaron et al. 1999) to demonstrate that the mutation operator implemented in the model is not neutral to the learning rate but introduces an upward bias.[13] A more positive example is provided by Izquierdo et al. (2007, 2008), who used Markov chain analysis and computer simulation to confirm and advance various insights on reinforcement learning put forward by Macy and Flache (2002) and Flache and Macy (2002).

13.8.2 Moves Towards Greater Mathematical Tractability: Simplifications

There are at least two types of simplifications that can help us to better understand the dynamics of a model. One consists in studying specific parameterisations of the original model that are thought to lead to particularly simple dynamics or to more tractable situations (Gilbert and Terna 2000; Gilbert 2007). Examples of this type of activity would be to run simulations without agents or with very few agents, explore the behaviour of the model using extreme parameter values, model very simple environments, etc. This activity is a common practice in the field (see, e.g. Gotts et al. 2003a, d).

A second type of simplification consists in creating an abstraction of the original model (i.e. a model of the model) which is mathematically tractable. An example of one possible abstraction would be to study the *expected* motion of the dynamic system (see the studies conducted by Galán and Izquierdo 2005; Edwards et al. 2003; Castellano et al. 2000; Huet et al. 2007; Mabrouk et al. 2007; Vilà 2008; Izquierdo et al. 2007, 2008 for illustrations of mean-field approximations). Since these mathematical abstractions do not correspond in a one-to-one way with the specifications of the formal model, any results obtained with them will not be conclusive in general, but they may give us insights suggesting areas of stability and basins of attraction, clarifying assumptions, assessing sensitivity to parameters or simply giving the option to illustrate graphically the expected dynamics of the

---

[13]This finding does not refute some of the most important conclusions obtained by the authors of the original model.

original model. This approach can also be used as a verification technique to detect potential errors and artefacts (Galán et al. 2009).

13.8.3 Moves Towards Greater Mathematical Complexity: Extensions

As argued before, understanding the dynamics of a model implies identifying the set of assumptions that are responsible for particular aspects of the obtained results. Naturally, to assess the relevance of any assumption in a model, it is useful to replace it with other alternatives, and this often leads to greater mathematical complexity.[14]

Ideally, the evaluation of the significance of an assumption is conducted by generalisation, i.e. by building a more general model that allows for a wide range of alternative competing assumptions, and contains the original assumption as a particular case. An example would be the introduction of arbitrary social networks of interaction in a model where every agent necessarily interacts with every other agent. In this case, the general model with arbitrary networks of interaction would correspond with the original model if the network is assumed to be complete, but any other network could also be studied within the same common framework. Another example is the introduction of noise in deterministic models.

Building models by generalisation is useful because it allows for a transparent, structured and systematic way of exploring the impact of various alternative assumptions that perform the same role in the model, but it often implies a loss in mathematical tractability (see, e.g. Izquierdo and Izquierdo 2006). Thus, it is often the case that a rigorous study of the impact of alternative assumptions in a model requires being prepared to slide up and down the tractability continuum depicted in Fig. 13.13 (Gotts et al. 2003). In fact, all the cases that are mentioned in the rest of this section involved greater complexity than the original models they considered, and computer simulation had to be employed to understand their dynamics.

In the literature there are many examples of the type of activity explained in this section. For example, Klemm et al. studied the relevance of various assumptions in Axelrod's model of dissemination of culture (Axelrod 1997b) by changing the network topology (Klemm et al. 2003a), investigating the role of dimensionality (Klemm et al. 2003b, 2005) and introducing noise (Klemm et al. 2003c). Another example is given by Izquierdo and Izquierdo (2007), who analysed the impact of using different structures of social networks in the efficiency of a market with quality variability.

---

[14]This is so because many assumptions we make in our models are, to some extent, for the sake of simplicity. As a matter of fact, in most cases the whole purpose of modelling is to build an abstraction of the world which is simpler than the world itself, so we can make inferences about the model that we cannot make directly from the real world (Edmonds 2001; Galán et al. 2009; Izquierdo et al. 2008a).

In the context of decision-making and learning, Flache and Hegselmann (1999) and Hegselmann and Flache (2000) compared two different decision-making algorithms that a set of players can use when confronting various types of social dilemmas. Similarly, Takadama et al. (2003) analysed the effect of three different learning algorithms within the same model.

Several authors, particularly in the literature of game theory, have investigated the effect of introducing noise in the decision-making of agents. This is useful not only to investigate the general effect of potential mistakes or experimentation but also to identify the stochastic stability of different outcomes (see Sect. 10 in Izquierdo et al. 2009). An illustrative example is given by Izquierdo et al. (2008), who investigate the reinforcement learning algorithm proposed by Bush and Mosteller (1955) using both mathematical analysis and simulation and find that the inclusion of small quantities of randomness in players' decisions can change the dynamics of the model dramatically.

Another assumption investigated in the literature is the effect of different spatial topologies (see, e.g. Flache and Hegselmann 2001, who generalised two of their cellular automata models by changing their—originally regular—grid structure). Finally, as mentioned in Sect. 13.5, it is increasingly common in the field of evolutionary game theory to assess the impact of various assumptions using computer simulation (see, e.g. Galán and Izquierdo 2005; Santos et al. 2006; Traulsen et al. 2006; Izquierdo and Izquierdo 2006).

13.9 Summary

In this chapter, we have provided a set of guidelines to understand the dynamics of computer models using both simulation and mathematical analysis. In doing so, it has become clear that mathematical analysis and computer simulation should not be regarded as alternative—or even opposed—approaches to the formal study of social systems but as complementary (Gotts et al. 2003, b). Not only can they provide fundamentally different insights on the same model, but they can also produce hints for solutions for each other. In short, there are plenty of synergies to be exploited by using the two techniques together, so the full potential of each technique cannot be reached unless they are used in conjunction.

To understand the dynamics of any particular computer model, we have seen that it is useful to see the computer model as the implementation of a function that transforms probability distributions over the set of possible inputs into probability distributions over the set of possible outputs. We refer to this function as the formal model that the computer model implements.

The mathematical approach to analyse formal models consists in examining the rules that define the model directly; the aim is to deduce the logical implications of these rules for any particular instance to which they can be applied. Our analysis of mathematical techniques to study formal models has been focused on the theory of

Markov chains. This theory is particularly useful for our purposes since many computer models can be meaningfully represented as time-homogenous Markov chains.

In contrast with mathematical analysis, the computer simulation approach does not look at the rules that define the formal model directly but instead tries to infer general properties of these rules by examining the outputs they produce when applied to particular instances of the input space. Thus, in the simulation approach, the data is produced by the computer using strict logical deduction, but the general patterns about how the rules transform the inputs into the outputs are inferred using generalisation by induction. Thus, in the general case—and in contrast with mathematical analysis—the inferences obtained using computer simulation will not be necessarily correct in a strict logical sense; but, on the other hand, computer simulation enables us to explore formal models beyond mathematical tractability, and the confidence we can place on the conclusions obtained with this approach can be rigorously assessed in statistical terms. Furthermore, as shown in this chapter, we can achieve any arbitrary level of accuracy in our computational approximations by running the model sufficiently many times.

Bearing in mind the relative strengths and limitations of both approaches, we have identified at least three different ways in which mathematical analysis and computer simulation can be usefully combined to produce a better understanding of the dynamics of computer models.

The first synergy appears at the time of characterising the dynamics of the formal model under study. To do that, we have shown how Markov chain analysis can be used to provide an overall picture of the dynamics of the model by categorising its different dynamic regimes and identifying when and how initial conditions are relevant. Having conducted such an analysis, one can then use computer simulation to design appropriate computational experiments with the aim of quantifying the probability distributions of the variables we are interested in. These probability distributions can always be approximated with any degree of accuracy by running the computer model several times.

The two other ways in which mathematical analysis and computer simulation can be combined derive from the fact that understanding the dynamics of a model involves not only characterising (i.e. describing) them but also finding out *why* such dynamics are being observed (i.e. discover causality). This often implies building supporting models that can be simpler or more complex than the original one. The rationale to move towards simplicity is to achieve greater mathematical tractability, and this often involves studying particularly simple parameterisations of the original model and creating abstractions which are amenable to mathematical analysis. The rationale to move towards complexity is to assess the relevance of specific assumptions, and it often involves building generalisations of the original model to explore the impact of competing assumptions that can perform the same role in the model but may lead to different results.

Let us conclude by encouraging the reader to put both mathematical analysis and computer simulation in their backpack and be happy to glide up and down the tractability spectrum where both simple and complex models lie. The benefits are out there.

Acknowledgements The authors have benefited from the financial support of the Spanish Ministry of Education and Science (projects DPI2005-05676 and TIN2008-06464-C03-02) and of the JCyL (projects VA006B09 and BU034A08). We are also very grateful to Nick Gotts, Bruce Edmonds and Gary Polhill for many extremely useful discussions.

Further Reading

Firstly, we suggest three things to read to learn more about Markov chain models. Grinstead and Snell (1997) provide an excellent introduction to the theory of finite Markov chains, with many examples and exercises. Häggström (2002) gives a clear and concise introduction to probability theory and Markov chain theory and then illustrates the usefulness of these theories by studying a range of stochastic algorithms with important applications in optimisation and other problems in computing. One of the algorithms covered is the Markov chain Monte Carlo method. Finally, Kulkarni (1995) provides a rigorous analysis of many types of useful stochastic processes, e.g. discrete and continuous time Markov chains, renewal processes, regenerative processes and Markov regenerative processes.

The reader may find three other papers helpful. Izquierdo et al. (2009) analyse the dynamics of ten well-known models in the social simulation literature using the theory of Markov chains, which is thus a good illustration of the approach in practice within the context of social simulation.[1] Epstein (2006) is a more general discussion, treating a variety of foundational and epistemological issues surrounding generative explanation in the social sciences and discussing the role of agent-based computational models in generative social science. Finally, Leombruni and Richiardi (2005) usefully discuss several issues surrounding the interpretation of simulation dynamics and the generalisation of the simulation results. For a different approach to analysing the dynamics of a simulation model, we refer the interested reader to Chap. 10 in this volume (Evans et al. 2017).

References

Arthur, W. B. (1989). Competing technologies, increasing returns, and lock-in by historical events. *Economic Journal, 99*(394), 116–131.

Arthur, W. B., Holland, J. H., LeBaron, B., Palmer, R., & Tayler, P. (1997). Asset pricing under endogenous expectations in an artificial stock market. In W. B. Arthur, S. Durlauf, & D. Lane (Eds.), *The economy as an evolving complex system II* (pp. 15–44). Reading, MA: Addison-Wesley Longman.

Axelrod, R. M. (1986). An evolutionary approach to norms. *American Political Science Review, 80*(4), 1095–1111.

---

[1]This comment is added by the editors as the authors are too modest to so describe their own work.

Axelrod, R. M. (1997a). Advancing the art of simulation in the social sciences. In R. Conte, R. Hegselmann, & P. Terna (Eds.), *Simulating social phenomena, Lecture notes in economics and mathematical systems* (Vol. 456, pp. 21–40). Berlin: Springer.

Axelrod, R. M. (1997b). The dissemination of culture: A model with local convergence and global polarization. *Journal of Conflict Resolution, 41*(2), 203–226.

Axelrod, R. M., & Bennett, D. S. (1993). A landscape theory of aggregation. *British Journal of Political Science, 23*(2), 211–233.

Axtell, R. L. (2000). Why agents? On the varied motivations for agent computing in the social sciences. In C. M. Macal & D. Sallach (Eds.), *Proceedings of the workshop on agent simulation: applications, models, and tools* (pp. 3–24). Argonne, IL: Argonne National Laboratory.

Axtell, R. L., & Epstein, J. M. (1994). Agent based modeling: Understanding our creations. *The Bulletin of the Santa Fe Institute, Winter, 1994*, 28–32.

Balzer, W., Brendel, K. R., & Hofmann, S. (2001). Bad arguments in the comparison of game theory and simulation in social studies. *Journal of Artificial Societies and Social Simulation, 4*(2). http://jasss.soc.surrey.ac.uk/4/2/1.html

Benveniste, A., Métivier, M., & Priouret, P. (1990). *Adaptive algorithms and stochastic approximations*. Berlin: Springer.

Böhm, C., & Jacopini, G. (1966). Flow diagrams, turing machines and languages with only two formation rules. *Communications of the ACM, 9*(5), 366–371.

Bush, R. R., & Mosteller, F. (1955). *Stochastic models for learning*. New York: Wiley.

Castellano, C., Marsili, M., & Vespignani, A. (2000). Nonequilibrium phase transition in a model for social influence. *Physical Review Letters, 85*(16), 3536–3539.

Cutland, N. (1980). *Computability: An introduction to recursive function theory*. Cambridge: Cambridge University Press.

Chung, K.L. (1960). Markov Chains with Stationary Transition Probabilities. Springer, Berlin.https://link.springer.com/book/10.1007%2F978-3-642-49686-8

Edmonds, B. (2001). The use of models: Making MABS actually work. In S. Moss & P. Davidsson (Eds.), *Multi-agent-based simulation, Lecture notes in artificial intelligence* (Vol. 1979, pp. 15–32). Berlin: Springer.

Edmonds, B. (2005). Simulation and complexity: How they can relate. In V. Feldmann & K. Mühlfeld (Eds.), *Virtual worlds of precision: Computer-based simulations in the sciences and social sciences* (pp. 5–32). Lit-Verlag: Münster, Germany.

Edwards, M., Huet, S., Goreaud, F., & Deffuant, G. (2003). Comparing an individual-based model of behaviour diffusion with its mean field aggregate approximation. *Journal of Artificial Societies and Social Simulation, 6*(4). http://jasss.soc.surrey.ac.uk/6/4/9.html

Ehrentreich, N. (2002) *The Santa Fe artificial stock market re-examined: Suggested corrections* (Betriebswirtschaftliche Diskussionsbeiträge Nr. 45/02). Halle/Saale, Germany: Wirtschaftswissenschaftliche Fakultät, Martin-Luther-Universität. http://econwpa.wustl.edu:80/eps/comp/papers/0209/0209001.pdf

Ehrentreich, N. (2006). Technical trading in the Santa Fe institute artificial stock market revisited. *Journal of Economic Behavior & Organization, 61*(4), 599–616.

Epstein, J. M. (2006). Remarks on the foundations of agent-based generative social science. In K. L. Judd & L. Tesfatsion (Eds.), *Handbook of computational economics, Agent-based computational economics* (Vol. 2, pp. 1585–1604). Amsterdam, The Netherlands: North-Holland.

Epstein, J. M., & Axtell, R. L. (1996). *Growing artificial societies: Social science from the bottom up*. Cambridge, MA: Brookings Institution Press/MIT Press.

Evans, A., Heppenstall, A., & Birkin, M. (2017). Understanding simulation results. In B. Edmonds & R. Meyer (Eds.), *Simulating social complexity: A handbook*. Cham: Springer.

Flache, A., & Hegselmann, R. (1999). Rationality vs. learning in the evolution of solidarity networks: A theoretical comparison. *Computational & Mathematical Organization Theory, 5*(2), 97–127.

Flache, A., & Hegselmann, R. (2001). Do irregular grids make a difference? Relaxing the spatial regularity assumption in cellular models of social dynamics. *Journal of Artificial Societies and Social Simulation, 4*(4). http://jasss.soc.surrey.ac.uk/4/4/6.html

Flache, A., & Macy, M. W. (2002). Stochastic collusion and the power law of learning. *Journal of Conflict Resolution, 46*(5), 629–653.

Galán, J. M., & Izquierdo, L. R. (2005). Appearances can be deceiving: Lessons learned re-implementing axelrod's 'evolutionary approach to norms'. *Journal of Artificial Societies and Social Simulation, 8*(3). http://jasss.soc.surrey.ac.uk/8/3/2.html

Galán, J. M., Izquierdo, L. R., Izquierdo, S. S., Santos, J. I., Del Olmo, R., López-Paredes, A., et al. (2009). Errors and artefacts in agent-based modelling. *Journal of Artificial Societies and Social Simulation, 12*(1). http://jasss.soc.surrey.ac.uk/12/1/1.html

Genz, A., & Kwong, K. S. (2000). Numerical evaluation of singular multivariate normal distributions. *Journal of Statistical Computation and Simulation, 68*(1), 1–21.

Gilbert, N. (1999). Simulation: A new way of doing social science. *The American Behavioral Scientist, 42*(10), 1485–1487.

Gilbert, N. (2007). *Agent-based models, Quantitative applications in the social sciences* (Vol. 153). London: Sage Publications.

Gilbert, N., & Terna, P. (2000). How to build and use agent-based models in social science. *Mind and Society, 1*(1), 57–72.

Gilbert, N., & Troitzsch, K. G. (1999). *Simulation for the social scientist*. Buckingham, UK: Open University Press.

Gotts, N. M., Polhill, J. G., & Adam, W. J. (2003, September 18–21). Simulation and analysis in agent-based modelling of land use change. In *Online proceedings of the first conference of the European social simulation association*, Groningen, The Netherlands. http://www.uni-koblenz.de/~essa/ESSA2003/proceedings.htm

Gotts, N. M., Polhill, J. G., & Law, A. N. R. (2003a). Agent-based simulation in the study of social dilemmas. *Artificial Intelligence Review, 19*(1), 3–92.

Gotts, N. M., Polhill, J. G., & Law, A. N. R. (2003b). Aspiration levels in a land-use simulation. *Cybernetics and Systems, 34*(8), 663–683.

Gotts, N. M., Polhill, J. G., Law, A. N. R, & Izquierdo, L. R. (2003, April 7–11). Dynamics of imitation in a land use simulation. In K. Dautenhahn & C. Nehaniv (Eds.), *Proceedings of the second international symposium on imitation in animals and artefacts* (pp. 39–46). Aberystwyth: University of Wales.

Grinstead, C. M., & Snell, J. L. (1997). Chapter 11: Markov chains. In C. M. Grinstead & J. L. Snell (Eds.), *Introduction to probability* (2nd Revised ed., pp. 405–470). Providence, RI: American Mathematical Society. http://www.dartmouth.edu/~chance/teaching_aids/books_articles/probability_book/book.html

Häggström, O. (2002). *Finite markov chains and algorithmic applications*. Cambridge: Cambridge University Press.

Harel, D. (1980). On folk theorems. *Communications of the ACM, 23*(7), 379–389.

Hauert, C., & Doebeli, M. (2004). Spatial structure often inhibits the evolution of cooperation in the snowdrift game. *Nature, 428*(6983), 643–646.

Hegselmann, R., & Flache, A. (2000). Rational and adaptive playing. *Analyse & Kritik, 22*(1), 75–97.

Holland, J. H., & Miller, J. H. (1991). Artificial adaptive agents in economic theory. *American Economic Review, 81*(2), 365–370.

Huet, S., Edwards, M., & Deffuant, G. (2007). Taking into account the variations of neighbourhood sizes in the mean-field approximation of the threshold model on a random network. *Journal of Artificial Societies and Social Simulation, 10*(1). http://jasss.soc.surrey.ac.uk/10/1/10.html

Imhof, L. A., Fudenberg, D., & Nowak, M. A. (2005). Evolutionary cycles of cooperation and defection. *Proceedings of the National Academy of Sciences of the United States of America, 102*(31), 10797–10800.

Izquierdo, L. R., Galán, J. M., Santos, J. I., & Olmo, R. (2008). Modelado de sistemas complejos mediante simulación basada en agentes y mediante dinámica de sistemas. *Empiria, 16*, 85–112.

Izquierdo, L. R., Izquierdo, S. S., Galán, J. M., & Santos, J. I. (2009). Techniques to understand computer simulations: Markov chain analysis. *Journal of Artificial Societies and Social Simulation, 12*(1). http://jasss.soc.surrey.ac.uk/12/1/6.html

Izquierdo, L. R., Izquierdo, S. S., Gotts, N. M., & Polhill, J. G. (2007). Transient and asymptotic dynamics of reinforcement learning in games. *Games and Economic Behavior, 61*(2), 259–276.

Izquierdo, L. R., & Polhill, J. G. (2006). Is your model susceptible to floating point errors? *Journal of Artificial Societies and Social Simulation, 9*(4). http://jasss.soc.surrey.ac.uk/9/4/4.html

Izquierdo, S. S., & Izquierdo, L. R. (2006). On the structural robustness of evolutionary models of cooperation. In E. Corchado, H. Yin, V. J. Botti, & C. Fyfe (Eds.), *Intelligent data engineering and automated learning - IDEAL 2006, Lecture notes in computer science* (Vol. 4224, pp. 172–182). Berlin: Springer.

Izquierdo, S. S., & Izquierdo, L. R. (2007). The impact on market efficiency of quality uncertainty without asymmetric information. *Journal of Business Research, 60*(8), 858–867.

Izquierdo, S. S., Izquierdo, L. R., & Gotts, N. M. (2008). Reinforcement learning dynamics in social dilemmas. *Journal of Artificial Societies and Social Simulation, 11*(2). http://jasss.soc.surrey.ac.uk/11/2/1.html

Janssen, J., & Manca, R. (2006). *Applied semi-markov processes.* New York, NY: Springer.

Karr, A. F. (1990). Markov processes. In D. P. Heyman & M. J. Sobel (Eds.), *Stochastic models, Handbooks in operations research and management science* (Vol. 2, pp. 95–123). Amsterdam: Elsevier.

Klemm, K., Eguíluz, V. M., Toral, R., & San Miguel, M. (2003a). Nonequilibrium transitions in complex networks: A model of social interaction. *Physical Review E, 67*(2), 026120.

Klemm, K., Eguíluz, V. M., Toral, R., & San Miguel, M. (2003b). Role of dimensionality in Axelrod's model for the dissemination of culture. *Physica A, 327*(1–2), 1–5.

Klemm, K., Eguíluz, V. M., Toral, R., & San Miguel, M. (2003c). Global culture: A noise-induced transition in finite systems. *Physical Review E, 67*(4), 045101.

Klemm, K., Eguíluz, V. M., Toral, R., & San Miguel, M. (2005). Globalization, polarization and cultural drift. *Journal of Economic Dynamics and Control, 29*(1–2), 321–334.

Kulkarni, V. G. (1995). *Modeling and analysis of stochastic systems.* Boca Raton, FL: Chapman & Hall/CRC.

Kulkarni, V. G. (1999). *Modeling, analysis, design, and control of stochastic systems.* New York: Springer.

Kushner, H. J., & Yin, G. G. (1997). *Stochastic approximation algorithms and applications.* New York, NY: Springer.

Lebaron, B., Arthur, W. B., & Palmer, R. (1999). Time series properties of an artificial stock market. *Journal of Economic Dynamics & Control, 23*(9–10), 1487–1516.

Leombruni, R., & Richiardi, M. (2005). Why are economists sceptical about agent-based simulations? *Physica A, 355*(1), 103–109.

Lieberman, E., Havlin, S., & Nowak, M. A. (2009). Evolutionary dynamics on graphs. *Nature, 433*(7023), 312–316.

Mabrouk, N., Deffuant, G., & Lobry, C. (2007, March 15–16). Confronting macro, meso and micro scale modelling of bacteria dynamics. In *M2M 2007: Third international model-to-model workshop*, Marseille, France. http://m2m2007.macaulay.ac.uk/M2M2007-Mabrouk.pdf

Macy, M. W., & Flache, A. (2002). Learning dynamics in social dilemmas. *Proceedings of the National Academy of Sciences of the United States of America, 99*(3), 7229–7236.

Miller, J. H., & Page, S. E. (2004). The standing ovation problem. *Complexity, 9*(5), 8–16.

Nowak, M. A., & May, R. M. (1992). Evolutionary games and spatial chaos. *Nature, 359*(6398), 826–829.

Nowak, M. A., & Sigmund, K. (1992). Tit for tat in heterogeneous populations. *Nature, 355*(6357), 250–253.

Nowak, M. A., & Sigmund, K. (1993). A strategy of win-stay, lose-shift that outperforms tit for tat in the Prisoner's Dilemma game. *Nature, 364*(6432), 56–58.

Ostrom, T. (1988). Computer simulation: The third symbol system. *Journal of Experimental Social Psychology, 24*(5), 381–392.

Polhill, J. G., Izquierdo, L. R., & Gotts, N. M. (2006). What every agent based modeller should know about floating point arithmetic. *Environmental Modelling & Software, 21*(3), 283–309.

Richiardi, M., Leombruni, R., Saam, N. J., & Sonnessa, M. (2006). A common protocol for agent-based social simulation. *Journal of Artificial Societies and Social Simulation, 9*(1). http://jasss.soc.surrey.ac.uk/9/1/15.html

Santos, F. C., Pacheco, J. M., & Lenaerts, T. (2006). Evolutionary dynamics of social dilemmas in structured heterogeneous populations. *Proceedings of the National Academy of Sciences of the United States of America, 103*(9), 3490–3494.

Suber, P. (2002). *Formal systems and machines: An isomorphism. Electronic hand-out for the course "Logical Systems"*. Richmond, IN: Earlham College. http://www.earlham.edu/~peters/courses/logsys/machines.htm

Takadama, K., Suematsu, Y. L., Sugimoto, N., Nawa, N. E., & Shimohara, K. (2003). Cross-element validation in multiagent-based simulation: Switching learning mechanisms in agents. *Journal of Artificial Societies and Social Simulation, 6*(4). http://jasss.soc.surrey.ac.uk/6/4/6.html

Takahashi, N. (2000). The emergence of generalized exchange. *American Journal of Sociology, 10*(4), 1105–1134.

Traulsen, A., Nowak, M. A., & Pacheco, J. M. (2006). Stochastic dynamics of invasion and fixation. *Physical Review E, 74*(1), 011909.

Vilà, X. (2008). A model-to-model analysis of bertrand competition. *Journal of Artificial Societies and Social Simulation, 11*(2). http://jasss.soc.surrey.ac.uk/11/2/11.html

Wikipedia. (2007). *Structured program theorem*. http://en.wikipedia.org/w/index.php?title=Structured_program_theorem&oldid=112885072

Wilensky, U. (1999). *NetLogo*. http://ccl.northwestern.edu/netlogo

Chapter 14
Interpreting and Understanding Simulations: The Philosophy of Social Simulation

R. Keith Sawyer

Abstract Simulations are usually directed at some version of the question: What is the relationship between the individual actor and the collective community? Among social scientists, this question generally falls under the topic of *emergence*. Sociological theorists and philosophers of science have developed sophisticated approaches to emergence, including the critical question: to what extent can emergent phenomena be reduced to explanations in terms of their components? Modelers often proceed without considering these issues; the risk is that one might develop a simulation that does not accurately reflect the observed empirical facts or one that implicitly sides with one side of a theoretical debate that remains unresolved. In this chapter, I provide some tips for those developing simulations, by drawing on a strong recent tradition of analyzing scientific explanation that is found primarily in the philosophy of science but also to some extent in sociology.

Why Read This Chapter?
To gain an overview of some key philosophical issues that underlie social simulation. Providing an awareness of them may help avoid the risk of presenting a very limited perspective on the social world in any simulations you develop.

14.1 Introduction

Researchers who develop multi-agent simulations often proceed without thinking about what the results will mean or how the results of the simulation might be used. After the simulation is completed, it is too late to begin to think about interpretation and understanding, because poorly designed simulations often turn out to be uninterpretable. An analysis of interpretation and understanding has to precede and inform the design. In this chapter, I provide some tips for those developing simulations, by drawing on a strong recent tradition of analyzing scientific explanation that is found

R.K. Sawyer (✉)
University of North Carolina, Chapel Hill, NC, 27599, USA
e-mail: rksawyer@email.unc.edu

© Springer International Publishing AG 2017
B. Edmonds, R. Meyer (eds.), *Simulating Social Complexity*,
Understanding Complex Systems, https://doi.org/10.1007/978-3-319-66948-9_14

primarily in the philosophy of science but also to some extent in sociology. My hope is that this exploration can help modelers by identifying hidden assumptions that often underlie simulation efforts and assumptions that sometimes can limit the explanatory power of the resulting simulation.

Modelers often proceed without considering these issues, in part, because of the historical roots of the approach. During the 1990s, two strands of work in computer science began to merge: *intelligent agents*, a tradition of studying autonomous agents that emerged from the artificial intelligence community, and *artificial life*, typically a graphically represented two-dimensional grid in which each cell's behavior is determined by its nearby neighbors. In the 1990s, a few members of both of these communities began to use the new technologies to simulate social phenomena. Several intelligent agent researchers began to explore systems with multiple agents in communication, which became known as *distributed artificial intelligence*, and artificial life researchers began to draw parallels between their two-dimensional grids and various real-world phenomena, such as ant colonies and urban traffic flows. The new field that emerged within computer science is generally known as *multi-agent systems*; the term refers to engineering computer systems with many independent and autonomous agents, which communicate with each other using a well-specified set of message types. When multi-agent systems are used specifically to simulate social phenomena, the simulations are known as multi-agent-based simulations (MABS), artificial societies, or social simulations.

Computer scientists had been working with multiple processor systems since the 1980s, when computer scientists began to experiment with breaking up a computational task into subtasks and then assigning the subtasks to separate, stand-alone computers. During this time, a specialized massively parallel computer called the *thinking machine* was famously built by the Thinking Machines Corporation, and computer scientists began developing formalisms to represent distributed computational algorithms. These early efforts at parallel computation were centrally managed and controlled, with the distribution of computation being used to speed up a task or to make it more efficient.

The multi-agent systems of the 1990s represented a significant shift from these earlier efforts, in that each computational agent was *autonomous*: capable of making its own decisions and choosing its own course of action. The shift to autonomous agents raised several interesting issues among computer scientists: if an agent were asked to execute a task, when and why would the agent agree to do it? Perhaps an agent would have a different understanding of how to execute a task than the requesting agent; how could two agents negotiate these understandings? Such questions would be unlikely to arise if one central agent had control over all of the distributed agents. But the rapid growth of the Internet—where there are, in fact, many autonomous computers and systems that communicate with each other every day—resulted in a real-world situation in which independent computational agents might choose not to respond to a request or might respond differently than the requester expected.

As a result of these historical developments, computer scientists found themselves grappling with questions that have long been central to sociology and economics. Why should a person take an action on behalf of the collective good? How to prevent free-riding and social loafing—situations where agents benefit from collective action but without contributing very much? What is the relationship between the individual actor and the collective community? What configurations of social network are best suited to different sorts of collective tasks?

Among social scientists, these questions generally fall under the topic of *emergence* (Sawyer 2005). Emergence refers to a certain kind of relation between system-level phenomena or properties and properties of the system components. The traditional method of understanding complex system phenomena has been to break the system up into its components, to analyze each of the components and the interactions among them. In this reductionist approach, the assumption is that at the end of this process, the complex system will be fully explained. In contrast, scholars of emergence are unified in arguing that this reductionist approach does not work for a certain class of complex system phenomena. There are different varieties of this argument; some argue that the reduction is not possible for epistemological reasons (it is simply too hard to explain using reduction although the whole is really nothing more than the sum of the parts), while others argue that it is not possible for ontological reasons (the whole is something more and different than the sum of the parts). Emergent phenomena have also been described as novel—not observed in any of the components—and unpredictable, even if one has already developed a full and complete explanation of the components and of their interactions.

Among sociologists, this debate is generally known as the individualism-collectivism debate. *Methodological individualists* are the reductionists, those who argue that all social phenomena can be fully explained by completely explaining the participating individuals and their interactions. *Collectivists* argue, in contrast, that some social phenomena cannot be explained by reduction to the analysis of individuals and their interactions. Several scholars have recently noted that agent-based simulations are an appropriate tool to explore these issues (Neumann 2006; Sawyer 2005; Schmid 2006).

In the mid-1990s, a few sociologists who were interested in the potential of computer simulation to address these questions began to join with computer scientists who were fascinated with the more theoretical dimensions of these very practical questions, and the field of multi-agent-based simulation (MABS) was born. Since that time, technology has rapidly advanced, and now there are several computer tools, relatively easy to use, that allow social scientists to develop multi-agent simulations of social phenomena.

After almost 15 years, this handbook provides an opportunity to look reflexively at this work and to ask: what do these simulations mean? How should scientists interpret them? Such questions have traditionally been associated with the philosophy of science. For over a century, philosophers of science have been exploring topics that are fundamental to science: understanding, explanation, perception, validation, and interpretation. In the following, I draw on concepts and arguments from within the philosophy of science to ask a question that is critical to scientists: What do these multi-agent based simulations mean? How should we interpret simulations?

In spite of the occasional participation of sociologists in social simulation projects, most modelers continue to proceed without an awareness of these important foundational questions. These are of more than simple theoretical importance; without an awareness of these somewhat complex issues of interpretation and understanding, the risk is that one might develop a simulation that does not accurately reflect the observed empirical facts or one that implicitly sides with one side of a theoretical debate that remains unresolved (thus clouding the interpretation of the simulation's results).

I address these questions by delving into the nature of *explanation*. Many developers of MABS believe that the simulations provide explanations of real phenomena—that the concrete specifics of what is going on in the world are revealed by examining the simulation. Interpreting and understanding the simulation thus result in an explanation of the target phenomenon.

Although simulation has many potential benefits to offer scientists, I argue that the meaning of a simulation is rarely obvious and unequivocal. For example, there are many different ways to understand and interpret a given simulation of a social phenomenon. These run roughly along a spectrum from a very narrowly focused and ungeneralizable simulation of a very specific instance of a social phenomenon to a grand-theoretical type of simulation that explains a very broad range of social emergence phenomena. The specific end of the spectrum results in better understanding of a single phenomenon, but not in any lawful regularities nor in any general knowledge about social life. The general end of the spectrum is something like the tradition of grand theory in sociology, with generalizable laws that explain a wide range of social phenomena. The center of the spectrum is associated with what Merton famously called "middle range theories"; this is what most mainstream sociologists today believe is the appropriate task of sociology, as the field has turned away from grand theorizing in recent decades.

All of these are valid forms of sociological explanation, and each has the potential to increase our understanding of social life. My concern is with the specific end of the spectrum: in some cases, it could be that a simulation explains only a very narrow single case, with no generalizability. This would be of limited usefulness to our understanding.

14.2 Interpreting Multi-agent Simulations

How should scientists outside of the simulation community interpret a multi-agent simulation of a real phenomenon? Although there has been almost no philosophical attention to these simulations, simulation developers themselves have often engaged in discussions of the scientific status of their simulations. Within this community, there is disagreement about the scientific status of the simulations. Opinions fall into two camps: the "simulation as theory" camp and the "simulation as experiment" camp.

Representing the first group, many of those developing computer simulations believe that in building them, they are engaged in a form of theory construction (Conte et al. 2001; Markovsky 1997; also see Ostrom 1988). They argue that social simulation, for example, is a more sophisticated and advanced form of social theory, because concepts and axioms must be rigorously specified to be implemented in a computer program, unlike the "discursive theorizing" of many sociologists, which is relatively vague and hard to empirically test (Conte et al. 2001; Turner 1993). As Markovsky (1997) noted, turning a (discursive) sociological theory into a simulation is not a transparent translation. A variable in the theory may turn out to be central to the simulation, or it may turn out not to matter very much; one cannot know which without going through the exercise of programming the simulation. Developing a simulation almost always reveals logical gaps in a theory, and these must be filled in before the simulation will work. As a result, simulations often introduce logical relationships that the original theory did not specify, and they contain gap-filling assumptions that the theory never made.

Representing the second group, other modelers have argued that a simulation is a *virtual experiment* (Carley and Gasser 1999). From this perspective, simulations cannot explain in and of themselves, but can only serve as tests of a theory—and the theory is what ultimately does the explaining. In a virtual experiment, a model is developed that simulates some real-world social phenomenon but with one or more features modified to create experimental conditions that can be contrasted. For example, the same business organization could be modeled multiple times but with a strong authority figure in one simulation and a diffuse authority structure in another (Carley and Gasser 1999). Whereas it would probably be impossible to implement such an experiment with real-world societies, a computer model readily allows such manipulation. When the model is started, the simulations that result behave in ways that are argued to be analogous to how the real-world organization would have behaved, in each of the different conditions. In this view, because the simulation plays the role of a data-generating experiment, it doesn't provide an explanation; rather, it provides raw data to aid in theorizing, and the theory ultimately provides the explanation.

14.3 Scientific Explanation

Explanations are attempts to account for *why* things happen—singular events or regular, repeatable patterns. In the philosophy of science, there is a long history of discussion surrounding scientific explanation, including the deductive-nomological (D-N) or covering law approach (Hempel 1965), the statistical relevance approach (Salmon 1971), and the mechanistic approach (Bechtel and Richardson 1993; Salmon 1984). Here I limit the term "explanation" to *causal* explanation (cf. Little 1998; Woodward 2003). The relation between causation and explanation is complex; some philosophers of science hold that all explanation must be causal, whereas others deny this. For example, in the deductive-nomological tradition of logical empiricism, laws are said to provide explanations, even though the status of causation is questionable—causation is thought to be nothing more than an observed regularity as captured by a covering law. In the more recent mechanistic approach, in contrast, causation is central to explanation. I take the mechanistic position that causal mechanism is central to explanation, but I first briefly summarize the covering law notion of explanation.

In the covering law approach, a phenomenon is said to be explained when salient properties of the event are shown to be consequents of general laws, where the antecedents can also be identified. The phenomenon is said to be explained by the combination of the antecedent conditions and the laws that then result in the phenomenon. A strength of the covering law approach is that laws both explain *and* predict; once a law is discovered, it can be used both to explain past phenomena and also to predict when similar phenomena will occur in the future.

Covering law models have always been problematic in the social sciences, primarily because of difficulty translating the notion of "law" to social reality. After all, advocates of the covering law model have had trouble adequately defining "law" even in the physical world (Hempel 1965). Candidates for social laws always have exceptions, and laws with exceptions are problematic in the D-N approach. There is a history of debate concerning whether social laws exist at all, with prominent social theorists such as Anthony Giddens arguing that there are no social laws (1984) and other prominent social theorists arguing that there are (e.g., Peter Blau 1977, 1983). Philosophers of social science have taken various positions on the status of social laws (Beed and Beed 2000; Kincaid 1990; Little 1993; McIntyre 1996). Much of this discussion centers on what constitutes a law: must it be invariant and universal (Davidson's 1980 "strict law"), or can it admit of some exceptions? Even the strongest advocates of lawful explanation admit that there are no strict laws in the social sciences; these laws will typically have exceptions, and the law cannot explain those exceptions.

In the last decade or so, philosophers of biology (Bechtel 2001; Bechtel and Richardson 1993; Craver 2001; Craver 2002; Glennan 1996; Machamer et al. 2000) and philosophers of social science (Elster 1989; Hedström 2005; Hedström and Swedberg 1998; Little 1991; Little 1998; Stinchcombe 1991) have begun to develop a different approach to explanation, one based on causal mechanisms rather than laws. In the mechanism approach, a phenomenon is said to be explained when the realizing mechanism that gave rise to the phenomenon is sufficiently described. Mechanistic accounts of explanation are centrally concerned with causation. For example, Salmon's (1984, 1994, 1997) causal mechanical model focuses on causal processes and their causal interactions; an explanation of an event traces the causal processes and interactions leading up to that event and also describes the processes and interactions that make up the event.

Hedström (2005) presented an account of how social simulations correspond to mechanistic explanations. Mechanistic explanations differ from covering law explanations by "specifying mechanisms that show how phenomena are brought about" (p. 24). Of course, how one defines "mechanism" is the crux of the approach. Some theorists believe that mechanisms provide causal explanations, whereas others do not. But what the approaches share is that "a mechanism explicates the details of how the regularities were brought about" (p. 24). Rather than explanation in terms of laws and regularities, a mechanism approach provides explanations by postulating the processes constituted by the operation of mechanisms that generate the observed phenomenon. For Hedström, "A social mechanism ... describes a constellation of entities and activities that are organized such that they regularly bring about a particular type of outcome" (p. 25). The explanation is provided by the specification of often unobservable causal mechanisms and the identification of the processes in which they are embedded. In other words, mechanists are willing to grant that macro-level regularities are observed; the covering law approach to sociology has, after all, resulted in the identification of lawlike regularities that are empirically supported. But for a mechanist, a covering law does not *explain*: "correlations and constant conjunctions do not explain but require explanation by reference to the entities and activities that brought them into existence" (p. 26).

14.4 MABS: Explaining by Simulating

Using MABS, researchers have begun to model the mechanisms whereby macroso-cial properties emerge from interacting networked agents. A MABS contains many autonomous computational agents that negotiate and collaborate with each other, in a distributed, self-organizing fashion. The parallels with causal mechanism approaches in the philosophy of science are striking (Sawyer 2004).

Hedström (2005) refers to his social simulation method as *empirically calibrated agent-based models (ECA)* to emphasize that the models should be grounded in quantitative empirical data. His recommended method is to (1) develop a stylized agent-based model "that explicates the logic of the mechanism assumed to be operative" (p. 143), (2) use relevant data to verify the mechanism actually works this way, and (3) run the model and modify it until it best matches relevant data. "Only when our explanatory account has passed all of these three stages can we claim to have an empirically verified mechanism-based explanation of a social outcome" (p. 144). Hedström provides an extended demonstration of the ECA method by modeling how social interactions might have given rise to the increase in youth unemployment in Stockholm in the 1990s. His model includes exactly as many computational agents as there were unemployed 20–24-year-olds in Stockholm during 1993 to 1999: 87,924. The demographic characteristics of these computational agents were an accurate reflection of their real-world counterparts. He then created a variety of different simulations as "virtual experiments" to see which simulation resulted in the best match to the observed empirical data.

The Stockholm simulation falls at the more specific end of the explanatory spectrum; even if it successfully simulates unemployment in Stockholm, it may not be helpful at understanding unemployment in any other city. Other MABS are designed to provide more general explanations. For example, many MABS have explored one of the most fundamental economic and sociological questions: What is the origin of social norms? For example, how do norms of cooperation and trust emerge? If autonomous agents seek to maximize personal utility, then under what conditions will agents cooperate with other agents? In game theory terms, this is a prisoner's dilemma problem. Many studies of cooperation in artificial societies have been implementations of the *iterated prisoner's dilemma (IPD)*, where agents interact in repeated trials of the game and agents can remember what other agents have done in the past (Axelrod 1997).

The sociologists Macy and Skvoretz (1998) developed an artificial society to explore the evolution of trust and cooperation between strangers. In prior simulations of the prisoner's dilemma, trust emerged in the iterated game with familiar neighbors, but trust did not emerge with strangers. Macy and Skvoretz hypothesized that if the agents were grouped into neighborhoods, norms of trust would emerge among neighbors within each neighborhood and that these norms would then extend to strangers. Their simulation contained 1000 agents that played the prisoner's dilemma game with both familiar neighbors and with strangers. To explore the effects of community on the evolution of PD strategy, the simulation defined neighborhoods that contained varying numbers of agents—from 9 agents per neighborhood to 50. Different runs of the simulation varied the *embeddedness* of interaction: the probability that in a given iteration, a player would be interacting with a neighbor or a stranger. These simulations showed that conventions for trusting strangers evolved in neighborhoods of all sizes, as long as agents interacted more with neighbors than strangers (embeddedness greater than 0.5). The rate of cooperation among strangers increased linearly as embeddedness was raised from 0.5 to 0.9. Simulations with smaller neighborhoods resulted in a higher rate of cooperation between strangers: at 0.9 embeddedness, the rate of cooperation between strangers was 0.62 in the 10-member neighborhood simulation and 0.45 in the 50-member neighborhood simulation (p. 655).

Macy and Skvoretz concluded that these neighborhoods—characterized by relatively dense interactions—allow conventions for trusting strangers to emerge and become stable and then diffuse to other neighborhoods via weak ties. If an epidemic of distrusting behavior evolves in one segment of the society, the large number of small neighborhoods facilitates the restoration of order (p. 657). This simulation demonstrates how social structure can influence micro- to macro-emergence processes; cooperation with strangers emerges when agents are grouped into neighborhoods, but not when they are ungrouped.

An advocate of the causal mechanist approach to explanation would argue that the Macy and Skvoretz simulation provides candidate explanations of several social phenomena. First, the simulation explains how norms of cooperation could emerge among friends in small communities—because exchanges are iterated, and agents can remember their past exchanges with each other, they learn that cooperation works to everyone's advantage. Second, the simulation explains how norms of cooperation with strangers could emerge—as local conventions diffuse through weak ties. And in addition, the simulation explains how several variables contribute to these effects—variables like the size of the neighborhood and the embeddedness of each agent.

Advocates of a covering law approach to explanation might prefer to think in terms of lawful generalizations. The above simulation suggests at least two: first, cooperation among strangers is greater when the neighborhoods are smaller, and second, cooperation among strangers increases linearly with embeddedness. In a D-N empiricist approach, such laws could be hypothesized and then tested through empirical study of existing human societies, and no understanding of the causal mechanism would be necessary. A mechanist like Hedström (2005) would counter that the identification of empirically supported lawful relations does not constitute an explanation. One has not identified a causal explanation until one has identified the underlying social mechanisms that realize the regularities captured by the law. The Macy and Skvoretz simulation helps to provide this form of causal explanation.

14.5 Potential Limitations of Simulations as Explanations

I am sympathetic to the mechanism approach. As Hedström points out, "It tends to produce more precise and intelligible explanations" (2005, p. 28); this is desirable for sociology, which I believe must work toward being an empirically grounded and theoretically rigorous science. It reduces theoretical fragmentation, because a single mechanistic account might potentially explain many different observed phenomena, from crime to social movements; think of Gladwell's best seller *The Tipping Point* (2000). And finally, knowing the underlying mechanism allows you to make a certain kind of causal claim, whereas covering law approaches give you essentially only correlations.

However, causal mechanist accounts of scientific explanation can be epistemically demanding. For example, many behaviors of a volume of gas can be explained by knowing a single number, its pressure; yet a mechanist account requires the identification of the locations and movements of all of the contained molecules. A strict focus on mechanistic explanation would hold that the ideal gas law does not explain the behavior of a volume of gas; only an account in terms of the individual trajectories of individual molecules would be explanatory. And even that would be an incomplete explanation, because the gas would manifest the same macroscopic behavior even if the individual molecules had each taken a different trajectory; certainly, an explanation should be able to account for these multiple realizations.

Many advocates of mechanism are unwilling to accept any place for the covering law approach: they argue that mechanisms are the *only* proper form of sociological explanation. In several publications, Sawyer (2003a, 2004, 2005) has described a class of social phenomena in which mechanism-based explanations would be of limited usefulness: when macro-level regularities are multiply realized in a wide range of radically different underlying mechanisms. In such a situation, an account of the mechanism underlying one realization of the regularity would not provide an explanation of the other instances of that regularity. Hedström's Stockholm simulation does not necessarily explain the rise in youth employment in any other city, for example; other cities might have different realizing mechanisms. But even though it might not be possible to develop a single simulation that accurately represents unemployment processes in all large cities, it might nonetheless be possible to develop a law (or set of laws) that was capable of explaining (in the deductive-nomological sense of the term) unemployment in a large number of cities.

A social mechanist account often requires information that is unavailable or that science is unable to provide. The Stockholm simulation was only possible because of the availability of detailed data gathered by the Swedish government. But covering law explanations can be developed with much less data or with data at a much larger grain size. For example, many behaviors of a society can be explained by knowing whether it is individualist or collectivist (Markus and Kitayama 1991; Triandis 1995). Such properties figure in lawful generalizations like "individualist societies are more likely to be concerned with ownership of creative products" (Sawyer 2006) and "collectivist societies are more likely to practice co-sleeping" (Morelli et al. 1992). In contrast to such simple and easy-to-understand regularities, a mechanist explanation of the same patterns requires quite a bit of knowledge about each participant in that society and their interactions with each other.

Even if a very good social simulation were developed, it might be very difficult to use that simulation to communicate to a broad, nontechnical audience what meaning or understanding to attribute to the phenomenon (or the simulation). And in extremely complex systems like human societies, it may be impossible to develop an explanation of macro phenomena in terms of individual actions and interactions, even though we may all agree that such processes nonetheless must exist at the individual level. The issue here is identifying the right level of description, and the mechanistic or realizing level is often too detailed to provide us with understanding. There are many cases in science where it seems that reduction is not the best strategy for scientific explanation. For example, higher-level events like mental events supervene on physical processes but do not seem to be reducible to a unique set of causal relationships in terms of them.

The most accurate simulation would come very close to replicating the natural phenomenon in all its particulars. After such a simulation has been successfully developed, the task remains to explain the simulation; and for a sufficiently detailed simulation, that could be just as difficult as the original task of explaining the data (Cilliers 1998). Computer programmers often have difficulty explaining exactly why their creations behave as they do, and artificial society developers are no different. Mechanistic accounts of explanation need to more directly address issues surrounding levels of explanation and epistemic and computational limits to human explanation and understanding (see Sawyer 2003a, 2004).

Social simulation unavoidably touches on the unresolved sociological issue of how explanation should proceed. Social simulations represent only individual agents and their interactions, and in this they are methodologically individualist (Conte et al. 2001; also see Drennan 2005). Methodological individualism is a sociological position that has its roots in the nineteenth-century origins of sociology; it argues that sociology should proceed by analyzing the individual participants in the social system, then their relations and the behaviors of bigger system components, and all the way up until we have an explanation of the social system. But if there are real emergent social properties, with downward causal powers over component individuals, then methodologically individualist simulation will fail to provide explanations of those social phenomena—for essentially the same reasons that philosophers of mind now believe that physicalism is inadequate to explain mental phenomena (see Sawyer 2002). Some social properties—such as the property of "being a collectivist society" or "being a church"—are multiply realized in widely different social systems. A simulation of a realizing mechanism of one instance of "being a church" would explain only one token instance but would fail to broadly explain the full range of mechanisms that could realize the social property. To return to the Macy and Skvoretz simulation, the norm of cooperation could emerge in many other realizing social systems, yet the norm might have the same downward causal effects regardless of its realizing mechanism. If so, then a simulation of one realization is only a partial explanation of a more general social phenomenon; it does not explain the other ways that human cooperative behavior could be realized.

Social simulations which contain only individual agents deny a sociological realism that accepts social properties as real. If macrosocial properties are real, then they have an ontological status distinct from their realizing mechanisms and may participate in causal relations (this point continues to be actively debated and the arguments are complex; see Sawyer 2003b). An accurate simulation of a social system that contains multiply realized macrosocial properties would have to represent not only individuals in interaction but also these higher-level system properties and entities (Sawyer 2003a).

The problem is that although a social simulation may provide a plausible account of how individual actions and interactions give rise to an emergent macro pattern, it is hard to know (1) whether or not that social simulation in fact captures the empirical reality of the emergence and, more critically, (2) even if all agree that the social simulation accurately captures an instance of the emergence of a macro pattern from a network of individual agents, there may be other networks and other emergence processes that could also give rise to the same macro pattern.

Issue (1) is the issue of validation and it is addressed in Chap. 9 in this same volume (David et al. 2017). My concern here is with issue (2), which Sawyer (2005) called *multiple realizability*: a social simulation may accurately represent the social mechanisms by which individual actions together give rise to an emergent macro phenomenon. But for a given macrosocial phenomenon, there could potentially be many different networks of people, acting in different combinations, that result in different emergence processes that lead to the same macrosocial phenomenon. If so, the social simulation would not provide a complete explanation of the macrosocial phenomenon but instead would only provide a limited and partial explanation.

The usual response to the multiple realizability issue is to argue that in many cases, the alternate realizations of the macro phenomena are not significantly different from each other. After all, every basketball team has five different players, but the fact that the five positions are occupied by different human beings does not substantially change the possible ways that the five can interact and the possible ways that plays emerge from the interactions of the five individuals. Some pairs of realization are quite similar to each other, so similar that understanding one realization is tantamount to understanding the other one of the pair—without necessarily developing an entirely distinct social simulation.

The problem with this response is that it fails in the face of *wild disjunction* (Fodor 1974): when a given macro phenomenon has multiple realizations and those realizations have no lawful relations with one another. If the multiple realizations are wildly disjunctive, then a social simulation of one of the realizations does not provide us with any understanding beyond that one realization. We are still left needing explanations of all of the other realizations. In many cases, wild disjunction is related to functionalist arguments (and the argument originated from functionalist perspectives in the philosophy of mind: Fodor 1974). "Being a church" is more likely to be multiply realized in wildly disjunctive fashion; if "church" is defined in terms of the functional needs, it satisfies for its society rather than in terms of structural features internal to the institution.

The mechanist could respond to wild disjunction concerns by empirically identifying all of the different ways that the macro phenomenon in question might emerge and then developing a suite of social simulations, each one of which would represent one of the realizing mechanisms. Then, could we say that the suite of social simulations, together, constituted a complete explanation of the macro phenomenon? I think so, although I would prefer to speak of a suite of "explanations" rather than to call the set a single explanation. The Stockholm unemployment simulation might only work for societies with generous social welfare systems; but then, another simulation could be developed for stingier governments, and two simulations together are not oppressively large given the outcome that unemployment everywhere is now fully explained.

The suite-of-simulation approach works fine as long as the number of realizing social simulations is manageable. But at some point, a suite of simulations would become so large that most sociologists would agree that it provided limited understanding of a phenomenon. Is 200 too many to be a meaningful explanation? Could as few as 20 still be too many? Even if it is computationally plausible and the number of person-hours required to develop all of the simulations is not excessive, it might nonetheless be of questionable value to undertake that work, because another path toward social explanation is available—the covering law model.

Many philosophical advocates of mechanism believe that mechanistic explanation is compatible with the existence of higher-level laws. Mechanisms are said to explain laws (Beed and Beed 2000; Bunge 2004; Elster 1998). Bunge (2004) and Little (1998) argued that causal mechanistic accounts are fully compatible with covering law explanations; the mechanisms do the explanatory work, and the covering laws provide a convenient shorthand that is often useful in scientific practice. However, it is possible that social laws may exist that are difficult to explain by identifying realizing mechanisms—in those cases where the laws relate wildly disjunctive, multiply realized social properties. If so, the scope of mechanistic explanation would be limited.

Many sociological theorists use the philosophical notion of emergence to argue that collective phenomena are collaboratively created by individuals yet are not reducible to individual action (Sawyer 2005). In the social sciences, emergence refers to processes and mechanisms of the micro- to macro-transition. Many of these accounts argue that although only individuals exist, collectives possess emergent properties that are irreducibly complex and thus cannot be reduced to individual properties. Thus they reject sociological realism and are methodologically collectivist. Other accounts argue that emergent properties are real.

The resolution to the apparent contradiction between mechanistic explanation and social emergence is to develop a sufficiently robust account of emergence so that mechanistic explanation and lawful explanation can be reconciled. Sawyer (2002, 2003b) proposed a version of emergence that he called *nonreductive individualism (NRI)*. Some emergent social properties may be real and may have autonomous causal powers, just like real properties at any other level of analysis. Nonreductive individualism argues that this is the case for social properties that are multiply realized in wildly disjunctive mechanisms. To the extent that social properties are multiply realized, artificial society simulations may be limited to the explanation of individual cases that do not generalize widely, resulting in a case study approach rather than a science of generalizable laws and theories. The emergentist nature of NRI is compatible with a more limited form of mechanism but one that is elaborated in a sociologically realist direction—with the mechanisms containing explicit models of social properties at levels of analysis above the individual.

If a social property is multiply realized in many different (methodologically individualist) mechanisms, a mechanistic explanation of any one realizing instance will have limited explanatory power—particularly if the social property participates in causal relations across its multiple realizations. A covering law approach might be necessary to capture generalizations of higher-level phenomena across different realizing mechanisms. Alternately, a mechanism could be proposed which explicitly models emergent social properties, in addition to individuals and their interactions. Although almost all artificial societies are currently individualist—with no representation of higher-level social properties—there is no reason why computer simulations could not be extended to model both individuals and macrosocial phenomena, apart from an implicit commitment to methodological individualism.

14.6 Conclusion

Social simulations are almost all methodologically individualist, in that they represent agents and their interactions, but not higher-level entities or properties. In other words, social simulations are representations of the realizing mechanisms of higher-level social properties. More generally, almost all agent-based simulations are representations of a realizing mechanism of some system-level phenomenon. Whether or not a complex system can be explained at the level of its realizing mechanisms, or requires explanation at the level of emergent macro properties, is an empirical question (Sawyer 2005). For example, it cannot be known a priori whether or not a given social property can be given a useful mechanistic explanation in terms of individuals—nor whether a given social property can be adequately simulated by representing only individuals and their interactions in the model.

If this "realizing mechanism" approach begins to seem limiting, then modelers could respond by incorporating system-level entities into their simulations. For example, sociologists could respond by developing simulations that contain the terms and properties of macro sociology, in addition to individual properties and relations. If macrosocial properties are indeed real, and have autonomous causal powers, then to be empirically accurate, any model would have to incorporate those properties. Although the social mechanism approach is commonly associated with methodological individualism—because its advocates assume that a social mechanism must be described in terms of individuals' intentional states and relations (e.g., Elster 1998; Hedström 2005)—there is no reason why social simulations cannot include systems and mechanisms at higher levels of analysis. The system dynamic models of an earlier era focused on macrosocial properties; but with the availability of multi-agent technology, new hybrid simulations could be developed that contain both societies of autonomous agents and explicit simulations of emergent macrosocial properties.

To explore how we should interpret and understand simulations, I have drawn on contemporary philosophical accounts of explanation and of causal mechanism. I conclude by cautioning against being overly confident that agent-based simulations, at least those that are based on mechanistic assumptions, provide complete explanations of a given system-level phenomenon. The explanation may not be complete even in those cases where the simulation is well conceived, is grounded in empirical observation and theory, and generates emergent processes that lead to empirically observed system outcomes. These successful simulations should be considered to be explanations of a given realizing instance of an emergence process, but not necessarily considered to be complete explanations of the target system phenomenon.

The question for sociologists is ultimately what path should sociology take? All sociologists define their goal to be the explanation of macrosocial phenomena—of groups and organizations, rather than of single individuals. Many sociologists believe that they can explain macrosocial phenomena without attending to the specific realizing mechanisms at the level of individuals and their interactions. These sociologists define sociology as the science of an autonomous social level of analysis. The social mechanists believe that this is the wrong approach; instead, the goal of sociology should be to identify and characterize these individual realizing mechanisms. Mechanists believe that there can be no autonomous science at the macrosocial level of analysis. These are the issues to be faced as we attempt to interpret and understand social simulations.

The question for complexity researchers more generally is the same: Can a mechanistic approach provide a complete explanation, or will scientific explanation need to incorporate some higher-level properties and entities? The answer to this question has direct implications for modelers. If the mechanistic approach is capable of providing a complete explanation of a given phenomenon, then a strict agent-based approach is appropriate. But if it is necessary to incorporate higher-level properties or entities, then agent-based simulations will need to include model entities that represent higher levels of organization.

Further Reading

Bechtel and Richardson (1993) provide a discussion of a range of philosophical issues related to the likely success of reductionist strategies in understanding and explaining complex systems, inspired by connectionist accounts of cognition, but relevant to complex systems at any level of analysis. Hedström (2005) makes a strong case for reductionist explanation of social systems, using mechanistic explanation and specifically multi-agent-based simulation in connection with empirical study.

For an examination of the philosophical accounts of mechanistic explanation and theories of emergence in sociology and philosophy, see Sawyer (2004). For an extensive review of historical and contemporary theories of emergence in the social sciences, primarily psychology and sociology, see Sawyer (2005). This advocates that sociology should be the science of social emergence. Conte et al. (2001) is a discussion between four different viewpoints specifically as they concern social simulation.

References

Axelrod, R. (1997). *The complexity of cooperation: Agent-based models of competition and collaboration.* Princeton, NJ: Princeton University Press.

Bechtel, W. (2001). The compatibility of complex systems and reduction: A case analysis of memory research. *Minds and Machines, 11*, 483–502.

Bechtel, W., & Richardson, R. C. (1993). *Discovering complexity: Decomposition and localization as strategies in scientific research.* Princeton, NJ: Princeton University Press.

Beed, C., & Beed, C. (2000). Is the case for social science laws strengthening? *Journal for the Theory of Social Behaviour, 30*(2), 131–153.

Blau, P. M. (1977). A macrosociological theory of social structure. *American Journal of Sociology, 83*(1), 26–54.

Blau, P. M. (1983). Comments on the prospects for a nomothetic theory of social structure. *Journal for the Theory of Social Behaviour, 13*(3), 265–271.

Bunge, M. (2004). How does it work?: The search for explanatory mechanisms. *Philosophy of the Social Sciences, 34*(2), 182–210.

Carley, K. M., & Gasser, L. (1999). Computational organization theory. In G. Weiss (Ed.), *Multiagent systems: A modern approach to distributed artificial intelligence* (pp. 299–330). Cambridge: MIT Press.

Cilliers, P. (1998). *Complexity and postmodernism: Understanding complex systems.* New York: Routledge.

Conte, R., Edmonds, B., Moss, S., & Sawyer, R. K. (2001). Sociology and social theory in agent based social simulation: A symposium. *Computational and Mathematical Organization Theory, 7*(3), 183–205.

Craver, C. (2001). Role functions, mechanisms and hierarchy. *Philosophy of Science, 68*, 31–55.

Craver, C. F. (2002). Interlevel experiments and multilevel mechanisms in the neuroscience of memory. *Philosophy of Science, 69*(Suppl), S83–S97.

David, N., Fachada, N., & Rosa, A. C. (2017). Verifying and validating simulations. doi:https://doi.org/10.1007/978-3-319-66948-9_9.

Davidson, D. (1980). *Essays on actions and events.* New York: Oxford University Press.

Drennan, M. (2005). The human science of simulation: A robust hermeneutics for artificial societies. *Journal of Artificial Societies and Social Simulation, 8*(1). http://jasss.soc.surrey.ac.uk/8/
1/3.html

Elster, J. (1989). *Nuts and bolts for the social sciences*. Cambridge: New York.

Elster, J. (1998). A plea for mechanism. In P. Hedström & R. Swedberg (Eds.), *Social mechanisms:
An analytical approach to social theory* (pp. 45–73). Cambridge, UK: Cambridge University
Press.

Fodor, J. A. (1974). Special sciences (or: The disunity of science as a working hypothesis).
Synthese, 28, 97–115.

Giddens, A. (1984). *The constitution of society: Outline of the theory of structuration*. Berkeley,
CA: University of California Press.

Gladwell, M. (2000). *The tipping point*. New York: Little, Brown.

Glennan, S. S. (1996). Mechanisms and the nature of causation. *Erkenntnis, 44*, 49–71.

Hedström, P. (2005). *Dissecting the social: On the principles of analytic sociology*. Cambridge,
UK: Cambridge University Press.

Hedström, P., & Swedberg, R. (Eds.). (1998). *Social mechanisms: An analytical approach to social
theory*. Cambridge, UK: Cambridge University Press.

Hempel, C. G. (1965). *Aspects of scientific explanation and other essays in the philosophy of
science*. New York: Free Press.

Kincaid, H. (1990). Defending laws in the social sciences. *Philosophy of the Social Sciences, 20*(1),
56–83.

Little, D. (1991). *Varieties of social explanation: An introduction to the philosophy of science*.
Boulder, CO: Westview Press.

Little, D. (1993). On the scope and limits of generalization in the social sciences. *Synthese, 97*(2),
183–207.

Little, D. (1998). *Microfoundations, method and causation: On the philosophy of the social
sciences*. New Brunswick, NJ: Transaction Publishers.

Machamer, P., Darden, L., & Craver, C. F. (2000). Thinking about mechanisms. *Philosophy of
Science, 67*, 1–25.

Macy, M. W., & Skvoretz, J. (1998). The evolution of trust and cooperation between strangers: A
computational model. *American Sociological Review, 63*, 638–660.

Markovsky, B. (1997). Building and testing multilevel theories. In J. Szmatka, J. Skvoretz, & J.
Berger (Eds.), *Status, network, and structure: Theory development in group processes* (pp. 13–
28). Stanford: Palo Alto, CA.

Markus, H. R., & Kitayama, S. (1991). Culture and the self: Implications for cognition, emotion,
and motivation. *Psychological Review, 98*(2), 224–253.

McIntyre, L. C. (1996). *Laws and explanation in the social sciences: Defending a science of human
behavior*. Boulder, CO: Westview.

Morelli, G. A., Rogoff, B., Oppenheim, D., & Goldsmith, D. (1992). Cultural variation in infants'
sleeping arrangements: Questions of independence. *Developmental Psychology, 28*(4), 604–
613.

Neumann, M. (2006, October 5–6). Emergence as an explanatory principle in artificial societies:
Reflections on the bottom-up approach to social theory. In F. Squazzoni (Ed.), *Epistemological
aspects of computer simulation in the social sciences: Second international workshop, EPOS
2006, Brescia, Italy, revised selected and invited papers, Lecture notes in computer science*
(Vol. 5466, pp. 69–88). Berlin: Springer.

Ostrom, T. (1988). Computer simulation: The third symbol system. *Journal of Experimental Social
Psychology, 24*, 381–392.

Salmon, W. (1971). Statistical explanation. In W. Salmon (Ed.), *Statistical explanation and
statistical relevance* (pp. 29–87). Pittsburgh: University of Pittsburgh Press.

Salmon, W. (1984). *Scientific explanation and the causal structure of the world*. Princeton, NJ:
Princeton University Press.

Salmon, W. (1994). Causality without counterfactuals. *Philosophy of Science, 61*, 297–312.

Salmon, W. (1997). Causality and explanation: A reply to two critiques. *Philosophy of Science, 64*, 461–477.

Sawyer, R. K. (2002). Nonreductive individualism, part 1: Supervenience and wild disjunction. *Philosophy of the Social Sciences, 32*(4), 537–559.

Sawyer, R. K. (2003a). Artificial societies: Multi agent systems and the micro-macro link in sociological theory. *Sociological Methods and Research, 31*(3), 37–75.

Sawyer, R. K. (2003b). Nonreductive individualism, part 2: Social causation. *Philosophy of the Social Sciences, 33*(2), 203–224.

Sawyer, R. K. (2004). The mechanisms of emergence. *Philosophy of the Social Sciences, 34*(2), 260–282.

Sawyer, R. K. (2005). *Social emergence: Societies as complex systems*. Cambridge: New York.

Sawyer, R. K. (2006). *Explaining creativity: The science of human innovation*. Oxford: New York.

Schmid, A. (2006, October 5–6). What does emerge in computer simulations? Simulation between epistemological and ontological emergence. In F. Squazzoni (Ed.), *Epistemological aspects of computer simulation in the social sciences: Second international workshop, EPOS 2006*, Brescia, Italy, revised selected and invited papers, *Lecture notes in computer science* (Vol. 5466, pp. 60–68). Berlin: Springer.

Stinchcombe, A. L. (1991). The conditions of fruitfulness of theorizing about mechanisms in social science. *Philosophy of the Social Sciences, 21*(3), 367–388.

Triandis, H. C. (1995). *Individualism & collectivism*. Boulder, CO: Westview Press.

Turner, J. H. (1993). *Classical sociological theory: A positivist perspective*. Chicago: Nelson-Hall.

Woodward, J. (2003). *Making things happen: A theory of causal explanation*. New York: Oxford University Press.

Chapter 15
Documenting Social Simulation Models: The ODD Protocol as a Standard

Volker Grimm, Gary Polhill, and Julia Touza

Abstract The clear documentation of simulations is important for their communication, replication, and comprehension. It is thus helpful for such documentation to follow minimum standards. The 'overview, design concepts, and details' document protocol (ODD) is specifically designed to guide the description of individual- and agent-based simulation models (ABMs) in journal articles. Popular among ecologists, it is also increasingly used in the social simulation community. Here, we describe the protocol and give an annotated example of its use, with a view in facilitating its wider adoption and encouraging higher standards in simulation description.

Why Read This Chapter?
To learn about the importance of documenting your simulation model and discover a lightweight and appropriate framework to guide you in doing this.

15.1 Introduction and History

A description protocol is a framework for guiding the description of something, in this case a social simulation model. It can be thought of as a checklist of things that need to be covered and rules that should be followed when specifying the details of a simulation (in a scholarly communication). Following such a protocol means that readers can become familiar with its form and that key elements are less likely to

V. Grimm (✉)
Department of Ecological Modelling, UFZ, Helmholtz Centre of Environmental Research – UFZ, Permoserstr. 15, D-04318, Leipzig, Germany
e-mail: volker.grimm@ufz.de

G. Polhill
The James Hutton Institute, Craigiebuckler, Aberdeen, AB15 8QH, UK

J. Touza
Environment Department, University of York, York, UK

© Springer International Publishing AG 2017
B. Edmonds, R. Meyer (eds.), *Simulating Social Complexity*,
Understanding Complex Systems, https://doi.org/10.1007/978-3-319-66948-9_15

be forgotten. This chapter describes a particular documentation protocol, the ODD (pronounced 'odd' or 'oh dee dee') protocol.

The ODD protocol (Grimm et al. 2006, 2010; Polhill et al. 2008; Polhill 2010; Müller et al. 2013) is a standard layout for describing individual- and agent-based simulation models (ABMs), especially for journal articles, conference papers, and other academic literature. It consists of seven elements which can be grouped into three blocks: overview, design concepts, and details (hence, 'ODD'; see Table 15.1). The purpose of ODD is to facilitate writing and reading of model descriptions, to better enable replication of model-based research, and to establish a set of design concepts that should be taken into account whilst developing an ABM. It does this in a relatively lightweight way, avoiding over formal approaches whilst ensuring that the essentials of a simulation are explicitly described in a flexible yet appropriate manner.

Originally, ODD was formulated by ecologists, where the proportion of ABMs described using ODD is increasingly fast and might cross the 50% margin in the near future. In social simulation, the acceptance of ODD has been slower. A first test, in which three existing descriptions of land use models were reformulated according to ODD, demonstrated the benefits of using ODD but also revealed that some refinements were needed to make it more suitable for social simulation (Polhill et al. 2008). In 2010, an update of ODD was released (Grimm et al. 2010), which is based on users' feedback and a review of more than 50 ODD-based model descriptions in the literature. In this update, ODD itself was only slightly modified, but the explanation of its elements was completely rewritten, with the specific intention of making it more suitable for social simulation.

Currently in social simulation, interest in ODD is also increasing (Polhill 2010). An indicator for this is the inclusion of ODD chapters in recent reference books (this volume; Heppenstall et al. 2012). ODD is also recommended by the Network for Computational Modelling for SocioEcological Science (CoMSES Net) and the Model Library of their node OpenABM.org. Moreover, a recent textbook of agent-based modelling uses ODD consistently (Railsback and Grimm 2012) so that the next generation of agent-based modellers is more likely to be familiar with ODD and hence to use it themselves.

15.2 The Purpose of ODD

Why is ODD (or a protocol very much like it) needed? There are a number of endeavours in agent-based social simulation that are facilitated through having a common approach to describing the models that is aimed at being readable and complete[1]:

---

[1]Many of these endeavours have been covered in submissions to the "model-to-model" series of workshops, organised by members of the social simulation community (Hales et al. 2003; Rouchier

- *Communication* is the most basic aim of anyone trying to publish their results. For agent-based modellers, this can pose a particular challenge, as our models can be complicated, with many components and submodels. As a critical mass of papers using ODD develops, so readers of agent-based modelling papers will find themselves sufficiently more familiar with papers structured using ODD than those using an arbitrary layout devised by the authors that they will find the former easier to read and understand than the latter.
- *Replication*, as we discuss later in this chapter, is a pillar of the scientific endeavour (Thiele and Grimm 2015). If our model descriptions are inadequate, our results are not repeatable, and the scientific value of our work commensurately reduced. ODD helps to encourage the adequacy of descriptions by saving authors having to 'reinvent the wheel' each time they describe a model, by providing a standard layout designed to ensure that all aspects of a model needed to replicate it are included in the account.
- *Comparing models* is likely to become increasingly important as work in agent-based modelling continues. If two or more research teams produce similar models with different outcomes, comparing the models will be essential to identify the cause of the variance in behaviour. Such comparisons will be much easier if all teams have used the same protocol to describe the models. At a conceptual level, the design concepts also enable comparison of models with greater differences and application domains.
- *Dialogue among disciplines* can be encouraged through a standard that is used by both the ecological and social simulation communities. This is especially useful for those developing coupled socio-ecosystem models (Polhill et al. 2008), which is a rapidly growing area of research (Polhill et al. 2011).

In the following, we briefly describe the rationale of ODD and how it is used, provide an example model description, and finally discuss benefits of ODD, current challenges, and its potential future development.

15.3   The ODD Protocol

A core principle of ODD is that first an 'overview' of a model's purpose, structure, and processes should be provided, *before* 'details' are presented. This allows readers to quickly get a comprehensive overview of what the model is, what it does, and for what purpose it was developed. This follows the journalistic 'inverted pyramid' style of writing, where a summary is provided in the first one or two paragraphs, and progressively further detail is added on the story the further on you read (see, e.g. Wheeler 2005). It allows the reader to easily access the information they are interested in at the level of detail they need. For experienced modellers, this

---

et al. 2008. The second workshop was held as a parallel session of the ESSA 2004 conference: see http://www.insisoc.org/ESSA04/M2M2.htm).

overview part is sufficient to understand what the model is for, to relate it to other models in the field, and to assess the overall design and complexity.

Before presenting the 'details', ODD requires a discussion of whether and how ten design concepts were taken into account whilst designing the model. This 'design concept' part of ODD does not describe the model itself but the principles and rationale underlying its design. 'Design concepts' are thus not needed for model replication but for making sure that important design decisions were made consciously and that readers are fully aware of these decisions. For example, it is important to be clear about what model output is designed to emerge from the behaviour the model's entities and their interactions and what, in contrast, is imposed by fixed rules and parameters. Ideally, key behaviours in a model emerge, whereas other elements might be imposed. If modellers are not fully aware of this difference, which is surprisingly often the case, they might impose too much so that model output is more or less hard-wired into its design, or they might get lost in a too complex model because too much emergence makes it hard to understand anything. Likewise, the design concept 'stochasticity' requires that modellers explicitly say what model processes include a stochastic component, why stochasticity was used, and how it was implemented. Note that, in contrast to the seven elements of ODD, the sequence in which design concepts are described can be changed, if needed, and design concepts that are not relevant for the model can be omitted.

The 'detail' part of ODD includes all details that are needed to re-implement the model. This includes information about the values of all model entities' state variables and attributes at the beginning of a simulation ('initialisation'), the external models, or data files that are possibly used as 'input data' describing the dynamics of one or more driving contextual or environmental variables (e.g. rainfall, market price, disturbance events) and 'details' where the submodels representing the processes listed in 'process overview and scheduling' are presented. Here, it is recommended for every submodel to start with the factual description of what the submodel is and then explain its rationale.

Model parameters should be presented in a table, referred to in the 'submodel' section of ODD, including parameter name, symbol, reference value, and—if the model refers to real systems—unit, range, and references, or sources for choosing parameter values. Note that the simulation experiments that were carried out to analyse the model, characterised by parameter settings, number of repeated runs, the set of observation variables used, and the statistical analyses of model output, are not part of ODD but ideally should be presented in a section 'simulation experiments' directly following the ODD-based model description.

15.4   How to Use ODD

To describe an ABM using ODD, the questions listed in Table 15.1 have to be answered. The identifiers of the three blocks of ODD elements—overview, design concepts, details—are not used themselves in ODD descriptions (except for 'design

Table 15.1 The seven elements of the ODD protocol. Descriptions of ABMs are compiled by answering the questions linked to each element

Overview	1. Purpose		What is the purpose of the model?
	2. Entities, state variables, and scales		What kind of entities are in the model? Do they represent managers, voters, landowners, firms, or something else? By what state variables or attributes are these entities characterised? What are the temporal and spatial resolutions and extents of the model?
	3. Process overview and scheduling		What entity does what, in what order? Is the order imposed or dynamic? When are state variables updated? How is time modelled: as discrete steps or as a continuum over which both continuous processes and discrete events can occur?
Design concepts	4. Design concepts	Basic principles	Which general concepts, theories, or hypotheses approaches are included in the model's design? How were they taken into account? Are they used at the level of submodels or at the system level?
		Emergence	What key results are emerging from the adaptive traits or behaviours of individuals? What results vary in complex/unpredictable ways when particular characteristics change? Are there other results that are more tightly imposed by model rules and hence less dependent on what individuals do?
		Adaptation	What adaptive traits do the individuals have? What rules do they have for making decisions or changing behaviour in response to changes in themselves or their environment? Do agents seek to increase some measure of success, or do they reproduce observed behaviours that they perceive as successful?
		Objectives	If agents (or groups) are explicitly programmed to meet some objective, what exactly is that, and how is it measured? When individuals make decisions by ranking alternatives, what criteria do they use? Note that the objective of such agents as group members may not refer to themselves but the group.
		Learning	May individuals change their adaptive traits over time as a consequence of their experience? If so, how?
		Prediction	Prediction can be part of decisionmaking; if an agent's learning procedures are based on estimating future consequences of decisions, how they do this? What internal models do agents use to estimate future conditions or consequences? What 'tacit' predictions are implied in these internal model's assumptions?

Table 15.1 (continued)

	Sensing	What aspects are individuals assumed to sense and consider? What aspects of which other entities can an individual perceive (e.g. displayed 'signals')? Is sensing local, through networks or global? Is the structure of networks imposed or emergent? Are the mechanisms by which agents obtain information modelled explicitly in a process, or is it simply 'known'?
	Interaction	What kinds of interactions among agents are assumed? Are there direct interactions where individuals encounter and affect others, or are interactions indirect, e.g. via competition for a mediating resource? If the interactions involve communication, how are such communications represented?
	Stochasticity	What processes are modelled by assuming they are random or partly random? Is stochasticity used, for example, to reproduce variability in processes for which it is unimportant to model the actual causes of the variability, or to cause model events or behaviours to occur with a specified frequency?
	Collectives	Do the individuals form or belong to aggregations that affect, and are affected by, the individuals? Such collectives can be an important intermediate level of organisation. How are collectives represented—as emergent properties of the individuals or as a separate kind of entity with its own state variables and traits?
	Observation	What data are collected from the ABM for testing, understanding, and analysing it, and how are they collected? Are all output data freely used, or are only certain data sampled and used, to imitate what can be observed in an empirical study?
Details	5. Initialisation	What is the initial state of the model world, i.e. at time t = 0? How many entities of what type are there initially, and what are the values of their state variables (or how were they set)? Is initialisation always the same, or is it varied? Are the initial values chosen arbitrarily or based on available data?
	6. Input data	Does the model use input from external sources such as data files or other models to represent processes that change over time?
	7. Sub-models	What are the submodels that represent the processes listed in 'process overview and scheduling'? What are the model parameters, their dimensions, and reference values? How were submodels designed or chosen, tested, and parameterised?

concepts', which is the only element of the corresponding block). Rather, the seven elements are used as numbered headlines in ODD-based model descriptions. For experienced ODD users, the questions in Table 15.1 are sufficient. For beginners, however, it is recommended to read the more detailed description of ODD in Grimm et al. (2010) and to use the template, which provides additional questions and examples and which is available via download.

15.5 An Example

In the supplementary material of Grimm et al. (2010), publications are listed which use ODD in a clear, comprehensive, and recommendable way. Many further examples are provided in the textbook by Railsback and Grimm (2012). In Grimm and Railsback (2012), Schelling's segregation model, as implemented in the model library of the software platform NetLogo (Wilensky 1999), is used as an example. Here, we demonstrate the process of model documentation using ODD by describing a model developed by Deffuant et al. (2002), which explores the emergence of extreme opinions in a population. We choose this model because it is simple but interesting, and opinion dynamic models are quite well known in the social simulation community. It is also one of the introductory examples in Gilbert (2007). The ODD for the Deffuant et al. model is interspersed with comments on the information included, with a view to provide some guidelines for those applying ODD to their own model. Clearly this is a very simple example, and many models would require more extensive description. The parts of ODD are set in italics and indented to distinguish them from comments. Normally the ODD description would simply form part of the text in the main body of a paper or in an appendix.[2]

15.5.1 Purpose

> The model's purpose is to study the evolution of the distribution of opinions in a population of interacting individuals, which is under the influence of extremists' views. Specifically, it aims to answer how marginal extreme opinions can manage to become the norm in large parts of a population. The central idea of the model is that people who have more extreme opinions are more confident than people with moderate views. More confident people are, however, assumed to more easily affect the opinion of others, who are less confident.

Comments The purpose section is deliberately brief. Even for more sophisticated models than this, we would not expect to see much more text here. This would otherwise repeat information in the rest of the paper. However, since the ODD,

---

[2]It is often the case that a substantial description needs to be included in the main text so readers can get an idea of what is being discussed, but maybe a more complete description might be added in an appendix.

to some extent, needs to stand alone and be comprehensive, the summary of the purpose is included as here. It is important to be specific here, for example, by using a statement starting with 'specifically'.

15.5.2   Entities, State Variables, and Scales

> The model includes only one type of entity: individuals. They are characterized by two continuous state variables, opinion x and uncertainty u. Opinions ranges from -1 to 1. Individuals with an opinion very close to $x = -1$ or $+1$ are referred to as 'extremists', all other individuals are 'moderates'. Uncertainty u defines an interval around an individuals' opinion and determines whether two individuals interact and, if they do, on the relative agreement of those two individuals which then determines how much opinion and uncertainty change in the interaction. One-time step of the model represents the time in which all individuals have randomly chosen another individual and possibly interacted with it. Simulations run until the distribution of opinions becomes stationary.

Comments For larger models, this section has the potential to get quite long if written in the same style as this example, which has only one type of entity, with two state variables. Other articles have taken the approach of using tables to express this information, one table per entity, with one row per state variable associated with that entity (see, e.g. Polhill et al. 2008); this row should include the variable's name, meaning, possible values, and physical units, if applicable. Other articles have used UML class diagrams (e.g. Bithel and Brasington 2009), as suggested in the original ODD article (Grimm et al. 2006); however, these do not provide a means for giving any description, however brief, of each state variable. Simply listing the entities and the data types of the state variables does not provide all the information that this element of ODD should provide. This, together with the fact that UML is focused on object-oriented design (which is used to implement the majority of ABMs but by no means all: NetLogo, for example, is not an object-oriented language, and many, particularly in agent-based social simulation, use declarative programming languages), meant that the recommendation to use UML was retracted in the recent ODD update (Grimm et al. 2010).

In declarative programming languages, the entities and their state variables may not be so explicitly represented in the program code as they are in object-oriented languages. For example, this information may be implicit in the arguments of rules. However, many declarative programs have a database of knowledge that the rules operate on. This database could be used to suggest entities and state variables. For example, a Prolog program might have a database containing the assertion person (Volker) and nationality (Volker, German). This suggests that a 'person' is an entity, and 'nationality' is a state variable. (It might be reasonable to suggest in general that assertions with one argument suggest entities and those with two state variables.)

15.5.3 Process Overview and Scheduling

In each time step, each individual chooses randomly one other individual to interact with, then the relative agreement between these two agents is evaluated, and the focal individual's opinion and uncertainty are immediately updated as a result of this opinion interaction. Updating of state variables is thus asynchronous. After all individuals have interacted, a convergence index is calculated which captures the level of convergence in the opinions of the population, additionally, and output updated (e.g. draw histogram of the population's opinions; write each individual's opinion to a file).

Comments This section briefly outlines the processes (or submodels) that the model runs through in every time step (ignoring initialisation) and in what order. Notice how each process is given an emphasised label, which corresponds to subsection headings in the submodel section; the same label should also be used in the program implementing the model. Whilst the ODD protocol does not make such precise stipulations as to formatting, there should be a clear one-to-one correspondence between the brief outlines of processes here and the details provided on each in the submodel section.

In describing larger models than Deffuant et al.'s, it may be appropriate to simply present the process overview as a list. Many models have a simple schedule structure consisting of a repeated sequence of actions; such a list would clearly show this schedule. However, others use more complicated scheduling arrangements (e.g. dynamic scheduling). In such cases, the rules determining when new events are added to the schedule would need to be described, as well as an (unordered) list of event types, each corresponding to a subsection of 'submodels'.

The 'schedule' in a declarative model may be even less clear, as it will depend on how the inference engine decides which rules to fire. However, declarative programs are at least asked a query to start the model, and this section would be an appropriate place to mention that. Some declarative programs also have an implied ordering to rule firing. For example, in Prolog, the rule a: x, y, z. will, in the event that the inference engine tries to prove a, try to prove x, then y, and then z. Suppose the model is started with the query. (a) In describing the model here, it might suffice simply to summarise how x, y, and z change the state of the model. Any subrules called by the inference engine trying to prove these could be given attention in the detail section.

The declarative programmer may also use language elements (such as cuts in Prolog) to manage the order of execution. In deciding which rules to describe here, a declarative modeller might focus on those changing the value of a state variable over time. The key point is that the program will do *something* to change the values of state variables over time in the course of its execution. Insofar as that can be described in a brief overview, it belongs here.

15.5.4 Design Concepts

Basic principles—This model extends earlier stylised models on opinion dynamics, which either used only binary opinions instead of a continuous range of opinions or where interactions only depended on whether opinion segments overlapped, but not on relative agreement (for references, see Deffuant et al. 2002).

Emergence—The distribution of opinions in the population emerges from interactions among the individuals.

Sensing—Individuals have complete information of their interaction partner's opinion and uncertainty.

Interaction—Pairs of individuals interact if their opinion segments, $[x - u, x + u]$, overlap.

Stochasticity—The interaction between individuals is a stochastic process because interaction partners are chosen randomly.

Observation—Two plots are used for observation: the histogram of opinions and the trajectories of each individual's opinion. Additionally, a convergence index is calculated.

Comments Note that the design concepts are only briefly addressed. This would be expected in larger models too. Note also that several design concepts have been omitted because they are not appropriate to the model. Specifically, adaptation, objectives, learning, prediction, and collectives have been left out here: individuals change their opinion after interaction, but this change is not adaptive since it is not linked to any objective; there are also no collectives since all individuals act on their own. Nevertheless, most models should be able to relate to some basic principles, emergence, interactions, observation, and most often also stochasticity. Small models might use the option of concatenating the design concepts into a single paragraph to save space.

15.5.5 Initialisation

Simulations are run with 1,000 individuals, of which a specified initial proportion, p_e, are extremists; p_+ denotes the proportion of 'positive' extremists, and p_- are the proportion of 'negative' extremists. Each moderate individual's initial opinion is drawn from a random uniform distribution between -1 and $+1$. Extremists have on opinion of either -1 or $+1$. Initially, individuals have a uniform uncertainty, which is larger for moderates than for extremists.

Comments This explains how the simulation is set up before the main schedule starts. In other models, this might include empirical data of various kinds from, for example, surveys. The key question to ask here, particularly given the potential for confusion with the next section ('input data'), is whether the data are used *only* to provide a value for a state variable before the schedule runs.

15.5.6 Input Data

The model does not include any input of external data.

Comments These are time-series data used to 'drive' the model. Some of these data *may* specify values for variables at time zero (i.e. during initialisation); however, if a data series specifies values for any time step other than during initialisation, then it is input data rather than initialisation. It is also important not to confuse 'input data' with parameter values.

15.5.7 Submodels

All model parameters are listed in the following table.

Parameter	Description
N	Number of individuals in population
U	Initial uncertainty of moderate individuals
μ	Speed of opinion dynamics
p_e	Initial proportion of extremists
$p+$	Initial proportion of positive extremists
$p-$	Initial proportion of negative extremists
u_e	Initial uncertainty of extremists

Opinion interaction—This is run for an agent j, whose 'opinion segment' s_j is defined in terms of its opinion x_j and uncertainty u_j as:

$$s_j = \left[x_j{-}u_j, \ x_j + u_j \right]$$

The length of the opinion segment is $2u_j$ and characterises an individual's overall uncertainty.

In opinion interaction, agent j (the influenced, focal, or 'calling' individual) is paired with a randomly chosen agent, i, the influencing individual. The 'overlap' of their opinion segments, h_{ij}, is then computed as:

$$h_{ij} = \min \left(x_i + u_i, \ x_j + u_j \right) - \max \left(x_i{-}u_i, \ x_j{-}u_j \right).$$

This overlap determines whether in opinion interaction will take place or not: Agent j will change its opinion if $h_{ij} > u_i$, which means that overlap of opinions is higher than the uncertainty of the influencing agent (see Fig. 15.1).

For opinion interactions, the relative agreement of the two agents' opinions, RA, is calculated by dividing the overlap of their opinion segments (h_{ij}) minus the length of the nonoverlapping part of influencing individual's opinion segment, ($2u_i - h_{ij}$), and this difference is divided by agent i's opinion segment length, $2u_i$ (Fig. 15.1 depicts these terms graphically):

$$RA = \left(h_{ij}{-} \left(2u_i{-}h_{ij} \right) \right) / 2u_i = 2 \left(h_{ij}{-}u_i \right) / 2u_i = \left(h_{ij}/u_i \right) {-}1$$

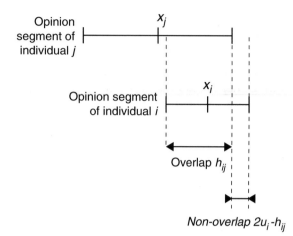

Fig. 15.1 Visualisation of the individual's opinions, uncertainties, and overlap in opinions in the model of (Deffuant et al. 2002)

The opinion and uncertainty of agent j are then updated as follows:

$$x_j = x_j + \mu \ RA \left(x_i + x_j\right)$$

$$u_j = u_j + \mu \ RA \left(u_i + u_j\right)$$

Thus, the new values are determined by the old values and the sum of the old values of both interacting individuals multiplied by the relative agreement, RA, and by parameter μ, which determines how fast opinions change.

The main features of this interaction model are, according to Deffuant et al. (2002):

- Individuals not only influence each other's opinions but also each other's uncertainties.
- Confident agents, who have low uncertainty, are more influential. This reflects the common observation that confident people more easily convince more uncertain people than the other way round—under the conditions that their opinions are not too different at the beginning.

Calculate convergence index—This index, y, is used as a summary model output for sensitivity analysis and an exploration of the model's parameter space. It is defined as:

$$y = q_+ + q_-$$

where q_+ and q_- are the proportions of initially moderate agents which become extremists in the positive extreme or negative extreme, respectively. If after reaching the steady state, none of the initially moderate agents became extremist, the index would take a value of zero. If half of them become positive extremists and the other half becomes negative extremists, the index would be 0.5. Finally, if all the initially moderate agents converge to only one extreme, the index would be one. Note that for calculating y, 'positive' or 'negative' extreme has to be defined via an interval close to the extreme, with a width of, for example, 0.15.

Comments Here, details on the two processes described in Sect. 3 are provided, in sufficient depth to enable replication, i.e. *opinion interaction* and *calculate convergence index*. Note how these names match with those used in the process overview in Sect. 3.

Authors describing larger models may find journal editors protesting at the length of the ODD if all submodels are described in the detail required. There are various ways such constraints can be handled. One is to include the submodels in an appendix or supplementary material to the paper. Another is to provide them as a technical report accessible separately (e.g. on a website) and referred to in the text. If space is not too limited, a summary of each submodel could be provided in the main text, longer than the brief description in the process overview but shorter than the full detail, the latter being provided separately. For very large models or where space is highly constrained, there may be little room for much more than the three overview sections in the journal article; again, making the full ODD available separately is a possible solution. Nevertheless, excluding the 'submodel' element entirely from the main text should be avoided because this would mean to ask readers to accept, in the main text of the article, the model as a black box. Description of the most important processes should therefore be included also in the main text.

15.6 Discussion

Since the example model by Deffuant et al. (2002) is very simple, using ODD here comes with the cost of making the model description longer than the original one, through requiring the ODD labels. The original model is actually relatively clear and easy to replicate (which might partly explain this model's success). However, easy replication is much more the exception than the rule (Hales et al. 2003; Rouchier et al. 2008; Thiele and Grimm 2015), and the more complex an ABM, the higher the risk that not all information is provided for unambiguous replication.

ODD facilitates writing comprehensive and clear documentations of ABMs. This does not only facilitate replication; it also makes writing and reading model documentations easier. Modellers no longer have to come up with their own format for describing their model, and readers know once they are familiar with the structure of ODD, *exactly* where to look for what kind of information.

Whether or not to use ODD as a standard format for model descriptions might look like a rather technical question, but it has fundamental consequences, which go far beyond the issue of replication. Once ODD is used as a standard, it will be become much easier to compare different models addressing similar questions. Even now, ODD can be used to review models in a certain field, by rewriting existing model descriptions according to ODD (Grimm et al. 2010). Building blocks of existing models, in particular specific submodels, which seem to be useful in general, will be much easier to identify and reuse in new models. This is particular so for ABMs representing human behaviour, and decisionmaking. Müller et al. (2013) found ODD too coarse for capturing all essential elements of the decision-making submodel in a systematic way; they therefore extended ODD to ODD + D, with the third 'D' standing for 'decisionmaking'. Most importantly, however, using ODD affects the way we design and formulate ABMs in the first place. After having used ODD for documenting two or three models, you start formulating ABMs by answering the ODD questions: What 'things' or entities do I need to represent in my model? What state variables and behavioural attributes do I need to characterise these entities? What processes do I want to represent explicitly, and how should they be scheduled? What are the spatial and temporal extent and resolution of my model, and why? What do I want to impose, and what to let emerge? What kind of interactions does the model include? For what purposes should I include stochasticity? How should the model world be initialised, what kinds of input data do I need, and how should I, in detail, formulate my submodels?

These questions do not impose any specific structure on simulation models, but they provide a clear checklist for both model developers and users. This helps avoiding 'ad hocery' in model design (Heine et al. 2005). Modellers can also more easily adopt designs of existing models and don't have to start from scratch all the time, as in most current social simulation models.

Criticisms of ODD include Amouroux et al. (2010), who, acknowledging its merits, find the protocol ambiguous and insufficiently specified to enable replication. This article pertained to the Grimm et al. (2006) first description of ODD. The update in Grimm et al. (2010) endeavoured to address issues such as these. However, the success of the latter article in so doing and indeed any future revisions of ODD can only be measured by comparing replication efforts based on ODD descriptions with those not conforming to any protocol—the norm prior to 2006 when ODD was first published. As suggested above, the record for articles not using ODD has not been particularly good: Rouchier et al. (2008) observe in their editorial to a special section of JASSS on the third Model-2-Model workshop that several researchers attempting replications have to approach the authors of the original articles to disambiguate model specifications. If the models were adequately described in the original articles, this should not be necessary.

Polhill et al. (2008) also observed that those used to object-oriented designs for modelling will find the separation of what will for them effectively amount to instance variables and methods (state variables and processes, respectively) counter-intuitive, if indeed not utterly opposed to encapsulation— one of the key principles of object orientation. For ODD, however, it is the reader who is important rather than programming principles intended to facilitate modularity and code reuse. It is also important that, as a documentation protocol, ODD does not tie itself to any particular ABM implementation environment. From the perspective of the human reader, it is illogical (to us at least) to discuss processes before being informed what it is the processes are operating on. Encapsulation is about hiding information; ODD has quite the opposite intention.

The main issue with ODD in social simulation circles as opposed to ecology, from which it originally grew, pertains to its use with declarative modelling environments. This matter has been raised in Polhill et al. (2008) and acknowledged in Grimm et al. (2010). Here we have tried to go further towards illustrating how a declarative modeller might prepare a description of their model that conforms to ODD. However, until researchers using declarative environments attempt to use ODD when writing an article and feedback on their findings, this matter cannot be properly addressed.

Certainly, ODD is not the silver bullet regarding standards for documenting ABMs. Nevertheless, even at the current stage, its benefits by far outweigh its limitations, and using it more widely is an important condition for further developments. Still, since ODD is a verbal format; not all ambiguities can be prevented. Whilst a more formal approach using, for example, XML or UML (e.g. Triebig and Klügl 2010 and for ABMs of land use/cover change, the MRPOTATOHEAD framework—Livermore 2010; Parker et al. 2008) might address such ambiguities, we consider it important that written, natural language formulations of ABMs exist (Grimm and Railsback 2005). This is the only way to make modelling, as a scientific activity, independent of technical aspects of mark-up or programming languages and operating systems. Further, verbal descriptions force us to *think* about a model, to try to understand what it is, what it does, and why it was designed in that way and not another (J. Everaars, *pers. comm.*). We doubt that a 'technical' standard for documenting ABMs—one that can be read by compilers or interpreters, would ever initiate and require this critical thinking about a model.

Nevertheless, it is already straightforward to translate ODD model description to NetLogo programs because much of the way models which are written in NetLogo correspond to the structure of ODD: the declaration of 'entities, state variables, and scales' is done via NetLogo's globals, turtles-own, and patches-own primitives. 'Initialisation' is done via the setup procedure, 'process overview and scheduling' which corresponds to the go procedure. 'Details' are implemented as NetLogo procedures, and 'design concepts' can be included, (as indeed can the entire ODD model description), on the 'information' tab of NetLogo's user interface.

15.7 Conclusion

Clearly describing simulations well so that other researchers can understand a simulation is important for the scientific development and use of complex simulations. It can help in the assessment and comprehension of simulation results by readers replicating simulations for checking and analysis by other researchers, transferring knowledge embedded within simulations from one domain to another, and allowing simulations to be better compared. It is thus an important factor for making use of simulations more rigorous and useful. A protocol such as ODD is useful in standardising such descriptions and encouraging minimum standards. As the field of social simulation matures, it is highly likely that the use of a protocol such as ODD will become standard practice.

The investment in learning and using ODD is minimal, but the benefits, both for its user and the scientific community, can be huge. We therefore recommend learning and testing ODD by rewriting the model description of an existing, moderately complex ABM, and, in particular, using ODD to formulate and document the next ABM you are going to develop.

Acknowledgments We are grateful to Bruce Edmonds for inviting us to contribute this chapter and for his helpful comments and suggested amendments to earlier drafts. Gary Polhill's contribution was funded by the Scottish Government.

Further Reading

Railsback and Grimm's (2012) textbook introduces to agent-based modelling with examples described using ODD. The OpenABM website (http://openabm.org) is a portal specifically designed to facilitate the dissemination of simulation code and descriptions of these using the ODD protocol. The original reference document for ODD is Grimm et al. (2006) with the most recent update being Grimm et al. (2010). Polhill (2010) is an overview of the 2010 update of ODD written specifically with the social simulation community in mind.

References

Amouroux, E., Gaudou, B., Desvaux, S., & Drogoul, A. (2010, November 1–4). O.D.D.: A promising but incomplete formalism for individual-based model specification. In T. B. Ho, D. N. Zuckerman, P. Kuonen, A. Demaille, & R.-D. Kutsche (Eds.), *2010 IEEE-RIVF international conference on computing and communication technologies: Research, innovation and vision for the future.* Hanoi, Vietnam: Vietnam National University.

Bithell, M., & Brasington, J. (2009). Coupling agent-based models of subsistence farming with individual-based forest models and dynamic models of water distribution. *Environmental Modelling & Software, 24,* 173–190.

Deffuant, G., Amblard, F., Weisbuch, G., & Faure, T. (2002). How can extremism prevail? A study based on the relative agreement interaction model. *Journal of Artificial Societies and Social Simulation, 5*(4.) http://jasss.soc.surrey.ac.uk/5/4/1.html

Gilbert, N. (2007). *Agent-based models.* London: Sage Publications.

Grimm, V., & Railsback, S. F. (2005). *Individual-based modeling and ecology.* Princeton, NJ: Princeton University Press.

Grimm, V., Berger, U., Bastiansen, F., Eliassen, S., Ginot, V., Giske, J., et al. (2006). A standard protocol for describing individual-based and agent-based models. *Ecological Modelling, 198,* 115–126.

Grimm, V., Berger, U., DeAngelis, D. L., Polhill, J. G., Giske, J., & Railsback, S. F. (2010). The ODD protocol: A review and first update. *Ecological Modelling, 221,* 2760–2768.

Grimm, V., & Railsback, S. F. (2012). Designing, formulating, and communicating agent-based models. In A. Heppenstall, A. Crooks, L. M. See, & M. Batty (Eds.), *Agent-based models of geographical systems* (pp. 361–377). Berlin: Springer.

Hales, D., Rouchier, J., & Edmonds, B. (2003). Model-to-model analysis. *Journal of Artificial Societies and Social Simulation, 6*(4). http://jasss.soc.surrey.ac.uk/6/4/5.html

Heine, B.-O., Meyer, M., & Strangfeld, O. (2005). Stylised facts and the contribution of simulation to the economic analysis of budgeting. *Journal of Artificial Societies and Social Simulation, 8*(4). http://jasss.soc.surrey.ac.uk/8/4/4.html

Heppenstall, A., Crooks, A., See, L. M., & Batty, M. (Eds.). (2012). *Agent-based models of geographical systems.* Berlin: Springer.

Livermore, M. (2010). MR POTATOHEAD framework: A software tool for collaborative land-use change modeling. In D. A. Swayne, W. Yang, A. A. Voinov, A. Rizzoli, & T. Filatova (Eds.), *Proceedings of the International environmental modelling and software society (iEMSs) 2010 international congress on environmental modelling and software: Modelling for environment's sake, fifth Biennial meeting*, Ottawa, Canada. http://www.iemss.org/iemss2010/index.php?n=Main

Müller, B., Bohn, F., Dreßler, G., Groeneveld, J., Klassert, C., Martin, R., et al. (2013). Describing human decisions in agent-based models–ODD+ D, an extension of the ODD protocol. *Environmental Modelling & Software, 48*, 37–48.

Parker, D. C., Entwisle, B., Rindfuss, R. R., Vanwey, L. K., Manson, S. M., Moran, E., et al. (2008). Case studies, cross-site comparisons, and the challenge of generalization: Comparing agent-based models of land-use change in frontier regions. *Journal of Land Use Science, 3*(1), 41–72.

Polhill, J. G., Gimona, A., & Aspinall, R. J. (2011). Agent-based modelling of land use effects on ecosystem processes and services. *Journal of Land Use Science, 6*(2–3), 75–81.

Polhill, J. G., Parker, D., Brown, D., & Grimm, V. (2008). Using the ODD protocol for describing three agent-based social simulation models of land use change. *Journal of Artificial Societies and Social Simulation, 11*(2). http://jasss.soc.surrey.ac.uk/11/2/3.html

Polhill, J. G. (2010). ODD updated. *Journal of Artificial Societies and Social Simulation, 13*(4). http://jasss.soc.surrey.ac.uk/13/4/9.html

Railsback, S. F., & Grimm, V. (2012). *Agent-based and individual-based modeling: A practical introduction*. Princeton, NJ: Princeton University Press.

Rouchier, J., Cioffi-Revilla, C., Polhill, J. G., & Takadama, K. (2008). Progress in model-to-model analysis. *Journal of Artificial Societies and Social Simulation, 11*(2). http://jasss.soc.surrey.ac.uk/11/2/8.html

Thiele, J. C., & Grimm, V. (2015). Replicating and breaking models: Good for you and good for ecology. *Oikos, 124*(6), 691–696.

Triebig, C., & Klügl, F. (2010). Elements of a documentation framework for agent-based simulation. *Cybernetics and Systems, 40*(5), 441–474.

Wheeler, S. (2005). Beyond the inverted pyramid: Developing news-writing skills. In R. Keeble (Ed.), *Print journalism: A critical introduction* (pp. 84–93). Abingdon, UK: Routledge.

Wilensky, U. (1999). NetLogo. http://ccl.northwestern.edu/netlogo

Part III
Mechanisms

Chapter 16
Utility, Games and Narratives

Guido Fioretti

Abstract This chapter provides a general overview of theories and tools to model decision-making. In particular, utility maximization and its application to collective decision-making, i.e. Game Theory, are discussed in detail. The most important exemplary games are presented, including the Prisoner's Dilemma, the Game of Chicken and the Minority Game, also known as the El Farol Bar Problem. After discussing the paradoxes and pitfalls of utility maximization, an alternative approach is introduced, which is based on seeking coherence between competing interpretations. An assessment of the pros and cons of competing approaches to modelling decision-making concludes the chapter.

Why Read This Chapter?
To appreciate how decision-making can be modelled in terms of utility maximization and game theory. To understand some of the paradoxes, limitations and major criticism of this approach and some of the alternatives.

16.1 Introduction

This chapter provides a general overview of theories and tools to model individual and collective decision-making. In particular, stress is laid on the interaction of several decision-makers.

A substantial part of this chapter is devoted to utility maximization and its application to collective decision-making, known as Game Theory. However, the pitfalls of utility maximization are thoroughly discussed, and the radically alternative approach of viewing decision-making as constructing narratives is presented with its emerging computational tools. In detail, the chapter is structured as follows.

Section 16.2 presents utility maximization and Game Theory with its Nash equilibria. The most important prototypical games are expounded in this section.

G. Fioretti (✉)
University of Bologna, Bologna, Italy
e-mail: guido.fioretti@unibo.it

© Springer International Publishing AG 2017   369
B. Edmonds, R. Meyer (eds.), *Simulating Social Complexity*,
Understanding Complex Systems, https://doi.org/10.1007/978-3-319-66948-9_16

Section 16.3 presents games that are not concerned with Nash equilibria. Section
16.4 illustrates the main paradoxes of utility maximization as well as the patches
that have been proposed to overcome them. Section 16.5 expounds the vision of
decision-making as constructing a narrative, supported by an empirical case study.
Section 16.6 aims at providing computational tools for this otherwise literary vision
of decision-making. Finally, Section 16.7 concludes by assessing the pros and cons
of competing approaches.

This chapter touches so many issues that a complete list of references to the
relevant literature would possibly be longer than the chapter itself. Instead of
references, a guide to the most relevant bibliography is provided at the end of the
chapter.

16.2 Utility and Games

Let $\{a_1, a_2, \ldots a_m\}$ be a set of *alternatives*. Let a_i denote a generic alternative,
henceforth called the i-th alternative where $i = 1, 2, \ldots m$.

By selecting an alternative, a decision-maker obtains one out of several possible
consequences. Let $\{c_{i1}, c_{i2}, \ldots c_{in}\}$ be the set of possible consequences of alternative a_i. Let c_{ij} denote a consequence of a_i, where $i = 1, 2, \ldots m$ and $j = 1, 2, \ldots n_i$.
The *expected utility* of alternative a_i is:

$$u\left(a_i\right) = \sum_{j=1}^{n_i} p\left(c_{ij}\right) u\left(c_{ij}\right) \tag{16.1}$$

where $p(c_{ij})$ is the probability of obtaining consequence c_{ij} and $u(c_{ij})$ is the utility of
consequence c_{ij}.

It is suggested that the one alternative should be chosen that maximizes expected
utility. Frank Ramsey, Bruno De Finetti and Leonard Savage demonstrated that this
is the only choice coherent with a set of postulates that they presented as self-
evident.

Among these postulates, the following ones have the strongest intuitive appeal:

Transitivity: transitivity of preferences means that if $a_i \succ a_j$ and $a_j \succ a_k$, then
$a_i \succ a_k$.
Independence: independence of irrelevant alternatives means that $a_i \succ a_j$ if
$a_i \cup a_k \succ a_j \cup a_k$, $\forall a_k$.
Completeness: completeness means that $\forall(a_i, a_j)$, a preference relation $\succ$ is
defined.

Utility maximization is neither concerned with conceiving alternatives nor with
the formation of preferences, which are assumed to be given and subsumed by the
utility function. Probabilities may eventually be updated by means of frequency
measurement, but at least their initial values are supposed to be given as well. Thus,
utility maximization takes as solved many of the problems with which its critics are
concerned.

Utility maximization takes a gambler playing dice or roulette as its prototypical setting. Indeed, in this setting the set of alternatives is given, utilities coincide with monetary prizes and probabilities can be assessed independently of utilities. By contrast, for some critics of utility maximization, gambling is not an adequate prototype of most real-life situations.

The interaction of several utility-maximizing decision-makers is covered by *Game Theory*. Game Theory assumes that collective decision-making is the combination of several individual decision processes, where each individual maximizes his utility depending on the alternatives selected by the other individuals. Since selecting an alternative implies considering what alternatives other players may select, alternatives are generally called *strategies* in this context.

Utility is called *payoff* in Game Theory. Games in which one player does better at another's expense are called *zero-sum games*. Games may be played once, or they may be repeated.

The bulk of Game Theory is concerned with equilibria. If each player knows the set of available strategies and no player can benefit by changing his or her strategy while the other players keep theirs unchanged, then the current choice of strategies and the corresponding payoffs constitute a *Nash equilibrium*. Since this implies stepping in another player's shoes in order to figure out what she would do if one selects a particular strategy, Nash equilibria are fixed points in self-referential loops of the kind "I think that you think that I think . . . ".

Note that being at a Nash equilibrium neither implies that each player reaches the highest possible payoff that she can attain nor that the sum of all payoffs of all players is the highest that can be attained. This is eventually a concern for economics, for it implies that individual interests may not produce the common good.

If a game is repeated, a Nash equilibrium may be realized either with *pure strategies*, meaning that players choose consistently one single alternative, or *mixed strategies*, meaning that players select one out of a set of available strategies according to a probability distribution. Accepting the idea of mixed strategies often allows to find Nash equilibria where there would be none if only pure strategies are allowed. However, the realism of random decision-makers choosing strategies according to a probability distribution is at least questionable.

Most of the games analysed by Game Theory involve two or in any case a very limited number of players. On the contrary, *evolutionary games* concern large populations of players playing different strategies that are subject to an evolutionary dynamics regulated by *replicator equations*. Successful strategies replicate and diffuse; unsuccessful strategies go extinct. Occasionally, new strategies may originate by random mutation.

The equilibrium concept of evolutionary games is that of *evolutionarily stable strategies*. An evolutionary stable strategy is such that, if almost every member of the population follows it, no mutant can successfully invade. Alternatively, evolutionary games may be played in order to observe typical dynamics, in which case they become akin to the influence games that will be handled in Sect. 16.3.

The following games propose prototypical modes of human interaction. Games used by experimental economics in order to evince human attitudes do not pertain to this list.

16.2.1 The Battle of Sexes

Imagine a couple where the husband would like to go to the football game, whereas the wife would like to go to the opera. Both would prefer to go to the same place rather than different ones.

The payoff matrix in Fig. 16.1 is an example of the Battle of Sexes, where the wife chooses a row and the husband chooses a column. Aside, a generic representation of the game where $L < M$.

This representation does not account for the additional harm that might come from going to different locations and going to the wrong one, i.e. the husband goes to the opera while the wife goes to the football game, satisfying neither. Taking account of this effect, this game would bear some similarity to the Game of Chicken of Sect. 16.2.7.

This game has two pure-strategy Nash equilibria, one where both go to the opera and another where both go to the football game. Furthermore, there is a Nash equilibrium in mixed strategies, where the players go to their preferred event more often than to the other one.

None of these equilibria is satisfactory. One possible resolution involves a commonly observed randomizing device, e.g. the couple may agree to flip a coin in order to decide where to go.

	Opera	Football		
Opera	3, 2	0, 0	M, L	0, 0
Football	0, 0	2, 3	0, 0	L, M

Fig. 16.1 A payoff matrix for the Battle of the Sexes (*left*) and its generic representation (*right*). The *left* number is the payoff of the row player (wife); the *right* number is the payoff of the column player (husband). In this generic representation, L is the payoff of the least preferred alternative, whereas M is the payoff of the most preferred alternative

16.2.2 The Stag Hunt

Rousseau described a situation where two individuals agree to hunt a stag, which none of them would be able to hunt alone. One hunter may eventually notice a hare and shoot at it. This would destroy the stag hunt so the other hunter would get nothing.

An example of the payoff matrix for the stag hunt is pictured in Fig. 16.2, along with its generic representation. The stag hunt requires that $C > B \geq D > S$.

This game has two pure-strategy Nash equilibria, one where both hunters hunt the stag and the other one where both hunters hunt a hare. The first equilibrium maximizes payoff, but the second equilibrium minimizes risk. There exists also a mixed-strategy Nash equilibrium, but no payoff matrix can make the hunters play "stag" with a probability higher than 1/2.

The stag hunt exemplifies the idea of society originating out of contracts between individuals. The examples of "social contract" provided by Hume are stag hunts:

- Two individuals must row a boat. If both choose to row, they can successfully move the boat, but if one does not, the other wastes his effort.
- Two neighbours wish to drain a meadow. If they both work to drain it, they will be successful, but if either fails to do his part, the meadow will not be drained.

Several animal behaviours have been described as stag hunts. For instance, orcas corral large schools of fish to the surface and stun them by hitting them with their tails. This works only if fishes do not have ways to escape, so it requires that all orcas collaborate to kill all fishes they caught rather than catching a few of them.

	Stag	Hare		
Stag	3, 3	0, 1	C, C	S, B
Hare	1, 0	1, 1	B, S	D, D

Fig. 16.2 A payoff matrix for the stag hunt (*left*) and its generic representation (*right*). The *left* number is the payoff of the row player; the *right* number is the payoff of the column player. In this generic representation, C is the payoff that accrues to both players if they cooperate, D is the payoff that accrues to both players if they defect from their agreement, S is the sucker's payoff and B is the betrayer's payoff

16.2.3 The Prisoner's Dilemma

The Prisoner's Dilemma is a central subject in economics, for it apparently contradicts its basic assumption that common good arises out of self-interested individuals. This difficulty is eventually overcome by repeating the game.

The basic formulation of the Prisoner's Dilemma is as follows. Two suspects, A and B, are arrested by the police. Having insufficient evidence for conviction, the police visits them separately, offering each prisoner the following deal: if one testifies for prosecution against the other and the other remains silent, the betrayer goes free, and the silent accomplice receives the full 10-year sentence. If both stay silent, both prisoners are sentenced to only 6 months in jail for a minor charge. If each betrays the other, each receives a 5-year sentence. Each prisoner must make the choice of whether to betray the other or to remain silent; unfortunately, neither prisoner knows what choice the other prisoner made.

The Prisoner's Dilemma describes any situation where individuals have an interest to be selfish, though if everyone cooperates, a better state would be attained. Examples may include unionizing, paying taxes, not polluting the environment or else. Figure 16.3 illustrates a payoff matrix for the Prisoner's Dilemma as well as its generic representation. The Prisoner's Dilemma requires that $B > C > D > S$.

The Prisoner's Dilemma has only one Nash equilibrium at (D, D). Notably, all individual incentives push towards this equilibrium. Nevertheless, this equilibrium is not socially optimal.

Eventually, the difficulty raised by the Prisoner's Dilemma can be overcome if players can repeat the game (which requires $2C > B + S$). In particular, by playing the Prisoner's Dilemma as an evolutionary game with large numbers of players and strategies, it is possible that islands of cooperation sustain themselves in a sea of selfish choices. One possibility for islands of cooperation to emerge is to allow reciprocity, e.g. with a "tit-for-tat" strategy: start with cooperating whenever you meet a new player, but defect if the other does. Another possibility is that players cooperate when they meet players that exhibit a randomly selected tag—e.g. a tie may be worn in order to inspire confidence—so that islands of cooperation emerge even if agents have no memory.

	Cooperate	Defect			
Cooperate	3, 3	0, 5		C, C	S, B
Defect	5, 0	1, 1		B, S	D, D

Fig. 16.3 A payoff matrix for the Prisoner's Dilemma (*left*) and its generic representation (*right*). The *left* number is the payoff of the row player; *right* number is the payoff of the column player. In this generic representation, C is the payoff if both players cooperate, D is the payoff if both defect from their agreement, S is the sucker's payoff and B is the betrayer's payoff

16.2.4   The Traveller's Dilemma

The Traveller's Dilemma is a non-zero-sum game in which two players attempt to maximize their own payoff, without any concern for the other player's payoff. It is a game that aims at highlighting a paradox of rationality. It is a thought experiment on the following problem.

An airline loses two suitcases belonging to two different travellers. The suitcases contain identical antiques. An airline manager tasked to settle the claims of both travellers explains that the airline is liable for a maximum of $100 per suitcase, and in order to determine a honest appraised value of the antiques, the manager separates both travellers and asks each of them to write down the amount of their value at no less than $2 and no more than $100. He also tells them that if both write down the same number, he will treat that number as the true value of both suitcases and reimburse both travellers that amount. However, if one writes down a smaller number than the other, this smaller number will be taken as the true value, and both travellers will receive that amount along with a bonus/malus: $2 extra will be paid to the traveller who wrote down the lower value and a $2 deduction will be taken from the person who wrote down the higher amount. The challenge is: what strategy should both travellers follow in order to decide what value they should write down?

If this game is actually played, nearly all the time, everyone chooses $100 and gets it. However, rational players should behave differently.

Rational players should value the antique slightly less than their fellow traveller, in order to get the bonus of $2. For instance, by pricing at $99, one would get $101, whereas the opponent would get $97. However, this triggers an infinite regression such that $2 is the only Nash equilibrium of this game. Thus, being rational does not pay.

The Traveller's Dilemma suggests that in reality people may coordinate and collaborate because of their bounded rationality, rather than in spite of it. If they would be smarter than they are, they would obtain less.

16.2.5   The Dollar Auction

The dollar auction is a non-zero-sum sequential game designed to illustrate a paradox brought about by rational choice theory. In this game, players with perfect information are compelled to make an ultimately irrational decision based on a sequence of rational choices.

The game involves an auctioneer who offers a one-dollar bill with the following rule: the dollar goes to the highest bidder, who pays the amount he bids. The second highest bidder must also pay the highest amount that he bids but gets nothing in return.

Suppose that the game begins with one of the players bidding 1 cent, hoping to make a $0.99 profit. He will be quickly outbid by another player bidding 2 cents, as a $0.98 profit is still desirable. Similarly, another bidder may bid 3 cents, making a $0.97 profit. At this point the first bidder may attempt to convert his loss of 1 cent into a gain of $0.97 by also bidding 3 cents. In this way, a series of bids is maintained.

One may expect that the bidders end up with offering $1.00 for a one-dollar bill, which is what the auction is for. However, a problem becomes evident as soon as the bidding reaches 99 cents. Suppose that one player bid 98 cents. The other players now have the choice of losing 98 cents or bidding one dollar, which would make their profit zero. After that, the original player has a choice of either losing 99 cents or bidding $1.01, losing only 1 cent. After this point these rational players continue to bid the value up well beyond the dollar, and neither makes a profit.

16.2.6 Pure Coordination Games

Pure coordination games are an empirical puzzle for Game Theory. Pure coordination games are one-shot games where players face a set of alternatives knowing that a positive payoff will only accrue to them if they coordinate on the same choice. For instance, two subjects may be shown a city map and asked, independently of one another, to select a meeting point. Or, subjects may be asked to select a positive integer. In the first case, they obtain a positive payoff if they select the same meeting point; in the second case, if they select the same integer.

The difficulty of pure coordination games derives from the fact that players cannot communicate and that the game is not repeated. The astonishing fact about pure coordination games is that, if they are actually played, players reach an agreement much more often than they would if they played randomly.

The commonly held explanation is that pure coordination games generally entail cues that single out one choice as more "salient" than others. For instance, subjects asked to select a meeting point generally end up with the railway station, whereas the majority of those asked to name a positive integer select the number 1.

Interestingly, this suggests that coordination may eventually be attained because of conventions, habits or values that do not enter the description of decision settings. People may not even be aware of what makes them coordinate with one another.

16.2.7 The Game of Chicken

The Game of Chicken models two drivers, both headed for a single-lane bridge from opposite directions. One must swerve, or both will die in the crash. However, if one driver swerves but the other does not, he will be called a "chicken". Figure 16.4 depicts a typical payoff matrix for the chicken game as well as its generic form.

Chicken is an anti-coordination game with two pure-strategy Nash equilibria where each player does the opposite of what the other does. Which equilibrium

	Swerve	Straight		
Swerve	0, 0	-1, +1	V/2, V/2	0, V
Straight	+1, -1	-10, -10	V, 0	$\frac{(V-C)}{2}, \frac{(V-C)}{2}$

Fig. 16.4 A payoff matrix for the Game of Chicken (*left*) and its generic representation (*right*). The *left* number is the payoff of the row player; the *right* number is the payoff of the column player. In this generic representation, V is the value of power, prestige, or of the available resource to be obtained; C is the cost if both players choose "straight"

is selected depends very much on the effectiveness in signalling precommitment before the game is played. For instance, a driver who disables the brakes and the steering wheel of his car and makes it known to the other driver may induce him to swerve.

Bertrand Russell remarked that the nuclear stalemate was much like the Game of Chicken[1]:

As played by irresponsible boys, this game is considered decadent and immoral, though only the lives of the players are risked. But when the game is played by eminent statesmen, who risk not only their own lives but those of many hundreds of millions of human beings, it is thought on both sides that the statesmen on one side are displaying a high degree of wisdom and courage, and only the statesmen on the other side are reprehensible. This, of course, is absurd. Both are to blame for playing such an incredibly dangerous game. The game may be played without misfortune a few times, but sooner or later it will come to be felt that loss of face is more dreadful than nuclear annihilation. The moment will come when neither side can face the derisive cry of "Chicken!" from the other side. When that moment is come, the statesmen of both sides will plunge the world into destruction.

The Game of Chicken has been reinterpreted in the context of animal behaviour. It is known as *Hawk-Dove game* among ethologists, where the Hawk-Dove game has the same payoff matrix as in Fig. 16.4. In the Hawk-Dove game, "swerve" and "straight" correspond to the following strategies, respectively:

Dove: retreat immediately if one's opponent initiates aggressive behaviour;

Hawk: initiate aggressive behaviour, not stopping until injured or until the opponent backs down.

While the original Game of Chicken assumes $C > V$ and cannot be repeated, the Hawk-Dove game lacks this requirement and is generally conceived as an evolutionary game.

The strategy "Dove" is not evolutionary stable, because it can be invaded by a "Hawk" mutant. If $V > C$, then the strategy "Hawk" is evolutionarily stable. If $V < C$,

---

[1] Bertrand W. Russell, *Common Sense and Nuclear Warfare*. London, George Allen and Unwin, 1959.

there is no evolutionarily stable strategy if individuals are restricted to following pure strategies, although there exists an evolutionarily stable strategy if players are allowed to use mixed strategies.

16.2.8 The War of Attrition

The war of attrition is a game of aggression where two contestants compete for a resource of value V by persisting with their intentions while constantly accumulating costs. Equivalently, this game can be seen as an auction in which the prize goes to the player with the highest bid B_h, and each player pays the loser's low bid B_l.

The war of attrition cannot be properly solved using its payoff matrix. In this game, the players' available resources are the only limit to the maximum value of bids. Since bids can be any number, if available resources are ignored, then the payoff matrix has infinite size. Nevertheless, its logic can be analysed.

Since players may bid any number, they may even exceed the value V that is contested over. Indeed, if both players bid higher than V, the high bidder does not so much win as lose less because $-B_l < V - B_h < 0$—a pyrrhic victory.

Since there is no value to bid which is beneficial in all cases, there is no dominant strategy. However, this does not preclude the existence of Nash equilibria. Any pair of strategies such that one player bids zero and the other player bids any value equal to V or higher, or mixes among any values V or higher, is a Nash equilibrium.

The war of attrition is akin to a Chicken or Hawk-Dove game—see Sect. 16.2.7— where if both players choose "swerve"/"Dove", they obtain 0 instead of $V/2$ as in Fig. 16.4.

The evolutionarily stable strategy when playing it as an evolutionary game is a probability density of random persistence times which cannot be predicted by the opponent in any particular contest. This result has led to the conclusion that, in this game, the optimal strategy is to behave in a completely unpredictable manner.

16.3 Influence Games

Contrary to those of Sect. 16.2, the games in this section are not concerned with Nash equilibria. Players are not assumed to figure out which alternatives the other players might choose, originating infinite regressions that can only stop at equilibrium points.

Rather, boundedly rational players are assumed to follow certain rules that may be quite simple but need not be necessarily so. The game then concerns what collective behaviours emerge out of mutual influence.

However, the games in this section are not so different from those of Sect. 16.2 when they are played as evolutionary games. Such is the case, for instance, of simulations where a large number of players iterate the Prisoner's Dilemma.

Two prototypical games will be expounded in this section. The *Ising model* (originally developed in physics, where it is also known as the *spin glass* model) is concerned with imitation. The *minority game*, also known as the *El Farol Bar Problem*, is concerned with the contrary of imitation. It is about doing the opposite of what others do.

16.3.1 The Ising Model

The Ising model was originally developed in physics in order to study the interaction between atoms in a ferromagnetic material. For this reason its agents can only take two states, or opinions in social applications, and are fixed in space.

The Ising model is an exceedingly stylized model of imitation dynamics. Clearly, many imitation models are more complex and more realistic than the Ising model. However, the closed-form solutions of the Ising model may guide the builder of more complex models in the process of understanding their behaviour.

In general, the Ising model is not presented as a game. It is done here in order to stress its symmetry with the minority game.

Let N players be denoted by means of an index $i = 1, 2, \ldots N$. Players must choose between an alternative $A = -1$ and an alternative $A = 1$.

The payoff of a player does not only depend on the alternative that she has chosen but also on the average of the alternatives chosen by the other players. Let m denote this average.

Since we want to reproduce situations where the individual follows the herd, the effect of m should be the stronger, the more homogeneous the group. Since $A \in \{-1; 1\}$ and consequently $m \in \{-1; 1\}$, we can reach this goal by requiring that the payoff depends on a term $A\,m$. This term may eventually be multiplied by a coefficient $J > 0$.

A stochastic term ε is necessary in order to understand our game as a system jumping between many equilibria. This term will disappear when expected values will be taken. In the end, the following functional form is chosen for the payoff of a player:

$$u(A) = v(A) + JAm + \varepsilon \tag{16.2}$$

where $u(A)$ is the total payoff of a player and $v(A)$ is its individual component. Furthermore, let us assume that this individual component takes the following form:

$$v(A) = \begin{cases} -h & \text{if } A = -1 \\ h & \text{if } A = 1 \end{cases} \tag{16.3}$$

where $h \in \Re$, $h > 0$.

By assuming that stochastic terms ε are Gumbel-distributed, we can apply the logit model. By combining Eqs. (16.2) and (16.3) we derive the following expressions for the probability that a player selects one of the two alternatives:

$$p\{A = -1\} = \frac{e^{\mu(-h-Jm+\varepsilon)}}{e^{\mu(-h-Jm+\varepsilon)} + e^{\mu(h+Jm+\varepsilon)}} \tag{16.4}$$

$$p\{A = 1\} = \frac{e^{\mu(h+Jm+\varepsilon)}}{e^{\mu(-h-Jm+\varepsilon)} + e^{\mu(h+Jm+\varepsilon)}} \tag{16.5}$$

The expected value of the selected alternative is $E\{A\} = -1 \bullet p\{A = -1\} + 1 \bullet p\{A = 1\}$. Since it is also $E\{A\} = m$, we obtain the following expression:

$$m = tanh\,(\mu h + \mu Jm) \tag{16.6}$$

where $tanh(x) = (e^x - e^{-x})/(e^x + e^{-x})$ is the hyperbolic tangent.

Equation (16.6) provides an analytic description of a game with herd behaviour on two alternatives described by means of a mean-field approximation. It admits a closed-form solution that provides the following findings:

- If $\mu J < 1$ and $h = 0$, there exists one single solution at $m = 0$. Consider that this is a discrete-time system, so its attractors are stable if all eigenvalues of the state transition function are in $(-1, 1)$. Intuitively, $\mu J < 1$ means that this system is globally stable. Furthermore, $h = 0$ means that the individual component of the payoff is zero, so the players have no incentive to choose one of the two alternatives. Consequently, the stochastic term makes $m = 0$ the only solution.
- If $\mu J < 1$ and $h \neq 0$, there exists one single solution with the same sign as h. As in the previous case, the system is globally stable so it admits one single solution. However, since in this case the players' payoff includes an individual component, it is this component that determines what equilibrium is reached. If most players prefer $A = -1$, the equilibrium will be $m \approx -1$; likewise, if most players prefer $A = 1$, then the equilibrium will be $m \approx 1$.
- If $\mu J \geq 1$ and $h = 0$, there exist two solutions: $m = 0$ and $m = \pm m(\mu J)$. In this case the system is globally unstable, but locally stable equilibria may exist. Since the individual component of the payoff is zero, the system may either tend towards $m = 0$ or $m \approx -1$ or $m \approx 1$.
- If $\mu J \geq 1$ and $h \neq 0$, the following subcases must be distinguished:

  - If, for any given μ and J, there exists a threshold $H(h) > 0$ such that $|h| \leq H$, then three solutions exist, one with the same sign as h and the other two with opposite sign. Condition $|h| \leq H$ means that the individual component of the payoff is limited even if not zero. Therefore, results are similar to the previous case.

- If, for any given μ and J, there exists a threshold $H(h) > 0$ such that $|h| > H$, then there exists one single solution with the same sign as h. Indeed if the individual component of the payoff can take any value, then the whole system is forced into its direction.

In the Ising model, each player observes the average behaviour of all other players. If each player observes only the behaviour of his neighbours, one obtains Schelling's model of racial segregation (according to Schelling's model of racial segregation, a city where Blacks and Whites are randomly distributed turns into a chessboard of homogeneous quarters if its inhabitants, although absolutely ready to accept the presence of the other colour, do not want to be a small minority in their own neighbourhoods).

16.3.2 The Minority Game

The minority game originates from a consideration inspired by the *El Farol* bar in Santa Fe, New Mexico (USA). The economist Brian Arthur remarked that people go to the bar in order to meet other people, but they do not want to go when all other people go because the bar is too crowded on such occasions. Thus, they want to do the opposite of what most people do—go to the bar when most people stay at home and stay at home when most people go to the bar. This is interesting, because the "El Farol Bar Problem" cannot have a stable equilibrium. This happens because once the majority observed what the minority did, it wants to imitate it, which turns the minority into majority, and so on endlessly.

Physicists Damien Challet and Yi-Cheng Zhang remarked that this is the essence of stock market dynamics. In the stock market, those traders gain, who buy when prices are low (because most traders are selling) and sell when prices are high (because most traders are buying). So all traders want to belong to the minority, which is clearly impossible, hence the inherent instability of this game. Among the physicists, the "El Farol Bar Problem" became the "Minority Game".[2]

Let us consider N players who either belong to a group denoted 0 or a group denoted 1. Players belonging to the minority group receive a positive payoff. Players belonging to the majority receive zero.

Strategies are functions that predict which will be the minority group in the next step given the minority group in the m previous steps. Thus, a strategy is a matrix with 2^m rows (dispositions with repetition of two elements of class m) and two columns. The first column entails all possible series of minority groups in the previous m steps, henceforth *histories*. The second column entails the group suggested to be minority in the next step. As an example, Fig. 16.5 illustrates a strategy with $m = 2$.

---

[2]The rest of this section has been extensively drawn from E. Moro, *The Minority Game: An Introductory Guide*, working paper available online.

Fig. 16.5 An example of a
strategy based on the two
previous steps of the minority
game. The *first column* lists
all possible stories. The
second column makes a
prediction depending on past
history

History	Prediction
0 0	0
0 1	1
1 0	1
1 1	1

Each player owns s strategies. If $s = 1$, the game is trivial because the time series of the minority group is periodical.

If $s > 1$, players choose the strategy that cumulated the greatest amount of payoffs. Thus, a number of feedbacks may arise between what strategies are chosen and their capability to predict the minority. The reason is that in this game players must adapt to an environment that they themselves create.

An important magnitude in this game is the variance of the time series of the number of players belonging to group 1 (or, equivalently, group 0). Henceforth, this magnitude will be denoted by σ^2.

The average of the number of players belonging to each group is generally close to $N/2$. If σ^2 is small, then the distribution of the number of players belonging to group 1 is concentrated around $N/2$. This implies that the minority is large, eventually close to its maximum ($N/2 - 1$). On the contrary, if σ^2 is large, the number of players belonging to group 1 tends to be either much smaller or much larger than $N/2$, implying that the minority is often very small.

Let us consider σ^2/N in order to normalize to the number of players. Let us define the *efficiency of coordination* $e_c = N/\sigma^2$ as the reciprocal of the extent to which players behave differently from one another.

Figure 16.6 depicts numerical simulations of e_c as a function of the number of histories in a strategy $2^m/N$. Graphs are shown for different values of s. The horizontal line marks the value that e_c attains if players would make a random choice among the strategies available to them.

With low m the efficiency of coordination is low. This happens because if memory is short, players have greater difficulties to adapt to the changing features of the game.

If only few strategies are available ($s = 2$, $s = 3$, $s = 4$), at intermediate values of m, many players guess the correct strategy, so e_c increases above the level that can be attained if strategies are chosen randomly. This threshold is marked by the dashed vertical line. However, this effect disappears if many strategies are available ($s = 8$, $s = 16$). In this case the decision process becomes similar to a random choice, so even at intermediate values of m, the efficiency of coordination is close to the level attained when strategies are chosen randomly.

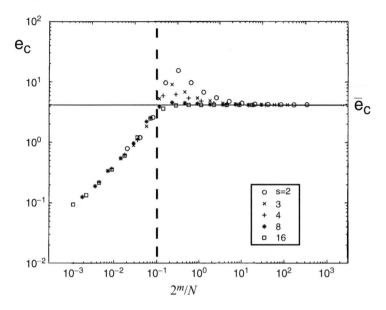

Fig. 16.6 Efficiency of coordination e_c as a function of the number of histories in a strategy $2^m/N$, for different values of the number of available strategies s. The *horizontal line* at $e_c = \bar{e}_c$ marks the efficiency level when players select a strategy at random. The *vertical dashed line* marks the point where e_c can be greater than $\bar{e}_c$.

Independently of the number of available strategies, with increasing m, the value of e_c tends to the level attained when strategies are chosen randomly. This happens because a history of length m occurs again after 2^m steps on average, so a strategy that is successful with a particular history needs 2^m steps in order to be successful again. With very high values of m, no strategy can present itself as particularly successful; therefore, a nearly random dynamics ensues.

Let us consider what information is available to players. The only information available to them is what group was the minority in previous time steps. Let this information be carried by a variable W_t, where $W_t = 0$ means that at time t the group 0 has been minority, $W_t = 1$ otherwise. The issue is whether this information is used efficiently; if it is not, there may exist arbitrage possibilities for players who utilize information more efficiently than their peers.

Let us consider W_t and W_{t+1} as distinct signals. Let us compute their mean mutual information $I(W_t, W_{t+1})$.[3]

---

[3]Given a source of binary symbols $\{a_1, a_2, \ldots a_M\}$ issued with probabilities $p_1, p_2, \ldots p_M$, the average information that they convey is defined as $H(A) = \sum_{i=1}^{M} p(a_i)\ log_2 1/p(a_i)$, and it is called *information entropy*. Suppose that there is a second source issuing symbols $\{b_1, b_2, \ldots b_N\}$ with information entropy $H(B)$. Let $H(A,B)$ denote the information entropy of the whole system. *Mean mutual information* $H(A) + H(B) - H(A,B)$ measures to what extent the two sources interact

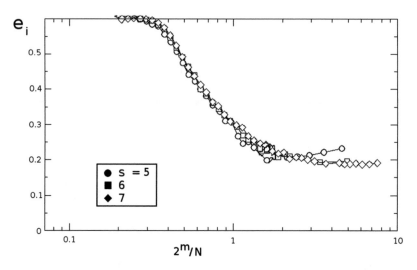

Fig. 16.7 Efficiency of information exploitation as a function of the number of stories in a strategy, normalized to the number of players

Mean mutual information measures whether the information entailed in the outcomes of two steps of the game, taken together, is greater than the sum of the information entailed in the outcomes of the two steps taken independently of one another. Thus, mean mutual information says whether a player, by observing the time series of the outcome of the game, could do better than his peers. Recalling the analogy with the stock market, $I(W_t, W_{t+1}) > 0$ means that a trader could gain from arbitrage.

Let us introduce *information efficiency* $e_i = 1/I(W_t, W_{t+1})$. Being the reciprocal of mean mutual information, information efficiency is high when mean mutual information is low, i.e. when information is efficiently exploited by the player, so there is little room for arbitrage.

Figure 16.7 depicts numerical simulations of e_i as a function of the number of stories in a strategy $2^m/N$. Graphs are shown for different values of s.

One may observe in Fig. 16.7 a sudden drop of e_i in the [0.3,1] interval. This interval is entailed in the interval [0.1,1] where e_c was observed to rise above the level corresponding to random choice in Fig. 16.6. Thus, we may subsume the behaviour of the minority game as in Table 16.1.

Table 16.1 shows that the minority game has two large behaviour modes, one inefficient in coordination but efficient in the exploitation of information and the other one efficient in coordination but inefficient in the exploitation of information.

---

to correlate their messages. Mean mutual information is zero if the two sources are independent of one another.

Table 16.1 Efficiency of coordination and efficiency of information exploitation in the minority game

$2^m/N < 0.1$	$2^m/N > 1$
Inefficient coordination	Efficient coordination
Low e_c	High e_c
Efficient information exploitation	Inefficient information exploitation
High e_i	Low e_i

In between is a tiny space where the efficiency of coordination and the efficiency of information exploitation may change dramatically depending on s and m.

Since the minority game is a stylized representation of stock markets, we may ask in which region stock markets operate. It is well known that traders are very many, so we may assume that N is very large. Human bounded rationality suggests that traders do not make use of complicated algorithms that take account of events far back in the past, so m should be in the order of a few units. Consequently, $2^m/N$ is likely to be very small.

This suggests that financial markets are characterized by low coordination, which implies irregular oscillations where large majorities and small minorities may appear. At the same time, financial markets are efficient in exploiting information. Thus, the observation of its time series offers few possibilities to extrapolate future courses.

16.4   Some Pitfalls of Utility Maximization

Utility maximization strikes its adepts for its elegance, simplicity and beauty. Unfortunately, empirical tests have shown that in many situations, decision-makers do not follow its prescriptions. Furthermore, there are cases where maximizing utility leads to paradoxical decisions.

Some of these paradoxes can be reduced to utility maximization by means of special additions to the basic theory. Others cannot, thereby suggesting that utility maximization, besides poor descriptive strength, may have poor normative value as well. In this section the main paradoxes will be discussed, together with their eventual resolution within the utility maximization framework.

16.4.1   Ellsberg's Paradox and Subadditive Probabilities

Suppose that a decision-maker is placed in front of two urns, henceforth denoted A and B. The decision-maker is informed that urn A entails white and black balls in equal proportion, e.g. urn A may contain ten white balls and ten black balls. Regarding urn B, the decision-maker knows only that it entails white and black balls. Suppose to ask the decision-maker to evaluate the probability to extract a white ball from urn A and the probability to extract a white ball from urn B.

Since urn A entails white and black balls in equal proportions, the probability to extract a white ball from urn A is 0.5. By contrast, nothing is known regarding the proportion of white to black balls in urn B. In cases like this, the so-called principle of insufficient reason—i.e. the fact that there is no reason to think otherwise—suggests to imagine that also urn B entails white and black balls in equal proportions. Thus, also in this case, the probability to extract a white ball is assessed at 0.5. And yet, something is not in order: intuitively, urn B should be characterized by a greater uncertainty than urn A!

Ellsberg's paradox actually deals with the size of the sample on which probabilities are evaluated. More precisely, Ellsberg's paradox places two extreme situations aside.

In the case of urn A, since we know that it entails white and black balls in equal proportions, we are able to compute probability with infinite precision. It is just like extracting a ball (and replacing it afterwards) infinite times. We are measuring probability on a sample of infinite size.

In the case of urn B, lack of knowledge on the proportion of white to black balls is equivalent to estimating the probability of extracting a white ball prior to any extraction. It means that the probability must be measured on a sample of size zero. We guess its value at 0.5, but the reliability of our estimate is very low.

One possibility for overcoming Ellsberg's paradox is that of representing uncertainty by means of two magnitudes. The first one is probability, whereas the second one is sample size. In statistics, sample size is expressed by *precision indicators*.

Another possibility is to resort to the theory of subadditive probabilities. While according to classical probability theory the sum of the probabilities of an exhaustive set of events must be equal to 1, according to the theory of subadditive probabilities, this holds only if probabilities are measured on a sample of infinite size. In all other cases, probabilities take values such that their sum is smaller than 1.

Let us consider the following example: we are playing dice in a clandestine gambling room. Since we fear that we are playing with unfair dice, we may not assign probability 1/6 to each face but rather something less, e.g. 1/8. Thus, the sum of the probabilities of all faces is $6 \times 1/8 = 3/4$, which is smaller than 1. Subsequently, if we have a possibility to throw a die many times—i.e. if we can increase the size of our sample—we may find out that that the die is unfair in the sense that, e.g. face "2" comes out with probability 1/3 while all other faces come out with probability 2/15. The sum of all these probabilities is $5 \times 2/15 + 1/3 = 2/3 + 1/3 = 1$.

Let us return to Ellsberg's paradox. In the case of urn A, the probability to extract a white ball is 0.5, and the probability to draw a black ball is 0.5. The sum of these probabilities is 1. In the case of urn B, the decision-maker may judge that the probability to extract a white ball is, for instance, 0.4 and that the probability of extracting a black ball is also 0.4. The sum of these probabilities is 0.8, and this does not constitute a problem for the theory of utility maximization.

By employing subadditive probabilities, utility maximization can be meaningfully applied to situations where probabilities are less than perfectly known. In the end, utility maximization can safely deal with the difficulty raised by Ellsberg's paradox.

16.4.2   Allais' Paradox and Prospect Theory

The following experiment was proposed by Maurice Allais. Subjects are asked to choose between the alternatives A and B reported on the rows of Table 16.2. It is empirically observed that most people choose alternative B.

Subsequently, the same subjects are confronted with the alternatives C and D reported on the rows of Table 16.3. It is empirically observed that most people choose alternative C.

Let us now examine the expected utilities of these two pairs of alternatives, namely, (A,B) and (C,D). Preferring (B) to (A) means that $u(2400) > 0.33 \times u(2500) + 0.66 \times u(2400)$, which can be written as $0.34 \times u(2400) > 0.33 \times u(2500)$. Unfortunately, preferring (C) to (D) implies just the opposite, i.e. $0.33 \times u(2500) > 0.34 \times u(2400)$. So it turns out that most people either do not behave rationally or do not maximize utility.

However, Allais' paradox is due to the presence of a tiny probability of not obtaining anything in alternative (A). Thus, it is due to aversion to risk.

Daniel Kahneman and Amos Tversky introduced non-linear transformations of utilities and probabilities in order to balance risk aversion. The transformed utilities and probabilities can describe the observed behaviour as expected utility maximization, and this is called prospect theory.

A prospect is a set of pairs $\{(c_1, p_1), (c_2, p_2), \ldots\}$, where c_j is a consequence that will obtain with probability p_j. As a preliminary step, prospects with identical consequences are summed, dominated prospects are eliminated and riskless components are ignored. Prospect theory prescribes that the utilities and the probabilities of the above prospects be transformed according to the following rules:

Table 16.2 The first choice in Allais' experiment

	Consequence 1	Consequence 2	Consequence 3
Alternative A	Receive \$2500 with probability 0.33	Receive \$2400 with probability 0.66	Receive nothing with probability 0.01
Alternative B	Receive \$2400 with probability 1.00		

Table 16.3 The second choice in Allais' experiment

	Consequence 1	Consequence 2
Alternative C	Receive \$2500 with probability 0.33	Receive nothing with probability 0.67
Alternative D	Receive \$2400 with probability 0.34	Receive nothing with probability 0.66

1. Utility is transformed by means of a non-linear function $v = f(u)$ such that $f'(u) > 0$ and $f''(u) < 0$ for $u > 0$, $f'(u) > 0$ and $f''(u) > 0$ for $u < 0$, with $|f''(u)|_{u<0} > |f''(u)|_{u>0}$.

2. Probabilities p are transformed into "weights" w by means of a non-linear function $w = g(p)$ such that $g(0) = 0$ and $g(1) = 1$ but $\exists \bar{p} \in (0,1)$ such that $\forall p < \bar{p}$ it is $g(p) \geq p$ and $\forall p > \bar{p}$ it is $g(p) \leq p$.

3. Weights w are transformed into coefficients q by means of the following rules:

$$q_{-h}^- = w^- (p_{-h}) \qquad\qquad \text{for } i = -h$$

$$q_i^- = w^- (p_{-h} + \cdots + p_i) - w^- (p_{-h} + \cdots + p_{i-1}) \qquad \text{for } -h < i \leq 0$$

$$q_i^+ = w^+ (p_i + \cdots + p_k) - w^+ (p_{i+1} + \cdots + p_k) \qquad \text{for } 0 \leq i < k$$

$$q_k^+ = w^+ (p_k) \qquad\qquad \text{for } i = k$$

where w^- and q^- refer to prospects with negative utility, denoted by an index $i \in [-h, 0]$, whereas w^+ and q^+ refer to prospects with positive utility, denoted by an index $i \in [0, k]$.

The v and q obtained at the end of this procedure can be used just like utilities and probabilities, respectively. However, note that prospect theory succeeds to eliminate the inconsistencies highlighted by Allais' paradox, but it does not explain why it works. It should be called a heuristic, rather than a theory.

In the end, utility maximization does succeed to cope with Allais' paradox but at the cost of a complicated patch that has the flavour of the epicycles that had to be added to the Ptolemaic system in order to support the idea that it was the Sun that was turning around the Earth. With such a patch, the elegance of the original theory is lost.

16.4.3 Preference Reversal in Slovic's Paradox

Let us consider a series of bets with different characteristics, for instance, a series of bets on different horses, or playing on a series of different slot machines, or a series of unfair dice that are different from one another. The game consists of choosing to bet on a specific horse, choosing to play on a specific slot machine or selecting a specific die to throw. In other words, the game consists of choosing one bet out of a series of bets.

Table 16.4 Slovic's experiment

	Consequence 1	Consequence 2
Pair of bets I		
Bet A_I	Win $4.00 with probability 0.99	Lose $1.00 with probability 0.01
Bet B_I	Win $16.00 with probability 0.33	Lose $2.00 with probability 0.67
Pair of bets II		
Bet A_{II}	Win $3.00 with probability 0.95	Lose $2.00 with probability 0.05
Bet B_{II}	Win $6.50 with probability 0.50	Lose $1.00 with probability 0.50
Pair of bets III		
Bet A_{III}	Win $2.00 with probability 0.80	Lose $1.00 with probability 0.20
Bet B_{III}	Win $9.00 with probability 0.20	Lose $0.50 with probability 0.80
Pair of bets IV		
Bet A_{IV}	Win $4.00 with probability 0.80	Lose $0.50 with probability 0.20
Bet B_{IV}	Win $40.00 with probability 0.10	Lose $1.00 with probability 0.90

In order to simplify matters, let us consider series composed by two bets. More specifically, let us consider the four pairs of bets illustrated in Table 16.4.

For any pair of bets, subjects are asked to select either bet A or bet B. On average, the number of subjects who prefer A to B is slightly greater than the number of subjects who prefer B to A.

At this point, a different game is played. Subjects are asked to imagine that they own a lottery ticket for each bet and that they have a possibility to sell it. That is, either they can wait for the outcome of each bet, where they may win or lose with a certain probability, or they can sell the ticket. In order to compare the willingness to play to the willingness to sell the ticket, subjects are asked to fix a minimum selling price for each bet.

In general, it is empirically observed that most people ask a higher price for bets B than for bets A.

However, for each pair of bets, bet A has the same expected (utility) value than bet B. Thus, utility maximizers should be indifferent between A and B. And yet the empirical evidence is that most subjects have a slight preference for A if they are asked to play one of the two bets but they definitely prefer B if they are asked to fix a selling price.

The distinguishing feature of bets A is that the first consequence has a much higher probability than the second one. Thus, one may assume that it is this difference of probability values that orientates decision-making.

The distinguishing feature of bets B is that the first consequence concerns a much larger amount of money than the second one. Probabilities, on the contrary, are sometimes very similar and sometimes very different from one another. Thus, one may assume that it is this difference of money values that orientates decision-making.

If subjects are asked to bet, their attention is caught by probabilities, so either they are indifferent or they prefer A to B. By contrast, if subjects are asked to sell lottery tickets, their attention is caught by money values, so they prefer B to A.

Slovic's paradox shows that preferences change if decision-makers focus on the probability of a consequence or, rather, on its utility (here, money value). This means that human beings are unable to evaluate probabilities and utilities independently of one another.

Slovic's paradox—often known as "preference reversal"—is destructive for utility maximization. Indeed, it undermines the assumption that a utility function and a probability function can be defined independently of one another. Ultimately, Slovic's paradox suggests that uncertain belief cannot be split into utilities and probabilities.

Obviously, several attempts have been made to reconcile preference reversal with the theory of rational choice. In particular, it has been found that preference reversal can be accommodated with the theory of rational choice if either violations of transitivity, or of independence, or of completeness of preferences are accepted. While the attempts to reconcile preference reversal with the theory of rational decision by relaxing transitivity or independence of preferences did not receive much attention because these properties are essential for our idea of rationality—see Sect. 16.2—the more recent idea of dropping completeness deserves some discussion. Indeed, allowing preferences to be incomplete amounts to accept the idea that utility functions can be defined, at most, for only *some* alternatives. Possibly, just the simplest and most repetitive ones.

16.4.4 Arrow's Paradox

The following paradox of social choice is due to Kenneth Arrow. Let A, B and C denote three alternatives, and let 1, 2 and 3 denote three individuals. Let us assume that:

- Individual 1 prefers alternative A to alternative B and alternative B to alternative C. Thus, he prefers alternative A to alternative C.
- Individual 2 prefers alternative B to alternative C and alternative C to alternative A. Thus, he prefers alternative B to alternative A.
- Individual 3 prefers alternative C to alternative A and alternative A to alternative B. Thus, he prefers alternative C to alternative B.

If these three individuals constitute a democratic community with a majority rule, then this community prefers A to B (individuals 1 and 3) and alternative B to alternative C (individuals 1 and 2). Thus, if the community wants to have transitive preferences, it must prefer A to C. But, the majority of its members (individuals 2 and 3) prefers C to A!

Arrow's paradox shows that there are conditions such that the aggregate outcome contradicts a basic assumption of utility maximization, even if individuals do not. It is not destructive for utility maximization as a theory of individual decision-making, but it impairs its extension to group or organizational decision-making.

Several proposals have been made in order to overcome Arrow's paradox. The most common solution is to allow individuals to have different preferences if all alternatives are presented to them, instead of being presented with pairs of alternatives. Or, one may limit voters to two alternatives presented in tournaments. In this way Arrow's paradox would disappear, and yet the final choice is not necessarily the one that would be preferred by the largest possible majority.

16.5 Seeking Coherence

As we have seen in Sect. 16.4, utility maximization is not a good descriptor of decision processes. Its proponents have objected that utility maximization is not meant to be a faithful description of what people actually do, but rather a prescription of what they should do. It pretends to be a normative theory, although it is not a descriptive theory.

However, the preference reversals highlighted by Slovic point to such a huge distance between theory and reality that even the normativeness of utility maximization might be questioned. If utilities do not exist prior to decision-making, it may make little sense to tell decision-makers that they should maximize them. Furthermore, if evolution shaped human reasoning along patterns that are different from utility maximization, we ought to be careful to declare these patterns "illogical" or "irrational". Rather, it may make sense to observe how human beings actually make their decisions, understand their rationales and eventually revise our theories.

Suppose to view human minds as coherence-seeking machines that make use of available information in order to construct a plausible interpretation of reality, be it social roles, scientific theories or else. By drawing causal relationships and by eliminating inconsistencies, a decision-maker tells herself a story that explains why certain facts are the way they are and why certain people did what they did. This story, a founding story that suggests a decision-maker what it is appropriate to do, is a *narrative*.

The construction of a narrative may require that issues that do not fit into the picture are ignored, downplayed or forgotten. It may require that opinions are changed even dramatically, and yet their purporters candidly claim that they have always been coherent throughout their lives or that they have been coherent in spite of having changed their opinion if their story is seen from a particular point of view.

Albeit disturbing for our idea of rationality, the extent and easiness with which human beings distort previous experiences have been proven by a number of experiments in psychology. It is easy to induce subjects to change opinion, while they are still convinced to have been coherent throughout the whole experiment. Experiments show that people remember past events to the extent that they fit their narratives and that they are ready to change their interpretation of the past if new evidence must be accommodated. On the whole, experimental evidence tells us that human beings are ready to lie to themselves in order to build coherent narratives.

This attitude is puzzling, because distorting reality in order to construct a coherent narrative is at odds with our idea of rationality. So either human nature is inherently irrational, or our idea of rationality is incorrect.

According to James March, reinventing the past is a crucial ability, for it enables decision-makers to conceive new goals and figure out a strategy in an uncertain future. Later, a similar argument has been made by Karl Weick under the label of "sense-making". In a nutshell, these authors suggest that in order to make decisions in the face of an uncertain future, it is good to have a narrative that explains the past as if previous decisions had been made along a coherent line. This line guides the decision-maker into the future, providing a rationale for action even if certainties are very few.

So here comes a straightforward argument for normativeness. If seeking coherence has the purpose of constructing a narrative, and if narratives are useful, then a decision theory based on constructing narratives should be regarded as rational and openly prescribed.

In business, politics and other fields, narratives may constitute the bulk of strategies. David Lane and Robert Maxfield have made a years-long field observation of the elaboration and modification of the narrative of a Silicon Valley firm. This study is worth reporting, because it is very clear in making us understand that narratives are useful precisely because they provide guidance in the face of an uncertain future and that their usefulness is not impaired by the fact that their coherence is based on arbitrary interpretations of reality.

16.5.1 A Real Story

In 1990 *E.* launched *LW*, an innovative technology for distributed control of electromechanical devices. Previously, control was centralized by one main processing unit that would command several peripheral devices. With *LW*, each device is endowed with a microprocessor and can communicate with all others, so the devices control one another. Distributed control is more resilient than centralized control and easily implements modular architectures to which additional devices can be added. Therefore, it is technically superior to centralized control.

Distributed control is particularly suited to the automation of office spaces in large buildings, post-Fordist productive plants as well as any setting where a large number of heterogeneous devices must coordinate their operations while retaining some flexibility. Thus, in its early days, *E.* focused on partnerships with large producers of the devices to be automated, e.g. a producer in the field of heating and air conditioning was offered a possibility to integrate a microchip in their devices, as well as on lifts, doors and windows in order to integrate all controls in a large building, from lighting to heating to theft protection.

With some disappointment, *E.* had to recognize that the *LW* technology was not exploited in its full potentialities. The problem was that each large producer was so specialized in its own industry that it had neither the power nor the capability to implement *LW* on all devices. For instance, a producer in the field of heating and air conditioning would not install *LW* on doors, windows, lights and lifts, for the production of these devices was covered by other firms. Indeed, the difficulty was that *E.* was attempting to create a single market for automation where the marketplace was covered by producers of several devices.

E. was aware of what huge difficulties were associated with the creation of a new market. Nevertheless, it deemed that long-term relations with a few specialized producers would pay in the long run. *E.* had a narrative, saying that large specialized producers would slowly but persistently adopt and impose *LW*. Consequently, it invested all of its resources in these relations.

By 1994, *E.* was losing confidence in this narrative. *E.* started to approach large system integrators of ICTs, such as *Olivetti* and *Ameritech*. However, the crucial move was that of hiring a person for this job, who did not come from Silicon Valley as all other executives did. Out of his suggestion, *E.* approached smaller companies which integrated devices that they bought from different producers. Technicians conceived the idea of embedding *LW* in a box that could be attached to any electromechanical device, of whatever producer.

Scholars of technological innovation know how difficult it is for visionary employees to convince their boss of the value of their idea. However, in the case of *E.*, the CEO embraced enthusiastically the new idea because it appeared to fit with his previous experience.

E.'s CEO had been the successful entrepreneur of a small firm that exploited digital technologies to produce private branch exchange systems (PBX) with innovative features. This firm had been able to displace giants such as *AT&T* by providing small independent installers with a superior product. When this CEO was confronted with the idea of addressing small independent integrators of electromechanical devices, he mapped this idea onto his previous experience.

In 1996, and within a few months, *E.* changed its narrative. *E.* presented itself as a provider of an innovative network technology designed for independent system integrators and based on a microchip that could be installed on any electromechanical device, of whatever producer.

Most importantly, *E.* told itself that it had always pursued this strategy. Nobody in the firm seemed to be aware that the firm's strategy had changed. According to the narrative that they had developed, they had always done what they were doing.

Moreover, when faced with evidence that the firm did change its strategy, management wished that the final publication would not stress this aspect (Lane, personal communication). This makes sense, for according to our idea of rationality, narratives should reflect "objective information", and decision-makers should stick to it. Thus, management did not want to appear irrational according to common wisdom.

However, the case of *E.* highlights that constructing a narrative by reinterpreting the past may be good and useful for decision-makers. Ultimately, the reported case reveals that precisely by reinterpreting its mission, *E.* was able to direct its investments. If the future is uncertain, as it is often the case, interpreting the past in order to find a direction for the future is a sensible activity.

So the trouble may rather lie with our idea of rationality. Since reinterpreting the past is regarded as irrational, then it must be done in secrecy. However, if reinterpreting the past may have positive effects, then it should be prescribed.

16.6 Tools for Modelling Coherence-Seeking

Although coherence-seeking cannot propose itself with a ready-made and ready-to-use formula such as utility maximization, there exist some tools that can be used to reproduce its building blocks. These are essentially classification tools, i.e. tools that form concepts out of information, and coherence tools, i.e. tools that arrange concepts into coherent stories.

In particular, the following tools will be reviewed in this section:

1. Unsupervised neural networks
2. Constraint satisfaction networks
3. Evidence Theory, also called belief functions theory

Unsupervised neural networks reproduce the formation of mental categories out of a flow of information. Constraint satisfaction networks arrange concepts into coherent maps. Finally, Evidence Theory assumes that actors receive information on possibilities and arrange them into coherent hypotheses. Although these tools have not been integrated with one another, they all concern the process of selecting some items from the flow of experiences, arranging them in a coherent whole and deciding accordingly.

Utility maximization makes sense in the restricted realm of games of chance, where it is possible to overview an exhaustive set of possibilities and enlist all of the consequences of any alternative. By contrast, reaching a decision by constructing coherence makes sense precisely because, quite often, such conditions do not hold. In order to stay within this focus, this review does not cover tools concerned with classification in a *given* set of categories, such as Case-Based Decision Theory and supervised neural networks.

16.6.1 Unsupervised Neural Networks

Human mental categories are not defined by prespecified similarity criteria that the objects to be classified should fulfil. Rather, mental categories are continuously constructed and modified according to the similarity of a just-received piece of information to the pieces of information that have already been stored in existing categories. For instance, a child observing house chairs may start with an idea of "chair" as an object having four legs and then observe an office chair with only one leg and yet sufficiently similar to house chairs to be added to their category, which from this time onwards does not have the number of legs as a common property of all the objects that it entails.

This is a radically different process from that of defining criteria in order to classify objects into given categories. In the case of mental categories, categories do not exist prior to the beginning of the classification process. Categories form out of similarity of certain input information to some information that has been previously received, so that these items are stored in the same "place" and concur to build up a mental category. There exist no criteria defined ex ante that control the classification of input information; since the only rule is similarity of information items, categories form depending on what items are received. For instance, the aforementioned mental category "chair" may depend on what chairs have been observed in the course of a life spent in Manhattan, rural China or tropical Africa, in modern times or the Middle Age.

By contrast, logical reasoning works with objects that have been clearly defined from the outset. The point here is that these definitions are actually made *once* mental categories have been formed. Eventually, a mental category formed around similarity judgements may suggest a definition for all the objects that it entails if they share some common feature, but this is not necessarily the case.

There are instances where definitions are not possible, simply because a mental category entails objects that do not have any common feature. For instance, the mental category expressed by the word "game" refers to children amusing themselves with toys and adults involved in a serious competition on a chessboard but also a set of wild animals. One may speculate that primitive man found some similarity between hunting and exercising for hunting and later on between sports and other leisures, including chess, and that the fact that chess was an amusement suggested some similarity to what children were supposed to do. So pairwise intersections of the meanings of the word "game" do exist, but this does not imply that all of their meanings have a common intersection. Nevertheless, human beings are perfectly at ease with this as well as many other concepts that cannot be defined.

Unsupervised neural networks (UNNs) are able to reproduce the idea that mental categories arise out of adding examples. Indeed, UNNs construct categories around the most frequent input patterns, mimicking the idea that a child creates a category "chair" upon observation of many objects similar to one another.

It is important to stress once again that the ensuing account deals with *unsupervised* neural networks only. Supervised neural networks (SNNs), with their neurons arranged in layers and training phases to teach them in what categories they must classify input information, do not fit into the present account, nor does Case-Based Decision Theory which, similarly to SNNs, is concerned with classifying information into previously defined categories. Directions to these tools will be provided in the bibliography, but they do not pertain to this chapter. Neural networks—of whatever sort—are composed by a set of *neurons* which produce an output $y \in \Re$ by summing inputs $x_1, x_2, \ldots x_N \in \Re$ by means of coefficients $a_1, a_2, \ldots a_N$:

$$y = \sum_{i=1}^{N} a_i x_i \tag{16.7}$$

For any set of coefficients a_i, this simple device is able to classify inputs in a category by yielding the same output y for several input vectors $\mathbf{x}$. This is due to the sheer fact that there exist several vectors $\mathbf{x}$ whose weighted sum yields the same y. For instance, if $N = 2$ and $\forall i$, it is $a_i = 1$, then e.g. $y = 10$, can arise out of $\mathbf{x}' = [9\ 1]$, $\mathbf{x}'' = [2.5\ 7.5]$ as well as many other vectors. In this sense, the neuron classifies the input vectors $[9\ 1]$ and $[2.5\ 7.5]$ in the same category.

Note that a neuron has no difficulty to classify input vectors that do not perfectly fit its categories. For instance, if there is a category $y = 10$ and a category $y = 11$, an input vector $\mathbf{x}''' = [2.1\ 8]$ is classified in the category $y = 10$ just as $\mathbf{x}'$ and $\mathbf{x}''$.

The shape of the categories implemented by a neuron depends on coefficients a_i. For instance, if $a_1 = 0.5$ and $a_2 = 20$, the input vector $\mathbf{x}' = [9\ 1]$ yields $y = 24.5$ and may not lie in the same category as $\mathbf{x}'' = [2.5\ 7.5]$, which yields $y = 151.25$.

Coefficients a_i may be chosen by the user of the network during a training phase, in which case we are dealing with a SNN. Alternatively, coefficients a_i may be initialized at random and subsequently changed by the network itself according to some endogenous mechanism. In this case we have a UNN, of which Kohonen networks are the most common instance. In UNNs, the ability of a neuron to change its categories stems from a feed-back from output y and a feed-forward from input $\mathbf{x}$, towards coefficients a_i:

$$\frac{da_i}{dt} = \varphi\,(\mathbf{a}, y)\ x_i - \gamma\,(\mathbf{a}, y)\ a_i \qquad \forall i \qquad (16.8)$$

where $\varphi(\mathbf{a},y)$ and $\gamma(\mathbf{a},y)$ may be linear or non-linear functions.

In Eq. (16.8), the term $\varphi\,(\mathbf{a}, y)\ x_i$ enables the neuron to learn input patterns. It entails both a feed-back (from y) and a feed-forward (from x_i). This *learning term* makes a_i increase when *both* y and x_i take high values, thereby enhancing those coefficients that happened to yield a high y when a particular x_i was high. Thus, the structure of coefficients vector $\mathbf{a}$ ultimately depends on which vectors $\mathbf{x}$ appeared most often as inputs.

The learning term is such that the neuron learns the patterns that it receives most often. This is sufficient to make the network work, but it also makes it unable to construct different categories if different patterns appear. Furthermore, since the learning term works by multiplying inputs and outputs, it may produce an explosive output that must be curbed in order to make the network work.

For both reasons, a *forgetting term* that makes the coefficients a_i decay towards zero is in order. In Eq. (16.8) the forgetting term is $\gamma(\mathbf{a}, y)a_i$. It depends on a feed-back from output y and, most importantly, on coefficient a_i itself.

Figure 16.8 illustrates the feed-backs and feed-forwards within a neuron of a UNN.

Simple, but non-trivial, examples of Eq. (16.8) are $\dot{\mathbf{a}} = \mu y\mathbf{x} - v\mathbf{a}$, $\dot{\mathbf{a}} = \mu\mathbf{x} - vy\mathbf{a}$, $\dot{\mathbf{a}} = \mu y\mathbf{x} - vy\mathbf{a}$ and $\dot{\mathbf{a}} = \mu y\mathbf{x} - vy^2\mathbf{a}$, where μ and v are constants. Each functional form corresponds to different strengths of the learning and forgetting terms.

In general, a network of neurons is able to discriminate input information according to much finer categories than a single neuron can do. As a rule, the greater the number of neurons, the finer the categories that a network constructs. However,

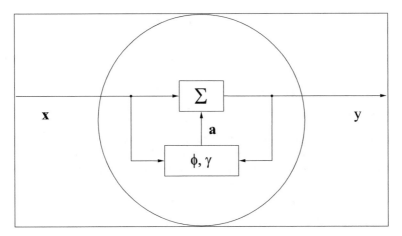

Fig. 16.8 The neuron of a UNN. The feed-backs and feed-forwards are responsible for the most notable properties of UNNs, including the absence of a training phase

a neural network is useful precisely because it is able to classify a huge amount of information into a few broad categories. If categories are so fine that they track input information exactly, then a neural net becomes useless. Thus, the number of neurons that a network should possess depends on the variability of the input as well as on user needs.

However, the behaviour of a neural network does not only depend on the number of its neurons but also on the structure of the connections between them. The fact is that just like the capabilities of neurons depend on feed-backs and feed-forwards, the capabilities of a neural network depend on linkages that eventually enable information to circulate in loops. If information can circulate within the network, then the whole network acquires a *memory*.

It is a *distributed* memory, fundamentally different in nature from the more usual *localized* memories. Localized memories such as books, disks, tapes, etc. store information at a particular point in space. This information can only be retrieved if one knows where its support is (e.g. the position of a book in a library or the address of a memory cell on a hard disk).

On the contrary, in a neural network, each neuron may be part of a number of information circuits where information is "memorized" as long as it does not stop to circulate. Although this *is* a memory, one cannot say that information is stored at any particular place, hence the name.

For obvious reasons, the information stored in a distributed memory cannot be retrieved by means of an address. However, a piece of information flowing in a particular loop can be retrieved by some other piece of information that is flowing close enough to it. Thus, in a distributed memory, information can be retrieved by means of *associations* of concepts, with a procedure that reminds of human "intuition". Intuition, according to this interpretation, would consist of associations between concepts that occur when information flowing in different but neighbouring circuits comes in touch.

16.6.2   Constraint Satisfaction Networks

Constraint satisfaction networks (CSNs) arrange concepts into coherent theories. CSNs are characterized by:

- Excitatory and inhibitory connections
- Feedbacks between neurons

Nodes represent possibilities, or concepts, or propositions. Connections represent inferences: an excitatory connection from node A to node B means "A implies B", whereas an inhibitory connection from node A to node B means "A implies B".

Let a_i denote the activation (the output) of node i, with $a_i \in \Re$. Let $w_{ij} \in \Re$ denote the weight by which node i multiplies the input arriving from node j. The net excitatory input to node i is:

$$E_i = \sum_j w_{ij}\, a_j \quad \text{if } w_{ij}a_j \geq 0 \tag{16.9}$$

The net inhibitory input to node i is:

$$I_i = \sum_j w_{ij}\, a_j \quad \text{if } w_{ij}a_j < 0 \tag{16.10}$$

At each time step, the activation of node i is increased by its excitatory inputs and decreased by its inhibitory inputs:

$$\Delta a_i = E_i\,(a_{\max} - a_i) + I_i\,(a_i - a_{\min}) \tag{16.11}$$

where, in general, $a_{max} = 1$ and $a_{min} = -1$. Feedbacks between neurons make the network maximize consonance:

$$C = \sum_{ij} w_{ij}a_i a_j \tag{16.12}$$

or, equivalently, minimize $Energy = -C$.

Consonance maximization means that those nodes are strengthened that represent possibilities, concepts or propositions that are coherent with one another. Thus, constraint satisfaction networks can be used to model any cognitive process characterized by a search for coherence. In particular, researchers have emphasized the ability of CSNs to construct narratives, much like humans actually do.

Notable applications of CSNs are the elaboration of scientific theories, which amounts to arrange empirical findings in a network of coherent causal relations, as well as the evaluation of guilt or innocence in a trial, which amounts to fitting testimonies in a coherent frame. Under this respect, CSNs share a common concern with Evidence Theory or, to be more precise, with its Dempster-Shafer's combination rule reported in Eq. (16.16) of Sect. 16.6.3.

When modelling decision-making, CSNs can be used to model the process of emphasizing the positive aspects of one alternative and the negative aspects of its competing alternatives until a coherent frame is available and a decision can be made. Notably, this oscillation between competing explanations reproduces at least one important aspect of *Gestalt* theories, namely, the idea that the human mind may shift among alternative interpretations of reality, as exemplified by Rubin's vase and other images where at least two interpretations are possible.[4] Many cues suggest that this is a fundamental pattern in decision-making.

A clear limitation of CSNs is that they work with given possibilities, concepts or propositions. In other words, CSN can reproduce the arrangement of possibilities and concepts into narratives, but not their arousal. By contrast, UNNs reproduce the arousal of concepts out of empirical experiences. Possibly, future models will be able to couple UNNs to CSNs in order to model both processes at a time.

16.6.3 Evidence Theory

Evidence Theory is a branch of the mathematics of uncertain reasoning that, unlike probability theory, does not assume that decision-makers know the set of all possible events. Rather than defining a "residual event" for anything that cannot be clearly expressed, Evidence Theory leaves a decision-maker's possibility set open to novelties.

This feature is implemented by assuming that the possibility set is not an algebra, in the sense that no operation is defined on it. A consequence of this assumption is that with no operation available, no residual event can be defined—simply because complementation does not exist. Therefore, the mathematical framework does not force any event into the system beyond those that a person has conceived. Novel possibilities can appear in the possibility set in the course of the calculations, and the possibility set is called *frame of discernment* in order to stress its cognitive nature.

Evidence Theory does not take a gambler as its prototypical subject but a judge or a detective. The reason is that a gambler playing with dice or throwing a coin knows what possibilities can occur. On the contrary, judges and detectives know that unexpected proves and testimonies may open up novel possibilities. Possibly, managers making investments, politicians steering their countries or just anyone in the important choices of her daily life is more akin to a judge or a detective looking for cues than a gambler looking for luck.

---

[4]The simplest picture of this kind is a cube depicted by its edges: it is up to the observer to choose which face stays in the front and which face stays in the rear. Rubin's vase is white and stands against a black background. The observer may see a white vase or two black profiles in front of one another.

Let us consider a frame of discernment Θ. Let us suppose that a person receives testimonies, or *bodies of evidence*, as numbers that to various extents support a set of possibilities $A_1, A_2, \ldots A_N$, where $A_1 \subseteq \Theta, A_2 \subseteq \Theta, \ldots A_N \subseteq \Theta$ and where the A_is are not necessarily disjoint sets.[5] Let us denote these numbers $\{m(A_1), m(A_2), \ldots m(\Theta)\}$, where $m(A_i)$ measures the amount of empirical evidence that supports possibility A_i.

Numbers m are exogenous to the person (the judge, the detective) who owns the frame of discernment. They are not subjective measures for this person, though they may be subjective evaluations of those who provide the testimonies. Numbers m are cardinal measures of the amount of empirical evidence supporting each possibility.

Since no operation is defined on the frame of discernment, the number m that has been assigned to Θ does not concern any specific possibility. Rather, it indicates how small the evidence is that supports the possibilities envisaged in the testimony or, in other words, how strongly a person fears that the possibilities that she is envisaging are not exhaustive. The greater the ignorance of a person on what possibilities exist, the greater $m(\Theta)$.

Note that $m(\Theta)$ can be smaller than any m of the A_is that Θ entails. Indeed, this applies to the A_is as well: if $A_i \supset A_j$, this does *not* imply that $m(A_i) > m(A_j)$. Although not strictly essential for Evidence Theory, numbers m are generally normalized by requiring that:

$$\sum_{i=1}^{N} m(A_i) + m(\Theta) = 1 \qquad (16.13)$$

For instance, if the original format of the testimony is:
$$\{5, 32, 12, 3\}$$

by applying Eq. (16.13) we obtain:

$$\{0.096, 0.615, 0.231, 0.058\}$$

where numbers sum up to one.

Let us suppose that a decision-maker wants to evaluate to what extent the available empirical evidence supports certain hypotheses that she is entertaining in her mind. Since a hypothesis concerns the truth of a possibility or a set of possibilities, hypotheses are subsets of the frame of discernment just as possibilities are. A body of evidence $\{m(A_1), m(A_2), \ldots m(\Theta)\}$ supports a hypothesis H to the extent that some A_is are included or at least intersect H.

Note that while the possibilities A_i entailed in the testimonies cannot be combined with one another (intersected, complemented, etc.) to form novel possibilities, a hypothesis H represents a construct of the owner of a frame of discernment (the judge, the detective, etc.). This person is absolutely free to conceive any hypothesis H as well as its opposite $\overline{H}$. Given a testimony $\{m(A_1), m(A_2), \ldots m(\Theta)\}$, the belief in hypothesis H is expressed by the following *belief function*:

---

[5]For simplicity, the theory is expounded with respect to a finite number of possibilities. No substantial change is needed if an infinite number of possibilities is considered.

$$Bel(H) = \sum_{A_i \subseteq H} m(A_i) \qquad (16.14)$$

By definition, $Bel(\oslash) = 0$ and $Bel(\Theta) = 1$. However, this last condition does not imply that any of the possibilities included in the frame of discernment must necessarily be realized. It simply means that any possibility must be conceived within the frame of discernment, independently of what possibilities are envisaged at a certain point in time. The belief function takes account of all evidence included in H. The *plausibility function* takes account of all evidence that intersect H:

$$Pl(H) = \sum_{H \cap A_i \neq \oslash} m(A_i) \qquad (16.15)$$

It can be shown that belief and plausibility are linked by the relation $Pl(H) = 1 - Bel(\overline{H})$, where $\overline{H}$ denotes the hypothesis opposite to H. If $m(\Theta) > 0$ these two measures are not equivalent, so both of them need to be considered. In general, $Bel(H) \leq Pl(H)$.

Let us suppose that some unexpected facts occur that are told by a new testimony. The new testimony must be combined with previous knowledge, confirming it to the extent that it is coherent with it. On the contrary, previous beliefs must be weakened if the new evidence disconfirms them.

Let $\{m(B_1), m(B_2), \dots m(\Theta)\}$ be the new testimony, which must be combined with $\{m(A_1), m(A_2), \dots m(\Theta)\}$. The new testimony may entail possibilities that are coherent with those of the previous testimony, possibilities that contradict those of the previous testimony and possibilities that partially support and partially contradict the previous testimony. Figure 16.9 illustrates contradictory, coherent and partially coherent/contradictory possibilities on the frame of discernment. Contradictory possibilities appear as disjoint sets. A possibility is coherent with another if it is included in it. Finally, two possibilities that are partially coherent, partially contradictory, intersect one another.

Let us suppose that two testimonies

$$\{m(A_1), m(A_2), \dots m(\Theta)\}$$

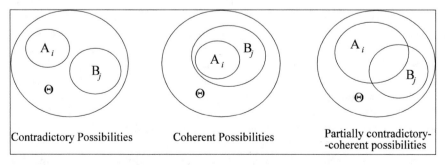

| Contradictory Possibilities | Coherent Possibilities | Partially contradictory- -coherent possibilities |

Fig. 16.9 *Left*, two contradictory possibilities. *Centre*, two coherent possibilities. *Right*, two partially coherent and partially contradictory possibilities

and

$$\{m\,(B_1)\,,m\,(B_2)\,,\ldots m\,(\Theta)\}$$

both of which satisfy Eq. (16.13) must be combined into a testimony

$$\{m\,(C_1)\,,m\,(C_2)\,,\ldots m\,(\Theta)\}$$

that also satisfies Eq. (16.13). Dempster-Shafer's combination rule yields a combined testimony $\{m(C_k)\}$ where the coherent possibilities between $\{m(A_i)\}$ and $\{m(B_j)\}$ have been stressed. According to Dempster-Shafer's combination rule, possibilities $\{C_k\}$ are defined by all intersections of each possibility in $\{A_1, A_2, \ldots \Theta\}$ with each possibility in $\{B_1, B_2, \ldots \Theta\}$. For any possibility C_k, the amount of empirical evidence is:

$$m\,(C_k) = \frac{\sum_{A_i \cap B_j = C_k} m\,(A_i)\ m\,(B_j)}{1 - \sum_{A_i \cap B_j = \varnothing} m\,(A_i)\ m\,(B_j)} \qquad (16.16)$$

The numerator of Eq. (16.16) measures the extent to which both the first and the second testimonies support the possibility C_k. Indeed, for each possible C_k, the sum extends to all pairs of possibilities from the two testimonies that are coherent on C_k (see Fig. 16.9). The more the intersections between the A_is and the B_js give rise to C_k and the greater their amounts of evidence, the larger the numerator.

The denominator is the complement to one of those elements of the second testimony that contradict the first one, for the complement to one is made on those A_is and B_js that are disjoint sets (see Fig. 16.9). The denominator represents a measure of the extent to which the two testimonies are coherent, in the sense that all evidence that supports contradictory possibilities is excluded.

Essentially, Dempster-Shafer's combination rule says that the evidence supporting possibility C_k is a fraction of the coherent evidence between $\{m(A_1), m(A_2), \ldots m(\Theta)\}$ and $\{m(B_1), m(B_2), \ldots m(\Theta)\}$. The amount of this fraction depends on the sum of all elements of the testimonies that support C_k.

Dempster-Shafer's rule can be iterated to combine any number of testimonies. The outcome of Dempster-Shafer's combination rule is independent of the order in which two testimonies are combined.

The above description stressed that Evidence Theory provides an algorithm for handling an exogenous flow of new, unexpected possibilities. However, its decision-makers are not supposed to conceive possibilities. They merely listen to exogenous testimonies that consist of possibilities and degrees of evidence supporting them and combine these testimonies into a coherent whole by means of Dempster-Shafer's theory. She does not conceive novel possibilities out of a creative effort. Rather, novel possibilities—the $\{C_k\}$—arise out of combination of exogenous inputs.

In this respect, Evidence Theory bears some similarity to UNNs, for both tools are open to novel arrangements of input information. These dynamical, open arrangements are eventually called "possibilities" in Evidence Theory or "categories" in UNNs. At the same time, Evidence Theory bears some similarity to CSNs when it comes to the combination of evidence, in the sense that both CSNs and Dempster-Shafer's combination rule Eq. (16.16) seek to improve the coherence of available information.

From one point of view, UNNs and CSNs take on the complementary tasks of modelling (a) the arousal of new concepts and (b) their combination into coherent narratives. However, Evidence Theory is a mathematical theory that may be developed into a framework able to encompass both CSNs and UNNs as simulation tools.

At present, this is pure speculation. However, it is also a possible narrative that may direct future research efforts.

16.7 Conclusions

This review presented tools to model decision-making according to two different paradigms, namely, utility maximization and coherence-seeking. The reader may feel unease because scientists do not provide a univocal answer to the demands of the modeller.

However, a pragmatic attitude may suggest that tools should be used depending on conditions. Utility maximization and Game Theory require that all available alternatives and all of their possible consequences can be listed. Thus, it may be sensible to make use of these tools when one such exhaustive list is available, eventually releasing the requirement of perfect rationality and the pursuit of Nash equilibria while assuming some form of bounded rationality as influence games do. By contrast, unsupervised neural networks, constraint satisfaction networks and Evidence Theory may be used when more challenging decision settings must be modelled. Modellers should remember that constructing narratives makes sense because decision-makers may be uncertain regarding what possibilities exist, so these tools become necessary only in this kind of decision settings.

The trouble, in this last case, is that the tools mentioned above have not been integrated into a unified framework. No simple formula is available to be readily used, so the modeller must resort to a higher degree of creativity and intuition. Or, from a more positive point of view, here is an exciting opportunity for modellers to participate to theory development.

Further Reading

Game Theory is a huge subject. Relevant handbooks are Aumann and Hart (1992, 1994, 2002) and, at a more introductory level, Rasmusen (2007). However, agent-based modellers should keep in mind that a substantial part of Game Theory has been developed around equilibrium states, which are generally not a main concern for them. Evolutionary games, thoroughly discussed in the above handbooks, are possibly closer to agent-based modelling. For other evolutionary mechanisms, see Chap. 21 in this volume (Chattoe-Brown and Edmonds 2012).

Neural networks are a huge subject as well. This field is currently split in two streams: on the one hand, research on neural networks as a model of cognitive processes in the brain and on the other hand, research on neural networks as an engineering tool for signal processing. A handbook oriented towards cognitive problems is Arbib (2002). Handbooks oriented towards engineering problems are Hu and Hwang (2002) and Graupe (2007). Specifically, unsupervised neural networks are often employed in pattern recognition. A comprehensive treatment of pattern recognition techniques is Ripley (1996).

All other tools and issues discussed in this chapter are in their infancy, so no generic reading can be mentioned. Interested scholars are better advised to start with the original papers mentioned in the bibliography, tracking developments on recent publications and working papers.

Arbib M. A. (Ed.). (2002). *The handbook of brain theory and neural networks* (2nd ed.). Cambridge, MA: MIT Press.

Aumann, R., & Hart, S. (Eds.). (1992). *Handbook of game theory with economic applications* (Vol. 1). Amsterdam: North-Holland.

Aumann, R., & Hart, S. (Eds.). (1994). *Handbook of game theory with economic applications* (Vol. 2). Amsterdam: North-Holland.

Aumann, R., & Hart, S. (Eds.). (2002). *Handbook of game theory with economic applications* (Vol. 3). Amsterdam: North-Holland.

Chattoe-Brown, E., & Edmonds, B. (2012). Evolutionary mechanisms. doi:https://doi.org/10.1007/978-3-319-66948-9_21.

Graupe, D. (2007). *Principles of artificial neural networks*. Singapore: World Scientific.

Hu, Y. H., & Hwang, J. N. (Eds.). (2002). *Handbook of neural network signal processing*. Boca Raton, FL: CRC Press.

Rasmusen E. (2007). *Games and information: An introduction to game theory* (4th ed.). Malden: Blackwell.

Ripley, B. D. (1996). *Pattern recognition and neural networks*. Cambridge: Cambridge University Press.

Weibull, J. W. (1997). *Evolutionary game theory*. Cambridge, MA: MIT Press.

Reasoned Bibliography

This chapter covered too many topics to be able to provide detailed references. Henceforth, a few basic publications will be listed that may be used by interested readers as a first orientation to each of the topics mentioned in this chapter.

Utility and Games

Utility maximization was pioneered by Frank Ramsay and Bruno De Finetti in the 1930s and subsequently refined by Leonard Savage in the 1950s. Savage still provides the most comprehensive explanation of this approach to uncertain reasoning.

Game Theory was initiated by John Von Neumann and Oskar Morgenstern in the 1940s. It subsequently developed into a huge research field within economics, with several specialized journals. Today, Game Theory is a field characterized by extreme mathematical sophistication and intricate conceptual constructions.

This chapter did not focus on the assumptions and methods of Game Theory but rather aimed at presenting the main prototypical games that have been devised hitherto. A classical treatise by Duncan Luce and Howard Raiffa may introduce the subject more easily than Von Neumann and Morgenstern did. Luce and Raiffa were first to present the Battle of the Sexes as well as the Prisoner's Dilemma, which they ascribed to Albert Tucker anyway.

Readers interested in evolutionary games may rather read the treatises written by Jörgen Weibull and Herbert Gintis, respectively. The former is more specific on evolutionary games and also more technical than the second one.

Robert Axelrod is the main reference so far it regards simulations of the iterated Prisoner's Dilemma with retaliation strategies. The idea that the iterated Prisoner's Dilemma could yield cooperation simply relying on tags is due to Rick Riolo.

The stag hunt and the Game of Chicken are classical, somehow commonsensical games. The Game of Chicken has been turned into the Hawk-Dove game by Maynard Smith and George Price. The Hawk-Dove game is not terribly different from the war of attrition, conceived by Maynard Smith and improved by Timothy Bishop and Chris Cannings.

The Traveller's Dilemma and the dollar auction are recent games invented by Kaushik Basu and Martin Shubik, respectively. Pure coordination games have been discovered by Thomas Schelling.

Axelrod, R. M. (1984). *The evolution of cooperation*. New York: Basic Books.

Basu, K. (1994). The traveller's dilemma: Paradoxes of rationality in game theory. *American Economic Review, 84*, 391–395.

Bishop, D. T., Cannings, C., & Smith, J. M. (1978). The war of attrition with random rewards. *Journal of Theoretical Biology, 74*, 377–389.

Gintis, H. (2000). *Game theory evolving*. Princeton, NJ: Princeton University Press.

Luce, R. D., & Raiffa, H. (1957) *Games and decision: Introduction and critical survey*. New York: Wiley.

Riolo, R. L., Cohen, M. D., & Axelrod, R. M. (2001). Evolution of cooperation without reciprocity. *Nature, 414*, 441–443.

Savage, L. (1954). *The foundations of statistics*. New York: Wiley.

Schelling, T. C. (1960). *The strategy of conflict*. Cambridge, MA: Harvard University Press.

Shubik, M. (1971). The dollar auction game: A paradox in noncooperative behavior and escalation. *Journal of Conflict Resolution, 15*, 109–111.

Smith, J. M., & Price, G. R. (1973) The logic of animal conflict. *Nature, 246*, 15–18.

Weibull, J. (1997). *Evolutionary game theory.* Cambridge, MA: The MIT Press.

Influence Games

Ernst Ising introduced his model in the 1920s. Since then, a huge literature appeared.

The Ising model is taught in most Physics courses around the world, so a number of good introductions are available on the Internet. A printed introduction by Barry Cipra is mentioned here for completeness.

Schelling's model of racial segregation was developed independently of the Ising model. However, it may be considered a variation of it.

The *El Farol Bar Problem* was conceived by Brian Arthur. Renamed *The Minority Game* and properly formalized, it was introduced to physicists by Damien Challet and Yi-Cheng Zhang.

A huge literature on the minority game has appeared on Physics journals. Good introductions have been proposed, among others, by Esteban Moro and Chi-Ho Yeung and Yi-Cheng Zhang.

Arthur, W. B. (1994). Inductive reasoning and bounded rationality. *The American Economic Review, 84*, 406–411.

Challet, D., & Zhang, Y. C. (1997). Emergence of cooperation and organization in an evolutionary game. *Physica A, 246*, 407–418.

Cipra, B. A. (1987). An introduction to the Ising model. *The American Mathematical Monthly, 94,* 937–959.

Moro, E. (2004). The minority game: An introductory guide. In E. Korutcheva & R. Cuerno (Eds.), *Advances in condensed matter and statistical physics* (pp. 263–286). New York: Nova Science Publishers. (Also available online as arXiv:cond-mat/0402651v1).

Schelling, T. C. (1971). Dynamic models of segregation. *Journal of Mathematical Sociology, 1,* 143–186.

Yeung, C. H., & Zhang, Y. C. (2009) Minority games. In R. A. Meyers (Ed.), *Encyclopedia of complexity and systems science.* Berlin, Springer. (Also available online as arXiv:0811.1479v2.).

Some Pitfalls of Utility Maximization

The idea that probabilities measured on samples of size zero are somewhat awkward is quite old and evidently linked to the frequentist view of probabilities. Daniel Ellsberg circulated this idea among economists, where in the meantime the subjectivist view of probability judgements had become dominant. Subadditive probabilities were conceived by Bernard Koopman in the 1940s and popularized among economists by David Schmeidler in the 1980s.

Maurice Allais submitted his decision problem to Leonard Savage, who did not behave according to his own axioms of rational choice. Since then, Savage presented utility maximization as a normative, not as a descriptive, theory. Prospect theory was advanced by Daniel Kahneman and Amos Tversky; it comes in a first version (1953) and a second version (1992).

The preference reversals highlighted by Paul Slovic have triggered a huge literature. A recent book edited by Sarah Lichtenstein and Paul Slovic gathers the most important contributions.

Kenneth Arrow originally devised his paradox as a logical difficulty to the idea of a Welfare State that would move the economy towards a socially desirable equilibrium. However, it may concern any form of group decision-making.

Michael Mandler is the main reference for a possible conciliation of Slovic's and Arrow's paradoxes with utility maximization, provided that preferences are incomplete.

Allais, M. (1953). Le comportement de l'homme rationnel devant le risque: critique des postulats et axiomes de l'école americaine. *Econometrica, 21,* 503–546.

Arrow, K. J. (1950) A difficulty in the concept of social welfare. *The Journal of Political Economy, 58,* 328–346. (Reprinted in *The Collected Papers of Kenneth J. Arrow.* Oxford: Blackwell, 1984).

Ellsberg, D. (1961). Risk, ambiguity, and the savage axioms. *The Quarterly Journal of Economics, 75,* 643–669.

Kahneman, D., & Tversky, A. (1953) Prospect theory: An analysis of decision under risk. *Econometrica, 21,* 503–546.

Koopman, B. O. (1940). The axioms and algebra of intuitive probability. *The Annals of Mathematics, 41,* 269–292.

Lichtenstein, S., & Slovic, P. (Eds.). (2006). *The construction of preference.* Cambridge: Cambridge University Press.

Mandler, M. (2005). Incomplete preferences and rational intransitivity of choice. *Games and Economic Behavior, 50,* 255–277.

Savage, L. (1967). Difficulties in the theory of personal probability. *Philosophy of Science, 34,* 305–310.

Schmeidler, D. (1989). Subjective probability and expected utility without additivity. *Econometrica, 57,* 571–587.

Tversky, A., & Kahneman, D. (1992). Advances in prospect theory: Cumulative representation of uncertainty. *Journal of Risk and Uncertainty, 5,* 297–323.

Decision-Making by Coherence-Seeking

In 1974 and 1976, James March was first to point to the fact that human beings distort their memories of the past in order to construct coherent stories that guide them into the future. This point has been also made by Karl Weick, who wrote a lengthy treatise on this subject a few decades later.

A number of psychological experiments confirm this idea. Interested readers may start with the works of Daryl Bem, Michael Conway and Michael Ross and a book edited by Ulric Neisser and Robyn Fivush.

However, the trouble with the idea of human beings reconstructing the past is that they are not willing to concede that they do so. Thus it is extremely difficult to find case studies. The one by David Lane and Robert Maxfield is possibly the only exception, though they were not allowed to publish all the materials they gathered during their investigation (Lane, personal communication).

A final remark on lack of communication in this stream of research. James March, Karl Weick and David Lane worked independently, possibly unaware of one another, focusing on the same issue but employing different expressions.

Bem, D. J. (1966). Inducing belief in false confessions. *Journal of Personality and Social Psychology, 3,* 707–710.

Bem, D. J. (1967). Self-perception: An alternative interpretation of cognitive dissonance phenomena. *Psychological Review, 74,* 183–200.

Cohen, M. D., & March, J. G. (1974) *Leadership and ambiguity: The American College President.* New York: McGraw-Hill.

Conway, M., & Ross, M. (1984). Getting what you want by revising what you had. *Journal of Personality and Social Psychology, 47,* 738–748.

Greenwald, A. (1980). The totalitarian ego: Fabrication and revision of personal history. *American Psychologist, 35,* 603–618.

Lane, D. A., & Maxfield, R. R. (2005) Ontological uncertainty and innovation. *Journal of Evolutionary Economics, 15,* 3–50.

March, J. G., & Olsen, J. P. (1976). Organizational learning and the ambiguity of the Past. In J. G. March & J. P. Olsen (Eds.), *Ambiguity and choice in organizations.* Bergen, Norway: Universitetsforlaget.

Neisser, U., & Fivush, R. (Eds.). (1994). *The remembering self: Construction and accuracy in the self-narrative.* Cambridge: Cambridge University Press.

Ross, M., & Newby-Clark, I. R. (1998). Constructing the past and future. *Social Cognition, 16,* 133–150.

Weick, K. E. (1979). *The Social psychology of organizing.* New York: Random House.

Weick, K. E. (1995). *Sensemaking in organizations.* Thousand Oaks, CA: Sage Publications.

Tools for Modelling Coherence-Seeking

This chapter did not deal with tools where categories pre-exist to the information that is being received, namely, supervised neural networks and Case-Based Decision Theory. Readers interested in supervised neural networks may start with the classical handbook by Rumelhart, McClelland and the PDP Research Group. Readers interested in Case-Based Decision Theory may refer to a series of articles by Itzhak Gilboa and David Schmeidler.

The earliest intuitions on the nature of mental categories date back to Ludwig Wittgenstein. A good explanation of the main features of mental categories, and why they are so different from our common idea of what a "category" is, is provided by George Lakoff in his *Women, Fire, and Dangerous Things*.

So far it regards unsupervised neural networks; the classic book by Teuvo Kohonen is still unrivalled for its combination of mathematical rigour and philosophical insight. Having been written at an early stage, it still keeps a strong link between artificial neural networks and the human brain.

Paul Thagard is the basic reference for constraint satisfaction networks. Constraint satisfaction networks appear in several contributions to the book *The Construction of Preference*, edited by Sarah Lichtenstein and Paul Slovic, mentioned in the section "Some Pitfalls of Utility Maximization". Regarding the importance of focussing on two alternatives in order to arrive at a decision, see Fioretti (2012).

Evidence Theory started with a book by Glenn Shafer in 1976 and triggered a small but continuous flow of mathematical works since then. An article by Guido Fioretti explains it to social scientists, along with examples of applications to decision problems.

Fioretti, G. (2009). Evidence theory as a procedure for handling novel events. *Metroeconomica*, *60*, 283–301.

Fioretti, G. (2012). *Either, or: Exploration of an emerging decision theory. IEEE Transactions on Systems, Man and Cybernetics C*, *42* (6): 854–864.

Gilboa, I., & Schmeidler, D. (1995). Case based decision theory. *Quarterly Journal of Economics*, *110*, 605–639.

Kohonen, T. (1989). *Self-organization and associative memory*. Berlin: Springer.

Lakoff, G. (1987). *Women, fire, and dangerous things*. Chicago: University of Chicago Press.

Rumelhart, D. E., McClelland, J. L., & the PDP Research Group. (1986). *Parallel distributed processing: explorations in the microstructure of cognition*. Cambridge, MA: MIT Press.

Shafer, G. (1976). *A mathematical theory of evidence*. Princeton, NJ: Princeton University Press.

Thagard, P. (2000). *Coherence in thought and action*. Cambridge, MA: MIT Press.

Chapter 17
Social Constraint

Martin Neumann

Abstract This chapter examines how a specific type of social constraint operates in Artificial Societies. The investigation concentrates on bottom-up behaviour regulation. Freedom of individual action selection is constraint by some kind of obligations that become operative in the individual decision-making process. This is the concept of norms. The *two-way dynamics* of norms is investigated in two main sections of the chapter: the effect of norms on a social macro-scale and the operation of social constraints in the individual agent. While normative modelling is becoming useful for a number of practical purposes, this chapter specifically addresses the benefits of this expanding research field to understand the dynamics of human societies. For this reason, both sections begin with an elaboration of the problem situation, derived from the empirical sciences. This enables to specify questions to agent-based modelling. Both sections then proceed with an evaluation of the state of the art in agent-based modelling. In the first case, sociology is consulted. Agent-based modelling promises an integrated view on the conception of norms in role theoretic and individualistic theories of society. A sample of existing models is examined. In the second case, socialisation research is consulted. In the process of socialisation, the obligatory force of norms becomes internalised by the individuals. A simulation of the feedback loop back into the mind of agents is only in the beginning. Research is predominantly on the level of the development of architectures. For this reason, a sample of architectures is evaluated.

Why Read This Chapter?
To understand social norms and their complexities, including how they can operate, how they can effectively constrain action and how such processes have been represented within simulations. The chapter also helps the reader to acquire an integrated view of norms and become aware of some of the relevant work simulating them using this framework.

M. Neumann (✉)
Jacobs University, Bremen, Germany
e-mail: ma.neumann@jacobs-university.de

© Springer International Publishing AG 2017   411
B. Edmonds, R. Meyer (eds.), *Simulating Social Complexity*,
Understanding Complex Systems, https://doi.org/10.1007/978-3-319-66948-9_17

17.1 Introduction

Some kind of mechanism for action selection has to be implemented in the agents. The decision-making process of isolated agents may be governed by BDI architectures. Developed by the philosopher Michael Bratman as a model of rational decision making—in particular to clarify the role of intentions in practical reasoning according to the norms of rationality—(Bratman 1987), the BDI framework has been adopted by Rao and Georgeff for the development of software technology (Rao and Georgeff 1991). BDI agents are data structures that represent beliefs about the environment and desires of the agents. Desires enable the goal-directed behaviour of an agent. Moreover, the decision-making process has to be managed, which is accomplished in two stages. To achieve a goal, a certain plan has to be selected, denoted as the intention of the agent. Secondly, the agent undertakes means-ends calculations, to estimate which actions are necessary to reach its goals. Intentions are crucial in mediating between the agent's beliefs and desires and the environment.

In groups, however, behaviour is more effective if agents orient their actions around other agents. The simplest example is of two robots moving towards each other. They have to decide how to pass. Hence, individual action selection has to be restricted by social constraints. In principle, there exist two options to model social constraints on the individual agent's action selection: top-down regulation by a central authority or bottom-up regulation of action selection. Top-down regulation can be exhibited computationally by a central processor or—in human societies—by coercion of a central authority. An evaluation of modelling approaches to the former kind of social regulation can be found in the chapter on Power and Authority. In bottom-up approaches, freedom of individual action selection is constraint by some kind of obligatory forces that become operative in the individual decision-making process even without a controlling authority. This is denoted by the concept of *norms*. Normative behaviour regulation can be enforced by some kind of generalised social influence such as sanctions and encouragement. Crucial for the normative behaviour regulation is that social constraints are internalised in the individual agent's decision-making process. This can range from more or less automatically executed habits to processes of normative reasoning and balancing competing goals. In contrast to top-down regulation, it is not based purely on coercion. Since bottom-up regulation is more flexible than predetermined constraints by a central authority, the past decades have witnessed a growing interest in the inclusion of norms in multi-agent simulation models. Numerous factors are responsible for attention being paid to norms in Artificial Societies, ranging from technical problems in the co-ordination of multi-agent system, e.g. with moving robots (Shoham and Tennenholtz 1992; Boman 1999), or practical problems such as e-commerce or electronic institutions (Lopez and Marquez 2004; Vazquez-Salceda et al. 2005) to philosophical interest in the foundation of morality (Axelrod 1986; Skyrms 1996, 2004) and the investigation of the wheels of social order (Conte and Castelfranchi 2001). The contribution of Artificial Societies to the latter problem is in the focus of this chapter: what is the potential contribution of Artificial Societies to the investigation of social order in human societies?

However, in the literature—not only, but *also* in the simulation literature—a great variety of concepts of norms exist. Typically, these differences are not much discussed in the AI literature. Nevertheless, some decisions are made, consciously or unconsciously. For this reason, this chapter aims to introduce the reader to the different concepts that are often only implicit in the different approaches to normative simulation models. This requires some background information about the different concepts of norms in the different empirical sciences. However, one reason for the variety of concepts of norms is that their investigation is scattered over a vast variety of different disciplines. For this reason also, this chapter has to concentrate on some empirical disciplines that are of particular importance for agent-based modelling. A first restriction is motivated by the decision to concentrate on bottom-up behaviour regulation in this chapter. This suggests to exclude the literature on norms that can be found in the political sciences or the theory of law, since these norm concepts are more closely related to top-down approaches. Secondly, agent-based modelling is of particular relevance for the study of social mechanisms. These mechanisms, however, are of some kind of generality. Typically, Artificial Societies investigate stylised facts. This suggests focussing the examination by excluding historical narratives that can be found in anthropological studies or historical sociology such as the work of Michel Foucault or Norbert Elias. Instead, the survey of the empirical sciences concentrates on the literature that investigates the theoretical foundation of the specific dynamics between the micro- and the macro-level that is the particular focus of agent-based modelling. Namely, how norms influence the operations and effects of social systems and how such a normative structure of the social system recursively affect the generating agent level. This calls for a survey of sociology and socialisation research in social psychology. They are most relevant sciences for an investigation of the contribution of norms to the wheels of social order.

Beside this introduction, the chapter contains three main sections. Sections 3 and 4 consist of two parts: one providing a broad overview of the empirical counterpart and a subsequent one about modelling. The sections can be read independently. A reader with a background in the empirical sciences who wants to get informed about simulation might concentrate on the parts evaluating simulation models. The sections provide the following information:

Section 2: first, a brief exposition of the *core concepts* of norms in the empirical sciences is provided. It is suggested to have a look at it, because here the core problems are exposed that are investigated in the following sections. These are the two main parts.

Section 3: an investigation of the dynamics of norm spreading from individuals to a social group and how this effects the operations of the social system is undertaken in this part. This refers to the sociological question of the operations and effects of normative behaviour regulation on the social *macro-level* (the emergence of a behaviour regularity). The section is divided into two parts: first, the sociological questions to agent-based modelling are developed. Then a sample of models is examined. The sample is divided into two categories of models: models inspired by a game theoretic problem description and cognitive models in an AI tradition. This section is particular relevant for modelling normative regulation in *Artificial Societies*.

Section 4: in this part, the recursive feedback loop is investigated. It is examined how social behaviour regulation is executed in the *individual agent*. This refers to the socio-psychological problem of norm internalisation. To represent this process, cognitively rich agents are necessary. Also this section is divided into two parts: first, a problem exposition will be provided. This refers to theories of socialisation. Then a sample of architectures is examined with regard to the question how norms become effective within the agent. This section might be the most interesting one for readers interested *cognitive modelling*.

Finally, the main findings of the chapter are summarised in concluding remarks and further reading suggested for those who want to investigate further.

17.2 The Concept of Norms

Before turning the attention to an examination of existing modelling approaches, a brief exposition of the core concepts of norms will be provided. In some way, some of these aspects have to be represented in a normative simulation model. Dependent on the research question, the developer of a normative simulation model may concentrate on only some aspects. In fact, no model includes *all* aspects. However, it might be useful to be aware of a more comprehensive perspective. Operationally, norms can be described as correlated structures. Agent behaviour exhibits regularities that can be found—to a certain degree—in an entire population. This can be described as a social constraint. From a social science perspective, norms are the most important concept of social constraints on behaviour regulation. Norms belong to the fundamental concepts in sociology to explain human behaviour. Summarising the individual and social aspects of norms, Gibbs provided the definition that 'a norm is a belief shared to some extent by members of a social unit as to what conduct ought to be in particular situations or circumstances' (Gibbs 1981, p. 7). Unfolding this concise definition reveals three essential components:

- An *individual* component: a belief.
- A *social* component: the belief is shared by other members of a social unit.
- A *deontic:* a conduct is obliged.

1. The capability to understand the meaning of a deontic language game is acquired in the childhood. The deontic prescribes individual behaviour. The capacity to play language games that are centred around words such as 'you ought' is the precondition for a normative orientation. It allows for the internalisation of norms. Although in the past decades consideration has been given to understanding processes of internalisation over an individual's entire life span, the central processes arguably take place during childhood. Indeed, the transmission

of cultural values has even been denoted as a 'second birth' (Claessens 1972). An individual may become accepted into a wider society through a variety of means. Norm internalisation is argued to represent one of the stronger mechanisms by which this process occurs. It is argued that in contrast to compliance under coercion or to simply copying other behaviour patterns, internalisation is coupled with an individual's intrinsic motivation and sense of identity. The mechanisms by which this occurs are the focus of socialisation research. Socialisation is the bridge between the individual and society (Krappmann 2006). Hence, socialisation research is at the border of psychology and sociology, and contributions from both disciplines can be found in the literature.

2. It is a central element of the concept of norms that it implies both a psychological component, namely, the mental state of a belief, as well as a social component. This becomes apparent by the fact that this belief is shared by a number of individuals. Norms are thus essential for a comprehension of the relation between structure and agency. This is often denoted as the micro-macro link. While the aspect of agency can be found in the psychological aspect of norms, structure is entailed in the social prescription. This makes norms a fundamental building block of the wheels of social order. An understanding of norms is thus crucial for an understanding of the social world. They are a central mechanism in the *two-way dynamics* creating of social order. On the one hand, individual interactions might produce social macro-patterns, namely, a normative order. On the other hand, once the agents on the individual level recognise patterns on the resultant social level, these patterns might have an effect on the mind of the individual agent, namely, the recognition of a deontic prescription. This complex dynamics can be embraced in the following schema (Fig. 17.1):

This schema can be illustrated by an example: interactions of individual actors might lead to a certain aggregate result, for instance, social stratification. This macro-property, however, might have an effect on how the actor cognitively

Fig. 17.1 The two-way dynamics of norms

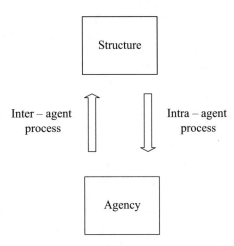

structures the perception of the world. This is an intra-agent process. For instance, actors might regard themselves as helpless losers or they might think that you can get what you want. This world perception, in turn, has an effect on how these actors behave in the interactions between the agents. Now the feedback loop is closed: the interactions result again in an aggregated macro-structure. The reader may think of other examples appropriate for his or her purpose. Presumably, it will turn out that the dynamics of the particular models can be described in terms of such a two-way dynamics between inter- and intra-agent processes.

17.2.1 The Sociological Perspective

In this section, process (1) is considered: the emergence and the properties of a normative macro-structure. Hence, this section will concentrate mainly on the social aspect of norms, i.e. the one-way dynamics of *inter*-agent processes. Before turning the attention to existing modelling approaches, the questions will be developed that need to be posed to normative multi-agent models for a comprehension of the effects and operations of norms in a social system. For this purpose, sociological theory will be utilised. Readers that are either familiar with this subject or with a purely practical interest might go on to Sect. 3.2). However, a more or less implicit reference to the problems posed by social theory can be discerned in most of the simulation models.

The *function* of normative behaviour regulation for social integration (i.e. a social structure) has been particularly emphasised by the classical role theoretic account within the sociological theory (Neumann 2008b). It can be traced back to the functionalist sociology of Durkheim and Parsons. This theoretical account was decisively influenced by anthropological research of the late 1920s when cultural anthropologists discovered the variety of the organisation of the social life and the correlation of the structure of personality and society (Malinowski 1927; Mead 1928). *Role theory* claims that action is guided by a normative orientation, insofar as social roles are described by social norms. The concept of norms is introduced in this account as a *structural constraint* of individual actors. This theory of action is often paraphrased as the 'homo sociologicus'. For a long time, this remained the dominant stream of sociological theory.

To specify where norms are placed in the theoretical architecture, let us consider the example of the famous Mr. Smith, introduced by Ralf Dahrendorf (1956) in his analysis of the 'homo sociologicus' to characterise key elements of sociological *role theory*. We first meet him at a cocktail party (in the early 1950s) and want to learn more about him. What is there to find out?

Mr. Smith is an adult male, ca. 35 years old. He holds a PhD and is an academic. Since he wears a wedding ring, we know that he is married. He lives in a middle-sized town in Germany and is a German citizen. Moreover, we discover that he is Protestant and that he arrived as a refugee after the Second World War in a town populated mostly by Catholics. We are told that this situation caused some difficulties for him. He is a lecturer by profession and he has two kids. Finally, we learn that he is the third chairmen of the local section of a political party, Y, a passionate and skilful card player and a similarly passionate though not very good driver. This approximates to what his friends would tell us.

In fact, we may have the feeling that we know him rather better now. After all, we have some expectations as to how a lecturer is likely to behave. As a lecturer, he stands in certain relations to colleagues and pupils. As a father, he will love and care for his children, and card playing is also typically associated with certain habits. If we know the party Y, we will know a lot more about his political values. However, all that we have found out represent *social facts*. There are a lot more lecturers, fathers and German citizens beside Mr. Smith. In fact, none of this information tell us anything about the unique identity of Mr. Smith. We simply discovered information about social positions, which can, of course, be occupied by varying persons. However, social positions are associated with specific *social roles*. Roles are defined by the specific attributes, behaviour and social relations required of them. Demands of society determine—to a certain degree—individual behaviour. Individuals are faced with obligations and expectations. This social demand is transmitted to the individual by *norms*. Norms are the 'casting mould' (Durkheim 1895) of individual action. They regulate how lecturers, fathers and members of political parties ought to act to fulfil the role expectations of society. However, nowadays, it is improbable that we will be told anything about driving competence. Thus, we have learned another lesson: norms may *change* over the course of time. In particular, Talcott Parsons (1937) emphasised that the ends of individual actions are not arbitrary but rather are prescribed by social norms. Thus, norms represent a key concept within sociological role theory.

To examine the explanatory account of this theoretical approach, the present investigation will abstract from a description of the content of concrete norms. We concentrate on the methodological characteristics of norms in general, rather than the content of specific norms. On closer inspection of this example, we find out that the concept of social norms is characterised by three *key elements*:

First, norms show some degree of generality. They are regarded as the 'casting mould' of individual action (Durkheim 1895). The very idea of role theory is that a social role must not be restricted to a unique individual. For instance, the roles of lecturer or chairman of a political party can be performed by different individuals. It might not, of course, be arbitrary as to who will play this role. In fact, it is a major focus of the empirical counterpart of role theory, statistical analysis of variables, to investigate the distribution of roles. For instance, monetary background might determine an individual's chances of securing an academic position. Nevertheless, academic positions are a feature of society, not the individual. In the classical account, the generality of norms is simply a given. However, agent-based models start from individual agents. Thus, in the individualistic approach norms have to *spread* in some way from one agent to another to gain generality. The explanation of norm spreading is essential for a reconstruction of social norms in terms of individual actors.

Secondly, the role set of father or lecturer encompasses a huge action repertoire. The choice of a concrete action cannot be determined only solely by an external force. The ends of an action have to be determined internally by the individual actor executing a specific role. This means that the ends of individual actions are (to a certain degree) determined by the society. For instance, it is a social norm how a caring father would look like and what kind of actions are to be undertaken. In fact, this varies between societies. This knowledge is often denoted as *internalisation*, even though the psychological mechanisms are not in the focus of sociological theory. Thus, already a comprehension of the inter-agent processes that constitute a social role calls for a related subjective element at the level of the individual agent.

Thirdly, this approach is characterised by a certain type of analysis: the normative integration of society (Parsons 1937; Davies and Moore 1945; Merton 1957). Hence, the question is to a lesser extent concerned with the origin of norms than with the *function* of norms for a society. For instance, the role of the father is to educate his child. The role of the lecturer is crucial for the socialisation of pupils. Thus, both roles are functionally relevant for the reproduction of the society.

However, in the past decades, this paradigm has been severely criticised: firstly, role theory has been criticised for sketching an *oversocialised* picture of man (Wrong 1961). Already in the 1960s, Homans (1964) claimed to 'bring man back in'. In fact, the individual actors of the role theory have often been regarded as more or less social automata. If they have properly internalised the norms, they execute the programme prescribed by their roles. Secondly, role theory is built on the claim that social phenomena should be explained with social factors (Durkheim 1895). Roles are a pregiven element in this theoretical architecture. Roles (and, thus, norms) emanate from society. However, the origin of both roles and society is left unexplained. In this so-called functionalist methodology of the role theoretic account, it is argued that to perform a social role is to perform a social function, and this social function is argued to be the source of the norm. However, this perspective lacks a description of a mechanism by which norms become effectively established. This deficit suggests to 'bring man back in'.

In fact, the explanatory deficit engendered an alternative. Others have suggested building sociology on the foundations of individual actors. This built on views advocated by John Stuart Mill. This is the programme of the so-called *method-ological individualism* (Boudon 1981; Raub and Voss 1981; Coleman 1990; Esser 1993). A shift 'from factors to actors' (Macy and Willer 2002) can be observed in the foundations of sociology in the recent decades. This approach regards the appearance of norms as an *aggregate product* of a sum of individual actions. This approach to norms has been particularly stressed by rational choice theories. The deficit of this approach, however, is a lack of a cognitive mechanism that could explain the deontic component of norms, i.e. how the social level may exhibit an obligatory force on the individual by norms. Agent-based modelling promises to overcome the complementary deficits of both theoretical accounts.

On the one hand, agent-based models contribute to the individualist theory build-ing strategy: obviously, agents are the fundamental building block of agent-based models. In agent-based simulation models (Artificial Societies), structures emerge from individual interaction. On the other hand, however, the great advantage of this methodology is that it allows to explicitly consider the complexity generated by individual interactions. Compared to purely analytical models, it is a great advantage of simulation approaches that they are not restricted to single or representative actors. This complexity generated by this particular feature enables the investigation of the feedback loop between individual interaction and collective dynamics. This has led to a growing awareness of its potential for investigating the building blocks of social structure. For this reason, it is even claimed that agent-based simulation allows us to 'discover the language in which the great book of social reality is written' (Deffuant et al. 2006), by constituting the promise to understand 'how actors produce, and are at the same time a product of social reality' (Deffuant et al. 2006). Since structures on the macro-level are a product of individual interaction, a causal understanding of the processes at work seems possible that might fill the gap between individual action and social structure with agent-based models.

This feature of agent-based modelling suggests that this methodology enables to bring together the role theoretic perspective that norms are a *structural constraints* and the individualistic perspective of norms as an *aggregated product* of a sum of individual actions. Hence, it might provide both a causal mechanism of how normative action constraints get established as well as how social forces exhibit cognitive power. For this reason, the focus will now shift to an evaluation of simulation models.

17.2.2   Simulation Models of Norms

After the theoretical problem exposition, the particular focus of this section is an investigation of the contribution of *simulation experiments* to sociological theory. The recourse to sociological theory in Sect. 3.1) revealed the following *questions*:

- Can they provide insights into the normative regulation of society, that is, do they also reproduce the findings of a functional analysis of the effects and operations of norms in a society (*focus of contribution*)?
- Moreover, do they allow for a causal reconstruction of the mechanisms that generate the functional interconnectedness on the social level? This implies that two further questions have to be addressed.
- What transforms the agents in such a way that they factually follow norms? That is, what are the causal mechanisms at work that enable an internalisation of norms (*transformation problem*)?
- By what mechanisms in the model can norm-abiding behaviour spread to or decay from one agent to another (*transmission problem*)?

These questions will be examined in this section. It has to be emphasised that the investigation concentrates on methodology, not on the contents of norms governing concrete roles such as father or lecturer. However, existing models are clustered around various intuitions about norms, conventions or standards of behaviour. The concrete research question differs from model to model. Some models concentrate on the emergence or spreading of norms. Others concentrate on functional aspects or the feedback of norms on individual agent's behaviour. A *multiplicity of concepts* is at hand. Hence, a comprehensive review of all models and accounts that may be in some way related to the study of norms would go beyond the scope of this investigation.

The overwhelming mass of models, however, can be traced back to (or is at least influenced by) *two* traditions in particular: first, *game theory* and, secondly, an architecture of cognitive agents with some roots in *artificial intelligence*. Tradition and theoretical background has a direct impact on the terminology used. Depending on their background, the models tend to be communicated in different scientific communities. Additionally, references in articles tend to depend on their authors' background. Under the perspective of content, the models in the AI tradition typically contain references to conceptual articles relating to agent architectures. Articles with models in a more game theoretical tradition typically refer to game theoretic literature for the characterisation of the interaction structures in which the authors are interested. Of course, this tradition-influenced framing, publishing and referencing is a *tendency*. It does *not* constitute a clear-cut disjunction without any intersection. It has to be emphasised that this is neither a very precise nor a disjunctive categorisation. To some degree, the distinction between game theory and DAI is a distinction in the mode of speech employed by the authors. Some problems of game theoretic models could also be formulated in a DAI language and vice versa. The categorisation of models as following the DAI tradition shall only indicate that the agents employed by these models are in some way cognitively richer than those in the so-called game theoretic models.

Nevertheless, this distinction gives a rough sketch of the line of thought followed by the models and also of the kind of problems, the concepts for their investigation and the mode of speech in which the paper is presented. Moreover, this categorisation provides hints to other areas of research that are closely related to the models considered in this article. For instance, simulation is only a small subdiscipline of game theory in general, and the distinction between analytical and simulation results is only gradual. Simulation models might describe problems in game theoretic terms, but the method of resolution is not that of analytical game theory (Binmore 1998). In fact, investigating norms with the means of *analytical* game theory is a highly active research field.

17.2.2.1 Models Using a Game Theoretic Problem Description

First, a closer examination of a sample of models applying a game theoretic problem description will be undertaken. The models investigated here build on the framework described in Chap. 16 on games and utility in this handbook (Fioretti 2017).

Axelrod (1986) studies the evolution of a standard not to cheat via a 'norms game' and a 'meta-norms game'. In the 'norms game', defectors may be punished by observers. In the 'meta-norms game', it is also the case that observers of a defection that do not punish the defector may be punished. Only the latter game leads to a widespread standard of not defecting.

Coleman (1987) investigates the effect of interaction structures on the evolution of cooperation in a prisoner's dilemma situation. Only small groups can prevent the exploitation of strangers.

Macy and Sato (2002) examine the effect of mobility on the emergence of trust among strangers in a trust game. While agents with low mobility trust only their neighbours, high mobility supports the evolution of trust among strangers.

Vieth (2003) investigates the evolution of fair division of a commodity in an ultimatum game. Including the ability to signal emotions leads to a perfectly fair share. If detection of emotions is costly the proposals even exceed fair share.

Bicchieri et al. (2003) present a model of a trust game. It demonstrates how a trust and reciprocate norm emerges in interactions among strangers. This is realised by several different conditional strategies.

Savarimuthu et al. (2007) study the convergence of different norms in the interactions of two different societies. Both societies play an ultimatum game against each other. Two mechanisms are examined: a normative advisor and a role model agent.

In the model by Sen and Airiau (2007), a co-ordination and a social dilemma game are examined. Agents learn norms in repeated interactions with *different* agents. This is denoted as social learning to distinguish this interaction type from repeated games with the same player. The whole population converges to a consistent norm.

Obviously, all models have been developed for differing concrete purposes. To examine the extent to which these models capture the explanatory problems of the *contribution* problem, *transformation* problem and *transmission* problem, the various accounts of the different models will be outlined in a table. Moreover, a short hint to the concrete implementation is provided. This will enable an evaluation inasmuch normative agent-based models have so far reached the goal to discover 'the language in which social reality is written'. These models are summarised in Table 17.1.

Lessons

The classical model employing a game theoretical approach for the problem description is Axelrod's model. The main contribution of this approach is a

Table 17.1 A sample of models using a game theoretic problem description

	Contribution	Transformation	Transmission	Implementation
Axelrod (1986)	Norm dynamics (*norms* broadly conceived!)	Sanctions	Social learning; replicator dynamics	Dynamical propensities
Coleman (1987)	Norm dynamics	Punishment by defections (memory restrictions for identifying defections as sanctions)	(a) group size (acquaintance) (b) additionally: Replicator dynamics	Conditional strategies
Macy and Sato (2002)	Norm dynamics	Losses by exclusion from interaction	Social learning	Dynamical propensities
Vieth (2003)	Norm dynamics	Losses by rejection	Social learning; replicator dynamics	Dynamical propensities
Bicchieri et al. (2003)	Norm dynamics	Sanctions by retaliating super game strategies	Strategy evolution; replicator dynamics	Conditional strategies
Savarimuthu et al. (2007)	Norm dynamics; functional analysis	Losses by rejection; advice	Advice updating based on collective experience	Dynamical propensities
Sen and Airiau (2007)	Norm dynamics	Experience	Social learning guiding behaviour convergence	Dynamical propensities

clear understanding of the emergence of (commonly shared) normative behaviour constraints. The starting point of the models is a dilemma situation. This is a consequence of the game theoretic problem description. Simulation allows for an evolutionary perspective by analysing repeated games. Typically, in the long run and under specific conditions (which vary from model to model), it is possible that behaviour conventions emerge, which are in the benefit of all agents or that represent some intuitions about fairness. The diffusion of a behavioural regularity is then regarded as norms. The subjective side of an obligatory force is not in the focus of this approach. Hence, what lessons can be gained from the investigation of this conception to model normative action constraints with regard to the specific questions?

Contribution

Already Axelrod's classical model provides a causal explanation for *norm spreading*. This includes a designation of mechanisms of norm transmission and normative transformation. An investigation of the functional effect of norms on the society is left aside. This orientation remained predominant in this line of research. Typically, models with a game theoretic background concentrate on the question of norm *dynamics*. They ask how a behaviour regularity emerges in an agent population. This is the problem of the rational choice tradition in sociological theory, namely, the perspective of norms as an aggregated product of a sum of individual actions.

Transformation

It is striking that, except for the model by Sen and Airiau, the transformation of individual behaviour in all models is driven by some kind of sanctions. However, also in the Sen and Airiau model, agents react to losses of utility values. This is the causal mechanism of norm spreading. The great advantage of this account is to shed light on the process of norm *change*. As it has become apparent in discussing Mr. Smith, this process can also be observed in human societies. However, norm change is only barely captured by the functional account of role theory. On the other hand, the models of this tradition only include a *restricted functional perspective*: on an individual level, the agents' choice of action is guided by the functional consideration of calculating the expected utility. However, a corresponding analysis on the social macro-level can be found only in Savarimuthu et al.'s model.

Transmission

With regard to the transmission of norms, it is striking that social learning is implemented in many game theoretic models by a *replicator dynamics*. Typically, this is *interpreted* as social learning by imitation. If applied in a context where *no real natural selection*, rather than some kind of *learning* is at work, then using a replicator dynamics amounts to saying: somehow the individuals learn in a way that—measured by the relative overall success of their type of behaviour—more successful types of behaviour become more frequent. As an effect, this may be true. However, no mechanism is indicated. In this dimension, the models struggle with the same kind of problem as functional analysis which the individualistic programme tries to resolve, namely, the lack of a causal explanation.

Implementation

From the perspective of the role theory of action, a weakness of this approach becomes apparent, one immediately related to the game theoretic problem description. Agents are faced with a strategic (binary) decision situation. Thus, they have a fixed set of behaviour (Moss 2001). For this reason, behaviour change can be implemented by dynamical propensies (i.e. the propensity to defect is dynamically updated). Faced with this situation, agents choose the alternative that maximises their expected utility. However, behaviour change goes not along with goal change. Agents can do no more than react to different environmental conditions. The agents' behaviour is guided strategic *adaptation*. An *active* element of normative orientation in the choice relating to the ends of action cannot be found in a game theoretic approach. This is simply due to the fact that agents do not possess any internal mechanism to reflect and eventually change their behaviour, other than the desire to maximise utility. This point has already been highlighted in Parsons' critique of 'utilitarian theories' of action (Parsons 1937), namely, that the ends of individual actions are in some way arbitrary. Even though the modelling of behaviour transformation is the strength of this kind of models, the ends of the action remain unchanged: the goal is to maximise utility. In this respect, the relation between the action and the ends of the action remains arbitrary.

However, the very idea of role theory is to provide an answer to the question: where do ends come from? Parsons' (and Durkheim's) answer was the internalisation of norms. A corresponding answer to this problem is not supplied in game theoretical models. This is due to the fact that agents do not act because they want to obey (or deviate from) a norm. They do not even 'know' norms. Even though the model provides a mechanism for the transformation of the agents, this is not identical with norm internalisation. This remains beyond the scope of this account. The agents' behaviour can only be *interpreted* as normative from the perspective of an external observer. Thus, transformation is not identical with internalisation. While the model provides a mechanism for behaviour transformation, it cannot capture the process of internalisation. Compared to the classical role theory, this is a principle limitation of a game theoretical description of the problem situation.

17.2.2.2 Models Utilising Cognitive Agents

This shortcoming calls for cognitively richer agents. For this reason, a sample of models in the AI tradition will be examined more closely.

Conte and Castelfranchi (1995b) investigate three different populations of food gathering agents: aggressive, strategic and normative agent populations. Aggressive agents attack 'eating' agents, strategic agents attack only weaker agents, and normative agents obey a finder-keeper norm. The aggregated performance of the normative population is the best with regard to the degree of aggression, welfare and equality.

In an extension of the above model, Castelfranchi et al. (1998) study the interaction of the different agent populations. Interaction leads to a breakdown of the beneficent effects of norms, which can only be preserved with the introduction of normative reputation and communication among agents.[1]

---

[1]For a more in-depth discussion of this model, the interested reader is referred to the chapter on reputation (Giardini et al. 2013).

Saam and Harrer (1999) present a further extension of Conte and Castelfranchi's model. They investigate the influence of social inequality and power relations on the effectiveness of a 'finder-keeper' norm.

Epstein (2000) examines the effect of norms on both the social macro- and the individual micro-level. On the macro-level, the model generates patterns of local conformity and global diversity. At the level of the individual agents, norms have the effect of relieving agents from individual thinking.

Flentge et al. (2001) study the emergence and effects of a possession norm by processes of memetic contagion. The norm is beneficent for the society but has short-term disadvantages for individual agents. Hence, the norm can only be retained in the presence of a sanctioning norm.

Verhagen (2001) tries to obtain predictability of social systems while preserving autonomy on the agent level through the introduction of norms. In the model, the degree of norm spreading and internalisation is studied.

Hales (2002) extends the Conte/Castelfranchi model by introducing stereotyping agents. Reputation is projected not on individual agents but on whole groups. This works effectively only when stereotyping is based on correct information. Even slight noise causes the norms to breakdown.

Burke et al. (2006) investigate the emergence of a spatial distribution of a binary norm. Patterns of local conformity and global diversity are generated by a decision process which is dependent on the local interactions with neighbouring agents.

The contribution of these models to the questions specified above can be summarised in Table 17.2, which also includes a brief remark on the implementation specification.

Lessons

The classical model of this kind of models is the one developed by Conte and Castelfranchi in 1995. It was the starting point for several extensions. While the scope of these models has been significantly extended in the past decade, the most significant contribution of this approach still can be regarded as to enhance the understanding of the operation and effects of normative behaviour constraints.

Epstein, Burke et al. and Verhagen do not study specific norms but examine mechanisms related to the operations of norms. In particular, the spreading of norms is studied by these authors. In this respect, they recover and refine (by the notion of local conformity and global diversity, a pattern that cannot be found in game theoretic models) the findings of game theoretic models with the means of cognitive agents. The other models concentrate mainly on studying the effects and operations of specific norms in the society. Analysed are questions such as the effect of possession norms, for instance, under the condition of social inequality and power relations. This is strongly influenced by Conte and Castelfranchi's problem exposition.

Table 17.2 A sample of models using cognitive agents

	Contribution	Transformation	Transmission	Implementation
Conte and Castelfranchi (1995b)	Functional analysis	—/—[a]	—/—[a]	Conditional strategies
Castelfranchi et al. (1998)	Functional analysis	Updating conditionals (of strategies) through knowledge	Updating knowledge by experience (and communication[b])	Conditional strategies
Saam and Harrer (1999)	Functional analysis	(a) —/—[a]	(a) —/—[a]	Conditional strategies
		(b) internalisation[b]	(b) obligation[b]	
Epstein (2000)	Norm dynamics; functional analysis	Observation	Social learning	Dynamical updating
Flentge et al. (2001)	Functional analysis	Memetic contagion	Contact	Conditional strategies
Verhagen (2001)	Norm dynamics	Internalisation	Communication	Decision tree
Hales (2002)	Functional analysis	Updating conditionals (of strategies) through knowledge	Updating knowledge by experience (and communication[b])	Conditional strategies
Burke et al. (2006)	Norm dynamics	Signals	Social learning guiding behaviour convergence	Dynamical propensities (threshold)

[a]The agents are/are not already moral agents
[b]Only in a second experiment

What lessons can be gain from the investigation of this conception to model normative action constraints with regard to the specific questions?

Contribution

A striking feature of these models is that they demonstrate how agent-based models are able to contribute to a *functional analysis* of norms. However, in contrast to the classical scheme of functional explanations in the social sciences, this result is reached by interactions of *individual agents*. Moreover, in these models, a much stronger notion of norms is deployed than typically in game theoretic models. Norms are not just reached by mutual agreement but are an explicitly prescribed action routine. The concept of norms in these models is in line with the role theoretical conception of norms as a *structural constraint* of individual actors. This conception of norms allows for a wider field of applications that could cover the role theoretic norm conception: these can be interpreted as internalised properties of the agents.

Transformation

There exist a wide range of varieties how agents change their behaviour. Behaviour transformation is not as straightforward as in game theoretic models. However, it has to be emphasised that the very first model of Conte and Castelfranchi did not include any behaviour transformation at all. Agents have no individual freedom in this model. As critics accuse the role theory, the action repertoire is also (depending on conditions) deterministic. Thus, even though the authors succeed in 'bringing man back in', the agents in the model are merely *normative automata*. Insofar as the norms are a pregiven element in the model, the approach can also be regarded as an 'over-socialised' conception of man.

This limitation has been overcome by the subsequent developments. With regard to the transformation problem, the agents have become more flexible than in the very first model. However, a key difference to game theoretic models still remains: while game theoretic models mostly concentrate on sanctioning, in models of cognitive agents, sanctions are only employed by Flentge et al. as the transformation mechanism.

However, while the norms in these models can be interpreted as internalised properties of the agents, an investigation of the process of internalisation is only in the beginning. So far no commonly accepted mechanism of internalisation has been identified. Memetic contagion is a candidate. In Verhagen's model, a quite sophisticated account is undertaken, including a self-model, a group model and a degree of autonomy. It is highly advanced in constructing a feedback loop between individual and collective dynamics. By the combination of a self-model and a group model, a representation of the (presumed) beliefs held in the society is integrated in the belief system of individual agents. Conceptually, this is quite close to Mr. Smith. However, it might be doubted whether the mechanisms applied are a theoretically valid representation of real processes.

Transmission

Complementary to the wide range of different mechanisms of agent transformation, also a variety of different transmission mechanisms are applied. Basically, agents apply some kind of knowledge updating process, if agent transformation takes place at all. Up to date, the transmission problem is no longer a blind spot of cognitive agents as it was the case in the (Conte and Castelfranchi 1995a, b) model. By comparison, communication plays a much more important role than in game theoretic models and is much more explicitly modelled in models within the AI tradition. The processes utilised are more realistic mechanisms than the replicator dynamics of game theoretic models. However, no consensus has been reached, what an appropriate mechanism would be. This is also due to the fact that a modelling of agent transformation and norm transmission is computationally more demanding than in game theoretic models. It has to be emphasised, however, that with regard to the transformation and the transmission problem, the borderlines of both approaches are no longer clear-cut. The models of Verhagen and Savarimuthu et al. include elements of the other line of thought.

Implementation

Since actions performed by cognitive agents cannot be reduced to the binary decision to cooperate or defect, more complex behaviour rules than dynamic propensities have to be applied. The dominant approach for the implementation of normative behaviour constraints in cognitive agents is based on a straightforward intuition, namely, to apply conditional strategies that are conditionally based on the agent's knowledge base. The strategy repertoire, however, depends on the concrete model. The overview of existing models has revealed that the focus of their contribution is mainly on the dynamics on the macro-level. The questions of norm transmission and the focus of contribution concentrate on the social macro-scale. The analysis is focused on the one-way dynamics of effects and operations of norms on the *social level*. Hence, the analysis is focused on the emergence of social structure out of individual interaction rather than on the relation between structure and agency. This is one aspect of the full dynamics, namely, the following process:

$$\text{inter-agent processes}: interaction \;\Rightarrow\; \text{macro-property}: structure$$

17.3   Socialisation: Norm Internalisation

It has been outlined, however, that the definition of a norm possesses social and *psychological* components. Norms are essential for a comprehension of the relation of structure and agency. While processes of emergence of aggregated behaviour standards from interaction among individual agents has been extensively studied, a comprehension of the reverse process how the aggregate level gets back into agents' minds is not as yet fully reached. A full comprehension of normative behaviour regulation, however, has also to include the reverse dynamics of the effect of social structure on the *individual agency*. Already the problem of agent transformation refers to the effect of structure on the level of the individual agent. This is the most problematic aspect in the agents' design. It would include the following dynamics:

$$structure \Rightarrow \text{Intra-agent processes}: agency$$

This would be a step towards closing the feedback loop of the *two-way* dynamics. Obviously, intra-agent processes are closely related to learning. The reader can find a more comprehensive examination of the state of the art in the chapter on evolution and learning. In particular, the effect of structure on (a transformation of) individual agency is particular relevant for studying the effects of norms, if agency is not restricted to a representative agent.

To represent such intra-agent processes, in particular the concept of *social learning* is well known in agent-based models. It is applied in a number of game theoretic models and can also be found in models of the AI tradition. The concepts of social learning, but also knowledge updating, can be traced back to *behaviouristic* psychological theories. Behaviourism is a theory of learning developed principally through experiments with animals. For instance, the conditioning experiments of Ivan Pavlov are well known: he demonstrated that dogs can be trained to exhibit a specific reaction such as salivation by presenting a specific stimulus such as the sound of a bell together with food (Pavlov 1927). Bandura (1962, 1969) extended the behaviouristic approach with a social dimension by developing a theory of social learning through *imitation*. From a behaviouristic perspective, norms constitute a learned behaviour and thus have to be explained using these theories. The dynamical propensities of models inspired by game theoretical concepts are a straightforward implementation of such a view on intra-agent processes. The propensity to cooperate or defect is updated in proportion to the propensity of sanctions. The propensity of sanctions, however, is a structural component resulting from inter-agent processes. Hence, agents learn to modify their behaviour according to structural conditions.

Here we find the feedback loop between social and individual components that are in fact essential for the concept of norms. However, the third component is missing: this approach does not include a concept of obligations. Deontics are out of the scope of this approach. This shortcoming can be traced back to the psychological theory that is represented in the agents: behaviourism is not capable of capturing mental processes. Indeed, it specifically avoids commenting on the mental processes involved. Under the influence of positivism, reference to unobservable entities such as the 'mind' has been regarded as not scientifically valid. Obligations are such unobservable entities. Hence, they cannot be represented by the means of behaviouristic learning theories that are applied in agent models.

In socialisation research, the complex cognitive processes necessary for grasping the meaning of an obligation is denoted as *internalisation*. It has already been shown that agent transformation is not the same as the internalisation of norms. This is also behaviourally important because normative behaviour, guided by deontics, need not be a statistical regularity, guided by propensities. In particular if moral reasoning is involved, deviant behaviour is not explained by chance variation, leading to some kind of normal distribution (where the mean value might be updated). There is a difference between norms and the normal.

To represent a complex cognitive concept such as norm internalisation calls for the cognitively rich agents of the AI tradition. However, the examination of current models has revealed that a comprehension of the cognitive mechanisms by which social behaviour regulation becomes effective in the individual mind is still in its fledgling stages. It has been shown that a multiplicity of concepts is at hand: while in the very beginning the agents were merely normative automata, there exist conceptualisations of normative agent transformation ranging from updating conditionals (of strategies) through knowledge to signalling and memetic contagion. However, no consensus is reached what are the most relevant mechanisms. It can be suspected that they remain effect generating rather than process representation mechanisms. As an agenda for the next decade, a closer examination of the processes by which normative obligation becomes accepted by humans might be useful. For this purpose, it is necessary to recall socialisation theory. This will help to clarify the problem situation. However, the results of socialisation research are not unequivocal. Hence, in the development of a normative architecture, some fundamental decisions have to be made.

17.3.1 The Perspective of Socialisation Research on Norms

To grasp an understanding of the decisions that have to be made in the cognitive design of an agent, first, some fundamental aspects of theories of socialisation are briefly highlighted. Subsequently, current architectures of normative agents are evaluated with regard to the questions posed by the empirical science.

Broadly speaking a conceptual dichotomy of two main approaches can be identified (Geulen 1991) in socialisation research concerning the relation between the individual and society: One position assumes a harmony between, or identity of, the individual and society. Philosophical precursors of this approach are Aristotle, Leibniz and Hegel. The second position stands in contrast and postulates an antagonism between the individual and society. Within this position, two further standpoints can be distinguished already in the philosophical tradition: Hobbes, for example, is representative of the argument that society should tame the individual. By contrast, the position of Rousseau is paradigmatic of an approach that advocates the need for releasing the individual from society. Both philosophers share the assumption that an antagonism exists between the individual and society, although they disagree about the implications.

As it has been outlined, socialisation research sits at the border of psychology and sociology, and contributions from both disciplines can be found in the literature. From a sociological perspective, the beginning of investigating socialisation processes cumulated in the work of Emil Durkheim, founding father of sociology and professor of pedagogy. Starting from a clinical and psychological perspective, Sigmund Freud developed a theory of socialisation, which in many aspects is surprisingly akin to Durkheim's approach.

The early theories of Freud and Durkheim agree in that they assume an *antagonism* between individual and society. Durkheim asserted that the individual consists of two parts: first, a private domain that is egoistic and guided purely by basic drives. The egoistic domain corresponds to that of the newborn child. The original human is a 'tabula rasa' in which social norms have to be implemented. Only through the process of socialisation do humans become socially and morally responsible persons. This is the second 'part' of the individual. Durkheim claimed that the best of us is of a social nature. Society, however, is coercive (Durkheim 1895) and can even compel individuals to commit suicide (Durkheim 1897). Norms are finally internalised once the individual no longer perceives this coercion (Durkheim 1907). Yet for Durkheim coercion nonetheless remains. As with Durkheim, Freud assumed the existence of an antagonism between individuals and society. This assumption can be discerned in his distinction between ego, id and superego. The id represents the drives of the child-like portion of the person. It is highly impulsive and takes into account only what it wants. It exclusively follows the pleasure principle (Freud 1932). The superego enables control of the primary drives: it represents the moral code of a society and involves feelings of shame and guilt (Freud 1955). It is the place where social norms can be found. It has been argued that the degree to which feelings of guilt are experienced is indicative of the degree of norm internalisation

(Kohlberg 1996). Finally, the ego is the controlling instance: it coordinates the demands of id, superego and outer world. According to Freud, the superego is the mirror of the society. Freud's theory of ego, superego and id, then, parallels Durkheim's assumption that the internalisation of norms involves social coercion (Geulen 1991). From the perspective of both, society is in radical conflict with human nature. Norms are given as an external fact. Both Durkheim and Freud regard the individual as *passive* and internalisation as a unidirectional process.

Building on G.H. Mead (1934), and the theories of cognitive and moral development of Piaget (1932, 1947) and Kohlberg (1996), in recent times identity theories have become influential in socialisation research. In contrast to an orientation based solely either on the individual subject or the society, identity theories emphasise the interaction of culture and individuals in the development of a morally responsible person (Bosma and Kunner 2001; Fuhrer and Trautner 2005; Keupp 1999).

Mead developed a concept of identity that in contrast to Durkheim did not reduce the individual to a private subject guided purely by drives. Instead, he developed a theory of the social constitution of individual identity. A crucial mechanism in the development of personality is the capability of role taking: to regard oneself from the perspective of the other. This ability enables individuals to anticipate the perspectives and expectations of others and thereby to come to accept social norms. In the process of role taking, the individual develops a consciousness whereby the individual is itself a stimulus for the reaction of the other in situations of social interaction. This is the distinction between the spontaneous 'I' and self-reflected 'me'. Together, they form what Mead denoted as identity: in other words, the 'self'. An abstraction of this process leads to the notion of the 'generalised other'. This is not a specific interaction partner but a placeholder for anybody. The notion of the 'generalised other' is the representation of society.

Identity theories follow Mead in seeing individual identity as the key link between person and culture. In contrast to the perspective to regard the social as constraining the individual, identity theories argue that socially embedded identity *enables action selection*. Action determination can be intrinsically or extrinsically motivated. The identity of individuals contributes to the development of their intrinsic motivation. There exist clear empirical evidence that sanctions and even incentives undermine intrinsic motivation (Deci and Ryan 2000). Norms, however, constitute a socially determined pattern of behaviour. Thus, norm obedience is always extrinsically motivated. However, at this point, *internalisation* comes into play. Extrinsic motivation can be internalised to different degrees, ranging from purely extrinsic behavioural regulation (e.g. sanctions) to motivations that are integrated into the self. Integration is attained when external guidelines have become part of personal identity. This is the highest degree of a transformation of external regulation into self-determination and is denoted as self-determined extrinsic motivation (Deci and Ryan 2000). In this case, a person is in line with itself if he or she orients behaviour around social norms. Integrated behaviour regulation is highly salient. Since norms are in full accordance with the personal values, action is regarded as autonomously motivated. Hence, the scale of internalisation from external regulation to integration is regarded as the scale from external control to *autonomy*. Internalisation of extrinsic motivation represents the bridge between psychological integrity and social cohesion. While intrinsic motivation is the paradigm of autonomous motivation, in the case of social behaviour regulation, autonomy can only be reached through a process of internalisation.

How is the concept of identity described? With reference to William James (1890), criteria for identity are formulated as consistency, continuity and effectiveness. Identity consists of an *inner* and *outer perspective*. Moreover, *personal* and *social identities* are differentiated (Tajfel 1970; Turner 1982; Turner and Onorato 1999). While the inner perspective is grounded on individual decisions, the outer perspective is based on ascription of others. Examples are ethnic or gender identity. However, the individual might decide to identify with these ascriptions. Then the ascription becomes part of the inner perspective. Examples can be found throughout the history. For instance, (beside other factors) elements of this psychological mechanism can be revealed in the black power movement in the 1960s or the raise of ethnic conflicts since the 1990s. Personal identity is the self-construction of a personal biography. Social identity is determined by peer and reference groups. This refers to social networks. While peers are the group to which the individual factually belongs, the individual need not belong to the reference group. It is sufficient to identify with the values of this group. For instance, this identification might constitute sympathy for a political party. The social identity is decisively responsible for the process by which social norms and values become part of individual goals. This is particularly dependent on the salience of group membership. Norm internalisation, however, is not a unidirectional process of the transmission of a given norm. While embedded in a social environment, the individual has an *active* role in the social group.

17.3.2 Normative Architectures

The brief overview of socialisation research suggests that for the design of normative agents in particular two main decisions have to be made:

Is an antagonism or an identity (respective harmony) between individual and society presumed? Hence, does the Artificial Society represent the theories of Durkheim and Freud, or identity theories that follow G. H. Mead?

How is the effect of normative behaviour regulation on the individual agent represented? Does the individual agent play an active or a passive role, i.e. has the individual agent something comparable to a personal identity?

The second question leads to a follow-up question, namely:

If agents play an active role, if and how can this represent a process of identity formation? In particular, are agents embedded in social networks of peer or reference groups?

It will now be examined how current architectures can be assessed from an empirical perspective. As the examination of current simulation models has revealed, a comprehension of the effects of the social level on individual agents is far from being sufficient so far. To provide an outlook of possible future modelling approaches of the effects of social norms on individual agents, a brief sample of normative agent *architectures* will be provided. In fact, the number of conceptually oriented articles on the architecture of normative agents exceeds the number of existing models. These architectures study how these processes could be modelled in principle. Typically, norms in concrete models are less sophisticated than concepts proposed in formal architectures (Conte and Dignum 2001). The development of architectures is a kind of requirement analysis: it specifies the essential components of normative agents. It can be expected that future implementations will be guided by deliberations that can be found in these architectures. For this reason, a sample of cases is selected (Neumann 2008a) for a closer examination with regard to the question of what decisions are made on how to represent effects of norms on individual agents.

Andrighetto et al. (2007) investigate the process of norm innovation. The behaviour of an agent may be interpreted by an observing agent as normative if it is marked as salient in the observer's normative board. Thus, norm instantiation is regarded as an inter-agent process.

Boella and van der Torre (2003) differentiate between three types of agents: agents who are the subject of norms, so-called defender agents, who are responsible for norm control and a normative authority that has legislative power and that monitors defender agents.

Boella and van der Torre (2006) rely on John Searle's notion of institutional facts (so-called 'counts-as' conditionals) to represent social reality in the agent architecture. A normative base and a 'counts-as' component transforms brute facts into obligations and permissions.

The Belief-Obligation-Intentions-Desire (BOID) architecture (Broersen et al. 2001) is the classical approach to represent norms in agent architectures. Obligations are added to the BDI architecture to represent social norms while preserving the agent's autonomy. Principles of the resolution of conflicts between the different components are investigated in the paper.

Boman (1999) proposes the use of supersoft decision theory to characterise real-time decision-making in the presence of risk and uncertainty. Moreover, agents can communicate with a normative decision module to act in accordance with social demands. Norms act as global constraints on individual behaviour.

Castelfranchi et al. (2000) explore the principles of deliberative normative reasoning. Agents are able to receive information about norms and society. The data is processed in a multi-level cognitive architecture. On this basis, norms can be adopted and used as meta-goals in the agent decision process.

Conte and Castelfranchi (1999) distinguish between a conventionalist (in rational philosophy) and a prescriptive (in philosophy of law) perspective on norms. A logical framework is introduced to preserve a weak intuition of the prescriptive perspective which is capable of integrating the conventionalist intuition.

Conte and Dignum (2001) argue that imitation is not sufficient to establish a cognitive representation of norms in an agent. Agents infer abstract standards from observed behaviour. This allows for normative reasoning and normative influence in accepting (or defeating) and defending norms.

Dignum et al. (2002) investigate the relations and possible conflicts between different components in an agent's decision process. The decision-making process of so-called B-doing agents is designed as a two-stage process, including norms as desires of society. The authors differentiate between abstract norms and concrete obligations.

Garcia-Camino et al. (2006) introduce norms as constraints for regulating the rules of interaction between agents in situations such as a Dutch auction protocol. These norms are regulated by an electronic institution (a virtual auctioneer) with an explicitly represented normative layer.

Lopez and Marquez (2004) explore the process of adopting or rejecting a normative goal in the BDI framework. Agents must recognise themselves as addressees of norms and must evaluate whether a normative goal has a higher or lower priority than those hindered by punishment for violating the norm.

Sadri et al. (2006) extend their concept of knowledge, goals and plan (KGP) agents by including norms based on the roles played by the agents. For this reason, the knowledge base KB of agents is upgraded by KB_{soc}, which caters for normative reasoning, and KB_{rev}, which resolves conflicts between personal and social goals.

Shoham and Tennenholtz (1992) propose building social laws into the action representation to guarantee the successful coexistence of multiple programmes (i.e. agents) and programmers. Norms are constraints on individual freedom. The authors investigate the problem of automatically deriving social laws that enable the execution of each agent's action plans in the agent system.

Vazquez-Salceda et al. (2005) provide a framework for the normative regulation of electronic institutions. Norms are instantiated and controlled by a central institution, which must consist of a means to detect norm violation and a means to sanction norm violators and repair the system.

How can these examples be evaluated with regard to the design decision suggested by socialisation research? The existing approaches can be regarded as a hierarchy of increasingly sophisticated accounts, ranging from mere constraints to abstract concepts. Broadly speaking, three concepts of norms can be differentiated: norms as constraints (the simplest choice), as obligations or as abstract concepts (the most sophisticated choice). This is summarised Table 17.3.

17.3.2.1 Constraints

The simplest and most straightforward way is to regard norms as mere constraints on the behaviour of individual agents. For example, the norm to drive on the right-hand side of the road restricts individual freedom. In this case, norms need not necessarily be recognised as such. They can be implemented off-line or can emerge

Table 17.3 A categorisation of approaches: constraints, obligations and abstract concepts

Constraints	Obligations	Abstract concepts
Garcia-Camino et al. (2006)	Sadri et al. (2005)	Dignum et al. (2002)
Boman (1999)	Broersen et al. (2001)	Andrighetto et al. (2007)
Shoham and Tennenholtz (1992)	Boella and van der Torre (2006)	Conte and Dignum (2001)
	Lopez and Marquez (2004)	
	Boella and van der Torre (2003)	
	Conte and Castelfranchi (1999)	
	Vazquez-Salceda et al. (2005)	
	Castelfranchi et al. (2000)	

in interaction processes. This may be sufficient for practical purposes. However, it follows that it is not possible to distinguish the norm from the normal. Hence, even though norms cannot be in contrast to individual desires in this account, the agents have no concept of obligations. They do not 'know' norms. Agents have a purely passive role. Since no decisions are possible, they remain merely normative automata.

17.3.2.2   Obligations

More sophisticated accounts treat norms as mental objects (Castelfranchi et al. 2000; Conte and Castelfranchi 1995a, b). This allows for deliberation about norms and, in particular, for the conscious violation of norms. Norms intervene in the process of goal generation, which might—or might not—lead to the revision of existing personal goals and the formation of normative goals. A number of accounts (such as the BOID architecture) rely on the notion of obligations. Obligations are explicit prescriptions that are always conditional to specific circumstances. One example of an obligation is not being permitted to smoke in restaurants. The rationale for including a separate obligation component next to a component of individual desires is geared towards ensuring an agent's *autonomy*: by explicitly separating individual and social desires, it is possible that the agent can deliberate over which component has priority. Conflicts may arise between different components. Compared to the literature on socialisation, a partial convergence with older theories can be observed. In particular, it is striking that Freud's architecture of the human psyche has some parallels to BOID agents: the Id, guided by egoistic drives taking into account only what it wants, can be found in the 'desires' component. Moreover, there is an obvious temptation to identify Freud's superego with the 'obligations' component. In fact, 'obligations' have been explicitly described as the desires of a society (Dignum et al. 2002). This conception is well supported by Freud's theory. With regard to current identity theories and the theory of self-determination, the situation is different: these theories emphasise that a full internalisation of norms is only realised when they have become part of one's identity. Thus, internalised norms form part of the person's own goals. According to identity theories, agents of this kind of architecture have not yet fully internalised norms. Norms, implemented in an 'obligations' component, do not represent complete external regulation, but by the same token, they are not part of the agent's own desires. In fact, the dichotomy between obligations and desires becomes only effective once conflicts between both components arise. This is explicitly wanted: the 'obligations' component is added to the architecture to enable norm compliance as well as violation. It is claimed that this process preserves the agent's autonomy. Hence, the dichotomy of obligations and desires refers to an *antagonism* between an individual and society. To represent the process of norm internalisation as described by modern theories, a dynamic relation between the components 'obligations' and 'desires' would be required: contingent on the salience of a norm, elements of the 'obligations' component should be imported to the 'desires' component.

17.3.2.3   Abstract Concepts

Agents may face several obligations that may contradict one another. For this reason, some authors differentiate between norms and obligations. Norms are regarded as more stable and abstract concepts than mere obligations (Dignum et al. 2002; Conte and Dignum 2001). One example of such an abstract norm is 'being altruistic': further inference processes are needed for the formation of concrete action goals from this abstract norm. The striking feature of this approach is to allow for normative reasoning. This calls for an *active* role of the agent. This conception of norms is a *precondition* for a modelling approach of social behaviour regulation based on identity conceptions.

In particular, the cognitive capacity of role taking constitutes a crucially important step in the development of goals from abstract concepts: that is, the ability to regard oneself from another's perspective. Interestingly, steps in this direction *can* be found in the AI literature. In Boella and van der Torre's architecture of a 'norm-governed system' (Boella and van der Torre 2003), the agent's decision-making process is governed by the belief that they are observed by other agents and by the belief that the other agents have expectancies with regard to how they ought to behave. This can be regarded as a first step in simulating identity theory. However, from the perspective of socio-psychological identity theories, it is a shortcoming of this architecture that the agents regard themselves only in terms of the question concerning whether they fulfil their—externally given—social role. Identity consists of an inner and outer perspective. The inner perspective is dependent on one's personal decisions. This is not the case in this architecture, which consists solely of an outer perspective. It can be questioned if and how an inner perspective can be modelled: among other things, the development of an inner perspective is correlated to a *social identity*. This social identity, however, is correlated to peer groups and reference groups. Hence, it refers to social networks, which can be simulated. In principle, a propensity to take over group norms could be simulated, dependent on the salience of group membership. To model such cognitive development, the agents thus need to be embedded in micro-social structures.

In conclusion, the concepts of norms as obligations and as an abstract concept are more closely related to concepts in empirical sciences than a mere constraint, which might be perfectly sufficient for practical purposes. It has to be noted, however, that they refer to *different* theories: the obligation concept of norms presumes an antagonism between the individual and the society, which is in line with Durkheim and Freud. The idea of norms as an abstract concept demands for a more active role of the agents. This is a precondition for modelling identity. There exist very first attempts that can be regarded as a modelling approach towards identity formation. Yet, it has to be emphasised that these are very first steps, and much is still not realised, such as to implement a correlation between network structures and salience of normative orientation. However, one principle deficiency of current models and architectures in attempting to represent the process of norm internalisation remains; namely, that agents do not have a childhood (Guerin 2008). However, socialisation theory describes childhood as the most important site for the internalisation of norms. Since agents have no childhood, the process of human *cognitive development* cannot be represented.

17.4 Conclusion

In conclusion, it can be retained that the interaction processes, resulting in macro-structural constraints, are quite well understood. In particular, the perspective to regard norms as an aggregated product of individual interactions is considerably elaborated. This is the view of sociological rational choice theories. In particular, the game theoretic paradigm has proved to be an effective means to study the dynamics of collective behaviour regularities. However, it lacks of an *active* element of normative orientation in the choice of the ends of action. The agents do not 'know' norms. Thus, these models do not capture the process of norm internalisation. Behaviour is merely guided by *adaptation* of agents to changing environmental conditions.

The role theoretic tradition emphasises that norms are structural constraints of individual behaviour. While models of cognitive agents in the AI tradition also have reached a substantial insights into norm dynamics, this aspect has been particularly studied these models. They have provided considerable insights into the effects of such structural constraints on a social macro-level. Hence, the inter-agent processes of interaction, leading to a macro-property of some kind of normatively structured social macro-level, are relatively good understood. There is, however, still a lot to do with regard to achieving a comprehensive understanding of how actors produce and are at the same time a product of social reality. While agent-based modelling *has reached* a substantial understanding of inter-agent processes, an investigation of the *recursive* impact on intra-agent processes is still in its fledgling stages.

This becomes apparent when considering socialisation theories of norm internalisation. An investigation of the effects of social behaviour regulation on individual agents is mostly at the level of conceptual considerations in the development of architectures. Here, the development of new models has to be aware of the decisions to be made. The philosophical orientation that implicitly underlies the BOID architecture is inspired by the classical accounts of Durkheim and Freud: by opposing obligations and desires, an antagonism between individuals and society is assumed. However, it has to be (and—implicitly—is) decided whether an antagonism individual and society is assumed or not. This is the question, whether the social macro-level is perceived as action constraint or as enabling action selection. Empirical research suggests that a social embedding in networks of peer and reference groups has a substantial impact on normative reasoning and thereby on the process of action selection (i.e. the agents' *desires*).

A comprehension of the two-way dynamics of the operations and effects of social behaviour regulation on a psychological as well as on the social level calls for interdisciplinary research. Agent-based modelling *is* an appropriate methodological tool for this effort. However, it has to be emphasised that developmental processes in the socialisation process are only barely captured by current simulations.

Further Reading

Even though they are quite old and some of their findings are out of date by now, it is still a good start (and not too much effort) to study the following two models to become familiar with the research field of normative agent-based models: Axelrod's (1986) evolutionary approach to norms and Conte and Castelfranchi's (1995a, b) paper on understanding the functions of norms in social groups (using simulation).

As an introduction into the design and logical foundations of normative architectures, the following anthologies are suggested: Boella et al. (2005) and Boella et al. (2007).

The relation of modelling and theory is particularly highlighted in the two anthologies (Conte and Dellarocas 2001; Lindemann et al. 2004). Here the reader will also find hints for further readings about the empirical and theoretical background.

For an overview of the theoretical background and developments in theorising norms, it is suggested to refer to Conte and Castelfranchi (1995a, b) and Therborn (2002).

References

Andrighetto, G., Campennì, M., Conte, R., & Paolucci, M. (2007). On the immergence of norms: A normative agent architecture. In G. P. Trajkovski & S. G. Collins (Eds.), *Emergent agents and socialities: Social and organizational aspects of intelligence. Proceedings of the 2007 Fall AAAI Symposium* (pp. 11–18). Menlo Park, CA: AAAI Press.

Axelrod, R. (1986). An evolutionary approach to norms. *American Political Science Review, 80,* 1095–1111.

Bandura, A. (1962). Social learning through imitation. In M. Jones (Ed.), *Nebraska symposium on motivation.* University of Nebraska Press: Lincoln.

Bandura, A. (1969). *Principles of behaviour modification.* New York: Holt.

Bicchieri, C., Duffy, J., & Tolle, G. (2003). Trust among strangers. *Philosophy of Science, 71,* 286–319.

Binmore, K. (1998). Review of the book "The complexity of cooperation: agent-based models of Competition and Collaboration" by R. Axelrod, Princeton, Princeton University Press. *Journal of Artificial Societies and Social Simulation, 1*(1). http://jasss.soc.surrey.ac.uk/1/1/review1.html.

Boella, G., & van der Torre, L. (2003). Norm governed multiagent systems: The delegation of control to autonomous agents. In *Proceedings of the IEEE/WIC International Conference on Intelligent Agent Technology 2003 (IAT 2003)* (pp. 329–335). IEEE Press.

Boella, G., & van der Torre, L. (2006). An architecture of a normative system: counts-as conditionals, obligations, and permissions. In *AAMAS '06: Proceedings of the Fifth International Joint Conference on Autonomous Agents and Multiagent Systems* (pp. 229–231). New York: ACM Press.

Boella, G., van der Torre, L., & Verhagen, H. (Eds.) (2005, April 12–15). *Proceedings of the Symposium on Normative Multi-Agent Systems (NORMAS 2005), AISB'05 Convention: Social Intelligence and Interaction in Animals, Robots and Agents,* University of Hertfordshire, Hatfield, UK. Hatfield: Society for the Study of Artificial Intelligence and the Simulation of Behaviour (SSAISB).

Boella, G., van der Torre, L., & Verhagen, H. (2007). Introduction to Normative Multiagent Systems. In G. Boella, L. van der Torre, & H. Verhagen (Eds.), *Normative multi-agent systems, Dagstuhl seminar proceedings 07122* (Vol. I, pp. 19–25). IBFI, Schloss Dagstuhl: Wadern.

Boman, M. (1999). Norms in artificial decision making. *Artificial Intelligence and Law, 7,* 17–35.

Bosma, H., & Kunner, E. (2001). Determinants and mechanisms in ego development: A review and synthesis. *Developmental Review, 21*(1), 39–66.

Boudon, R. (1981). *The logic of social action.* London: Routledge.

Bratman, M. (1987). *Intentions, plans and practical reasoning.* Stanford: CSLI.

Broersen, J., Dastani, M., Huang, Z., & van der Torre, L. (2001). The BOID architecture: Conflicts between beliefs, obligations, intentions, and desires. In E. André, S. Sen, C. Frasson, & J. P. Müller (Eds.), *Proceedings of the 5th International Conference on Autonomous Agents* (pp. 9–16). New York: ACM Press.

Burke, M., Fournier, G., & Prasad, K. (2006). The emergence of local norms in networks. *Complexity, 11,* 65–83.

Castelfranchi, C., Conte, R., & Paolucci, M. (1998). Normative reputation and the costs of compliance. *Journal of Artificial Societies and Social Simulation, 1*(3). http://www.soc.surrey.ac.uk/JASSS/1/3/3.html

Castelfranchi, C., Dignum, F., & Treur, J. (2000) Deliberative normative agents: Principles and architecture. In: N. R. Jennings & Y. Lesperance (Eds.), *Intelligent Agents VI, Agent Theories, Architectures, and Languages (ATAL), 6th International Workshop, ATAL '99, Orlando, Florida, USA, July 15–17, 1999, Proceedings (LNCS, 1757)* (pp. 364–378). Berlin: Springer.

Claessens, D. (1972). *Familie und wertsystem: Eine studie zur zweiten sozio-kulturellen geburt des menschen.* Berlin: Duncker & Humblot.

Coleman, J. (1987). The emergence of norms in varying social structures. *Angewandte Sozialforschung, 14,* 17–30.

Coleman, J. (1990). *Foundations of social theory.* Cambridge: Belknap Press of Harvard University Press.

Conte, R., & Castelfranchi, C. (1995a). *Cognitive and social action.* London: UCL Press.

Conte, R., & Castelfranchi, C. (1995b). Understanding the functions of norms in social groups through simulation. In N. Gilbert & R. Conte (Eds.), *Artificial societies: The computer simulation of social life* (pp. 252–267). London: UCL Press.

Conte, R., & Castelfranchi, C. (1999). From conventions to prescriptions: Towards an integrated view of norms. *Artificial Intelligence and Law, 7,* 323–340.

Conte, R., & Castelfranchi, C. (2001). Are incentives good enough to achieve (info) social order? In R. Conte & C. Dellarocas (Eds.), *Social order in multiagent systems.* Dordrecht: Kluwer.

Conte, R., & Dellarocas, C. (Eds.). (2001). *Social order in multiagent systems.* Dordrecht: Kluwer.

Conte, R., & Dignum, F. (2001). From social monitoring to normative influence. *Journal of Artificial Societies and Social Simulation, 4*(2). http://www.soc.surrey.ac.uk/JASSS/4/2/7.html

Dahrendorf, R. (1956). *Homo sociologicus. Ein versuch zu geschichte, bedeutung und kritik der kategorie der sozialen rolle.* Opladen: Westdeutscher Verlag.

Davies, K., & Moore, W. (1945). Some principles of stratification. *American Sociological Review, 10,* 242–249.

Deci, E., & Ryan, R. (2000). The "what" and "why" of goal pursuits: Human needs and the self-determination of behavior. *Psychological Inquiry, 11*(4), 227–268.

Deffuant, G., Moss, S., & Jager, W. (2006). Dialogues concerning a (possibly) new science. *Journal of Artificial Societies and Social Simulation, 9*(1). http://jasss.soc.surrey.ac.uk/9/1/1.html

Dignum, F., Kinny, D., & Sonenberg, L. (2002). From desires, obligations and norms to goals. *Cognitive Science Quarterly, 2,* 407. http://www.cs.uu.nl/people/dignum/papers/CSQ.pdf

Durkheim, E. ([1895] 1970). *Regeln der soziologischen methode.* Neuwied: Luchterhand.

Durkheim, E. ([1897] 2006). *Der selbstmord.* Frankfurt am Main.: Suhrkamp.

Durkheim, E. ([1907] 1972). *Erziehung und soziologie.* Düsseldorf: Schwann.

Epstein, J. (2000). *Learning to be thoughtless: Social norms and individual computation* (Working Paper No. 6). Center on Social and Economic Dynamics. http://www.santafe.edu/media/workingpapers/00-03-022.pdf

Esser, H. (1993). *Soziologie: Allgemeine grundlagen.* Frankfurt am Main: Campus.

Fioretti, G. (2017). Utility, games, and narratives. doi:https://doi.org/10.1007/978-3-319-66948-9_16.

Flentge, F., Polani, D., & Uthmann, T. (2001). Modelling the emergence of possession norms using memes. *Journal of Artificial Societies and Social Simulation, 4*(4). http://jasss.soc.surrey.ac.uk/4/4/3.html

Freud, S. (1932). Neue vorlesungen zur einführung in die psychoanalyse. In S. Freud (Ed.), *Gesammelte Werke* (Vol. 15). London: Imago.

Freud, S. (1955). *Abriss der psychoanalyse.* Frankfurt am Main: Fischer.

Fuhrer, U., & Trautner, N. (2005). Entwicklung und identität. In J. Asendorpf (Ed.), *Enzyklopädie der pschologie—Entwicklungspsychologie* (Vol. 3). Göttingen: Hofgrebe.

Garcia-Camino, A., Rodriguez-Aguilar, J. A., Sierra, C., & Vasconcelos, W. (2006, May 08–12). Norm-oriented programming of electronic institutions: A rule-based approach. In H. Nakashima, M. Wellman, G. Weiss, & P. Stone (Eds.), *Proceedings of the 5th International Joint Conference on Autonomous Agents and Multi-agent Systems 2006,* Hakodate, Japan (pp. 670–672). New York: ACM Press.

Geulen, D. (1991). Die historische entwicklung sozialisationstheoretischer ansätze. In K. Hurrelmann & D. Ulich (Eds.), *Neues handbuch der sozialisationsforschung.* Beltz: Weinheim.

Giardini, F., Conte, R., & Paolucci, M. (2013). In B. Edmonds & R. Meyer (Eds.), *Simulating social complexity: A handbook.* Berlin: Springer-Verlag.

Gibbs, J. P. (1981). *Norms, deviance and social control: Conceptual matters.* New York: Elsevier.

Guerin, F. (2008, April 1–4). Constructivism in AI: Prospects, progress and challenges. In *Proceedings of the AISB Convention 2008, Volume 12: Computing and Philosophy,* Aberdeen, Scotland (pp. 20–27). London: The Society for the Study of Artificial Intelligence and Simulation of Behaviour.

Hales, D. (2002). Group reputation supports beneficent norms. *Journal of Artificial Societies and Social Simulation, 5*(4). http://jasss.soc.surrey.ac.uk/5/4/4.html

Homans, G. (1964). Bringing man back in. *American Sociological Review, 29,* 809–818.

James, W. (1890). *The principles of psychology*. New York: Holt.

Keupp, H. (1999). *Identitätskontruktionen*. Hamburg: Rowohlt.

Kohlberg, L. (1996). *Die psychologie der moralentwicklung*. Frankfurt am Main: Suhrkamp.

Krappmann, L. (2006). Sozialisationsforschung im spannungsfeld zwischen gesellschaftlicher reproduktion und entstehender handlungsfähigkeit. In W. Schneider & F. Wilkening (Eds.), *Enzyklopädie der psychologie—Entwicklungspsychologie* (Vol. 1). Göttingen: Hofgrebe.

Lindemann, G., Moldt, D., & Paolucci, M. (Eds.). (2004). *Regulated agent-based social systems: First international workshop, RASTA 2002, Lecture Notes in Artificial Intelligence* (Vol. 2934). Berlin: Springer.

Lopez, F., & Marquez, A. (2004). An architecture for autonomous normative agents. In *Proceedings of the 5th Mexican International Conference in Computer Science, ENC'04* (pp. 96–103). Los Alamitos: IEEE Computer Society.

Macy, M., & Sato, Y. (2002). Trust, cooperation, and market formation in the U.S. and Japan. *Proceedings of the National Academy of Sciences, 99*, 7214–7220.

Macy, M., & Willer, R. (2002). From factors to actors: Computational sociology and agent-based modelling. *American Review of Sociology, 28*, 143–166.

Malinowski, B. (1927). *Sex and repression in primitive society*. New York: Humanities Press.

Mead, M. (1928). *Coming of age in Samoa*. New York: William Morrow.

Mead, G. H. ([1934] 1968). *Geist, identität und gesellschaft*. Frankfurt am Main.: Suhrkamp.

Merton, R. (1957). *Social theory and social structure* (2nd ed.). Glencoe: The Free Press.

Moss, S. (2001). Game theory: Limitations and an alternative. *Journal of Artificial Societies and Social Simulation, 4*(2). http://www.soc.surrey.ac.uk/JASSS/4/2/2.html

Neumann, M. (2008a, July 14–17). A classification of normative architectures. In G. Deffuant, & C. Cioffi-Revilla (Eds.), *Proceedings of the Second World Congress of Social Simulation (WCSS'08)*, Fairfax, VA.

Neumann, M. (2008b). Homo socionicus: A case study of normative agents. *Journal of Artificial Societies and Social Simulation, 11*(4). http://jasss.soc.surrey.ac.uk/11/4/6.html

Parsons, T. ([1937] 1968). *The structure of social action: A study in social theory with special reference to a group of recent European writers*. New York: Free Press.

Pavlov, I. P. (1927). *Conditioned reflexes: An investigation of the physiological activity of the cerebral cortex*. London: Oxford University Press.

Piaget, J. ([1932] 1983). *Das moralische urteil beim kinde*. Stuttgart: Klett-Cotta.

Piaget, J. ([1947] 1955). *Psychologie der intelligenz*. Zürich: Rascher.

Rao, A., & M. Georgeff (1991, April 22–25). Modeling rational agents within a BDI architecture. In J. F. Allen, R. Fikes, & E. Sandewall (Eds.), *Proceedings of the 2nd International Conference on Principles of Knowledge Representation and Reasoning (KR'91)*, Cambridge, MA (pp. 473–484). San Francisco: Morgan Kaufmann.

Raub, W., & Voss, T. (1981). *Individuelles handeln und gesellschaftliche folgen: Das individualistische programm in den sozialwissenschaften*. Darmstadt: Luchterhand.

Saam, N., & Harrer, A. (1999). Simulating norms, social inequality, and functional change in artificial societies. *Journal of Artificial Societies and Social Simulation, 2*(1). http://jasss.soc.surrey.ac.uk/2/1/2.html

Sadri, F., Stathis, K., & Toni, F. (2006). Normative KGP agents. *Computational and Mathematical Organization Theory, 12*, 101–126.

Savarimuthu, B., Purvis, M., Cranefield, S. & Purvis, M. A. (2007). *How do norms emerge in multi-agent societies? Mechanism design* (The Information Science Discussion Paper Series, 2007/01). Dunedin: University of Otago.

Sen, S., & Airiau, S. (2007, January 6–12). Emergence of norms through social learning. In M. Veloso (Ed.), *Proceedings of the Twentieth Joint Conference on Artificial Intelligence (IJCAI-07)*, Hyderabad, India (pp. 1507–1512).

Shoham, Y., & Tennenholtz, M. (1992, July 12–16). On the synthesis of useful social laws for artificial agent societies (preliminary report). In W. R. Swartout (Ed.), *Proceedings of the 10th National Conference on Artificial Intelligence (AAAI'92)*, San Jose, CA (pp. 276–281). Cambridge: AAAI Press/The MIT Press.

Skyrms, B. (1996). *Evolution of the social contract*. Cambridge: Cambridge University Press.

Skyrms, B. (2004). *The stage hunt and the evolution of social structure*. Cambridge: Cambridge University Press.

Tajfel, H. (1970). Experiments in intergroup discrimination. *Scientific American, 223*(5), 96–102.

Therborn, G. (2002). Back to norms! On the scope and dynamics of normative action. *Current Sociology, 50*(6), 863–880.

Turner, J. C. (1982). Towards a cognitive redefinition of the social group. In H. Tajfel (Ed.), *Social identity and intergroup relations*. Cambridge: Cambridge University Press.

Turner, J. C., & Onorato, R. S. (1999). Social identity, personality, and the self-concept: A self categorising perspective. In T. Tylor et al. (Eds.), *The psychology of the social group*. Mahwah, NJ: Erlbaum.

Vazquez-Salceda, J., Aldewereld, H., & Dignum, F. (2005). Norms in multiagent systems: From theory to practice. *International Journal of Computer Systems and Engineering, 20*, 225–236.

Verhagen, H. (2001). Simulation of the learning of norms. *Social Science Computer Review, 19*, 296–306.

Vieth, M. (2003). Die evolution von fairnessnormen im ultimatumspiel: Eine spieltheoretische modellierung. *Zeitschrift für Soziologie, 32*, 346–367.

Wrong, D. (1961). The oversocialised conception of man. *American Sociological Review, 26*, 183–193.

Chapter 18
Reputation for Complex Societies

Francesca Giardini, Rosaria Conte, and Mario Paolucci

Abstract Reputation, the germ of gossip, is addressed in this chapter as a distributed instrument for social order. In literature, reputation is shown to promote (a) social control in cooperative contexts—like social groups and subgroups—and (b) partner selection in competitive ones, like (e-) markets and industrial districts. Current technology that affects, employs and extends reputation, applied to electronic markets or multi-agent systems, is discussed in light of its theoretical background. In order to compare reputation systems with their original analogue, a social cognitive model of reputation is presented. The application of the model to the theoretical study of norm-abiding behaviour and partner selection are discussed, as well as the refinement and improvement of current reputation technology. The chapter concludes with remarks and ideas for future research.

18.1   Reputation in Social Systems: A General Introduction

Ever since hominid settlements started to grow, human societies have needed to cope with problems of social order. How to avoid fraud and cheating in wider, unfamiliar groups? How to choose trustworthy partners when the likelihood of re-encounter is low? How to isolate cheaters and establish worthwhile alliances with the "good guys"?

Social knowledge like reputation and its transmission (i.e. gossip) play a fundamental role in creating and maintaining social order, adding at the same time cohesiveness to social groups and allowing for distributed social control and sanctioning (plus a number of other functionalities; see Boehm 1999). Reputation is a property that even unwilling and unaware individuals derive from the generation, transmission and manipulation of a special type of social belief, which

F. Giardini (✉)
Faculty of Behavioral and Social Sciences, Department of Sociology,
University of Groningen, Groningen, The Netherlands (NL)
e-mail: f.giardini@rug.nl

R. Conte • M. Paolucci
Laboratory for Agent-Based Social Simulation (LABSS), Institute of Cognitive Sciences
and Technologies, National Research Council, Rome, Italy

© Springer International Publishing AG 2017
B. Edmonds, R. Meyer (eds.), *Simulating Social Complexity*,
Understanding Complex Systems, https://doi.org/10.1007/978-3-319-66948-9_18

has contributed to the regulation of natural societies from the dawn of mankind (Dunbar 1996). People use reputational information for many things, including: to make decisions about possible interactions, to evaluate candidate partners and to understand and predict their behaviours (Alexander 1987).

It has long been known that reputation is a fundamental generator, vehicle and manipulator of social knowledge for enforcing reciprocity and other social norms (Conte and Paolucci 2002). In particular, in the study of cooperation and social dilemmas, the role of reputation as a partner selection mechanism started to be appreciated since the early 1980s (Kreps and Wilson 1982). However, at that stage, there was little understanding of its dynamic and cognitive underpinnings. Despite its critical role in the enforcement of altruism, cooperation and social exchange, the socio-cognitive study of reputation is relatively new. Hence, how this critical type of knowledge is manipulated in the minds of agents, how social structures and infrastructures generate, transmit and transform it, has not yet been fully clarified. Consequently, the full picture of how it affects agents' behaviour is also unclear. Partly, this is because reputation extends beyond the boundaries of academic disciplines, emerging as a prototypical cross-disciplinary topic (Paolucci and Sichman 2014).

The aim of this chapter is to guide the reader through the multiplicity of computational approaches concerned with the reputation mechanism and its dynamics. Reputation is a complex social phenomenon that cannot be treated as a static attribute of agenthood, with no regard for the underlying process of transmission. We claim that reputation is both the process and the effect of transmitting information and that further specifications about the process and its mechanisms are needed. We will follow this with three different applications of the cognitive theory of reputation to model social phenomena: the Sim-Norm model, the Socrate framework and the Repage architecture.

This introduction will be followed. In order to lay the ground for understanding the multiplicity of reputation, we will present by an outline of reputation research in some different domains, namely, social psychology, management and experimental economics and agent-based simulation. This will show the variety of viewpoints that can be used to describe and explore this complex phenomenon. We will then focus on some of the work in electronic markets and multi-agent simulations that include reputation mechanisms. Electronic markets are a typical example of a complex environment where centralized control is not possible and decentralized solutions are far from being effective. In recent years, the Internet has contributed to a growing number of auction sites that facilitate the exchange of goods between individual consumers, without guaranteeing either transparency or the safety of the transactions. On the other hand, multi-agent applications are concerned with the problem of assessing the reliability of single agents and of social networks.

In Sect. 18.6 we propose a cognitive model of reputation, which aims to solve some of the problems left open by existing systems, starting from a theoretical analysis of cognitive underpinnings of reputation formation and spreading. This model will be tested in the following section, where a description of three different implementations is of the model and their results are then provided. We also describe a set of simulation studies on gossip, in which private transmission of unverified information is able to support cooperation in a public goods game. Moving from the observation that reputation and punishment are considered the most important mechanisms for social control, a systematic comparison of their effects will show that their combination represents a powerful way of detecting and deterring cheaters. Finally, we draw some conclusions about some future directions for research in this area.

18.2   State of the Art: An Overview on Reputation in Natural and Artificial Societies

According to Frith and Frith (2006), there are three ways to learn about other people: through direct experience, through observation and through "cultural information". When the first two modalities are not available, reputational information becomes essential in order to obtain some knowledge about potential partner(s) in an interaction and thus to form expectations about their behaviour. Reputation allows people to predict, at least partially or approximately, what kind of social interaction they can expect and how that interaction may possibly develop. Reputation is therefore a coordination device whose predictive power is essential in social interactions (Paolucci and Conte 2009).

Reputation and its transmission (gossip) have an extraordinary preventive power: it substitutes personal experience in (a) identifying cheaters and isolating them and in (b) finding trustful partners. It makes available most of the benefits of evaluating someone, without the costs of direct interaction.

Furthermore, in human societies gossip facilitates the formation of groups (Gluckman 1963): gossipers share and transmit relevant social information about members within the group (Barkow 1996) while, at the same time, isolating those in out-groups. Gossip contributes to stratification and social control, since it works as a tool for sanctioning deviant behaviours and for promoting those behaviours that are functional with respect to the group's goals and objectives (e.g. via a learning process). Reputation is also considered as a means for sustaining and promoting the diffusion of norms and norm conformity (Wilson et al. 2000). On the other hand, reputation can be used to pursuit self-interest, either by promoting one's achievements or by spreading negative information about others (Paine 1967; Noon and Delbridge 1993).

Reputation plays a key role in evolutionary theories of cooperation, supporting indirect reciprocity (Nowak and Sigmund 1998a, b, 2005). Theories of indirect reciprocity explain large-scale human cooperation in terms of conditional helping by individuals who want to uphold a reputation and then to be included in future cooperation (Panchanathan and Boyd 2004). By means of computer simulations, Nowak and Sigmund (1998a, b) showed that reputation can sustain the emergence of indirect reciprocity—getting people to cooperate (even with strangers) in order to receive cooperation, without the necessity of any kind of contract or keeping track of contributions. Theories of indirect reciprocity explain large-scale human cooperation in terms of conditional helping by individuals who want to uphold a reputation and then to be included in future cooperation (Panchanathan and Boyd 2004). In a "market for cooperators" (Noë and Hammerstein 1994), or in partner choice, building a positive reputation for generosity can be seen as a long-term investment. Here, individuals may compete for the most altruistic partners, leading non-altruists to become ostracized (Roberts 1998).

As Alexander (1987) pointed out "indirect reciprocity involves reputation and status, and results in everyone in the group continually being assessed and reassessed". In the last few years, attention to reputation has grown both within single disciplines and in interdisciplinary contexts (Milinski 2016; Wu et al. 2016). This has involved a variety of methodologies, going from online large-scale experimental studies using dynamic networks (Rand et al. 2011; Wang et al. 2012) to economic laboratory experiments, and has included important advances in the study of reputation as a means to support cooperation in a variety of contexts (Beersma and Van Kleef 2011; Piazza and Bering 2008; Sommerfeld et al. 2008).

Reputation and gossip are also crucial in other fields of the social sciences like management and organization science, governance and business ethics, where the importance of reputation in branding became apparent (Fombrun and Shanley 1990). The economic interest in the subject matter came from the fact that reputation can be applied at the super-individual level; corporate reputation is considered as an external and intangible asset tied to the history of a firm and coming from stakeholders' and consumers' perceptions (Fombrun 1996). Rose and Thomsen (2004) claim that a good reputation and a good financial performance are mutually dependent—a good reputation may influence the financial assets of a firm and vice versa. Several researchers have tried to create a corporate reputation index containing the most relevant dimensions to take into account when dealing with corporate reputation. Cravens et al. (2003) interviewed 650 CEOs in order to create a reliable index, but their index has so many entries, ranging from global strategy to employees' attributes, that it is not easy to foresee how such a tool could be used. Gray and Balmer (1998) distinguish between corporate image and corporate reputation. Corporate image is the mental picture consumers hold about a firm, and is thus similar to individual perception, whereas the reputation results more from the firm's communication and long-term strategy. Generally speaking, corporate reputation is treated as an aggregate evaluation that stakeholders, consumers, managers, employees and institutions form about a firm. However, the mechanisms leading to the final result are not well defined.

If social order is a constant of human evolution, it is particularly crucial in an e-society where the boundaries of interaction are widening. The increasingly fast development of ICT technologies dramatically enlarges the range of interaction among users, generating new types of aggregation, from civic communities to electronic markets and from professional networking to e-citizenship. What is the effect of this widening of social boundaries? Communication and interaction technologies modify the range, structures and modalities of interaction, with consequences that are only partially explored, often only to resume the stereotype of technological unfriendliness (e.g. the negative impact of computer terminals, as opposed to face-to-face interaction, on subjects' cooperativeness in experimental studies of collective and social dilemmas (Sell and Wilson 1991; Rocco and Warglien 1995). Detailed studies of the effects of technological infrastructures on interaction styles and modes are lacking. Perhaps, an exception to this is represented by the research on the effects of asymmetry of information within the markets. Asymmetry of information is known to encourage fraud and low-quality production in many situations. As exemplified by Akerlof (1970), asymmetry of information can drive honest traders and high-quality goods out of the market. The result is a market where only "lemons", or fraudulent commodities, are available—often to the detriment of both sellers and buyers. The classical example of such a market is the used car market, where only sellers have information about problems with the cars they are selling, and most consumers are incapable of discerning these problems. This phenomenon is an intrinsic feature of e-markets, but goes back to eleventh-century Maghribi traders moving along the coast of the Mediterranean Sea (Greif 1993). Contemporary online traders such as users of Internet auction sites face the same problem of mediaeval traders: online buyers can learn about the quality (or condition) of the good only once they have already paid for it.

Auction sites vary from the very generic, concerning the products being offered and operated on a global scale (e.g. eBay), to those that focus on specific products on a national scale (many car auction sites). Buying through auction sites offers less control to the buyers than even online retailers, as the sellers are not visible and have not made major investments. Consumers who purchase through auction sites must rely on the accuracy and reliability of the seller. Sellers on the Internet may actively try to communicate their reputation to potential buyers, increasing the expected impact of reputation on buying decisions. Melnik and Alm (2002) investigated whether an e-seller's reputation matters. Their results indicated that reputation had a positive—albeit relatively small—impact on the price levels consumers were willing to pay. Moreover, Yamagishi et al. (2004) show that reputation has a significant positive effect on the quality of products. In any case, the strength of reputational mechanisms does not seem to be diminished by the spread of anonymous contexts in which interactions take place at a distance and are mediated by a computer (as happens online). In this sense, the new technologies allow for more information. A greater number of people can interact due to overcoming spatial limitations. These new opportunities for large-scale interaction, as well as the chance to make opinions accessible to the community of Internet users (i.e. bidirectionality), have allowed the development of systems based on online feedback mechanisms (Dellarocas 2003). We are witnessing the proliferation of services that rely on reputation systems (e.g. eBay, Amazon, TripAdvisor), and experimental studies show that even in online anonymous contexts, where the way in which reputation is assigned is opaque, reputation is used to actively avoid defectors (Capraro et al. 2016). Technical challenges for large-scale systems can be met with the use of simple reputation systems, as in the case of collaborative filtering algorithms (Petroni et al. 2016).

Despite the role of reputation in economic transactions, online reputation systems are only moderately efficient (Bolton et al. 2004; Resnick and Zeckhauser 2001), showing a stronger effect of negative feedbacks on price reduction than a positive one on price increase (Diekmann et al. 2014).

In all of these cases, the notion of reputation is weak and essentially reduced to centralized image: no direct exchange of information takes place among participants but only reports to a central authority, which calculates the resultant reputation score. This mechanism is debatable alone and can be insufficient, but it can be complemented by detailed comments or forums. For example, when forums are available, this is the solution chosen by TripAdvisor, whose users can provide detailed comments about hotels, restaurants, tourist attractions and services. These comments are displayed along with real reputation exchanges that are performed in parallel, thus offering interested users as much information as possible. Moreover, many people do not bother to provide reputational feedback (under-provision), and if they do, they lean on providing only positive reports (overscoring).

Agent-based social simulation has taught us some lessons: (1) what matters about reputation is its transmission (Castelfranchi et al. 1998), since by this means agents acquire-cost; (2) reputation has more impact than directly acquired information. In a simulation study, Pinyol et al. (2008) showed that if agents transmitted only their own evaluations about one another (image), the circulation of social knowledge ceases quickly. To exchange information about reputation, agents need to participate in circulating reputation whether they believe it or not (gossip), and, to preserve their autonomy, they must decide how, when and about whom to gossip. In a simulation study, Pinyol et al. (2008) showed that if agents transmitted only their own evaluations about one another (image), the circulation of social knowledge ceases quickly. To exchange information about reputation, agents need to participate in circulating reputation whether they believe it or not (gossip) and, to preserve their autonomy, they must decide how, when and about whom to gossip. What is missing in the study of reputation is the merging of these separate directions in an interdisciplinary integrated approach, which accounts for both its social cognitive mechanisms and structures.

18.3 Simulating Reputation: Current Systems

So far, the simulation-based study of reputation has been undertaken for the sake of social theory, namely, in the account of prosocial behaviour—be it cooperative, altruistic or norm abiding—among autonomous, i.e. self-interested agents. Thanks to computational methods, social simulation has contributed to our understanding of reputation as a means to promote norm-abiding behaviour in social groups and as a tool for improving partner selection in electronic markets and computational settings.

Several attempts have been made to model and use reputation in artificial societies, especially in two subfields of information technology: computerized interaction (with a special reference to electronic marketplaces) and agent-mediated interaction. It is worth emphasizing that in these domains trust and reputation are actually treated as the same phenomenon, and often the fundamentals of reputation mechanisms are derived from trust algorithms (Moukas et al. 1999; Zacharia 1999; Zacharia et al. 1999). We will review some of the main contributions in online reputation reporting systems and in multi-agent systems, in order to achieve a better understanding of the complex issue of implementing and effectively using reputation in artificial societies.

18.3.1 Online Reputation Reporting Systems

The continuously growing volume of transactions on the World Wide Web and the growing number of frauds that appears to entail[1] have led scholars from different disciplines to develop new online reputation reporting systems. These systems are intended to provide a reliable way to deal with reputation scores or feedbacks, allowing agents to find cooperative partners and avoid cheaters.

The existing systems can be roughly divided into two subsets, agent-oriented individual approaches and agent-oriented social approaches, depending on how agents acquire reputational information about other agents.

The *agent-oriented individual approach* has been dominated by Marsh's ideas on trust (Marsh 1992, 1994a, b), on which many further developments and algorithms are based. This kind of approach is characterized by two attributes: (1) any one agent may seek potential cooperation partners, and (2) the agent only relies on its experiences from earlier transactions. When a potential partner proposes a transaction, the recipient calculates the "situational reputation" by weighing the reputation of his potential trading partner against other factors, such as potential output and the importance of the transaction. If the resulting value is higher than a certain "cooperation threshold", the transaction takes place and the agent updates the reputation value according to the outcomes of the transaction. If the threshold is not reached, the agent rejects the transaction offer, an action that may be punished by a "reputation decline". These individual-based models (Bachmann 1998; Marsh 1994a; Ripperger 1998) differ with regard to their memory span. Agents may forget their experiences slowly, fast or never, and this has important consequences for the dynamics of the overall level of trust in the system.

---

[1] According to the US-based Internet Crime Complaint Center (IC3), losses as a result of auto-auction fraud exceeded $8.2 million dollars in 2011.

In agent-oriented social approaches, agents not only rely on their direct experience but are also allowed to consider third-party information (Abdul-Rahman and Hailes 1997; Rasmusson 1996; Rasmusson and Janson 1996; Yu and Singh 2000). Although these approaches share the same basic idea—i.e. experiences of other agents in the network can be used when searching for the right transaction partner—they use upon different methods to weigh the third-party information and to deal with "friends of friends". Thus the question arises as to how to react to information from agents who do not seem to be very trustworthy. A similar problem arises with the storage and distribution of information. To form a complete picture of its potential trading partners, each agent needs both direct (its own) and indirect (third-party) evaluations in order to be able to estimate the validity and the informational content of such a picture.

Regan and Cohen (2005) propose a system for computing indirect and direct reputation within a computer-mediated market. Buyers rely on reputation information about sellers when choosing from whom to buy a product. If they do not have direct experience from previous transactions with a particular seller, they take indirect reputation into account by asking other buyers for their evaluations of the potential sellers. The received information is then combined to mitigate effects of deception. The objective of this system is to propose a mechanism which reduces reputation in the face of undesirable practices in online applications, especially on the part of sellers, and to prevent the market from turning into a "lemons market" where only low-quality goods are listed for sale.

One serious problem with the model by Regan and Cohen and similar other models concerns the transmission of reputation. In these kinds of models, agents only react to reputation requests, while proactive, spontaneous delivery of reputation information to selected recipients is not considered. However, this simple solution is quite effective. On the other hand, despite its simplicity, these types of model tackle the problem of collusion between rating agents, because by keeping the evaluation of sellers remains among buyers (i.e. not disclosing it to the sellers). Therefore sellers cannot influence their own scores.

Turning to electronic marketplaces, classic systems like eBay show a characteristic bias towards positive evaluations (Resnick and Zeckhauser 2002). This suggests that factual cooperation among users at the information level may lead to a "courtesy" equilibrium (Conte and Paolucci 2003). As Cabral and Hortaçsu (2010) formally prove, negative feedbacks trigger a decline in sale price that drives the targeted sellers out of the market. Good sellers, however, can gain from "buying a reputation" by building up a record of favourable feedback through purchases rather than sales. Thus those who suffer a bad reputation stay out—at least until they decide to change identity—while those who stay in can but enjoy a good reputation: after a good start, they will hardly receive negative feedback and even if they do, it will not get to the point of spoiling their good name. Under such conditions, even good sellers may have an incentive to sell lemons, considering that it takes time for their reputation scores to go down.

Intuitively, the courtesy equilibrium reduces the deterrent effect of reputation. If a reputation system is meant to reduce frauds and improve the quality of products, it needs to be constructed in such a way as to avoid the emergence of a courtesy equilibrium. It is not by chance that among the possible remedies to ameliorate eBay, Dellarocas (2003) suggested a short-memory system, erasing all feedbacks but the very last one.

18.3.2 MAS Applications

Models of trust and reputation for multi-agent systems applications (e.g. Yu and Singh 2000; Carbo et al. 2002; Sabater and Sierra 2002; Schillo et al. 2000; Huynh et al. 2004; for exhaustive reviews see Ramchurn et al. 2004; Sabater and Sierra 2004; Pinyol and Sabater-Mir 2013) present interesting ideas and advances over conventional online reputation systems, with their notion of a distributed reputation.

Yu and Singh (2000) proposed an agent-oriented model for social reputation and trust management, which focuses on electronic societies and MAS. Their model introduces a gossip mechanism for informing neighbours of defective transaction partners, in which the gossip is transferred link-by-link through the network of agents. It also has a mechanism to allow agents to include other agents' testimonies in its reputation calculations. Agents store information about the outcome of every transaction they ever had and recall this information in case they are planning to bargain with the same agent again (direct evaluation). If the agent meets an agent it has not traded with before, the reputation mechanism comes into play. In this mechanism, so-called referral chains are generated that can make third-party information available across several intermediate stations. An agent is thus able to gain reputation information with the help of other agents in the network. Since a referral chain represents only a small part of the whole network, the information delivered will most likely be a partial view instead of global score as in centralized systems like eBay.

In the context of several extensive experiments, Yu and Singh showed that the implementation of their mechanism results in a stable system, in which the reputation of cheaters decreases rapidly while cooperating agents experienced a slow, almost linear increase in reputation. However, some problems remain. The model does not allow agents to combine their own experience with the network information. Thus, it might take unnecessarily long to react to a suddenly defecting agent that cooperated before. In addition, Yu and Singh do not give an explanation of how their agent-centred storage of social knowledge (e.g. the referral chains) is supposed to be organized. Consequently, no analysis of network load and storage intensity can be done.

ReGreT (Sabater 2004) is another MAS application in which the link between trust and reputation is very strong. In this, reputation is only one of the dimensions an agent resort to in order to evaluate the trustworthiness of another agent. In ReGreT, reputational information and direct experience have different values, and the former is considered less reliable than the latter.

A system called Liar Identification for Agent Reputation (LIAR) has been proposed by Muller and Vercouter (2008), based on three levels of reputation: direct, indirect and recommendation based. To implement those elements, LIAR explicitly models a social commitment mechanism, social norms and the operations over them.

SOARI (Service Oriented Architecture for Reputation Interaction) is a reputation ontology that has been proposed by Nardin et al. (2008). SOARI is a service-oriented architecture that provides support to the semantic interoperability among agents that implement heterogeneous reputation models. The main contribute of SOARI is to provide a mapping among different reputation models, represented by a common reputation ontology especially designed for agents' interaction, in the form of a service that can be executed externally to agents and is available online as an on-demand service for agents.

As these example shows, the "agentized environment" produces interesting solutions that may apply also to online communities. This is for two main reasons. *Firstly*, in this environment two problems of order arise: meeting users' expectations (external efficiency) and promoting agents' performance (internal efficiency). Internal efficiency is instrumental to the external one, but it re-proposes the problem of social control at the level of the agent. In order to promote the former, agents must be in an environment where they evaluate and act upon each other's behaviours. *Secondly*, agent systems can be used to help determine (a) what type of agents, (b) what type of beliefs and (c) what type of processes among agents are required to achieve useful social control. More specifically, they can be used to map out what type of agent and processes are needed for *which* desirable result, including better efficiency, encouraging equity (and hence users' trust), discouraging discrimination and fostering collaboration at the information level or object level (or both).

However, in models of Internet systems, the notion of reputation is weak and essentially reduced to centralized image: participants do not exchange information directly but only report their evaluations to a central authority, which calculates their global reputation value. The solutions proposed for MAS systems are interesting, but these are insufficient to meet the problems left open by online contexts. There is a tendency to consider reputation as an external attribute of agents without taking into account the processes of creation and transmission of that reputation. Is there an alternative? How can we understand the effects of reputation on transactions if we do not model the process of reputation creation and transmission?

18.4   An Alternative Approach: The Social Cognitive Process of Reputation

Current models operate with a highly simplified model of reputation, in which different experiences and items of information are reduced to a single accumulator. In this section, we will model reputation as a social cognitive process and briefly discuss advantages and disadvantages of this approach.

A social cognitive process involves symbolic mental representations (e.g. social beliefs and goals[2]) that are manipulated by individuals and agents in the process of social reasoning by means of the operations that agents perform upon them (social reasoning).

Social cognitive processes are aimed at modelling (and possibly implementing) systems acting in a social—be it natural or artificial—environment. These processes employ explicit representations of a variety of mental states (including social goals, motivations, obligations) and operations (such as social reasoning and decision-making) necessary for an intelligent social system to act in some domain and influence other agents thus triggering the processes of (social learning, influence, and control). To represent reputation as a social cognitive process, two different constructs are needed, namely, image and reputation. After giving the definition of those constructs, we will show how agents can behave when evaluating someone and transmitting these evaluations. Thus playing one of three different roles: evaluator, beneficiary and target.

18.4.1 Image and Reputation

An *image* consists of a set of evaluative beliefs (Miceli and Castelfranchi 2000) held by an agent (the "evaluator") concerning the characteristics of another agent (the "target"). It is an assessment of its positive or negative qualities with regard to a norm or competence. The *image* relevant for social reputation may concern a subset of the target's characteristics, e.g. its willingness to comply with socially accepted norms and customs.

An agent's *reputation* we argue is distinct from, although closely related to, its image. More precisely, we define *reputation* as consisting of three distinct but interrelated objects: (1) a cognitive representation, i.e. a believed evaluation of another agent; (2) a population object, i.e. an evaluation that is propagated to others; and (3) an objective emergent property at the agent level, i.e. what the agent is believed to be. As an illustration, when we say that "John has a very good reputation as a dentist", we are implicitly assuming that (1) someone believes that he is good at his job, (2) an indefinite number of people share that belief, and (3) he actually possesses some skills; therefore his reputation is grounded in some objective properties.

Reputation is a highly dynamic phenomenon in two distinct senses: it is subject to change, especially due to the effect of corruption, errors and deception, and it emerges as an effect of a multilevel process within the society of agents (Conte and Paolucci 2002). This involves emergence both from agents to society and from

---

[2]A belief or a goal is social when it mentions another agent and possibly one or more of his or her mental states (for an in-depth discussion of these notions, see Conte and Castelfranchi 1995; Conte 1999).

society back to the individual agents. In particular, it proceeds from the level of individual cognition to the level of social propagation (population level) and from there back to individual cognition. Once it reaches the population level, it gives rise to an additional property at the agent level. Reputation is the immaterial, more powerful equivalent of the scarlet letter sewn to one's clothes Nathaniel Hawthorne described in his masterpiece. It is more powerful because it may not be perceived by the individual to whom it is attached and therefore harder for an individual (him/her) to control or manipulate. The objective nature of reputation (in our sense) also makes it impersonal, and therefore, spreading reputation can carry less responsibility than spreading image.

To formalize these concepts, we will begin by defining the building blocks of "image". An agent has made an evaluation when he or she believes that a given entity, be it another agent, an organization, a firm, etc., can achieve a specific goal of some agent who is (often, but not always) the same as the evaluator. An agent has made a *social evaluation when his or her belief concerns another agent as a means for achieving this goal. Thus, E* targets *T* and benefits *B*. Evaluations may concern physical, mental and social properties of targets; agents may evaluate a target with regard to both capacity and willingness to achieve a shared goal. The latter, willingness to achieve a goal or interest, is particular to social evaluations. Formally, e (with $e \in E$) may evaluate t ($t \in T$) with regard to a state of the world that is in b's ($b \in B$) interest, but of which b may not be aware.

To make this analysis more concrete, we will start with an example in which we consider a classic multi-agent situation in which a set of agents fight for access to a scarce resource (food). Assume that a norm of "precedence"—a proscription against attacking agents who are consuming their "own" resources—is applied to reduce conflicts. The norm is disadvantageous for the norm follower in the short run, but is advantageous for the community and thus eventually for the individual followers. We will call N the set of norm followers, or normative agents, and C the set of cheaters, or violators of the norm. With regard to social evaluations (image), the targets coincide with the set of all agents; $T = N \cup C$ (all are evaluated). For reasons of simplicity, the agents carrying out the evaluation are restricted to the norm followers: $E = N = N \cup C$: indeed, if normative agents benefit globally from the presence of the norm, cheaters in this simple setting benefit even more; they can attack the weaker while they themselves are safe from attacks by the gullible normative.

It is very easy to find examples where all three sets (E, T and B) coincide. General behavioural norms, such as "Do not commit murder", apply to, benefit, and are evaluated by all agents. However, there are also situations in which beneficiaries, targets and evaluators are separate, for example, when norms safeguard the interests of a subset of the population. Consider the quality of TV programmes for children, broadcast in the afternoon. Here, we can identify three more or less distinct sets. The children are the beneficiaries, while adults entrusted with taking care of children are the evaluators. It could be argued that B and E still overlap, since E may be said to adopt B's interests. The targets of evaluation are the writers of programmes and the decision-makers at the broadcast stations. There may be a non-empty intersection between E and T but no full overlap. If the target of evaluation is the broadcaster itself (a supra-individual entity), the intersection can be considered to be empty.

Extending this formalization to include reputation, we have to differentiate it further. To assume that a target t is assigned, a given reputation implies assuming that t is believed to be "good" or "bad", but it does not imply sharing either evaluation. While image is based on direct experience or observation, therefore an evaluator is assumed to believe his/her own evaluation; reputation therefore involves one more set of agents: in addition to evaluators E, targets T and beneficiaries B, we have a set M of memetic agents who share the meta-belief. This means that they simply believe that some other agents had a positive experience with John, therefore they hold the meta-belief that John has a positive reputation ("I believe that others believe that he is a good dentist"). It is important to stress the fact that a memetic agent does not need to hold the evaluation belief, but she simply need to transmit it. If they contribute to the diffusion of reputation, the memetic agents can also be labelled as gossipers G. Often, E can be taken as a subset of M; the evaluators are aware of the effect of evaluation. In most situations, the intersection between the two sets is at least non-empty.

18.4.2   Identifying Reputational Roles

We have seen that agents may play more than one role simultaneously: evaluator, beneficiary, target and memetic/gossiper. In order to implement a socio-cognitive model of reputation, we need to describe the characteristics of the four roles in more detail.

18.4.2.1   Evaluator

Autonomous agents continually asses their environment and form evaluations as effect of interaction and perception. Social evaluations are formed when agents evaluate one another with regard to their goals (Castelfranchi 1998).

This image, based on direct experience, drives future actions: it serves to identify friends and to avoid enemies or cheating partners. Agents also observe interactions between third parties and evaluate them with regard to the goals or interests of a given set of agents (the beneficiaries). Information thus obtained may be used to draw inferences about the target's likelihood to violate other rights in the future. Agents evaluate one another with regard to their own goals and the goals they adopt from either other individual agents (e.g. their children) or supra-individual agents, such as groups, organizations or abstract social entities.

18.4.2.2 Beneficiary

A beneficiary is the entity that benefits from the action with regard to which targets are evaluated. Beneficiaries can be individual agents, groups and organizations or even abstract social entities like social values and institutions. Beneficiaries may be aware of their goals and interests, and of the evaluations, but this is not necessarily the case. In principle, their goals might simply be adopted by the evaluators—as it happens, for example, when members of the majority support norms protecting minorities. Evaluators often are a subset of the beneficiaries.

Beneficiaries may be implicit in the evaluation. This is particularly the case when it refers to a social value (honesty, altruism, etc.); the benefit itself and those who take advantage of it are left implicit and may coincide with the whole society. The beneficiary of the behaviour under evaluation is also a beneficiary of this evaluation: the more an (accurate) evaluation spreads, the likelier the execution of the positively evaluated behaviour.

18.4.2.3 Target

The target of social evaluation is the entity that is evaluated. Targets of reputation (targets) should be autonomous agents endowed with mental states, possibly with an explicit decision-making or deliberative capacity. Consequently, they are a locus of social responsibility: they hold the power to prevent social harm and possibly to respond for it, in case any harm occurs.

Other than beneficiaries, targets are always explicit. They may be individual entities or supra-individual like a group, a collective, an abstract entity or a social artefact, such as an institution.

18.4.2.4 Gossiper (Memetic Agent)

An agent is a (potential) memetic agent if she transmits (is in position to transmit) reputation information about a target to another agent or set of agents. Although sharing awareness of a given target reputation, memetic agents do not necessarily share the corresponding image (social evaluation) of the target. That is, they do not necessarily believe it to be true.

Memetic agents (if they are also targets) may deserve a negative evaluation; they may actually convey information that they hold to be false in order to enjoy the advantages of sharing reputation information. By sharing reputation, the agent will be considered as part of the in-group by other evaluators, and therefore gain a good reputation without sustaining the costs of its acquisition.

18.5   Implementing the Social and Cognitive Processes of Reputation: Sim-Norm and Repage

Sim-Norm was the first attempt to implement the social and cognitive theory of reputation. The model was developed to examine the effect of reputation on the efficiency of a norm of precedence (Conte et al. 1998; Conte and Paolucci 1999; Paolucci 2000) in reducing aggression, measured both at the global (i.e. societal) and local (i.e. individual) level. In particular, Sim-Norm was designed to explore *why* self-interested agents exercise social control, and its results confirmed that reputation can have a positive impact on social control.

Sim-Norm revolved around the question of which ingredients are necessary for social order to be established in a society of agents. The role of norms as aggression controllers in artificial populations living under conditions of resource scarcity was addressed. We set out to explore two hypotheses:

1. Norm-based social order can be maintained, and its costs reduced via distributed social control.
2. Social cognitive mechanisms are needed to account for distributed social control. In particular, the propagation of social beliefs plays a decisive role in distributing social control at low or zero individual costs and high global benefit. More precisely, while individually acquired evaluation of other agents gave norm executors no significant advantage, the transmission of these evaluations among norm executors proved decisive in levelling the outcomes of norm-abiders and cheaters (if numerically balanced).

The model defines agents as objects moving in a two-dimensional environment (a 10×10 grid) with randomly scattered food. At the beginning of each run, agents and food items are assigned locations at random. A location is a cell in the grid. The same cell cannot contain more than one object at a time (except when an agent is eating). The agents move through the grid in search of food, stopping to eat to build up their strength when they find it. The agents can be attacked only when eating; no other type of aggression is allowed. At the beginning of each step of the simulation, every agent selects an action from the six available routines: *eat, move-to-food-seen, move-to-food-smelled, attack, move-random* and *pause*. Actions are supposed to be simultaneous and time consuming.

To investigate the role of norms in the control of aggression, we compared scenarios in which agents follow a norm—implemented as a restriction on attacks— with identical scenarios, in which they follow utilitarian rules. In all scenarios, each agent can perform only one of three strategies:

- Blind aggression, or control condition, in which aggression is not constrained. If the agent can perform no better move (eating, moving to food seen or smelled), then it will attack without further considerations. Blind agents have access to neither their own strength nor the eater's strength; these parameters never enter their decision-making process.

- Utilitarian, in which aggression is constrained by strategic reasoning. Agents will only attack those eaters whose strength is lower than their own. An eater's strength is "visible", that is, one step away from the agent's current location. While blind agents observe no rule at all, utilitarian agents observe a rule of personal utility, which does not qualify as a norm.
- Normative (*N*), in which aggression is constrained by a norm. We introduced a finder-keeper precept, assigning a "moral right" to food items to finders, who become possessors of the food. Possession of food is ascribed to an agent on the grounds of spatial vicinity; food owned is flagged, and every player knows to whom it belongs. Each food unit may have up to five owners, decided on the basis of proximity at the time of creation. The norm then prescribes that agents cannot attack other agents who are eating their own food.

The strategies can also be characterized by the kind of agents they allow to attack: while blind agents attack anybody, the utilitarian agents attack only the weaker, and the normative agents, respecting a norm of private property, will not attack agents who are eating their own food.

These strategies were compared (Castelfranchi et al. 1998) using an efficiency measure (the average strength of the population after *n* periods of simulation) and a fairness measure (the individual deviation from the average strength). The first two series of experiments showed that normative agents perform less well than nonnormative agents in mixed populations, as they alone bear the costs of social control and are exploited by utilitarian agents.

In a following series of experiments, *image* was added to the preceding experimental picture. In this model, useful knowledge can be drawn from personal experience, but therefore still at one's own cost. To reduce cost differences among subpopulations, image is insufficient. Henceforth, we provided the cooperative agents with the capacity to exchange with their (believed-to-be) respectful neighbours at distance one from them images of other agents. With the implementation of a mechanism of transmission of information, we can speak of a reputation system. We ran the experiments again with normative agents exchanging information about cheaters. The results suggest that circulating knowledge about others' behaviours significantly improves normative agents' outcomes in a mixed population.

The spreading of reputation can then be interpreted as a mechanism of cost redistribution for the normative population. Communication allows compliant agents to easily acquire preventive information, sparing them the costs of direct confrontations with cheaters. By spreading the news that some "guys" cheat, the good guys (1) protect themselves, (2) at the same time punish the cheaters and possibly (3) exercise an indirect influence on the bad guys to obey the norm. Social control is therefore explained as an indirect effect of a "reciprocal altruism" of knowledge. The model inspired further research in the social simulation community: Saam and Harrer (1999) used the same model to explore the interaction between normative control and power, whereas Hales (2002) applied an extended version of Sim-Norm to investigate the effects of group reputation. In his model, agents are given the cognitive capacity to categorize other agents as members of a group and project reputation onto whole groups instead of individual agents (a form of stereotyping).

Repage (Sabater et al. 2006) is a computational system for reputation management. Based on a model of reputation, image and their interplay, Repage provides evaluations of potential partners and is fed with information transmitted from others plus outcomes from direct experience. This is fundamental to account for (and to design) limited autonomous agents as exchange partners. To select good partners, agents need to form and update own social evaluations; hence, they must exchange evaluations with one another.

In order to preserve their autonomy, agents need to *decide* whether to share others' evaluations of a given target. If agents would automatically accept reported evaluations and transmit them as their own, they would not be autonomous anymore. In addition, in order to exchange information about reputation, agents need to participate in circulating it, whether they believe it or not; but again to preserve their autonomy, they must *decide* how, when and about whom to gossip.

In sum, the distinction between image and reputation suggests a way out from the paradox of sociality, i.e. the trade-off between agents' autonomy and their need to adapt to social environment. On one hand, agents are autonomous if they select partners based on their social evaluations (images). On the other, they need to update evaluations by taking into account others' evaluations. Hence, social evaluations must circulate and be represented as "reported evaluations" (reputation), before and in order for agents to decide whether to accept them or not. To represent this level of cognitive detail in artificial agents' design, there is a need for a specialized subsystem. This is what Repage provides.

Repage is a sophisticated cognitive architecture that operates on a subset of the predicates that constitute the memory of the agent, that is, of those predicates that are relevant for dealing with image and reputation. Predicates about reputation, as discussed above, must contain an evaluation about a target which contains three aspects: the type of the evaluation (either personal experience or image or third party image), the role of the target (either informant or seller) and the actual content. To store the content, a simple number is used, as in eBay and in most reputation systems. This sharp representation, however, is quite implausible in inter-agent communication, which is one of the central aspects of Repage; in real life no one tells that "People are saying that Jane is 0.234 good". To capture the lack of precision coming from vague utterances, e.g. "I believe that agent X is good, I mean, very good — good, that is", and from noise in the communication or in the recollection from memory, the actual value of an evaluation is represented in a fuzzy way, by a n-tuple of positive real values that sum to one.

Finally, each predicate has a strength value associated to it. This value is a function of the strength of its antecedents and of some special characteristics intrinsic to that type of predicate. The network of dependencies specifies which predicates contribute to the values of other predicates. In fact, each predicate in the Repage memory has a set of antecedents and a set of consequences. If an antecedent changes its value or is removed, the predicate is notified, thanks to the work of the detectors. Then the predicate recalculates its value and notifies the change to its consequences. Aggregation and other interesting properties of these representations are detailed in Sabater and Paolucci (2007). An example of Repage in action can be found in Quattrociocchi et al. (2008).

To illustrate the behaviour of Repage, let us consider an example about a potential purchase. The scenario is the following: agent X is a buyer who knows that agent Y sells what he needs but knows nothing about the quality of agent Y (the target of the evaluations) as a seller. Therefore, he turns to other agents in search for information—the kind of behaviour that can be found, for example, in Internet forums, auctions and in most agent systems. Then, agent X receives a communication from agent Z saying that his image of agent Y as a seller is very good. Since agent X does not yet have an image about agent Z as an informer, he resorts to a default image (i.e. usually quite low). The uncertain image as an informer adds uncertainty to the value of the communication, resulting in a decision to look for more information.

Later on, agent X has received six communications from different agents containing their image of agent Z as an informer. Three of them give a good report and three a bad one. This information is enough for agent X now to build an image about agent Z as an informer, so this new image substitutes the default candidate image that was used so far. However, the newly formed image is insufficient to take any strategic decision—the target seems to show an irregular behaviour.

At this point, agent X decides to try a direct interaction with agent Y. Because he is not sure about agent Y, he resorts to a low-risk interaction. The result of this interaction is completely satisfactory and has important effects in the Repage memory. The candidate image about agent Y as a seller becomes a full image, in this case a positive one.

Moreover, this positive image is compared (via the fuzzy metric presented above) with the information provided by agent Z (which was a positive evaluation of agent Y as a seller); since the comparison shows that the evaluations are coherent, a positive confirmation of the image of agent Z as an informer is generated. This reinforcement of the image of agent Z as a good informer at the same time reinforces the image of agent Y as a good seller. Consequently, there is a positive feedback between the image of agent Y as a good seller and the image of agent Z as a good informer. As a final wave of feedback, the image of the three agents who gave a good evaluation of Z as an informer is increased, while the image of the other three is decreased. This feedback is a necessary and relevant part of the Repage model.

Taking into account the correlations between different reputation attributes, Nardin et al. (2014) compare Repage with other architectures via a multivariate statistical approach. Their analysis shows that, in most cases, there is a benefit in using a more expressive communication language.

18.5.1   A Simulation Model of Reputation Spreading in an Industrial District: SOCRATE

SOCRATE is an attempt to test the cognitive theory of reputation in an ideal-typical economic setting, modelled after an industrial district in which firms exchange goods and information (Giardini et al. 2008; Di Tosto et al. 2010). In this model, the focus is on the interplay between the market structure and the social relationships among agents. Social links and the resulting social structure, usually informal, are defining features of industrial clusters (Porter 1998; Fioretti 2005; Squazzoni and Boero 2002), in which trust and reputation play a crucial role. Social evaluations are the building blocks of social and economic relationships inside the cluster; they are used to select trustworthy partners, to create and enlarge the social network and to exert social control on cheaters. We designed an artificial environment in which agents can choose among several potential suppliers by relying either on their own evaluations or on other agents' evaluations. In the latter case, the availability of truthful information could help agents to find reliable partners without bearing the costs of potentially harmful interactions with bad suppliers. Moreover, evaluations can be transmitted either as image (with an explicit source and the consequent risk of retaliation) or as reputation.

This model was developed with the aim of answering the following questions: How does false information affect the quality of the cluster? What are the effects of image and reputation, respectively, on the economic performance of firms?

There are two different kinds of interactions among agents in the model: material exchange and evaluation exchange. The former refers to the exchange of products between leader firms and their suppliers, and it leads to the creation of a supply chain network. The latter consists in the flow of social evaluations among the firms, which is of paramount importance in this setting, where agents can transmit true or false evaluations in order to either help or hamper their fellows searching for a good partner.

Agents are firms organized into different layers, in line with their role in the production cycle. The number of layers can vary according to the characteristics of the cluster, but a minimum of two layers is required. We implemented three layers: Layer 0 (*L0*) is represented by leader firms that supply the final product and are supplied by firms on Layer 1 (*L1*). On Layer 2 (*L2*), there are firms providing raw material to firms in *L1*.

Reputation and image transmission are exchanged within layers, so for instance firms on *L0* and *L1* are not allowed to talk each other. Agents in *L0* have to select suppliers that produce with a quality above the average among all *L1* agents. Suppliers can be directly tested or they can be chosen, thanks to the information received by other *L0* firms acting as informers. Buying products from *L1* and asking for information to *L0* fellows are competing activities that cannot be performed contemporaneously. In turn, once received an order for a product, *L1* firms should select a good supplier (above the average quality) among those in *L2*. After each interaction with a supplier, both *L0* and *L1* agents create an evaluation, i.e. an image, of it, comparing the quality of the product they bought with the quality threshold value. Agents are endowed with a table in which all the values of the tested partners are recorded and stored for future selections. Under the reputation condition, evaluations are exchanged without revealing their source, thus injecting the cluster with untested information. In this condition, retaliation against untrustful informers is unattainable.

Our results showed that the quality of products was higher in the cluster with reputational information, compared to the cluster with image, for the same percentages of cheating. We also replicated the results by varying the distribution of firms on the three layers, thus designing a market with harsh competition for good partners, and we found that the exchange of reputational information also allows the whole cluster to obtain higher profits (Di Tosto et al. 2010).

SOCRATE results provided further support to the hypotheses about the importance of reputation for social control, showing again that social evaluations and their features have consequences also in economic terms.

18.6 Gossip as Reputation Transmission and Its Effect on Cooperation in Social Dilemmas

Gossip is a multifaceted social phenomenon, widespread in human societies and serving several functions: it is a valuable source of information about community and its members, but it is also essential to map the social environment, to promote membership and to sanction deviant behaviours in a public way (Giardini and Conte 2012). In human groups, exchanging evaluations serves as a means to create and maintain relationships between individuals, and it might be pivotal to either the creation or the enforcement of other kinds of relationships (friendship, acquaintances, business, etc.).

When cooperation is framed as a public goods game (Hardin 1968; Gardner, Ostrom, Walker 1990), cooperation can emerge only if individuals sacrifice short-term gains in favour of the long-term collective good. In large groups, this translates into a high probability that individuals will tend to interact with complete strangers with little or no opportunities for positive reciprocity. Simulation data and lab experiments show that cooperation can hardly be sustained in groups, unless costly punishment is provided (Carpenter 2007; Fehr and Gachter 2000). Although effective in many contexts, costly punishment increases the amount of cooperation but not the average pay-off of the group (Dreber et al. 2008). Those who punish pay a cost for that. In repeated games, cooperators who do not bear the costs of punishing defectors are better off than cooperators who punish (Ohtsuki et al. 2009). Evidence from different kinds of communities show that an essential mechanism for supporting cooperation is gossip and reputational threats can effectively promote trust (Greif 1993).

Giardini et al. (2014) developed a computational model in which they tested whether different reputation-based strategies may have an effect on cooperation rates in mixed populations. An essential element in the functioning of reputation is the action linked to it, although sometimes this action is implicit, for example, when that reputation is used to avoid cheaters. When partner selection is available, cheaters are avoided because of their reputations (Roberts 2008), but in indirect reciprocity models where cooperators cannot choose with whom to interact, players with bad image scores do not receive donations (Nowak and Sigmund 1998a).

In order to understand how the action linked to reputation might affect the overall cooperation levels, three different "reputation-based strategies" were defined, *as follows*:

- *Refuse* means that gossipers can refuse to contribute to the group when they know (on the basis of direct experience and gossip) that there is a majority of defectors in the group.
- *Compare* refers to the action of comparing between groups and actively looking for a better group.
- *Leader* is a refined form of partner choice in which group formation is delegated to a single agent, randomly selected to act as a leader and then allowed to choose its group mates. When the leader belongs to the population of "gossipers", it can use information received about others in order to select the best partners. A remarkable feature of this model is that information is privately transmitted among gossipers; therefore, it can become redundant and unreliable.

The results show that cooperation rates are higher when agents can compare their present situation and switch to a better one, i.e. when they can avoid free-riders, and this solution allows gossipers to get the highest scores in large groups of 25 agents. Moreover, the combination of punishment and gossip can make cooperation increase to its maximum in large groups, irrespective of the specific gossip strategy.

Group size can be a crucial factor, as showed by Suzuki and Akiyama (2005), who implemented a simulation model in which players in a PGG game can know other players' image score. In their work, cooperation can emerge in groups of four individuals, but increasing the size of groups inevitably leads to a decrease in the frequency of cooperation. The authors explain this result in terms of the limited observability of reputations in large communities with many individuals. In order to test whether this group size limitation also holds when agents are arranged on different networks, Vilone et al. (2016) compared two different network topologies, a small-world network and a bipartite graph. When reputation-based partner selection was available in a population distributed on a bipartite graph, full cooperation was reached after ten generations, also for larger groups of 20 individuals. This result has been replicated also with private gossip and errors in transmissions (Giardini and Vilone 2016) showing the importance of reputation in promoting informal social control and sustaining cooperation.

18.7 Conclusion and Future Work

Over the last two decades, there has been a significant increase in research on reputation and gossip. There is growing evidence that the presence of reputation can strongly promote cooperation and represents an effective way to maintain social control (Milinski 2016). Since reputation is a social coordination device emerging from the interplay of different information flowing in the social space, it could be difficult to test for emerging dynamics in a laboratory. This is especially true if we want to verify the difference in the usage of information with and without an explicit source, and we want to measure such a difference emerging from multiple interactions. Using artificial agents, i.e. computer programmes that behave according to some rules defined by the experimenter, we are able to investigate the complex interplay between the micro level of agents' motivations and the macro level of collective behaviours.

In this chapter, we discussed current studies of reputation as a distributed instrument for social order. After a critical review of current technologies of reputation in electronic institutions and agentized environments, a theory of reputation as a social cognitive artefact was presented. In this view, reputation allows agents to cooperate at a social meta-level, exchanging information for partner selection in competitive settings like markets and for cheater isolation and punishment in cooperative settings like teamwork and grouping.

To exemplify both functionalities, we introduced two major simulation models of reputation in artificial societies. Both have been used mainly as a theory-building tool. The first, *Sim-Norm*, is a reputation-based model for norm compliance. The main findings from simulations show that, if circulated among norm-abiders only, reputation allows for the costs of compliance to be redistributed between two balanced subpopulations of norm-abiders and cheaters. In such a way, it contributes to the fitness of the former, neutralizing the advantage of cheaters. However, results also show that as soon as the latter start to bluff and optimistic errors begin to spread in the population, things worsen for norm-abiders, to the point that the advantage produced by reputation is nullified.

Repage, a much more complex computational model than Sim-Norm, was developed to test the impact of image, reputation and their interaction on the market. Based on our social cognitive theory, it allows the distinction between image and reputation to be made and the trade-off between agents' autonomy and their liability to social influence to be coped with. Repage allows the circulation of reputation whether or not third parties accept it as true. Socrate is an attempt to combine complex agents (endowed with a memory and able to manage different kinds of evaluations) with a market in which agents must protect themselves from both informational and material cheating. In this context, reputation has been proven useful to punish cheaters, but it also prevented the social network from collapse. We also discussed agent-based models of the evolution of cooperation in which gossip and punishment were compared as tools for social control, showing the importance of the former as an informal way of sanctioning non-cooperators.

These results clearly show that differentiating image from reputation provides a means for coping with informational cheating and that further work is needed to achieve a better understanding of this complex phenomenon. The long-term results of these studies are expected to do several things, as follows:

(a) Answer the question as to how to cope with informational cheating (by testing the above hypothesis)
(b) Provide guidelines about how to realize technologies of reputation that achieve specified objectives (e.g. promoting respect of contracts vs. increasing volume of transactions)
(c) Show the impact of reputation on the competitiveness of firms within and between districts

Acknowledgements The authors would like to thank Jordi Sabater and Samuele Marmo for their helpful collaboration. This work was partially supported by the Italian Ministry of University and Scientific Research under the Firb programme (Socrate project, contract number RBNE03Y338) and by the European Community under the FP6 programme (eRep project, contract number CIT5-028575; EMIL project, contract number IST-FP6-33841).

Further Reading

For a more in-depth treatment of the contents of this chapter, we refer the reader to the monograph *Reputation in Artificial Societies* (Conte and Paolucci 2002). For more on the same line of research, with an easier presentation aimed to dissemination, we suggest the booklet published as the result of the eRep project (Paolucci et al. 2009). More recently, Hendrikx, Bubendorfer and Chard (2014) published a review of existing reputation systems, and the book by Bertino and Matei (2014) illustrated a project for the study of reputation in Wikipedia.

Due to the focus on the theoretical background of reputation, only a narrow selection of simulation models of reputation could be discussed in this chapter. Sabater and Sierra (2004) give a detailed and well-informed overview of current models of trust and reputation using a variety of mechanisms. Another good starting point for the reader interested in different models and mechanisms is the review by Ramchurn and colleagues (Ramchurn et al. 2004).

Further advanced issues for specialized reputation subfields can be found in Jøsang et al. (2007), a review of online trust and reputation systems, and in Koenig et al. (2008), regarding the Internet of Services approach to Grid Computing.

References

Abdul-Rahman, A., & Hailes, S. (1997). A distributed trust model. In *Proceedings of the 1997 Workshop on New Security Paradigms*. Langdale: ACM.
Akerlof, G. (1970). The market for lemons: Quality uncertainty and the market mechanisms. *The Quarterly Journal of Economics, 84*, 488–500.

Alexander, R. (1987). *The biology of moral systems*. New York: De Gruyter.

Bachmann, R. (1998). In T. Malsch (Ed.), *Kooperation, vertrauen und macht in systemen verteilter künstlicher intelligenz. Eine vorstudie zum verhältnis von soziologischer theorie und technischer modellierung* (Sozionik ed., pp. 197–234). Berlin: Sigma.

Barkow, J. H. (1996). Beneath new culture is old psychology: Gossip and social stratification. In J. H. Barkow, L. Cosmides, & J. Tooby (Eds.), *The adapted mind. Evolutionary psychology and the generation of culture*. New York: Oxford University Press.

Beersma, B., & Van Kleef, G. A. (2011). How the grapevine keeps you in line: Gossip increases contributions to the group. *Social Psychological and Personality Science, 2*, 642–649.

Bertino, E., & Matei, S. A. (Eds.). (2014). *Roles, trust, and reputation in social media knowledge markets: theory and methods (computational social sciences)*. Berlin: Springer-Verlag.

Boehm, C. (1999). *Hierarchy in the forest: The evolution of egalitarian behavior*. Cambridge: Harvard University Press.

Bolton, G., Katok, E., & Ockenfels, A. (2004). How Effective Are Electronic Reputation Mechanisms? An Experimental Investigation. *Management Science, 50*(11), 1587–1602.

Cabral, L., & Hortaçsu, A. (2010). The dynamics of seller reputation: Evidence from ebay. *The Journal of Industrial Economics, 58*(1), 54–78.

Capraro, V., Giardini, F., Vilone, D., & Paolucci, M. (2016). Partner selection supported by opaque reputation promotes cooperative behavior. *Judgement and Decision Making, 11*(6), 589–600.

Carbo, J., Molina, J. M., & Davila, J. (2002). Comparing predictions of SPORAS vs. a fuzzy reputation agent system. In *3rd International Conference on Fuzzy Sets and Fuzzy Systems* (pp. 147–153). Switzerland: Interlaken.

Carpenter, J. P. (2007). Punishing free-riders: How group size affects mutual monitoring and the provision of public goods. *Games and Economic Behavior, 60*(1), 31–51.

Castelfranchi, C. (1998). Modelling social action for AI agents. *Artificial Intelligence, 103*, 157–182.

Castelfranchi, C., Conte, R., & Paolucci, M. (1998). Normative reputation and the costs of compliance. *Journal of Artificial Societies and Social Simulation, 1*(3). http://jasss.soc.surrey.ac.uk/1/3/3.html

Conte, R. (1999). Social intelligence among autonomous agents. *Computational and Mathematical Organization Theory, 5*, 202–228.

Conte, R., & Castelfranchi, C. (1995). *Cognitive and social action*. London: London University College of London Press.

Conte, R., & Paolucci, M. (1999). Reproduction of normative agents: A simulation study. *Adaptive Behaviour*, special issue on Simulation Models of Social Agents, *7*, 301–322.

Conte, R., & Paolucci, M. (2002). *Reputation in artificial societies: Social beliefs for social order*. Norwell: Kluwer Academic.

Conte, R., & Paolucci, M. (2003). Social cognitive factors of unfair ratings in reputation reporting systems. In J. Liu, C. Liu, M. Klush, N. Zhong, & N. Cercone (Eds.), *Proceedings of the IEEE/WIC International Conference on Web Intelligence-WI 2003* (pp. 316–322). Halifax: IEEE Computer Press.

Conte, R., Castelfranchi, C., & Dignum, F. (1998). Autonomous norm-acceptance. In *Intelligent agents*. Berlin: Springer.

Cravens, K., Goad Oliver, E., & Ramamoorti, S. (2003). The reputation index: Measuring and managing corporate reputation. *European Management Journal, 21*(2), 201–212.

Dellarocas, C. N. (2003). *The digitalization of word-of-mouth: Promise and challenges of online feedback mechanisms* (MIT Sloan Working Paper 4296-03).

Di Tosto, G., Giardini, F., & Conte, R. (2010). Reputation and economic performance in industrial districts: Modelling social complexity through multi-agent systems. In K. Takadama, C. Cioffi-Revilla, & G. Deffuant (Eds.), *Simulating interacting agents and social phenomena* (pp. 165–176). Tokyo: Springer.

Diekmann, A., Jann, B., Przepiorka, W., & Wehrli, S. (2014). Reputation formation and the evolution of cooperation in anonymous online markets. *American Sociological Review, 79*(1), 65–85.

Dreber, A., Rand, D., Fudenberg, D., & Nowak, M. (2008). Winners don't punish. *Nature, 452*, 348–351.

Dunbar, R. (1996). *Grooming, gossip, and the evolution of language.* London: Faber and Faber.

Fehr, E., & Gachter, S. (2000). Fairness and retaliation: The economics of reciprocity. *Journal of Economics Perspectives, 14*, 159–181.

Fioretti, G. (2005). Agent based models of industrial clusters and districts. In *Contemporary issues in urban and regional economics.* New York: Nova Science.

Fombrun, C. (1996). *Reputation*, in Wiley Encyclopedia of Management.

Fombrun, C., & Shanley, M. (1990). What's in a name? Reputation building and corporate strategy. *Academy of Management Journal, 33*(2), 233–258.

Frith, C. D., & Frith, U. (2006). How we predict what other people are going to do. *Brain Research, 1079*(1), 36–46.

Gardner, R., Ostrom, E., & Walker, J. (1990). The Nature of Common-Pool Resource Problems. *Rationality and Society, 2*, 335–358.

Giardini, F., & Conte, R. (2012). Gossip for social control in natural and artificial societies. *Simulation, 88*(1), 18–32.

Giardini, F., & Vilone, D. (2016). Evolution of gossip-based indirect reciprocity on a bipartite network. *Scientific Reports, 6*, 37931.

Giardini, F., Di Tosto, G., & Conte, R. (2008). A model for simulating reputation dynamics in industrial districts. *Simulation Modelling Practice and Theory, 16*, 231–241.

Giardini, F., Paolucci, M., Adamatti, D., & Conte, R. (2014). Group size and gossip strategies: An ABM tool for investigating reputation-based cooperation. In *MABS 2014: Multi-agent-based simulation XV* (pp. 104–118). Cham: Springer-Verlag.

Gluckman, M. (1963). Gossip and scandal. *Current Anthropology, 4*, 307–316.

Gray, E. R., & Balmer, J. M. T. (1998). Managing corporate image and corporate reputation. *Long Rage Planning, 31*(5), 695–702.

Greif, A. (1993). Contract enforceability and economic institutions in early trade: the Maghribi traders' coalition. *American Economic Review, 83*(3), 525–548.

Hales, D. (2002). Group reputation supports beneficent norms. *Journal of Artificial Societies and Social Simulation. 5*(4). http://jasss.soc.surrey.ac.uk/5/4/4.html

Hardin, G. (1968). The tragedy of the commons. *Science, 162*, 1243–1248.

Hendrikx, F., Bubendorfer, K., & Chard, R. (2015). Reputation systems: a survey and taxonomy. *Journal of Parallel and Distributed Computing, 75*, 184–197.

Huynh, D., Jennings, N. R., & Shadbolt, N. R. (2004). Developing an integrated trust and reputation model for open multi-agent systems. In *Proceedings of the Workshop on Trust in Agent Societies (AAMAS-04)* (pp. 65–74). New York: ACM.

Jøsang, A., Ismail, R., & Boyd, C. (2007). A survey of trust and reputation systems for online service provision. *Decision Support Systems, 43*(2), 618–644.

Koenig, S., Hudert, S., Eymann, T., & Paolucci, M. (2008). Towards reputation enhanced electronic negotiations for service oriented computing. In R. Falcone, S. K. Barber, J. Sabater-Mir, & M. P. Singh (Eds.), *Trust in agent societies—11th International Workshop, TRUST 2008*, Estoril, Portugal, May 12–13, 2008. Revised selected and invited papers. Berlin: Springer.

Kreps, D., & Wilson, R. (1982). Reputation and imperfect information. *Journal of Economic Theory, 27*, 253–279.

Marsh, S. (1992). Trust and reliance in multi-agent systems: A preliminary report. MAAMAW'92, 4th European Workshop on Modelling Autonomous Agents in a Multi-Agent World, Rome.

Marsh, S. (1994a). *Formalising trust as a computational concept.* Stirling: Department of Computing Science and Mathematics, University of Stirling.

Marsh, S. (1994b). Optimism and pessimism in trust. In *Proceedings of the Ibero-American Conference on Artificial Intelligence (IBERAMIA-94).* Caracas: McGraw-Hill.

Melnik, M. I., & Alm, J. (2002). Does a seller's ecommerce reputation matter? Evidence from ebay auctions. *The Journal of Industrial Economics, 50*(3), 337–349.

Miceli, M., & Castelfranchi, C. (2000). The role of evaluation in cognition and social interaction. In K. Dautenhahn (Ed.), *Human cognition and social agent technology.* Amsterdam: Benjamins.

Milinski, M. (2016). Reputation, a universal currency for human social interactions. *Philosophical Transactions of the Royal Society B: Biological Sciences, 371*(1687), 20150100.

Moukas, A., Zacharia, G., & Maes, P. (1999). Amalthaea and histos: multiagent systems for WWW sites and reputation recommendations. In M. Klusch (Ed.), *Intelligent information agents. Agent-based information discovery and management on the internet* (pp. 292–322). Berlin: Springer.

Muller, G., & Vercouter, L. (2008). *L.I.A.R. achieving social control in open and decentralised multi-agent systems* (Technical report). Saint-Étienne: École Nationale SupÉrieure des Mines de Saint-Étienne.

Nardin, L. G., Brandão, A. A., Sichman, J. S., & Vercouter, L. (2008) SOARI: A service oriented architecture to support agent reputation models interoperability. In R. Falcone, S. K. Barber, J. Sabater-Mir, & M. P. Singh (Eds.), *Trust in agent societies: 11th International Workshop, TRUST 2008, Estoril, Portugal, May 12–13, 2008. Revised selected and invited papers, no. 5396 in Lecture notes in artificial intelligence.* (pp. 292–307). Berlin: Springer. doi:https://doi.org/10.1007/978-3-540-92803-4_15.

Nardin, L. G., Brandão, A., Kira, E., & Sichman, J. S. (2014). Effects of reputation communication expressiveness in virtual societies. *Computational and Mathematical Organization Theory, 20*(2), 113–132.

Noë, A., & Hammerstein, P. (1994). Biological markets: Supply and demand determine the effect of partner choice in cooperation, mutualism and mating. *Behavioral Ecology and Sociobiology, 35*, 1–11.

Noon, M., & Delbridge, R. (1993). News from behind my hand: Gossip in organizations. *Organization Studies, 14*, 23–36.

Nowak, M. A., & Sigmund, K. (1998a). Evolution of indirect reciprocity by image scoring. *Nature, 393*, 573–577.

Nowak, M. A., & Sigmund, K. (1998b). The dynamics of indirect reciprocity. *Journal of Theoretical Biology, 194*, 561–574.

Nowak, M. A., & Sigmund, K. (2005). Evolution of indirect reciprocity. *Nature, 437*, 1291–1298.

Ohtsuki, H., Iwasa, Y., & Nowak, M. A. (2009). Indirect reciprocity provides only a narrow margin of efficiency for costly punishment. *Nature, 457*(7225), 79–82.

Paine, R. (1967). What is gossip about? An alternative hypothesis. *Man, 2*, 278–285.

Panchanathan, K., & Boyd, R. (2004). Indirect reciprocity can stabilize cooperation without the second-order free rider problem. *Nature, 432*, 499–502.

Paolucci, M. (2000). False reputation in social control. *Advances in Complex Systems, 3*(1-4), 39–51.

Paolucci, M., & Conte, R. (2009). Reputation: Social transmission for partner selection. In G. P. Trajkovski & S. G. Collins (Eds.), *Handbook of research on agent-based societies: Social and cultural interactions* (pp. 243–260). Hershey: IGI.

Paolucci, M., & Sichman, J. S. (2014). Reputation to understand society. *Computational and Mathematical Organization Theory, 20*(2), 211–217.

Paolucci, M., Eymann, T., Jager, W., Sabater-Mir, J., Conte, R., Marmo, S., et al. (2009). *Social knowledge for e-governance: Theory and technology of reputation.* Rome: ISTC-CNR.

Petroni, F., Querzoni, L., Beraldi, R., & Paolucci, M. (2016). LCBM: A fast and lightweight collaborative filtering algorithm for binary ratings. *Journal of Systems and Software, 117*, 583–594.

Piazza, J., & Bering, J. M. (2008). Concerns about reputation via gossip promote generous allocations in an economic game. *Evolution and Human Behavior, 29*, 172–178.

Pinyol, I., & Sabater-Mir, J. (2013). Computational trust and reputation models for open multi-agent systems: A review. *Artificial Intelligence Review, 40*(1), 1–25.

Pinyol, I., Paolucci, M., Sabater-Mir, J., & Conte, R. (2008). Beyond accuracy. Reputation for partner selection with lies and retaliation. In L. Antunes, M. Paolucci, & E. Norling (Eds.), *Multi-agent-based simulation VIII* (pp. 128–140). Berlin: Springer.

Porter, M. (1998). Clusters and the new economics of competition. *Harvard Business Review, 76*, 77–90.

Quattrociocchi, W., Paolucci, M., & Conte, R. (2008). Reputation and uncertainty reduction: Simulating partner selection. In *Proceedings of the International Workshop on Trust in Agent Societies* (pp. 308–325). Berlin: Springer.

Ramchurn, S. D., Huynh, D., & Jennings, N. R. (2004). Trust in multiagent systems. *The Knowledge Engineering Review, 19*(1), 1–25.

Rand, D. G., Arbesman, S., & Christakis, N. A. (2011). Dynamic social networks promote cooperation in experiments with humans. *Proceedings of the National Academy of Sciences of the United States of America, 108*, 19193–19198.

Rasmusson, L. (1996). *Socially controlled global agent systems* (Working Paper). Department of Computer and Systems Science, Royal Institute of Technology, Stockholm.

Rasmusson, L., & Janson, S. (1996). Simulated social control for secure internet commerce. New security paradigms workshop, Lake Arrowhead, California, United States. In *Proceedings of the 1996 Workshop on New Security Paradigms* (pp. 18–25). New York: ACM.

Regan, K., & Cohen, R. (2005). Indirect reputation assessment for adaptive buying agents in electronic markets. In *Baseweb workshop at AI 2005 Conference*.

Resnick, P., & Zeckhauser, R. (2001). *Trust among strangers in internet transactions: Empirical analysis of ebay's reputation system* (Working Paper for the NBER Workshop on Empirical Studies of Electronic Commerce).

Resnick, P., & Zeckhauser, R. (2002). Trust among strangers in internet transactions: empirical analysis of eBay's reputation system. In M. R. Baye (Ed.), *The economics of the internet and E-commerce, Volume 11 of Advances in Applied Microeconomics*. Amsterdam: Elsevier Science.

Ripperger, T. (1998). *Ökonomik des vertrauens: Analyse eines organisationsprinzips*. Tübingen: Mohr Siebeck.

Roberts, G. (1998). Competitive altruism: From reciprocity to the handicap principle. *Proceedings of the Royal Society of London B, 265*(1394), 427–431.

Roberts G. (2008). Evolution of direct and indirect reciprocity. Proceedings of the Royal Society B: Biological Sciences, *275*(1631), 173–179.

Rocco, E., & Warglien, M. (1995). La comunicazione mediata da computer e l'emergere dell'opportunismo elettronico. *Sistemi Intelligenti, 7*(3), 393–420.

Rose, C., & Thomsen, S. (2004). The impact of corporate reputation on performance: Some danish evidence. *European Mangement Journal, 22*(2), 201–210.

Saam, N., & Harrer, A. (1999). Simulating norms, social inequality, and functional change in artificial societies. *Journal of Artificial Societies and Social Simulation, 2*(1). http://jasss.soc.surrey.ac.uk/2/1/2.html

Sabater, J. (2004). Evaluating the ReGreT system. *Applied Artificial Intelligence, 18*(9–10), 797–813.

Sabater, J., & Paolucci, M. (2007). On representation and aggregation of social evaluations in computational trust and reputation models. *International Journal of Approximate Reasoning, 46*(3), 458–483.

Sabater, J., & Sierra, C. (2002). Reputation and social network analysis in multi-agent systems. In *Proceedings of the First International Joint Conference on Autonomous Agents & Multiagent Systems (AAMAS)*, Bologna, Italy. New York: ACM.

Sabater, J., & Sierra, C. (2004). Review on computational trust and reputation models. *Artificial Intelligence Review, 24*(1), 33–60.

Sabater, J., Paolucci, M., & Conte, R. (2006). Repage: REPutation and ImAGE among limited autonomous partners. *Journal of Artificial Societies and Social Simulation, 9*(2). http://jasss.soc.surrey.ac.uk/9/2/3.html

Schillo, M., Funk, P., & Rovatsos, M. (2000). Using trust for detecting deceitful agents in artificial societies. *Applied Artificial Intelligence, 14*, 825–848.

Sell, J., & Wilson, R. (1991). Levels of information and contributions to public goods. *Social Forces, 70*, 107–124.

Sommerfeld, R. D., Krambeck, H., & Milinski, M. (2008). Multiple gossip statements and their effects on reputation and trustworthiness. *Proceedings of the Royal Society B: Biological Sciences, 275*, 2529–2536.

Squazzoni, F., & Boero, R. (2002). Economic performance, inter-firm relations and local institutional engineering in a computational prototype of industrial districts, *Journal of Artificial Societies and Social Simulation, 5*(1). http://jasss.soc.surrey.ac.uk/5/1/1.html

Suzuki, S., & Akiyama, E. (2005). Reputation and the evolution of cooperation in sizable groups. *Proceedings of the Royal Society of London B: Biological Sciences, 272*(1570), 1373–1377.

Vilone, D., Giardini, F., & Paolucci, M. (2016). Exploring reputation-based cooperation: Reputation-based partner selection and network topology support the emergence of cooperation in groups. In F. Cecconi (Ed.), *New frontiers in the study of social phenomena* (pp. 101–114). Cham: Springer.

Wang, J., Suri, S., & Watts, D. J. (2012). Cooperation and assortativity with dynamic partner updating. *Proceedings of the National Academy of Sciences of the United States of America, 109*, 14363–14368.

Wilson, D. S., Wilczynski, C., Wells, A., & Weiser, L. (2000). Gossip and other aspects of language as group-level adaptations. In C. Heyes & L. Huber (Eds.), *The evolution of cognition*. Cambridge: MIT Press.

Wu, J., Balliet, D., & Van Lange, P. A. M. (2016). Gossip versus punishment: The efficiency of reputation to promote and maintain cooperation. *Scientific Reports, 6*, 23919.

Yamagishi, T., Matsuda, M., Yoshikai, N., Takahashi, H., & Usui, Y. (2004). Solving the lemons problem with reputation. An experimental study of online trading. In *eTrust: Forming relationships in the online world*. New York: Russell Sage Foundation.

Yu, B., & Singh, M. P. (2000). A social mechanism of reputation management in electronic communities. In *Proceedings of the 4th International Workshop on Cooperative Information Agents IV: The Future of Information Agents in Cyberspace*. Berlin: Springer-Verlag.

Zacharia, G. (1999). Trust management through reputation mechanisms. In C. Castelfranchi, R. Falcone, & B. S. Firozabadi (Eds.). *Proceedings of the Workshop on Deception, Fraud and Trust in Agent Societies, Seattle* (pp.163–167).

Zacharia, G., Moukas, A., & Maes, P. (1999). Collaborative Reputation Mechanisms in Electronic Marketplaces. In *Proceedings of the 32nd Hawaii International Conference on System Sciences*, Wailea, Maui.

Chapter 19
Social Networks and Spatial Distribution

Frédéric Amblard and Walter Quattrociocchi

Abstract In most agent-based social simulation models, the issue of the organisation of the agents' population matters. The topology, in which agents interact,
be it spatially structured or a social network, can have important impacts on the
obtained results in social simulation. Unfortunately, the necessary data about the
target system is often lacking; therefore, you have to use models in order to
reproduce realistic spatial distributions of the population and/or realistic social
networks among the agents. In this chapter, we identify the main issues concerning
this point and describe several models of social networks or of spatial distribution
that can be integrated in agent-based simulation to go a step forwards from the use
of a purely random model. In each case, we identify several output measures that
allow quantifying their impacts.

Why Read This Chapter?
To learn about interaction topologies for agents, from social networks to structures
representing geographical space and the main questions and options an agent-based
modeller has to face when developing and initialising a model.

19.1 Introduction

Independent of the methodology followed to build a model, any agent-based
modeller has to face not only the design of the agents and their behaviour but also the
design of the topology in which the agents interact. This can be a spatial structure,
so that the agents are distributed within a representation of geographical space, or a
social structure, linking agents as nodes in a network, or even both.

F. Amblard (✉)
Institut de Recherche en Informatique de Toulouse (IRIT), Université Toulouse 1 Capitole,
Toulouse, France
e-mail: frederic.amblard@ut-capitole.fr

W. Quattrociocchi
IMT School for Advanced Studies, Lucca, Italy

© Springer International Publishing AG 2017   471
B. Edmonds, R. Meyer (eds.), *Simulating Social Complexity*,
Understanding Complex Systems, https://doi.org/10.1007/978-3-319-66948-9_19

Interaction topologies can be determined explicitly or implicitly. They are *explicit* when they are specified as modelling hypotheses and thus clearly defined within the model. Conversely, topologies are *implicit* when they are inferred from other processes and thus not a definite part of the model. We will come back to the consequences of such a classification in the next section.

There are three major issues to solve when dealing with social and spatial structures:

Implementation: How to represent the structure in the model, e.g. continuous versus discrete representation of space or which data structure to choose for the social network.

Initialisation: How to initialise the chosen structure, e.g. which initial shape of the network should be chosen and how should the population of agents be distributed on this network.

Observation: How to characterise a given structure and/or its evolution, potentially taking into account agents' states related to their place in the structure. This latter point raises the question of the indicators to observe during the simulation in order to follow changes in either spatial or social structures.

Since the answers to these questions differ quite substantially for spatial and social structures, we will discuss them separately in Sects. 3 and 4. The following section, while focussing on the topic of explicitly versus implicitly defined interaction topologies, will also discuss the situation where both social and spatial structures have to be taken into account at the same time, leading to either spatially embedded networks or graph-like representations of spatial structures.

19.2　Explicit and Implicit Structures

As mentioned in the introduction, there is a difference between explicit and implicit structures. We define explicit structures as clearly implemented modelling hypotheses, which can therefore be identified in the model. Implicit structures, on the other hand, are not directly defined in the model and are rather determined as a result of other processes and/or hypotheses in the model. We will demonstrate the difference with the help of two examples and then go on to explore the consequences of ex- versus implicitness with regard to the three issues of implementation, initialisation and observation introduced above.

19.2.1　Example 1: Schelling's Segregation Model

To present it briefly, the segregation model of Thomas Schelling (1971) is composed of a set of agents, some red, the others green. Each agent is positioned on an empty square of a chessboard (representing the environment). If the proportion of

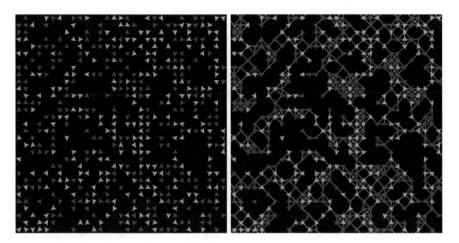

Fig. 19.1 Schelling's segregation model in NetLogo, the spatial structure is represented on the *left* and the inferred social structure represented as a graph on the *right* figure

neighbours of the same colour falls below the value of the agent's tolerance variable, the agent moves to a randomly chosen empty square (originally: the nearest empty square at which it would be satisfied with the neighbourhood); otherwise, it stays where it is.

The spatial structure of this model is explicit and is represented explicitly as a grid. As such, it is discrete, regular and static (the distribution of the agents on this structure evolves, not the structure itself). The social structure, on the other hand, is implicit in this model. During the simulation the agents take into account the type (green or red) of their neighbours on the grid, but the corresponding social structure is not defined as such and is inferred from the spatial distribution of the agents (cf. Fig. 19.1). As discussed later, it can be interesting in such a case to characterise the implicit social structure of the model, as this is the one that drives the model, whereas the spatial structure merely acts as a constraint (in particular concerning the maximal number of neighbours) on the evolution of the social structure.

19.2.2   Example 2: IMAGES Innovation Dynamics Model

In the innovation dynamics model developed in the FAIR-IMAGES project (IMAGES 2004), agents, representing farmers, are linked via a social network. This defines the paths, over which information is diffused, and controls which agents influence each other. The social network is determined, at least in part, from the geographical locations of the agents to account for the fact that geographical neighbours tend to know each other. In this example, the social structure is explicit as it is built into the model by the hypothesis (Fig. 19.2).

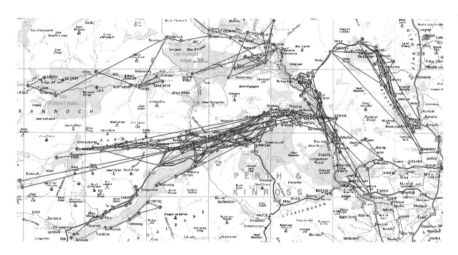

Fig. 19.2 Reconstruction of the social network among agents incorporating the geographical distance (IMAGES 2004)

19.2.3 Questions Linked to the Implicit/Explicit Property of the Structures

The property of being explicit or implicit enables us to narrow down the range of possible answers to the three questions raised in the introduction. To begin with, the question of implementation can only be asked when dealing with explicit structures; the same is true for the question of initialisation. However, characterising the implicit social structure in spatial models, i.e. identifying at a given time step the whole set of interactions among agents that could or do take place, can give useful hints for understanding the underlying dynamics. Identifying, for instance, separate components in the implicit social network inferred from a spatial model is more informative than solely identifying spatial clusters as it confirms that there is effectively no connection among the different groups.

In the next two sections, we will discuss the conditions in which social and spatial structures can be considered as independent features of social analysis and can therefore be presented independently. This is generally the case but, as we will detail in the last section, there are some exceptions.

19.3 Social Networks

Social networks have been analysed extensively during the last decade. From the social network of scientists (co-authorship network or co-citation network) (Newman 2001; Jeong et al. 2002; Newman 2004; Meyer et al. 2009) to the social network of dolphins (Lusseau 2003), many empirical studies on large graphs popularised this fascinating subject: "social" links between social beings. Neither empirical analysis nor theoretical modelling is new in this field. From the formalisation of graphs by Euler in the eighteenth century in order to represent paths in Königsberg, which led to the now well-established graph theory, to social network analysis in sociology, originating from the use of the socio-matrix of Moreno, social networks are now quite commonly used in agent-based simulation to explicitly represent the topology of interactions among a population of agents.

In order to clarify the kind of modelling issues you are dealing with, we can divide modelling of social networks into three categories: (a) static networks, (b) dynamic networks with the dynamics independent of the agents' states (for instance, random rewiring process) and (c) dynamic networks evolving dependent on the agents' states. In this chapter we will concentrate on the first case since it is the most common, although the use of the second case has recently started to grow rapidly, while the third case is still in its incipient stage.

In each case, the same three questions arise with regard to implementation, initialisation and observation:

- Which data structure is best suited to represent the network?
- Which initial (and in the static case, only) network configuration to use?
- How to identify something interesting from my simulation including the network (e.g. a social network effect)?

19.3.1 Which Data Structure to Use?

Although it could seem trivial, especially when you use a high-level modelling platform such as NetLogo or Repast, this issue is important concerning the execution efficiency of your model. More important is that even, depending on your choice, biases are linked to some data structures when using particular classes of networks such as scale-free networks.

Basically, using an object-oriented approach, you have two choices: either to embed social links within the agent as pointers to other agents or to externalise the whole set of links as a global collection (called SocialNetwork, for instance). The former is more practical when having 1–n interactions rather than 1–1 interactions, i.e. taking into account all neighbours' states to determine the new state of the agent rather than picking one agent at random in the neighbourhood. The difference between the two solutions is mainly related to the scheduling you will use. You can choose either to first schedule the agents, picking an agent at random from the population and then selecting one (or more) of its social links, or you can choose to pick a random link from the global collection and then execute the corresponding interaction. While this choice will depend a lot on the kind of model you are implementing, it is crucial when using scale-free networks since both options may produce a bias, and you will have to choose the solution that is more relevant for the purpose of your model.

To be able to explain this further, we need to quickly introduce scale-free networks (they will be presented in more detail in Sect. 3.2.4). Their main property is that the distribution of the number of links per agent (node) follows a power law, meaning that very few agents—the so-called hubs—have a lot of links, while the majority of agents have only a few.

If you now choose to schedule the agents first, you effectively apply an egalitarian rule on the individuals, which however results in the links involving the hubs being less frequently scheduled than the links involving agents with few social relations. Take for example a population of 100 agents where one agent has 30 links and another has only one link. Each of these two agents has the same probability to be scheduled (0.01), but if you then proceed to select a random link from the scheduled agent, the links of the hub agent each have a 0.01/30 probability to be chosen, while the one link of the other agent still has a 0.01 probability to be picked.

Conversely, if you schedule the collection of links first, i.e. apply an egalitarian rule on the individual links, the influence of the hubs in the global dynamics will be strengthened, as they are involved in more links. Therefore, the initial states of the hub agents are also interesting with respect to the whole dynamics of the model.

19.3.2   Which Initial Network Configuration to Use?

This question mainly arises when choosing which (initial) social structure to implement. There is a large choice of models—each with some advantages and some drawbacks—that can be distinguished into four categories: (a) regular graphs, lattices or grids being the most long-standing structure used in social simulation (inherited from the cellular automata approaches), (b) random graphs, (c) small-world networks and (d) scale-free networks. Concerning the latter three categories, Newman et al. (2006) regroup an important set of articles that can be useful as advanced material on this point.

19.3.2.1   Lattices

The field of social modelling inherited many tools from mathematics and physics and in particular cellular automata (Wolfram 1986). The corresponding underlying interaction structure is then in general a grid and in many cases a torus. The cells of the automata represent the agents, and their social neighbourhood is defined from the regular grid with a von Neumann or a Moore neighbourhood. The von Neumann neighbourhood links a cell to its four adjacent cells (north, east, south, west), while a Moore neighbourhood adds four more neighbours (NE, SE, SW, NW; cf. Fig. 19.3).

The main advantage of using regular grids stems from visualisation, regular grids enabling very efficient visualisations of diffusion processes or clustering process (cf. Fig. 19.4).

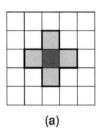 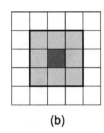 

(a)   (b)

Fig. 19.3 von Neumann (**a**) and 3 × 3 Moore neighbourhood (**b**) on a regular grid; the illustration on the *right* shows a torus, i.e. the result of linking the borders (north with south, east with west) of a grid (Flache and Hegselmann 2001)

Fig. 19.4 Opinion dynamics model on a regular grid (Jager and Amblard 2005)

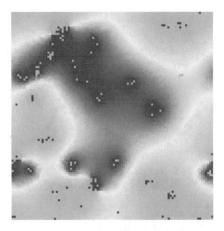

The first important point concerning the regular structures deals with connectivity. In contrast to other kinds of networks (random ones, for instance; see next section), using regular networks makes it difficult to change the connectivity of the structure, i.e. the number of links per agent. The exploration of connectivity effects on the model behaviour is limited in this case to specific values (4, 8, 12, 24, ... in the case of a two-dimensional regular grid).

The second point deals with the dimension of the regular structure. A one-dimensional lattice corresponds to a circle (see Fig. 19.5), two-dimensional structures to grids (chessboard) and three-dimensional structures to cubic graphs. However, we have to notice that only 2D regular structures benefit from a visualisation advantage, higher dimensions suffering from the classical disadvantage associated with the visualisation of dynamics on complex graphs.

The presence or absence of borders is important in regular graphs. A classic example is the 2D grid, which—if not implemented as a torus—is not a regular structure anymore, since agents localised at the borders have fewer connections. Moreover, these agents being linked to each other create a bias in the simulation (Chopard and Droz 1998). This bias is sometimes needed for any dynamics to

Fig. 19.5 Regular 1D
structure with
connectivity $= 4$

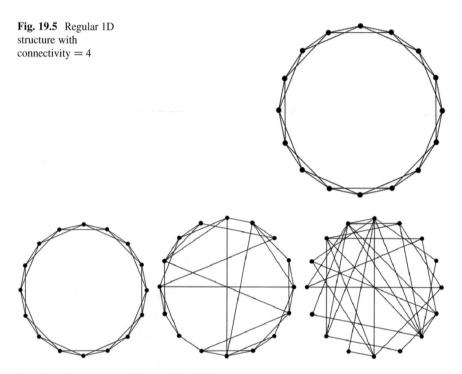

Fig. 19.6 The β-model of Watts (1999) enables to go from regular graphs (on the *left*) to random
graphs (on the *right*) using rewiring of edges

happen and could either correspond to a modelling hypothesis or to an unwanted
artefact. This point is probably far clearer on one-dimensional graphs, where
if you do not close the circle, the diameter[1] is approximately the size of the
population, whereas if you close it, the diameter is half the population size. This
issue corresponds to border effects identified on cellular automata.

19.3.2.2 Random Graphs

Another kind of model that can be used to generate social structures is the random
graph model (Solomonoff and Rapoport 1951; Erdös and Rényi 1960); see Fig. 19.6
on the right for an illustration. As told by Newman et al. (2006), there are two ways
to build random graphs containing n vertices: one (denoted as $G_{n,m}$) is to specify the
number m of edges between vertex pairs chosen at random, and the other (denoted
as $G_{n,p}$) is to specify a probability p for an edge to link any two vertices. Both of
them correspond to graphs that have on average the same properties when they are

---

[1]The diameter of a graph is defined as the length of the longest-shortest path in the graph.

big enough. The only difference is that in $G_{n,m}$ the number of edges is fixed, while in $G_{n,p}$ it may fluctuate from one instance to the other; but on average it is also fixed.

The first important property of random graphs is that they show a phase transition when the average degree of a vertex is 1. Below this transition, we obtain a number of small components, while above this threshold, the model exhibits a giant component with some isolated nodes. The giant component is a subset of the graph vertices, each of which is reachable from any of the others along some path(s). The random graphs that are used most often in agent-based social simulation are slightly above the threshold, in the giant-component phase with some isolated nodes.

Another important point concerns the degree distribution. The properties and behaviour of a network are affected in many ways by its degree distribution (Albert et al. 2000; Cohen et al. 2000; Callaway et al. 2000). $G_{n,p}$ has a binomial degree distribution (Poisson distribution for large n), which is sharply peaked and has a tail that decays quicker than any exponential distribution.

It is possible to define random graphs with any desired degree distribution (Bender and Canfield 1978; Luczak 1992; Molloy and Reed 1995). In this case, one considers graphs with a given degree sequence rather than with a degree distribution. A degree sequence is a set of degrees k_1, k_2, k_3, ... for each of the corresponding vertices 1, 2, 3, ... Molloy and Reed (1995) suggest the following algorithm:

- Create a list in which the label i of each vertex appears exactly k_i times.
- Pair up elements from this list uniformly at random until none remain.
- Add an edge to the graph joining the two vertices of each pair.

According to Molloy and Reed (1995), these graphs possess a phase transition at which a giant component appears, just as in the standard Poisson random graph.

In the context of agent-based social simulation, a great advantage of random graphs over regular graphs is that you can easily change and precisely tune the average connectivity of the graph and—applying Molloy and Reed's algorithm—the distribution of the edges among vertices.

Replacing regular graphs with random graphs, several scientists experienced a "social network" effect, i.e. models having different macroscopic behaviours depending on the chosen interaction structure (Stocker et al. 2001; Stocker et al. 2002; Holme and Grönlund 2005; Huet et al. 2007; Deffuant 2006; Gong and Xiao 2007; Kottonau and Pahl-Wostl 2004; Pujol et al. 2005). The fact is that these two classes of networks have very different characteristics. In terms of clustering, regular graphs exhibit more clustering or local redundancy than random graphs. On the other hand, random graphs lead to a shorter diameter and average path length among the pairs of individuals than a regular graph. The mean path length for a random graph scales logarithmically with graph size. For more details concerning random graphs models, we refer the interested reader to Bollobas (2001).

19.3.2.3   Small-World Networks

The question arising at this stage is are there classes of graphs between these two extremes (random and regular graphs) that may have other characteristics? Watts and Strogatz (1998) introduced such a model, motivated by the observation that many real-world graphs share two main properties:

- *The small-world effect*, i.e. most vertices are connected via a short path in the network.[2]
- *High clustering*, corresponding to the phenomenon that the more neighbours two individuals have in common, the more likely they are to be connected themselves.

Watts and Strogatz defined a network to be a small-world network if it exhibits both of these properties, that is, if the mean vertex-vertex distance l is comparable to that of a random graph and the clustering coefficient is comparable to that of a regular lattice.

To construct such a network, Watts and Strogatz found the following algorithm. Starting from a regular lattice with the desired degree, each link has a probability p to be rewired, i.e. to be disconnected from one of its vertices and reconnected with another vertex chosen uniformly at random. The result is the creation of shortcuts in the regular structure. Watts and Strogatz (1998) imposed additional constraints on the rewiring process: a vertex cannot be linked to itself, and any two vertices cannot be linked by more than one edge. Moreover, the rewiring process only rewired one end and not both ends of the link. These added conditions prevent the resulting network from being a random graph even in the limit $p = 1$.

A simplified version of this model was proposed by Newman and Watts (1999). Starting with a lattice and taking each link one by one, add another link between a pair of vertices chosen at random with the probability p without removing the existing one. This corresponds to the addition of Lkp new links on average to the starting lattice, Lk being the initial number of links in the graph (Fig. 19.7).

Fig. 19.7 Average path length (*red*) and clustering coefficient (*green*) represented as the probability of rewiring evolves

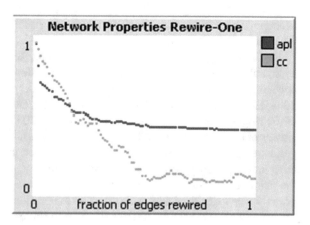

[2]Short path being defined by Watts and Strogatz as comparable to those found in random graphs of the same size and average degree.

A number of other models have been proposed to achieve the combination of short path length with high clustering coefficient. The oldest one is the random biased net of Rapoport (1957), in which clustering is added to a random graph by *triadic closure*, i.e. the deliberate completion of connected triples of vertices to form triangles in the network, thereby increasing the clustering coefficient. Another attempt to model clustering was made in the 1980s by Holland and Leinhardt (1981), using the class of network models known as *exponential random graphs*. Another method for generating clustering in networks could be membership of individuals in groups. This has been investigated by Newman et al. (2001; 2003b) using so-called *bipartite graph* models.

19.3.2.4 Scale-Free Networks

Even with the small-world effect, the hypothesis that complex systems such as cells or social systems are based upon components—i.e. molecules or individuals—randomly wired together has proven to be incomplete.

In fact, several empirical data analyses of real networks found out that for many systems, including citation networks, the World Wide Web, the Internet, and metabolic networks, the degree distribution approximates a power law (de Solla Price 1965; Albert et al. 1999; Faloutsos et al. 1999; Broder et al. 2000).

This corresponds to a new class of network since neither of the previously discussed networks such as random graphs and small-world models have a power-law degree distribution. Barabási and Albert (1999) called these graphs *scale-free networks* and proposed that power laws could potentially be a generic property of many networks and that the properties of these networks can be explained by having the graph grow dynamically rather than being static. Their paper proposes a specific model of a growing network that generates power-law degree distributions similar to those seen in the World Wide Web and other networks.

Their suggested mechanism has two components: (1) the network is growing, i.e. vertices are added continuously to it, and (2) vertices gain new edges in proportion to the number they already have, a process that Barabási and Albert call *preferential attachment*.[3] Therefore, the network grows by addition of a single new vertex at each time step, with m edges connected to it. The other end of the edge is chosen at random with probability proportional to degree.

$$P(k_i) = \frac{k_i}{\sum_j k_j} \tag{19.1}$$

---

[3]The preferential attachment mechanism has appeared in several different fields under different names. In information science, it is known as cumulative advantage (de Solla Price 1976), in sociology as the Matthew effect (Merton 1968) and in economics as the Gibrat principle (Simon 1955).

An extension to this model proposed by the same authors (Albert and Barabási 2000) follows another way of building the graph. In this model one of three events occurs at each time step:

- With probability p, m new edges are added to the network. One end of each new edge is connected to a node selected uniformly at random from the network and the other to a node chosen using the preferential attachment process according to the probability given just before.
- With probability q, m edges are rewired, meaning that a vertex i is selected at random, and one of its edges chosen at random is removed and replaced with a new edge whose other end is connected to a vertex chosen again according to the preferential attachment process.
- With probability $1 - p - q$, a new node is added to the network. The new node has m new edges that are connected to nodes already present in the system via preferential attachment in the normal fashion.

This model produces a degree distribution that again has a power-law tail, with an exponent γ that depends on the parameters p, q and m, and can vary anywhere in the range from 2 to ∞.

Dorogovtsev et al. (2000) consider a variation on the preceding model applied to directed graphs, in which preferential attachment takes place with respect only to the incoming edges at each vertex. Each vertex has m outgoing edges, which attach to other pre-existing vertices with attachment probability proportional only to those vertices' in-degree.

Another extension proposed by Krapivsky et al. (2000) explores what happens when $\Pi(k)$, the probability distribution of connectivity per node, is not linear in k. They studied the case in which $\Pi(k)$ takes the power-law form $\Pi(k) \propto k^\alpha$.

In a subsequent model, Bianconi and Barabási (2001), motivated by the Google effect (emergence of a new hub from an existing scale-free network), introduced the idea that some nodes are intrinsically "better" or faster growing than others. In the model, each vertex in the network has a fixed fitness value η_i, chosen from some distribution $\rho(\eta)$, that corresponds to an intrinsic ability to compete for edges at the expense of other vertices. Each such vertex connects to m others, the probability of connecting to vertex i being proportional to the product of the degree and the fitness of i:

$$\Pi = \frac{\eta_i k_i}{\sum_j \eta_j k_j} \qquad (19.2)$$

The use of models like preferential attachment in an agent-based social simulation context follows two motivations that are relatively distinct. On the one hand, such models are used as initial configurations for the simulation: therefore the construction of the scale-free network, even if it could be considered as a model of a growing network, should rather be envisaged as an algorithm to build the initial state of the simulation. On the other hand, it can also be seen as a subject of research, the focus of a model, for instance, in the field of modelling social network dynamics. In this case, preferential attachment mechanisms could be included in the model of social network evolution (if such a mechanism is relevant, of course) (Fig. 19.8).

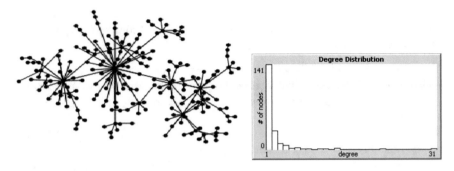

Fig. 19.8 Scale-free network generated using the preferential attachment model implemented in NetLogo (*left*) and the distribution of number of nodes per degree (*right*), which follows a scale-free distribution

As the structure and the evolution of a scale-free network cannot be considered as separate concepts, the same holds true for the topology and the overlying inter-actions among components. For instance, Pastor-Satorras and Vespignani (2001) discovered that the epidemic threshold on top of scale-free networks converges towards 0, meaning that in scale-free topologies, even weak viruses diffuse fast within the network. In particular, the dynamic aspects of real phenomena occurring on top of scale-free networks would allow investigating several properties of real social networks, for instance, how changes in the network structure trigger emergent phenomena by local interaction.

19.3.3   How to Distribute Agents on the Network?

This question deals with the efficiency of the model, because each link between nodes (agents) represents an interaction, and the shape of the topology, i.e. the interaction space that we are interested in, characterising its emergent properties. To exemplify, let us take the example of an opinion dynamics model including extremist agents that have a significant impact on the global dynamics. What should the initial state of the population be? Are the extremists homogeneously distributed in the population, i.e. over the social network, or are they organised in dense communities, or any case in between? This question could have an important impact on the emerging phenomenon and is the consequence of two choices: a modelling choice, i.e. what is the initial state of your model, and a more technical choice, i.e. the possibility to generate the distribution wanted. With regard to the latter, you can always generate an ad-hoc situation either using real data or building them artificially; however, it is often the case that you wish to use a generic way to distribute agents among a network. Usually, the starting point would be to use a random distribution. You first generate your network, and you distribute the different agents uniformly over this network assuming that each agent's state is independent from its location in the network. However, you do not have to believe strongly in any social determinism to question this independence and to wish to explore other kinds of distributions. Two cases occur then:

You have modelling hypotheses concerning the relation between the existence of a link between two agents and their actual state. In this case, you would probably proceed using a stochastic algorithm that enables some flexibility applying this relation.

You are only able to describe the global state you want to attain. In this case, you probably have to operate iterative permutations among the agents. For each permutation, you have to compute the result. If it is better (i.e. closer to the state wanted), then you keep the permutation, and if not, you reject this permutation and keep the previous state.

Note that the latter solution can become very expensive with regard to the evaluation of a given state and is not guaranteed to obtain the wanted final state in a finite time. Therefore, the most prudent approach would be to take the agents' characteristics into account when deciding on the presence of a link between two agents. Such a solution has been proposed by Thiriot and Kant (2008) explicitly drawing on agents' attributes and link types for the generation of interaction networks. In this perspective, the generated social network is a part of the modelling hypotheses, and it has been shown that the corresponding graphs differ a lot from the ones obtained with abstract models like the ones described in this section.

19.3.4 Which Measures to Use?

Once you have your model running, especially when it deals with agents changing their state over a (possibly evolving, irregular) social network, the first image is usually a messy network with overlapping links and nodes changing colours over time. Without specific efforts dedicated to the network layout (which is not a trivial task), the global picture is not understandable at first glance. Therefore, you would want to add some indicators enabling you to better understand what happens in the model. There are two main categories: the first one—which is currently under development—concerns indicators linking the states of nodes with their position in the network, while the more classical indicators of the second category help to characterise the structure of the network only.

Some important properties associated with graphs influence the characteristics of the dynamics taking place on the graph. This is mainly the case for the diameter, especially dealing with diffusion processes, and the clustering coefficient, dealing, for instance, with influence processes. In the particular case of a regular network, where all nodes are equal, it can be determined depending on the connectivity and the dimension of the graph. The average path length gives information that is equivalent. The clustering coefficient is defined locally as, for a given node, the rate of existing links among its neighbours compared to the number of possible links. Local redundancy of links is therefore important as an inertial process that can reinforce or go against an influence dynamics. In the following paragraphs, we will briefly describe the main indicators you can use to gain some insight about the phenomena emerging in the model.

The first question you could ask is, looking at a particular dimension of the agents' state vector, do they have a tendency to regroup or cluster according to this state or not. Or phrased in a different way: do connected agents tend to become similar? A useful indicator for this would be an averaged similarity measure over the network, calculating the mean distance among all connected agents of the population.

In order to characterise the two main features of small-world networks (i.e. small-world effect and high clustering), several indicators are used. The small-world effect is measured simply by averaging the distance among any pair of vertices in the graph. This results in the average path length index.

Concerning the clustering, a number of indicators have been proposed. Watts and Strogatz (1998) suggested the *clustering coefficient*, used classically in social network analysis and consisting in averaging a local clustering coefficient over all vertices of the network. The local clustering coefficient C_i of the vertex i is defined as the number of existing links among the neighbours of the vertex i, divided by the number of possible links among these neighbours. This quantity is 1 for a fully connected graph but tends towards 0 for a random graph as the graph becomes large.

The problem is that this indicator is heavily biased in favour of vertices with a low degree due to the small number of possible links (denominator) they have. When averaging the local clustering coefficient over all nodes without additional weighting, this could make a huge difference in the value of C. A better way to calculate the average probability of a pair of neighbours being connected is to count the total number of pairs of vertices on the entire graph that have a common neighbour and the total number of such pairs that are themselves connected, and divide the one by the other. Newman et al. (2001) have expressed this, i.e. the fraction of transitive triples, as:

$$C = 3 * \text{(number of triangles on the graph)}/\text{(number of connected triples of vertices)}$$

Another definition for clustering coefficient proposed by Newman (2003a) consists in calculating the probability that the friend of your friend is also your friend, resulting in:

$$C = 6 * \text{(number of triangles on the graph)}/\text{(number of paths of length 2)}$$

Dealing with small-world networks, Watts and Strogatz (1998) defined a network to have high clustering if $C \gg C_{rg}$ (this latter being the clustering coefficient for random graphs).

Watts and Strogatz defined a network to be a small-world network if it shows both of those properties, that is, if the mean vertex-vertex distance l is comparable with that on a random graph (l_{rg}), and the clustering coefficient is much greater than that for a random graph. Walsh (1999) used this idea to define the *proximity ratio*:

$$\mu = \frac{C/C_{rg}}{l/l_{rg}} \tag{19.3}$$

which is of order 1 on a random graph but much greater than 1 on a network obeying the Watts-Strogatz definition of a small-world network.

One of the most important properties of social networks is the so-called notion of *power*. As a shared definition of power is still object of debate, the design of metrics able to characterise its causes and consequences is a pressing challenge. In particular, the social network approach emphasises the concept of power as inherently relational, i.e. determined by the network topology. Hence, the focus must be put on the relative positions of nodes. In order to characterise such a property, the concept of *centrality* has emerged. The simplest centrality metric, namely, the degree centrality, measures the number of edges that connect a node to other nodes in a network. Over the years, many more complex centrality metrics have been proposed and studied, including status score (Katz 1953), α-centrality (Bonacich and Lloyd 2001), betweenness centrality (Freeman 1979) and several others based on the random walk, the most famous of which is the eigenvector centrality used by Google's PageRank algorithm (Page et al. 1999). The temporal declination of these concepts is meaningful, and Kossinets et al. (2008) have shown that nodes that are topologically more central are not necessarily central from a temporal point of view, hence the concept of temporal centrality. The temporalisation of network metrics is currently a pressing scientific challenge (Casteigts et al. 2010; Santoro et al. 2011).

19.3.5   What Kind of Network Effect Can You Anticipate?

As mentioned by Newman et al. (2006), the behaviour of dynamical systems on networks is one of the most challenging areas of research. At the end of their paper, Watts and Strogatz (1998) present one of the most tractable problems in the field that is the spread of disease using the SIR (susceptible/infective/removed) epidemic model. They measure the position of the epidemic threshold—the critical value of the probability r of infection of a susceptible individual by an infective one at which the disease first becomes an epidemic, spreading through the population rather than dying out. They found a clear decline in the critical r with increasing p, the probability of links rewiring, indicating that the small-world topology makes it easier for the disease to spread. They also found that the typical time for the disease to spread through the population decreases while increasing p.

One of the simplest models of the spread of non-conserved information on a network is one in which an idea or a disease starts at a single vertex and spreads first to all its neighbouring vertices. From these, it spreads to all of their neighbours, and so forth, until there are no accessible vertices left that have not yet been infected. This process is known in computer science as breadth-first search. Depending on the type of network model chosen, it could happen (this is mostly the case on real networks) that you obtain a large number of small components plus, optionally, a single giant component. If we simulate the spread of a rumour or disease on such a network using breadth-first search, either we will see a small outbreak corresponding to one of the small components or, if there is a giant component and we happen to start a breadth-first search within it, we will see a giant outbreak that fills a large portion of the system. The latter is precisely what we call an epidemic, when we are talking about disease, and the phase transition at which the giant component appears in a graph is also a phase transition between a regime in which epidemics are not possible and a regime in which they are. In fact, if the disease starts its breadth-first search at a randomly chosen vertex in the network, then the probability of seeing an epidemic is precisely equal to the fraction of the graph occupied by the giant component. By studying, either analytically or numerically, the behaviour of various epidemiological models on networks, the authors hope to get a better idea of how real diseases will behave, and in some cases, they have found entirely new behaviours that had not previously been observed in epidemiological models.

Ball et al. (1997) consider networks with two levels of mixing, meaning that each vertex in the network belongs both to the network as a whole and to one of a specified set of subgroups (e.g. family) with different properties of spread within the network. Disease spreading is again modelled using the SIR model. In the model, people can be in one of three states: susceptible (S), meaning they can catch the disease but haven't yet; infective (I), meaning they have caught the disease and can pass it on to others; and removed (R), meaning they have recovered from the disease and can neither pass it on nor catch it again, or they have died. Ball et al. (1997) found that the rapid spread of a disease within groups such as families can lead to epidemic outbreaks in the population as a whole, even when the probability of interfamily communication of the disease is low enough that epidemic outbreaks normally would not be possible. The reason for this is the following: if transmission between family members takes place readily enough that most members of a family will contract the disease once one of them does, then we can regard the disease as spreading on a super-network in which vertices are the families, not individuals. Roughly speaking, the spread of the disease between families will take place with n^2 times the normal person-to-person probability, where n is the number of people in a family.

An alternative approach to calculating the effect of clustering on SIR epidemics has been presented by Keeling (1999). What Keeling's method does is to include, in approximate form, the effect of the short-scale structure of the network—the clustering—but treat everything else using a standard fully mixed approximation. Thus, things like the effect of degree distributions is absent from his calculations. But the effect of clustering is made very clear. Keeling found that a lower fraction of the population needs to be vaccinated against a disease to prevent an epidemic if the clustering coefficient is high.

In another paper, Pastor-Satorras and Vespignani (2001) address the behaviour of an endemic disease model on networks with scale-free degree distributions. Their motivation for the work was an interest in the dynamics of computer virus infections, which is why they look at scale-free networks; computer viruses spread over the Internet and the Internet has a scale-free form, as demonstrated by Faloutsos et al. (1999). They used a derivation of the SIR model, the SIS (susceptible/infected/susceptible) model, which simply considers that individuals recover with no immunity to the disease and are thus immediately susceptible once they have recovered. In their work, Pastor-Satorras and Vespignani grow networks according to the scale-free model of Barabási and Albert (1999) and then simulate the SIS model on this network starting with some fixed initial number of infective computers. Pastor-Satorras and Vespignani do not find oscillations in the number of infected individuals for any value of the independent parameter of the model. No matter what value the parameter takes, the system is always in the epidemic regime; there is no epidemic threshold in this system. No matter how short a time computers spend in the infective state or how little they pass on the virus, the virus remains endemic. Moreover, the average fraction of the population infected at any one time decreases exponentially with the infectiousness of the disease.

Watts (2002) has looked at the behaviour of cascading processes. Unlike disease, the spread of some kinds of information, such as rumours, fashions, or opinion, depends not only on susceptible individuals having contacts with infective ones but also on their having contact with such individuals in sufficient numbers to persuade them to change their position or beliefs on an issue. People have a threshold for adoption of trends. Each individual has a threshold t for adoption of the trend being modelled, which is chosen at random from a specified distribution. When the proportion of a person's contacts that have already adopted the trend rises above this threshold, the person will also adopt it. This model is similar to the rioting model of Granovetter (1978). Watts gives an exact solution for his model on random graphs for the case where initially a low density of vertices has adopted the trend. The solution depends crucially on the presence of individuals who have very low thresholds t. In particular, there must exist a sufficient density of individuals in the network whose thresholds are so low that they will adopt the trend if only a single one of their neighbours does. As Watts argues, the trend will only propagate and cause a cascade if the density of these individuals is high enough to form a percolating subgraph in the network. The fundamental result of this analysis is that, as a function of the average degree z of a vertex in the graph, there are two different thresholds for cascading the spread of information. Below $z = 1$, no cascades happen because the network itself has no giant component. Cascades also cease occurring when z is quite large, the exact value depending on the chosen distribution of t. The reason for this upper threshold is that as z becomes large, the value of t for vertices that adopt the trend when only a single neighbour does so becomes small, and hence there are fewer of such vertices. For large enough z, these vertices fail to percolate and so cascades stop happening.

Another field of application deals with the robustness of networks. If we have information or disease propagating through a network, how robust is the propagation to failure or removal of vertices? The Internet, for example, is a highly robust network because there are many different paths by which information can get from any vertex to any other. The question can also be rephrased in terms of disease. If a certain fraction of all the people in a network are removed in some way from a network—by immunisation against disease, for instance—what effect will this have on the spread of the disease?

Albert et al. (2000) discuss network resilience for two specific types of model networks, random graphs and scale-free networks. The principal conclusion of the paper is that scale-free networks are substantially more robust to the random deletion of vertices than Erdös-Rényi random graphs, but substantially less robust to deletion specifically targeting the vertices with the highest degrees. The mean vertex-vertex distance in the scale-free network increases as vertices are deleted, but it does so much more slowly than in the random graph. Similarly, the size of the largest component goes down substantially more slowly as vertices are deleted in the scale-free network than in the random graph. By contrast, the random graph's behaviour is almost identical whether one deletes the highest-degree vertices or vertices chosen at random. Thus scale-free networks are highly robust to random failure, but highly fragile to targeted attacks. More recent work has shown that there are networks with even higher resilience than scale-free networks (Costa 2004; Rozenfeld and Ben-Avraham 2004; Tanizawa et al. 2005).

As a conclusion on network effects, the existence of isolated components and the average degree of the graph are the first important factors that play a role in the dynamics occurring in agent-based social simulation. After that, depending on the kind of phenomenon studied, social influence or diffusion dynamics, for instance, clustering coefficient and average path length should be considered.

19.4 Spatial Structure

19.4.1 Which Spatial Structure to Use?

Despite the increased use of social networks in agent-based simulation, in many cases, one has to take into account the spatial dimension. In this section, we will first present some issues dealing with this point, then some models that allow the distribution of a population of agents in a geographical space, as well as some measures to try and qualify such distributions and their effects on the simulation.

19.4.1.1 Torus or Not?

There may also be arbitrary effects introduced by the spatial bounds or limits placed on the phenomenon or study area. This occurs since spatial phenomena may be unbounded or have ambiguous transition zones in reality. In the model, ignoring spatial dependency or interaction outside the study area may create edge effects. For example, in Schelling's segregation model, there is a higher probability to stabilise when you have fewer neighbours (e.g. three in the corners or five on a border) for a particular density.

The choice of spatial bounds also imposes artificial shapes on the study area that can affect apparent spatial patterns such as the degree of clustering. A possible solution is similar to the sensitivity analysis strategy for the modifiable areal unit problem or MAUP: change the limits of the study area and compare the results of the analysis under each realisation. Another possible solution is to overbound the study area, i.e. to deliberately model an area that encompasses the actual study area. It is also feasible to eliminate edge effects in spatial modelling and simulation by mapping the region to a boundless object such as a torus or sphere.

19.4.2   How to Distribute Agents on the Space?

To begin with the simplest case, let us present the Poisson process in a 2D continuous space. Such a distribution can be used in particular to define the null hypothesis of a totally random structure for spatial distribution that will enable us to compare other kinds of spatial structures. In order to simulate a Poisson process of intensity λ on a domain of surface D, we first define the number of points N to be distributed, picking it at random from a Poisson law of parameter λD. For each point A_i, the coordinates x_i and y_i are therefore taken at random from a uniform law. For a rectangular domain, it is sufficient to bound the values of x and y depending on the studied domain. For more complex zones, the method can be the same, deleting the points in a larger rectangle that are not part of the specific zone. Studying a population where the number of agents N is known, we can use the same method without picking the number N at random and using the known value instead. The corresponding process is called a binomial process (Tomppo 1986). Moreover, note that because in a Poisson process random points can be infinitely close, the space occupied by the represented agents is not taken into account. Some processes have been developed to deal with this (Ripley 1977; Cressie 1993), corresponding to a random repartition of the points under the constraint that two points have to be at least $2R$ apart, where R is the specified radius of an agent. It can also be interpreted as the random repartition of nonoverlapping disks of radius R.

Practically, spatial structures are rarely totally random, and it is quite frequent to have more aggregated structures. The Neyman-Scott cluster process can be used to simulate such more elaborated structures (Ripley 1977; Cressie 1993). The Neyman-Scott process is built from a master Poisson process whose N_{ag} (number of aggregates) points are used as centres for the aggregates. Each aggregate is then composed of a random number of points whose positions are independent and distributed within a radius R around the centre of the aggregate (Fig. 19.9).

In order to simulate complex interaction, the Gibbs process can be useful. The general class of Gibbs processes allows obtaining various complex structures (Tomppo 1986). The main idea of the Gibbs process is to distribute the points of the pattern according to attraction/repulsion relations with a range. Such a process can be defined using a cost function $f(r)$ which represents the cost associated to the presence of two points in the pattern separated by a distance r. For a given point pattern, we can calculate a total cost, equal to the sum of the costs associated to each couple of points (see Eq. (19.4) below). By definition, the lower the total cost

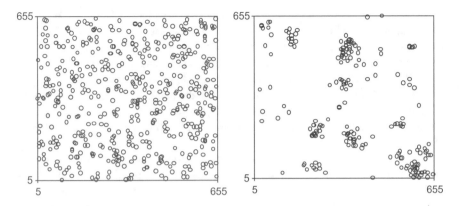

Fig. 19.9 Two different spatial distributions, random (*left*) and aggregated (*right*) from Goreaud (2000)

obtained on a pattern, the stronger is the probability for this pattern to be the result of the Gibbs process considered. For a given r, if the cost function $f(r)$ is positive, the probability is low to have two points at a distance r (we could say that there is repulsion at distance r). Conversely, if the cost function $f(r)$ is negative, it is highly probable to find a couple of points separated by a distance r (we could say there is attraction at a distance r):

$$\text{totalCost} = \sum_{i,j} f\left(d\left(A_i, A_j\right)\right) \qquad (19.4)$$

where d is the distance between points A_i and A_j.

To obtain the most probable realisation of the process, we aim to simulate patterns whose total costs are low enough, i.e. get closer to a minimum of the cost function. Such patterns can be obtained using an algorithm of depletions and replacements: starting, for instance, from a Poisson pattern, we iterate a big number of times a modification of this pattern, which consists in moving randomly an arbitrarily chosen point. If this move increases the total cost, the previous configuration is kept unchanged, else the new position is kept. Then, the attractions and repulsions among points lead to a progressive reorganisation of the original pattern towards a pattern that is more aggregated or more regular. Following Tomppo (1986), this algorithm converges quickly enough towards a realisation of the process.

In the preceding paragraphs, we were discussing the spatial distribution of homogeneous points. What if the points are heterogeneous? In the case of heterogeneous patterns that do not correspond to the superposition of independent distributions of different homogeneous groups, the preceding methodology can be used independently for each class of points.

In order to consider a more difficult case, let us consider that now each point can have a radius as well as a colour. More generally speaking, the variables associated to the points are called "marks", and the following approach can be applied either to qualitative (the colour) or quantitative (the radius) marks. We can then define stochastic processes whose realisations are marked point patterns. These are marked punctual processes, some mathematical objects that can generate an infinite number of marked point patterns, all different, but sharing some common properties, in particular the spatial structure of points and marks (Stoyan 1987; Goulard et al. 1995).

For a marked pattern, one can consider the position of points (the pattern as such) and the attribution of marks separately. The positioning of points can be done using tools and methods from the preceding part. Rather than to consider probabilities of having points separated by a distance r, we now have to reason about the joint probability P (see Eq. (19.5)) that two points at a distance r have marks m and m'. This function expresses neighbourhood relationships among values of a mark. The function $g_M(r, m, m')$ can moreover be linked to the number of neighbours having a mark m' at a distance r of a point marked m:

$$P\left(M\left(A_1\right) = m \text{ and } M\left(A_2\right) = m'\right) = \lambda_M(m)\, \lambda_M\left(m'\right) g_M\left(r, m, m'\right) \qquad (19.5)$$

The hypothesis of a totally random distribution of marks can be used as null hypothesis for marked stochastic processes. Then if, for a given case, the points of a subpopulation SP_2 are less often near the points of a subpopulation SP_1 than in the null hypothesis, we can speak about repulsion between the two subpopulations considered. Conversely, if the points from SP_2 are more numerous in the vicinity of SP_1 than in the null hypothesis, we can talk about attraction of SP_2 by SP_1.

One can easily use a "depletion and replacement" algorithm with an associated cost function in this particular case to generate marked point patterns of the wanted properties with regard to inter- and intra-attraction/repulsion among the elements of the different groups.

This situation can be seen as two different cases that could be treated using nearly the same methods: the case where mark values are drawn at random from a known statistical law and the case where mark values are fixed from an existing distribution (histogram, for instance) and where we have to distribute these mark values spatially. This corresponds to a random distribution of marks and to a list of given marks, respectively.

19.4.3 Which Measures to Use?

To analyse a spatial structure, there are several established methods from spatial statistics to characterise the spatial structure of point patterns (Ripley 1981; Diggle 1983; Cressie 1993).

One can distinguish between methods based on quadrants, the required data for which are the number of individuals in the quadrants, having variable position and

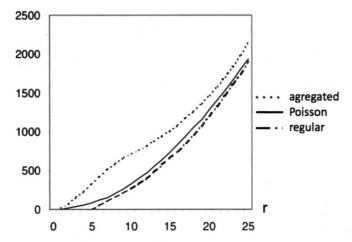

Fig. 19.10 Estimation of the $K(r)$ function for different repartitions (aggregated, Poisson and regular) from Goreaud (2000)

size (Chessel 1978), and methods based on distance, for which the distances among points or individuals or positions are used as input. Indicators à la "Clark and Evans" (Clark and Evans 1954) are classical examples of methods based on distance. They calculate for each point an index of spatial structure considering the n (fixed) closest neighbours. The average value of this index can then be compared to the theoretical value calculated for a null hypothesis (for instance, a Poisson process) in order to characterise the spatial structure with regard to aggregates or regularity, attraction or repulsion, etc. (Pretzsch 1993; Füldner 1995).

For a homogeneous punctual process having a density λ, Ripley (1976, 1977) showed that it is possible to characterise the second-order propriety using a function $K(r)$ such that $\lambda K(r)$ is the expected value of the number of neighbours at a distance r of a point chosen at random from the pattern. This function is linked directly to the function of density of a couple of points $g(r)$ defined previously:

$$\lambda K(r) = E \text{ (number of neighbours at distance} < r \text{ from } A_i)$$

$$K(r) = \int_{s=0}^{r} g(s)2\pi \ sds \tag{19.6}$$

For the Poisson process, which serves as null hypothesis, the expected value of the number of neighbours at distance r of a given point from the pattern is $\lambda \pi r^2$, and then $K(r) = \pi r^2$. For an aggregated process, the points have on average more neighbours than in the null hypothesis and thus $K(r) > \pi r^2$. Conversely, for a regular process, the points have on average fewer neighbours than in the null hypothesis and $K(r) < \pi r^2$ (Fig. 19.10). In the most frequent case where we do not know the process, the function $K(r)$ has to be estimated with the unique known realisation: the real point pattern. We then approach the expected value of the number of neighbours around a given point using its average on the whole set of individuals belonging to the pattern. We then obtain a first approximated indicator of $K(r)$, noted as $\widehat{K}(r)$ defined as follows:

$$\widehat{K}(r) = \frac{1}{\lambda} \frac{1}{N} \sum_{i=1}^{N} \sum_{j \neq i} k_{ij} \qquad (19.7)$$

N is the number of points in the studied domain of surface D; $\lambda = N/D$ is the estimator of the process density; and k_{ij} takes the value 1 if the distance between points i and j is below r and 0 otherwise.

This first estimator is unfortunately biased as it underestimates the values of $K(r)$ because of the points situated close to the border of the studied domain having fewer neighbours than a given point of the process. This general problem, known as border or edge effect and discussed in Sect. 4.1.1, is met in a lot of methods for the analysis of spatial structure.

To correct the border effect, two methods are known from the literature:

- Ohser and Stoyan (Stoyan et al. 1995) propose a global correction method, which is based on the geometrical properties of the studied domain. This method has the advantage of being quite simple, so quick enough to calculate.
- Ripley (1977) proposes a local correction method that is based on the contribution of each point situated nearby the border. This local correction has the advantage of being equally usable to calculate the individual indices relative to each point of the pattern, just as those proposed by Getis and Franklin (1987). Some works seem to show that this class of estimators is more robust than the estimators of the Ohser and Stoyan type (Kiêu and Mora 1999).

This latter solution consists in replacing the coefficient k_{ij} with the inverse of the proportion of the perimeter of the circle C_{ij} (centred on A_i and passing through A_j) for the points situated near the border of the domain studied. It corresponds to an estimation of the number of points situated at the same distance that would be outside of this domain. Ripley (1977) shows this estimator is not biased:

$$k_{ij} = \frac{\text{totalPerimeter}}{\text{PerimeterInsideTheZone}} = \frac{2\pi r}{C_{\text{inside}}} \geq 1 \qquad (19.8)$$

An alternative to the function $K(r)$ is the function $L(r)$ proposed by Besag (1977) which is easier to interpret. For a Poisson process, for any distance r, $L(r) = 0$. Aggregated processes and regular ones are situated under and above the x-axis, respectively:

$$L(r) = \sqrt{\frac{K(r)}{\pi}} - r \quad \text{estimated by} \quad \widehat{L}(r) = \sqrt{\frac{\widehat{K}(r)}{\pi}} - r \qquad (19.9)$$

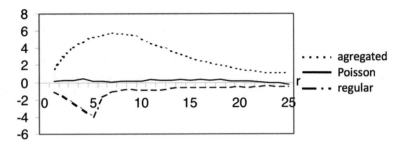

Fig. 19.11 Estimation of the $L(r)$ function for different repartitions (aggregated, Poisson and regular) from Goreaud (2000)

However, appropriate such methods of punctual processes are to deal with agent-based simulations and allow initialising and measuring and therefore characterising such systems, agent-based simulations have an important aspect that is not addressed by this literature: the dynamics. In many cases, the agents in such systems will move in the environment, and the spatial properties of the system are represented more accurately considering a set of trajectories of the agents rather than a succession of static spatial repartition. Even though punctual processes enable to capture spatial clustering effects, when considering, for example, the Boids model (Reynolds 1987), the dynamics of the flock will be overlooked. Therefore, there is a need for statistical tools that enable to characterise the interrelations between sets of trajectories to analyse some models properly (Fig. 19.11).

19.4.4   What Effects Can You Anticipate?

In spatial agent-based simulation, and without aiming at being exhaustive, the main phenomena you can look at and therefore characterise depend greatly on the population of agents in the model. Dealing with a homogeneous population of agents, you may observe spatial clustering, i.e. the formation of groups in space, that could be characterised by the methods presented beforehand. Introducing heterogeneity into the population, the main question deals with the spatial mixing of subpopulations, as present, for instance, in Schelling's segregation model. Dealing with a population of agents which may change their states, methods to characterise diffusion processes could be used. However, one of the most interesting points in this case does not really deal with the efficiency of the diffusion (evolution of the number of infected agents in an epidemiological model, for instance) but rather with the characterisation of the shape of such a diffusion. In this case, considering a diffusion process that takes place, you aim at characterising the shape of the interface rather than the global phenomenon that takes place. To this aim, fractal analysis gives interesting hints but is not appropriate for many phenomena, especially dynamical ones.

Finally, an important phenomenon that can occur at the macro-level of any spatial agent-based simulation concerns the density dependence of the results. For an identified phenomenon (if we take, for instance, the segregation model of Schelling), there will be an impact of the density of agents (either locally or globally) on the appearance of the phenomenon. This is exactly the case of segregation in Schelling's model, where a very low number of empty places (thus, a high density of agents) can freeze the system.

19.5   Conclusion

This chapter aims at presenting ways to deal with the distribution of agents in social simulation. Two kinds of distributions considered are distribution over a graph or social network on the one hand and spatial distribution on the other hand. While these two cases are very common in social simulation, too little effort is spent on either the characterisation or the investigation of the impact of the distribution on the final results. The methods presented in this chapter are certainly not exhaustive and not even pertinent to all cases, but they do present a first step towards pointing the curiosity of agent-based modellers at techniques that would definitely be useful for social simulation.

Further Reading

The literature on dynamic aspects of social networks is rapidly developing. For social network models and their analysis, we currently recommend Newman et al. (2006). For spatial aspects, we recommend Diggle (1983). For more details concerning random graphs models, we refer the interested reader to Bollobas (2001).

References

Albert, R., & Barabási, A.-L. (2000). Topology of evolving networks: Local events and universality. *Physical Review Letters, 85*, 5234–5237.
Albert, R., Jeong, H., & Barabási, A.-L. (1999). Diameter of the world-wide web. *Nature, 401*, 130–131.
Albert, R., Jeong, H., & Barabási, A.-L. (2000). Error and attack tolerance of complex networks. *Nature, 406*, 379–381.
Ball, F., Mollisson, D., & Scalia-Tomba, G. (1997). Epidemics with two levels of mixing. *Annals of Applied Probability, 7*, 46–89.
Barabási, A.-L., & Albert, R. (1999). Emergence of scaling in random networks. *Science, 286*, 509–512.

Bender, E. A., & Canfield, E. R. (1978). The asymptotic number of labeled graphs with given degree sequences. *Journal of Combinatorial Theory, Series A, 24*, 296–307.

Besag, J. (1977). Contribution to the discussion of Dr. Ripley's paper. *Journal of the Royal Statistical Society: Series B, 39*, 193–195.

Bianconi, G., & Barabási, A.-L. (2001). Competition and multiscaling in evolving networks. *Europhysics Letters, 90*, 436–442.

Bollobas, B. (2001). *Random graphs* (2nd ed.). New York: Academic.

Bonacich, P., & Lloyd, P. (2001). Eigenvector-like measures of centrality for asymmetric relations. *Social Networks, 23*(3), 191–201.

Broder, A., et al. (2000). Graph structure in the web. *Computer Networks, 33*, 309–320.

Callaway, D. S., Newman, M. E. J., Strogatz, S. H., & Watts, D. J. (2000). Network robustness and fragility: Percolation on random graphs. *Physical Review Letters, 85*, 5468–5471.

Casteigts, A., Flocchini, P., Quattrociocchi, W., & Santoro, N. (2010). Time-varying graphs and dynamic networks. arXiv:1012.0009v3.

Chessel, D. (1978). Description non paramétrique de la dispersion spatiale des individus d'une espèce. In J. M. Legay & R. Tomassone (Eds.), *Biométrie et écologie* (pp. 45–133). Paris: Société Française de Biométrie.

Chopard, B., & Droz, M. (1998). *Cellular automata modeling of physical systems*. Cambridge: Cambridge University Press.

Clark, P. J., & Evans, F. C. (1954). Distance to nearest neighbor as a measure of spatial relationships in populations. *Ecology, 35*(4), 445–453.

Cohen, R., Erez, K., Ben-Avraham, D., & Havlin, S. (2000). Resilience of the Internet to random breakdowns. *Physical Review Letters, 85*, 4626–4628.

Costa, L. D. (2004). Reinforcing the resilience of complex networks. *Physical Review E, 69*, 066127.

Cressie, N. A. C. (1993). *Statistics for spatial data*. New York: Wiley.

de Solla Price, D. J. (1965). Networks of scientific papers. *Science, 149*, 510–515.

de Solla Price, D. J. (1976). A general theory of bibliometric and other cumulative advantage processes. *Journal of the American Society for Information Science, 27*, 292–306.

Deffuant, G. (2006). Comparing extremism propagation patterns in continuous opinion models. *Journal of Artificial Societies and Social Simulation, 9*(3). http://jasss.soc.surrey.ac.uk/9/3/8.html

Diggle, P. J. (1983). *Statistical Analysis of Spatial Point Patterns*. New York: Academic.

Dorogovtsev, S. N., Mendes, J. F. F., & Samukhin, A. N. (2000). Structure of growing networks with preferential linking. *Physical Review Letters, 85*, 4633–4636.

Erdös, P., & Rényi, A. (1960). On the evolution of random graphs. *Publications of the Mathematics Institute of Hungarian Academy of Science, 5*, 17–61.

Faloutsos, M., Faloutsos, O., & Faloutsos, C. (1999). On power-law relationships of the internet topology. *Computer Communications Review, 29*, 251–262.

Flache, A., & Hegselmann, R. (2001). Do irregular grids make a difference? Relaxing the spatial regularity assumption in cellular models of social dynamics. *Journal of Artificial Societies and Social Simulation, 4*(4). http://www.soc.surrey.ac.uk/JASSS/4/4/6.html

Freeman, L. C. (1979). Centrality in social networks: Conceptual clarification. *Social Networks, 1*, 215–239.

Füldner, K. (1995). *Strukturbeschreibung von Buchen-Edellaubholz-Mischwäldern*. Göttingen: Cuvillier Verlag.

Getis, A., & Franklin, J. (1987). Second order neighborhood analysis of mapped point patterns. *Ecology, 68*(3), 473–477.

Gong, X., & Xiao, R. (2007). Research on multi-agent simulation of epidemic news spread characteristics. *Journal of Artificial Societies and Social Simulation, 10*(3). http://jasss.soc.surrey.ac.uk/10/3/1.html

Goreaud, F. (2000). *Apports de l'analyse de la structure spatiale en forêt tempérée à l'étude et la modélisation des peuplements complexes*. Diploma thesis, ENGREF (École nationale du génie rural des eaux et forêts), Nancy, France.

Goulard, M., Pages, L., & Cabanettes, A. (1995). Marked point process: Using correlation functions to explore a spatial data set. *Biometrical Journal, 37*(7), 837–853.

Granovetter, M. (1978). Threshold models of collective behaviour. *American Journal of Sociology, 83*(6), 1420–1443.

Holland, P. W., & Leinhardt, S. (1981). An exponential family of probability distributions for directed graphs. *Journal of the American Statistical Association, 76*, 33–65.

Holme, P., & Grönlund, A. (2005). Modelling the dynamics of youth subcultures. *Journal of Artificial Societies and Social Simulation, 8*(3). http://jasss.soc.surrey.ac.uk/8/3/3.html

Huet, S., Edwards, M., & Deffuant, G. (2007). Taking into account the variations of neighbourhood sizes in the mean-field approximation of the threshold model on a random network. *Journal of Artificial Societies and Social Simulation, 10*(1). http://jasss.soc.surrey.ac.uk/10/1/10.html

IMAGES. (2004). *Improving agri-environmental policies: A simulation approach to the role of the cognitive properties of farmers and institutions* (Final report of the FAIR3 CT 2092 - IMAGES project). http://wwwlisc.clermont.cemagref.fr/ImagesProject/FinalReport/final%20report%20V2last.pdf.

Jager, W., & Amblard, F. (2005). Uniformity, bipolarisation and pluriformity captured as generic stylised behaviour with an agent-based simulation model of attitude change. *Computational and Mathematical Organization Theory, 10*, 295–303.

Jeong, H., et al. (2002). Evolution of the social network of scientific collaborations. *Physica A, 311*, 590–614.

Katz, L. (1953). A new status derived from sociometric analysis. *Psychometrika, 18*, 39–43.

Keeling, M. J. (1999). The effects of local spatial structure on epidemiological invasion. *Proceedings of the Royal Society B: Biological Sciences, 266*, 859–867.

Kiêu, K., & Mora, M. (1999). Estimating the reduced moments of a random measure. *Advances in Applied Probability, 31*(1), 48–62.

Kossinets, G., Kleinberg, J., & Watts, D. (2008). The structure of information pathways in a social communication network. In Y. Li, B. Liu, & S. Sarawagi (Eds.), *Proceedings of the 14th ACM SIGKDD International Conference on Knowledge Discovery and Data Mining (KDD'08)* (pp. 435–443). New York: ACM.

Kottonau, J., & Pahl-Wostl, C. (2004). Simulating political attitudes and voting behaviour. *Journal of Artificial Societies and Social Simulation, 7*(4). http://jasss.soc.surrey.ac.uk/7/4/6.html

Krapivsky, P. L., Redner, S., & Leyvraz, F. (2000). Connectivity of growing random networks. *Physical Review Letters, 85*, 4629–4632.

Luczak, T. (1992). Sparse random graphs with a given degree sequence. In A. M. Frieze & T. Luczak (Eds.), *Proceedings of the Symposium on Random Graphs, Poznan 1989* (pp. 165–182). New York: Wiley.

Lusseau, D. (2003). The emergent properties of a dolphin social network. *Proceedings of the Royal Society B: Biological Sciences, 270*(2), S186–S188.

Merton, R. K. (1968). The Matthew effect in science. *Science, 159*, 56–63.

Meyer, M., Lorscheid, I., & Troitzsch, K. (2009). The development of social simulation as reflected in the first ten years of JASSS: A citation and co-citation analysis. *Journal of Artificial Societies and Social Simulation, 12*(4). http://jasss.soc.surrey.ac.uk/12/4/12.html

Molloy, M., & Reed, B. (1995). A critical point for random graphs with a given degree sequence. *Random Structures & Algorithms, 6*, 161–179.

Newman, M. E. J. (2001). The structure of scientific collaboration networks. *Proceedings of the National Academy of Sciences of the United States of America, 98*(2), 404–409.

Newman, M. E. J. (2003a). Ego-centered networks and the ripple effect. *Social Networks, 25*, 83–95.

Newman, M. E. J. (2003b). Properties of highly clustered networks. *Physical Review E, 68*, 016128.

Newman, M. E. J. (2004). Co-authorship networks and patterns of scientific collaboration. *Proceedings of the National Academy of Sciences of the United States of America, 101*, 5200–5205.

Newman, M. E. J., & Watts, D. J. (1999). Scaling and percolation in the small-world network model. *Physical Review E, 60*, 7332–7342.

Newman, M. E. J., Strogatz, S. H., & Watts, D. J. (2001). Random graphs with arbitrary degree distributions and their applications. *Physical Review E, 64*, 026118.

Newman, M. E. J., Barabási, A.-L., & Watts, D. J. (2006). *The structure and dynamics of networks*. Princeton, NJ: Princeton University Press.

Page, L., Brin, S., Motwani, R., & Winograd T. (1999). *The PageRank citation ranking: Bringing order to the web* (Technical Report, 1999–66. Stanford InfoLab, Stanford University). http://ilpubs.stanford.edu:8090/422/1/1999-66.pdf

Pastor-Satorras, R., & Vespignani, A. (2001). Epidemic spreading in scale-free networks. *Physical Review Letters, 86*, 3200–3203.

Pretzsch, H. (1993). *Analyse und Reproduktion räumlicher Bestandesstrukturen: Versuche mit dem Strukturgenerator STRUGEN* (Schriften aus der forstlichen Fakultät der Universität Göttingen und der N.F.V., 114). Universität Göttingen.

Pujol, J. M., Flache, A., Delgado, J., & Sanguesa, R. (2005). How can social networks ever become complex? Modelling the emergence of complex networks from local social exchanges. *Journal of Artificial Societies and Social Simulation, 8*(4). http://jasss.soc.surrey.ac.uk/8/4/12.html

Rapoport, A. (1957). Contribution to the theory of random and biased nets. *Bulletin of Mathematical Biophysics, 19*, 257–277.

Reynolds, C. W. (1987). Flocks, herds, and schools: A distributed behavioral model. *Computer Graphics, 21*(4), 25–34.

Ripley, B. D. (1976). The second order analysis of stationary point process. *Journal of Applied Probability, 13*, 255–266.

Ripley, B. D. (1977). Modelling spatial patterns. *Journal of the Royal Statistical Society. Series B, 39*, 172–212.

Ripley, B. D. (1981). *Spatial statistics*. New York: Wiley.

Rozenfeld, H. D., & Ben-Avraham, D. (2004). Designer nets from local strategies. *Physical Review E, 70*, 056107.

Santoro, N., Quattrociocchi, W., Flocchini, P., Casteigts, A., & Amblard, F. (2011). Time-varying graphs and social network analysis: Temporal indicators and metrics. In *Proceedings of SNAMAS 2011* (pp. 33–38).

Schelling, T. (1971). Dynamic Models of Segregation. *Journal of Mathematical Sociology, 1*(1), 143–186.

Simon, H. A. (1955). On a class of skew distribution functions. *Biometrika, 42*, 425–440.

Solomonoff, R., & Rapoport, A. (1951). Connectivity of random nets. *Bulletin of Mathematical Biophysics, 13*, 107–117.

Stocker, R., Green, D. G., & Newth, D. (2001). Consensus and cohesion in simulated social networks. *Journal of Artificial Societies and Social Simulation, 4*(4). http://jasss.soc.surrey.ac.uk/4/4/5.html

Stocker, R., Cornforth, D., & Bossomaier, T. R. J. (2002). Network structures and agreement in social network simulations. *Journal of Artificial Societies and Social Simulation, 5*(4). http://jasss.soc.surrey.ac.uk/5/4/3.html

Stoyan, D. (1987). Statistical analysis of spatial point process: A soft-core model and cross correlations of marks. *Biometrical Journal, 29*, 971–980.

Stoyan, D., Kendall, W. S., & Mecke, J. (1995). *Stochastic geometry and its applications*. New York: Wiley.

Tanizawa, T., Paul, G., Cohen, R., Havlin, S., & Stanley, H. E. (2005). Optimization of network robustness to waves of targeted and random attacks. *Physical Review E, 71*, 047101.

Thiriot, S., & Kant, J-D (2008). Generate country-scale networks of interaction from scattered statistics. In F. Squazzoni (Ed.), *Proceedings of the 5th Conference of the European Social Simulation Association, ESSA '08*, Brescia, Italy.

Tomppo, E. (1986). *Models and methods for analysing spatial patterns of trees, Communicationes Instituti Forestalis Fenniae* (Vol. 138). Helsinki: The Finnish Forest Research Institute.

Walsh, T. (1999). Search in a small world. In T. Dean (Ed.), *Proceedings of the 16th International Joint Conference on Artificial Intelligence, IJCAI '99* (pp. 1172–1177). San Francisco: Morgan Kauffman.

Watts, D. J. (1999). *Small worlds: The dynamics of networks between order and randomness.* Princeton, NJ: Princeton University Press.

Watts, D. J. (2002). A simple model of global cascades on random networks. *Proceedings of the National Academy of Sciences of the United States of America, 99*, 5766–5771.

Watts, D. J., & Strogatz, S. H. (1998). Collective dynamics of small world networks. *Nature, 393*, 440–442.

Wolfram, S. (1986). *Theory and applications of cellular automata.* Singapore: World Scientific.

Chapter 20
Learning

Michael W. Macy, Steve Benard, and Andreas Flache

Abstract Learning and evolution are adaptive or "backward-looking" models of social and biological systems. Learning changes the probability distribution of traits within an individual through direct and vicarious reinforcement, while evolution changes the probability distribution of traits within a population through reproduction and selection. Compared to forward-looking models of rational calculation that identify equilibrium outcomes, adaptive models pose fewer cognitive requirements and reveal both equilibrium and out-of-equilibrium dynamics. However, they are also less general than analytical models and require relatively stable environments. In this chapter, we review the conceptual and practical foundations of several approaches to models of learning that offer powerful tools for modeling social processes. These include the Bush-Mosteller stochastic learning model, the Roth-Erev matching model, feed-forward and attractor neural networks, and belief learning. Evolutionary approaches include replicator dynamics and genetic algorithms. A unifying theme is showing how complex patterns can arise from relatively simple adaptive rules.

Why Read This Chapter?
To understand the properties of various individual or collective learning algorithms and be able to implement them within an agent (where evolution is considered as a particular kind of collective learning).

M.W. Macy (✉)
Department of Information Science, Cornell University, Ithaca, NY 14853-7601, USA
e-mail: mwmacy@cornell.edu

S. Benard
Department of Sociology, Indiana University, Bloomington, IN, USA

A. Flache
Department of Sociology, ICS, University of Groningen, Groningen, The Netherlands

© Springer International Publishing AG 2017
B. Edmonds, R. Meyer (eds.), *Simulating Social Complexity*,
Understanding Complex Systems, https://doi.org/10.1007/978-3-319-66948-9_20

20.1 Introduction

Evolution and learning are basic explanatory mechanisms for consequentialist theories of adaptive self-organization in complex systems.[1] These theories are consequentialist in that behavioral traits are selected by their outcomes. Positive outcomes increase the probability that the associated trait will be repeated (in learning theory) or reproduced (in evolutionary theory), while negative outcomes reduce it. Explanatory outcomes might be rewards and punishments (in learning theory), survival and reproduction (in evolutionary models), systemic requisites (in functionalism), equilibrium payoffs (in game theory), or the interests of a dominant class (in conflict theory).

An obvious problem in consequentialist models is that the explanatory logic runs in the opposite direction from the temporal ordering of events. Behavioral traits are the explanandum and their outcomes the explanans. This explanatory strategy collapses into teleology unless mechanisms can be identified that bridge the temporal gap. While expected utility theory and game theory posit a forward-looking and analytic causal mechanism, learning and evolution provide a backward-looking and experiential link. In everyday life, decisions are often highly routine, with little conscious deliberation. These routines can take the form of social norms, protocols, habits, traditions, and rituals. Learning and evolution explain how these routines emerge, proliferate, and change in the course of consequential social interaction, based on *experience* instead of calculation. In these models, *repetition*, not prediction, brings the future to bear on the present, by recycling the lessons of the past. Through repeated exposure to a recurrent problem, the consequences of alternative courses of action can be iteratively explored, by the individual actor (learning) or by a population (evolution).

Backward-looking rationality is based on rules rather than choices (Vanberg 1994). A choice is an instrumental, case-specific comparison of alternative courses of action, while rules are behavioral routines that provide standard solutions to recurrent problems. Rules can take the form of strategies, norms, customs, habits, morals, conventions, traditions, rituals, or heuristics. Rule-based decision-making is backward-looking in that the link between outcomes and the actions that produce them runs backward in time. The outcomes that explain the actions are not those the action will produce in the future; they are the outcomes that were previously experienced when the rule was followed in the past.

Learning alters the probability distribution of behavioral traits within a given individual, through processes of direct and vicarious reinforcement. Evolution alters the frequency distribution of traits within a population, through processes of reproduction and selection. Whether selection operates at the individual or population level, the units of adaptation need not be limited to human actors but may include larger entities such as firms or organizations that adapt their behavior

---

[1]Much of the material in this chapter has been previously published in Macy (1996, 1997, 1998, 2004) Macy and Flache (2002), and Flache and Macy (2002).

in response to environmental feedback. Nor is evolutionary adaptation limited to genetic propagation. In cultural evolution, norms, customs, conventions, and rituals propagate via role modeling, occupational training, social influence, imitation, and persuasion. For example, a firm's problem-solving strategies improve over time through exposure to recurrent choices, under the relentless selection pressure of market competition. Suboptimal routines are removed from the repertoires of actors by learning and imitation, and any residuals are removed from the population by bankruptcy and takeover. The outcomes may not be optimal, but we are often left with well-crafted routines that make their bearers look much more calculating than they really are (or need to be), like a veteran outfielder who catches a fly ball as if she had computed its trajectory.

20.2 Evolution

Selection pressures influence the probability that particular traits will be replicated, in the course of competition for scarce resources (ecological selection) or competition for a mate (sexual selection). Although evolution is often equated with ecological selection, sexual selection is at least as important. By building on partial solutions rather than discarding them, genetic crossover in sexual reproduction can exponentially increase the rate at which a species can explore an adaptive landscape, compared to reliance on trial and error. Paradoxically, sexual selection can sometimes inhibit ecological adaptation, especially among males. Gender differences in parental investment cause females to be choosier about mates and thus sexual selection to be more pronounced in males. An example is the peacock's large and cumbersome tail, which attracts the attention of peahens (who are relatively drab) as well as predators. Sexually selected traits tend to become exaggerated as males trap one another in an arms race to see who can have the largest antlers or to be bravest in battle.

Selection pressures can operate at multiple levels in a nested hierarchy, from groups of individuals with similar traits down to individual carriers of those traits, down to the traits themselves. Evolution through group selection was advanced by Wynne-Edwards (1962, 1986) as a solution to one of evolution's persistent puzzles—the viability of altruism in the face of egoistic ecological counterpressures. Pro-social in-group behavior confers a collective advantage over rival groups of rugged individualists. However, the theory was later dismissed by Williams in *Adaptation and Natural Selection* (Williams 1966), which showed that between-group variation gets swamped by within-group variation as group size increases. Moreover, group selection relies entirely on differential rates of extinction, with no plausible mechanism for the whole-cloth replication of successful groups.

Sexual selection suggests a more plausible explanation for the persistence of altruistic behaviors that reduce the chances of ecological selection. Contrary to Herbert Spencer's infamous view of evolution as "survival of the fittest," generosity can flourish even when these traits are ecologically disadvantageous, by attracting females who have evolved a preference for "romantic" males who are ready to sacrifice for their partner. Traits that reduce the ecological fitness of an individual carrier can also flourish if the trait increases the selection chances of other individuals with that trait. Hamilton (1964) introduced this gene-centric theory of kin altruism, later popularized by Dawkins' in the *Selfish Gene* (Dawkins 1976).

Allison (1992) extended the theory to benevolence based on cultural relatedness, such as geographical proximity or a shared cultural marker. This may explain why gene-culture coevolution seems to favor a tendency to associate with those who are similar, to differentiate from "outsiders," and to defend the in-group against social trespass with the emotional ferocity of parents defending their offspring.

This model also shows how evolutionary principles initially developed to explain biological adaptation can be extended to explain social and cultural change. Prominent examples include the evolution of languages, religions, laws, organizations, and institutions. This approach has a long and checkered history. Social Darwinism is a discredited nineteenth-century theory that used biological principles as analogs for social processes such as market competition and colonial domination. Many sociologists still reject all theories of social or cultural evolution, along with biological explanations of human behavior, which they associate with racist and elitist theories of "survival of the fittest." Others, like the sociobiologist E. O. Wilson (1988, p. 167), believe "genes hold culture on a leash," leaving little room for cultural evolution to modify the products of natural selection. Similarly, evolutionary psychologists like Cosmides and Tooby search for the historical origins of human behavior as the product of ancestral natural selection rather than ongoing social or cultural evolution.

In contrast, a growing number of sociologists and economists are exploring the possibility that human behaviors and institutions may be heavily influenced by processes of social and cultural selection that are independent of biological imperatives. These include DiMaggio and Powell (the new institutional sociology), Nelson and Winter (evolutionary economics), and Hannan and Freeman (organizational ecology).

One particularly compelling application is the explanation of cultural diversity. In biological evolution, speciation occurs when geographic separation allows populations to evolve in different directions to the point that individuals from each group can no longer mate. Speciation implies that all life has evolved from a very small number of common ancestors, perhaps only one. The theory has been applied to the evolution of myriad Indo-European languages that are mutually incomprehensible despite having a common ancestor. In sociocultural models, speciation operates through homophily (attraction to those who are similar), xenophobia (aversion to those who are different), and influence (the tendency to become more similar to those to whom we are attracted and to differentiate from those we despise).

Critics counter that sociocultural evolutionists have failed to identify any underlying replicative device equivalent to the gene. Dawkins has proposed the "meme" as the unit of cultural evolution, but there is as yet no evidence that these exist. Yet Charles Darwin developed the theory of natural selection without knowing that phenotypes are coded genetically in DNA. Perhaps the secrets of cultural evolution are waiting to be unlocked by impending breakthroughs in cognitive psychology.

The boundary between learning and evolution becomes muddied by a hybrid mechanism, often characterized as "cultural evolution." In cultural evolution, norms, customs, conventions, and rituals propagate via role modeling, occupational training, social influence, and imitation. Cultural evolution resembles learning in that the rules are soft wired and can therefore be changed without replacing the carrier. Cultural evolution also resembles biological evolution in that rules used by successful carriers are more likely to propagate to other members of the population. However, because cultural rules are soft wired, the rules can propagate without replacing the carriers. For example, norms can jump from one organism to another by imitation (Dawkins 1976; Durham 1992; Boyd and Richerson 1985; Lopreato 1990). A successful norm is one that can cause its carrier to act in ways that increase the chances that the norm will be adopted by others. Cultural evolution can also be driven by social learning (Bandura 1977) in which individuals respond to the effects of vicarious reinforcement. Social learning and role modeling can provide an efficient shortcut past the hard lessons of direct experience.

Imitation of successful role models is the principal rationale for modeling cultural evolution as an analog of natural selection (Boyd and Richerson 1985; Dawkins 1976). However, social influence differs decisively from sociobiological adaptation. Softwired rules can spread without replacement of their carriers, which means that reproductive fitness loses its privileged position as the criteria for replication. While "imitation of the fittest" is a reasonable specification of cultural selection pressures, it is clearly not the only possibility. Replication of hardwired rules may be a misleading model for cultural evolution, and researchers need to be cautious in using Darwinian analogs as templates for modeling the diffusion of cultural rules. In cultural models of "imitation of the fittest," actors must not only know which actor is most successful; they must also know the underlying strategy that is responsible for that success. Yet successful actors may not be willing to share this information. For very simple strategies, it may be sufficient to observe successful behaviors. However, conditional strategies based on "if-then" rules cannot always be deduced from systematic observation. Researchers should therefore exercise caution in using biological models based on Darwinian principles to model cultural evolution, which is a hybrid of the ideal types of evolution and learning.

20.3 Learning

The most elementary principle of learning is simple reinforcement. Thorndike (1898) first formulated the theory of reinforcement as the "law of effect," based on the principle that "pleasure stamps in, pain stamps out." If a behavioral response has a favorable outcome, the neural pathways that triggered the behavior are strengthened, which "loads the dice in favor of those of its performances which make for the most permanent interests of the brain's owner" (James 1981, p. 143). This connectionist theory anticipates the error back propagation used in contemporary neural networks (Rumelhart and McClelland 1988). These models show how highly complex behavioral responses can be acquired through repeated exposure to a problem.

Reinforcement theory relaxes three key behavioral assumptions in models of forward-looking rationality:

1. Propinquity replaces causality as the link between choices and payoffs.
2. Reward and punishment replace utility as the motivation for choice.
3. Melioration replaces optimization as the basis for the distribution of choices over time.

We consider each of these in turn.

1. *Propinquity, not causality.* Compared to forward-looking calculation, the law of effect imposes a lighter cognitive load on decision makers by assuming experiential induction rather than logical deduction. Players explore the likely consequences of alternative choices and develop preferences for those associated with better outcomes, even though the association may be coincident, "super-stitious," or causally spurious. The outcomes that matter are those that have already occurred, not those that an analytical actor might predict in the future. Anticipated outcomes are but the consciously projected distillations of prior exposure to a recurring problem. Research using fMRI supports the view that purposive assessment of means and ends can take place *after* decisions are made, suggesting that "rational choice" may be not so much a theory of decision but a theory of how decisions are rationalized to self and others.

   Reinforcement learning applies to both intended and unintended conse-quences of action. Because repetition, not foresight, links payoffs back to the choices that produce them, learning models need not assume that the payoffs are the intended consequences of action. Thus, the models can be applied to expressive behaviors that lack a deliberate or instrumental motive. Frank's (1988) evolutionary model of trust and commitment formalizes the backward-looking rationality of emotions like vengeance and sympathy. An angry or frightened actor may not be capable of deliberate and sober optimization of self-interest, yet the response to the stimulus has consequences for the individual, and these in turn can modify the probability that the associated behavior will be repeated.

2. *Reward and punishment, not utility.* Learning theory differs from expected utility theory in positing two distinct cognitive mechanisms that guide decisions toward better outcomes, *approach* (driven by reward) and *avoidance* (driven by punishment). The distinction means that aspiration levels are very important for learning theory. The effect of an outcome depends decisively on whether it is coded as gain or loss, satisfactory or unsatisfactory, pleasant or aversive.

3. *Melioration, not optimization.* Melioration refers to suboptimal gradient climb-ing when confronted with what Herrnstein and Drazin (1991) call "distributed choice" across recurrent decisions. A good example of distributed choice is the decision whether to cooperate in an iterated prisoner's dilemma game.

Suppose each side is satisfied when the partner cooperates and dissatisfied when the partner defects. Melioration implies a tendency to repeat choices with satisfactory outcomes even if other choices have higher utility, a behavioral tendency March and Simon (1958) call "satisficing." In contrast, unsatisfactory outcomes induce searching for alternative outcomes, including a tendency to revisit alternative choices whose outcomes are even worse, a pattern we call "dissatisficing." While satisficing is suboptimal when judged by conventional game-theoretic criteria, it may be more effective in leading actors out of a suboptimal equilibrium than if they were to use more sophisticated decision rules, such as "testing the waters" to see if they could occasionally get away with cheating. Gradient search is highly path dependent and not very good at backing out of evolutionary cul-de-sacs. Course correction can sometimes steer adaptive individuals to globally optimal solutions, making simple gradient climbers look much smarter than they need to be. Often, however, adaptive actors get stuck in local optima. Both reinforcement and reproduction are biased toward *better* strategies, but they carry no guarantee of finding the highest peak on the adaptive landscape, however relentless the search. Thus, learning theory can be usefully applied to the equilibrium selection problem in game theory. In repeated games (such as an ongoing prisoner's dilemma), there is often an indefinitely large number of analytic equilibria. However, not all these equilibria are learnable, either by individuals (via reinforcement) or by populations (via evolution). Learning theory has also been used to identify a fundamental solution concept for these games—stochastic collusion—based on a random walk from a self-limiting noncooperative equilibrium into a self-reinforcing cooperative equilibrium (Macy and Flache 2002).

20.4   Modeling Evolution

Replicator dynamics are the most widely used model of evolutionary selection (Taylor and Jonker 1978). In these models, the frequency of a strategy changes from generation to generation as a monotonic function of its "payoff advantage," defined in terms of the difference between the average payoff of that strategy and the average payoff in the population as a whole. The more successful a strategy is on average, the more frequent it tends to be in the next generation.

Replicator dynamics typically assume that in every generation, every population member encounters every other member exactly once, and replication is based on the outcome of this interaction relative to the payoff earned by all other members of the population. However, in natural settings, actors are unlikely to interact with or have information about the relative success of every member of a large population. The mechanism can also be implemented based on local interaction (limited to network "neighbors") and local replication (neighbors compete only with one another for offspring).

The outcomes of replicator dynamics depend on the initial distribution of strategies, since the performance of any given strategy will depend on its effectiveness in interaction with other strategies. For example, aggressive strategies perform much better in a population that is accommodating than one that is equally aggressive. It is also not possible for replicator dynamics to invent new strategies that were not present at the outset.

These limitations are minimized by using genetic algorithms. The genetic algorithm was proposed by Holland (1975) as a problem-solving device, modeled after the recursive system in natural ecologies. The algorithm provides a simple but elegant way to write a computer program that can improve through experience. The program consists of a string of symbols that carry machine instructions. The symbols are often binary digits called "bits" with values of 0 and 1. The string is analogous to a chromosome containing multiple genes. The analog of the gene is a bit or combination of bits that comprises a specific instruction. The values of the bits and bit combinations are analogous to the alleles of the gene. A one-bit gene has two alleles (0 and 1), a two-bit gene has four alleles (00, 01, 10, and 11), and so on. The number of bits in a gene depends on the instruction. An instruction to go left or right requires only a single bit. However, an instruction to go left, right, up, or down requires two bits. When the gene's instructions are followed, there is some performance evaluation that measures the program's reproductive fitness relative to other programs in a computational ecology. Relative fitness determines the probability that each strategy will propagate. Propagation occurs when two mated programs recombine through processes like "crossover" and "inversion." In crossover, the mated programs (or strings) are randomly split, and the "left" half of one string is combined with the "right" half of the other, and vice versa, creating two new strings. If two different protocols are each effective, but in different ways, crossover allows them to create an entirely new strategy that may combine the best abilities of each parent, making it superior to either. If so, then the new rule may go on to eventually displace both parent rules in the population of strategies. In addition, the new strings contain random copying errors. These mutations continually refresh the heterogeneity of the population, in the face of selection pressures that tend to reduce it.

To illustrate, consider the eight-bit string **10011010** mated with *11000101*. (The typefaces might represent gender, although the algorithm does not require sexual reproduction.) Each bit could be a specific gene, such as whether to trust a partner under eight different conditions (Macy and Skvoretz 1998). In mating, the two parent strings are randomly broken, say after the third gene. The two offspring would then be **100***00101* and *110***11010**. However, a chance copying error on the last gene might make the second child a mutant, with *110***11011**. At the end of each generation, each individual's probability of mating is a monotonic (often linear) function of relative performance during that generation, based on stochastic sampling (Goldberg 1989):

$$P_{ij} = \frac{F_i}{\sum\limits_{n=1}^{N} F_n} \text{ for } j = 1 \text{ to } N, \ j \neq i \qquad (20.1)$$

where P_{ij} is the probability that j is mated with i, F_i is i's "fitness" (or cumulative payoff over all previous rounds in that generation), and N is the size of the population. If the best strategy had only a small performance edge over the worst, it had only a small edge in the race to reproduce. With stochastic sampling, each individual, even the least fit, selects a mate from the fitness-weighted pool of eligibles. In each pairing, the two parents combined their chromosomes to create a single offspring that replaces the less fit parent. The two chromosomes are combined through crossover.

20.5   Learning Models

The need for a cognitive alternative to evolutionary models is reflected in a growing number of formal learning-theoretic models of behavior (Macy 1991; Roth and Erev 1995; Fudenberg and Levine 1998; Young 1998; Cohen et al. 2001). In general form, learning models consist of a probabilistic decision rule and a learning algorithm in which outcomes are evaluated relative to an aspiration level, and the corresponding decision rules are updated accordingly.

All stochastic learning models share two important principles, the law of effect and probabilistic decision-making (Macy 1989, 1991; Börgers & Sarin 1997; Roth and Erev 1995; Erev and Roth 1998, Erev et al. 1999; for more references cf. Erev et al. 1999). The law of effect implies that the propensity of an action increases if it is associated with a positively evaluated outcome, and it declines if the outcome is negatively evaluated. Probabilistic choice means that actors hold a propensity qX for every action X. The probability pX to choose action X then increases in the magnitude of the propensity for X relative to the propensities for the other actions.

Whether an outcome is evaluated as positive or negative depends on the evaluation function. An outcome is positive if it exceeds the actor's aspirations. There are basically three substantively different approaches for modeling the aspiration level, fixed interior aspiration, fixed zero aspiration, and moving average aspiration. Fixed interior aspiration assumes that some payoffs are below the aspiration level and are evaluated negatively, while other payoffs are above the aspiration level and are evaluated positively (e.g., Macy 1989, 1991; Fudenberg and Levine 1998). The fixed zero aspiration approach also fixes the aspiration level, but it does so at the minimum possible payoff (Roth and Erev 1995; Börgers and Sarin 1997; Erev and Roth 1998). In other words, in the fixed zero aspiration model, every payoff is deemed "good enough" to increase or at least not reduce the corresponding propensity, but higher payoffs increase propensities more than lower payoffs do. Finally, moving average aspiration models assume that the aspiration level approaches the average of the payoffs experienced recently, so that players get used to whatever outcome they may experience often enough (Macy and Flache 2002; Börgers and Sarin 1997; Erev and Rapoport 1998; Erev et al. 1999). Clearly, these assumptions have profound effects on model dynamics.

20.5.1 Bush-Mosteller Stochastic Learning Model

One of the simplest models of reinforcement learning is the Bush-Mosteller model (Bush and Mosteller 1950). The Bush-Mosteller stochastic learning algorithm updates probabilities following an action a as follows:

$$
p_{a,t+1} = \begin{cases} p_{a,t} + (1 - p_{a,t}) \; l \, s_{a,t}, & s_{a,t} \geq 0 \\[2mm] p_{a,t} + p_{a,t} \; l \; s_{a,t}, & s_{a,t} < 0 \end{cases} \quad , \quad a \in \{C, D\} \tag{20.2}
$$

In Eq. (20.2), $p_{a,t}$ is the probability of action a at time t, and $s_{a,t}$ is a positive or negative stimulus ($0 \leq s_{a,t} \leq 1$). The change in the probability for the action not taken, b, obtains from the constraint that probabilities always sum to one, i.e., $p_{b,t+1} = 1 - p_{a,t+1}$. The parameter l is a constant ($0 < l < 1$) that scales the learning rate. With $l \approx 0$, learning is very slow, and with $l \approx 1$, the model approximates a "win-stay, lose-shift" strategy (Catania 1992).

For any value of l, Eq. (20.2) implies a decreasing effect of reward as the associated propensity approaches unity, but an increasing effect of punishment. Similarly, as the propensity approaches zero, there is a decreasing effect of punishment and a growing effect of reward. This constrains probabilities to approach asymptotically their natural limits.

20.5.2 The Roth-Erev Matching Model

Roth and Erev (Roth and Erev 1995; Erev and Roth 1998; Erev et al. 1999) have proposed a learning-theoretic alternative to the Bush-Mosteller formulation. Their model draws on the "matching law" which holds that adaptive actors will choose between alternatives in a ratio that matches the ratio of reward. Like the Bush-Mosteller model, the Roth-Erev payoff matching model implements the three basic principles that distinguish learning from utility theory—experiential induction (vs. logical deduction), reward and punishment (vs. utility), and melioration (vs. optimization). The similarity in substantive assumptions makes it tempting to assume that the two models are mathematically equivalent or, if not, that they nevertheless give equivalent solutions.

On closer inspection, however, we find important differences, identified by Flache and Macy (2002). Each specification implements reinforcement learning in different ways and with different results. Roth and Erev (1995, Erev & Roth 1998) propose a baseline model of reinforcement learning with a fixed zero reference point. The law of effect is implemented such that the propensity for action X is simply the sum of all payoffs a player ever experienced when playing X. The probability to choose action X at time t is then the propensity for X divided by the sum of all action propensities at time t. The sum of the propensities increases over time, such that payoffs have decreasing effects on choice probabilities. However, this also undermines the law of effect. Suppose, after some time, a new action is carried out and yields a higher payoff than every other action experienced before. The probability of repetition of this action will nevertheless be negligible, because its recent payoff is small in comparison with the accumulated payoffs stored in the propensities for the other actions. As a consequence, the baseline model of Roth and Erev (1995) fails to identify particular results, because it has the tendency to lock the learning dynamics into any outcome that occurs sufficiently often early on. Roth and Erev amend this problem by introducing a "forgetting parameter" that keeps propensities low relative to recent payoffs. With this, they increase the sensitivity of the model to recent reinforcement. Roth and Erev used a variant of this baseline model to estimate globally applicable parameters from data collected across a variety of human subject experiments. They concluded that "low rationality" models of reinforcement learning may often provide a more accurate prediction than forward-looking models. Like the Bush-Mosteller, the Roth-Erev model is stochastic, but the probabilities are not equivalent to propensities. The propensity q for action a at time T is the sum of all stimuli s_a a player has ever received when playing a:

$$q_{a,T} = \sum_{t=1}^{T} s_{a,t}, \quad a \in \{C, D\}. \tag{20.3}$$

Roth and Erev then use a "probabilistic choice rule" to translate propensities into probabilities. The probability p_a of action a at time $t+1$ is the propensity for a divided by the sum of the propensities at time t:

$$p_{a,t+1} = \frac{q_{a,t}}{q_{a,t} + q_{b,t}}, \quad (a, b) \in \{C, D\}, a \neq b \tag{20.4}$$

where a and b represent binary choices. Following action a, the associated propensity q_a increases if the payoff is positive relative to aspirations (by increasing the numerator in Eq. (20.4)) and decreases if negative. The propensity for b remains constant, but the probability of b declines (by increasing the denominator in the equivalent expression for $p_{b,t+1}$).

20.5.3 Artificial Neural Networks

Bush-Mosteller and Roth-Erev are very simple learning models that allow an actor to identify strategies that generally have more satisfactory outcomes. However, the actor cannot learn the conditions in which a strategy is more or less effective. Artificial neural nets add perception to reinforcement, so that actors can learn conditional strategies.

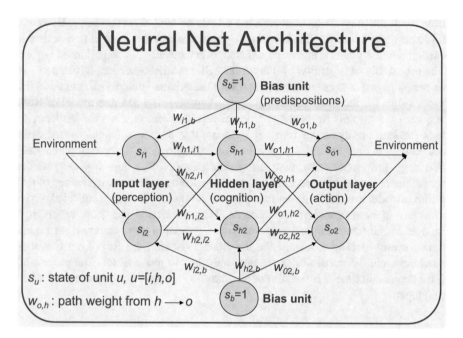

Fig. 20.1 Simple example of a feed-forward network with one hidden layer

An artificial neural network is a simple type of self-programmable learning device based on parallel distributed processing (Rumelhart and McClelland 1988). Like genetic algorithms, neural nets have a biological analog, in this case, the nerve systems of living organisms. In elementary form, the device consists of a web of neuron-like nodes (or neurodes) that fire when triggered by impulses of sufficient strength and in turn stimulate other nodes when fired. The magnitude of an impulse depends on the strength of the connection (or "synapses") between the two neurodes. The network learns by modifying these path coefficients, usually in response to environmental feedback about its performance.

There are two broad classes of neural networks that are most relevant to social scientists, feed-forward networks, and attractor networks. Feed-forward networks consist of four types of nodes, usually arranged in layers, as illustrated in Fig. 20.1. The most straightforward are input (sensory) and output (response) nodes. Input nodes are triggered by stimuli from the environment. In Fig. 20.1, there are two input nodes. I1 has been activated by the environment ($+1$), while I2 has not (-1). Output nodes, in turn, trigger action by the organism on the environment. In Fig. 20.1, output node O1 has been triggered, as indicated by the output value of $+1$.

The other two types of nodes are less intuitive. Intermediate (or "hidden") nodes link sensory and response nodes so as to increase the combinations of multiple stimuli that can be differentiated. Figure 20.1 shows a network with a single layer containing two hidden nodes, H1 and H2. The number of hidden layers and the number of hidden nodes in each layer vary with the complexity of the stimulus patterns the network must learn to recognize and the complexity of the responses the network must learn to perform. Unlike sensory and response nodes, hidden nodes have no direct contact with the environment, hence their name.

Bias nodes are a type of hidden node that has no inputs. Instead, a bias node continuously fires, creating a predisposition toward excitation or inhibition in the nodes it stimulates, depending on the size and sign of the weights on the pathway from the bias node to the nodes it influences. If the path weight is positive, the bias is toward excitation, and if the path is negative, the bias is toward inhibition. The weighted paths from the bias node to other hidden and output nodes correspond to the activation thresholds for these nodes.

A feed-forward network involves two processes—action (firing of the output nodes) and learning (error correction). The action phase consists of the forward propagation of influence (either excitation or inhibition) from the input nodes to the hidden nodes to the output nodes. The influence of a node depends on the state of the node and the weight of the neural pathway to a node further forward in the network. The learning phase consists of the backward propagation of error from the output nodes to the hidden nodes to the input nodes, followed by the adjustment of the weights so as to reduce the error in the output nodes.

The action phase begins with the input nodes. Each node has a "state" which can be binary (e.g., 0 or 1 to indicate whether the node "fires") or continuous (e.g., 9 to indicate that the node did not fire as strongly as one whose state is 1.0). The states of the input nodes are controlled entirely by the environment and correspond to a pattern that the network perceives. The input nodes can influence the output nodes directly as well as via hidden nodes that are connected to the input nodes.

To illustrate, consider a neural network in which the input nodes are numbered from $i = 1$ to I. The ith input is selected, and the input register to all nodes j influenced by i is then updated by multiplying the state of i times the weight w_{ij} on the ij path. This updating is repeated for each input node. The pathways that link the nodes are weighted with values that determine the strength of the signals moving along the path. A low absolute value means that an input has little influence on the output. A large positive weight makes the input operate as an excitor. When it fires, the input excites an otherwise inhibited output. A large negative path weight makes the input operate as an inhibitor. When it fires, the input inhibits an otherwise excited output.

Next, the nodes whose input registers have been fully updated (e.g., the nodes in the first layer of hidden nodes) must update their states. The state of a node is updated by aggregating the values in the node's input register, including the input from its bias node (which always fires, e.g., $B_i = 1$).

Updating states is based on the activation function. Three activation functions are commonly used. Hard-limit functions fire the node iff the aggregate input exceeds zero. Sigmoid stochastic functions fire the node with a probability given by the aggregate input. Sigmoid deterministic functions fire the node with a magnitude given by the aggregate input.[2] For example, if node k is influenced by two input nodes, i and j and a bias node b, then k sums the influence $ik = i*w_{ij} + k*w_{ik} + b*w_{ib}$. Positive weights cause i, j, and b to activate k and negative weights inhibit. If j is hard limited, then if $ik > 0$, $k = 1$, else $k = 0$. If the activation function is stochastic, k is activated with a probability $p = 1/(1 + e(-ik))$. If the sigmoid function is deterministic, k is activated with magnitude p.

Once the states have been activated for all nodes whose input registers have been fully updated (e.g., the first layer of hidden nodes and/or one or more output nodes that are directly connected to an input node), these nodes in turn update the input registers of nodes they influence going forward (e.g., the second layer of hidden nodes and/or some or all of the output nodes). Once this updating is complete, all nodes whose input registers have been fully updated aggregate across their inputs and update their states, and so on until the states of the output nodes have been updated. This completes the action phase.

The network learns by modifying the path weights linking the neurodes. Learning only occurs when the response to a given sensory input pattern is unsatisfactory. The paths are then adjusted so as to reduce the probability of repeating the mistake the next time this pattern is encountered. In many applications, neural nets are trained to recognize certain patterns or combinations of inputs. For example, suppose we want to train a neural net to predict stock prices from a set of market indicators. We first train the net to correctly predict known prices. The net begins with random path coefficients. These generate a prediction. The error is then used to adjust the weights in an iterative process that improves the predictions. These weights are analogous to those in a linear regression, and like a regression, the weights can then be applied to new data to predict the unknown.

20.5.3.1 Back Propagation of Error

A feed-forward neural network is trained by adjusting the weights on the paths between the nodes. These weights correspond to the influence that a node will have in causing nodes further forward in the network to fire. When those nodes fire incorrectly, the weights responsible for the error must be adjusted accordingly. Just as the influence process propagates forward, from the input nodes to the hidden layer to the output nodes, the attribution of responsibility for errors propagates backward, from the output nodes to the hidden layers. The back propagation of error begins with the output nodes and works back through the network to the input nodes, in the opposite direction from the influence process that "feeds forward" from input nodes to hidden nodes to output nodes. First, the output error is calculated for each output node. This is simply the difference between the expected state of the ith output node $(\widehat{S_i})$, and the state that was observed (S_i). For an output node, error refers to

---

[2]A multilayer neural net requires a nonlinear activation function (such as a sigmoid). If the functions are linear, the multilayer net reduces to a single-layer I-O network.

the difference between the expected output for a given pattern of inputs and the observed output:

$$e_o = s_o \left(1 - s_o\right) \left(\widehat{s}_o - s_o\right) \tag{20.5}$$

where the term $s_o \left(1 - s_o\right)$ limits the error to the unit interval. If the initial weights were randomly assigned, it is unlikely that the output will be correct. For example, if we observe an output value of 0.37 and we expected 1, then the error is -0.53.

Once the error has been updated for each output node, these errors are used to update the error for the nodes that influenced the output nodes. These nodes are usually located in the last layer of hidden nodes, but they can be anywhere and can even include input nodes that are wired directly to an output.[3] Then the errors of the nodes in the last hidden layer are used to update error back further still to the hidden nodes that influenced the last layer of hidden nodes, and so on, back to the first layer of hidden nodes, until the errors for all hidden nodes have been updated. Input nodes cannot have error, since they simply represent an exogenous pattern that the network is asked to learn. Back propagation means that the error observed in an output node o $\left(\widehat{s}_o - s_o\right)$ is allocated not only to o but to all the hidden nodes that influenced o, based on the strength of their influence. The total error of a hidden node h is then simply the summation over all n_h allocated errors from the n nodes i that h influenced, including output nodes as well as other hidden nodes:

$$e_h = s_h \left(1 - s_h\right) \sum_{i=1}^{n} w_{hi} e_i \tag{20.6}$$

Once the errors for all hidden, bias, and output nodes have been back propagated, the weight on the path from i to j is updated:

$$w'_{ij} = w_{ij} + \lambda s_i e_j \tag{20.7}$$

where λ is a fractional learning rate. The learning algorithm means that the influence of i on j increases if j's error was positive (i.e., the expected output exceeded the observed) and decreases if j's influence was negative.

Note that the Bush-Mosteller model is equivalent to a neural net with only a single bias unit and an output, but with no sensory inputs or hidden units. Such a device is capable of learning *how often* to act, but not *when* to act, that is, it is incapable of learning conditional strategies. In contrast, a feed-forward network can learn to differentiate environmental cues and respond using more sophisticated protocols for contingent strategies.

---

[3]However, if an input node is wired to hidden nodes as well as output nodes, the error for this node cannot be updated until the errors for all hidden nodes that it influenced have been updated.

20.5.3.2 Attractor Neural Network

Feed-forward networks are the most widely used but not the only type of artificial neural network. An alternative design is the attractor neural network (Churchland and Sejnowski 1994; Quinlan 1991), originally developed and investigated by Hopfield (1982; Hopfield and Tank 1985). In a recent article, Nowak and Vallacher (1998) note the potential of these computational networks for modeling group dynamics. This approach promises to provide a fertile expansion to social network analysis, which has often assumed that social ties are binary and static. A neural network provides a way to dynamically model a social network in which learning occurs at both the individual and structural levels, as relations evolve in response to the behaviors they constrain.

Unlike feed-forward neural networks (Rumelhart and McClelland 1988), which are organized into hierarchies of input, hidden, and output nodes, attractor (or "feed-lateral") networks are internally undifferentiated. Nodes differ only in their states and in their relational alignments, but they are functionally identical. Without input units to receive directed feedback from the environment, these models are "unsupervised" and thus have no centralized mechanism to coordinate learning of efficient solutions. In the absence of formal training, each node operates using a set of behavioral rules or functions that compute changes of state ("decisions") in light of available information. Zeggelink (1994) calls these "object-oriented models," where each agent receives input from other agents and may transform these inputs into a change of state, which in turn serves as input for other agents.

An important agent-level rule that characterizes attractor networks is that individual nodes seek to minimize "energy" (also "stress" or "dissonance") across all relations with other nodes. As with feed-forward networks, this adaptation occurs in two discrete stages. In the action phase, nodes change their states to maximize similarity with nodes to which they are strongly connected. In the learning phase, they update their weights to strengthen ties to similar nodes. Thus, beginning with some (perhaps random) configuration, the network proceeds to search over an optimization landscape as nodes repeatedly cycle through these changes of weights and states.

In addition to variation in path strength, neural networks typically have paths that inhibit as well as excite. That is, nodes may be connected with negative as well as positive weights. In a social network application, agents connected by negative ties might correspond to "negative referents" (Schwartz and Ames 1977), who provoke differentiation rather than imitation.

Ultimately, these systems are able to locate stable configurations (called "attractors"), for which any change of state or weight would result in a net increase in stress for the affected nodes. Hopfield (1982) compares these equilibria to memories and shows that these systems of undifferentiated nodes can learn to implement higher-order cognitive functions. However, although the system may converge at a stable equilibrium that allows all nodes to be locally satisfied (i.e., a "local optimum"), this does not guarantee that the converged pattern will minimize overall dissonance (a "global optimum").

This class of models generally uses complete networks, with each node characterized by one or more binary or continuous states and linked to other nodes through endogenous weights. Like other neural networks, attractor networks learn stable configurations by iteratively adjusting the weights between individual nodes, without any global coordination. In this case, the weights change over time through a Hebbian learning rule: the weight w_{ij} is a function of the correlation of states for nodes i and j over time. Specifically, Hebbian learning implies the following rules:

- To the extent that nodes i and j adopt the same states at the same time, the weight of their common tie will increase until it approaches some upper limit (e.g., 1.0).
- To the extent that nodes i and j simultaneously adopt different states, the weight of their common tie will decrease until it approaches some lower limit (e.g., 0.0).

Although Hebbian learning was developed to study memory in cognitive systems, it corresponds to the homophily principle in social psychology (Homans 1951) and social network theory (McPherson and Smith-Lovin 1987), which holds that agents tend to be attracted to those whom they more closely resemble. This hypothesis is also consistent with structural balance theory (Cartwright and Harary 1956; Heider 1958) and has been widely supported in studies of interpersonal attraction and interaction, where it has been called "the law of attraction" (Byrne 1971; Byrne and Griffitt 1966).

An important property of attractor networks is that individual nodes seek to minimize "energy" (or dissonance) across all relations with other nodes—a process that parallels but differs from the pursuit of balanced relations in structural balance theory. These networks also posit self-reinforcing dynamics of attraction and influence as well as repulsion and differentiation.

Following Nowak and Vallacher (1998), Macy et al. (2003) apply the Hopfield model of dynamic attraction to the study of polarization in social networks. In this application, observed similarity/difference between states determines the strength and valence of the tie to a given referent. This attraction and repulsion may be described anecdotally in terms of liking, respect, or credibility and their opposites. In their application of the Hopfield model, each node has $N - 1$ undirected ties to other nodes. These ties include weights, which determine the strength and valence of influence between agents. Formally, social pressure on agent i to adopt a binary state s (where $s = \pm 1$) is the sum of the states of all other agents j, where influence from each agent is conditioned by the weight (w_{ij}) of the dyadic tie between i and j ($-1.0 < w_{ij} < 1.0$):

$$P_{is} = \frac{\sum_{j=1}^{N} w_{ij} s_j}{N - 1}, j \neq i \tag{20.8}$$

Thus, social pressure ($-1 < P_{is} < 1$) to adopt s becomes increasingly positive as i's "friends" adopt s ($s = 1$) and i's "enemies" reject s ($s = -1$). The pressure can also become negative in the opposite circumstances. The model extends to multiple states in a straightforward way, where Eq. (20.8) independently determines the pressure on agent i for each binary state s. Strong positive or negative social pressure does not guarantee that an agent will accommodate, however. It is effective only if i is willing and able to respond to peer influence. If i is closed-minded or if a given trait is not under i's control (e.g., ethnicity or gender), then no change to s will occur. The probability π that agent i will change state s is a cumulative logistic function of social pressure:

$$\pi_{is} = \frac{1}{1 + e^{-10P_{is}}} \qquad (20.9)$$

Agent i adopts s if $\pi > C + \chi$, where C is the inflection point of the sigmoid, χ is a random number drawn from a uniform distribution in the interval $[C - \varepsilon, C + \varepsilon]$, and ε is an exogenous error parameter ($0 \leq \varepsilon \leq 1$). At one extreme, $\varepsilon = 0$ produces highly deterministic behavior, such that any social pressure above the trigger value always leads to conformity and pressures below the trigger value entail differentiation. Following Harsanyi (1973), $\varepsilon > 0$ allows for a "smoothed best reply" in which pressure levels near the trigger point leave the agent relatively indifferent and thus likely to explore behaviors on either side of the threshold. In the Hopfield model, the path weight w_{ij} changes as a function of similarity in the states of node i and j. Weights begin with uniformly distributed random values, subject to the constraints that weights are symmetric ($w_{ij} = w_{ji}$). Across a vector of K distinct states s_{ik} (or the position of agent i on issue k), agent i compares its own states to the observed states of another agent j and adjusts the weight upward or downward corresponding to their aggregated level of agreement or disagreement. Based on the correspondence of states for agents i and j, their weight will change at each discrete time point t in proportion to a parameter λ, which defines the rate of structural learning ($0 < \lambda < 1$):

$$w_{ij,t+1} = w_{ijt}(1 - \lambda) + \frac{\lambda}{K} \sum_{k=1}^{K} s_{jkt} s_{ikt}, j \neq i \qquad (20.10)$$

As correspondence of states can be positive (agreement) or negative (disagreement), ties can grow positive or negative over time, with weights between any two agents always symmetric.

Note one significant departure from structural balance theory. Although the agents in this model are clearly designed to maintain balance in their behaviors with both positive and negative referents, this assumption is not "wired in" to the relations themselves. That is, two agents i and j feel no direct need for consistency in their relations with a third agent h. Indeed, i has no knowledge of the jh relationship and thus no ability to adjust the ij relation so as to balance the triad.

Given an initially random configuration of states and weights, these agents will search for a profile that minimizes dissonance across their relations. Structural balance theory predicts that system-level stability can only occur when the group either has become uniform or has polarized into two (Cartwright and Harary 1956) or more (Davis 1967) internally cohesive and mutually antipathetic cliques. However, there is no guarantee in this model that they will achieve a globally optimal state in structural balance.

20.5.4 Belief Learning

Actors learn not only what is useful for obtaining rewards and avoiding punishments; they also update their beliefs about what is true and what is false. There are two main models of belief learning in the literature, Bayesian belief learning and fictitious play[4] (cf. Offerman 1997; Fudenberg and Levine 1998). These models differ in their assumptions about how players learn from observations. Both models assume that players believe that something is true with some fixed unknown probability p. In Bayesian learning, players then use Bayes' learning rule to rationally update over time their beliefs about p. In a nutshell, Bayes' learning rule implies that actors' assessment of the true value of p converges in the long run on the relative frequency of events that they observe. However, in the short and medium term, Bayesian learners remain suspicious in the sense that they take into account that observed events are an imperfect indication of p (Offerman 1997).

Fudenberg and Levine note that fictitious play is a special case of Bayesian learning. Fictitious play is a form of Bayesian learning that always puts full weight on the belief that the true value of p corresponds to the relative frequency observed in past events. Fudenberg and Levine (1998) note that it is an implausible property of fictitious play that a slight change in beliefs may radically alter behavior. The reason is that the best reply function always is a step function. As a remedy, Fudenberg and Levine introduce a smooth reply curve. The reply curve assigns a probability distribution that corresponds to the relative frequency of events. With strict best reply, the reply curve is a step function. Instead, a smooth reply curve assigns some probability to the action that is not strict best reply. This probability decreases in the difference between the expected payoffs. Specifically, when expected payoffs are equal, actors choose with equal probability, whereas their choice probabilities converge on pure strategies when the difference in expected payoffs approaches the maximum value.

The strict best reply function corresponds to the rule, "play X if the expected payoff for X is better than the expected payoff for Y, given your belief p. Otherwise play Y." Smooth best reply is then introduced with the modification to play the strict best reply strategy only with a probability of $1 - \eta$, whereas the alternative is played

---

[4]The Cournot rule may be considered as a third degenerate model of belief learning. According to the Cournot rule, players assume that the behavior of the opponent in the previous round will always occur again in the present round.

with probability η. The probability η, in turn, decreases in the absolute difference between expected payoffs $|uX(p) - uY(p)|$, where $\eta = 0.5$ if players are indifferent.

Belief learning generally converges with the predictions of evolutionary selection models. The approaches are also similar in the predicted effects of initial conditions on end results. Broadly, the initial distribution of strategies in independent populations in evolutionary selection corresponds to the initial beliefs players' hold about their opponent. For example, when two pure strategy Nash equilibria are feasible, then the one tends to be selected toward which the initial strategy distribution in evolutionary selection or initial beliefs in belief learning are biased.

20.6 Conclusion

Evolution and learning are examples of backward-looking consequentialist models, in which outcomes of agents' past actions influence their future choices, either through selection (in the case of evolution) or reinforcement (in the case of learning). Backward-looking models make weaker assumptions about agents' cognitive capacities than forward-looking models and thus may be appropriate for settings in which agents lack the ability, resources, information, and motivation to engage in intensive cognitive processing, as in most everyday instances of collective action. Backward-looking models may also be useful for understanding behavior driven by affect, rather than calculation. Forward-looking models may be more appropriate in applications such as investment decisions, international diplomacy, or military strategy, where the stakes are high enough to warrant collection of all relevant information and the actors are highly skilled strategists. However, even where the cognitive assumptions of the models are plausible, forward-looking models are generally limited to the identification of static equilibria but not necessarily whether and how agents will reach those equilibria.

When implemented computationally, backward-looking models can show how likely agents are to reach particular equilibria, as well as the paths by which those equilibria may be reached. However, computational models are also less general than analytical models. Furthermore, backward-looking models will be of little help if change in the environment outpaces the rate of adaptation. These limitations underscore the importance of robustness testing over a range of parameter values.

Evolutionary models are most appropriate for theoretical questions in which adaptation takes place at the population level, through processes of selection. Biological evolution is the most obvious analog, but social and cultural evolution are likely to be more important for social scientists. However, as we note above, researchers must be cautious about drawing analogies from the biological to social/cultural dynamics.

Learning models based on reinforcement and Bayesian updating are useful in applications that do not require conditional strategies based on pattern recognition. When agents must learn more complex conditional strategies, feed-forward neural networks may be employed. Furthermore, attractor neural networks are useful for modeling structural influence, such as conformity pressures from peers.

Models of evolution and learning are powerful tools for modeling social processes. Both show how complex patterns can arise when agents rely on relatively simple, experience-driven decision rules. This chapter seeks to provide researchers with an overview of promising research in this area and the tools necessary to further develop this research.

Further Reading

We refer readers interested in particular learning models and their application in agent-based simulation to Macy and Flache (2002), which gives a brief introduction into principles of reinforcement learning and discusses by means of simulation models how reinforcement learning affects behavior in social dilemma situations, whereas Macy (1996) compares two different approaches of modeling learning behavior by means of computer simulations. Fudenberg and Levine (1998) give a very good overview on how various learning rules relate to game-theoretic rationality and equilibrium concepts.

For some wider background reading, we recommend Macy (2004), which introduces the basic principles of learning theory applied to social behavior; Holland et al. (1986), which presents a framework in terms of rule-based mental models for understanding inductive reasoning and learning; and Sun (2008), which is a handbook of computational cognitive modeling.

References

Allison, P. (1992). The cultural evolution of beneficent norms. *Social Forces, 71*, 279–301.

Bandura, A. (1977). *Social learning theory*. Englewood Cliffs, NJ: Prentice Hall.

Börgers, T., & Sarin, R. (1997). Learning through reinforcement and replicator dynamics. *Journal of Economic Theory, 77*, 1–14.

Boyd, R., & Richerson, P. J. (1985). *Culture and the evolutionary process*. Chicago, IL: University of Chicago Press.

Bush, R. R., & Mosteller, F. (1950). *Stochastic models for learning*. New York: Wiley.

Byrne, D. E. (1971). *The attraction paradigm*. New York: Academic.

Byrne, D. E., & Griffitt, D. (1966). A development investigation of the law of attraction. *Journal of Personality and Social Psychology, 4*, 699–702.

Cartwright, D., & Harary, F. (1956). Structural balance: A generalization of Heider's theory. *Psychological Review, 63*, 277–293.

Catania, A. C. (1992). *Learning* (3rd ed.). Englewood Cliffs, NJ: Prentice Hall.

Churchland, P. S., & Sejnowski, T. J. (1994). *The computational brain*. Cambridge, MA: MIT Press.

Cohen, M. D., Riolo, R., & Axelrod, R. (2001). The role of social structure in the maintenance of cooperative regimes. *Rationality and Society, 13*(1), 5–32.

Davis, J. A. (1967). Clustering and structural balance in graphs. *Human Relations, 20*, 181–187.

Dawkins, R. (1976). *The selfish gene*. Oxford: Oxford University Press.

Durham, W. H. (1992). *Coevolution: genes, culture and human diversity*. Stanford, CA: Stanford University Press.

Erev, I., & Rapoport, A. (1998). Coordination, "magic", and reinforcement learning in a market entry game. *Games and Economic Behavior, 23*, 146–175.

Erev, I., & Roth, A. E. (1998). Predicting how people play games: Reinforcement learning in experimental games with unique, mixed strategy equilibria. *American Economic Review, 88*(4), 848–879.

Erev, I., Bereby-Meyer, Y., & Roth, A. E. (1999). The effect of adding a constant to all payoffs: Experimental investigation, and implications for reinforcement learning models. *Journal of Economic Behavior and Organizations, 39*(1), 111–128.

Flache, A., & Macy, M. W. (2002). Stochastic collusion and the power law of learning: A general reinforcement learning model of cooperation. *Journal of Conflict Resolution, 46*(5), 629–653.

Frank, R. (1988). *Passions within reason: The strategic role of the emotions*. New York: Norton.

Fudenberg, D., & Levine, D. (1998). *The theory of learning in games*. Boston: MIT Press.

Goldberg, D. (1989). *Genetic algorithms in search, optimization, and machine learning*. New York: Addison-Wesley.

Hamilton, W. (1964). The genetic evolution of social behaviour. *Journal of Theoretical Biology, 17*, 1–54.

Harsanyi, J. (1973). Games with randomly disturbed payoffs. *International Journal of Game Theory, 2*, 1–23.

Heider, F. (1958). *The psychology of interpersonal relations*. New York: Wiley.

Herrnstein, R. J., & Drazin, P. (1991). Meliorization: A theory of distributed choice. *Journal of Economic Perspectives, 5*(3), 137–156.

Holland, J. (1975). *Adaptation in natural and artificial systems*. Ann Arbor: University of Michigan Press.

Holland, J. H., Holyoak, K. J., Nisbett, R. E., & Thagard, P. R. (1986). *Induction: Processes of inference, learning, and discovery*. Cambridge, MA: MIT Press.

Homans, G. C. (1951). *The human group*. New York: Harcourt Brace.

Hopfield, J. J. (1982). Neural networks and physical systems with emergent collective computational abilities. *Proceedings of the National Academy of Sciences, 79*, 2554–2558.

Hopfield, J. J., & Tank, D. W. (1985). "Neural" computation of decisions in optimization problems. *Biological Cybernetics, 52*, 141–152.

James, W. (1981). *Principles of psychology*. Cambridge, MA: Harvard University Press.

Lopreato, J. (1990). From social evolutionism to biocultural evolutionism. *Sociological Forum, 5*, 187–212.

Macy, M. (1989). Walking out of social traps: A stochastic learning model for the prisoner's dilemma. *Rationality and Society, 2*, 197–219.

Macy, M. (1991). Learning to cooperate: Stochastic and tacit collusion in social exchange. *American Journal of Sociology, 97*, 808–843.

Macy, M. (1996). Natural selection and social learning in prisoner's dilemma: Co-adaptation with genetic algorithms and artificial neural networks. *Sociological Methods and Research, 25*, 103–137.

Macy, M. (1997). Identity, interest, and emergent rationality. *Rationality and Society, 9*, 427–448.

Macy, M. (1998). Social order in an artificial world. *Journal of Artificial Societies and Social Simulation, 1*(1). http://jasss.soc.surrey.ac.uk/1/1/4.html

Macy, M. (2004). Learning theory. In G. Ritzer (Ed.), *Encyclopedia of social theory*. Thousand Oaks, CA: Sage.

Macy, M., & Flache, A. (2002). Learning dynamics in social dilemmas. *Proceedings of the National Academy of Sciences, 99*, 7229–7236.

Macy, M., & Skvoretz, J. (1998). The evolution of trust and cooperation between strangers: A computational model. *American Sociological Review, 63*, 638–660.

Macy, M., Kitts, J., Flache, A., & Benard, S. (2003). Polarization in dynamic networks: A Hopfield model of emergent structure. In R. Breiger & K. Carley (Eds.), *Dynamic social network modeling and analysis: Workshop summary and papers* (pp. 162–173). Washington, DC: National Academy Press.

March, J. G., & Simon, H. A. (1958). *Organizations*. New York: Wiley.

McPherson, J. M., & Smith-Lovin, L. (1987). Homophily in voluntary organizations: status distance and the composition of face to face groups. *American Sociological Review, 52*, 370–379.

Nowak, A., & Vallacher, R. R. (1998). Toward computational social psychology: Cellular automata and neural network models of interpersonal dynamics. In S. J. Read & L. C. Miller (Eds.), *Connectionist models of social reasoning and social behavior* (pp. 277–311). Mahwah, NJ: Lawrence Erlbaum.

Offerman, T. (1997). *Beliefs and decision rules in public good games*. Dordrecht: Kluwer.

Quinlan, P. T. (1991). *Connectionism and psychology: A psychological perspective on new connectionist research*. Chicago, IL: University of Chicago Press.

Roth, A. E., & Erev, I. (1995). Learning in extensive-form games: Experimental data and simple dynamic models in intermediate term. *Games and Economic Behavior, 8*, 164–212.

Rumelhart, D. E., & McClelland, J. L. (1988). *Parallel distributed processing: Explorations in the microstructure of cognition*. Cambridge, MA: MIT Press.

Schwartz, S. H., & Ames, R. E. (1977). Positive and negative referent others as sources of influence: A case of helping. *Sociometry, 40*, 12–21.

Sun, R. (Ed.). (2008). *The Cambridge handbook of computational psychology*. Cambridge: Cambridge University Press.

Taylor, P. D., & Jonker, L. (1978). Evolutionary stable strategies and game dynamics. *Mathematical Biosciences, 40*, 145–156.

Thorndike, E. L. (1898). *Animal intelligence: An experimental study of the associative processes in animals, Psychological Review, Monograph Supplements, No. 8*. New York: Macmillan.

Vanberg, V. (1994). *Rules and choice in economics*. London: Routledge.

Williams, G. C. (1966). *Adaptation and natural selection*. Princeton, NJ: Princeton University Press.

Wilson, E. O. (1988). *On human nature*. Cambridge, MA: Harvard University Press.

Wynne-Edwards, V. C. (1962). *Animal dispersion in relation to social behaviour*. Edinburgh: Oliver & Boyd.

Wynne-Edwards, V. C. (1986). *Evolution through group selection*. Oxford: Blackwell Scientific.

Young, H. P. (1998). *Individual strategy and social structure. An evolutionary theory of institutions*. Princeton, NJ: Princeton University Press.

Zeggelink, E. (1994). Dynamics of structure: An individual oriented approach. *Social Networks, 16*, 295–333.

Chapter 21
Evolutionary Mechanisms

Edmund Chattoe-Brown and Bruce Edmonds

Abstract After an introduction, the abstract idea of evolution is analysed into four processes which are illustrated with respect to a simple evolutionary game. A brief history of evolutionary ideas in the social sciences is given, illustrating the different ways in which the idea of evolution has been used. The technique of Genetic Algorithms (GA) is then described and discussed including the representation of the problem and the composition of the initial population, the Fitness Function, the reproduction process, the Genetic Operators, issues of convergence and some generalisations of the approach including endogenising the evolutionary process. Genetic Programming (GP) and Classifier Systems (CS) are also briefly introduced as potential developments of GA. Four detailed examples of social science applications of evolutionary techniques are then presented: the use of GA in the Arifovic "cobweb" model, using CS in a model of price setting developed by Moss, the role of GP in understanding decision-making processes in a stock market model and relating evolutionary ideas to social science in a model of survival for "strict" churches. The chapter concludes with a discussion of the prospects and difficulties of using the idea of biological evolution in the social sciences.

Why Read This Chapter?

To learn about techniques that may be useful in designing simulations of adaptive systems including Genetic Algorithms (GA), Classifier Systems (CS) and Genetic Programming (GP). The chapter will also tell you about simulations that have a fundamentally evolutionary structure—those with variation, selection and replications of entities—showing how this might be made relevant to social science problems.

E. Chattoe-Brown (✉)
Department of Sociology, University of Leicester, Leicester, UK
e-mail: ecb18@le.ac.uk

B. Edmonds
Centre for Policy Modelling, Manchester Metropolitan University, All Saints Campus,
Oxford Road, Manchester, M1 6BH, UK

© Springer International Publishing AG 2017   525
B. Edmonds, R. Meyer (eds.), *Simulating Social Complexity*,
Understanding Complex Systems, https://doi.org/10.1007/978-3-319-66948-9_21

21.1   Introduction

There are now many simulations of complex social phenomena that have structures or component processes analogous to biological evolution (see Arifovic (1994), Chattoe (2006a), Dosi et al. (1999), Lomborg (1996), Nelson and Winter (1982), Oliphant (1996), Windrum and Birchenhall (1998) to get a flavour of the diversity of approach and applications). Clearly the process of biological evolution is complex and has resulted in the development of complex (and in several cases social) systems. However, biological evolution follows very specific mechanisms and is clearly not strictly isomorphic with social processes. For a start, biological evolution occurs over larger time spans than most social processes. Further, it is unlikely, as sociobiology (Wilson 1975) and evolutionary psychology (Buss 2015) are sometimes supposed to imply, that the domain of social behaviour will actually prove reducible to genetics. Thus, it is not immediately apparent *why* evolutionary ideas have had such an influence upon the modelling of social processes. Nevertheless, simulations of social phenomena have been strongly influenced by our understanding of biological evolution, and this has occurred via two main routes: through *analogies* with biological evolution and through computer science approaches.

In the first case, conceptions of evolution have been used as a way of understanding social processes, and then simulations have been made using these conceptions. For example, Nelson and Winter (1982) modelled growth and change in firms using the idea of random variation (new products or production processes) and selective retention (whether these novelties in fact sustain profitability—the survival requirement for firms—in an environment defined by what other firms are currently doing).

In the second case, computer science has taken up the ideas of evolution and applied it to engineering problems. Most importantly in machine learning, ideas from biological evolution have inspired whole families of techniques in what has become known as "evolutionary computation". The most famous of these techniques are Genetic Algorithms (Holland 1975; Mitchell 1996) and Genetic Programming (Koza 1992a, 1994) discussed below. These algorithms have then been applied in social simulations with different degrees of modification, from using them unchanged as "off the shelf" plug-ins (e.g. to model learning processes) to specifying simulation processes that use the core evolutionary idea but are completely re-engineered for a particular modelling purpose or domain. There is no a priori reason to suppose that a particular technique from computer science will be the most appropriate algorithm in a social simulation (including those with a biological inspiration) as we shall see below, but it certainly presents a wealth of evolutionary ideas and results that are potentially applicable in some form. Like any theory, the trick is to use good judgement and a clear specification in applying an algorithm to a particular social domain (Chattoe 1998, 2006b).

What is certainly the case is that biological evolution offers an example of how complex and self-organised phenomena can emerge from randomness, so it is natural to look to this as a possible conceptual framework with which to understand social phenomena with similar properties. (In particular, while it may be reasonable to assume deliberation and rationality in some social contexts, it is extremely unlikely to apply to all social structures and phenomena. As such, some kind of blind variation and retention—evolution—is probably the only well-defined theoretical alternative). The extent to which evolution-like processes are generally applicable to social phenomena is unknown (largely because this foundational issue has not received much attention to date), but these processes certainly are a rich source of ideas, and it may be that there are some aspects of social complexity that will prove to be explicable by models thus inspired. It is already the case that many social simulation models have taken this path and thus have the potential to play a part in helping us to understand social complexity (even if they only serve as horrible examples).

This chapter looks at some of the most widely used approaches to this kind of modelling, discusses others, gives examples and critically discusses the field along with areas of potential development.

21.2 An Abstract Description of Biological Evolution

We will not provide full details of biological evolution as currently understood in the neo-Darwinian synthesis.[1] Rather we will take from this a generalised model of evolution that will potentially cover a variety of social processes. This description will then be used to discuss an example from evolutionary game theory (Vega-Redondo 1996). This will unpack the implications of the abstract description and demonstrate its generality. This generalisation is a preliminary to discussing evolutionary simulations of social phenomena based on the abstract description as a framework.

21.2.1 The Four Process Description

The basic components in the biological theory are the genotype (the set of instructions or genome) and the phenotype (the "body" which the genotype specifies) in which these instructions are embedded. The phenotype is constructed using "instructions" encoded in the genotype. The phenotype has various capabilities including reproduction. Maintenance of the phenotype (and the embedded genotype) requires a number of potentially scarce inputs (food, water). The phenotypic capabilities include management of inputs and outputs to the organism. Poor adaptation of these capabilities with respect to either external or internal environment will

---

[1]For details about this, see any good textbook on biology (e.g. Dobzhansky et al. 1977).

result in malfunction and consequent death. The death of a particular phenotype also ends its reproductive activity and removes the corresponding genotype from the population. Variation occurs by mutation during reproduction, giving rise to novel genotypes (and hence subsequent phenotypes) in the resulting offspring. Genotypic variations are not selected directly by the environment but according to the overall capabilities of their phenotypes. In biology, phenotype alterations cannot be transmitted to the genotype for physiological reasons, but in social systems, this "Lamarckian" adjunct to evolution (which is not, however, adequate to explain change in its own right) is both possible and plausible. In particular, it allows for combinations of evolutionary learning at the social level and deliberate action at the individual level (Chattoe 2006a).

A full specification of an evolutionary model requires descriptions of the following processes:

1. *Generation of phenotypes*: A specification of the genotypes and the phenotypes these correspond to. This may not specify a 1-1 mapping between genotypes and phenotypes but describe the process by which phenotypes are actually constructed from genotypes. This is necessary when genotypes cannot be enumerated.
2. *Capabilities of the phenotypes*: A specification of ways in which phenotypes may use their capabilities to affect the internal and external environment, including the behaviour and numbers of other phenotypes. Lamarckian systems include the capability to modify the genotype using environmental feedback during the lifetime of the phenotype.
3. *Mechanisms of reproduction and variation*: A specification of the process by which phenotypes reproduce including possible differences between ancestor and successor genotypes resulting from reproduction. Reproduction may involve a single ancestor genotype (parthenogenesis) or a pair (sexual reproduction). In principle, multiple parents could be modelled if appropriate for particular social domains (like policies decided by committees) though this approach has not been used so far.
4. *Mechanism of selection*: A specification of all the processes impinging on the phenotype and their effects. This is the converse of the second mechanism; the capabilities of one phenotype form part of the selection process for the others. Some processes, such as fighting to the death, can be seen as directly selective. However, even indirect processes like global warming may interact with phenotypic capabilities in ways that affect fitness.

In these process specifications, it may be convenient to distinguish (and model separately) the "environment" as the subset of objects impinging on phenotypes which display no processes of the first three types. Whether a separate representation of the environment is useful depends on the process being modelled. At one extreme, a person in a desert is almost exclusively dealing with the environment. At the other, rats in an otherwise empty cage interact almost entirely with each other.

Obviously, some of these specifications could be extremely complex depending on the system being modelled. The division into system components is necessarily imprecise but not arbitrary. It is based on the considerable observed integrity of organisms relative to their environment. (This integrity is also observed in social "organisms" like firms which have clearly—and often legally—defined boundaries). The first and third specifications involve processes internal to the organism, while the second and fourth represent the organism's effect on the external world and the converse.

Of course, social processes, even "evolutionary social processes", are not constrained by the above specification. For example, what most closely corresponds to the genotype might not be separable from what corresponds to the phenotype. Nevertheless, however, for a very broad class of evolutionary simulations, it will be necessary to implement something very similar to the above four categories.

21.2.2 Illustrative Example: A Simple Evolutionary Game

Despite the potential complexity of specifying complete models for biological systems, this description can also be used to clarify and analyse relatively simple evolutionary systems. In this section, we shall provide a description for an evolutionary game. The purpose is not to comment on evolutionary game theory per se but to show how the description raises issues relevant to our understanding of evolutionary models.

For each agent, the genotype is one of a set of finite state automata producing a single action in each period, for example, the complete set of one- and two-state automata leading to the actions "co-operate" (C) and "defect" (D) in a prisoner's dilemma (see, e.g. Lomborg 1996). The action is compared with the action of a co-player (another agent), and the result is an adjustment to the "energy level" for each agent depending on the game payoffs and chosen strategies. If agents reach a certain energy level, they produce an exact copy. (This model dispenses with variation and involves asexual reproduction.) If the energy level of any agent reaches zero, it dies and is removed from the environment. Reproduction reduces the energy level considerably. Merely existing also does so but at a much lower rate.

With some qualifications, this is an example of a complete description discussed in the last section. It reveals some interesting things about the process of constructing such descriptions.

Firstly, this model involves a very attenuated environment compared to real social systems. Agents have a single external capability involving one of two actions and thus affecting the energy levels of their co-players. The effect of these actions is also the only environmental process that impinges on agents. The model of the environment just consists of mechanisms for deciding when and which agents will play, administering actions and energy changes, producing copies and removing dead agents. In real social systems, exogenous events (both social and environmental) are likely to be very important.

Secondly, the discussion of energy levels still sounds biological, but this is simply to make the interpretation of the example more straightforward in the light of the recent discussion. As we shall show subsequently, the survival criterion can just as easily be profitability or organisation membership levels.

Thirdly (and perhaps most importantly), there has been sleight of hand in the description of the model. We have already described Lamarckism (genotype modification by the phenotype during the organism's lifetime), and the construction of the phenotype by the genotype during gestation is a fundamental part of the evolutionary process. But in this example, the genotype is effectively "reconstructing" the phenotype every time the finite state automaton generates an action. There is nothing observable about a particular agent, given the description above, except the sequence of actions they choose. There is no way for an agent to establish that another is actually the same when it plays D on one occasion and C on another or that two plays of D in successive periods actually come from two different agents. In fact, this is the point at which models of social evolution develop intuitively from the simple description of biological evolution used so far. The capabilities of social agents (such as consumers, families, churches and firms) include the "senses" that give them the ability to record actions and reactions in memory. Furthermore, they have mental capabilities that permit the processing of sense data in various ways, some subset of which we might call rationality. The simple automata described above are reactive, in that their actions depend in systematic ways on external stimuli, but they can hardly be said to be rational or reflective in that their "decision process" involves no choice points, internal representations of the world or "deliberation". Such distinctions shelve into the deep waters of defining intelligence, but the important point is that we can make useful distinctions between different kinds of adaptability based on the specifications of process we use in our models without compromising the evolutionary framework we have set up. It is in this way that the complex relationship between selection and reasoned action may begin to be addressed.

Thus, even in this simple example, one can see not only the general evolutionary structure of the simulation but also that the conception differs in significant ways from the corresponding biological process.

21.3 Evolutionary Ideas in the Social Sciences

From early on, since the publication of *On the Origin of Species* (Darwin 1859), Darwin's ideas of evolution have influenced those who have studied social phenomena. For example, Tarde (1884) published a paper discussing "natural" and "social" Darwinism. This marked a shift from looking at the social organisation of individuals to the patterns of social products (fashions, ideas, tunes, laws and so on). Tarde (1903, p.74) put it like this:

> but self-propagation and not self-organisation is the prime demand of the social as well as of the vital thing. Organisation is but the means of which propagation, of which *generative* or *imitative* imitation, is the end.

However, it was from the latter half of the twentieth century that the full force of the analogy with biological evolution (as understood in the neo-Darwinian synthesis) was felt in the social sciences. There were those who sought to understand the continuous change in cultural behaviours over long time scales in this way, for example (Boyd and Richerson 1985; Campbell 1965; Cavalli-Sforza and Feldman 1973; Cloak 1975; Csányi 1989). Richard Dawkins coined the term "meme" as a discrete and identifiable unit of cultural inheritance corresponding to the biological gene (Dawkins 1976, 1982), an idea which has influenced a multitude of thinkers including Costall (1991), Lynch (1996), Dennett (1990), Heyes and Plotkin (1989), Hull (1982, 1988) and Westoby (1994). Another stream of influence has been the philosophy of science via the idea that truth might result from the evolution of competing hypotheses (Popper 1979), a position known as evolutionary epistemology since (Campbell 1974). The ultimate reflection of the shift described by Tarde above is that the human mind is "merely" the niche where memes survive (Blackmore 1999) or which they exploit (as "viruses of the mind", Dawkins 1993)— the human brain is programmed by the memes, rather than using them (Dennett 1990). This fits in with the idea of the social intelligence hypothesis (Kummer et al. 1997) that the biological reason the brain evolved is because it allows specific cultures to develop in groups giving specific survival value with respect to the ecological niches they inhabit (Reader 1970). All of these ideas hinge on the importance of imitation (Dautenhahn and Nehaniv 2002), since without this process, individual memes, ideas or cultural patterns would be quickly lost.

Evolutionary theories are applied in a wide variety of disciplines. As mentioned above, evolutionary theories are applied to culture and anthropology, as in the work of Boyd and Richerson, Cavalli-Sforza and Feldman and Csányi. The evolution of language can be seen as an analogy to biological evolution, as described by Hoenigswald and Wiener (1987). In computer science, Genetic Programming and Genetic Algorithms (as well as the more rarely used Classifier Systems) are descendants of the evolutionary view as well, for example, in the work of several individuals at the Santa Fe Institute (Holland 1975; Kauffman 1993). Learning theories of humans, applied to individuals, groups and society, can be tied to evolutionary theory, as shown in the work of Campbell (1965, 1974). The work of several philosophers of science also shows an evolutionary perspective on knowledge, as in the work of Popper (1979) and Kuhn (1970). Such theories have been used to account for brain development by Gerald Edelman (1992) and extended to the milliseconds-to-minutes time scale of thought and action by William Calvin (1996a, 1996b). Evolutionary theory (and in some cases, explicit modelling) is present in economics, often tied to the development of technology, as in the work of Nelson and Winter (1982), or to the evolution of institutions and practices as in the work of Dosi et al. (1999), Hodgson (1993) and North (1990). Sociology too has used evolutionary ideas and simulations to understand the evolution of social order (Lomborg 1996; Macy 1996), changing populations of organisations (Hannan and Freeman 1993) and the survival of so-called strict churches (Chattoe 2006a).

Interestingly, however, against these creative approaches must be set forces in particular social sciences that have slowed or marginalised their adoption. In sociology, the conversion of functionalism (potentially a form of social evolution) into a virtual religion was followed by a huge backlash against untestable grand theory which made these ideas virtually beyond the pale for 20 years or so (Chattoe 2002; Runciman 1998). It is quite likely that confused associations with social Darwinism, eugenics and even Nazism have not helped the use of biological analogies in social science from the 1940s until quite recently. In economics, the focus on deliberate rationality and well-defined equilibria has meant that evolutionary approaches are judged ad hoc unless they can be reinterpreted to support the core assumptions of economics. (This can be observed, e.g. in evolutionary approaches to equilibrium selection where the object is not to understand the dynamics of the system but to support the claim that particular equilibria are robust.) In psychology, while there appears to be no overt objection to evolutionary approaches, it seems to be the case (perhaps for historical reasons) that the main interest in these ideas is to explain behaviour using genetic accounts of cognitive structure rather than using evolutionary analogies.

In subsequent sections, having shown that interest in evolutionary ideas is widespread, we turn to technical details of various kinds of evolutionary algorithm, their strengths, weaknesses and social applicability, so the reader is able to evaluate their use and consider applications in their own areas of research interest. We start with the Genetic Algorithm, which is easiest to describe, and then move to Genetic Programming and the (more rarely used but in some sense more satisfactory as an analogy) Classifier Systems. The final example doesn't rely directly on the use of an evolutionary algorithm but clearly attempts to model a social process using a biological analogy.

21.4   The Basic Genetic Algorithm

This section describes the basic operation and limitations of the Genetic Algorithm. This leads to a description of ways in which the Genetic Algorithm can be generalised and a detailed discussion of one specific way of generalising it (Genetic Programming) in the subsequent section.

21.4.1   What Is a Genetic Algorithm?

The Genetic Algorithm is actually a family of programmes developed by John Holland (1975) and his coworkers at the University of Michigan. The following algorithm describes the structure of a typical Genetic Algorithm. It is the different ways in which various parts of the algorithm can be implemented which produces the wide variety of Genetic Algorithms available. Each part of the algorithm will be discussed in more detail in a subsequent section. For the purposes of illustration, consider an attempt to solve the notorious Travelling Salesman Problem that involves producing the shortest tour of a set of cities at known distances visiting each once only (Grefenstette et al. 1985).

1. Represent potential solutions to the problem as data structures.
2. Generate a number of these solutions/structures and store them in a composite data structure called the Solution Pool.
3. Evaluate the "fitness" of each solution in the Solution Pool using a Fitness Function.
4. Make copies of each solution in the Solution Pool, the number of copies depending positively on its fitness according to a Reproduction Function. These copies are stored in a second (temporary) composite data structure called the Breeding Pool.
5. Apply Genetic Operators to copies in the Breeding Pool chosen as "parents" and return one or more of the resulting "offspring" to the Solution Pool, randomly overwriting solutions which are already there. Repeat this step until some proportion of the Solution Pool has been replaced.
6. Repeat steps 3, 4 and 5 until the population of the Solution Pool satisfies a stopping condition. One such condition is that the Solution Pool should be within a certain distance of homogeneity.

There is an obvious parallel between this algorithm and the process of biological evolution that inspired it. The string representing a solution to a problem corresponds to the genotype and each element to a gene. The Fitness Function represents the environment that selects whole genotypes on the basis of their relative performance. The Genetic Operators correspond to the processes causing genetic variation in biology that allow better genes to propagate, while poorer ones are selected out. This class of Genetic Algorithms has a number of interesting properties (for further discussion, see Goldberg 1989).

1. It is evolutionary. Genetic Operators combine and modify solutions directly to generate new ones. Non-evolutionary search algorithms typically generate solutions "from scratch" even if the location of these solutions is determined by the current location of the search process. The common Genetic Operators are based on biological processes of variation. Genetic Operators permit short subsections of parent solutions to be propagated unchanged in their offspring. These subsections (called schemata) are selected through their effect on the overall fitness of solutions. Schemata that produce high fitness for the solutions in which they occur continue to be propagated, while those producing lower fitness tend to die out. (Note that while it is not possible to assign a meaningful fitness to single genes, it is possible to talk about the relative fitness of whole genotypes differing by one or more genes. By extension, this permits talk about successful "combinations" of genes.) The Genetic Operators also mix "genetic material" (different solutions in the Breeding Pool) and thus help to ensure that all the promising areas of the Problem Space are explored continuously. These ideas clearly resonate with the social production of knowledge, in science, for example.

2. It is non-local. Each solution is potentially exploring a different area of the Problem Space although solutions can "cluster" in promising areas to explore them more thoroughly. This allows for societies to be "smarter" than their members.

3. It is probabilistic. The Fitness Function ensures that fitter solutions participate in Genetic Operators more often because they have more copies in the Breeding Pool and are thus more likely to propagate their useful schemata. However, it sometimes happens that a solution of low overall fitness contains useful schemata. The probabilistic replacement of only a proportion of the Solution Pool with new solutions means that a small number of poor solutions will survive for sufficient generations that these schemata have a good chance of being incorporated into fitter solutions. This probabilistic approach to survival (when coupled with non-locality and the use of Genetic Operators) means that the Genetic Algorithm avoids getting stuck on nonoptimal peaks in the Problem Space. Consider a Problem Space with two peaks, one higher than the other. A simple hill-climbing algorithm, if it happens to start "near" the lower peak, will climb up it and then be stuck at a nonoptimal position. By contrast, there is nothing to prevent the Genetic Operators from producing a new solution somewhere on the higher peak. Once this happens, there is a possibility of solutions fitter than those at the top of the lower peak and these will come to dominate the population. The search process can thus "jump" from one peak to another which most variants of hill climbing don't do.

4. It is implicitly parallel. In contrast with the behaviour of serial search algorithms that operate on a single best solution and improve it further, the Genetic Algorithm uses a population of solutions and simultaneously explores the area each occupies in the Problem Space. The results of these explorations are repeatedly used to modify the direction taken by each solution. The parallelism arises because the "side effects" of exploring the area surrounding each solution affect all the other solutions through the functioning of Genetic Operators. The whole is thus greater than the sum of its parts.

5. It is highly general. The Genetic Algorithm makes relatively few assumptions about the Problem Space in advance. Instead, it tries to extract the maximum amount of information from the process of traversing it. For example, non-evolutionary heuristic search algorithms use features like the gradient (first differential) which may not be calculable for highly irregular Problem Spaces. By contrast, in the Genetic Algorithm, all operations take place directly on a representation of the potential solution. The Fitness Function also evaluates fitness directly from solutions rather than using derived measures. Although no search technique escapes the fact that all such techniques exploit *some* properties of the problem space they are applied upon, in practice, Genetic Algorithms are good at finding acceptable solutions to hard problems (which, in some cases, defeat other methods), albeit not always the best solution. Ironically, social evolutionary learning may be *better* at finding the solutions to difficult problems than rationality which struggles without high levels of knowledge about environmental structure.

21.4.2 The Problem Representation and Initial Population

The most important step in developing a Genetic Algorithm also requires the most human ingenuity. A good representation for solutions to the problem is vital to efficient convergence. Some solutions have more obvious representations than others do. In the Travelling Salesman Problem, for example, the obvious representation is an ordered list of numbers representing cities. For example, the solution (1 4 3 2) involves starting at city 1, then going to city 4 and so on. Once a representation has been developed, a number of solutions are generated and form the initial population in the Solution Pool. These solutions can be generated randomly, or they may make use of some other (perhaps "quick and dirty") algorithm producing better than random fitness. The optimum size of the initial population depends on the size of the Problem Space. A population of almost any size will ultimately converge. But the efficiency of the Genetic Algorithm relies on the availability of useful genetic material that can be propagated and developed by the Genetic Operators. The larger the initial population, the greater the likelihood that it will already contain schemata of an arbitrary quality. This must be set against the increased computational cost of manipulating the larger Solution Pool. The initial population must also be sufficiently large that it covers the Problem Space adequately. One natural criterion is that any given point in the Problem Space should not be more than a certain "distance" from some initial solution. A final requirement for a good solution representation is that all "genes" should be similarly important to overall fitness, rather than some "genes" being much more important than others. Equivalent variations at different positions should have a broadly similar effect on overall fitness. In the Travelling Salesman Problem, all the positions in the list are equivalent. They all represent cities. The efficiency of the Genetic Algorithms relies on the exponential propagation of successful schemata, and this efficiency is impaired if schemata differ too much in importance as the system then becomes "bottlenecked" on certain genes.

21.4.3 The Fitness Function

The Fitness Function is at least as important as the solution representation for the efficiency of the Genetic Algorithm. It assigns fitness to each solution by reference to the problem that solution is designed to solve. The main requirement for the Fitness Function is that it must generate a fitness for any syntactically correct solution. (These are commonly referred to as "legal" solutions.) In the Travelling Salesman Problem, an obvious Fitness Function satisfying this requirement would be the reciprocal of the tour length. The reciprocal is used because the definition of the problem involves finding the shortest tour. Given this goal, we should regard shorter tours as fitter. More complicated problems like constrained optimisation can also be handled using the Fitness Function. One approach is simply to reject all solutions that do not satisfy the constraints. This involves assigning them a fitness of 0. However, where solutions satisfying the constraints are sparse, a more efficient method is to add terms to the Fitness Function reflecting the extent of constraint satisfaction. These "penalty terms" lower the fitness of solutions that fail to satisfy the constraints but do not necessarily reduce it to zero.

21.4.4   The Process of Reproduction

Reproduction is sometimes classified as a Genetic Operator in that it takes a number of solutions (the Solution Pool) and produces a new set (the Breeding Pool). However, it is a Genetic Operator of a special type in that it uses additional information (the fitness of solutions and the Reproduction Function) in generating that population. The Reproduction Function links the fitness of individual solutions and the number of copies they produce. This process mimics the reproductive success of fitter organisms in biological systems. The number of copies depends on the type of Genetic Algorithm. Typically, the fittest solutions in the Solution Pool may produce two or three copies, while the worst may produce none. In order that potentially useful "genetic material" be retained, it is important that fitter solutions do not proliferate too rapidly, nor less fit solutions die out too fast. Despite their low fitness, poor solutions may contain useful schemata that need to be incorporated into better solutions. Ensuring "adequate" survival for instrumental efficiency is a matter of trial and error and depends on the problem and the type of Genetic Algorithm being used. There are two main types of reproduction strategies.

In the first, the Holland-type algorithm (Holland 1975), the copies of each solution make up the Breeding Pool as described above. The Breeding Pool thus contains more copies of fitter solutions. There are two main kinds of Reproduction Function. The first is proportional fitness: Here the number of copies produced for each solution is equal to the size of the Solution Pool normalised by some function according to the "share of fitness" accruing to each particular solution. Fitter solutions, responsible for a larger share of total fitness, produce more copies. This system is similar to that used in replicator dynamics (Vega-Redondo 1996): It is performance relative to the average that determines the number of offspring. The second possibility is rank-based fitness. In this case, the number of copies depends on fitness rank. For example, the fittest five solutions may receive two copies each, the least fit receive no copies, and all others receive one. Both types of function have probabilistic equivalents. Instead of determining the actual number of copies, the function can determine the probability of drawing each type. The reproduction operator is then applied repeatedly, drawing from the probability distribution until the Breeding Pool is full. Clearly, this will still result in a greater proportion of fitter solutions in the Breeding Pool. The Reproduction Function can be linear or arbitrarily complex. In practice, the "shape" of the Reproduction Function is chosen on the basis of experience to optimise the performance of the Genetic Algorithm.

The second reproduction strategy, the GENITOR algorithm (Whitley 1989), does not involve a Breeding Pool. Instead of copying solutions into the Breeding Pool and then copying the results of Genetic Operators back again, the GENITOR takes parent solutions sequentially from the Solution Pool, applies Genetic Operators and returns the offspring immediately to the Solution Pool. The Solution Pool is kept sorted by rank, and new solutions are appropriately placed according to fitness. A new solution either overwrites the solution with fitness nearest to its own or is inserted into the Solution Pool so that all solutions with lower fitness move down one place and the solution with the lowest fitness is removed altogether. The GENITOR algorithm ensures that fitter solutions are more likely to become parents by using a skewed distribution to select them.

The differences between these strategies are instructive. The GENITOR algorithm is more similar to the interaction of biological organisms. The parents produce offspring that are introduced into a population that probably still contains at least one parent. Fitness affects which parents will mate, rather than generating offspring from all individuals in the Solution Pool. Even the "pecking order" interpretation of the introduction of offspring seems relatively intelligible. By contrast, the Breeding Pool in the Holland-type algorithm seems to be an abstraction with little descriptive plausibility. The Holland-type algorithm effectively splits the process of reproduction into two parts: the proliferation of fitter individuals and the subsequent generation of variation in their offspring. In biological systems, both processes result from the "same" act of reproduction. Furthermore, the differential production of offspring emerges from the relative fitness of parents. It is not explicitly designed into the system. In functional terms, both types of algorithm promote the survival of the fittest through variation and selective retention. In instrumental terms, one is sometimes more suitable than the other for a particular Problem Space. In descriptive terms, the GENITOR algorithm seems more appropriate to biological systems. (It can also be given a more plausible behavioural interpretation in social contexts).

21.4.5   The Genetic Operators

There are two main types of Genetic Operator that correspond to the biological phenomena of recombination and mutation. These are the original Genetic Operators developed by Holland (1975). Recombination Genetic Operators involve more than one solution and the exchange of genetic material to produce offspring. The commonest example is the Crossover Operator. Two solutions are broken at the same randomly selected point (n), and the "head" of each solution is joined to the "tail" of the other to produce two new solutions as shown in Fig. 21.1. Here a_i identifies an ordered set of k genes from one parent, and b_i identifies those from another.

One of the two new solutions is then chosen with equal probability as the offspring to be placed in the Solution Pool.

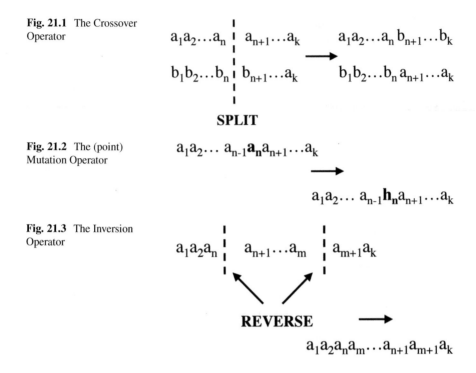

Fig. 21.1 The Crossover Operator

Fig. 21.2 The (point) Mutation Operator

Fig. 21.3 The Inversion Operator

Mutation Genetic Operators involve a single solution and introduce new genetic possibilities. The two main kinds of mutation Genetic Operators used in Genetic Algorithms correspond to so-called large scale and point mutation. In the Mutation Operator (realising point mutation), one gene is altered to another value from the legal range (selected with equal probability) and shown in Fig. 21.2 as h_n.

One commonly used Genetic Operator (corresponding to large-scale chromosomal mutation) is the Inversion Operator (see Fig. 21.3) which involves reversing the order of a set of genes between two randomly selected points (n and m) in the genotype.

The Inversion Operator provides an opportunity to discuss positional effects in solution representations although these can arise in all Genetic Operators except point mutation. It is not problematic to invert (reverse) the order in which a section of a city tour takes place in the Travelling Salesman Problem. However, there may be problem representations for which we have no reason to expect that the Inversion Operator will generate solutions that are even syntactically correct let alone fit. There are two solutions to the production of illegal solutions by Genetic Operators. One is to use the penalty method. The other is simply to avoid unsuitable Genetic Operators by design. Positional effects pose particular problems if genes have different meanings, some representing one sort of object and some another. In this case, inverted solutions are almost certain not to be legal. This difficulty will be addressed further in the section on developing the Genetic Algorithm.

In biological systems, recombination ensures that the genes of sexually reproduced offspring are different from those of both parents. Various forms of mutation
guarantee that entirely new genetic possibilities are also being introduced continuously into the gene pool. In the Genetic Algorithm, the Genetic Operators perform
the same function, but the probability with which each is applied has to be tuned
to ensure that useful genetic material can be properly incorporated before it is lost.
Typically, the Crossover Operator is applied with a high probability to each solution
and the Mutation Operator with a low probability to each gene leading to a moderate
probability of some mutation occurring in each solution. Other Genetic Operators
are applied with intermediate probability. These probabilities are intended to reflect
very approximately the relative importance of each process in biological systems.

For instrumental uses of the Genetic Algorithm, the setting of probabilities is a
matter of experience. If the probabilities of application are too low, especially for the
Mutation Operator, there is a danger of premature convergence on a local optimum
followed by inefficient "mutation only" search. (In such cases, the advantages of
parallel search are lost and the Genetic Algorithm effectively reverts to undirected
serial search.) By contrast, if the probabilities of application are too high, excessive
mixing destroys useful schemata before they can be combined into fit solutions.

There is a wide variety of other Genetic Operators discussed in the literature
(Goldberg 1989; Mitchell 1996), some developed descriptively from biological
systems, and others designed instrumentally to work on particular problems. The
descriptive use of Genetic Operators in the example provided here means that
although it is important to bear the instrumental examples in mind, they should not
be regarded as definitive. The processes of variation that affect the social analogues
of genotypes should be established empirically just as they were for biological
genes.

21.4.6 Convergence

Because the Genetic Algorithm is a powerful technique, many of the problems it
is used to solve are very hard to tackle by other means. Although it is possible
to test the Genetic Algorithm by comparison with other techniques for simple
problems, there is a danger that conclusions about performance will not scale
to more complex cases. One consequence of this is the difficulty of defining
satisfactory conditions for convergence. Provided the Problem Space is suitable,
a non-evolutionary algorithm will find the best solution within a certain time. In the
same time, the Genetic Algorithm is only statistically likely to converge (though, in
practice, it will actually do so for a far larger class of problems). As a result, unlike
some iterative procedures, the Genetic Algorithm cannot simply be stopped after
a fixed number of generations. Instead, the properties of the Solution Pool must
be analysed to determine when the programme should stop. The simplest method
involves stopping when the fittest solution is "good enough". Clearly, this involves
a value judgement external to the definition of the problem. Another possibility is

to stop the programme when the rate of change in best solution fitness drops below a specified level. Unfortunately, the behaviour of the Genetic Algorithm means that improvements in fitness are often "stepped" as the Genetic Operators give rise to whole ranges of new possibilities to be explored. For this reason more sophisticated approaches analyse the Solution Pool continuously and measure fitness in the whole population. Another advantage of this technique is that it allows for the fact that convergence is never total because of the Mutation Operator. There is always a certain amount of "mutation noise" in the Solution Pool even when it has converged.

21.4.7   Developing the Genetic Algorithm

The previous section was intended to provide a summary of the main aspects of design and a feel for the operation of a typical instrumental Genetic Algorithm (one that is supposed to solve a predefined problem as efficiently as possible). In the next two subsections, we describe a variety of generalisations that move the Genetic Algorithm away from the instrumental interpretation and towards the possibility of realistic *description* of certain social processes. This involves enriching the syntax for solution representations, developing formal techniques for analysing the behaviour of evolutionary models and making various aspects of the evolutionary process endogenous. The fact that these generalisations develop naturally from previous discussions suggests that a suitably sophisticated Genetic Algorithm might serve as a framework for evolutionary models of (carefully chosen) social phenomena. We shall try to show that Genetic Programming (as an extension of Genetic Algorithms) is particularly suitable for this purpose.

21.4.7.1   Generalising the Solution Representation

In the simplest Genetic Algorithm, the solution representation is just a list of numbers with a fixed length. Each gene (number) in the genotype (list) represents an object like a city in the Travelling Salesman Problem. But there is no reason why the Genetic Algorithm should be limited to solving problems using such a restricted representation. The enrichment of the syntax for solution representations has proceeded in three overlapping domains: the computational improvement of programmes implementing Genetic Algorithms, the incorporation of useful insights from biology and the study of theoretical requirements for the use of different solution representations.

Developments of the first sort are those which broaden the capabilities of the Genetic Algorithm itself. Instead of solutions of fixed length "hard coded" by the programmer, Goldberg et al. (1990) have developed a "messy" Genetic Algorithm. This evolves an encoding of optimal length by varying the lengths of potential solutions as well as their encoding interpretations. Schraudolph and Belew (1992) have also addressed this problem, developing a technique called Dynamic Parameter Encoding that changes the solution encoding in response to an analysis of the current Solution Pool. (This technique avoids the loss of efficiency that results from premature convergence and the consequent failure of parallel search.) Finally, Harvey (1993) has stressed the importance of variable length genotypes in systems that are to display genuine increases in behavioural complexity.

Developments of the second sort have arisen from the study of biological systems. Smith et al. (1992) have developed a Genetic Algorithm that produces a diverse coexistent population of solutions in "equilibrium" rather than one dominated by a single "optimal" solution. In this way, the coexistent population is capable of generalisation. This approach also forms the basis of the Classifier Systems discussed in Forrest (1991). Here groups of "if [condition]-then [action]" rules form coexistent data structures that can jointly perform computational tasks. Belew (1989, 1990) has developed this notion further by considering models in which the solutions themselves take in information from the environment and carry out a simple form of learning. Koza (1992b, 1992c) considers the possibility of co-evolution. This is a process in which the fitness of a solution population is not defined relative to a fixed environment or Fitness Function but rather in terms of another population. He applies this technique to game strategy learning by Genetic Programmes. Clearly this development is important to models of social systems where we can seldom define, let alone agree, a clear objective ranking of alternative social arrangements. In a sense, it is the existence of a Fitness Function that identifies instrumental (rather than descriptive) applications of evolutionary algorithms. The exception might be a model in which different solutions to a problem were created "subconsciously" in the style of a Genetic Algorithm but were then evaluated "rationally" by an agent. For an example, see Chattoe and Gilbert (1997).

Developments of the third sort involve the adaptation of formal systems such as grammars to serve as solution representations. Antoinisse has developed a representation and set of Genetic Operators that can be used for any problem in which legal solutions can be expressed as statements in a formal grammar (Antoinisse 1991). Koza (1992a, 1994) has developed a similar though far more general representation involving the syntax of computer languages. This approach (called Genetic Programming) will receive detailed discussion in its own section shortly.

21.4.7.2 Making the Process of Evolution Endogenous

So far, most of the Genetic Algorithm generalisations discussed have been instrumental in their motivation and use. The abstractions and limitations in the simple Genetic Algorithm have not been viewed as unrealistic but merely unhelpful (since they are engineering solutions rather than attempts to describe and understand complex social behaviour). The interesting question from the perspective of this chapter is how it is possible to develop simulations based on evolutionary algorithms which are not just instrumentally effective (e.g. allowing firms to survive by learning about their market situation) but actually provide a convincing ("descriptive") insight into their decision processes and the complexity of the resulting system. At the same time, the powerful self-organising capabilities of evolutionary algorithms may serve to provide an alternative explanation of observed stability (and instability) in social systems which do not (or cannot) involve a high level of individual rationality. Despite the instrumental nature of most current developments in Genetic Algorithms, the trend of these developments suggests an important issue for the design of descriptive models.

Most of the developments discussed above can be characterised as making various aspects of the process of evolution endogenous. Instead of exogenous system level parameters that are externally "tuned" by the programmer for instrumental purposes, various parts of the evolutionary process become internalised attributes of the individual solutions. They need not be represented in the solution explicitly as numerical parameters. They are parameters in the more general sense that they alter the process of evolution and may be adjusted by the programmer. For example, the level of mutation may emerge from some other process (such as endogenous copying of information through imitation) rather than being "applied" to the solutions. Co-evolution provides a good example of this approach. In the instrumental Genetic Algorithm, the Fitness Function is specified by the programmer and applied equally to all solutions, producing an answer to some question of interest. To follow an old Darwinian example, this is equivalent to the deliberate breeding of particular dog breeds. In co-evolving Genetic Algorithms, as in biological evolution, there is no fixed Fitness Function. Fitness can only be measured relative to the behaviour of other agents that constitute an important part of the environment. This is equivalent to the production of the dog species by biological evolution. Another example is provided by the Classifier Systems briefly discussed above. The simple Genetic Algorithm assumes that the fitness of an individual solution is independent of the fitness of other solutions. In practice, the fitness of one solution may depend on the existence and behaviour of other solutions. In biology, this is acknowledged in the treatment of altruism (Becker 1976; Boorman and Levitt 1980) and of group selection (Hughes 1988).

The use of solutions that are syntactically identical also abstracts from another important feature of evolution. Because the solutions only differ semantically, there is no sense in measuring the relative "cost" of each. By contrast, when solutions differ syntactically, selection pressure may operate to produce shorter solutions as well as better ones. In descriptive models, "fitness" no longer measures an abstract quantity but describes the efficient scheduling of all scarce resources used including time. The less time is spent making decisions (provided they are sensible), the more time can be spent on other things. To put this point in its most general terms, organisms (and firms) are dynamic solutions to a dynamic environment, while the simple Genetic Algorithm is a static solution to a static environment. Since social environments are dynamic, one way in which social agents can evolve or adapt is by evolving or adapting their models of that environment. Thus, an important way in which descriptive models can make the evolutionary process endogenous is by simulating agents that develop and test their own interpretations of the world in an evolutionary manner rather than being "gifted" with a fixed set of interpretations or decision processes by the modeller (Dosi et al. 1999).

The value of making parts of the process specification endogenous can only be assessed in specific cases using descriptive plausibility as the main criterion. For example, if the rate of mutation can realistically be treated as fixed over the lifetime of a given evolutionary process, it makes little practical difference whether it is represented as an extra global parameter or as part of the representation for each solution. In such cases, instrumental considerations such as computational efficiency may as well decide the matter. By contrast, making fitness endogenous will probably have a major effect on the behaviour of the system. In particular, there will be a tension between the descriptive plausibility of this change and the "instrumental" desirability of convergence to a unique optimum facilitated by an external Fitness Function.

This aspect of Genetic Algorithm design provides a new insight into the distinction between instrumental and descriptive models. Instrumental models are those that allow the programmer to achieve her goals whatever they are. By contrast, the only goal that is permitted to shape a descriptive model is that of effective description as determined by empirical evidence. What determines the extent to which mutation should be modelled as a process inhering in agents is the extent to which the mutation process inheres in agents. Only once it has been shown that the mutation rate does not vary significantly across agents should it be represented as an environmental variable.

To sum up then, Genetic Algorithms constitute a broad class of powerful evolutionary search mechanisms with an active research agenda. Some (but not all) of the subsequent developments to the basic Genetic Algorithm are valuable to the descriptive modelling of social systems. (In addition, some developments may have value in the characterisation of models. In the long term, it may be possible to prove formal convergence results for descriptively realistic systems.) We now turn to a discussion of Genetic Programming, a significant variant of the Genetic Algorithm based on the idea of "evolving" computer programmes which can both solve instrumental problems and represent sets of practices agents use to address the problems their environment creates for them.

21.5 Genetic Programming

The fundamental insight of Genetic Programming (Koza 1992a, 1994) is that evolutionary algorithms do not need to be limited to static representations or adaptation in a static environment. The approach originated in an instrumental concern, the possibility of evolving efficient computer programmes rather than having to design them explicitly (Koza 1991). However, it rapidly became clear that the power of the technique could be extended to any process which could be represented as an algorithm provided the fitness of different solutions could be measured (Koza 1992d). The possibility of developing descriptive models of agents was also considered early on Koza (1992c). In most models of this kind, however, the fitness of the programme representing an agent is assessed by its ability to fulfil exogenous goals. Agents typically "compete" against the environment on an equal footing rather than constituting that environment.

The potential of such an approach is tremendous. It involves the possibility of an evolutionary process that operates on the richest representation language we can envisage: the set of computable functions. These functions can model the capability to collect, abstract, store and process data from the environment, transfer it between agents and use it to determine action. Furthermore, we know that (in principle at least) languages within the class of computable functions can also represent important features of human consciousness like self-awareness and self-modification of complex mental representations (Kampis 1991; Metcalfe 1994; Fagin et al. 1995).

A simple example illustrates the most common solution representation used in Genetic Programming. This can be visualised as a tree structure and translates exactly into the set of "S-expressions" available in the LISP programming language (Friedman and Felleisen 1987). This is convenient for programming purposes because LISP comes already equipped to perform operations on S-expressions and can therefore easily and efficiently implement suitable Genetic Operators. The tree structure in Fig. 21.4 is equivalent to the S-expression (OR (AND (NOT D0) (NOT D1))) (AND D0 D1)). This is the definition of the XOR (exclusive or) function. For obvious reasons, D0 and D1 are referred to as terminals, and the set {AND, OR, NOT} are referred to as functions. The choice of a suitable set of functions and terminals (the equivalent of the solution representation in Genetic Algorithms) is a key part of Genetic Programming. Functions are by no means limited to the logical operators. They can also include mathematical operators and programming language

Fig. 21.4 An S-expression

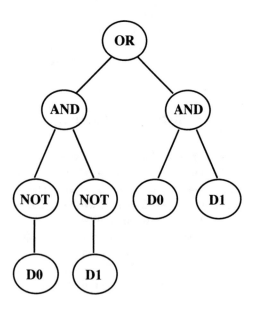

instructions. Similarly, terminals can represent numerical (or physical) constants, a variety of "sensor" inputs from the environment (including the observable actions of other agents) and "symbolic" variables like "true" and "false".

The instrumental measurement of fitness involves providing the S-expressions with different "inputs" (in this case truth values for D0 and D1) and assessing the extent to which the desired "output" results. For example, in Koza (1991), a programme to generate random numbers was tested by measuring the statistical properties of the number sequences it generated and rewarding such features as uncorrelated residuals. If S-expressions represent agents that are capable of action in an environment, success can be measured by the ability to modify the relationship between the agent and the environment in a certain way, for example, by following a trail successfully. (The further along the trail an agent gets, the fitter its programme.) It should be noted that the instrumental measurement of fitness requires a fairly precisely defined problem and solution grammar. On the other hand, the descriptive modelling of interaction need not. In order to do "well enough" in the market, a firm only needs to make some profit in every period sufficient to cover its costs. It may or may not have an internal goal to do better than this or even to make as much profit as it possibly can, but this goal is not required for its survival (and may, in some cases, actually be counterproductive).

This discussion raises several potential difficulties with the descriptive use of Genetic Programming. However, these appear to recede on further consideration of the corresponding solutions to these problems in instrumental applications. The first difficulty is designing Genetic Operators that are guaranteed to produce meaningful offspring. In the S-expression representation, it is clear that a cut can be made at any point on the tree and the crossing of two such fragmented parents will always result in two legal offspring. However, the price to be paid for this advantage is that solutions must have a hierarchical form. More complicated function sets, mixing numerical and logical functions, for example, must restrict Crossover to prevent such outcomes as (+4 TRUE) or (NOT 6).

However, given the descriptive interpretation of Genetic Operators, it is plausible that agents should know the syntactic rules of combination for the set of terminals and operators they possess. As such, the relevant "descriptive" Genetic Operators may execute rather more slowly than the simple instrumental ones, but it is not unreasonable to suppose that only syntactically correct trees will result. However, this raises another interesting possibility for using Genetic Operators. A good illustration is provided by a second difficulty with Genetic Programming, that of "bloating". This occurs because Genetic Programmes sometimes grow very large and contain substantial amounts of syntactically redundant material. (If a tree is trying to converge on a specific numerical value, for example any sub-trees evaluating to 0 are syntactically redundant.) Bloating produces a number of difficulties. Firstly, it slows down the evaluation of trees. Secondly, it becomes harder to interpret the trees and assess their behavioural plausibility. Finally, it is descriptively unsatisfactory. We do not expect real human decision processes to contain pointless operations (although bureaucratically specified production processes might for example). Unfortunately, the obvious solution (the exogenous penalisation of long solutions) lacks precision. It is not possible to establish how long solutions to a particular problem "ought" to be without making arbitrary assumptions. The result is an ungrounded trade-off between length and quality. An interesting alternative is to introduce "purely syntactic" Genetic Operators. These take no account of tree fitness but simply look for redundant material within trees. For example, a Genetic Operator which replaced instances of the pattern (* constant 0) with 0 would be very simple to implement.

This approach allows firms, for example, to apply plausible syntactic knowledge to the structure of their decision processes ("rationalisation" in the non-pejorative sense) without compromising the assumption (opposed to extreme economic rationality) that they cannot evaluate the fitness of a strategy without trying it in the market.

It also suggests a possible solution to another persistent problem with Genetic Programmes, that of interpretation. Even quite small trees are often hard to interpret and thus to evaluate behaviourally. Application of syntactic Genetic Operators may reduce the tree to a form in which it can be more easily interpreted. Another approach might be to use a Genetic Programming instrumentally to interpret trees, giving the greatest fitness to the shortest tree which can predict the output of a decision process tree to within a certain degree of accuracy. Thus, in principle at least, the Genetic Programming approach can be extended to include processes that are behaviourally similar to abstraction and refinement of the decision process itself.

As in the discussion of Genetic Algorithms above, we have kept this discussion of Genetic Programming relatively technical with some digressions about its general relevance to modelling social behaviour. In the final section of this chapter, we will present some evolutionary models in social science specifically based on evolutionary algorithms. This discussion allows us to move from general to specific issues about the applicability of biological analogies to social systems. In particular, we will try to show why models based on Genetic Programming and some Classifier Systems are more behaviourally plausible than those based on Genetic Algorithms.

21.6   Example Applications of Evolutionary Algorithms

21.6.1   Example Using Genetic Algorithms: The Arifovic "Cobweb" Model

Arifovic (1994) is probably responsible for the best-known simulation of this type representing the quantity setting decisions of firms to show convergence in a cobweb model.[2] She argues that the Genetic Algorithm both produces convergence over a

---

[2]A cobweb model is one in which the amount produced in a market must be chosen before market prices are observed. It is intended to explain why prices might be subject to periodic fluctuations in certain types of markets.

wider range of model parameters than various forms of rational and adaptive learning but also that it mimics the convergence behaviour of humans in experimental cobweb markets. Arifovic draws attention to two different interpretations of the Genetic Algorithm and explores the behaviour of both. In the "single population interpretation", each firm constitutes a single genotype, and the Genetic Algorithm operates over the whole market. In the "multiple population interpretation", each firm has a number of genotypes representing alternate solutions to the quantity setting decision and operates its own "internal" Genetic Algorithm to choose between them.

She shows that using a basic Holland-type Genetic Algorithm, neither interpretation leads to convergence on the rational expectations equilibrium for the cobweb market. When she adds her "Election" Genetic Operator, however, both interpretations do so. The Election Operator involves using Crossover but then evaluating the offspring for profitability on the basis of the price prevailing in the previous period. The effect of this is to add some "direction" to the application of Genetic Operators, in fact a hill-climbing component. An offspring is only added to the population if it would have performed better than its parents did in the previous period. This approach does not require any implausible knowledge as it is based on past events. However, it appears that the motivation for introducing the Election Operator is instrumental, namely, to ensure perfect convergence to the rational expectations equilibrium (a goal of economic theory rather than a property of real markets necessarily). Interestingly, the graphs shown in the paper suggest that the Genetic Algorithm has done very well in converging to a stable (if mutation noise augmented) price fairly close to the rational expectations equilibrium. In fact, Arifovic shows how the Election Operator endogenously reduces the effective mutation rate to zero as the system approaches the theoretical equilibrium. She also points out that the Election Operator does not harm the ability of the Genetic Algorithm to learn a new equilibrium if the parameters of the cobweb model change. What she doesn't explain is why the goal of the model should be to produce the theoretical equilibrium.

In fact, there are problems with both of her models that serve as instructive examples in the application of evolutionary ideas. The single population interpretation seems to involve a standard Holland-type Genetic Algorithm even down to a Breeding Pool that has no behavioural interpretation in real systems. There is also a problem with the use of Genetic Operators that is general in Genetic Algorithms. The way in which the bit strings are interpreted is very precise. If one firm uses Crossover involving the price strategy of another, it is necessary to "gift" a common representation to all firms and assume that firms know precisely where bit strings should "fit" in their own strategies. Given the encoding Arifovic uses, inserting a bit string one position to the left by mistake doubles the price it produces. In descriptive terms, this seems to be the worst of both worlds. It is easy to see how one firm could charge the same price as another or (with more difficulty) acquire a "narrative" strategy fragment like "keep investment in a fixed ratio to profit" but not how firms could come to share a very precise arbitrary representation and copy instances around exactly. More generally, encoding price in this way is just behaviourally odd.

It is hard to imagine what a firm would think it was doing if it took a "bit" of one of its prices and "inserted it" into another. Of course, the effect would be to raise or lower the price, but the way of going about it is very bizarre.

We think the reason for this is that an encoding is not a procedure that is endogenously evolved. A Genetic Programme that calculates price by taking the previous price of another firm, adding unit cost and then adding 2 is telling a firm behaviourally how to determine price. These are "real" procedures given by the ontology of what a firm knows: the set of operators and terminals. By contrast there has to be reason why a firm would bother to encode its price as a bit string rather than just operating on it directly. Unless this encoding is "gifted", it is not clear how (or why) the firm would develop it.

The multiple population interpretation is much more plausible in behavioural terms since the problem representation only needs to be the same within a firm, although the strangeness of "combining" prices remains. A firm applying Genetic Operators to its own strategies can reasonably be assumed to know how they are encoded, however.

However, both interpretations come up against a serious empirical problem noted by Olivetti (1994). Because the Election Operator is effectively a hill-climbing algorithm, it fails to converge under quite small changes to the assumptions of the model. In particular, Olivetti shows that the system doesn't converge when a white noise stochastic disturbance is added to the demand function. This suggests that Arifovic has not understood the main advantage of the Genetic Algorithm, and her pursuit of instrumental convergence at the expense of behavioural plausibility is actually counterproductive. In a sense, this is just a reprise of the previous instrumental insight. Genetic Algorithms perform better on difficult problems precisely because they do not "hill climb" (as the Election Operator does) and can thus "jump" from one optimum to another through parallel search.

21.6.2   Example Using Classifier Systems: The Moss Price Setting Model

As discussed briefly above, Classifier Systems consist of sets of "if [condition]-then [action]" rules that can collectively solve problems. They are evolutionary because new rules are typically generated using a Genetic Algorithm to select, recombine and mutate the most effective rules in the population. However, there is one significant (and potentially problematic) difference between Classifier Systems and Genetic Algorithms or Genetic Programming. This is the allocation of fitness to the individual rules, frequently using the so-called bucket brigade algorithm. This allows individual rules to "bid" fitness in order to take part in the set that is used to solve the problem in a particular instance. Rules taking part in a successful outcome then receive "recompense" also in terms of fitness. Unfortunately, the behavioural interpretation for this algorithm is not clear. In addition, the system is required to make decisions about how to "allocate" fitness between participating rules. This is the "credit assignment problem" recognised in artificial intelligence, and it is hard to produce effective general solutions. In particular, rules that are only used occasionally may nonetheless be essential under specific circumstances. (It is possible that an instrumental approach and lack of biological awareness have created this problem but that it is not actually intrinsic to this kind of modelling. In biological evolution, there is no credit assignment. Phenotypic traits stand and fall together.)

That said, the Classifier System has one definite advantage over both Genetic Programming and Genetic Algorithms assuming these difficulties can be overcome. This is that the individual rules may be much simpler (and hence more easily interpreted behaviourally) than Genetic Programmes. This ease of interpretation also makes it more plausible that individual rules (rather than sub-trees from Genetic Programmes or very precisely encoded bit strings from Genetic Algorithms) might be transferred meaningfully between firms either by interpretation of observable actions or "gossip". Interestingly, despite their advantages, Classifier Systems are easily the least applied evolutionary algorithms for understanding social behaviour, and this lacuna offers real opportunities for new research.

In what appears to be one of the earliest applications to firm decision-making, Moss (1992) compares a Classifier System and a (non-evolutionary) algorithm of his own design on the task of price setting in a monopoly. His algorithm hypothesises specific relationships between variables in the market and then tests these. For example, if an inverse relationship between price and profit is postulated, the firm experiments by raising price and seeing whether profit actually falls. If not, the hypothesis is rejected and another generated. If it works, but only over a range, then the hypothesis is progressively refined. The conclusion that Moss draws from this approach illustrates an important advantage of Genetic Programming over Genetic Algorithms and (some) Classifier Systems—that its solutions are explicitly based on process and therefore explanatory. Moss points out that the simple Classifier System simply evolves a price, while his algorithm shows how the firm evolves a representation of the world that allows it to set a price. Although not doing it quite as explicitly as his algorithm, a Genetic Programme may incorporate a stylised representation of market relationships into the encoding of the decision process. (Of course, in certain circumstances, the firm may lack the operators and terminals to deduce these relationships adequately, or they may not form a reliable basis for action. In this case, simpler strategies like "price following"—simply setting the same price as another firm—are likely to result.)

To return to the point made by Moss, all the Classifier System models so far developed to study firm behaviour seem to be "flat" and "hard coded". By "flat" we mean that only a single rule is needed to bridge the gap between information received and action taken. In practice, the Classifier System paradigm is capable of representing sets of rules that may trigger each other in complex patterns to generate the final output. This set of rules may also encapsulate evolved knowledge of the environment although "hard coding" prevents this. For example, we might model the production process as a Classifier System in which the rules describe the microstructure of the factory floor: where each worker went to get raw materials, what sequence of actions they performed to transform them and where they put the results. In such a model, events (the arrival of a partially assembled computer at your position on the production line) trigger actions (the insertion of a particular component). However, running out of "your" component would trigger a whole other set of actions like stopping the production line and calling the warehouse. The construction of such "thick" Classifier Systems is a task for future research.

"Hard coding" implies that each rule bridges the gap between input and output in the same way, suggesting the common representation of Genetic Algorithms with its attendant behavioural implausibility. In the models described above, decision-makers do not have the option to add to the set of conditions or to change the mappings between conditions and actions: changing price on the basis of customer loyalty rather than costs, for example. There is nothing in the Classifier System architecture to prevent this, but all the current models seem to implement the architecture in a simplified and behaviourally implausible way that makes it more like a Genetic Algorithm than Genetic Programming in terms of "hard coding" of representations and decision processes.

21.6.3   Example Using Genetic Programming: An Artificial Stock Market

In this example (Edmonds 2002), there is a simulated market for a limited number of stocks, with a fixed number of simulated traders and a single "market maker". Each trader starts off with an amount of cash and can, in each trading period, seek to buy or sell each of the kinds of stock. Thus, at any time, a trader might have a mixture of cash and amounts of each stock. A single market maker sets the prices of each stock at the beginning of each trading period depending on the last price and the previous amount of buying and selling of it. The "fundamental" is the dividend paid on each stock, which for each stock is modelled as a slowly moving random walk. There is a transaction cost for each buy or sell action by the traders. Thus, there is some incentive to buy and hold stocks and not trade too much, but in general, more money can be made (or lost) in short-term speculation. The market is endogenous except for the slowly changing dividend rate so that the prices depend on the buy and sell actions and a trader's success depends on "outsmarting" the other traders.

In the original artificial stock market model (Arthur et al. 1997), each artificial trader had a fixed set of price prediction strategies. At each time interval, they would see which of these strategies was most successful at predicting the price in the recent past (fixed number of time cycles) and rely on the best of these to predict the immediate future price movements. Depending on its prediction using this best strategy, it would either buy or sell.

In the model presented here, each trader has a small population of action strategies for each stock, encoded as a GP tree. In each time period, each artificial trader evaluates each of these strategies for each stock. The strategies are evaluated against the recent past (a fixed number of time cycles) to calculate how much value (current value based on cash plus stock holdings at current market prices) the trader would have had if they had used this strategy (taking into account transactions costs and dividends gained), assuming that the prices were as in the recent past. The trader then picks the best strategy for each stock and (given constraints of cash and holdings) tries to apply this strategy in their next buy and sell (or hold) actions.

At the end of each trading period, the set of action strategy trees are slightly evolved using the GP algorithm. That is to say that they are probabilistically "remembered" in the next trading round depending on their evaluated success, with a few of them crossed in a GP manner to produce new variations on the old strategies and very few utterly new random strategies introduced. As a result of this, a lot of evolution of small populations is occurring, namely, a population for *each* trader and *each* stock. Here, each GP tree represents a possible strategy that the trader could think of for that stock. The Genetic Programming algorithm represents the trader's learning process for each stock, thinking up new variations of remembered strategies, discarding strategies that are currently unsuccessful and occasionally thinking up completely novel strategies. This is a direct implementation of Campbell's model of creative thinking known as "Blind Variation and Selective Attention" (Campbell 1965). Further, it introduces notions of analogy and expertise into the model. A strategy that is good for one stock is a priori likely to be good for another similar stock. Thus, if a new stock is introduced, agents may use existing strategies to decide what to do about it. A new trader will have relatively poor strategies generally and will not necessarily have the feedback to choose the most appropriate strategy for a new stock. By contrast, an expert will have both a good set of strategies to choose from and better judgement of which to choose. These aspects of social (evolutionary) learning are clearly important in domains where there is genuine novelty which many traditional approaches do not handle well (or in some cases at all).

The nodes of the strategy trees can be any mixture of appropriate nodes and types. This model uses a relatively rich set of nodes, allowing arithmetic, logic, conditionals, branching, averaging, statistical market indices, random numbers, comparisons, time lags and the past observed actions of other traders. With a certain amount of extra programming, the trees can be strongly typed (Haynes et al. 1996), i.e. certain nodes can take inputs that are only a specific type (say numeric) and output a different type (say Boolean)—for example, the comparison "greater than". This complicates the programming of the Genetic Operators but can result in richer and more specific trees.

Below are a couple of examples of strategies in this version of the stock market model. The output of the expression is ultimately a numeric value which indicates buy or sell (for positive or negative numbers), but only if that buy or sell is of a greater magnitude than a minimal threshold (which is a parameter, allowing for the "do nothing"—hold—option):

- [minus [priceNow 'stock-1'] [maxHistoricalPrice 'stock-1']]—*Sell if price is greater than the maximum historical price otherwise buy.*
- [lagNumeric [2] [divide [doneByLast 'trader-2' 'stock-3'] [indexNow]]]—*Buy or sell according to what was done by trader-2 for stock-3 divided by the price index three time periods ago.*

The field of evolutionary computation is primarily concerned with the efficiency and effectiveness of its algorithms in solving explicitly posed problems. However, efficiency is not the primary consideration here but rather how to make such algorithms correspond to the behaviour of observed social actors. In this model, a large population of strategies within each individual would correspond to a very powerful ability in a human to find near-optimal strategies, which is clearly unrealistic. Thus, a relatively small population of strategies is "better" since it does mean that particular traders get "locked in" to a narrow range of strategies for a period of time (maybe they all do so badly that a random, novel strategy does better eventually). This reflects the existence of "group think" and trading "styles" that can reasonably be anticipated in real markets.

Other relevant issues might be that traders are unlikely to ever completely discard a strategy that has worked well in the past. (Many evolutionary models fail to take account of the fact that humans are much better at recall from structured memory than they are at reasoning. Such a model might thus "file" all past strategies but only have a very small subset of the currently most effective ones in live memory. However, if things started going very badly, it would be easy to choose not from randomly generated strategies but from "past successes". It is an interesting question whether this would be a more effective strategy.) Clearly however, the only ultimate tests are whether the resulting learning behaviour sufficiently matches that of observed markets and whether the set of operators and terminals can be grounded in (or at least abstracted from) the strategies used by real traders. (Either test taken alone is insufficient. Simply matching behaviour may be a coincidence, while "realistic" trading strategies that don't match behaviour have either been abstracted inappropriately or don't really capture what traders do. It is quite possible that what they are able to report doing is only part of what they actually do.)

Given such a market and trader structure, what transpires is a sort of learning "arms-race" where each trader is trying to "outlearn" the others, detecting the patterns in their actions and exploiting them. The fact that all agents are following some strategy at all times ensures that (potentially) there are patterns in existence to be outlearned. Under a wide range of conditions and parameter settings, one readily observes many of the qualitative patterns observed in real stock markets—speculative bubbles and crashes, clustered volatility, long-term inflation of prices and so on. Based on the simulation methodology proposed by Gilbert and Troitzsch (2005) and the idea of generative social science put forward by Epstein (2007), this outcome shows how a set of assumptions about individual actions (how traders implement and evolve their strategies) can potentially be falsified against aggregate properties of the system such as price trends across the range of stocks. Such models are an active area of research; a recent PhD, which surveys these, is Martinez-Jaramillo (2007).

21.6.4   Example: The Functional Survival of "Strict" Churches

There are clear advantages to using existing evolutionary algorithms to understand complex social processes as we hope we have shown through the examples above. Apart from an opportunity to discuss the "technicalities" of evolutionary algorithms through looking at simple cases, it is valuable to have programmes that can be used "off the shelf" (rather than needing to be developed from scratch) and for which there is an active research agenda of technical developments and formal analysis which can be drawn on. However, the major downside of the approach has also been hinted at (and will be discussed in more detail in the conclusion). Great care must be exercised in choosing a domain of application for evolutionary algorithms in understanding complex social systems. The more an evolutionary algorithm is used "as is", the smaller its potential domain of social application is likely to be. Furthermore, while it is possible, by careful choice of the exact algorithm, to relax some of the more socially unhelpful assumptions of evolutionary algorithms (the example of an external Fitness Function and a separate Breeding Pool have already been discussed), the danger is that some domains will simply require too much modification of the basic evolutionary algorithm to the point where the result becomes awkward or the algorithm incoherent. (A major problem with existing models has been the inability of their interpretations to stand up to scrutiny. In some cases, such as the Election Operator proposed by Arifovic, it appears that even the designers of these models are not fully aware of the implications of biological evolution.)

As suggested at the beginning of the chapter, the other approach, formerly rare but now increasingly popular, is to start not with an evolutionary algorithm but with a social system and build a simulation that is nonetheless evolutionary based on the structure of that. The challenge of choosing domains with a clear analogy to biological evolution remains but is not further complicated by the need to unpick and redesign the assumptions of an evolutionary algorithm. Such an example of a "bespoke" evolutionary simulation is provided in this section.

Iannaccone (1994) puts forward an interesting argument to explain the potentially counter-intuitive finding that "strict churches are strong". It might seem that a church that asked a lot of you, in terms of money, time and appropriate behaviour, would be less robust (in this consumerist era) than one that simply allowed you to attend on "high days and holidays" (choosing your own level of participation). However, the evidence suggests that it is the liberal churches that are losing members fastest. Iannaccone proposes that this can be explained by reflecting on the nature of religious experience. The satisfaction that people get out of an act of worship depends not just on their own level of involvement but also that of all other participants. This creates a free rider problem for "rational" worshippers.

Each would like to derive the social benefit while minimising their individual contribution. Churches are thus constantly at the mercy of those who want to turn up at Christmas to a full and lively church but don't want to take part in the everyday work (like learning to sing the hymns together) that makes this possible.

Iannaccone then argues that an interesting social process can potentially deal with this problem. If we suppose that churches do things like demanding time, money and appropriate behaviour from their worshippers, this affects the satisfaction that worshippers can derive from certain patterns of activity. If the church can somehow make non-religious activities less possible and less comfortable, it shifts the time allocations of a "rational" worshipper towards the religious activities and can simultaneously reward him or her with the greater social benefit that comes from the church "guiding" its members in this way. To take a mildly contrived example, Muslims don't drink alcohol. They also dress distinctively. A Muslim who wanted to drink couldn't safely ask his friends to join him, could easily be seen entering or leaving a pub by other Muslims and would probably feel out of place and uncomfortable once inside (quite apart from any guilt the church had managed to instil). The net effect is that Muslims do not spend much time in pubs (while many others in the UK do) and have more time for religious activity. Of course, it is easy to pick holes in the specifics of Iannaccone's argument. Why would the Muslim not dress up in other clothes? (That itself might need explanation though). Why not engage in another non-religious activity that was not forbidden? Why assume that only religious activities are club goods? (Isn't a good night at the pub just as much a result of collective effort?)

However, regardless of the details, the basic evolutionary point is this. Religious groups set up relatively fixed "creeds" that tell members when and how to worship, what to wear and eat, how much money must be given to the church and so on. Given these creeds, worshippers join and leave churches. To survive, churches need worshippers and a certain amount of "labour" and income to maintain buildings, pay religious leaders and so on. Is it in fact the case as Iannaccone argues that the dynamics of this system will result in the differential survival of strict churches at the expense of liberal ones? This is, in fact, a very general framework for looking at social change. Organisations like firms depend on the ability to sell their product and recruit workers in a way that generates profit. Organisations like hospitals are simultaneously required to meet external goals set by their funders and honour their commitments to their "customers": On one hand, the budget for surgery may be exhausted. On the other, you can't turn away someone who is nearly dead from a car crash knowing they will never survive to the next nearest accident and emergency department. This evolutionary interplay between organisations facing external constraints and their members is ubiquitous in social systems.

Before reporting the results and discussing their implications, two issues must be dealt with. Because this is a "two-sided" process (involving worshippers and churches), we must attend to the assumptions made about the behaviour of these groups. In the model discussed here, it was assumed that churches were simply defined by a fixed set of practices and did not adapt themselves. This is clearly a simplification but not a foolish one. Although creeds do adapt, they often do so over very long periods, and this is a risky process. If worshippers feel that a creed is just being changed for expedience (rather than in a way consistent with doctrine), they may lose faith just as fast as in a church whose creed is clearly irrelevant to changed circumstances. Speculatively, the great religions are those that have homed in on the unchanging challenges and solutions that people face in all times and all places, while the ephemeral ones are those that are particular to a place or set of circumstances. Conversely, the model assumes that worshippers are strictly rational in choosing the allocations of time to different activities that maximise their satisfaction. Again, this assumption isn't as artificial as it may seem. Although we do not choose religions like we choose baked beans, there is still a sense in which a religion must strike a chord in us (or come to do so). It is hard to imagine that a religion that someone hated and disbelieved in could be followed for long merely out of a sense of duty. Thus, here, satisfaction is being used in a strictly subjective sense without inquiring into any potential objective correlates. This life, for me, is better than that life. In terms of predicting individual behaviour, this renders satisfaction a truism, but in the context of the model (and explaining the survival of different kinds of churches), what matters is not what people happen to like but the fact that they pursue it. To sum up, we could have represented the churches as more adaptive and the worshippers as less adaptive, but since we are interested in the *interplay* of their behaviours (and, incidentally, this novel approach reveals a shortage of social science data about how creeds change and worshippers participate in detail), there is no definite advantage to doing so.

In a nutshell, the model works as follows (more details can be found in Chattoe (2006a)). Each agent allocates their time to activities generating satisfaction (and different agents like different things to different extents). They can generate new time allocations in two main ways. One is by introspection, simply reflecting that a bit more of this and a bit less of that might be nicer. The other is by meeting other agents and seeing if their time allocations would work better. This means, for example, that an agnostic who meets a worshipper from church A may suddenly realise that leading their life in faith A would actually be much more satisfying than anything they have come up with themselves. Conversely, someone "brought up in" church B (and thus mainly getting ideas from other B worshippers about "the good life") may suddenly realise that a life involving no churchgoing at all is much better for him or her (after meeting an agnostic). Of course, who you meet will depend on which church you are in and how big the churches (and agnostic populations) are. It may be hard to meet agnostics if you are in a big strict church, and similarly, there are those whom a more unusual religion might suit very well who will simply not encounter its creed. Churches are created at a low rate, and each one comes with a creed that specifies how much time and money members must contribute and how many non-religious activities are "forbidden". Members can only have time allocations that are compatible with the creed of the church. These allocations determine the social benefits of membership discussed above. If a church cannot meet minimum membership and money constraints, it disappears. Thus, over time, churches come and go, differing in their "strictness", and their survival is decided by their ability to attract worshippers and contributions. Worshippers make decisions that are reasonable (but not strictly rational in that they are not able instantaneously to choose the best time allocation and church for them—which may include no church—for any state of the environment). This system reproduces some stylised facts about religion. New churches start small and are often (but not always) slow to grow. Churches can appear to fade and then experience resurgences. There are a lot of small churches and very few large ones.

What happens? In fact, there is almost no difference between the lifetimes of liberal churches and mildly strict ones. What is clear however is that very strict churches (and especially cults—which proscribe all non-religious activities) do not last very long at all. It is important to be clear about this as people often confuse membership with longevity. It is true that strict churches can grow very fast and (for a while) very large, but the issue at stake here is whether they will survive in the long term. To the extent that the assumptions of the simulation are realistic, the answer would appear to be no. Thus, we have seen how it is possible to implement a reasonably coherent biological analogy in a social context without using a pre-existing evolutionary algorithm.

21.7 Conclusion: Using Biological Analogies to Understand Social Systems

Having presented a number of case studies of evolutionary algorithms in different application areas, we are now in a position to draw some general conclusions about the design and use of evolutionary simulations. Despite the fact that some have claimed that a generalised version of evolution (Blind Variation and Selective Attention) *is* the basic template for human creativity (Campbell 1965) and that it is plausible that some processes similar to biological evolution do occur in human societies, it is unlikely that these processes will be direct translations of biological evolution in all its details. For this reason, we would propose that research into evolutionary models proceeds as follows (although it is inevitable that there will be some backward and forward interplay between the stages for reasons discussed below):

1. Start with your substantive research domain of interest (linguistics, stock markets, the rise and fall of religious groups) and consider the general arguments for representing these (or parts of them) in evolutionary terms. While it is seldom spelt out explicitly, there are actually rather few candidate "general social theories" to explain the dynamic interaction of choice and change. Unless one believes that individuals have the power and knowledge required for rational action to benefit them (and note that this condition isn't met in situations as simple as the two-person one-shot prisoner's dilemma), evolution is really the only coherent and completely specified theory available.[3] Thus (and obviously

---

[3] In fact, it might be argued that it is the *only* one. Rational choice cannot contend with novelty or the origin of social order. By focusing on *relative* performance, no matter how absolutely poor, evolution can produce order from randomness.

the authors are biassed in this regard) if you believe that agents act on imperfect knowledge in an independently operating[4] environment (such that there often is a gap between what you expect to happen and what happens however effectively you collect and process data about your environment), it is worth considering an evolutionary approach. We would argue that these conditions are met in most social settings, but economists would disagree.

2. Consider the explicit specification of an evolutionary process for your particular domain of research (perhaps using the four process specification above as a guide). The key choice made in this context is a "coherent" object of selection (OOS) whose presence or absence is empirically accessible. This makes organisations and firms with any kind of formal status particularly suitable. For informal groups like families, for example, it is much less clear what constitutes a "unit". (Is it, in a traditional society setting, that they physically survive or, in a modern setting, that they still cohabit or are still on speaking terms? The problems here are evident). Interestingly, individuals (while obviously "physically" coherent) are still problematic as objects of selection. Unless the model involves "bare" survival, it is less obvious what happens when an agent is "selected". However, examples still exist, such as who is trading in particular markets. Most of the rest of the evolutionary process specification follows naturally from the choice of an OOS. It then becomes fairly clear what the resource driving selection is (food for tribal groups, profit for firms, membership for voluntary organisations, attention for memes), what causes the birth and death of OOS (sexual reproduction, merger, religious inspiration, bankruptcy, lack of interest or memorability and so on) and what variation occurs between OOS.

This last is an interesting area and one where it is very important to have a clearly specified domain of application. For example, consider industrial organisation. Textbook economic theory creates in the mind an image of the archetypal kettle factory (of variable size), selling kettles "at the factory gates" directly to customers and ploughing profits straight back into growth and better technology. In such a world, a firm that is successful early on can make lots of poor judgements later because it has efficient technology, market dominance, retained profit and so on. As such, evolutionary pressure rapidly ceases to operate. Further, this kind of firm does not "reproduce" (it merely gets larger), and even imitation of its strategy by other firms (that are smaller and poorer) may not cause the effective "spread" of social practices required by an evolutionary approach. (What works for the dominant firm may actually be harmful to smaller "followers".)

By contrast, we can see the more modern forms of competition by franchises and chains (Chattoe 1999) or the more realistic detail of "supply chain production" as much more naturally modelled in evolutionary terms. In the first case, firms do directly "reproduce" a set of practices (and style of product, décor,

---

[4]This independence comes *both* from other social actors and physical processes like climate and erosion.

amount of choice and so on) from branch to branch. More successful chains have more branches. Furthermore, the "scale" of competition is determined by the number of branches, and it is thus reasonable to say that successful business practices proliferate. Wimpy may drive out "Joe Smith's Diner" from a particular town, but Joe Smith is never a real competitor with the Wimpy organisation even if he deters them from setting up a branch in that town. This means that selection pressure continues to operate with chains at any scale competing with other chains at similar scales. Short of outright monopoly, there is never a dominant market position that is stable.[5]

In the second case, we can see how open-ended evolution may create new opportunities for business and that supply chains as a whole constitute "ecologies" (Chattoe-Brown 2009). Initially, each firm may transport its own goods to market, but once markets are sufficiently distant and numerous, there may be economies of scale in offering specialist transport and logistics services (e.g. all goods going from Bristol to Cardiff in one week may be carried by a single carter, or a firm may create a distribution infrastructure, so not all goods are transported directly from origin to destination but via cost-saving looped routes). Again, it is clear how the organisations here must operate successful practices that satisfy both suppliers (those who want to deliver goods) and customers (those who want to receive them) and, further, how the nature of the business environment may change continuously as a consequence of innovation (whether technical or social). The creation of the refrigerated ship or the internal combustion engine may foreclose some business opportunities (like raising animals in the city or harness making) and give rise to others which may or may not be taken up (spot markets, garages).

These examples show several things. Firstly, it is necessary to be very clear what you are trying to understand as only then can the fitness of the evolutionary analogy be assessed. Secondly, it is useful to have a systematic way (Chattoe 1998, 2006b) of specifying evolutionary models since these stand or fall on their most implausible assumptions (particularly in social sciences which aren't very keen on this approach).[6] Thirdly, there are a lot more opportunities for evolutionary modelling than are visible to the "naked eye", particularly to those who take the trouble to develop both domain knowledge and a broad evolutionary perspective. Considering the ubiquity of branch competition and intermediate production in the real world, the economic literature is amazingly distorted towards the "autonomous kettle factory view", and simulation models of realistic market structures are scarcer still (though this is just starting to change). The price of adopting a novel method is scepticism by one's peers

---

[5]This is probably because the market is spatially distributed, and the only way of making additional profits is by opening more branches (with associated costs). There are no major economies of scale to be exploited as when the kettle factory simply gets bigger and bigger with all customers continuing to bear the transport costs.

[6]More informally, "the assumptions you don't realise you are making are the ones that will do you in".

(and associated difficulties in "routine" academic advancement), but the rewards are large domains of unexplored research opportunities and the consequent possibility for real innovation. Finally, don't forget that it is always possible to use an evolutionary algorithm as a "black box learning system" within the "mind" of an agent or organisation, although there is a design issue about interpreting this kind of model discussed previously. Further, even as a "black box", the learning algorithm can make a crucial difference in simulations (Edmonds and Moss 2001), and one cannot simply assume that any learning algorithm will do.

3. Explore whether data for your chosen domain is available (or can readily be got using standard social science methods).[7] If it is available, does it exist at both the individual level and in aggregate? For example, is there observational data about firm price setting practices (e.g. in board meetings) and long-term historical data about the birth, death and merger of firms in a particular industry and their prices over time? Because simulation is a relatively new method, it is still possible to build and publish exploratory (or less flatteringly "toy") models of evolutionary processes, but it is likely to get harder and may become impossible unless the evolutionary model or the application domain is novel. It is almost certainly good scientific practice to make the accessibility of data part of the research design, but it does not follow from this that only models based on available data are scientific. The requirement of falsifiability is served by the data being collectable "in principle", not already collected. The best argument to support claims of scientific status for a simulation is to consider (as a design principle) how each parameter could be calibrated using existing data or existing research methods. (The case is obviously weaker if someone has to come up with a new data collection method first although it helps if its approach or requirements can be sketched out a priori.)

This aspect of research design also feeds into the decision about whether to use an existing evolutionary algorithm and, if so, which one. The emerging methodology of social simulation (Gilbert and Troitzsch 2005, pp. 15–18; Epstein 2007) is to make a set of empirically grounded hypotheses at the micro level (firms set prices thus) and then to falsify this ensemble at the macro level. (The real distribution of survival times for firms is thus: It does or does not match the simulated distribution of survival times produced by the model.) A problem will arise if it is hard to interpret the simulated price setting practices. Suppose, for example, we use GP to model the evolution of trading strategies in stock markets. We may use interviews or observation of real traders to decide what terminals and operators are appropriate but, having let the simulation run and observed plausible aggregate properties, we may still not know (and find it extremely hard to work out because of the interpretation issue) whether the

---

[7]In a way, it is a black mark against simulation that this needs to be said. Nobody would dream of designing a piece of statistical or ethnographic work without reference to the availability or accessibility of data!

evolved strategies used are actually anything like those which traders would (or could) use.

Equating the empirical validation of the GP grammar with the validation of the strategies evolved from it is bit like assuming that, because we have derived a Swahili grammar from listening to native speakers, we are then qualified to decide when Swahili speakers are telling the truth (rather than simply talking intelligibly). It muddles syntax and semantics. The design principle here is then to consider how the chosen evolutionary algorithm will be interpreted to establish the validity of evolved practices. (Creative approaches may be possible here like getting real traders to design or choose GP trees to trade for them or getting them to "critique" what are effectively verbal "translations" of strategies derived from apparently successful GP trees as if they were from real traders.) In this regard, it is their potential ease of interpretation that makes the relative neglect of CS models seem more surprising in evolutionary modelling.

4. Having first got a clear sense of what needs to be modelled, it is then possible to choose a modelling technique in a principled way. As the analysis of case studies suggests, the danger with a "method-led" approach is that the social domain will be stylised (or simply falsified) to fit the method. A subsidiary difficulty with the method-led approach is that even if the researcher is wise enough to use a modified evolutionary algorithm to mirror a social process accurately (rather than distorting or abstracting the domain to fit the method), inadequate technical understanding may render the modified algorithm incoherent or ineffective. It is thus very important to understand fully any methods you plan to apply particularly with regard to any instrumental assumptions they contain. (In making convergence her goal for the GA cobweb model, Arifovic introduced an election operator which actually rendered the GA *less* effective in solving hard problems. This issue would probably have been foreseen in advance by a competent instrumental user of the GA technique. The muddle arose from the *interface* between social description and the GA as a highly effective instrumental optimisation device.)

Having chosen a modelling technique, all its supporting assumptions must also be examined in the light of the application domain. For example, it is very important not to confuse single and multiple population interpretations of a GA: Do firms each have multiple candidate pricing strategies and choose them by an evolutionary process, or is there one overall evolutionary process, in which single pricing strategies succeed and fail with the associated firms "carrying" them? Each model (or some combination) might be justified on empirical grounds but only if the difference in interpretation is kept clearly in mind. Although we are sceptical that systems of realistic social complexity would allow this, the principled choice of methods means that it is even possible that some domains would not require simulation at all but could be handled by mathematical models of evolution like replicator dynamics (Weibull 1995) or stochastic models (Moran 1962).

By contrast, however, if the properties of the chosen social domain are too far from a standard evolutionary algorithm (such that it can neither be used wholesale or deconstructed without collapsing into incoherence), the best solution is to build a bespoke evolutionary model as was done for the "strict churches" case study. (At the end of the day, evolutionary algorithms were themselves "evolved" in a completely different engineering environment, and we would not therefore expect them to apply widely in social systems. Thus, great care needs to be taken to use them only where they clearly do apply and thus have real value.) With free and widely used agent-based modelling packages like NetLogo[8] and associated teaching materials (Gilbert and Troitzsch 2005, Gilbert 2007), this is now much easier than it was. Ten years ago, one reason to use an existing algorithm was simply the significant cost of building your own from scratch. To sum up this strategy of research, the decision to use, modify or build an evolutionary algorithm from scratch should be a conscious and principled one based on a clear understanding of the domain and existing social science data about it.

The final piece of advice is not technical or methodological but presentational. In applying a novel method, be prepared to suffer equally at the hands of those who don't understand it and those who do! One of the hardest things to do in academia is to strike a balance between rejecting ill-founded criticisms or those that translate to "I just don't like this" without also rejecting real objections that may devalue months (or even years) of your effort (and still, frustratingly for you, be part of "good science"). To judge criticisms in a novel area, you must be especially well informed and thus confident of your ground. For example, there is no clear-cut evidence for Lamarckism (modification of the genotype by the phenotype during the life of the organism in a way that can then be transmitted by reproduction) in biology, but in social systems, such processes are ubiquitous. (Someone discovers a good way to discipline children. Those who were thus disciplined do the same thing to their children. This is an acid test because, with hindsight, the "victims" have to see it as beneficial and thus not have been warped by it, even if it was hateful at the time. Punishments so nasty that the victims won't inflict them, or so ineffective that the parents stop bothering, will die out.) Failure to understand this issue may either set you on the path of non-Lamarckian (and thus quite possibly implausible) evolutionary models of social systems or of apologising mistakenly for building Lamarckian models which don't "truly" reflect biological evolution (when that was never the design criterion for using biological analogies in social science anyway).

The best way to address these issues is hopefully to follow the systematic procedure outlined above. This minimises the chances that you will miss things which critics can use to reject your models (and if they are hostile enough, your whole approach) and ensures that by justifying the models to yourself, you can actually justify them to others. In popular scientific folklore, Darwin (still the greatest evolutionist) spent a considerable period trying to anticipate all possible objections to his theory and see how valid they were (and what counters he could

---

[8]http://ccl.northwestern.edu/netlogo/

provide) before he presented his work. Given how fraught the acceptance of his theory has been anyway, imagine if he had not troubled to take that step!

We hope we have shown, by the use of diverse case studies and different evolutionary modelling techniques, both the considerable advantages and (potentially avoidable) limitations of this approach and encourage interested readers to take these ideas forward both in developing new kinds of models and applying evolutionary models to novel domains. The field is still wide open, and we are always pleased to hear from potential students, coworkers, collaborators, supporters or funders!

Acknowledgements Edmund Chattoe-Brown acknowledges the financial support of the Economic and Social Research Council as part of the SIMIAN (http://www.simian.ac.uk) node of the National Centre for Research Methods (http://www.ncrm.ac.uk).

Further Reading

Gilbert and Troitzsch (2005) is a good general introduction to social science simulation and deals with evolutionary techniques explicitly, while Gilbert (2007) is recommended as an introduction to this kind of simulation for studying evolution in social systems. For deeper introductions to the basic techniques, see Goldberg (1989), which is still an excellent introduction to GA despite its age (for a more up-to-date introduction, see Mitchell (1996), and Koza (1992a, 1994)) for a very accessible explanation of GP with lots of examples. Forrest (1991) is a good introduction to techniques in Classifier Systems.

More details about the four example models are given in the following: Chattoe (2006a) shows how a simulation using an evolutionary approach can be related to mainstream social science issues, Edmonds (2002) gives an example of the application of a GP-based simulation to an economic case, and Moss (1992) is a relatively rare example of a classifier-based model.

References

Antoinisse, H. (1991). A grammar based Genetic Algorithm. In G. Rawlins (Ed.), *Foundations of Genetic Algorithms: Proceedings of the first workshop on the foundations of Genetic Algorithms and Classifier Systems, Indiana University, 15-18 July 1990* (pp. 193–204). San Mateo, CA: Morgan Kaufmann.

Arifovic, J. (1994). Genetic Algorithm learning and the cobweb model. *Journal of Economic Dynamics and Control, 18*, 3–28.

Arthur, W.B., Holland, J.H., LeBaron, B., Palmer, R., & Tayler, P. (1997). Asset pricing under endogenous expectations in an artificial stock market. In W. B. Arthur, S. N. Durlauf, & D. A. Lane (Eds.), *The economy as a complex evolving system II, Santa Fe Institute Studies in the Science of Complexity, proceedings* (Vol. XXVII, pp. 15–44). Reading, MA: Addison-Wesley.

Becker, G. (1976). Altruism, egoism and genetic fitness: Economics and sociobiology. *Journal of Economic Literature, 14*, 817–826.

Belew, R. (1989). When both individuals and populations search: Adding simple learning to the Genetic Algorithm. In J. Schaffer (Ed.), *Proceedings of the third international conference on Genetic Algorithms, George Mason University, 4-7 June 1989* (pp. 34–41). San Francisco, CA: Morgan Kaufmann.

Belew, R. (1990). Evolution, learning and culture: Computational metaphors for adaptive search. *Complex Systems, 4*, 11–49.

Blackmore, S. (1999). *The meme machine*. Oxford: Oxford University Press.

Boorman, S., & Levitt, P. (1980). *The genetics of altruism*. St Louis, MO: Academic Press.

Boyd, R., & Richerson, P. J. (1985). *Culture and the evolutionary process*. Chicago: University of Chicago Press.

Buss, D. M. (2015). *Evolutionary psychology: The new science of the mind* (5th ed.). London: Psychology Press.

Calvin, W. (1996a). *How brains think: Evolving intelligence, then and now*. New York: Basic Books.

Calvin, W. (1996b). *The cerebral code: Thinking a thought in the mosaics of the mind*. Cambridge, MA: MIT Press.

Campbell, D. T. (1965). Variation and selective retention in socio-cultural evolution. In H. R. Barringer, G. I. Blanksten, & R. W. Mack (Eds.), *Social change in developing areas: A reinterpretation of evolutionary theory* (pp. 19–49). Cambridge, MA: Schenkman.

Campbell, D. T. (1974). Evolutionary epistemology. In P. A. Schlipp (Ed.), *The philosophy of Karl R. Popper, The library of living philosophers* (Vol. XIV, pp. 412–463). LaSalle, IL: Open Court.

Cavalli-Sforza, L., & Feldman, M. (1973). Cultural versus biological inheritance: Phenotypic transmission from parents to children. *Human Genetics, 25*, 618–637.

Chattoe, E., & Gilbert, N. (1997). A simulation of adaptation mechanisms in budgetary decision making. In R. Conte, R. Hegselmann, & P. Terna (Eds.), *Simulating social phenomena, Lecture notes in economics and mathematical systems* (Vol. 456, pp. 401–418). Berlin: Springer.

Chattoe, E. (1998). Just how (un)realistic are evolutionary algorithms as representations of social processes? *Journal of Artificial Societies and Social Simulation, 1*(3), 2. http://www.soc.surrey.ac.uk/JASSS/1/3/2.html

Chattoe, E. (1999, June 7–9). *A co-evolutionary simulation of multi-branch enterprises*. Paper presented at the European Meeting on Applied Evolutionary Economics, Grenoble. http://webu2.upmf-grenoble.fr/iepe/textes/chatoe2.PDF

Chattoe, E. (2002). Developing the selectionist paradigm in sociology. *Sociology, 36*, 817–833.

Chattoe, E. (2006a). Using simulation to develop and test functionalist explanations: A case study of dynamic church membership. *British Journal of Sociology, 57*, 379–397.

Chattoe, E. (2006b). Using evolutionary analogies in social science: Two case studies. In A. Wimmer & R. Kössler (Eds.), *Understanding change: Models, methodologies and metaphors* (pp. 89–98). Basingstoke: Palgrave Macmillan.

Chattoe-Brown, E. (2009). *The implications of different analogies between biology and society for effective functionalist analysis* (Draft paper). Department of Sociology, University of Leicester, Leicester.

Cloak, F. T. (1975). Is a cultural ethology possible? *Human Ecology, 3*, 161–182.

Costall, A. (1991). The meme meme. *Cultural Dynamics, 4*, 321–335.

Csányi, V. (1989). *Evolutionary systems and society: A general theory of life, mind and culture*. Durham, NC: Duke University Press.

Darwin, C. R. (1859). *On the origin of species by means of natural selection*. London: John Murray.

Dautenhahn, K., & Nehaniv, C. L. (Eds.). (2002). *Imitation in animals and artifacts*. Cambridge, MA: MIT Press.

Dawkins, R. (1976). *The selfish gene*. Oxford: Oxford University Press.

Dawkins, R. (1982). Organisms, groups and memes: Replicators or vehicles? In R. Dawkins (Ed.), *The extended phenotype*. Oxford: Oxford University Press.

Dawkins, R. (1993). Viruses of the mind. In B. Dahlbohm (Ed.), *Dennett and his critics* (pp. 13–27). Malden, MA: Blackwell Publishers.

Dennett, D. (1990). Memes and the exploitation of imagination. *Journal of Aesthetics and Art Criticism, 48*, 127–135.

Dobzhansky, T., Ayala, F. J., Stebbins, G. L., & Valentine, J. W. (1977). *Evolution*. San Francisco: W.H. Freeman.

Dosi, G., Marengo, L., Bassanini, A., & Valente, M. (1999). Norms as emergent properties of adaptive learning: The case of economic routines. *Journal of Evolutionary Economics, 9*, 5–26.

Edelman, G. (1992). *Bright air, brilliant fire: On the matter of the mind*. New York: Basic Books.

Edmonds, B. (2002). Exploring the value of prediction in an artificial stock market. In V. M. Butz, O. Sigaud, & P. Gérard (Eds.), *Anticipatory behaviour in adaptive learning systems, Lecture notes in computer science* (Vol. 2684, pp. 285–296). Berlin: Springer.

Edmonds, B., & Moss, S. (2001). The importance of representing cognitive processes in multi-agent models. In G. Dorffner, H. Bischof, & K. Hornik (Eds.), *Artificial neural networks: ICANN 2001, International Conference Vienna, Austria, August 21–25, 2001, Proceedings, Lecture notes in computer science* (Vol. 2130, pp. 759–766). Berlin: Springer.

Epstein, J. M. (2007). *Generative social science: Studies in agent-based computational modelling*. Princeton, NJ: Princeton University Press.

Fagin, R., Halpern, J., Moses, Y., & Vardi, M. (1995). *Reasoning about knowledge*. Cambridge, MA: MIT Press.

Forrest, S. (1991). *Parallelism and programming in Classifier Systems*. London: Pitman.

Friedman, D. P., & Felleisen, M. (1987). *The little LISPER*. Cambridge, MA: MIT Press.

Gilbert, N., & Troitzsch, K. G. (2005). *Simulation for the social scientist* (2nd ed.). Buckingham, UK: Open University Press.

Gilbert, N. (2007). *Agent-based models, Quantitative applications in the social sciences* (Vol. 153). London: Sage Publications.

Goldberg, D. E., Deb, K., & Korb, B. (1990). Messy Genetic Algorithms revisited: Studies in mixed size and scale. *Complex Systems, 4*, 415–444.

Goldberg, D. E. (1989). *Genetic Algorithms in search, optimization and machine learning*. Boston, MA: Addison-Wesley.

Grefenstette, J., Gopal, R., Rosmaita, B., & Van Gucht, D. (1985). Genetic Algorithms for the travelling salesman problem. In J. Grefenstette (Ed.), *Proceedings of the first international conference on Genetic Algorithms and Their Applications, Carnegie Mellon University, Pittsburgh, PA, 24-26 July 1985* (pp. 160–168). Hillsdale, NJ: Lawrence Erlbaum.

Hannan, M. T., & Freeman, J. (1993). *Organizational ecology*. Cambridge, MA: Harvard University Press.

Harvey, I. (1993). Evolutionary robotics and SAGA: The case for hill crawling and tournament selection. In C. G. Langton (Ed.), *Artificial Life III: Proceedings of the workshop on Artificial Life, Santa Fe, New Mexico, June 1992* (pp. 299–326). Boston: Addison-Wesley.

Haynes, T., Schoenefeld, D., & Wainwright, R. (1996). Type inheritance in strongly typed Genetic Programming. In P. J. Angeline & J. E. Kinnear (Eds.), *Advances in Genetic Programming 2* (pp. 359–376). Boston, MA: MIT Press.

Heyes, C. M., & Plotkin, H. C. (1989). Replicators and interactors in cultural evolution. In M. Ruse (Ed.), *What the philosophy of biology is: Essays dedicated to David Hull*. Amsterdam: Kluwer Academic Publishers.

Hodgson, G. (1993). *Economics and evolution: Bringing life back into economics*. Cambridge, MA: Polity Press.

Hoenigswald, H. M., & Wiener, L. S. (1987). *Biological metaphor and cladistics classification*. London: Francis Pinter.

Holland, J. H. (1975). *Adaptation in natural and artificial systems*. Ann Arbor, MI: University of Michigan Press.

Hughes, A. (1988). *Evolution and human kinship*. Oxford: Oxford University Press.

Hull, D. L. (1982). The naked meme. In H. C. Plotkin (Ed.), *Learning development and culture: Essays in evolutionary epistemology*. New York: Wiley.

Hull, D. L. (1988). Interactors versus vehicles. In H. C. Plotkin (Ed.), *The role of behaviour in evolution*. Cambridge, MA: MIT Press.

Iannaccone, L. (1994). Why strict churches are strong. *American Journal of Sociology, 99*, 1180–1211.

Kampis, G. (1991). *Self-modifying systems in biology: A new framework for dynamics, information and complexity*. Oxford: Pergamon Press.

Kauffman, S. A. (1993). *The origins of order, self-organization and selection in evolution*. Oxford: Oxford University Press.

Koza, J. R. (1991). Evolving a computer program to generate random numbers using the Genetic Programming paradigm. In R. Belew & L. Booker (Eds.), *Proceedings of the fourth international conference on Genetic Algorithms, UCSD, San Diego 13-16 July 1991* (pp. 37–44). San Francisco: Morgan Kaufmann.

Koza, J. R. (1992a). *Genetic Programming: On the Programming of Computers by Means of Natural Selection*. Cambridge, MA: A Bradford Book/MIT Press.

Koza, J. R. (1992b). Genetic evolution and co-evolution of computer programmes. In C. Langton, C. Taylor, J. Farmer, & S. Rassmussen (Eds.), *Artificial Life II: Proceedings of the workshop on Artificial Life, Santa Fe, New Mexico, February 1990* (pp. 603–629). Redwood City, CA: Addison-Wesley.

Koza, J. R. (1992c). Evolution and co-evolution of computer programs to control independently acting agents. In J.-A. Meyer & S. Wilson (Eds.), *From animals to animats: Proceedings of the first international conference on Simulation of Adaptive Behaviour (SAB 90), Paris, 24-28 September 1990* (pp. 366–375). Cambridge, MA: A Bradford Book/MIT Press.

Koza, J. R. (1992d). A genetic approach to econometric modelling. In P. Bourgine & B. Walliser (Eds.), *Economics and cognitive science: Selected papers from the second international conference on Economics and Artificial Intelligence, Paris, 4-6 July 1990* (pp. 57–75). Oxford: Pergamon Press.

Koza, J. R. (1994). *Genetic Programming II: Automatic discovery of reusable programs*. Cambridge, MA: A Bradford Book/MIT Press.

Kuhn, T. S. (1970). *The structure of scientific revolutions*. Chicago: University of Chicago Press.

Kummer, H., Daston, L., Gigerenzer, G., & Silk, J. (1997). The social intelligence hypothesis. In P. Weingart, P. Richerson, S. D. Mitchell, & S. Maasen (Eds.), *Human by nature: Between biology and the social sciences* (pp. 157–179). Hillsdale, NJ: Lawrence Erlbaum.

Lomborg, B. (1996). Nucleus and shield: The evolution of social structure in the Iterated Prisoner's Dilemma. *American Sociological Review, 61*, 278–307.

Lynch, A. (1996). *Thought contagion, How belief spreads through society: The new science of memes*. New York: Basic Books.

Macy, M. (1996). Natural selection and social learning in the Prisoner's Dilemma: Co-adaptation with Genetic Algorithms and Artificial Neural Networks. *Sociological Methods and Research, 25*, 103–137.

Martinez-Jaramillo, S. (2007). *Artificial financial markets: An agent based approach to reproduce stylized facts and to study the Red Queen Effect* (PhD Thesis). Centre for Computational Finance and Economic Agents (CCFEA), University of Essex, UK.

Metcalfe, J. (Ed.). (1994). *Metacognition: Knowing about knowing*. Cambridge, MA: A Bradford Book/MIT Press.

Mitchell, M. (1996). *An introduction to Genetic Algorithms*. Cambridge, MA: A Bradford Book/MIT Press.

Moran, P. A. P. (1962). *The statistical processes of evolutionary theory*. Oxford: Clarendon Press.

Moss, S. (1992). Artificial Intelligence models of complex economic systems. In S. Moss & J. Rae (Eds.), *Artificial Intelligence and economic analysis: Prospects and problems* (pp. 25–42). Cheltenham, UK: Edward Elgar.

Nelson, R. R., & Winter Jr., S. G. (1982). *An evolutionary theory of economic change*. Cambridge, MA: Belknap Press of Harvard University Press.

North, D. C. (1990). *Institutions, institutional change and economic performance*. Cambridge: Cambridge University Press.

Oliphant, M. (1996). The development of Saussurean communication. *BioSystems, 37*, 31–38.

Olivetti, C. (1994). *Do Genetic Algorithms converge to economic equilibria?* (Discussion Paper, 24). Department of Economics, University of Rome "La Sapienza", Rome, Italy.

Popper, K. R. (1979). *Objective knowledge: An evolutionary approach.* Oxford: Clarendon Press.

Reader, J. (1970). *Man on Earth.* London: Collins.

Runciman, W. G. (1998). The selectionist paradigm and its implications for sociology. *Sociology, 32,* 163–188.

Schraudolph, N., & Belew, R. (1992). Dynamic parameter encoding for Genetic Algorithms. *Machine Learning, 9,* 9–21.

Smith, R., Forrest, S., & Perelson, A. (1992). *Searching for diverse co-operative populations with Genetic Algorithms* (TCGA Report, 92002). The Clearinghouse for Genetic Algorithms, Department of Engineering Mechanics, University of Alabama, Tuscaloosa.

Tarde, G. (1884). Darwinisme naturel et Darwinisme social. *Revue Philosophique, XVII,* 607–637.

Tarde, G. (1903). *The laws of imitation.* New York: Henry Holt.

Vega-Redondo, F. (1996). *Evolution, games and economic behaviour.* Oxford: Oxford University Press.

Weibull, J. (1995). *Evolutionary game theory.* Cambridge, MA: MIT Press.

Westoby, A. (1994) *The ecology of intentions: How to make memes and influence people: Culturology.* http://ase.tufts.edu/cogstud/papers/ecointen.htm.

Whitley, D. (1989). The GENITOR algorithm and selection pressure: Why rank-based allocation of reproductive trials is best. In J. Schaffer (Ed.), *Proceedings of the third international conference on Genetic Algorithms, George Mason University, 4-7 June 1989* (pp. 116–121). San Francisco, CA: Morgan Kaufmann.

Wilson, E. O. (1975). *Sociobiology: The new synthesis.* Cambridge, MA: Harvard University Press.

Windrum, P., & Birchenhall, C. (1998). Developing simulation models with policy relevance: Getting to grips with UK science policy. In P. Ahrweiler & N. Gilbert (Eds.), *Computer simulations in science and technology studies* (pp. 183–206). Berlin: Springer.

Part IV
Applications

Chapter 22
Agent-Based Modelling and Simulation Applied to Environmental Management

Christophe Le Page, Didier Bazile, Nicolas Becu, Pierre Bommel, François Bousquet, Michel Etienne, Raphael Mathevet, Véronique Souchère, Guy Trébuil, and Jacques Weber

Abstract The purpose of this chapter is to summarize how agent-based modelling and simulation (ABMS) is being used in the area of environmental management. With the science of complex systems now being widely recognized as an appropriate one to tackle the main issues of ecological management, ABMS is emerging as one of the most promising approaches. To avoid any confusion and disbelief about the actual usefulness of ABMS, the objectives of the modelling process have to be unambiguously made explicit. It is still quite common to consider ABMS as mostly useful to deliver recommendations to a lone decision-maker, yet a variety of different purposes have progressively emerged, from gaining understanding through raising awareness, facilitating communication, promoting coordination or mitigating conflicts. Whatever the goal, the description of an agent-based model remains challenging. Some standard protocols have been recently proposed, but

C. Le Page (✉) • F. Bousquet
CIRAD, UPR47 Green, Montpellier, 34398, France
e-mail: christophe.le_page@cirad.fr; francois.bousquet@cirad.fr

D. Bazile
CIRAD, UPR47 Green, Montpellier, 34398, France

Instituto de Geografía, PUCV, Valparaiso, Chile
e-mail: didier.bazile@cirad.fr

N. Becu
CNRS, Laboratoire de geographie PRODIG 2, Paris, France
e-mail: nicolas.becu@univ-paris1.fr

P. Bomme
CIRAD, UPR47 Green, Montpellier, 34398, France

Brasilia University, UnB-CDS, 70070-914, Brasilia DF, Brazil
e-mail: pierre.bommel@cirad.fr

M. Etienne
Ecodevelopment Unit, National Institute for Agronomic Research, Avignon, France
e-mail: etienne@avignon.inra.fr

© Springer International Publishing AG 2017
B. Edmonds, R. Meyer (eds.), *Simulating Social Complexity*,
Understanding Complex Systems, https://doi.org/10.1007/978-3-319-66948-9_22

still a comprehensive description requires a lot of space, often too much for the maximum length of a paper authorized by a scientific journal. To account for the diversity and the swelling of ABMS in the field of ecological management, a review of recent publications based on a lightened descriptive framework is proposed. The objective of the descriptions is not to allow the replication of the models but rather to characterize the types of spatial representation, the properties of the agents and the features of the scenarios that have been explored and also to mention which simulation platforms were used to implement them (if any). This chapter concludes with a discussion of recurrent questions and stimulating challenges currently faced by ABMS for environmental management.

Why Read This Chapter?
To understand the recent shift of paradigms prevailing in both environmental modelling and renewable resource management that led to the emerging rise in the application of ABMS. Also, to learn about a practical way to characterize applications of ABMS to environmental management and to see this framework applied to review a selection of recent applications of ABMS from various fields related to environmental management including the dynamics of land use changes, water, forest and wildlife management, agriculture, livestock productions and epidemiology.

22.1 Introduction

In this chapter, we state that there is a combined shift in the way of thinking in both ecosystem management and ecological modelling fields. For the last 20 years, the status of computer simulation in the field of renewable resource management has changed. This chapter investigates how agent-based modelling and simulation

R. Mathevet
Centre d'Ecologie Fonctionnelle et Evolutive, UMR 5175, CNRS 1919 Route de Mende, 34293, Montpellier, France; raphael.mathevet@cefe.cnrs.fr

V. Souchère
UMR SADAPT INRA AgroParisTech, B.P. 1, 78850, Thiverval Grignon, France
e-mail: souchere@grignon.inra.fr

G. Trébuil
CIRAD, UPR47 Green, Montpellier, 34398, France

CU-CIRAD ComMod Project, Chulalongkorn University, Bangkok, Thailand
e-mail: guy.trebuil@cirad.fr

J. Weber
CIRAD, Direction R'egionale Ile de France, 42 Rue Scheffer, 75116, Paris, France
e-mail: jaques.weber@cirad.fr

(ABMS) may have contributed to this evolution and what are the challenges it has to face for such a combination to remain fruitful.

Biosphere 2, an artificial closed ecological system built in Arizona (USA) in the late 1980s, was supposed to test if and how people could live and work in a closed biosphere. It proved to be sustainable for eight humans for 2 years, when low oxygen level and wild fluctuations in carbon dioxide led to the end of the experience. Biosphere 2 represents the quest for "engineering nature" that has fascinated a lot of people (including a non-scientific audience) during the second part of the last century. The human aspect of this "adventure" mainly dealt with the psychological impact on a few people living in enclosed environments. In the real world, the relationships between human beings and the biosphere are based on tight linkages between cultural and biological diversity. Launched around 20 years before the Biosphere 2 project, the Man and Biosphere Programme (MAB) of UNESCO is seeking to improve the global relationship between people and their environment. This is now the kind of approach—in line with the Millennium Development Goal #7 from the United Nations—that is attracting more and more interest.

In ecological management, the place of people directly involved in the management scheme is now widely recognized as central, and the impact of their activities has both to be considered as promoting and endangering different types of biodiversity. At the same time, ABMS has progressively demonstrated its ability to explicitly represent the way people are using resources, the impact of this management on plant and animal dynamics and the way ecosystems adapt to it. The next section discusses how both trends have been reinforcing each other in more detail.

The third section of this chapter gives a review of recent applications of ABMS in the field of environmental management. To avoid confusion due to the coexistence of multiple terms not clearly distinguishable, we use ABMS here as an umbrella term to refer indifferently to what authors may have denominated "agent-based modelling", "multi-agent simulation" or even "multi-agent-based simulation" (also the name of an international workshop where applications dealing with environmental management are regularly presented). Our review is covering the dynamics of land use changes; water, forest and wildlife management; but also agriculture, livestock productions and epidemiology. We are focusing here on models with explicit consideration of the stakeholders (in this chapter, this is how we will denominate people directly concerned by the local environmental management system). Bousquet and Le Page (2004) proposed a more extensive review of ABMS in ecological modelling. For a specific review of ABMS dealing with animal social behaviour, see Chap. 24 in this handbook (Hemelrijk 2017).

22.2 A Shift in Intertwined Paradigms

During the last two decades, evidences accumulate that the interlinked fields of ecosystem management and environmental modelling are changing from one way of thinking to another. This is a kind of paired dynamics where agents of change from one field are fostering the evolution of conceptual views in the other one. A survey of the articles published in *Journal of Environmental Management* and *Ecological Modelling*—just to refer to a couple of authoritative journals in those fields—clearly reveals this combined shift of paradigms. Another indication from the scientific literature was given when the former *Conservation Ecology* journal was renamed *Ecology and Society* in June 1997.

Among ecologists, it has become well accepted that classical equilibrium theories are inadequate and that ecosystems are facing cycles of adaptive change made of persistence and novelty (Holling 1986). Concepts from the sciences of complexity are now widely adopted in ecology (Levin 1998), and the perception of ecosystems as complex adaptive systems, in which patterns at higher levels emerge from localized interactions and selection processes acting at lower levels, has begun to affect the management of renewable resources (Levin 1999).

Beyond the standard concept of "integrated renewable resource management", the challenge is now to develop a new "integrative science for resilience and sustainability" focusing on the interactions between ecological and social components and taking into account the heterogeneity and interdependent dynamics of these components (Berkes and Folke 1998). The relationships between stakeholders dealing with the access and use of renewable resources are the core of these intertwined ecological and social dynamics that are driving the changes observed in many ecosystems.

Panarchy is a useful concept to understand how renewable resource management is affected by this new paradigm in ecology. It has been formalized as the process by which ecological and social systems grow, adapt, transform and abruptly collapse (Gunderson and Holling 2002). The back loop of such changes is a critical time when uncertainties arise and when resilience is tested and established (Holling 2004). This new theoretical background is making sense to social scientists working on renewable resource management (Abel 1998) and to interdisciplinary groups expanding ecological regime shifts theory to dynamics in social and economic systems (Kinzig et al. 2006).

For a long period, the mainstream postulate in ecological modelling has been that science should first help to understand the "natural functioning" of a given ecosystem, so that the impacts of external shocks due to human activities ("anthropic pressures") could be monitored. Models were mainly predictive, oriented towards decision-makers who were supposed to be supported by powerful tools (expert systems, decision support systems) in selecting the "best", "optimal" management option. Nowadays, command-and-control approaches are seen as "being worse than inadequate" (Levin 1999).

Evidently, there is a growing need for more flexible (usable and understandable by diverse participants) and adaptive (easily modified to accommodate unforeseen situations and new ideas) models that should allow any involved stakeholders (ecosystem and resource managers among others) to gain insights through exploration of simulation scenarios that mimic the challenges they face. Similar to the role of metaphor in narratives, such simulation models do not strive for prediction anymore, but rather aim at sparking creativity, facilitating discussion, clarifying communication and contributing to collective understanding of problems and potential solutions (Carpenter et al. 1999). To underline the change of status of simulation models used in such a way, the term "companion modelling" has been proposed (ComMod 2003; Etienne 2011).

In recent years, ABMS has attracted more and more attention in the field of environmental management (Bousquet and Le Page 2004; Hare and Deadman 2004). Recent compilations of experiences have been edited (Gimblett 2002; Janssen 2002; Bousquet et al. 2005; Perez and Batten 2006). We propose to review recent ABMS applications in the field of environmental management based on a simplified framework presented in the next section.

22.3 A Framework for Characterizing Applications of ABMS to Environmental Management

To standardize the description of ecological models based on the interactions between elementary entities (individual-based models and agent-based models), Grimm et al. (2006) have recently proposed a protocol based on a three-block sequence: overview, design concepts and details (ODD). It is a kind of guideline for authors wishing to publish their model whose fulfilment corresponds to an entire article devoted to communicating the details of the model. Hare and Deadman (2004) also proposed a first classification scheme from the analysis of 11 case studies. Revisiting some elements from these two contributions, we propose here to successively give some insights about (1) the purpose of the model, (2) the way the environment is represented, (3) the architecture of the different agents, (4) the implementation (translation of the conceptual model into a computer programme) and (5) the simulation scenarios.

22.3.1 What Is the Model's Purpose?

As recommended by Grimm and the 27 other participants to the collective design of the ODD protocol (2006), a concise formulation of the model's purpose has to be stated first: it is crucial to understand why some aspects of reality are included while others are ignored. The reasons leading to start a modelling process are not always clearly given. Is it mainly to gain understanding and increase scientific knowledge? Is it more about raising awareness of stakeholders who do not have a clear picture of a complex system? Does it aim at facilitating communication or supporting decision? The more the information about the model's purpose will be precise, the less confusion and disbelieving about its real usefulness will remain.

22.3.2   How Is the Environment Represented?

Applications of ABMS to investigate environmental management issues are relying on a fundamental principle: they represent interacting social and ecological dynamics. On the one hand, agents represent some sort of stakeholders: individual people at the micro level and/or at more aggregated levels some groups of individuals defining (*lato sensu*) institutions (social groups such as families, economic groups such as farmers' organizations, political groups such as nongovernmental organizations). On the other hand, the environment, holding some sort of renewable resources, stands for the landscape. The renewable resources are contributing to define the landscape, and in turn the way the landscape is structured and developed influences the renewable resources. Typically, the resources are modified by direct actions of agents on their environment, whereas the resources also exhibit some intrinsic natural dynamics (growth, dispersal, etc.). At the same time, agents' decisions are somehow modified by the environment as the state of resources evolves. In such situations, the implementation of the social dynamics is performed through defining the behaviours of agents, whereas the implementation of the natural dynamics is commonly ascribed to the spatial entities defining the environment. Furthermore, spatial entities viewed as "management entities" can support the reification of the specific relationships between a stakeholder using the renewable resource and the renewable resource itself.

Yet some applications of ABMS to environmental management do not represent any spatially explicit natural dynamics. The data related to the environmental conditions (i.e. the overall quantity of available resource), used by the agents to make their decisions, are managed just like any other kind of information. But even when the environmental conditions are not spatially explicit, the explicit representation of space can help to structure the interactions among the agents. For instance, in the simulation of land use changes, the cognitive reasoning of agents can be embedded in cellular automata (CA) where each cell (space portion) represents a decisional entity that considers its neighbourhood to evaluate the transition function determining the next land use. Typically, acquaintances are then straightforwardly set from direct geographical proximity. It is still possible to stick on CA with more flexible "social-oriented" ways to define the acquaintances, but as soon as decisional entities control more than a single space portion, they have to be disembodied from the spatial entities. The FEARLUS model proposed by Polhill et al. (2001) and described in Sect. 4 is a good illustration of such a situation.

Applications of ABMS explicitly representing some renewable resources have to deal with the fact that renewable resources are usually heterogeneously scattered over the landscape being shaped by their patterns. Any irregularities or specificities in the topological properties of the environment legitimate to incorporate a spatially explicit representation of the environment. The combined use of geographic information systems (GIS) and ABMS is a promising approach to implement such integration (Gimblett 2002), particularly when there is a need to refer explicitly to an actual landscape in a realistic way. More generally, the relationship between an artificial landscape and a real one can be pinpointed by referring to three levels of proximity: (1) none, in case of theoretical, abstract landscape; (2) intermediate, when the reference to a given landscape is implicit; and (3) high, when the reference to a given landscape is explicit. Theoretical ABMS applications frequently use purely abstract landscapes, such as in the well-known sugar and spice virtual world of Sugarscape (Epstein and Axtell 1996). In the intermediate case, the implicit reference to a given landscape may exist through matching proportions in the composition of the landscape and similar patterns in its spatial configuration. When the reference to an actual landscape is explicit, the use of GIS is required to design the environment. An example of such realistic representation of a given landscape is given by Etienne et al. (2003). Characterizing the relationship between the simulated environment and the reality is a good way to estimate to what extent the model may provide a wide scope: the rule "the more realistic, the less generic" is hardly refutable.

From a technical point of view, the representation of space in ABMS applications with spatially explicit representation of the environment could be either continuous or discrete. Most of the time, the representation of space is based on a raster grid (the space is regularly dissected into a matrix of similar elementary components), and less frequently it is made of a collection of vector polygons. The continuous mode is quite uncommon in ABMS. This is related to the standard scheduling of ABMS that relies on either a discrete-time approach or on a discrete-event approach. Therefore, dealing with time (regular or irregular) intervals, the spatial resolution of the virtual landscape can be chosen so that the elementary spatial entity (defined as the smallest homogeneous portion of space) can be used as the unit to characterize distances or neighbourhood. Using a discrete mode to represent the space allows to easily define aggregated spatial entities that directly refer to different ecological scales relevant to specific natural or social dynamics, as well as to the management units specifically handled by the stakeholders. The corresponding spatial entities are interrelated according to a hierarchical organization, through aggregations.

22.3.3 How Are the Agents Modelled?

As we restrict our study to the sole applications with explicit consideration of the stakeholders, by "agent" we mean a computer entity representing a kind of stakeholder (individual) or a group of stakeholders. We stick with that operational definition even when the authors opt for another terminology and propose to characterize each kind of agent by considering two aspects: internal reasoning leading to decision-making and interactions with the other agents (coordination).

22.3.3.1 Internal Reasoning

Decision-making is the internal process that specifies how an agent behaves. It encompasses two dimensions: sophistication (from reactive to cognitive) and adaptiveness (through evaluation). What is called the "behaviour" of an agent refers to a wide range of notions. In some situations, the behaviour of an agent is simply and straightforwardly characterized by the value of a key parameter, as in the theoretical exploration of the tragedy of the commons by Pepper and Smuts (2000) where agents are either restrained (intake rate of resource is set to 50%) or unrestrained (intake rate of resource is set to 99%) in their foraging activity, when all the other biological functions (perception, movement, reproduction, mortality) are the same. In some other cases, the behaviour of a given agent does not only depend on internal characteristics, like the driftwood collector agents proposed by Thébaud and Locatelli (2001) who are stealing wood collected by other agents only when their attitude is still disrespectful (internal property) and when they cannot be observed (no peer pressure). Whatever the factors determining the behaviour of an agent are, this behaviour may or may not change over time. When the behaviour is simply characterized by a value of a key parameter, the adaptiveness can be taken into account without any particular architectural design. For more sophisticated behaviours, it becomes necessary to use a design pattern linking the agent to its behavioural attitude. With such a design pattern, the different behavioural attitudes are made explicit through corresponding subclasses, as shown in Fig. 22.1 with an example taken from the Dricol model (Thébaud and Locatelli 2001). The adoption of a particular attitude is updated according to some evaluation function.

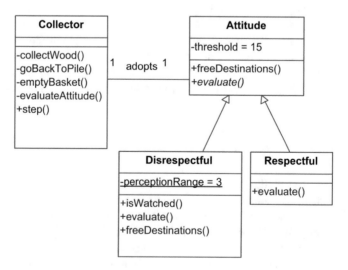

Fig. 22.1 Design pattern of the driftwood collector agents (Thébaud and Locatelli 2001)

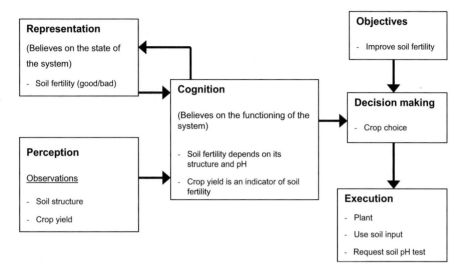

Fig. 22.2 Architecture of farmer agents from the CATCHSCAPE model (Becu et al. 2003)

Regarding the degree of sophistication of the decision-making process, the so-called *reactive* agents implement a direct coupling between perception (often related to little instantaneous information) and action. The forager agents of Pepper and Smuts (2000) mentioned above are typical reactive agents. On the opposite side, *cognitive* agents implement more complex decision-making processes by explicitly deliberating about different possibilities of action and by referring to specific representations of their environment, which is of particular importance for applications of ABMS to environmental management (Bousquet and Le Page 2004). An example of such agents is given by Becu et al. (2003): farmer agents evaluate direct observations and messages received from others (social network), update their knowledge base and evaluate options according to their objectives (see the corresponding architecture in Fig. 22.2).

22.3.3.2 Interactions with Other Agents (Coordination)

Bousquet (2001) synthesized his general approach of multi-agent systems to study environmental management issues with a diagram (see Fig. 22.3). We will refer here to the three kinds of interactions depicted in Fig. 22.3 to describe the types of agents implemented in applications of ABMS to environmental management.

The deliberative process of one agent is quite often influenced by some other closely related agents. The proximity may be either spatial (local neighbourhood) or social (acquaintances). In situations like the Dricol model (Thébaud and Locatelli 2001), what matters is just the presence of other agents in the surroundings. When

Fig. 22.3 Interactions
between agents via the
environment (Ie), through
peer-to-peer communication
(Ii) and via the collective
level (Ic) (Bousquet 2001)

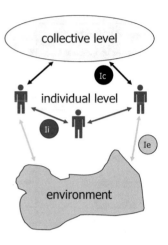

more information about the related agents is needed, then the rules to access
this information have to be specified. It is often assumed that the information is
directly accessible through browsing the agents belonging to a given network. This
corresponds to "Ie" in Fig. 22.3: other agents perceived through the environment are
considered as part of the environment of one agent.

Agents may strictly control the access to their internal information unless they
intentionally decide to communicate it ("Ii" in Fig. 22.3). Then the sharing of
information has to go through direct exchanges of peer-to-peer messages, with a
specified protocol of communication.

Relating agents directly to the collective level ("Ic" in Fig. 22.3) is most often
achieved via the notion of *groups* to which they can belong and for which they
can be representative to outsiders. Inspired by the Aalaadin metamodel proposed
by Ferber and Gutknecht (1998), recent applications of ABMS to agricultural
water (Abrami 2004) and waste water (Courdier et al. 2002, described in the next
section) management as well as to epidemiology (Muller et al. 2004, described
in the next section) illustrate how both notions of *group* and *role* are useful to
handle levels of organization and behaviours within levels of organization. Even
when "Ic" are not implemented through specific features of the agents' architecture,
the mutual influence of both collective and individual levels is fundamental in
renewable resource management. On one hand, individuals' behaviours are driven
by collective norms and rules; on the other hand, these norms and rules are evolving
through agents' interactions. This individual-society dynamic linkage, introducing
the notion of institution, relies on the representation of common referents (Bousquet
2001). Such "mediatory objects" are, for instance, the water temples involved in the
coordination of a complex rice terraces system in Bali (Lansing and Kremer 1993;
Janssen 2007).

22.3.4 Implementation

We believe it is useful to indicate whether a simulation platform was used or not to implement the model. Nowadays, some established generic tools such as Ascape, Cormas, Mason, (Net)(Star)Logo, Repast or Swarm are being used by large communities of users. Intending to release researchers from low-level technical-operational issues, their development is boosted by their comparisons performed through the implementation of simple benchmark models (Railsback et al. 2006) and the analysis of their abilities to fulfil identified requirements (Marietto et al. 2003). The maintainers and developers of such generic platforms have also taken into consideration some sensitive technical aspects (floating point arithmetic, random numbers generators, etc.) recently pointed out (Polhill et al. 2006) and provide some elements to help users to escape these numerical traps. Additionally to the benefit of not having to re-implement basic functionalities from scratch, it may also happen that a previous model, made available online in the library of existing models, presents some similarities with the new model to be developed.

Nevertheless, using a generic platform is not a panacea. It may incidentally lead to poorer presentation of the developed models if the authors (wrongly) assume that any reader is aware of the platform's general principles. A new research stream (model-to-model comparison) recently emerged from the fact that it is very difficult to replicate simulation models from what is reported in publications (Hales et al. 2003). Reproducing results, however, is a sine qua non condition for making ABMS a more rigorous tool for science. It may be achieved through a better description of individual models, but also through the maintenance and development of strong communities of users sharing the same tools for implementation. This kind of stimulating diversity of the simulation platforms may contribute to identify some generic "shorthand" conventions that could minimize the effort to describe the model rigorously and completely (Grimm et al. 2006).

22.3.5 Simulation Scenarios

Some ABMS platforms like NetLogo, for simplicity purposes, merge in the same (unique) implementation file the definition of the domain entities with the specification of the simulation scenario. In their ODD protocol, Grimm et al. (2006) suggest to state the elements pertaining to the scheduling of a "standard" simulation scenario in the first part (overview) of the sequential protocol, then come back to some specific design concepts characterizing the domain entities (discussed here in Sect. 22.3.3) in the second part (design) of the sequential protocol, and finally to describe the initialization of a standard scenario in the third and last part (details). Yet, a clear separation between model and simulation should be promoted when seeking genericity. At the level of the agents, focusing on the description of their internal structure and potential behaviour may help to identify some modules of their architecture that could be reused in other contexts. At the level of the initialization and scheduling of the simulation, the same benefit can be expected: for instance, generating parameter values for a population of agents from a statistical distribution or creating an initial landscape fitting some schematic patterns (Berger and Schreinemachers 2006).

The notion of "standard" scenario is not always very easily recognizable. Some authors prefer to start by presenting what they call "reference" scenarios that correspond to "extreme" situations. For instance, whenever the structure of a given model makes sense to mention it, a "no agents" simulation scenario should be available without any modifications, i.e. just initializing the number of agents to zero. These scenarios can be used either as a verification step in the modelling process (to test that the implementation is a faithful translation of the conceptual model) or as a reference to compare the outputs obtained from more plausible simulation scenarios. More generally, simulation scenarios have to address validation by questioning the results through looking back at the system under study. In ABMS, validation is a multidimensional notion. Depending on the purpose assigned to the model, the focus will be mainly set on (1) checking if the simulated data are fitting available real datasets, (2) looking for comparable processes observed in other case studies and (3) evaluating to what extent the stakeholders accept the model and its outputs as a fair representation of their system. For a more detailed discussion of validation, see Chap. 9 in this volume (David et al. 2017).

Another essential dimension of simulation scenarios relates to the model output used to observe them. Confronting the interpretations of the same simulated results built from specific stakeholders' viewpoints may be an effective way to share the different opinions and then highlight the need to improve the agents' coordination mechanisms or even to achieve a compromise (Etienne et al. 2003).

22.4   A Review of Recent Applications of ABMS to Environmental Management

To classify the recent applications of ABMS in environmental management is not an easy task, as the range of covered topics is wide: dynamics of land use changes; water, forest and wildlife management; but also agriculture, livestock productions and epidemiology. Some topics like epidemiology can easily be treated separately. Some others are likely to be appearing simultaneously in some case studies, especially for those dealing with multiple uses of the same renewable resource and/or representing landscape with several land-use types. This latter situation frames an entire research field in human geography that is focusing on the dynamics of land-use/cover changes (LUCC). In the classification proposed below, some applications clearly related to LUCC are listed in other subsections (mainly in "agriculture" and in "forest"). Conversely, the "LUCC" subsection contains applications that are related to some other topics specifically addressed later on.

Finally, some other topics like biodiversity are multidimensional, and thus the related case studies can be split into several other topics (for instance, biodiversity related to endangered species is reversed into "wildlife"). Whenever it undoubtedly exists for an application, we are mentioning the relevance to other topics. A specific category has been added to group the examples of ABMS addressing theoretical issues in ecological management.

As the number of publications related to the use of ABMS in ecological management is booming, it is almost impossible to analyse all of them. So for each of the categories presented above, we had to select only a few representative case studies to be briefly described by referring as much as possible to the elements discussed in previous sections (the case studies that were not selected to be analysed are just mentioned in the introduction paragraph of each category). Following Hare and Deadman (2004) who proposed a taxonomy of ABMS in environmental management as a first step to provoke discussion and feedback, our purpose here is to contribute to the framing of a practical bibliographic survey by proposing some key characteristics useful for comparing applications of ABMS in ecological management. See "Appendix" for a table recording the key characteristics of the selected case studies.

22.4.1 Theoretical Issues in Environmental Management

Thébaud and Locatelli (2001) have designed a simple model of driftwood collection to study the emergence of resource-sharing conventions; Pepper and Smuts (2000) have investigated the evolution of cooperation in an ecological context with simple reactive agents foraging either almost everything or just half of a renewable resource. Schreinemachers and Berger (2006) have compared respective advantages of heuristic and optimizing agent decision architectures; Rouchier et al. (2001) have compared economic and social rationales of nomad herdsmen securing their access to rangelands; Evans and Kelley (2004) have compared experimental economics and ABMS results to explore land-use decision-making dynamics; Soulié and Thébaud (2006) represent a virtual fishery targeting different species in different areas to analyse the effects of spatial fishing bans as management tools.

22.4.2 Dynamics of Land-Use/Cover Changes

Parker et al. (2003) have recently reviewed the application of multi-agent systems to better understand the forces driving land-use/cover change (MAS/LUCC). Their detailed state of the art presents a wide range of explanatory and descriptive applications. Since this authoritative paper has been published, new applications related to LUCC have continued to flourish. For instance, Caplat et al. (2006) have simulated pine encroachment in a Mediterranean upland, Matthews (2006) has proposed a generic tool called PALM (People and Landscape Model) for simulating resource flows in a rural livestock-based subsistence community; LUCITA (Lim et al. 2002), an ABM representing colonist household decision-making and land-use change in the Amazon rainforest, has been developed further (Deadman et al. 2004); Bonaudo et al. (2005) have designed an ABM to simulate the pioneer fronts in the same Transamazon highway region, but at a lower scale; Manson (2005, 2006) has continued to explore scenarios of population and institutional changes in the Southern Yucatan Peninsular Region of Mexico; Huigen (2004) and Huigen et al. (2006) have developed MameLuke to simulate settling decisions and behaviours in the San Mariano watershed, Philippines. Below we describe in more detail a selection of applications that are also characterized in the overview table presented in the "Appendix".

22.4.2.1   FEARLUS, Land-Use and Land Ownership Dynamics (Polhill et al. 2001)

FEARLUS, an abstract model of land use and land ownership implemented with Swarm, has been developed to improve the understanding of LUCC in rural Scotland by simulating the relative success of imitative versus nonimitative process of land-use selection in different kinds of environment. An abstract regional environment is defined as a toroidal raster grid made out of 8-connex land parcels, each being characterized by fixed biophysical conditions. The same external conditions that vary over time apply in the same way to all land parcels. These two factors are determining the economic return of a given land use at a particular time and place. The land manager agents decide about the land uses of the land parcels they own (initially a single one) according to a specific selection algorithm. During the simulation, they can buy and sell land parcels (landless managers leave the simulation; new ones may enter it by buying a land parcel). Simulation scenarios were defined on several grids by pairing selection algorithms from the predefined sets of five imitative and five nonimitative selection algorithms.

22.4.2.2   Greenbelt to Control Residential Development (Brown et al. 2004)

This ABM, the simplest version of which being strictly equivalent to a mathematical model, has been developed to investigate the effectiveness of a greenbelt located beside a developed area for delaying residential development outside the greenbelt. The environment is represented as an abstract cellular lattice where each cell is characterized by two values: a constant one to account for aesthetic quality and a variable one to denote the proximity to service centres. Service centres are called agents but actually they are more passive entities as they do not exhibit any decision-making. Residential agents, all equipped with the same aesthetic and service centre preferences, decide their location among a set of randomly selected cells according to a given utility function. The Swarm platform was used for implementation. Scenarios, scheduled with periodic introductions of new agents, are based on the values of residential agents' preferences and on the spatial distribution of aesthetic quality.

22.4.2.3 LUCC in the Northern Mountains of Vietnam (Castella et al. 2005, 2005; Castella and Verburg 2007)

Castella et al. (2005) developed the SAMBA model under the Cormas simulation platform to simulate the land-use changes during the transition period of decollectivization in a commune of the northern Vietnam uplands. This simple and adaptable model with heuristic value represented the diversity of land-use systems during the 1980s as a function of household demographic composition and paddy field endowment in the lowland areas. The environment in which agents make decisions was made of a 2500 cell grid, and 6 different land-use types could be attributed to each cell, representing a plot of 1000 m^2, also characterized by its distance to the village. While there was no coordination among farmer agents with reactive behaviour in the early version of the model, interactions among them were added later and the model coupled to a GIS to extrapolate the dynamics to the regional landscape level (Castella et al. 2005). The simulated scenarios tested the effects of the size of the environment, the overall population and household composition and the rules for the allocation of the paddy fields on the agricultural dynamics and differentiation among farming households. More recently, this process-oriented model was compared to a spatially explicit statistical regression-based pattern-oriented model (CLUE-s) implemented at the same site. While SAMBA better represented the land-use structure related to villages, CLUE-s captured the overall pattern better. Such complementarity supports a pattern-to-process modelling approach to add knowledge of the area to empirically calibrated models (Castella and Verburg 2007).

22.4.2.4 Competing Rangeland and Rice Cropping Land Uses in Senegal (D'Aquino et al. 2003)

To test the direct design and use of role-playing games (RPG) and ABMS with farmers and herders competing for land use in the Senegal River Valley, participatory simulation workshops were organized in several villages. The ABM used during the last day of the workshops was straightforwardly implemented with the Cormas platform from the characteristics and rules collectively agreed the day before when crafting and testing a RPG representing stakeholders' activities related to agriculture and cattle raising. The environment is set as a raster grid incorporating soil, vegetation and water properties of the village landscape as stated by the stakeholders (a GIS was used only to clear ambiguities). The same crude rules defined and applied during the RPG were used to implement the autonomous reactive farmer agents. After displaying the scenario identified during the RPG, new questions emerged and were investigated by running the corresponding simulation scenarios. The hot debates that emerged demonstrate the potential of these tools for the improvement of collective processes about renewable resource management.

22.4.2.5 Landscape Dynamics in the Méjan Plateau, Massif Central, France (Etienne et al. 2003)

Etienne et al. (2003) developed a multi-agent system in order to support a companion modelling approach on landscape dynamics in the Méjan plateau of the Massif Central, the mountain range of central France. The purpose of the model is to support the coordination process among stakeholders concerned with pine encroachment. The environment is a cellular automaton coming from the rasterization of a vector map. Several procedures account for vegetation changes due to pine encroachment according to natural succession trends and range, timber or conservation management decisions. The three agent types (sheep farmers, foresters and the National Park) are concerned by this global biological process, but it affects their management goals in a very different way (sheep production, timber production, nature conservation). The model is used to simulate and compare collectively contrasting management scenarios arising from different agreements. Simulation results were used to support the emergence of collective projects leading to a jointly agreed management plan.

22.4.2.6 GEMACE: Multiple Uses of the Rhone River Delta, Southern France (Mathevet et al. 2003)

This ABM developed with the Cormas platform simulates the socio-economic dynamic between hunting managers and farmers in the Camargue (Rhone river delta, southern France), through the market of the wildfowling leasing system, in interaction with ecological and spatial dynamics. A CA represents an archetypal region based on a spatial representation of the main types of estates, distributed around a nature reserve. Each cell is characterized by water and salt levels through land relief, land-use history, infrastructure, spatial neighbourhood and current land use. A wintering duck population, heterogeneously distributed in its habitats, is affected by various factors such as land-use changes, wetland management, hunting harvest and disturbance. Land-use decisions are made at farmland level by farmers and hunting managers that are communicating agents. Their strategy, farming or hunting oriented, is based on crop rotation, allocation of land use and water management and may change according to some specific representations and values related to farming and hunting. Scenario runs allowed discussing the structuring of the waterfowl hunting area resulting from the individual functioning of farms in conjunction with a nature reserve and other hunting units and the conservation policy.

22.4.3   Water Management

In the field of sustainable development, water management resources are an issue of major importance. ABMS dealing with water management is used to simulate the management of irrigated ecosystems, to represent the interactions among stakeholders by capturing their views and formalizing the decision-making mechanisms (especially negotiation processes), to capture the socio-economic aspects of potable water management and evaluate scenarios based on alternative control measures, etc. For instance, Haffner and Gramel (2001) have investigated strategies for water supply companies to deal with nitrate pollution; Janssen (2001) has simulated the effects of tax rates related to the intensive use of phosphorus on lake eutrophication; Becu et al. (2003) have developed CATCHSCAPE to simulate the impact of upstream irrigation management on downstream agricultural viability in a small catchment of Northern Thailand; Krywkow et al. (2002) have simulated the effects of river engineering alternatives on the water balance of the Meuse river in the Netherlands and have related this hydrological module to stakeholders' negotiations and decisions. Below we describe in more detail a selection of applications that are also characterized in the table presented in the "Appendix".

22.4.3.1   SHADOC: Viability of Irrigated Systems in the Senegal River Valley (Barreteau and Bousquet 2000; Barreteau et al. 2004)

To examine how existing social networks affect the viability of irrigated systems in the Senegal River Valley, the SHADOC ABM focuses on rules used for credit assignment, water allocation and cropping season assessment, as well as on organization and coordination of farmers in an irrigation scheme represented as a place of acquisition and distribution of two resources: water and credit. The model used a spatially non-explicit representation: all plots are subject to the same hydrological cycle regardless of their exact geographical position. The societal model is structured with three types of group agents in charge of credit management, watercourse and pumping station management. As far as individual agents (farmers) are concerned, the model employs a four-level social categorization with different types of farmers according to their own cultivation objective. Each agent acts according to a set of rules local to him. Each agent also has its own point of view about the state of the system and especially its potential relations with other agents. SHADOC was first designed as a tool for simulating scenarios of collective rules and individual behaviours.

22.4.3.2   MANGA: Collective Rules of Water Allocation in a Watershed (Le Bars et al. 2005)

MANGA has been developed to test the economics, environmental and ethical consequences of particular water rules in order to improve the collective management of water resources according to agricultural constraints, different actors' behaviours and confrontation of decision rules of each actor. Their modelling approach takes into account cognitive agents (farmers or water supplier) trying to obtain the water they need via negotiation with the others as a result of its individual preferences, rationality and objectives. The MANGA model used a spatially non-explicit representation for coupling social and environmental models. To implement the decision-making process of the cognitive agents, the authors used the BDI formalism and more particularly the PRS architecture. During simulations, MANGA allows to test several water allocation rules based on water request, irrigated corn area or behaviour evolution.

22.4.3.3   Sinuse: Water Demand Management in Tunisia (Feuillette et al. 2003)

Sinuse is a simulator conceived to simulate the interactions between a water table and the decisions of farmers in Tunisia. The farmers' decisions are driven by economic objectives, but the dynamics of the system is mainly dependent on the interactions among agents. The agents interact through message sending to exchange land and to team up to build wells. They also interact through imitation and influence on the land price. They interact through the environment as they share a common resource, the water table which has its own dynamics and depends on the number of active wells. The model was developed with Cormas platform. Simulations study the influence of various policies such as subsidies for improved irrigation equipment.

22.4.3.4   Water Management and Water Temple Networks in Bali (Janssen 2007; Lansing and Kremer 1993)

Do irrigation systems necessarily need a centralized authority to solve complex coordination problems? An ancestral Balinese system of coordination based on villages of organized rice farmers (subaks) linked via irrigation canals has served as a case study to investigate this question. Actions to be done on each specific date for each subak are traditionally related to offerings to temples. The original model was recently re-implemented to deeper investigate why the temple level would be the best level for coordination. The environment is set as a network of 172 subaks, together with a network of 12 dams allocating the water to the subaks. Each subak has up to four neighbouring subaks. It selects one cropping plan out of 49 predefined ones. The corresponding water demand is affecting the runoff between dams. Harvests are affected by water stresses and pest outbreaks. The densities of pest in subaks are changing due to local growth (related to the presence of rice) and migration (based on a diffusion process). Six simulation scenarios based on the level of social coordination were explored by Lansing and Kremer. Additionally, to the two extreme scenarios defined with a single group of all 172 subaks (full synchronization) and 172 separate groups (no synchronization), four intermediate scenarios were tested, based on groups defined from the existing system of temples.

22.4.3.5 Sharing Irrigation Water in the Lingmuteychu Watershed, Bhutan (Gurung et al. 2006)

Raj Gurung and colleagues used ABMS, following the companion modelling approach, to facilitate water management negotiations in Bhutan. A conceptual model was first implemented as a role-playing game to validate the proposed environment, the behavioural rules and the emergent properties of the game. It was then translated into a computerized multi-agent system under the Cormas platform, which allowed different scenarios to be explored. Communicating farmer-agent exchanged water and labour, either within a kinship network or among an acquaintance network. Different modes of communication (intra-village and intervillage) were simulated, and a communication observer displayed the exchange of water among farmers.

22.4.4 Forestry

Applications of ABMS in forestry are either focusing on LUCC issues or on management issues. For instance, Moreno et al. (2007) have simulated social and environmental aspects of deforestation in the Caparo Forest Reserve of Venezuela; Nute et al. (2004) have developed NED-2, an agent-based decision support system that integrates vegetation growth, wildlife and silviculture modules to simulate forest ecosystem management plans and perform goal analysis on different views of the management unit. Below we describe in more detail a selection of applications that are also characterized in the table presented in the "Appendix".

22.4.4.1 Deforestation and Afforestation in South-Central Indiana (Hoffmann et al. 2002)

Hoffmann et al. (2002) propose an original way of using ABMS to improve scientific knowledge on the interactions between human activities and forest patterns in Indiana, during the last 200 years. The environment is a raster artificial landscape randomly generated from the 1820s land-cover ratio between crops, fallows and forests and randomly calculated slopes. Farmer is the only type of agent identified, but they can behave differently according to two potential goals (utility maximizing or learning reinforcement) and two actions: deforestation or afforestation. Simulations are used to check through statistical analysis of a high number of runs the impact of ecological (slope), social (stakeholders goals) or economic (agricultural prices, returns) factors in changing land-use patterns.

22.4.4.2   Forest Plantation Co-management (Purnomo and Guizol 2006; Purnomo et al. 2005)

This ABMS modelling approach links social, economic and biophysical dynamics to explore scenarios of co-management of forest resources in Indonesia. The purpose is to create a common dynamic representation to facilitate negotiations between stakeholders for growing trees. The environment is a simplified forest landscape (forest plots, road, agricultural land) represented on a cellular automaton. Each stakeholder has explicit communication capacities, behaviours and rationales from which emerge specific actions that impact landscape dynamics. The model is used to simulate different types of collaboration between stakeholders, and both biophysical and economic indicators are provided to measure the impact of each scenario on forest landscape and smallholder incomes. Simulation results are supposed to support the selection of the system of governance providing the best pathway to accelerate plantation development, local community poverty alleviation and forest landscape improvement.

22.4.5   Wildlife

Understanding how human activities impact on the population of animals in the wild is a concern shared by conservationists, by external harvesters (hunters, fishermen) and by local people. Viewed as a source of food or as an emblem of biodiversity, management schemes first have to ensure the viability of the population. Viewed as competitors for the living space of local people, management schemes have to control the population. For instance, Zunga et al. (1998) have simulated conflicts between elephants and people in the Mid-Zambezi Valley; Galvin et al. (2006) have used ABMS to analyse how the situation in the Ngorongoro Conservation Area (NCA) in northern Tanzania could be modified to improve human welfare without compromising wildlife conservation value; Jepsen et al. (2005) have investigated the ecological impacts of pesticide use in Denmark. Below we describe in more detail a selection of applications that are also characterized in the table presented in the "Appendix".

22.4.5.1 Water Management in Mediterranean Reedbeds (Mathevet et al. 2003)

Using the Cormas platform, the authors have developed an ABM to be used in environmental planning and to support collective decision-making by allowing evaluation of the long-term impact of several water management scenarios on the habitat and its fauna of large Mediterranean reedbeds. A hydro-ecological module (water level, reedbed, fish, common and rare bird populations) is linked to a socio-economic module (reed market and management). Each cell was assigned a type of land use, land tenure and topography from a GIS to create a virtual reedbed similar to the studied wetland. Five types of interacting agents represent the users of the wetland. They are characterized by specific attributes (satisfaction, cash amount, estates, etc.) and exploit several hydro-functional units. The behaviour of the agents depends on their utility function based on their evaluation of the access cost to the reedbed and on their beliefs. The ecological and socio-economic consequences of individual management decisions go beyond the estates and relate to the whole system at a different timescale.

22.4.5.2 Giant Pandas in China (An et al. 2005)

Using data from Wolong Nature Reserve for giant pandas (China), this ABM simulates the impact of the growing rural population on the forests and panda habitat. The model was implemented using Java-Swarm 2.1.1 and IMSHED that provides a graphical interface to set parameters and run the program. It has three major components: household development, fuelwood demand and fuelwood growth and harvesting. The simulated landscape was built from GIS data. Two resolutions were identified for sub-models requiring extensive human demographic factors and for landscape sub-models. Both person and household are cognitive agents that were defined from socio-economic survey. They allowed simulating the demographic and household dynamics. Agents interact with each other and their environment through their activities according to a set of rules. The main interaction between humans and the environment is realized through fuelwood collection according to demand. This model was used to test several scenarios and particular features of complexity, to understand the roles of socio-economic and demographic factors, identifying particular areas of special concern, and conservation policy.

22.4.5.3 Traditional Hunting of Small Antelopes in Cameroon (Bousquet et al. 2001)

To investigate the viability of populations of blue duikers, a small antelope traditionally hunted by villagers in the forests of Eastern Cameroon, an ABM has been developed with the Cormas platform. The raster spatial grid was defined by reading data from a GIS map corresponding to a village that was surveyed during several months. Each cell represents 4 ha (the size of the blue duiker habitat) and is characterized by a cover (river, road or village) and a reference to a hunting locality. The population dynamics of the blue duiker is simulated through the implementation of biological functions (growth, age-dependent natural mortality, migration, reproduction) applied to all the individual antelope agents. Hunter agents decide on the location of their traps by selecting one hunting locality out of the four they use. A first set of simulation scenarios was based on unilateral decisions of hunter agents, all of them following the same general rule. Coordination among kinship groups of hunters was introduced in a second set of experiences.

22.4.5.4   Whale Watching in Canada (Anwar et al. 2007)

To investigate the interactions between whale-watching boats and marine mammals in the Saguenay St. Lawrence Marine Park and the adjacent Marine Protected Area in the St. Lawrence estuary, in Quebec, this ABM was implemented with the Repast platform. A raster grid defined from a GIS database represents the landscape of the studied area. The boats are cognitive agents and whales are simple reactive agents. Several simulations were run to explore various decision strategies of the boat agents and how these strategies can impact on whales. For each simulation, the happiness factor was used as an indicator of how successful the boat agents were in achieving their goals. Results showed that cooperative behaviour that involves a combination of innovator and imitator strategies based on information sharing yields a higher average happiness factor over non-cooperative and purely innovator behaviours. However, this cooperative behaviour creates increased risk for the whale population in the estuary.

22.4.6   Agriculture

ABMS applied to agriculture is mainly focussing on decision-making processes at the farm level (typically, agents represent households). Economic aspects usually play a pivotal role, and standard procedures like linear programming are often used to represent individual choices among available production, investment, marketing alternatives, etc. This economic module is then embedded into a more integrated framework to explicitly represent spatial and social aspects. For instance, to investigate technology diffusion, resource use changes and policy analysis in a Chilean region, Berger (2001) has connected an economic sub-model based on recursive linear programming to an hydrological sub-model; Ziervogel et al. (2005) have used ABMS to assess the impact of using seasonal forecasts among smallholder farmers in Lesotho; Sulistyawati et al. (2005) have analysed the consequence at the landscape level of swidden cultivation of rice and the planting and tapping of rubber by Indonesian households whose demography and economic welfare are simulated. Below we describe in more detail a selection of applications that are also characterized in the table presented in the "Appendix".

22.4.6.1 Agricultural Pest and Disease Incursions in Australia (Elliston and Beare 2006)

To analyse the effectiveness and regional economic implications of alternative management strategies for a range of different scenarios of a disease incursion in the agricultural sector, an ABM has been developed with the Cormas platform and applied to the case of the wheat disease Karnal bunt in a region of South East Queensland, Australia. A cellular spatial grid allows representing the spread of the pest across neighbouring paddocks and a range of potential transmission pathways including the wind, farm inputs and agents (farmers, contractors and quarantine officers) through their movement over the spatial grid. Farmers make cropping decisions about planting, spraying for weeds, harvesting and the use of contract labour. They can directly identify and report signs of a Karnal bunt incursion on their property. The incursion can also be detected from quality inspection when farm production reaches the collective storage unit. Then a quarantine response, based on a recursive checking in the neighbourhood of infected farms, is implemented by officer agents. Simulation scenarios are based on one hand on levels of farmer detection and reporting and on the other hand on the way the disease was first introduced into the system (limited and slowly expanding incursion versus potentially rapid expansion from a wide use of contaminated fertilizer).

22.4.6.2 Agripolis: Policy Impact on Farms' Structural Changes in Western Europe (Happe et al. 2006)

Agripolis is the evolution of a model developed by Balmann (1997). It describes the dynamics of an agricultural region composed of farms managed by farmers (an agent represents both of these concepts). The landscape and the market are the other agents. The farm agent has a cognitive decision-making process: this process corresponds to the traditional modelling in agricultural economics, where agents try to maximize their income. The land market is the central interaction institution between agents in Agripolis. Farm agents extend their land by renting land from farm landowners. The allocation of land is done through auctions. The Agripolis model was used to study a region in southwest Germany. A sensitivity analysis is done to analyse the relationship between policy change and determinants of structural changes such as the interest rate, managerial abilities and technical change.

22.4.6.3 Adaptive Watershed Management in the Mountains of Northern Thailand (Barnaud et al. 2007)

This companion modelling experiment aims at facilitating a learning process among Akha highlanders about the socio-economic aspects (i.e. allocation of formal and informal credit, on- and off-farm employment) of the expansion of plantation crops to mitigate soil erosion risk on steep land. Farmers' individual decision-making regarding investment in perennial crops, assignment of family labour to off-farm wage-earning activities and search for credit were modelled. The simulated scenarios looked at the effects of the duration of the grace period of the loans, the distribution of formal credit among the three types of farms and different structures of the networks of acquaintances for informal credit. Two main indicators were used to analyse the results of the simulations for each type of farm: (1) the total area under plantation crops (ecological indicator) and (2) the proportion of bankrupt farms leaving the agricultural sector (socio-economic indicator).

22.4.7 Livestock Management

Janssen et al. (2000) have used adaptive agents to study the co-evolution of management and policies in a complex rangeland system; another work lead by Janssen (Janssen et al. 2002) has investigated the implications of spatial heterogeneity of grazing pressure on the resilience of rangelands; Bah et al. (2006) have simulated the multiple uses of land and resources around drillings in Sahel under variable rainfall patterns; Milner-Gulland et al. (2006) have built an ABM of livestock owners' decision-making, based on data collected over 2 years in five villages in southeast Kazakhstan. Below we describe in more detail a selection of applications that are also characterized in the table presented in the "Appendix".

22.4.7.1 Rangeland Patterns in Australia (Gross et al. 2006)

To evaluate general behaviours of rangeland systems in Australia, this ABM represents a landscape made of enterprises, which are cognitive agents that represent a commercial grazing property. Each property is defined by an area, the quality of land in each patch and its livestock. Behaviours of the enterprise agents are defined by a strategy set comprised of a set of rules, which evolves over time to represent learning. A government agent has an institutional strategy set that also varies through time. Its main roles are to collect taxes and deliver drought relief in the form of interest payment subsidies. The biophysical sub-models allow simulating plant and livestock dynamics. Pastoral decisions are made by the enterprise agents according to a set of rules. The variation in the level of financial weakness leads to the adoption of a new strategy by an enterprise. Each one is randomly associated with a rate of learning. Implemented in the C++ programming language, the model is fed by inputs of historical data. The simulations emphasize consequences of interactions between environmental heterogeneity and learning rate.

22.4.7.2 Collective Management of Animal Wastes in La Reunion (Courdier et al. 2002)

To investigate the collective management of livestock farming wastes on La Reunion Island, an ABM called Biomas has been developed with the Geamas platform. Biomas simulates the organization of transfers of organic materials between two kinds of agents: surplus-producing farms (i.e. farms where livestock-raising activity dominates) and deficit farms (i.e. predominantly crop production). The environment is represented as a network of "situated objects". Their association with the Geamas agents enables the agents to act on the environment (e.g. "crop" agents are linked to "plot" situated objects). Some situated objects like "road sections" are only related to other situated objects. Graphs of connected situated objects allow representing itineraries. The agents in Biomas are interacting through direct exchanges of messages; they are also linked to a "group" agent through a membership process. The "group" agent is responsible for imposing the management constraints on all its members or individual agents by means of contracts and implements a penalty system in the case of disregard of the regulations. The simulation scenarios are based on the constraints and regulations defined at this "group" level.

22.4.8 Epidemiology

Models developed for the spread of infectious diseases in human populations are typically implemented assuming homogeneous population mixing, without a spatial dimension, social (and network) dimension or symptom-based behaviour. ABMS offers a great potential to challenge these assumptions. Recently Ling Bian (2004) proposed a conceptual framework for individual-based spatially explicit epidemiological modelling, discussing four aspects: (1) population segments or unique individuals as the modelling unit, (2) continuous process or discrete events for disease development through time, (3) travelling wave or network dispersion for transmission of diseases in space and (4) interactions within and between night-time (at home) and daytime (at work) groups. As an illustration, she compares a simple population-based model of influenza to an equivalent schematic individual-based one. This abstract model has been utilized by Dunham (2005) to develop a generic ABMS tool. Recently, the fear of bioterrorism has also stimulated intensive studies in the USA; see, for instance, BioWar, developed by Carley et al. (2006) to simulate anthrax and smallpox attacks on the scale of a city.

22.4.8.1 Bovine Leukaemia (Bagni et al. 2002)

Two methodologies (system dynamics and agent based) used to simulate the spread of a viral disease (bovine leukaemia) are compared. The purpose is, through "what-if" analysis, to assess the system's behaviour under various conditions and to evaluate alternative sanitary policies. Based from the same set of Unified Modelling Language (UML) diagrams, Vensim and Swarm are the two platforms that have been used to implement the conceptual model. The environment represents in an abstract way a dairy farm segmented into sectors. "Cow" and "farm sector" are the two types of autonomous agents in this model. The integration at the farm level is directly achieved through the "model swarm". Scenarios focus particularly on the number of cows detected as positive at sanitary controls (as opposed to the total number of infected cows).

22.4.8.2 Malaria in Haiti (Rateb et al. 2005)

To assess the impact of education on malaria healthcare in Haiti, an ABM with a realistic representation of Haiti has been designed. The environment is set as a raster grid with cells characterized by land covers (sea, road, land, mountain, city, school and hospital) associated with specific contamination probabilities (this is how mosquitoes are represented in the model). Apart from an epidemiological status, autonomous agents (representing individual people) are characterized by a mobility capability and an education score which value corresponds to the time agents take to attribute existing symptoms to malaria and therefore to go to a hospital. Implemented in StarLogo, three scenarios based on the number of schools and hospitals have been discussed.

22.4.8.3 Sleeping Sickness in Cameroon (Muller et al. 2004)

To understand the spread of human African trypanosomiasis, and ultimately to elaborate a tool to evaluate risk and test control strategies, an ABM has been developed with the MadKit platform and tested with data from one village in Southern Cameroon. The space is not explicitly represented in this model. This is due to the metamodel associated with the MadKit platform: the system under study has to be described through "agent-group-role" interactions (Ferber and Gutknecht 1998). Hence, surprisingly, locations can only be depicted as agents here (see Fig. 22.4). They are characterized by a proportional surface area and a number of animals.

Location agents, as "groups", are responsible for "enrolling" tsetse and human agents that will, as members of the same group, be able to interact through the sending of "bite" messages. The probability for a human agent to be bitten is

Fig. 22.4 The representation of space as clustering of three kinds of "location" agents: village (pentagons), cocoa plantations (hexagons) and forest (rectangles). (Muller et al. 2004)

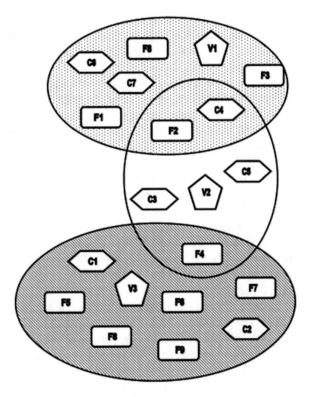

inversely proportional to the number of animals. Simulation scenarios are based on the organization of space, to investigate the effect of the size and number of transmission areas.

22.4.9   General Considerations

To fill in the table presented in the "Appendix" from the description of the models found in publications was not always easy. This is partly due to the fact that the elements to be detailed in the columns of the table require further refinements and more precise definitions. But this can also be attributed to the heterogeneity in the way model contents are detailed by the authors. The lack of a general framework to document such kind of models is patent, and all designers of ABM should become aware and refer to the framework proposed by Grimm et al. (2006). Even when the code of the implemented model is published (in appendices of articles or on a website), it is quite challenging and time-consuming to dive into it to retrieve specific information. This difficulty has triggered a bias: we have tended to select the applications we know better. As the co-authors of this chapter all belong to the same scientific network, the representativeness of the selected case studies may be questioned. This kind of task—a systematic survey based on a set of unambiguously defined characteristics—should be undertaken at the whole scientific community level, in a continuous way. Ideally, it should use an effective tool for mass collaborative authoring like wiki.

The environment, abstract or realistic, is most often represented as a raster grid. The spatial resolution, when it makes sense to define it precisely (for realistic simulated landscapes), is always clearly related to a key characteristic of one of the model's components.

The number of applications with interactions involving the collective level is rather low. This does not necessarily imply cognitive agents. In the model of Bousquet et al. (2001), for instance, the collective level is related to the kinship structure of the small antelopes' population; when a young individual becomes mature, it leaves the parental habitat territory and starts to move around to look for a potential partner to establish a new family group in an unoccupied habitat. The group left by the young adult is affected in such a way that the reproduction can be activated again.

In our review, theoretical case studies are less numerous than empirical case studies. The prevalence of theoretical case studies is only significant for the LUCC category. It suggests that the proportion of empirical applications of ABMS is gaining ground compared to theoretical and abstract contributions. As analysed by Janssen and Ostrom (2006), this could be explained by the fact that theoretical models, more frequent at the beginning, have demonstrated that ABMS can provide novel insights to scientific inquiry. The increased availability of more and more relevant ecological and socio-economics data then paved the way to the rise of empirically based ABMS.

22.5   Why ABMS Is More and More Applied to Environmental Management

If ABMS is becoming more and more popular in environmental modelling, it is mainly because it demonstrates a potential to overcome the limitations of other kinds of models to take into account elements and processes that can hardly be ignored to consider the underlying research questions. Another aspect has to be stressed: ABMS is structurally an integrative modelling approach. It can easily be expressed with other modelling formalisms and tools. Additionally, to the evidential use of CA as a way to represent the space in ABMS applications dealing with environmental management, several other fruitful associations with complementary tools (GIS to handle spatial requests, linear programming modules directly used by agents to perform maximization of utility functions, etc.) have already been explored. Beyond technical aspects, ABMS can also be seen as a methodological step of a wider approach, like in companion modelling when it is jointly used with role-playing games to allow stakeholders' participation in the design of the tools used during participatory simulation workshops (Bousquet et al. 2002).

22.5.1   Getting Rid of Empirically Implausible Assumptions

In ecology, traditional general population models are assuming that (1) individuals are identical, (2) the interaction between individuals is global and (3) the spatial distribution of individuals is uniform. Required to ensure analytical tractability, these overly simplified assumptions significantly limit the usefulness of such population-based approaches. The assumption of "perfect mixing" on which population-based modelling approaches rely (two individuals randomly picked can be interchanged) is only valid when the environment is homogeneous or when all individuals facing the same environmental conditions react in exactly the same way. One way to account for heterogeneity is to define subpopulations as classes of similar individuals (for instance, based on their age). Then a distribution function of the individual states is sufficient. But when interactions between individuals are depending on the local configuration of the environment (including the other individuals), the spatial heterogeneity and the interindividual variability (two key drivers of evolution) cannot be left out anymore. Spatially explicit individual-based models (IBM) allow representing any kind of details critical to the system under study, thus relaxing assumptions distorting the reality in an unacceptable manner. This is the main reason why for more than two decades now (DeAngelis and Gross 1992; Huston et al. 1988; Judson 1994; Kawata and Toquenaga 1994), IBM has been more and more widely used in ecological modelling (for a recent guideline to make IBM more coherent and effective, see Grimm and Railsback 2005).

When it comes to include human decision-making processes into models, the standard way consists in assuming that all individuals equally informed (perfect information sharing) exhibit a standard behaviour based on rationality to achieve optimization. It is well known that renewable resource management addresses self-referential situations. The success of an individual strategy highly depends on the ability to "best guess" what the other individuals may do over time (Batten 2007). This is closely related to the notion of "representation" defined by Rouchier et al. (2000) as the understanding an agent has of what it perceives and that enables it to evaluate and then to choose the actions it can undertake on its environment. Do all agents agree, or do they have very different approaches to the same object or agent? One of the main interests of ABMS is to offer the possibility to explore the second option.

More generally, ABMS is a valid technical methodology to take into account heterogeneity in parameter values and in behaviours. In abstract and theoretical ABMS, standard statistical distribution functions can be used to assign particular parameter values to the different instances of agents created from the same class. Railsback et al. (2006) have compared how the main generic ABMS simulation platforms handle the initialization of one attribute of a population of agents from a random normal distribution (see version #14 of their benchmark "StupidModel"). For more realistic applications of ABMS, Berger and Schreinemachers (2006) recently introduced a straightforward approach to empirical parameterization using a common sampling frame to randomly select observation units for both biophysical measurements and socio-economic surveys. The heterogeneity in behaviours is usually considered with each agent having to select one behavioural module from a set of existing ones. From a conceptual design point of view, heterogeneity of behaviours is easier to represent with a hierarchy of classes. Subclasses of a generic agent class are a proper design when a given agent does not update its behaviour over time. To account for such an adaptive ability, the agent class has to be linked to a hierarchy of behaviours, as shown in Fig. 22.1. Beyond selecting an alternative out of a predefined set of options, it is even possible to define innovative agents equipped with some evolutionary programming to drive the creation of new behavioural patterns by recombining elementary behavioural components.

22.5.2   Dealing with Multiple Nested Levels

The seminal paper of Simon (1973) envisions hierarchical organizations as adaptive structures and not only as top-down sequences of authoritative control. This view was instilled in ecology by Allen and Starr (1982), who promoted the idea that biotic and abiotic processes at work in ecosystems are developing mutually reinforcing relationships over distinct ranges of scales. Each level, made from components interacting at the same timescale, communicates some information to the next higher and slower level. Reciprocally, any level can contribute to maintain the stability of faster and smaller levels. In the field of environmental management, both social and biophysical systems are characterized by hierarchical, nested structures. For example, family members interact to form a household, which may interact with other households in a village through political and economic institutions. Populations formed of individual species members aggregate to form communities, which, in turn, collectively define ecosystems. Holling (2001) nicely illustrates this with two mirroring examples: on one side the components of the boreal forest represented over time and space scales (from needle to landscape) and on the other side the institutional hierarchy of rule sets (from the decisions of small groups of individuals to constitution and culture) represented along dimensions of the number of people involved and the turnover times.

These ideas have been conceptualized to frame the emerging paradigm of "panarchy" (Gunderson and Holling 2002): the hierarchical structure of socio-ecological systems is exhibiting never-ending cycles of growth, accumulation, restructuring and renewal. The key concepts of this heuristic model are undoubtedly expanding the theoretical understanding of environmental management. What is their concrete contribution to the evolution of ecological modelling? To what extent can ABMS claim to represent them in a better way than other kinds of models?

The aggregation between hierarchical levels is very difficult to model in a purely analytical or statistical framework. In ecology, aggregation methods are applicable for models involving two levels of organization (individual and population) and their corresponding timescales (fast and slow) to reduce the dimension of the initial dynamical system to an aggregated one governing few global variables evolving at the slow timescale (Auger et al. 2000). The reverse way is much more difficult to integrate to models. How to account for the influence of changes at the global level on transitions at the microscopic level? Moreover, how to simulate both ways simultaneously at work? The main challenge deals with the coordination and scheduling of the different processes running at different levels: at the collective level, explicit decisions about temporarily giving back the control to lower-level component entities and conversely decisions from lower-level entities to create a group and to give the control to it. In the scientific community of ABMS, these ideas have directly inspired the production of conceptual organizational metamodels like Aalaadin (Ferber and Gutknecht 1998), specific features in generic simulation platforms like the threaded scheduling of agents in Swarm as well as applications like simulating hydrological processes (runoff, erosion and infiltration on heterogeneous soil surfaces) with "waterball", pond and river agents (Servat et al. 1998).

22.5.3 Beyond Decision Support Systems: Exploring New Dimensions in the Way to Use Models

As they represent complex adaptive systems which are unpredictable as a whole, ABMS applied to environmental management should caution about the large uncertainties related to their predictive abilities (Bradbury 2002). Still, empirically based ABMS can be used as a decision support system, for instance, to assist policymakers in prioritizing and targeting alternative policy interventions, as Berger et al. (2006) did in Uganda and Chile. Nevertheless, when multiple perceptions of the reality coexist, the statement "everything is defined by the reality of the observed phenomena" can be questioned. Therefore, ABMS in the field of ecological management should take some distance with the positivist posture that designates the scientific knowledge as the only authentic one. Relating empirical observations of phenomena to each other, in a way which is consistent with fundamental theory, phenomenological modelling is a means to represent the phenomena in a formalized and synthetic way. Descriptive rather than explanatory, this approach does not truly gain understanding of the phenomena, but can claim, in simple cases, to predict them. Parker et al. (2003) refer to these two distinct explanatory and descriptive approaches to clarify the potential roles of ABMS in LUCC. Anyway, the general rule "the more realistic the application, the more descriptive the approach" may not necessarily always apply. Explanatory goals can be assigned to models closely related to a real situation as well.

In contrast to the positivist approach, the constructivist approach refers to "constructed" knowledge, contingent on human perception and social experience and not necessarily reflecting any external "transcendent" realities. Starting "from scratch" to collectively design a model is a straightforward implementation of the constructivist approach. Among scientists, it will integrate within and between disciplines. By involving stakeholders, instead of showing them a simplification of their knowledge, the collective design of the model is seeking a mutual recognition of everyone's representation. In such a context, ABMS is more a communication platform to facilitate collective learning than a turnkey itinerary for piloting renewable resource management (Bousquet et al. 1999; ComMod 2003; Etienne et al. 2003; Gurung et al. 2006).

22.6 Drawbacks, Pitfalls and Remaining Challenges

22.6.1 Verification and Validation of ABMS

This is a problem challenging ABMS in general that is addressed in Chaps. 7 (Galán et al. 2017) and 9 (David et al. 2017) of this book. In the field of ecological management, as in other fields of applications, some authors claim to intentionally bridle the development of their agent-based model to design a strict equivalent to a mathematical equation-based model, as a means to verify it (Brown et al. 2004). The same process has been tested with mathematical representations of discrete distributed systems like Petri nets (Bakam et al. 2001).

22.6.2   Capturing the Metarules Governing the Adoption of Alternative Strategies

Nowadays, a set of tested and reliable tools and methods is available to better understand decision rules of actors and integrate them in computer agents (Janssen and Ostrom 2006). What remains much more challenging is to capture the rules governing the changes in agents' behaviours. An example of such a kind of "metarule" is the "evaluateAttitude" of the collector agent (see Fig. 22.1) defined by Thébaud and Locatelli (2001). In such a stylized model, the metarule is simply based on a threshold value of a specific parameter (the size of the pile of collected driftwood). When it comes to making the rules explicit to governing the changes of behavioural rules of human beings in real situations, methods are still weak. The metarules, if they exist, that control changes of strategies are difficult to grasp and elicit and by consequence to implement in an empirical-based ABM. One reason is the timescale which is greater for these metarules than for decision rules and which makes direct observation and verification harder to carry.

22.6.3   Improving the Representation of Space

Representing space with a CA, by far the most frequent way in current applications of ABMS in environmental management, is easy. But, as recently pointed out by Bithell and Macmillan (2007), imposition of a fixed grid upon the dynamics may cause important phenomena to be misrepresented when interactions between individuals are mediated by their size and may become too consumed by computer resources when the system scale exceeds the size of individuals by a large factor. How to handle discrete spatial data that is potentially completely unstructured and how to discover patterns of neighbourhood relationships between the discrete individuals within it? New directions like particle in cell are suggested.

In the next few years, we can also expect more applications based on autonomous agents moving over a GIS-based model of the landscape, with rendering algorithms determining what an individual agent is able to "see". Already used to simulate recreational activities (see, for instance, Bishop and Gimblett, 2000), behavioural responses to 3D virtual landscape may become more common.

Appendix

Publications	Model name	Topic	Issue	Environment	Agents	Software
Polhill et al. (2001)	FEARLUS	LUCC	Land-use and land ownership dynamics	Raster 7,7	Land manager (49) HeB(10) Ie	Swarm
				Region		
Brown et al. (2004)		LUCC	Greenbelt to control residential development	Raster X,80	Resident (10) Ho Ie R	Swarm
				Suburb		
Castella et al. (2005)	SAMBA (Generic)	LUCC	LU changes under decollectivization in north Vietnam	Raster 50,50 (25 Ha)	Household (50) HeP Ie R	Cormas
				Commune level (Abstract)		
Castella et al. (2005)	SAMBA-GIS (More realistic)	LUCC	LU changes under decollectivization in north Vietnam	Raster Regional level (More realistic)	Household (x) HeP Ie R (x = nb household in each village)	Cormas
						ArcView
D'Aquino et al. (2003)	SelfCormas	LUCC; agriculture; livestock	Competing rangeland and rice cropping land uses in Senegal	Raster 20,20 (5 Ha)	Farmer (20) HeB(2) Ie R	Cormas
				Village		
Etienne et al. (2003)	Mejan	LUCC; livestock; forestry; wildlife	Pine encroachment on original landscapes in France	Raster 23793 (1 Ha)	Farmer (40) HeB Ie,Ii,Ic C	Cormas
						MapInfo
					Forester (2) HeB Ie,Ii,Ic R	
					National Park (1) Ho Ie,Ii,Ic C	
				Plateau		

Reference	Model	Domain	Application	Scale	Agents	Platform
Mathevet et al. (2003)	GEMACE	LUCC; agriculture; wildlife	Competing hunting and hunting activities in Camargue (south of France)	Raster	Farmer Ie, Ii	Cormas
				Region	Hunting manager Ie, Ii	
Barreteau and Bousquet (2000) and Barreteau et al. (2004)	SHADOC	Water; agriculture	Viability of irrigated systems in the Senegal River Valley		Farmer	
Le Bars et al. (2005)	MANGA	Water			Farmer (*n*) Ie,Ii C	
					Water supplier (1) Ie C	
Feuillette et al. (2003)	Sinuse	Water; agriculture	Water table level	Raster 2400 (1 Ha)	Farmer HeP Ie,Ii,Ic	Cormas
				Watershed		
Lansing and Kremer (1993) and Janssen (2007)		Water; agriculture	Water management and water temple networks in Bali	Network 172	Village (172) HeB(49) Ie,Ic R	
				Watershed		
Raj Gurung et al. (2006)	Limbukha	Water management	Negotiation of irrigation water sharing between two Bhutanese communities	Grid 8,13 (10.4 Ha)	Farmer (12) *HeB* Ii	Cormas
				(Abstract) Village		
Hoffmann et al. (2002)	LUCIM	Forestry; LUCC	Deforestation and afforestation in Indiana, USA	Raster 100	Farmer (10) HeB Ie R	
				State		

Publications	Model name	Topic	Issue	Environment	Agents	Software
Purnomo and Guizol (2006)		Forestry	Forest plantation co-management in Indonesia	Raster 50,50	Developer Ie,Ii,Ic	
					Smallholder Ie,Ii,Ic C	
					Broker Ie,Ii,Ic	
					Government Ie,Ii,Ic	
				Forest massif		
Mathevet et al. (2003)	ReedSim	Wildlife	Water management in Mediterranean reedbeds			Cormas
An et al. (2005)		Wildlife	Impact of the growing rural population on the forests and panda habitat in China			
Bousquet et al. (2001)	Djemiong	Wildlife	Traditional hunting of small antelopes in Cameroon	Raster 2042 (4 Ha)	Hunter (90) HeP Ie,Ic R	Cormas
					Antelop (7350) HeB(3) Ie,Ic R	
				Village		
Anwar et al. (2007)		Wildlife	Interactions between whale-watching boats and whales in the St. Lawrence estuary	Raster	Boat C	Repast
					Whale R	
Elliston and Beare (2006)		Agriculture; epidemiology	Agricultural pest and disease incursions in Australia			Cormas
Happe et al. 2006	Agripolis	Agriculture; livestock	Policy impact on structural changes of W. Europe farms	Raster 73439 (1 Ha)	Farms (2869) HeP Ic	
				Region		

				Raster (300 Ha) (Realistic)	Farmer (12) HeP Ii R	Cormas
Barnaud et al. (2007)	MaeSalaep 2.2	Agriculture; LUCC	LU strategies in transitional swidden agricultural systems, Thailand	Village catchment	Credit sources (3)	
Gross et al. (2006)		Livestock	Evaluation of behaviours of rangeland systems in Australia		Enterprise C	
					Government	
Courdier et al. (2002)	Biomas	Livestock; agriculture	Collective management of animal wastes in La Reunion	Network	Livestock farm (48) HeP Ie,Ii,Ic	Geamas
				Region	Crop farm (59) HeP Ie,Ii,Ic	
					Shipping agent (34) HeP Ii	
Bagni et al. (2002)		Epidemiology; livestock	Evaluation of sanitary policies to control the spread of a viral pathology of bovines	Abstract	Farm sector	Swarm
				Farm	Cow	
Rateb et al. (2005)		Epidemiology	Impact of education on malaria healthcare in Haiti	Raster	Inhabitant *HeP* R	StarLogo
				country		
Muller et al. (2004)		Epidemiology; agriculture; livestock	Risk and control strategies of African trypanosomiasis in Southern Cameroon	Network	Villager Ic R	MadKit
				Village		

Topic and Issue

When multiple topics are covered by a case study, the first in the list indicates the one we used to classify it. Within each topic we have tried to order the case studies from the more abstract and theoretical ones to the more realistic ones. This information can be retrieved from the issue: only case studies representing a real system mention a geographical location.

Environment

- First line: mode of representation, with the general following pattern:
  [none, network, raster, vector] $N(x)$
  N indicates the number of elementary spatial entities (nodes of network, cells or polygons), when raster mode, N, is given as number of lines x number of columns, unless some cells have been discarded from the rectangular grid because they were out of bound (then only the total number is given), and (x) indicates the spatial resolution.
- Second line: level of organization at which the issue is considered (for instance, village, biophysical entity (watershed, forest massif, plateau, etc.), city, conurbation, province, country, etc.)

Agents

One line per type of agent (the practical definition given in this paper applies, regardless of the terminology used by the authors). The general pattern of information looks like:
  name(x) [Ho;HeC;HeB(y)] [Ie;Ii;Ic] [R;C]

- (x) indicates the number of instances defined when initializing a standard scenario, italic mentions that this initial number change during simulation.
- When $x > 1$, to account for the heterogeneity of the population of agents, we propose the following coding: "Ho" stands for a homogeneous population (identical agents), and "He" stands for a heterogeneous population. "HeP" indicates that the heterogeneity lies only in parameter values, while "HeB" indicates that the heterogeneity lies in behaviours. In such a case, each agent is equipped with one behavioural module selected from a set of (y) existing ones. Italic points out adaptive agents updating either parameter value (*HeP*) or behaviour (*HeB*) during simulation.
- [Ie, Ii, Ic] indicates the nature of relationships as defined in the text and shown in Fig. 22.3.
- [R; C] indicates if agents are clearly either reactive or cognitive.

Further Reading

1. The special issue of JASSS in 2001[1] on "ABM, Game Theory and Natural Resource Management issues" presents a set of papers selected from a workshop held in Montpellier in March 2000, most of them dealing with collective decision-making processes in the field of natural resource management and environment.
2. Gimblett (2002) is a book on integrating GIS and ABM, derived from a workshop held in March 1998 at the Santa Fe Institute. It provides contributions from computer scientists, geographers, landscape architects, biologists, anthropologists, social scientists and ecologists focusing on spatially explicit simulation modelling with agents.
3. Janssen (2002) provides a state-of-the-art review of the theory and application of multi-agent systems for ecosystem management and addresses a number of important topics including the participatory use of models. For a detailed review of this book, see Terna (2005).
4. López Paredes and Hernández Iglesias (2008) advocate why agent-based simulations provide a new and exciting avenue for natural resource planning and management: researches and advisers can compare and explore alternative scenarios and institutional arrangements to evaluate the consequences of policy actions in terms of economic, social and ecological impacts. But as a new field it demands from the modellers a great deal of creativeness, expertise and "wise choice", as the papers collected in this book show.

References

Abel, T. (1998). Complex adaptive systems, evolutionism, and ecology within anthropology: Interdisciplinary research for understanding cultural and ecological dynamics. *Georgia Journal of Ecological Anthropology, 2*, 6–29.

Abrami, G. (2004). *Niveaux d'organisation dans la modélisation multi-agent pour la gestion de ressources renouvelables. Application à la mise en oeuvre de règles collectives de gestion de l'eau agricole dans la basse-vallée de la Drôme.* Montpellier: Ecole Nationale du Génie Rural, des Eaux et Forêts.

Allen, T. F. H., & Starr, T. B. (1982). *Hierarchy: Perspectives for ecological complexity.* Chicago: University of Chicago Press.

An, L., Linderman, M., Qi, J., Shortridge, A., & Liu, J. (2005). Exploring complexity in a human–environment system: An agent-based spatial model for multidisciplinary and multiscale integration. *Annals of the Association of American Geographers, 95*(1), 54–79.

Anwar, S. M., Jeanneret, C., Parrott, L., & Marceau, D. (2007). Conceptualization and implementation of a multi-agent model to simulate whale-watching tours in the St. Lawrence estuary in Quebec, Canada. *Environmental Modelling and Software, 22*(12), 1775–1787.

---

[1] http://jasss.soc.surrey.ac.uk/4/2/contents.html

Auger, P., Charles, S., Viala, M., & Poggiale, J.-C. (2000). Aggregation and emergence in ecological modelling: Integration of ecological levels. *Ecological Modelling, 127*, 11–20.

Bagni, R., Berchi, R., & Cariello, P. (2002). A comparison of simulation models applied to epidemics. *Journal of Artificial Societies and Social Simulation, 5*(3), 5.

Bah, A., Touré, I., Le Page, C., Ickowicz, A., & Diop, A. T. (2006). An agent-based model to understand the multiple uses of land and resources around drilling in Sahel. *Mathematical and Computer Modelling, 44*(5–6), 513–534.

Bakam, I., Kordon, F., Le Page, C., & Bousquet, F. (2001). Formalization of a multi-agent model using colored petri nets for the study of an hunting management system. In *Lecture notes in computer science* (Vol. 1871, pp. 123–132). Berlin: Springer.

Balmann, A. (1997). Farm based modelling of regional structural change: A cellular automata approach. *European Review of Agricultural Economics, 24*(1), 85–108.

Barnaud, C., Promburom, T., Trébuil, G., & Bousquet, F. (2007). An evolving simulation and gaming process for adaptive resource management in the highlands of northern Thailand. *Simulation and Gaming, 38*(3), 398–420.

Barreteau, O., & Bousquet, F. (2000). SHADOC: A multi-agent model to tackle viability of irrigated systems. *Annals of Operations Research, 94*, 139–162.

Barreteau, O., Bousquet, F., Millier, C., & Weber, J. (2004). Suitability of multi-agent simulations to study irrigated system viability: Application to case studies in the Senegal River Valley. *Agricultural Systems, 80*, 255–275.

Batten, D. (2007). Are some human ecosystems self-defeating? *Environmental Modelling and Software, 22*, 649–655.

Becu, N., Perez, P., Walker, A., Barreteau, O., & Le Page, C. (2003). Agent based simulation of a small catchment water management in northern Thailand: Description of the CATCHSCAPE model. *Ecological Modelling, 170*(2-3), 319–331.

Berger, T. (2001). Agent-based spatial models applied to agriculture: A simulation tool for technology diffusion, resource use changes and policy analysis. *Agricultural Economics, 25*, 245–260.

Berger, T., & Schreinemachers, P. (2006). Creating agents and landscapes for multiagent systems from random samples. *Ecology and Society, 11*(2), 19. http://www.ecologyandsociety.org/vol11/iss2/art19/

Berger, T., Schreinemachers, P., & Woelcke, J. (2006). Multi-agent simulation for the targeting of development policies in less-favored areas. *Agricultural Systems, 88*, 28–43.

Berkes, F., & Folke, C. (Eds.). (1998). *Linking ecological and social systems: Management practices and social mechanisms for building resilience*. Cambridge, UK: Cambridge University Press.

Bian, L. (2004). A conceptual framework for an individual-based spatially explicit epidemiological model. *Environment and Planning B: Planning and Design, 31*(3), 381–395.

Bishop, I., & Gimblett, R. (2000). Management of recreational areas: GIS, autonomous agents, and virtual reality. *Environment and Planning B: Planning and Design, 27*(3), 423–435.

Bithell, M., & Macmillan, W. D. (2007). Escape from the cell: Spatially explicit modelling with and without grids. *Ecological Modelling, 200*, 59–78.

Bonaudo, T., Bommel, P., & Tourrand, J. F. (2005). Modelling the pioneers fronts of the transamazon highway region. In *Proceedings of CABM-HEMA-SMAGET 2005, joint conference on multi-agent modeling for environmental management, Bourg St Maurice, Les Arcs, France, 21-25 March 2005*. http://smaget.lyon.cemagref.fr/contenu/SMAGET%20proc/PAPERS/BonaudoBommelTourrand.pdf

Bousquet, F. (2001). *Modélisation d'accompagnement. Simulations multi-agents et gestion des ressources naturelles et renouvelables*. Unpublished Habilitation à Diriger des Recherches, Lyon I University, Lyon, France.

Bousquet, F., Barreteau, O., d'Aquino, P., Etienne, M., Boissau, S., Aubert, S., et al. (2002). Multi-agent systems and role games: Collective learning processes for ecosystem management. In M. A. Janssen (Ed.), *Complexity and ecosystem management: The theory and practice of multi-agent systems* (pp. 248–285). Cheltenham: Edward Elgar Publishing.

Bousquet, F., Barreteau, O., Le Page, C., Mullon, C., & Weber, J. (1999). An environmental modelling approach: The use of multi-agents simulations. In F. Blasco & A. Weill (Eds.), *Advances in environmental and ecological modelling* (pp. 113–122). Paris: Elsevier.

Bousquet, F., & Le Page, C. (2004). Multi-agent simulations and ecosystem management: A review. *Ecological Modelling, 176*, 313–332.

Bousquet, F., Le Page, C., Bakam, I., & Takforyan, A. (2001). Multiagent simulations of hunting wild meat in a village in eastern Cameroon. *Ecological Modelling, 138*, 331–346.

Bousquet, F., Trébuil, G., & Hardy, B. (Eds.). (2005). *Companion modeling and multi-agent systems for integrated natural resources management in Asia*. Los Baños, Philippines: International Rice Research Institute.

Bradbury, R. (2002). Futures, prediction and other foolishness. In M. A. Janssen (Ed.), *Complexity and ecosystem management: The theory and practice of multi-agent systems* (pp. 48–62). Cheltenham: Edward Elgar Publishing.

Brown, D. G., Page, S. E., Riolo, R., & Rand, W. (2004). Agent-based and analytical modeling to evaluate the effectiveness of greenbelts. *Environmental Modelling and Software, 19*, 1097–1109.

Caplat, P., Lepart, J., & Marty, P. (2006). Landscape patterns and agriculture: Modelling the long-term effects of human practices on Pinus sylvestris spatial dynamics (Causse Mejean, France). *Landscape Ecology, 21*(5), 657–670.

Carley, K. M., et al. (2006). BioWar: Scalable agent-based model of bioattacks. *IEEE Transactions on Systems, Man, and Cybernetics, Part A: Systems and Humans, 36*(2), 252–265.

Carpenter, S., Brock, W., & Hanson, P. (1999). Ecological and social dynamics in simple models of ecosystem management. *Conservation Ecology, 3*(2), 4.

Castella, J.-C., Boissau, S., Trung, T. N., & Quang, D. D. (2005). Agrarian transition and lowland–upland interactions in mountain areas in northern Vietnam: Application of a multi-agent simulation model. *Agricultural Systems, 86*, 312–332.

Castella, J.-C., Trung, T. N., & Boissau, S. (2005). Participatory simulation of land-use changes in the Northern Mountains of Vietnam: The combined use of an agent-based model, a role-playing game, and a geographic information system. *Ecology and Society, 10*(1), 27.

Castella, J.-C., & Verburg, P. H. (2007). Combination of process-oriented and pattern-oriented models of land use change in a mountain area of Vietnam. *Ecological Modelling, 202*(3-4), 410–420.

ComMod. (2003). Our companion modelling approach. *Journal of Artificial Societies and Social Simulation, 6*(2), http://jasss.soc.surrey.ac.uk/6/2/1.html

Courdier, R., Guerrin, F., Andriamasinoro, F. H., & Paillat, J. M. (2002). Agent-based simulation of complex systems: Application to collective management of animal wastes. *Journal of Artificial Societies and Social Simulation, 5*(3), http://jasss.soc.surrey.ac.uk/5/3/4.html

D'Aquino, P., Le Page, C., Bousquet, F., & Bah, A. (2003). Using self-designed role-playing games and a multi-agent system to empower a local decision-making process for land use management: The SelfCormas experiment in Senegal. *Journal of Artificial Societies and Social Simulation, 6*(3), http://jasss.soc.surrey.ac.uk/6/3/5.html

David, N., Fachada, N., & Rosa, A. C. (2017). Verifying and validating simulations. doi:https://doi.org/10.1007/978-3-319-66948-9_9.

Deadman, P., Robinson, D., Moran, E., & Brondízio, E. (2004). Colonist household decisionmaking and land-use change in the Amazon rainforest: An agent-based simulation. *Environment and Planning B: Planning and Design, 31*(5), 693–709.

DeAngelis, D. L., & Gross, L. J. (Eds.). (1992). *Individual-based models and approaches in ecology*. New York: Chapman and Hall.

Dunham, J. B. (2005). An agent-based spatially explicit epidemiological model in MASON. *Journal of Artificial Societies and Social Simulation, 9*(1), http://jasss.soc.surrey.ac.uk/9/1/3.html

Elliston, L., & Beare, S. (2006). Managing agricultural pest and disease incursions: An application of agent-based modelling. In P. Perez & D. Batten (Eds.), *Complex science for a complex world: Exploring human ecosystems with agents* (pp. 177–189). Canberra: ANU E Press.

Epstein, J. M., & Axtell, R. L. (1996). *Growing artificial societies: Social science from the bottom up*. Washington, D.C.: Brookings Institution Press.

Etienne, M. (Ed.). (2011). *Companion modelling: A participatory approach to support sustainable development*. Versailles: Quae.

Etienne, M., Le Page, C., & Cohen, M. (2003). A Step-by-step approach to building land management scenarios based on multiple viewpoints on multi-agent system simulations. *Journal of Artificial Societies and Social Simulation, 6*(2), http://jasss.soc.surrey.ac.uk/6/2/2.html

Evans, T. P., & Kelley, H. (2004). Multi-scale analysis of a household level agent-based model of landcover change. *Journal of Environmental Management, 72*, 57–72.

Ferber, J., & Gutknecht, O. (1998). A meta-model for the analysis and design of organizations in multi-agent systems. In Y. Demazeau (Ed.), *Proceedings of the 3rd International Conference on Multi-Agent Systems (ICMAS '98), Cité des Sciences – La Villette, Paris, France, July 4-7, 1998* (pp. 128–135). IEEE: Los Alamos, NM.

Feuillette, S., Bousquet, F., & Le Goulven, P. (2003). SINUSE: A multi-agent model to negotiate water demand management on a free access water table. *Environmental Modelling and Software, 18*, 413–427.

Galán, J. M., Izquierdo, L. R., Izquierdo, S. S., Santos, J. L., del Olmo, R., & López-Paredes, A. (2017). doi:https://doi.org/10.1007/978-3-319-66948-9_7.

Galvin, K. A., Thornton, P. K., Roque de Pinho, J., Sunderland, J., & Boone, R. B. (2006). Integrated modeling and its potential for resolving conflicts between conservation and people in the rangelands of East Africa. *Human Ecology, 34*(2), 155–183.

Gimblett, R. (Ed.). (2002). *Integrating geographic information systems and agent-based modeling techniques for understanding social and ecological processes, Santa Fe Institute Studies in the Sciences of Complexity*. Oxford: Oxford University Press.

Grimm, V., Berger, U., Bastiansen, F., Eliassen, S., Ginot, V., Giske, J., et al. (2006). A standard protocol for describing individual-based and agent-based models. *Ecological Modelling, 198*, 115–126.

Grimm, V., & Railsback, S. F. (2005). *Individual-based modeling and ecology*. Princeton, NJ: Princeton University Press.

Gross, J. E., McAllister, R. R. J., Abel, N., Stafford Smith, D. M., & Maru, Y. (2006). Australian rangelands as complex adaptive systems: A conceptual model and preliminary results. *Environmental Modelling and Software, 21*(9), 1264–1272.

Gunderson, L. H., & Holling, C. S. (2002). *Panarchy: Understanding transformations in human and natural systems*. Washington, D.C.: Island Press.

Gurung, T. R., Bousquet, F., & Trébuil, G. (2006). Companion modeling, conflict resolution, and institution building: Sharing irrigation water in the Lingmuteychu watershed, Bhutan. *Ecology and Society, 11*(2), 36.

Haffner, Y., & Gramel, S. (2001). Modelling strategies for water supply companies to deal with nitrate pollution. *Journal of Artificial Societies and Social Simulation, 4*(3), http://jasss.soc.surrey.ac.uk/4/3/11.html

Hales, D., Rouchier, J., & Edmonds, B. (2003). Model-to-model analysis. *Journal of Artificial Societies and Social Simulation, 6*(4), http://jasss.soc.surrey.ac.uk/6/4/5.html

Happe, K., Kellermann, K., & Balmann, A. (2006). Agent-based analysis of agricultural policies: An Illustration of the agricultural policy simulator AgriPoliS, its adaptation and behavior. *Ecology and Society, 11*(1), 49.

Hare, M., & Deadman, P. (2004). Further towards a taxonomy of agent-based simulation models in environmental management. *Mathematics and Computers in Simulation, 64*, 25–40.

Hemelrijk, C. (2017). Simulating complexity of animal social behaviour. doi:https://doi.org/10.1007/978-3-319-66948-9_24.

Hoffmann, M., Kelley, H., & Evans, T. (2002). Simulating land-cover change in South-Central Indiana: An agent-based model of deforestation and afforestation. In M. A. Janssen (Ed.), *Complexity and ecosystem management: The theory and practice of multi-agent systems* (pp. 218–247). Cheltenham: Edward Elgar Publishing.

Holling, C. S. (1986). The resilience of terrestrial ecosystems; local surprise and global change. In W. C. Clark & R. E. Munn (Eds.), *Sustainable development of the biosphere* (pp. 292–317). Cambridge: Cambridge University Press.

Holling, C. S. (2001). Understanding the complexity of economic, ecological, and social systems. *Ecosystems, 4*, 390–405.

Holling, C. S. (2004). From complex regions to complex worlds. *Ecology and Society, 9*(1), 11.

Huigen, M. G. A. (2004). First principles of the MameLuke multi-actor modelling framework for land use change, illustrated with a Philippine case study. *Journal of Environmental Management, 72*, 5–21.

Huigen, M. G. A., Overmars, K. P., & de Groot, W. T. (2006). Multiactor modeling of settling decisions and behavior in the San Mariano watershed, the Philippines: A first application with the MameLuke framework. *Ecology and Society, 11*(2), 33.

Huston, M., DeAngelis, D. L., & Post, W. (1988). New computer models unify ecological theory. *BioScience, 38*(10), 682–691.

Janssen, M. A. (2001). An exploratory integrated model to assess management of lake eutrophication. *Ecological Modelling, 140*, 111–124.

Janssen, M. A. (Ed.). (2002). *Complexity and ecosystem management: The theory and practice of multi-agent systems*. Cheltenham: Edward Elgar Publishing.

Janssen, M. A. (2007). Coordination in irrigation systems: An analysis of the Lansing–Kremer model of Bali. *Agricultural Systems, 93*, 170–190.

Janssen, M. A., Anderies, J. M., Stafford Smith, M., & Walker, B. H. (2002). Implications of spatial heterogeneity of grazing pressure on the resilience of rangelands. In M. A. Janssen (Ed.), *Complexity and ecosystem management: The theory and practice of multi-agent systems* (pp. 103–126). Cheltenham: Edward Elgar Publishing.

Janssen, M. A., & Ostrom, E. (2006). Empirically based, agent-based models. *Ecology and Society, 11*(2).

Janssen, M. A., Walker, B. H., Langridge, J., & Abel, N. (2000). An adaptive agent model for analysing co-evolution of management and policies in a complex rangeland system. *Ecological Modelling, 131*(2-3), 249–268.

Jepsen, J. U., Topping, C. J., Odderskær, P., & Andersen, P. N. (2005). Evaluating consequences of land-use strategies on wildlife populations using multiple-species predictive scenarios. *Agriculture, Ecosystems and Environment, 105*, 581–594.

Judson, O. P. (1994). The rise of the individual-based model in ecology. *Trends in Ecology and Evolution, 9*(1), 9–14.

Kawata, M., & Toquenaga, Y. (1994). From artificial individuals to global patterns. *Trends in Ecology and Evolution, 9*(11), 417–421.

Kinzig, A. P., et al. (2006). Resilience and regime shifts: Assessing cascading effects. *Ecology and Society, 11*(1), 20.

Krywkow, J., Valkering, P., Rotmans, J., & van der Veen, A. (2002, June 24–27). Agent-based and integrated assessment modelling for incorporating social dynamics in the management of the Meuse in the Dutch Province of Limburg. In A.E. Rizzoli & A.J. Jakeman (Eds.), *Integrated assessment and decision support: Proceedings of the First Biennial Meeting of the International Environmental Modelling and Software Society iEMSs, University of Lugano, Switzerland* (Vol. 2, pp. 263–268). Manno: iEMSs.

Lansing, J. S., & Kremer, J. N. (1993). Emergent properties of Balinese water temple networks: Coadaptation on a rugged fitness landscape. *American Anthropologist, 95*(1), 97–114.

Le Bars, M., Attonaty, J.-M., Ferrand, N., & Pinson, S. (2005). An agent-based simulation testing the impact of water allocation on farmers' collective behaviors. *Simulation, 81*(3), 223–235.

Levin, S. A. (1998). Ecosystems and the biosphere as complex adaptive systems. *Ecosystems, 1*, 431–436.

Levin, S. A. (1999). Towards a science of ecological management. *Conservation Ecology, 3*(2), 6.

Lim, K., Deadman, P. J., Moran, E., Brondizio, E., & McCracken, S. (2002). Agent-based simulations of household decision making and land use change near Altamira, Brazil. In R. Gimblett (Ed.), *Integrating geographic information systems and agent-based modeling techniques for understanding social and ecological processes, Santa Fe Institute Studies in the Sciences of Complexity* (pp. 277–310). Oxford: Oxford University Press.

López Paredes, A., & Hernández Iglesias, C. (Eds.). (2008). *Agent-based modelling in natural resource management*. Valladolid: INSISOC.

Manson, S. (2005). Agent-based modeling and genetic programming for modeling land change in the Southern Yucatan Peninsular Region of Mexico. *Agriculture, Ecosystems and Environment, 111*, 47–62.

Manson, S. (2006). Land use in the southern Yucatan peninsular region of Mexico: Scenarios of population and institutional change. *Computers, Environment and Urban Systems, 30*, 230–253.

Marietto, M. B., David, N., Sichman, J. S., & Coelho, H. (2003). Requirements analysis of agent-based simulation platforms: State of the art and new prospects. In J. S. Sichman, F. Bousquet, & P. Davidsson (Eds.), *MABS'02, Proceedings of the 3rd international conference on Multi-Agent-Based Simulation, Lecture notes in artificial intelligence* (Vol. 2581, pp. 125–141). Berlin: Springer.

Mathevet, R., Bousquet, F., Le Page, C., & Antona, M. (2003). Agent-based simulations of interactions between duck population, farming decisions and leasing of hunting rights in the Camargue (Southern France). *Ecological Modelling, 165*(2/3), 107–126.

Mathevet, R., Mauchamp, A., Lifran, R., Poulin, B., & Lefebvre, G. (2003). ReedSim: Simulating ecological and economical dynamics of Mediterranean reedbeds. In *Proceedings of MODSIM03, International Congress on Modelling and simulation, integrative modelling of biophysical, social, and economic systems for resource management solutions, Townsville, Australia, 14–17 July 2003*. Modelling & Simulation Society of Australia & New Zealand. http://mssanz.org.au/MODSIM03/Volume_03/B02/01_Mathevet.pdf

Matthews, R. (2006). The people and landscape model (PALM): Towards full integration of human decision-making and biophysical simulation models. *Ecological Modelling, 194*, 329–343.

Milner-Gulland, E. J., Kerven, C., Behnke, R., Wright, I. A., & Smailov, A. (2006). A multi-agent system model of pastoralist behaviour in Kazakhstan. *Ecological Complexity, 3*(1), 23–36.

Moreno, N., Quintero, R., Ablan, M., Barros, R., Dávila, J., Ramírez, H., et al. (2007). Biocomplexity of deforestation in the Caparo tropical forest reserve in Venezuela: An integrated multi-agent and cellular automata model. *Environmental Modelling and Software, 22*, 664–673.

Muller, G., Grébaut, P., & Gouteux, J.-P. (2004). An agent-based model of sleeping sickness: Simulation trials of a forest focus in southern Cameroon. *C.R. Biologies, 327*, 1–11.

Nute, D., et al. (2004). NED-2: An agent-based decision support system for forest ecosystem management. *Environmental Modelling and Software, 19*, 831–843.

Parker, D. C., Manson, S. M., Janssen, M. A., Hoffmann, M. J., & Deadman, P. (2003). Multi-agent systems for the simulation of land-use and land-cover change: A review. *Annals of the Association of American Geographers, 93*(2), 314–337.

Pepper, J. W., & Smuts, B. B. (2000). The evolution of cooperation in an ecological context: An agent-based model. In T. A. Kohler & G. J. Gumerman (Eds.), *Dynamics in human and primate societies, Santa Fe Institute Studies in the Sciences of Complexity* (pp. 45–76). New York: Oxford University Press.

Perez, P., & Batten, D. (Eds.). (2006). *Complex science for a complex world: Exploring human ecosystems with agents*. Canberra: ANU E Press.

Polhill, J. G., Gotts, N. M., & Law, A. N. R. (2001). Imitative versus non-imitative strategies in a land use simulation. *Cybernetics & Systems, 32*(1–2), 285–307.

Polhill, J. G., Izquierdo, L. R., & Gotts, N. M. (2006). What every agent-based modeller should know about floating point arithmetic. *Environmental Modelling and Software, 21*, 283–309.

Purnomo, H., & Guizol, P. (2006). Simulating forest plantation co-management with a multi-agent system. *Mathematical and Computer Modelling, 44*, 535–552.

Purnomo, H., Mendoza, G. A., Prabhu, R., & Yasmi, Y. (2005). Developing multi-stakeholder forest management scenarios: A multi-agent system simulation approach applied in Indonesia. *Forest Policy and Economics, 7*(4), 475–491.

Railsback, S. F., Lytinen, S. L., & Jackson, S. K. (2006). Agent-based simulation platforms: Review and development recommendations. *Simulation, 82*(9), 609–623.

Rateb, F., Pavard, B., Bellamine-BenSaoud, N., Merelo, J. J., & Arenas, M. G. (2005). Modeling malaria with multi-agent systems. *International Journal of Intelligent Information Technologies, 1*(2), 17–27.

Rouchier, J., Bousquet, F., Barreteau, O., Le Page, C., & Bonnefoy, J.-L. (2000). Multi-agent modelling and renewable resources issues: The relevance of shared representation for interacting agents. In S. Moss & P. Davidsson (Eds.), *Multi-agent-based simulation, Second International Workshop, MABS 2000, Boston, MA, USA, July; Revised and additional papers, Lecture notes in artificial intelligence* (Vol. 1979, pp. 341–348). Berlin: Springer.

Rouchier, J., Bousquet, F., Requier-Desjardins, M., & Antona, M. (2001). A multi-agent model for describing transhumance in North Cameroon: Comparison of different rationality to develop a routine. *Journal of Economic Dynamics & Control, 25*, 527–559.

Schreinemachers, P., & Berger, T. (2006). Land-use decisions in developing countries and their representation in multi-agent systems. *Journal of Land Use Science, 1*(1), 29–44.

Servat, D., Perrier, E., Treuil, J.-P., & Drogoul, A. (1998). When agents emerge from agents: Introducing multi-scale viewpoints in multi-agent simulations. In J. S. Sichman, R. Conte, & N. Gilbert (Eds.), *Proceedings of the first international workshop on multi-agent systems and agent-based simulation, Lecture notes in artificial intelligence* (Vol. 1534, pp. 183–198). London: Springer.

Simon, H. A. (1973). The organization of complex systems. In H. H. Pattee (Ed.), *Hierarchy theory: The challenge of complex systems* (pp. 1–27). New York: Braziller.

Soulié, J.-C., & Thébaud, O. (2006). Modeling fleet response in regulated fisheries: An agent-based approach. *Mathematical and Computer Modelling, 44*, 553–554.

Sulistyawati, E., Noble, I. R., & Roderick, M. L. (2005). A simulation model to study land use strategies in swidden agriculture systems. *Agricultural Systems, 85*, 271–288.

Terna, P. (2005). Review of "Complexity and ecosystem management: The theory and practice of multi-agent systems". *Journal of Artificial Societies and Social Simulation, 8*(2), http://jasss.soc.surrey.ac.uk/8/2/reviews/terna.html

Thébaud, O., & Locatelli, B. (2001). Modelling the emergence of resource-sharing conventions: An agent-based approach. *Journal of Artificial Societies and Social Simulation, 4*(2), http://jasss.soc.surrey.ac.uk/4/2/3.html

Ziervogel, G., Bithell, M., Washington, R., & Downing, T. (2005). Agent-based social simulation: A method for assessing the impact of seasonal climate forecast applications among smallholder farmers. *Agricultural Systems, 83*, 1–26.

Zunga, Q., Vagnini, A., Le Page, C., Touré, I., Lieurain, E., & Bousquet, F. (1998). Coupler Systèmes d'Information Géographiques et Systèmes Multi-Agents pour modéliser les dynamiques de transfotmation des paysages. Le cas des dynamiques foncières de la Moyenne Vallée du Zambèze (Zimbabwe). In N. Ferrand (Ed.), *Modèles et systèmes multi-agents pour la gestion de l'environnement et des territoires* (pp. 193–206). Clermont-Ferrand: Cemagref Editions.

Chapter 23
Distributed Computer Systems

David Hales

Abstract Ideas derived from social simulation models can directly inform the design of distributed computer systems. This is particularly the case when systems are "open", in the sense of having no centralised control, where traditional design approaches struggle. In this chapter, we indicate the key features of social simulation work that are valuable for distributed systems design. We also discuss the differences between social and biological models in this respect. We give examples of socially inspired systems from the currently active area of peer-to-peer systems, and finally we discuss open areas for future research in the field.

Why Read This Chapter?
To understand how simulating social complexity might be used in the process of designing distributed computer systems.

23.1 Introduction

Massive and open distributed computer systems provide a major application area for ideas and techniques developed within social simulation and complex systems modelling. In the early years of the twenty-first century, there has been an explosion in global networking infrastructure in the form of wired and wireless broadband connections to the Internet encompassing both traditional general-purpose computer systems, mobile devices and specialist appliances and services. The challenge is to utilise such diverse infrastructure to provide novel services that satisfy user needs reliably. Traditional methods of software design and testing are not always applicable to this challenge. Why is this? And what can social simulation and complexity perspectives bring to addressing the challenge? This chapter answers these questions by providing a general overview of some of the major benefits of approaching design from socially inspired perspectives in addition to examples of

D. Hales (✉)
Centre for Policy Modelling, Manchester Metropolitan University, Manchester, UK
e-mail: dave@davidhales.com

© Springer International Publishing AG 2017
B. Edmonds, R. Meyer (eds.), *Simulating Social Complexity*,
Understanding Complex Systems, https://doi.org/10.1007/978-3-319-66948-9_23

applications in the area of peer-to-peer (P2P) systems and protocols. Finally, we will speculate on possible future directions in the area.

This chapter is not an exhaustive survey of the area, for example, we have not discussed the application of social network analysis techniques to web graphs and social networks that are constructed within, or facilitated by, distributed software systems (Staab et al. 2005). Both these are active areas. Further, we have not discussed the active research area based on randomised "Gossiping" approaches, where information is diffused over networks through randomised copying of information between adjacent nodes (Wang et al. 2007).

23.2 What Is Wrong with Traditional Design Approaches?

Traditional design approaches to systems and software often assume that systems are essentially "closed"—meaning they are under the control of some administrative authority that can control access, authenticate users and manage system resources such as issuing software components and updates. Consider the simplest situation in which we have a single computer system, which is not connected to a network, that is required to solve some task. Here, design follows a traditional process of analysis of requirements, specification of requirements then design and iterative refinement, until the system meets the specified requirements. User requirements are assumed to be discoverable and translatable into specifications at a level of detail that can inform a design. The designer has, generally, freedom to dictate how the system should achieve the required tasks via the coordination of various software components. This coordination generally follows an elaborate sequencing of events where the output from one component becomes the input to others. The design task is to get that sequence right.

In "open systems", it is assumed that there are multiple authorities. This means that the components that comprise the system cannot be assumed to be under central control. An extreme example of this might be an open peer-to-peer (P2P) system in which each node in the system executes on a user machine under the control of a different user. Popular file sharing systems operate in this way, allowing each user to run any variant of client software they choose. These kinds of systems function because the client software implements publicly available peer communication protocols, allowing the nodes to interconnect and provide functionality. However, due to the open nature of such systems, it is not possible for the designer, a priori, to control the sequence of processing in each node. Hence the designer needs to consider what kinds of protocols will produce *acceptable* system level behaviours under *plausible* assumptions of node behaviour. This requires a radically different design perspective in which systems need to be designed as self-repairing and self-organising systems in which behaviour emerges bottom-up rather than being centrally controlled.

One term for this approach, based on self-organisation and emergence, is so-called self-star (or self-*) systems (Babaoglu et al. 2005). The term is a broad expression that aims to capture all kinds of self-organising computer systems that continue to function acceptably under realistic conditions.

But what kinds of design approach can be employed? Currently there is no accepted general theory of self-organisation or emergence—rather there are some interesting models at different levels of abstraction that capture certain phenomena. Many such models have been produced within the biological sciences to explain complex self-organising biological phenomena. Biological systems, particularly co-evolving systems, appear to evidence many of the desirable properties required by self-* computer systems. Hence, several proposed self-* techniques have drawn on biological inspiration (Babaoglu et al. 2006).

23.3 Social Versus Biological Inspiration

It is useful to ask in what way social organisation differs from the biological level. In this section we briefly consider this question with regard to desirable properties for information systems. An important aspect of human social systems (HSS) is their ability (like biological systems) to both preserve structures—with organisations and institutions persisting over time—and adapt to changing environments and needs. The evolution of HSS is not based on DNA, but rather on a complex interplay between behaviour, learning and individual goals. Here we present some distinguishing aspects of HSS.

23.3.1 Rapid Change

A feature of HSS is the speed at which reorganisations can occur. Revolutions in social organisation can take place within the lifetime of a single individual. Hence, although HSS often show stable patterns over long periods, rapid change is possible. The ability to respond rapidly would appear to be a desirable property in rapidly changing information system environments; however, for engineering purposes, one must ensure that such fast changes (unlike revolutions!) can be both predicted and controlled.

23.3.2 Non-Darwinian Evolution

HSS do not evolve in a Darwinian fashion. Cultural and social evolution is not mediated by random mutations and selection of some base replicator over vast time periods, but rather follows a kind of collective learning process. That is, the information storage media supporting the change by learning—and hence (as noted above), both the mechanisms for change and their time scale—are very different from those of Darwinian evolution. Individuals within HSS can learn both directly from their own parents (vertical transmission), from other members of the older generation (diagonal transmission) or from their peers (horizontal transmission). Hence, new cultural traits (behaviours, beliefs, skills) can be propagated quickly through a HSS. This can be contrasted with simple Darwinian transmission in which, typically, only vertical transmission of genetic information is possible. Although it is possible to characterise certain processes of cultural evolution based on the fitness of cultural replicators (Boyd and Richerson 1985) or memes (Dawkins 1976), it is important to realise such replicators are not physical, like DNA, but part of a socio-cognitive process—passing through human minds—and may follow many kinds of selective process (Lumsden and Wilson 1981). The problems of using the idea of biological evolution in the social sciences are discussed in more detail in Chap. 21 (Chattoe-Brown and Edmonds 2017) of this volume.

23.3.3   Stable Under Internal Conflict

HSS exist because individuals need others to achieve their aims and goals. Production in all HSS is collective, involving some specialisation of roles. In large modern and post-modern HSS, roles are highly specialised, requiring large and complex coordination and redistribution methods. However, although HSS may sometime appear well integrated, they also embody inherent conflicts and tensions between individual and group goals. What may be in the interests of one individual or group may be in direct opposition to another. Hence, HSS embody and mediate conflict on many levels.

This aspect is highly relevant to distributed and open information systems. A major shift from the closed monolithic design approach is the need to deal with and tolerate inevitable conflicts between subcomponents of a system. For example, different users may have different goals that directly conflict. Some components may want to destroy the entire system. In open systems this behaviour cannot be eliminated and hence needs in some way to be tolerated.

23.3.4   Only Partial Views and Controversy

Although HSS are composed of goal-directed intelligent agents, there is little evidence that individuals or groups within them have a full view or understanding of the HSS. Each individual tends to have a partial view often resulting from specialisation within, and complexity of, the system. Such partial views, often dictated by immediate material goals, may have a normative (how things "should" be) character rather than a more scientific descriptive one (how things "are"). Consequently, the ideas that circulate within HSS concerning the HSS itself tend to take on an "ideological" form. Given this, social theories are rarely as consensual as those from biological sciences. Thus, social theories include a high degree of controversy, and they lack the generally accepted foundational structure found in our understanding of biology. However, from an information systems perspective, such controversy is not problematic: we do not care if a given social theory is true for HSS or not; we only care if the ideas and mechanisms in the theory can be usefully applied in information systems. This last point, of course, also holds for controversial theories from biology as well (e.g. Lamarckian evolution).

23.3.5   Trust and Socially Beneficial Norms

In trying to understand the stability of socially functional behaviour, much work within the social sciences has focused on the formation and fixation of "norms" of behaviour. Many researchers working with multi-agent systems (MAS) have attempted to create artificial versions of norms to regulate MAS behaviours— although much of these have not been based on theories from HSS (although see Conte and Paolucci 2002). Certainly the establishment and stability of beneficial norms (such as not cheating one's neighbour) is a desirable property visible in all stable HSS (Hales 2002). This point (the existence and power of norms) is of course closely related to the previous point, which notes that norms can influence understanding and perception.

It is widely agreed that, in HSS, many observed behaviours do not follow the same pattern as would be expected from simple Darwinian evolution or individual "rational" behaviour—in the sense of maximising the chance of achieving individual goals. Behaviour is often more socially beneficial and cooperative or altruistic, generally directed towards the good of the group or organisation within which the individual is embedded. (We note the widespread appearance of altruistic behaviour among many species of social mammals—such that, once again, we speak here of a difference in degree between HSS and other social animals.) Many theories and mechanisms have been proposed by social scientists for this kind of behaviour (Axelrod 1984), with many of these formalised as computer algorithms; furthermore, several of these have already been translated for use in information systems (Cohen 2003; Hales and Edmonds 2005).

23.3.6   Generalised Exchange and Economics

Almost all HSS evidence some kind of generalised exchange mechanisms (GEM) —i.e. some kind of money. The emergence of GEM allows for coordination through trade and markets. That is, collective coordination can occur where individual entities (individuals or firms) behave to achieve their own goals. It is an open (and perhaps overly simplified) question whether certain norms are required to support GEM or, rather, most norms are created via economic behaviour within GEM (Edmonds and Hales 2005). Certainly, the formation and maintenance of GEM would be an essential feature of any self-organised economic behaviour within information systems—currently many information systems work by assuming an existing GEM a priori, i.e. they are parasitic on HSS supplying the trust and norms required. Such systems require trusted and centralised nodes before they can operate because they do not emerge such nodes in on-going interaction. However, given that GEM exist, a huge amount of economic theory, including new evolutionary economics and game theory, can be applied to information systems.

23.4   What Can Social Simulation Offer the Designer?

Social simulation work has the potential to offer a number of insights that can be applied to aid design of distributed computer systems. Social simulators have had no choice but to start from the assumption of open systems composed of autonomous agents—since most social systems embody these aspects. In addition, much social simulation work is concerned with key aspects of self-* systems such as:

- *Emergence and self-organisation*: understanding the micro-to-macro and the macro-to-micro link. Phenomena of interest often result from bottom-up processes that create emergent structures that then constrain or guide (top-down) the dynamics of the system.
- *Cooperation and trust*: getting disparate components to "hang together" even with bad guys around. In order for socially integrated cooperation to emerge, it is generally necessary to employ distributed mechanisms to control selfish and free-riding behaviour. One mechanism for this is to use markets (see Chap. 25 in this volume, Rouchier 2017) but there are other methods.
- *Evolving robust network structure*: constructing and maintaining functional topologies robustly. Distributed systems often form dynamic networks in which the maintenance of certain topological structures improves system level performance.
- *Constructing adaptive/evolutionary heuristics* rather than rational action models. Models of both software and user behaviour in distributed systems are based on implicit or explicit models. Traditional approaches in game theory and economics have assumed rational action, but these are rarely applicable in distributed systems.

These key aspects have import into two broad areas of systems design. Firstly, simulation models that produce desirable properties can be adapted into distributed system protocols that attempt to reproduce those properties. Secondly, models of agent behaviour, other than rational action approaches, can be borrowed as models of user behaviour in order to test existing and new protocols.

Currently, however, it is an open question as to how results obtained from social simulation models can be productively applied to the design of distributed information systems. There is currently no general method whereby desirable results from a social simulation model can be imported into a distributed system. It is certainly not currently the case that techniques and models can be simply "slotted into" distributed systems. Extensive experimentation and modification is required. Hence, in this chapter we give specific examples from P2P systems where such an approach has been applied.

23.5   What About Agent-Orientated Design Approaches?

Multi-agent system (MAS) design approaches have previously been proposed (Wooldridge and Jennings 1995), which attempt to address some of the design issues raised by open systems. Those approaches start with a "blank sheet" design approach rather than looking for biological or social inspiration. The focus therefore has tended to be on logical foundations, proof, agent languages and communication protocols. For example, the BDI agent framework starts from the assumption that agents within a system follow a particular logical architecture based on "folk psychological" cognitive objects—such as beliefs or intentions (Rao and Georgeff 1991). However, such approaches have difficulty scaling to large societies with complex interactions particularly where the effects of emergence and self-organisation are important. A more recent approach within MAS work has been to look towards self-organising approaches using simulation to capture processes of emergence (Brueckner et al. 2006). In this work heavy use has been made of biological and socially inspired approaches.

23.6   Examples of Socially Inspired P2P Systems

Here we give very brief outlines of some P2P protocols that have been directly inspired by social simulation models. While P2P systems have not been the only distributed systems that benefited from social inspiration, we have focused on this particular technology because it is currently, at the time of writing, a very active research area and increasingly widely deployed on the Internet.

23.6.1   Reciprocity-Based BitTorrent P2P System

BitTorrent (Cohen 2003) is an open P2P file sharing system that draws directly from the social simulation work of Robert Axelrod (1984) on the evolution of cooperation. The protocol is based on a form of the famous Tit-For-Tat (TFT) strategy popularised by Axelrod computer simulation tournaments. Strategies were compared by having agents play the canonical prisoner's dilemma (PD) game.

The PD game captures a social dilemma in the form of a minimal game in which two players each select a move from two alternatives (either to cooperate or defect) and then each player receives a score (or pay-off). If both players cooperate, then both get a reward pay-off (R). If both defect they are punished, both obtaining pay-off P. If one player selects defect and the other selects cooperate, then the defector gets T (the "temptation"), and the other receives S (the "sucker"). When these pay-offs, which are numbers representing some kind of desirable utility (e.g. money), obey the following constraints $T > R > P > S$ and $2R > T + S$, then we say the game represents a prisoner's dilemma. When both players cooperate, this maximises the collective good, but when one player defects and another cooperates, this represents a form of free riding with the defector gaining a higher score (T) at the expense of the cooperator (S).

Axelrod asked researchers to submit computer programs to a "tournament" where they repeatedly played the PD against each other accumulating pay-offs. The result of the tournaments was that a simple strategy, TFT, did remarkably well against the majority of other submitted programs—although other strategies can also survive within the complex ecology that occurs when there is a population of competing strategies.

TFT operates in environments where the PD is played repeatedly with the same partners for a number of rounds. The basic strategy is simple: a player starts by cooperating and then in subsequent rounds copies the move made in the previous round by its opponent. This means defectors are punished in the future: the strategy relies on future reciprocity. To put it another way, the "shadow" of future interactions motivates cooperative behaviour in the present. In many situations this simple strategy can outperform pure defection.

In the context of BitTorrent, the basic mechanism is simple: files are split into small chunks (about 1 MB each) and downloaded by peers, initially, from a single hosting source. Peers then effectively "trade" chunks with each other using a TFT-like strategy—i.e. if two peers offer each other a required chunk, then this equates to mutual cooperation. However, if either does not reciprocate, then this is analogous to a defect, and the suckered peer will retaliate in future interactions.

The process is actually a little more subtle because each peer is constantly looking at the upload rate/download rate from each connected peer in time—so it does not work just by file chunk but by time unit within each file chunk. While a file is being downloaded between peers, each peer maintains a rolling average of the download rate from each of the peers it is connected to. It then tries to match its uploading rate accordingly. If a peer determines that another is not downloading fast enough, then it may "choke" (stop uploading) to that other. Figure 23.1 shows a schematic diagram of the way the BitTorrent protocol structures population interactions.

Additionally, peers periodically try connecting to new peers randomly by uploading to them—testing for better rates. This means that if a peer does not

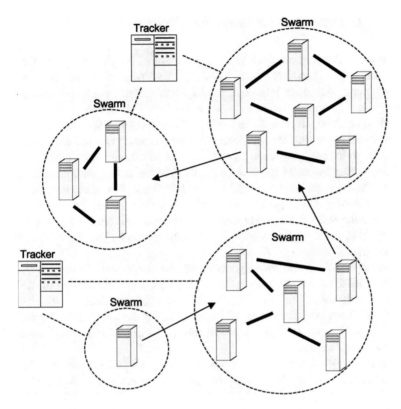

Fig. 23.1 A schematic of a portion of a BitTorrent system. The trackers support swarms of peers each downloading the same file from each other. Thick lines indicate file data. They are constantly in flux due to the application of the TFT-like "choking" protocol. Trackers store references to each peer in each supported swarms. It is not unknown for trackers to support thousands of swarms and for swarms to contain hundreds of peers. Arrows show how peers might, through user intervention, move between swarms. Generally at least one peer in each swarm would be a "seeder" that holds a full copy of the file being shared

upload data to other peers (a kind of defecting strategy), then it is punished by other peers in the future (by not sharing file chunks)—hence, a TFT-like strategy based on punishment in future interactions is used.

Axelrod used the TFT result to justify sociological hypotheses such as understanding how fraternisation broke out between enemies across the trenches of WW1. Cohen has applied a modified form of TFT to produce a file sharing system resistant to free riding. However, TFT has certain limitations, it requires future interactions with the same individuals, and each has to keep records of the last move made by each opponent. Without fixed identities it is possible for hacked clients to cheat BitTorrent. Although it appears that widespread cheating has not actually spread in the population of clients. It is an open question as to why this might be (but see (Hales and Patarin 2006) for a hypothesis).

23.6.2 Group Selection-Based P2P Systems

Recent work, drawing on agent-based simulations of cooperative group formation based on "tags" (social labels or cues) and dynamic social networks suggests a mechanism that does not require reciprocal arrangements but can produce cooperation and specialisation between nodes in a P2P (Riolo et al. 2001; Hales and Edmonds 2005). It is based on the idea of cultural group selection and the well-known social psychological phenomena that people tend to favour those believed to be similar to themselves—even when this is based on seemingly arbitrary criteria (e.g. supporting the same football team). Despite the rather complex lineage, like TFT, the mechanism is refreshingly simple. Individuals interact in cliques (subsets of the population). Periodically, if they find another individual who is getting higher utility than themselves, they copy them—changing to their clique and adopting their strategy. Also, periodically, individuals form new cliques by joining with a randomly selected other.

Defectors can do well initially, suckering the cooperators in their clique—but ultimately all the cooperators leave the clique for pastures new, leaving the defectors all alone with nobody to free ride on. Those copying a defector (who does well initially) will also copy their strategy, further reducing the free-riding potential in the clique. So a clique containing any free riders quickly dissolves, but those containing only cooperators grow.

Given an open system of autonomous agents, all cliques will eventually be invaded by a free rider who will exploit and dissolve the clique. However, so long as other new cooperative cliques are being created, cooperation will persist in the overall population. In the context of social labels or "tags", cliques are defined as those individuals sharing particular labels (e.g. supporting the same football team). In the context of P2P systems, the clique is defined as all the other peers each peer is connected to (its neighbourhood), and movement between cliques follows a process of network "rewiring".

Through agent-based simulation, the formation and maintenance of high levels of cooperation in the single round PD and in a P2P file sharing scenario has been demonstrated (Hales and Edmonds 2005). The mechanism appears to be highly scalable with zero scaling cost—i.e. it does not take longer to establish cooperation in bigger populations. Figure 23.2 shows the evolution of cooperative clusters within a simulated network of peer nodes. A similar approach was presented by Hales and Arteconi (2006) that produced small-world connected cooperative networks rather than disconnected components.

In addition to maintaining cooperation between nodes in P2P, the same group selection approach has been applied to other areas such as the coordination of robots in a simulated warehouse scenario and to support specialisation between nodes in a P2P job sharing system (Hales and Edmonds 2003; Hales 2006).

(a) Grouping before cooperation   (b) Cooperation spreading

(c) Giant cooperative component breaks apart   (d) Cooperative groups are formed

Fig. 23.2 Evolution of the connection network between nodes playing the prisoner's dilemma. From an initially random topology composed of all nodes playing the defect strategy (*dark shaded nodes*), components quickly evolve, still containing all defect nodes (**a**). Then a large cooperative component emerges in which all nodes cooperate (**b**). Subsequently the large component begins to break apart as defect nodes invade the large cooperative component and make it less desirable for cooperative nodes (**c**). Finally an ecology of cooperative components dynamically persists as new components form and old components die (**d**). Note: the cooperative status of a node is indicated by a light shade

23.6.3   Segregation-Based P2P Systems

The model of segregation proposed by Thomas Schelling is well known within social simulation (Schelling 1969, 1971). The model demonstrates how a macro-structure of segregated clusters or regions robustly emerges from simple local behaviour rules. Schelling's original model consists of agents on a two-dimensional

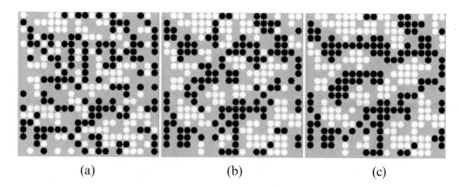

(a)	(b)	(c)

Fig. 23.3 Example run of the Schelling segregation model. From an initially random distribution of agents (**a**) clusters emerge over successive iterations (**b**) and (**c**). In this example the satisfaction proportion is 0.5, meaning that agents are unsatisfied if less than 50% of neighbours in the grid are a different colour (taken from Edmonds and Hales (2005))

grid. Each grid location can hold a single agent or may be empty. Each agent maintains a periodically updated satisfaction function. Agents take one of two colours that are fixed. An agent is said to be satisfied if at least some proportion of adjacent agents on the grid have the same colour; otherwise, the agent is said to be not satisfied. Unsatisfied agents move randomly in the grid to a free location. The main finding of the model is that even if the satisfaction proportion is very low, this still leads to high levels of segregation by colour—i.e. large clusters of agents emerge with the same colour. Figure 23.3 shows an example run of the model in which clusters of similar colours emerge over time.

The results of the segregation model are robust even when nodes randomly leave and enter the system—the clusters are maintained. Also agents in the segregation model only require local information in order to decide on their actions. These properties are highly desirable for producing distributed information systems, and therefore it is not surprising that designs based on the model have been proposed.

Sing and Haahr (2006) propose a general framework for applying a modified form of Schelling's model to topology adoption in a P2P network. They show how a simple protocol can be derived that maintains a "hub-based" backbone topology within unstructured networks. Hubs are nodes in a network that maintain many links to other nodes. By maintaining certain proportions of these within networks, it is possible to improve system performance for certain tasks. For many tasks linking the hubs to form a backbone within the network can further increase performance. For example, the Gnutella[1] file sharing network maintains high-bandwidth hubs called "super-peers" that speed file queries and data transfer between nodes. Figure 23.4 shows an example of a small network maintaining a hub backbone.

In the P2P model nodes represent agents, and neighbours are represented by explicit lists of neighbour links (a so-called view in P2P terminology). Nodes adapt

---

[1]http://www.gnutella.com.

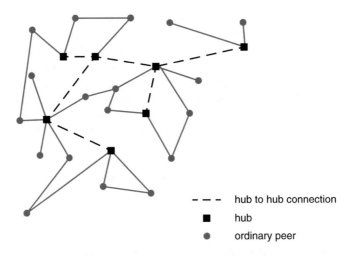

Fig. 23.4 An example of a hub-based peer-to-peer topology. Hubs are linked and perform specialist functions that improve system performance (taken from Singh and Haahr (2006))

their links based on a satisfaction function. Hub nodes (in the minority) are only satisfied if they have at least some number of other hubs in their view. Normal nodes are only satisfied if they have at least one hub node in their view. Hence different node types use different satisfaction functions and exist in a network rather than lattice. It is demonstrated via simulation that the network maintains a stable and connected topology supporting a hub backbone under a number of conditions, including dynamic networks in which nodes enter and leave the network over time.

The approach presented by Sing and Haahr is given as a general approach (a template design pattern) that may be specialised to other P2P application areas rather than just self-organising hub topologies. For example, they apply the same pattern to decrease bandwidth bottlenecks and increase system performance of a P2P by clustering similar nodes based on bandwidth capacity (Singh and Haahr 2004).

23.7   Possible Future Research

In the following sections, we give a brief outline of some promising possible areas related to socially inspired distributed systems research.

23.7.1   Design Patterns

Social simulators and distributed systems researchers currently constitute very different communities with different backgrounds and goals. A major problem for moving knowledge between these disciplines is the different language, assumptions and outlets used by them. One promising approach for communicating techniques from social simulation to distributed systems designers is to develop so-called design patterns which provide general templates of application for given techniques. This approach has been influential within object-oriented programming, and recently biologically inspired approaches have been cast as design patterns (Gamma et al. 1995; Babaoglu et al. 2006). Design patterns are not formal and need not be tied to a specific computer language but rather provide a consistent framework and nomenclature in which to describe techniques that solve recurrent problems the designer may encounter. At the time of writing, few, if any, detailed attempts have been made to present techniques from social simulation within a consistent framework of design patterns.

23.7.2 The Human in the Loop: Techno-Social Systems

Most distributed and open systems function via human user behaviour being embedded within them. In order to understand and design such systems, some model of user behaviour is required. This is particularly important when certain kinds of user intervention are required for the system to operate effectively. For example, for current file sharing systems (e.g. BitTorrent) to operate, users are required to perform certain kinds of altruistic actions such as initially uploading new files and maintaining sharing of files after they have been downloaded (so-called seeding). Web2.0 systems often require users to create, upload and maintain content (e.g. Wikipedia). It seems that classical notions of rational action are not appropriate models of user behaviour in these contexts. Hence, explicitly or implicitly, such distributed systems require models of user behaviour which capture, at some level, realistic behaviour. Such systems can be viewed as techno-social systems—social systems that are highly technologically mediated.

One promising method for understanding and modelling such systems is to make use of the participatory modelling approach discussed in Chap. 12 (Barreteau et al. 2017). In such a system, user behaviour is monitored within simulations of the technical infrastructure that mediates their interactions. Such an approach can generate empirically informed and experimentally derived behaviour models derived from situated social interactions. This is currently, at the time of writing, an underdeveloped research area.

Interestingly, from the perspective of distributed systems, if it is possible to model user behaviour at a sufficient level of detail based on experimental result, then certain aspects of that behaviour could be incorporated into the technological infrastructure itself as protocols.

23.7.3   *Power, Leadership and Hierarchy*

A major area of interest to social scientists is the concept of power—what kinds of process can lead to some individuals and groups becoming more powerful than others? Most explanations are tightly related to theories of inequality and economic relationships; hence, this is a vast and complex area.

Here we give just a brief very speculative sketch of recent computational work, motivated by sociological questions, that could have significant import into understanding and engineering certain kinds of properties (e.g. in peer-to-peer systems), in which differential power relationships emerge and may, perhaps, be utilised in a functional way. See Chap. 27 in this volume (Geller and Moss 2017) for a detailed overview of modelling power and authority in social simulation.

Interactions in human society are increasingly seen as being situated within formal and informal networks (Kirman and Vriend 2001). These interactions are often modelled using the abstraction of a game capturing interaction possibilities between linked agents (Zimmermann et al. 2001). When agents have the ability to change their networks based on past experience and some goals or predisposition, then, over time, networks evolve and change.

Interestingly, even if agents start with more-or-less equal endowments and freedom to act, and follow the same rules, vastly unequal outcomes can be produced. This can lead to a situation in which some nodes become objectively more powerful than other nodes through topological location (within the evolved network) and exploitative game interactions over time.

Zimmermann et al. (2001) found this in their simulations of agents playing a version of the prisoner's dilemma on an evolving network. Their motivation and interpretation is socio-economic: agents accumulate "wealth" from the pay-offs of playing games with neighbours and make or break connections to neighbours based on a simple satisfaction heuristic similar to a rule discussed in Kirman (1993).

Figure 23.5 shows an example of an emergent stable hierarchical network structure. Interestingly, it was found that, over time, some nodes accumulate large amounts of "wealth" (through exploitative game behaviour) and other nodes become "leaders" by being at the top of a hierarchy. These unequal topological and wealth distributions emerge from simple self-interested behaviour within the network. Essentially, leaders, through their own actions, can rearrange the topology of the network significantly, whereas those on the bottom of the hierarchy have little "topological power".

The idea of explicitly recognising the possibility of differential power between subunits in self-* systems and harnessing this is an idea rarely discussed in engineering contexts but could offer new ways to solve difficult coordination problems.

Considering P2P applications, one can envisage certain kinds of task in which differential power would be required for efficient operation. Consider, e.g. two nodes negotiating an exchange on behalf of their group or follower nodes. This might

Fig. 23.5 Forms of "hierarchy", "leadership" and unequal wealth distribution have been observed to emerge in simulated interaction networks (from Zimmermann et al. (2001)). Nodes play PD-like games with neighbours and break connections based on a simple satisfaction rule. Hierarchies are produced in which some nodes are more connected and hence can affect the network dramatically by their individual actions—a form of "topological power"

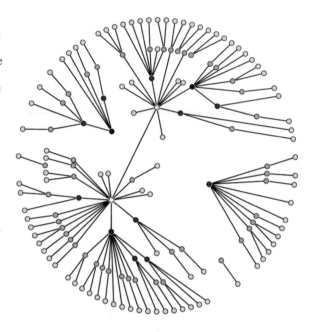

be more efficient than individual nodes having to negotiate with each other every time they wished to interact. Or consider a node reducing intra-group conflict by imposing a central plan of action.

We mention the notion of engineering emergent power structures, briefly and speculatively here, because we consider power to be an under-explored phenomenon within evolving information systems. Agents, units or nodes are often assumed to have equal power. It is rare for human societies to possess such egalitarian properties, and perhaps many self-*-like properties are facilitated by the application of unequal power relationships. We consider this a fascinating area for future work.

23.8   Summary

We have introduced the idea of social inspiration for distributed systems design and given some specific examples from P2P systems. We have argued that social simulation work can directly inform protocol designs. We have identified some of the current open issues and problem areas within this research space and pointed out promising areas for future research.

Increasingly, distributed systems designers are looking at self-organisation as a way to address their difficult design problems. In addition, there has been an explosive growth in the use of such systems over the Internet, particularly with the high take-up of peer-to-peer systems. The idea of "social software" and the so-called "Web2.0" approach also indicate that social processes are becoming increasingly central to systems design. We believe that the wealth of models and methods produced with social simulation should have a major impact in this area over the coming decade.

Further Reading

The interested reader could look at the recent series of the IEEE Self-Adaptive and Self-Organising systems (SASO) conference proceedings, which started in 2007 and have been organised annually (http://www.saso-conference.org/). To get an idea of current work in social simulation, a good place to start is the open access online Journal Artificial Societies and Social Simulation (JASSS); see http://jasss.soc.surrey.ac.uk/JASSS.html.

References

Axelrod, R. (1984). *The evolution of cooperation*. New York: Basic Books.

Babaoglu, O., et al. (2006). Design patterns from biology for distributed computing. *ACM Transactions on Autonomous and Adaptive Systems, 1*(1), 26–66.

Babaoglu, O., et al. (Eds.). (2005). *Self-star properties in complex information systems: Conceptual and practical foundations, Lecture Notes in Computer Science, 3460*. Berlin: Springer.

Barreteau, O. et al. (2017). *Participatory approaches*, Chap. 11 in this volume.

Boyd, R., & Richerson, P. (1985). *Culture and the evolutionary process*. Chicago: University of Chicago Press.

Engineering self-organising systems. In Brueckner, S., Di Marzo Serugendo, G., Hales, D., & Zambonelli, F. (Eds.) (2006). *Proceedings of the 3rd Workshop on Engineering Self-Organising Applications (EOSA'05), Lecture Notes in Artificial Intelligence, 3910*. Berlin: Springer.

Chattoe-Brown, E., & Edmonds, B. (2017). *Evolutionary mechanisms*, Chap. 21 in this volume.

Cohen, B. (2003, June 5–6). Incentives build robustness in BitTorrent. In J. Chuang, & R. Krishnan (Eds.), *Proceedings of the First Workshop on the Economics of Peer-2-Peer Systems*, 2003, Berkley, CA.http://www2.sims.berkeley.edu/research/conferences/p2pecon/papers/s4-cohen.pdf

Conte, R., & Paolucci, M. (2002). *Reputation in artificial societies: Social beliefs for social order*. Amsterdam: Kluwer.

Dawkins, R. (1976). *The selfish gene*. Oxford: Oxford University Press.

Edmonds, B., & Hales, D. (2005). Computational simulation as theoretical experiment. *Journal of Mathematical Sociology, 29*(3), 209–232.

Gamma, E., Helm, R., Johnson, R., & Vlissides, J. (1995). *Design patterns: Elements of reusable object-oriented software*. Reading, MA: Addison-Wesley.

Geller, A., & Moss, S. (2017). *Modeling power and authority: An emergentist view from Afghanistan*, Chap. 27 in this volume.

Hales, D., & Edmonds, B. (2003, July 2003). Evolving social rationality for MAS using "Tags". In J. S. Rosenchein et al. (Eds.), *Proceedings of the 2nd International Conference on Autonomous Agents and Multi-agent Systems, (AAMAS 2003)*, Melbourne (pp. 497–503). New York: ACM Press.

Hales, D., & Patarin, S. (2006). How to cheat BitTorrent and why nobody does. In J. Jost, F. Reed-Tsochas, & P. Schuster (Eds.), *ECCS2006, Proceedings of the European Conference on Complex Systems, Towards a Science of Complex Systems*.http://www.cabdyn.ox.ac.uk/complexity_PDFs/ECCS06/Conference_Proceedings/PDF/p19.pdf

Hales, D. (2002). Group reputation supports beneficent norms. *Journal of Artificial Societies and Social Simulation, 5*(4). http://jasss.soc.surrey.ac.uk/5/4/4.html

Hales, D. (2006). Emergent group-level selection in a peer-to-peer network. *Complexus, 3*, 108–118.

Hales, D., & Arteconi, S. (2006). SLACER: A self-organizing protocol for coordination in P2P networks. *IEEE Intelligent Systems, 21*(2), 29–35.

Hales, D., & Edmonds, B. (2005). Applying a socially-inspired technique (tags) to improve cooperation in P2P networks. *IEEE Transactions in Systems, Man and Cybernetics - Part A: Systems and Humans, 35*(3), 385–395.

Kirman, A. (1993). Ants, rationality and recruitment. *Quarterly Journal of Economics, 108*, 137–156.

Kirman, A. P., & Vriend, N. J. (2001). Evolving market structure: An ACE model of price dispersion and loyalty. *Journal of Economic Dynamics and Control, 25*(3/4), 459–502.

Lumsden, C., & Wilson, E. (1981). *Genes, mind and culture*. London: Harvard University Press.

Rao, A. S., & Georgeff, M. P. (1991). Modeling rational agents within a BDI-architecture. In R. Fikes, & E. Sandewall (Eds.), *Proceedings of Knowledge Representation and Reasoning* (KR&R-91) (pp. 473–484). San Mateo, CA: Morgan Kaufmann.

Riolo, R., Cohen, M., & Axelrod, R. (2001). Evolution of cooperation without reciprocity. *Nature, 414*, 441–443.

Rouchier, J. (2017). *Agent-Based Simulation as a Useful Tool for the Simulation of Markets*, Chap. 25 in this volume.

Schelling, T. (1969). Models of segregation. *American Economic Review, 59*, 488–493.

Schelling, T. (1971). Dynamic models of segregation. *Journal of Mathematical Sociology, 1*(1), 143–186.

Singh, A., & Haahr, M. (2004, 18–22 September). Topology adaptation in P2P networks using Schelling's model. In J. C. Oh, & D. Mosse (Eds.), *Proceedings of the Workshop on Games and Emergent Behaviours in Distributed Computing Environments, co-located with PPSN VIII*, Birmingham, UK.

Singh, A., & Haahr, M. (2006). Creating an adaptive network of hubs using Schelling's model. *Communications of the ACM, 49*(3), 69–73.

Staab, S., et al. (2005). Social networks applied. *IEEE Intelligent Systems, 20*(1), 80–93.

Wang, F.-Y., Carley, K. M., Zeng, D., & Mao, W. (2007). Social computing: From social informatics to social intelligence. *IEEE Intelligent Systems, 22*(2), 79–83.

Wooldridge, M., & Jennings, N. R. (1995). Intelligent agents: Theory and practice. *The Knowledge Engineering Review, 10*(2), 115–152.

Zimmermann, M. G., Eguiluz, V. M., & San Miguel, M. (2001). Cooperation, adaptation and the emergence of leadership. In A. Kirman & J. B. Zimmermann (Eds.), *Economics with heterogeneous interacting agents* (pp. 73–86). Berlin: Springer.

Chapter 24
Simulating Complexity of Animal Social Behaviour

Charlotte Hemelrijk

Abstract Complex social phenomena occur not only among humans, but also throughout the animal kingdom, from bacteria and amoebae to non-human primates. At a lower complexity they concern phenomena such as the formation of groups and their coordination (during travelling, foraging, and nest choice) and at a higher complexity they deal with individuals that develop individual differences that affect the social structure of a group (such as its dominance hierarchy, dominance style, social relationships and task division). In this chapter, we survey models that give insight into the way in which such complex social phenomena may originate by self-organisation in groups of beetle larvae, in colonies of ants and bumblebees, in groups of fish, and groups of primates. We confine ourselves to simulations and models within the framework of complexity science. These models show that the interactions of an individual with others and with its environment lead to patterns at a group level that are emergent and are not coded in the individual (genetically or physiologically), such as the oblong shape of a fish school, variable shape in bird flocks, specific swarming pattern in ants, the centrality of dominants in primates, patterns of exchange and of 'reconciliation' and the task division among bumble bees. The hypotheses provided by these models appear to be more parsimonious than usual in the number of adaptive traits and the degree of cognitive sophistication involved. With regard to the usefulness of these simulations, we discuss for each model what kind of insight it provides, whether it is biologically relevant, and if so, whether it is specific to the species and environment and to what extent it delivers testable hypotheses.

Why Read This Chapter?

To get an overview of simulation models aimed at understanding animal social behaviour, such as travelling, foraging, dominance or task division. The chapter also provides an analysis of the kinds of insight each simulation provides, how specific these insights are and whether they are testable.

C. Hemelrijk (✉)
Centre for Life Sciences, Groningen Institute of Evolutionary Life Sciences, University of Groningen, Nijenborgh 7, Groningen, 9747AG, The Netherlands
e-mail: c.k.hemelrijk@rug.nl

© Springer International Publishing AG 2017
B. Edmonds, R. Meyer (eds.), *Simulating Social Complexity*,
Understanding Complex Systems, https://doi.org/10.1007/978-3-319-66948-9_24

24.1   Introduction

Many complex social phenomena of human behaviour are also observed in animals. For instance, humans coordinate their movement while searching for the right restaurant. They also make and follow paths on lawns in between different university buildings in order to cross between them. This resembles the path marking and following behaviour of ants as they forage and select different food sources. Furthermore, a division of tasks is found in both human social organisations and in large colonies of social insects, such as honeybees. Human social relationships are diverse, and so are the social relationships of primates. In groups of humans and animals, competition results in stable relationships in which one individual consistently beats the other. Within a group, individuals can be ordered in a dominance hierarchy. Furthermore, dominant relationships are also found between groups and between classes of individuals, such as between the sexes. Further, societies may differ in their dominance style, such as whether they are egalitarian or despotic. In summary, despite their hugely inferior cognitive capacities, animals show a number of complex social phenomena that resemble those of humans. From this the question arises whether or not these complex phenomena originate by self-organisation in the same way for both humans and animals. Therefore, it is important to survey complex social behaviour in animals in addition to that of humans.

We divide the chapter in subsections dealing with different (groups of) complex phenomena ordered in increasing complexity: From group formation (which is simple in animals) and coordination under various circumstances (such as during travelling, foraging, selection of shelter and nest site) we continue to the social organisation of the group (its dominance hierarchy, dominance style, dominance classes, social relationships, personality style and task division).

We mainly confine ourselves to individual-based models that are spatially explicit. Individuals in the model are steered by behavioural rules that are fixed or that are based on parameters that change (compare self-organisation with(out) structural changes, Pfeifer and Scheier 1999); they react to their local environment only. The environment may contain only other individuals or may also contain food. Food may be continuously abundant, or it may be depleted and may regrow or not after being eaten.

In general, we discuss for each model whether it has led to a new perspective in the study of the phenomenon; whether this perspective is more 'parsimonious' than previous explanations in terms of cognitive sophistication and the number of specific behavioural adaptations; whether the new explanation concerns a general principle or is specific to a certain species or environment; and whether it produces hypotheses that can easily be tested in real animals. In the evaluation we also provide literature for further reading and indicate important areas for future modelling.

24.2   Group Formation and Coordination

Everywhere in nature groups are formed. They are formed among individuals of all kinds of species in all stages of their lives, and in a few cases groups contain several species. Groups may differ in size, longevity, the degree of heterogeneity and also in the way they are formed. We distinguish two processes of group formation, namely social and environmental processes. Social processes concern forms of social attraction, and environmental processes relate to attraction to food resources, shelter and the like. Below we discuss models for each process separately and for their combination.

24.2.1   Social Attraction

Attraction is often mediated through visual and chemical cues. Visual attraction is important in many species of fish and birds. Chemical attraction (through pheromones) occurs among single-celled individuals (such as bacteria and amoebae), ants and beetle larvae.

One of the simplest aggregation processes has been studied in certain experiments and models of cluster formation of beetle larvae (Camazine et al. 2001). In their natural habitat, these beetle larvae live in oak trees, and grouping helps them to overcome the production of toxic resin by the host tree. They emit a pheromone and move towards higher concentrations of this chemical. In the experimental setup, larvae are kept in a Petri dish. In the model, the behavioural rules of following a gradient are represented in individuals that roam in an environment that resembles a Petri dish. It appears that both in the model and in the experiment the speed of cluster growth and the final number and distribution of clusters depend on the initial distribution and density of the larvae. By means of both model and experimental data it has been shown that cluster growth can be explained by a positive feedback. A larger group emits more pheromone and therefore attracts more members. Consequently, its size increases in turn emitting more pheromone, etc. This process is faster at a higher density, because individuals meet each other more frequently. Thus clusters appear sooner. Furthermore, growth is faster at higher densities, because more individuals are available for growing.

The location of the clusters has been studied. After starting from a random distribution of individuals, a single cluster remains in the centre of the Petri dish. This location is due to the higher frequency with which individuals visit the centre than the periphery. Starting from a peripheral cluster, there is an attraction to both the peripheral cluster and the centre. Thus, there is a kind of competition between clusters to attract additional individuals. The final distribution of clusters may consist of one cluster (at the centre, at the periphery or at an intermediate location) or of two clusters (in the periphery and in the centre). The final distribution depends on self-organisation and on three factors, namely, the initial density of individuals, initial distribution of clusters, and whether there is social attraction. Although the model is devised for the specific experimental setup and species, the mechanisms of group formation and growth shown in the model are found in many animal species. For instance, similar patterns are observed in models and experiments of cockroaches.

Cockroaches aggregate by sensing the relative humidity. They tend to move towards lower humidity. Here larger groups grow faster, because individuals have a higher tendency to stop and form a group if they collide with a larger number of others. Besides, they rest longer if more individuals are resting close by. They use the relative humidity as a cue to estimate the number of individuals. A lower relative humidity correlates with a higher number of individuals close by Jeanson et al. (2005). Despite the different underlying process of social attraction (pheromonal, visual or based on relative humidity), pattern formation is similar to that of beetle larvae.

24.2.2 Foraging

A model that deals with group formation solely through environmental causes concerns the splitting and merging of groups as it is found in the so-called fission-fusion system of primates, in particular that of spider monkeys (Ramos-Fernández et al. 2006). It relates the pattern of group formation to various distributions of food, because particularly in these fission-fusion societies, the food distribution may have an effect on grouping. In the model, the food distribution is based on a distribution of resources in forests that follows an inverse power law as described by Enquist and co-authors (Enquist and Niklas 2001; Enquist et al. 1999). The foragers maximise their food intake by moving to larger food patches that are closer by (they minimise the factor distance divided by patch size). Further, they do not visit patches that they have visited before. Individuals have no social behaviour, and they meet in groups only because they accidentally visit the same food patch. For a distribution with both large and small trees (patches) in roughly equal numbers the model leads to a frequency distribution of subgroup sizes that resembles that of spider monkeys. Results hold, however, only if foragers have complete knowledge of the environment in terms of patch size and distance. The resemblance is not found if individuals only know (a random) part of the environment. Furthermore, if they have complete knowledge, individuals meet with certain partners more often than expected if the choice of the food patch is a random choice. In this way, certain networks of encounters develop. Another serious shortcoming of the model is that food can only be depleted: there is no food renewal, and individuals do not return to former food patches. The main conclusion is that the ecological environment influences grouping: Ecologies that differ in the variation of their patch sizes (high, medium or low variance) also differ in the distribution of subgroup sizes. This model delivers mainly a proof of principle.

24.2.3   Combination of Social and Environmental Processes

The relation between ecology and subgroup formation is also shown in a simulation study of orang-utans in the Ketambe forest in Indonesia (te Boekhorst and Hogeweg 1994a). The predictions for and resemblance to empirical data that are delivered in this model (te Boekhorst and Hogeweg 1994a) are more specific than those in the model of spider monkeys discussed above (Ramos-Fernández et al. 2006). Here, realistic patterns of grouping arise from simple foraging rules in interaction with the structure of the forest and in the presence of two social rules. The environment and population are built to resemble the Ketambe area in the composition of the community and the size of the habitat. Two main categories of food trees are distinguished, figs (very large and non-seasonal) and fruit trees (small and seasonal). There are 480,000 trees of which 1200 bear fruit at the same time, once a year for ten consecutive days unless depleted earlier by the individuals. The crop size and spatial distribution of trees are specified: fig trees are clustered and fruit trees distributed randomly. Furthermore there are randomly distributed sources of protein. These are renewed immediately after being eaten. In the fruiting season extra individuals migrate into the area. With regard to the behavioural rules, individuals first search for a fig tree. If this is not found while moving approximately in the direction of others, because other individuals may indicate the presence of food on a distant fig tree, individuals look for fruits close by. Upon entering a tree, the individual feeds until satiated or the tree is emptied. Next the individual rests close by and starts all over again later on. A further social rule causes adult males to avoid each other.

The resulting grouping patterns in the model resemble those of real orang-utans in many respects: Temporary aggregations are found in the enormous fig trees. This happens because in these big trees, individuals may feed until satiated and then leave separately. However, when feeding in the much smaller fruit trees, food is insufficient, and therefore individuals move to the next tree. This causes them to travel together. Thus, travelling bands develop mainly in the fruit season. In this season parties are larger than when fruit is scarce. Groupings emerge as a consequence of simple foraging rules in interaction with the forest structure. Thus, the model makes clear that differences between forest structures will lead to different grouping patterns in apes. This is a parsimonious explanation, because it is not necessary to think in terms of costs and benefits of sociality to explain these grouping patterns, rather they arise as a side effect of feeding behaviour. The empirical data analysis and ideas for the model of orang-utans were inspired by findings in another model on grouping in chimpanzees.

This model of chimpanzees concerns both foraging and social attraction and is meant to explain group formation, in particular the fission-fusion society (te Boekhorst and Hogeweg 1994b). It offers new explanations for the relatively solitary life of females and the numerous and large subgroups of males. Subgroups of males had so far been supposed to form in order to join in defending the community. To explain such cooperation, males were believed to be genetically related to each other. Further, the solitary life of females was attributed to competition for food.

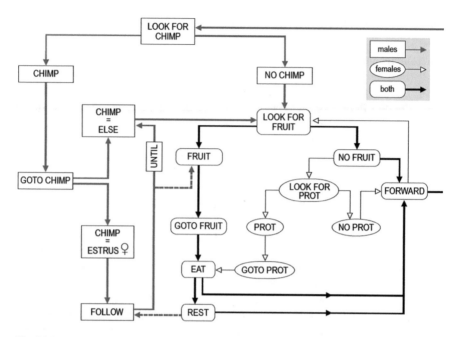

Fig. 24.1 Behavioural rules of artificial chimpanzees (te Boekhorst and Hogeweg 1994b)

However, the model shows that their patterns of grouping may arise in an entirely different way: they may result from a difference between the sexes in diet and in the priority of foraging versus reproduction. The model resembles a certain community of chimpanzees in Gombe in its community composition (number of males and females, 11 and 14, respectively, and the number of females that are synchronously in oestrus) and its habitat size (4 km²), its number of trees, the number of trees that bear fruit at the same time, the length of their fruit bearing, their crop size and the speed of their depletion. The number of protein resources is modelled more arbitrarily, approximating such forests. With regard to their behavioural rules, individuals of both sexes look for fruit and when satiated have a rest close to the tree (Fig. 24.1). If not satiated, they move towards the next fruit tree. They continue to do this until they are satiated. If there is no fruit on a specific tree, females (but not males) look for proteins before searching another fruit tree. Males have one priority over finding food: finding females. Whenever they see a chimpanzee, they approach it to investigate whether it is a female in oestrus. If this is the case, males in the model follow her until she is no longer in oestrus.

In the model, patterns of grouping resemble those of real chimpanzees. Male groups emerge because all males have the same diet, which differs from that of females, and because they will follow the same female in oestrus together. Furthermore, in the model, as in real chimpanzees, males appear to travel over longer distances per day than females. They do so because they are travelling in larger groups, and this leads to faster depletion of food sources, and therefore they have to visit a larger number of trees in order to become satiated. This interconnection between the number of trees visited, group size and distance covered led to the hypothesis for orang-utans that different kinds of groupings (aggregation and travel bands) originate in different trees (fig and fruit). Note that these results were robust for a large range of different ecological variables and different compositions of the community. Here, a difference between the sexes in their diet and their priorities for sex and food appeared essential. With regard to the chimpanzee community, the authors conclude that, to explain its fission-fusion structure, the genetically based theory that kin-related males are jointly defending the community is not needed. In fact, in subsequent DNA studies no special genetic relatedness was found among cooperating male chimpanzees in Kibale (Goldberg and Wrangham 1997). Instead, chimpanzee-like party structures may emerge by self-organisation if chimpanzees search for food and for mates in a forest. Besides, the model can be used to explain the frequent bisexual groups observed in bonobos as being caused by their prolonged period of oestrus. Whereas these models have been controversial among primatologists for a long time, their usefulness is slowly becoming accepted (Aureli et al. 2008).

24.2.4 Group Coordination and Foraging

Groups of social insects, for instance ants, are remarkably efficient in foraging. They collectively choose a food source that is closer rather than one (of the same quality) that is further away (Deneubourg and Goss 1989). Their choice is made without comparing the distance to different food sources. Experiments and models show that ants use trail pheromones as a mechanism of collective 'decision-making': Individuals mark their path with pheromone and at crossings follow the path that is more strongly marked. As they return to the nest sooner when the food source is close by, they obviously imprint the shorter path more often with pheromones. This results in a positive feedback: as the shorter path receives stronger markings, it also receives more ants, etc. Thus, the interaction between ants and their environment results in the adaptive and efficient exploitation of food sources. The 'preference' for a food source nearby rather than one further away is a side effect of pheromonal marking. This marking also helps a single ant to find its way and may initially have been an adaptation to cause the ant to return to its nest. It is actually more than just that, because its intensity is adapted to the quality of the food source. The process of marking and following may also lead to mistakes. For instance, if a path is initially developed to a food source of low quality, and later on a food source of high quality is introduced elsewhere, the ants may remain fixated on the source of low quality even if it is further from the nest due to 'historical constraint'.

Army ants live in colonies of 200,000 individuals; they are virtually blind; they travel in a swarm with about 100,000 workers back and forth to collect food. Different species of army ants display highly structured patterns of swarming that may be species specific (Fig. 24.2a, b). For example, *Eciton burchellii* has a more

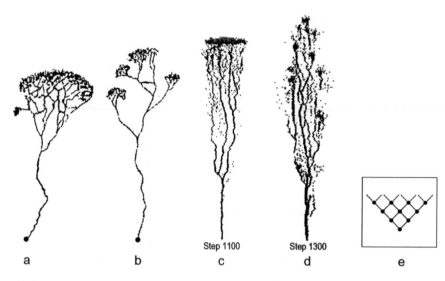

Fig. 24.2 Foraging patterns of two species of army ants, *E. burchellii* and *E. rapax*, empirical data and models. Empirical data: *E. burchellii* (**a**) and *E. rapax* (**b**); models: (**c**) a few food clumps, (**d**) frequent occurrence of single food items, and (**e**) network of nodes (after Deneubourg and Goss 1989)

dispersed swarm than *Eciton rapax*. Such species-specific differences in swarming are usually regarded as a separate adaptation, which is assumed to be based on corresponding differences in the underlying behavioural tendencies of coordination. Deneubourg and co-authors, however, give a simpler explanation (Deneubourg and Goss 1989; Deneubourg et al. 1998). In a model, they have shown that such markedly different swarming patterns may arise from a single-rule system of laying and following pheromone trails when ants are raiding food sources with different spatial distributions (Fig. 24.2c, d). The authors have shown this in a model in which 'artificial ants' move in a network of points (Fig. 24.2e) and mark their path with pheromones. When choosing between left and right, they prefer the more strongly marked direction. By introducing different distributions of food in the model (either uniformly distributed single insects or sparsely distributed colonies of insects), different swarming patterns arise from the interaction between the flow of ants heading away from the nest to collect food and the spatial distribution of the foragers returning with the food. These different swarm types are remarkably similar to those of the two species of army ants mentioned above (for empirical confirmation see Franks et al. 1991). Therefore, these different swarm forms reflect the variation in diet of the different species. Thus, the explanation of the model is more parsimonious than if we assume the different swarm forms to arise from a specific adaptation in rules of swarming. In summary, this model teaches us the effects of the environment on swarm coordination.

With regard to its evolution, natural selection shapes the necessary traits for the successful marking and following of trails depending on the size of the food source and other environmental characteristics (for an evolutionary model of this, see Solé et al. 2001).

24.2.5 Group Coordination in a Homogeneous Environment

Even in environments (such as the open sea, savannah and sky) that are virtually uniform without environmental structure, remarkable coordination is observed in the swarms of many animal species, e.g. of insects, fish, birds and ungulates. Coordination appears flexible even in swarms of a very large size (for instance, of up to ten million individuals in certain species of fish). Swarming behaviour has been modelled in several ways. The most simplistic representations of emergent phenomena have used partial differential equations. Slightly more complex behaviour has been obtained using particle-based models derived from statistical mechanics in physics (Vicsek et al. 1995). These models have been used to explain the phase transition of unordered to ordered swarms in locusts (Buhl et al. 2006). Yet the biologically most relevant results come from models wherein individuals coordinate with their local neighbours by following only three rules based on zones of perception (Fig. 24.3): They avoid neighbours that are close by (separation), align to others up to an intermediate distance (alignment) and approach those further away (cohesion). These models have been applied to describe herds of ungulates (Gueron et al. 1996), schools of fish (Couzin et al. 2002; Hemelrijk and Kunz 2005; Huth and Wissel 1992; Huth and Wissel 1994; Kunz and Hemelrijk 2003; for a review see Parrish and Viscido 2005) and swarms of birds (Hildenbrandt et al. 2010; Reynolds 1987).

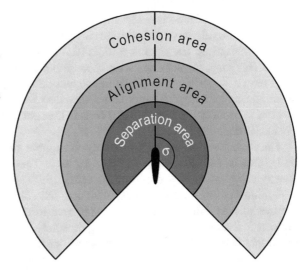

Fig. 24.3 Behavioural areas of avoidance (separation), alignment and attraction (cohesion) with a dead (blind) angle at the back (from Hemelrijk and Hildenbrandt 2008). The separation angle is indicated as σ

Through these models we obtain insight into a number of important biological aspects of swarming, which have mainly been related to schools of fish.

Firstly, we get insight into the *coordination* of schools. Schools coordinate with remarkable flexibility even into the absence of a *leader* and without a directional preference. The direction of their movement is merely the consequence of the location and heading of others in the school. With regard to the question whether there is a leader in a swarm, such a leader fish is supposed to be located at the front (Bumann and Krause 1993). However, neither in such models of fish schools nor in real swarms individuals appear to have a fixed location. Instead frontal locations are continuously switched in the model on average every 2 s, in real fish every 1.4 s (Gnatapogon elongates, Huth and Wissel 1994). Thus, there cannot be consistent leaders.

Furthermore, if a number of individuals have a directional preference (for instance, for certain food sources or breeding locations), but most of them do not, those with such a preference will automatically lead the school. Huse et al. (2002) showed that even if the percentage of individuals with a certain preferred direction is very small (though above 7%), this may influence the direction of the entire school.

If individuals prefer a different direction, for instance, because they aim to go to different food locations, the school may react differently depending on the degree to which two preferred directions differ (Couzin et al. 2005): If the directions differ little, the group will follow the average direction between the two ('a compromise'). If the angle between both directions is large, the group will either follow the direction that is preferred by the majority or in the absence of a 'convincing majority' it will randomly choose one of the two. In these examples of swarming the models help us to understand the processes that determine the direction in which a school is heading.

Secondly, we obtain insight into the *segregation* of individuals that differ in a certain trait. In a school, individuals may be segregated, for example, according to size. Usually this is attributed to an intentional or genetic preference for being near individuals of the same size or body form. Models show, however, that this segregation may also arise directly as a side effect of differences in body size without any preference (e.g. see Couzin et al. 2002; Hemelrijk 2005; Kunz and Hemelrijk 2003). This may, for instance, arise because larger individuals due to their larger body have a larger range at which they avoid others who are too close. Thus, by avoiding smaller individuals more often than the reverse, large individuals may end up at the periphery leaving the small ones in the centre (as has been found in water insects (Romey 1995)).

Thirdly, natural schools of fish show a number of traits that are believed to be helpful in the *protection against predators*: Their shape is oblong and their density is highest at the front. Bumann et al. (1997) argue that an *oblong form* and *high frontal density* protect against predation: the oblong shape reduces the size of the frontal area, where predators are supposed to attack and high frontal density protects individuals against approaching predators. Hemelrijk and co-authors have noted that it is unlikely that individual fish actively organise themselves so as to create these two patterns (Hemelrijk and Hildenbrandt 2008; Hemelrijk and Kunz

2005). Therefore, the authors studied in a model whether these patterns might arise by self-organisation as a side effect of their coordinated movements. This indeed appeared to be the case, and their emergence appeared to be robust (independent of school size and speed). These patterns come about because, during travelling, individuals avoid collisions mostly by slowing down (as they do not perceive others to be directly behind them, in the so-called blind angle, Fig. 24.3). Consequently, they leave a gap between their former forward neighbours, and subsequently these former neighbours move inwards to be closer together. Thus, the school becomes oblong.

Furthermore, when individuals fall back, a loose tail builds up, and this automatically leaves the highest density at the front. In the model it appears that larger schools are relatively more oblong because they are denser, and so more individuals fall back to avoid collision. Faster schools appear to be less oblong. This arises because fast individuals have greater difficulty to turn; thus, the path of the school is straighter, the school is more aligned (polarised) and therefore, fewer individuals fall back. Consequently, the core of a faster school is denser, and the tail is looser than they are in slower schools. Recent tests in real fish (mullets) confirm the specific relationships between group shape and its size, density and polarisation as found in the model (Hemelrijk et al. 2010). Although this indicates that the shape in real schools develops in a similar way, it is still necessary to investigate shape and frontal density in more species and to study the effects of different speeds on these traits.

Fourth, these models help us to understand why flocks of birds are more variable in their shapes than schools of fish (when they are not under attack). The cause of the variable shapes lies in their difference in locomotion, birds fly and fish swim (Hemelrijk and Hildenbrandt 2012). Flying behaviour implies that during turns, birds do neither speed up in the outer corner nor slowdown in the inner corner. Instead all individuals make a turn at the same constant speed and with the same curvature (Fig. 24.4, *top left*; Hemelrijk and Hildenbrandt 2012). This causes the shape of the flock to change relative to the movement direction with each turn. If a flock is wide in its shape and it turns under 90 degrees to the right, it will subsequently be oblong. Here, an individual that is situated at the left of the flock before the turn will be located at the rear of the flock after the turn. Thus individuals change location in the flock during turns. Furthermore, during turns individuals need to roll over their shoulder if they want to make the turn sharp (Fig. 24.4, *top right*). As a side effect they lose lift and consequently, move downwards a bit. The flock thus loses altitude temporarily and may change shape vertically (Fig. 24.4, *bottom*; Hemelrijk and Hildenbrandt 2012).

Fifth, in real animals predation and attacks on swarms result in a spectacular range of behavioural *patterns of evasion by schooling prey*. These patterns are supposed to confuse the predator. They have been labelled 'tight ball', 'bend', 'hourglass', 'fountain effect', 'vacuole', 'split', 'join' and 'herd'. They have been described for schools of several prey species and predators (Axelsen et al. 2001; Lee 2006; Nottestad and Axelsen 1999; Parrish 1993; Pitcher and Wyche 1983). Most of these patterns may be obtained in a model by simple behavioural rules of prey and predator (see e.g. Inada and Kawachi 2002). Many of the different

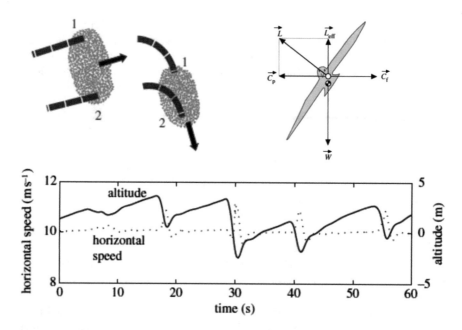

Fig. 24.4 Flock turning in *StarDisplay* (Hemelrijk and Hildenbrandt 2012). (*Top left*) Turning of a flock by 90° at almost fixed speed. (*Top right*) Loss of effective lift during turning, with L = lift, W = weight, L_{eff} = effective lift, C_p = centripetal force, C_f = centrifugal force. (*Bottom*) Loss of altitude while turning approximately each 10s during an interval of 60s

patterns of evasion result in self-organisation in models of schooling, which are built in such a way that upon detecting the predator, individuals compromise between their tendency to avoid the predator and to coordinate with their group members. Though these models do not exactly fit real data, they give us insight into how specific collective evasion patterns may arise.

One pattern of collective escape was not obtained in these models, the 'wave of agitation'. In flocks of starlings, it is observed as a dark band moving away from the predator, usually a Peregrine falcon (Procaccini et al. 2011). It was unknown what individual manoeuvre caused the wave. It could be a density wave, where individuals flee from danger and come temporarily closer together to other flock members, or an orientation wave, where individuals make a skitter movement, rolling sideward and back again, temporarily displaying a larger surface of their wing area. This cannot be distinguished in video recordings, because birds are too far away, at a distance of about 1 km. When trying out waves of each escape manoeuvre in a computational model, *StarDisplay*, it became apparent that the observation of a dark band moving over the flock surface happened only when a skittering escape movement was displayed, not when individuals were fleeing from danger by moving closer to other group members (Hemelrijk et al. 2015). Thus, the model delivers the hypothesis that in starling flocks the dark, moving bands

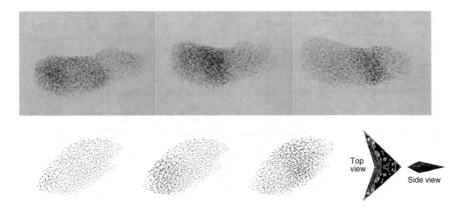

Fig. 24.5 Agitation waves in starling flocks. (*Top*) Agitation wave of real starlings (photo by Carere, Procaccini et al. 2011). (*Bottom, left*) Wave of agitation due to the zigzag-like manoeuvre in *Stardisplay* (black band moving to the right) (*Bottom, right*) Shape of bird when observing wing area from above and from the side (Hemelrijk et al. 2015)

of agitation waves are reflecting a skittering escape motion, because it causes us to temporarily see a larger wing surface due to the associated rolling behaviour (Fig. 24.5).

24.3 Social Organisation

Although groups may be beneficial for their members in so far as they provide protection against predation, they also result in competition for food, mates and space. If individuals meet for the first time, such competitive interactions may initially have a random outcome. Over time, however, a dominance hierarchy develops, whereby certain individuals are consistently victorious over others and are said to have a higher dominance value than others (Drews 1993). Individuals may perceive the dominance value of the other from the body posture of the other (in primates) or from their pheromone composition (in insects).

With regard to the question which individual becomes dominant, there are two extremely opposing views: dominance as a fixed trait by inherited predisposition and dominance by chance and self-organisation. While some argue for the importance of predisposition of dominance by its (genetic) inheritance (Ellis 1991), others reject this for the following reasons: Experimental results show that the dominance of an individual depends on the order of its introduction in a group (Bernstein and Gordon 1980). Dominance changes with experience, because the effects of victory and defeat in conflicts are self-reinforcing, the so-called winner-loser effect. This implies that winning a fight increases the probability of victory in the next fight and losing a fight increases the probability of defeat the next time. This effect has

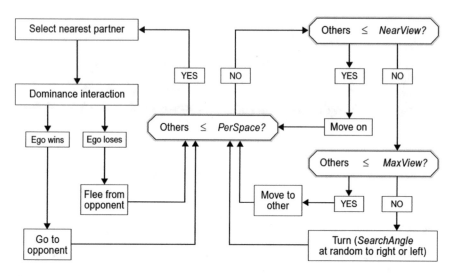

Fig. 24.6 Flowchart of the behavioural rules of individuals in DomWorld. *PerSpace, NearView* and *MaxView* indicate close, intermediate and long distances

been established empirically in many animal species and is accompanied by psychological and physiological changes, such as hormonal fluctuations (Bonabeau et al. 1996a; see Chase et al. 1994; Hemelrijk 2000; for a recent review, see Hsu et al. 2006; Mazur 1985).

The self-reinforcing effects of fighting are the core of a model called *DomWorld*, which concerns individuals that group and compete (Hemelrijk 1996, 1999b; Hemelrijk and Wantia 2005). It leads to a number of emergent phenomena that have relevance for many species, in particular for primates and more specifically macaques.

24.3.1   The Basic DomWorld Model

The model *DomWorld* consists of a homogeneous world in which individuals are grouping and competing. It does not specify what they compete about. Grouping is implemented by making individuals react to others in different manners depending on the distance to the other (Fig. 24.6). An individual is attracted to others far away (within *MaxView*), it continues to follow its direction if it perceives others at an intermediate distance (within *NearView*), and it decides whether or not to perform a competitive dominance interaction if it encounters others close by (within *PerSpace*) (Hemelrijk 2000). After winning a dominance interaction, it chases away the other, and after losing a fight, it flees from it.

Dominance interactions are implemented after the *DoDom* interactions by Hogeweg and Hesper (1985), which were extended to reflect differences in the intensity of aggression between species and between the sexes and with the decision whether or not to attack depending on risks involved (Hemelrijk 1998, 1999b). Each individual has a dominance value which indicates the individual's capacity to win. At the beginning of a competitive interaction, both individuals display their dominance value and observe that of the other. The outcome of the fight depends on the relative dominance of both partners and on chance. The probability of winning is higher the higher the dominance value of an individual in relation to that of the other. Initially, the dominance values are the same for all individuals. Thus, during the first encounter, chance decides who wins. After winning, the dominance of the winner increases and that of the loser decreases. Consequently, the winner has a greater chance to win again (and vice versa) which reflects the self-reinforcing effects of the victories (and defeats) in conflicts of real animals.

We allow for rank reversals; when, unexpectedly, a lower-ranking individual defeats a higher-ranking opponent; this outcome has a greater impact on the dominance values of both opponents, which change with a greater amount than when, as we would expect, the same individual conquers a lower-ranking opponent (conform to detailed behavioural studies on bumble bees by Honk and Hogeweg 1981). Furthermore, in their decision whether or not to attack, we made individuals sensitive to the risks, i.e. the 'will' to undertake an aggressive interaction (instead of remaining nonaggressively close by) increases with the chance to defeat the opponent, which depends on the relative dominance ranks of both opponents (Hemelrijk 1998).

We also represented the intensity of aggression in which primate societies differ (in some species individuals bite, in other species they merely approach, threaten and slap) as a scaling factor (called *StepDom*) that weighs the changes in dominance value after a fight more heavily if the fight was intense (such as biting) than if the fight was mild (involving threats and slaps or merely approaches and retreats) (Hemelrijk 1999b). In several models, we distinguished two types of individuals representing males and females (Hemelrijk et al. 2003). We gave males a higher initial dominance value and a higher intensity of aggression (reflecting their larger body size, stronger muscular structure and larger canines than those of females). In fights between both types of individuals, the intensity of the fight was determined by the initiator (the attacker).

24.3.2 Spatial Structure

The major advantage of group life is supposedly to be protection against predators. Central positions are supposed to be safest, because here individuals are shielded by other group members from predators approaching from the outside. Therefore, according to the well-known 'selfish herd' theory of Hamilton (1971), individuals have evolved a preference for a position in the centre, the so-called 'centripetal instinct'. If competition for this location is won by dominants, dominants will end up in the centre. This is thought to be the main reason why in many animal

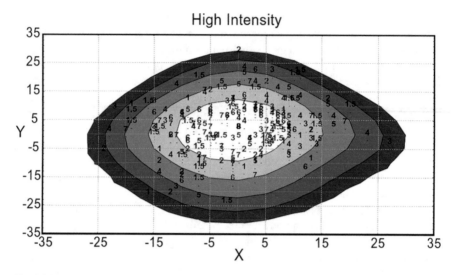

Fig. 24.7 Spatial-social structure. Darker shading indicates areas with individuals of decreasing dominance rank

species dominants are seen to occupy the centre. However, in *DomWorld* this spatial structure emerges, even though such a preference is lacking, and there is no 'centripetal instinct' nor threat of predation (Hemelrijk 2000).

The spatial configuration, with dominant individuals in the centre and subordinates at the periphery of the group (Fig. 24.7), emerges in the model due to a feedback between the dominance hierarchy and the spatial location of the individuals of different rank. During the development of the hierarchy, some individuals become permanent losers. Such low-ranking individuals must end up at the periphery by being constantly chased away. This automatically leaves dominants in the centre. Also, in real animals, such a spatial structure may occur although its members have no centripetal instinct nor experience a threat of predation. For instance, in the elegant experiments with fish by Krause (1993), central dominants were observed, although no centre-oriented locomotion appeared (Krause and Tegeder 1994). Furthermore, this spatial structure has been described in hammerhead sharks in spite of the absence of any predatory threat (Klimley 1985). Thus, the model provides a new way of understanding spatial structure.

24.3.3 Dominance Style: Egalitarian and Despotic Societies

High dominance ranking is supposed to be associated with benefits such as priority of access to mates, food and safe locations. If benefits are strongly biased towards higher-ranking individuals, the society is called 'despotic', whereas if access to resources is more equally distributed, it is called 'egalitarian'. These terms have been used to classify social systems of many animal species (such as insects, birds and primates). Egalitarian and despotic species of primates, such as macaques, appear to differ in many other traits too, such as in group density, their intensity and frequency of aggression and in their frequency and patterns of affiliation (grooming). Usually these differences are explained from the perspective of optimisation of single traits by natural selection. However, Thierry (1990b) suggests that in macaques the many behavioural differences can be traced back to two inherited differences, namely degree of nepotism (i.e. cooperation among kin) and intensity of aggression. Note that despotic macaques display aggression of a higher intensity, i.e. they bite more often, whereas egalitarian macaques display aggression that is milder, they only threaten and slap.

The model *DomWorld* presents an even simpler hypothesis (Hemelrijk 1999b), namely that a mere difference in intensity of aggression produces both types of societies. By increasing the value of one parameter, namely that of intensity of aggression, the artificial society switches from a typically egalitarian society to a despotic one. For instance, compared to egalitarian artificial societies, despotic ones are more loosely grouped, showing a higher frequency of attack, their behaviour is more rank-related, aggression is more asymmetric, spatial centrality of dominants is clearer and female dominance over males is greater. All these differences between despotic and egalitarian societies arise via feedback between the development of the hierarchy and spatial structure, and this happens only at a high intensity of aggression. The steep hierarchy develops from the high aggression intensity, because each outcome has a stronger impact than at a low intensity and it is strengthened further via a mutual feedback between the hierarchy and the spatial structure with dominants in the centre and subordinates at the periphery (Hemelrijk 1999b, 2000).

Pronounced rank-development causes low-ranking individuals to be continuously chased away by others, and thus the group spreads out (1 in Fig. 24.8). Consequently, the frequency of attack diminishes among the individuals (2 in Fig. 24.8), and therefore the hierarchy stabilises (3 in Fig. 24.8). While low-ranking individuals flee from everyone, this automatically leaves dominants in the centre, and thus a spatial-social structure develops (Fig. 24.7). Since individuals of similar dominance are treated by others in more or less the same way, similar individuals remain close together; therefore, they interact mainly with others of similar rank; thus, if a rank reversal between two opponents occurs, it is only a minor one because opponents are often similar in dominance. In this way the spatial structure stabilises the hierarchy, and it maintains the hierarchical differentiation (4 and 5 in Fig. 24.8). Also, the hierarchical differentiation and the hierarchical stability mutually strengthen each other (6 in Fig. 24.8).

In short, the model (Hemelrijk 1999b) makes it clear that changing a single parameter representing the intensity of aggression may cause a switch from an egalitarian to a despotic society. Since all the differences resemble those found between egalitarian and despotic societies of macaques, this implies that in real macaques these differences may also be entirely due to a single trait, intensity of aggression. Apart from intense aggression, such as biting, however, a high frequency

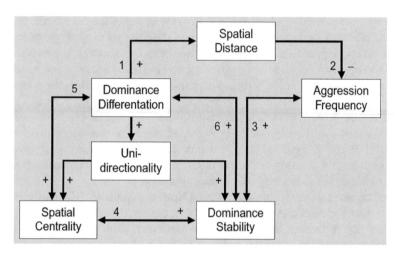

Fig. 24.8 Interconnection between variables causing spatial-social structuring at a high aggression intensity

of aggression can also cause this switch (Hemelrijk 1999a). A higher frequency of aggression also leads to a steeper hierarchy. This in turn results in a clearer spatial structure, which again strengthens the development of the hierarchy, and this has the cascade of consequences as described for aggression intensity.

The lack of individual recognition among group members may be considered as one of the shortcomings of the model. In a subsequent model, this was corrected by having each individual keep a record of the dominance value of each group member. This value was updated depending on its experiences gained with other group members (Hemelrijk 1996, 2000). With regard to the development of spatial structure, hierarchy and dominance style remained similar, but patterns were weaker than in the case of a direct perception of dominance without individual recognition. Weaker patterns arise due to the contradictory experiences that different individuals have with specific others. This will impede a hierarchical development and thus weaken the accompanying consequences. Even though this may be more realistic for certain modelling purposes, it is more useful to have clearer patterns in a simpler model. Such a caricature is more helpful for building upon understanding and developing new ideas.

Dominance style is usually considered to be species specific, but Preuschoft et al. (1998) raised the question whether competitive regimes (egalitarian versus despotic) should not rather be considered as sex specific. In their study of the competitive regime of both sexes of Barbary macaques, they found an unexpected sex difference: males behave in an egalitarian way, whereas females are despotic. It is unexpected that the sex with the larger body size and fiercer aggression evolved a more egalitarian dominance style. Therefore, it seems to be a separate adaptation. However, the same difference in dominance style between the sexes was also found in *DomWorld*: males appear to be more egalitarian than females. The unexpectedly stronger egalitarianism of males in the model is due to yet another difference between the sexes, the higher initial dominance of males compared to females (which relates to differences in body size). Consequently, single events of victory and defeat have less impact on their overall power or dominance. Therefore, they lead to less hierarchical differentiation than among females, who are much smaller and weaker and on whom each victory and defeat therefore has more impact. The greater the sexual difference in initial dominance between the sexes, the more egalitarian the males behave among themselves compared to the behaviour of the females among themselves. The conclusion of this study is that the degree of sexual dimorphism may influence the competitive regime of each sex, in the model and in real primates. Further empirical studies are needed to investigate whether the degree of sexual dimorphism is directly proportional to the steepness of the gradient of the hierarchy of females compared to males.

With regard to its evolution, dominance style is supposed to be a consequence of different degrees of competition within and between groups (van Schaik 1989). According to this theory, when competition within groups is high and that between groups is low, despotic societies evolve and if it is reversed, and competition between groups is high, egalitarian groups emerge. In line with this, *DomWorld* already shows that high competition within a group leads to a despotic society (Hemelrijk 1999b). However, in a subsequent study, competition between groups appears to favour despotic rather than egalitarian groups (Wantia 2007). To study effects of competition between groups, the *DomWorld* model was extended to several groups (*GroupWorld*, Wantia 2007). Here, as in real primates, usually individuals of high rank participate in encounters between groups (see e.g. Cooper 2004).

The model generates a number of unexpected results. Firstly, among groups of the same dominance style, competition between groups does not affect dominance style, since it happens at a very low frequency compared to competition within groups. However, in competition between groups of different dominance style, remarkable results came to light. Unexpectedly, under most conditions groups with a more despotic dominance style were victorious over others with a more egalitarian style. This arose due to the greater power of individuals of the highest rank of the despotic group compared to that of the egalitarian group. In the model, this is a consequence of the stronger hierarchy in despotic groups compared to that in egalitarian groups. In reality this effect may be even stronger, because higher-ranking individuals may also obtain relatively more resources in a despotic group than in an egalitarian one. The outcome of fights between groups depends, however, on the details of the fights between groups and the composition of the group. When participants of intergroup fights fought in dyads or in coalitions of equal size, the despotic group out-competed the egalitarian one. If, however, individuals of egalitarian groups, for one reason or another, fought in coalitions of a larger size or if their coalitions included more males than those of the despotic groups, the egalitarian group had a chance to win. Thus, the main conclusion of the study is that it depends on a number of factors simultaneously, which dominance style will be favoured. Therefore, this model suggests that group composition and details of what happens in fights between groups should be studied in order to increase our understanding of the origination of dominance style.

24.3.4 Distribution of Affiliation, Grooming

Grooming (to clean the pelage of dirt and parasites) is supposed to be altruistic and therefore it should be rewarded in return (Trivers 1971). Using the theory of social exchange, Seyfarth (1977) argues that female primates try to exchange grooming for receipt of support in fights. For this, they direct the supposedly altruistic grooming behaviour more towards higher- than towards lower-ranking partners. As a consequence of this supposed competition for grooming partners of high rank, and by being defeated by dominant competitors, females will end up grooming close-ranking partners most frequently and are groomed themselves most often by females ranking just below them. According to him, this explains the following grooming patterns that are apparent among female primates: (a) high-ranking individuals receive more grooming than others and (b) most grooming takes place between individuals that are adjacent in rank.

DomWorld presents a simpler alternative to the explanation by Seyfarth (Hemel-rijk 2002b) in which a mental mechanism of social exchange for future social benefits is absent. This is important, because it is doubtful whether grooming involves any real costs at all (see Wilkinson 1988).

The core of the argument is that individuals more often groom those whom they encounter more frequently. In the case of spatial structure with dominants in the centre and individuals in closer proximity to those that are closer in dominance (as described for primates (e.g. see Itani 1954), these patterns will follow automatically. Individuals more often groom those that are nearby in rank (as found by Seyfarth 1977). Further, because dominants are more often in the centre, they are frequently surrounded on all sides by others. Subordinates, however, are at the edge of the group, and therefore have nobody on one side. Consequently, dominants simply meet others more frequently than subordinates do (as is shown in *DomWorld*, Hemelrijk 2000). Therefore, dominants are more often involved in grooming than subordinates (as found by Seyfarth).

In support of the model, grooming patterns in despotic macaques appear to be more dominance oriented, and in egalitarian species grooming is more often directed at anyone (Thierry et al. 1990). To establish the relevance of this model-based hypothesis for real primates, it should be tested further in more species and groups, whether or not the patterns of grooming among individuals of similar rank and of the receipt of grooming by individuals of higher rank occur especially in groups with centrally located dominants, and less so in those with a weak spatial structure.

Also the generally noted pattern of reciprocation of grooming in primates can be explained as a side effect of spatial-social structuring rather than being 'calculated' in the sense that individuals keep track of the number of acts received and tune the acts they give correspondingly (as has been the standard explanation). Instead, the model suggests that reciprocation emerges by self-organisation. Since individuals are more often close to some than to others (Fig. 24.7), they will groom them more often than that they groom others and be groomed by them more often than by others (Puga-Gonzalez et al. 2009).

In primates it is often observed that when two individuals are fighting, they also groom each other soon after a fight. This has been labelled 'reconciliation' (Aureli et al. 2002). In primates about 10% of the fights are followed (within 10 minutes) by such 'immediate' grooming or reconciliation. To explain this, it has been assumed that primates understand the importance of reconciliation in order for maintaining 'good relationships'. This has been used for explaining that they reconcile fights especially with those with whom they groom more frequently, their 'friends', because these friends are their most important relationships. A simpler explanation for these patterns in terms of the cognition involved by the individuals performing it is given when grooming behaviour is added to the model *DomWorld*, called *GroofiWorld* (Puga-Gonzalez et al. 2009). In *GroofiWorld*, upon meeting someone close by (in *PerSpace*), individuals groom the other if they think they will be defeated by it. This model generates patterns of 'reconciliation' that are statistically similar to empirical data but without individuals having any intention to maintain good relationships with others or especially with friends. Instead, the patterns of 'reconciliation' emerge because individuals have a higher number of opportunities to groom with the former opponent soon after a fight than at another randomly chosen point in time. After their fight, the former opponents are closer together in space than they are at another, randomly chosen point in time. Indeed in studies of reconciliation in animals, be it primates, horses or goats, former opponents are closer in space after a fight than at other, randomly chosen points in time and thus have more opportunities to groom. The higher percentage of fights that individuals reconcile with their 'friends' (frequent grooming partners) rather than with others is simpler explained by the model as well: Compared to others, 'friends' are more often groomed anyhow, thus also after a fight. This is probably due to their closer spatial proximity compared to non-friends, which in the model is a side effect of dominance interactions. Similar explanations are also given by the model for the supposedly cognitively sophisticated pattern of 'consolation' (Puga-Gonzalez et al. 2014).

24.3.5 *Dominance Relationships Between the Sexes*

Most primate species live in bisexual groups. Apart from the order of Lemuriformes, males are usually dominant. However, variation occurs and even if males are larger than females, females are sometimes dominant over a number of them (Smuts 1987). To study whether this may arise through chance and the self-reinforcing effects of dominance interactions, the sexes are modelled in *DomWorld*. When, for the sake of simplicity, the sexes in *DomWorld* are distinguished only in terms of an inferior fighting capacity of females compared to that of males, surprisingly, males

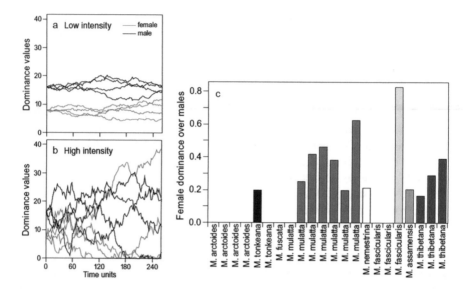

Fig. 24.9 Female dominance over males or degree of rank overlap: (**a**, **b**) Female dominance over males over time and aggression intensity (Hemelrijk 1999a, b), (**c**) Degree of rank overlap of females with males (i.e. female dominance over males) in groups of egalitarian macaques (in *black*) and in groups of despotic macaques (in lighter shades of *grey*) (Hemelrijk et al. 2008)

appear to become less dominant over females at a high intensity of aggression than at a low intensity (Hemelrijk 1999b; Hemelrijk et al. 2003). This is due to the stronger hierarchical differentiation, which causes the hierarchy to develop in a more pronounced way. This implies that the hierarchy is also more strongly developed for each sex separately. Since this differentiation increases the overlap in dominance between the sexes, it causes more females to be dominant over males (Fig. 24.9b) than in the case of a hierarchy that has been less developed (Fig. 24.9a).

Similarly to females, adolescent males of despotic macaques have more ease than those of egalitarian species in outranking adult females (Thierry 1990a). Thierry explains this as a consequence of the stronger cooperation to suppress males among related females of despotic macaques than of egalitarian ones, and van Schaik (1989) emphasises the greater benefits associated with coalitions for females of despotic species than egalitarian ones. However, *DomWorld* explains greater female dominance, as we have seen, simply as a side effect of more pronounced hierarchical differentiation.

Differences in overlap of ranks between the sexes may affect sexual behaviour. Males of certain species, such as Bonnet macaques, have difficulty in mating with females that outrank them (Rosenblum and Nadler 1971). Therefore, following the model, we expect that despotic females (because they outrank more males) have fewer males to mate with than egalitarian ones. In line with this, in macaques despotic females are observed to mate with fewer partners and almost exclusively with males of the highest ranks (Caldecott 1986); this observation is attributed to the evolution of a more pronounced female preference in despotic than in egalitarian species of macaques. The explanation derived from the model, however, is simpler. In a subsequent study, in support of this hypothesis, the relative dominance position of both sexes in egalitarian and despotic macaque species indeed appeared to differ in macaques as expected: females of despotic species were dominant over a significantly higher percentage of the males in their group than females of egalitarian species. (Fig. 24.9c).

In a similar way as high aggression intensity, a high frequency of aggression in the model also results in more female dominance. A higher frequency of aggression in the model can be obtained by increasing the *SearchAngle* over which individuals search for others, when no one is perceived as dominant (Fig. 24.6). Due to a greater *SearchAngle* they return to the others sooner, the group becomes denser, and thus the frequency of aggression is higher. A difference in group density may be of relevance to the difference in female dominance between common chimpanzees and bonobos (also known as pygmy chimpanzees (Stanford 1998)). Despite their similar sexual dimorphism, female dominance in pygmy chimpanzees is higher than among common chimpanzees. This is usually attributed to more intensive coalition formation among pygmy females against males. However, in line with *DomWorld*, we may also explain it as a side effect of the difference in density between both species (Hemelrijk 2002a; Hemelrijk et al. 2003). Density is high in groups of pygmy chimpanzees. Due to the higher density there is a higher frequency of aggression and according to *DomWorld* this may result in more female dominance over males. This hypothesis should be tested by comparing different groups of bonobos and by studying the relationship between female dominance and frequency of aggression.

Sexual attraction in real animals is usually thought to be accompanied by strategies of exchange. For instance, chimpanzee males are described as exchanging sex for food with females (Goodall 1986; Stanford 1996; Tutin 1980). Yet, in spite of detailed statistical studies, we have found no evidence that males obtain more copulations with or more offspring from those females whom they allow to share their food more often (Hemelrijk et al. 1999, 2001, 1992; Meier et al. 2000). Male tolerance of females seems to increase during the females' period of oestrus even without noticeable benefits. Thus, we need another explanation for male tolerance of females. *DomWorld* provides us with such an alternative hypothesis. 'Sexual attraction' of males to females is implemented in such a way that males have a greater inclination to approach females than individuals of their own sex. In the model (and in the preceding models and empirical studies of Fig. 24.9), we measure the relative dominance position of females compared to males by counting the number of males ranking below each female and calculating this figure (Mann-Whitney U-value, Fig. 24.10a). It appears that this value of relative female dominance to males increases with sexual attraction as an automatic consequence of the more frequent encounters between the sexes (Fig. 24.10b, synchronous). This result is in line with the observation that female dominance in chimpanzees increases when males are sexually attracted to the females (Yerkes 1940). The question of whether female dominance over males also increases during sexual attraction in other species should be studied in the future.

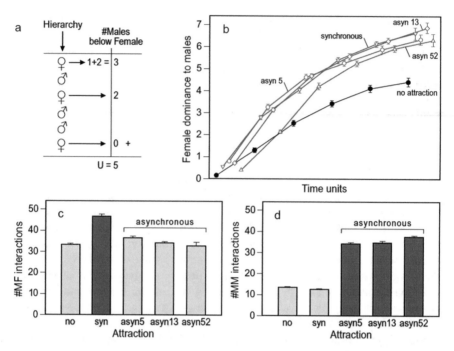

Fig. 24.10 Female dominance over males and sexual attraction. (**a**) Measurement of female dominance; (**b**) Female dominance over males over time without attraction (control) and when attracted to females that cycle synchronously and asynchronously; (**c**) number of interactions between sexes with(out) attraction during (a)synchronous cycling; (**d**) number of interactions among males with(out) attraction during (a)synchronous cycling. *Asyn* asynchronous cycling, *syn* synchronous cycling; 5, 13, 52 are durations of oestrus period

Whereas the examples mentioned above concern species in which females are synchronously sexually attractive (tumescent), in other species they cycle asynchronously. In the model, however, female dominance over males is relatively similar regardless of whether they are attractive synchronously or asynchronously (Fig. 24.10b, syn, asyn). The process leading to increased female dominance differs, however, for the two conditions. If single females are attractive in turn, many males cluster close to a single female. Consequently, in contrast to synchronous tumescence, the frequency of interaction between the sexes remains similar to that when females are not attractive to males, but the frequency of male-male interactions is increased markedly (Fig. 24.10c, d). Due to the higher frequency of interactions among males, the differentiation of the male hierarchy is stronger than without attraction and this causes certain males to become subordinate to some females (Fig. 24.10b, asyn).

Furthermore, the adult sex ratio (or percentage of males) in the group influences the relative dominance of females compared to that of males (Hemelrijk et al. 2008). Female dominance appears to be higher when there are more males in the group. This arises from a shift in the relative number of intra- and inter-sexual interactions.

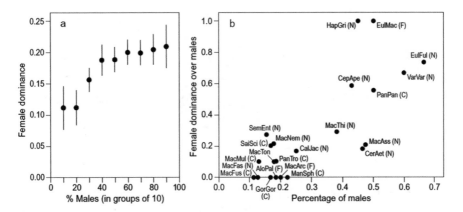

Fig. 24.11 Percentage of males in the group and female dominance over males. (**a**) Model (average and SE), (**b**) Real data of primates. Six-letter codes indicate species. Environmental conditions: *N* natural, *F* free-ranging, *C* captive condition

A higher proportion of males causes both sexes to interact more often with males. Due to the males' higher intensity of aggression, this causes a greater differentiation of the dominance values of both females and males. Consequently, at a high intensity of aggression, the hierarchy of females overlaps more with that of males, and thus, the dominance position of females is higher in relation to males than if there are fewer males in the group. Subsequent analysis of these patterns in real primates has confirmed that female dominance increases with a higher percentage of males in the group (Fig. 24.11b). It appeared that in line with the preceding modelling results, in groups of a despotic species (rhesus macaques) a higher percentage of males appeared to be correlated with greater female dominance over them, whereas such a correlation was absent among groups of an egalitarian species (stump-tailed macaques). Similarly, when studying the influence of women (students) on decisions taken in a group with a high level of conflict, women were shown to have more influence the higher the percentage of males in the group (Stroebe et al. 2016).

24.3.6 Strategies of Attack

When real animals are brought together for the first time, they perform dominance interactions only during a limited period. This has been empirically established in several animal species, e.g. chickens (Guhl 1968) and primates (Kummer 1974). The interpretation is that individuals fight to reduce the ambiguity of their relationships (Pagel & Dawkins 1997); once these are clear, energy should be saved. On the other hand, it has also been suggested that individuals should continuously strive for a higher ranking and therefore always attack, unless an opponent is clearly believed to be superior (e.g. see Datta & Beauchamp 1991).

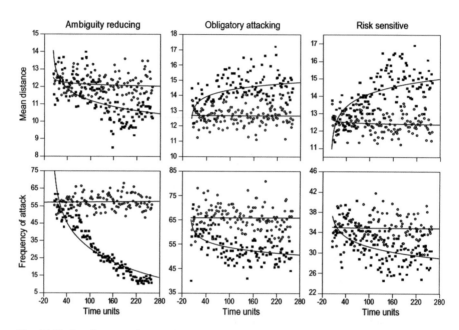

Fig. 24.12 Development of mean distance (*top*) and frequency of aggressive interactions (*bottom*) among individuals for different attack strategies and intensities of aggression (logarithmic line fitting). Open circles represent *StepDom* of 0.1, closed blocks of 1.0

In the *DomWorld* model, we compare these popular ethological views with each other and a control strategy, in which individuals invariably attack others upon meeting them (this is called the 'Obligate' attack strategy). Here, the 'Ambiguity-Reducing' strategy is a symmetrical rule in which individuals are more likely to attack opponents closer in rank to themselves. In the so-called 'Risk Sensitive' strategy, the probability of an attack is higher when the opponent is of a lower rank (Hemelrijk 1998). Remarkably, it appears that, with time, the frequency of aggression decreases in all three attack strategies, at least when groups are cohesive and the intensity of aggression is sufficiently high (Fig. 24.12).

This decrease of aggression is a direct consequence of the 'Ambiguity-Reducing' strategy, but unexpectedly it also develops in the other two (see Fig. 24.12). Due to the high intensity of aggression, each interaction has a strong impact on the rank of both partners; thus a steep hierarchy develops. This automatically implies that some individuals become permanent losers, and that, by fleeing repeatedly, they move further and further away from others. The increased distance among individuals in turn results in a decrease of the frequency of encounters and hence aggression. This provides a test for the real world: it has to be examined whether the development of the dominance hierarchy is accompanied not only by a reduction of aggression but also by an increase in interindividual distances.

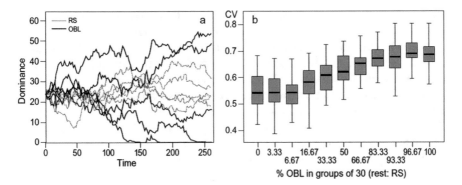

Fig. 24.13 Dominance distribution and personality types: (**a**) example of hierarchical development of mixed group (*fat lines*: obligate attackers, *dotted lines*: risk-sensitive individuals). $N = 10$, 5 of each type. (**b**) Hierarchical differentiation in mixed groups with different ratios of obligatorily attacking (OBL) and risk-sensitive (RS) individuals. *CV* Mean coefficient of variation of *DomValues*. Box = SE, whiskers = SD

24.3.7   Personality Types

Two of these attack strategies, i.e. 'risk-sensitive' and 'obligate' attack (Hemelrijk 2000) resemble the attack strategies of individuals of different personality types, namely those of the cautious and the bold personality, respectively (Koolhaas et al. 2001). When groups in the model consist of both types (mixed groups), the differentiation of dominance values appears to be greater among individuals that are attacking obligatorily than risk sensitively due to their higher frequency of attack (Hemelrijk 2000). Consequently, obligate attackers rise very high and descend very low in the hierarchy (resulting in a bimodal distribution of dominance values, Fig. 24.13a), whereas risk-sensitive attack leads to less variation, a unimodal distribution of values (Fig. 24.13a) and therefore to more intermediate dominance positions. Further, among risk-sensitive individuals, the average dominance is slightly higher than among those that always attack. This is due to higher 'intelligence' of the risk-sensitive attack strategy, because these individuals preferably attack when the risk of losing the fight is minimal.

This resembles the distribution of dominance in mixed groups of a bird species, the great tits (Verbeek et al. 1999). Here, bold individuals ascend very high up in the dominance hierarchy or descend very low, whereas cautious individuals have intermediate ranks that on average are above those of bold individuals. Differences in high and low rank of individuals were explained by the different stages of moulting of their feathers, a difference in tendency to attack from a familiar territory or an unfamiliar one, and a difference in speed of recovery from defeats. *DomWorld* shows that there is no need to add causes based on different stages of moulting or on familiarity with a territory or to different speed of recovery. Thus, the model produces a far simpler explanation for the distribution of dominance values in these groups of great tits. To verify these results empirically, differences in risk sensitivity of both types of personality need to be confirmed empirically.

Secondly, the model provides us with an alternative explanation for the associations between dominance behaviour and personality type in great tits found by Dingemanse and Goede (2004). This association appears to differ among individuals who own a territory and those who do not; whereas among territory owners bold ones were dominant over cautious ones, the reverse held for those without a territory. To explain this, the authors use a context-specific argument in which they need an additional trait, namely speed of recovery from defeat (Carere et al. 2001). They argue that among those individuals without a territory, bolds have more difficulty in recovering from defeat than cautious ones and therefore become low in rank, whereas territory owners do not suffer this setback, and therefore, they become higher in rank.

Alternatively, a simpler explanation in line with our model may apply. Here, we start from existing dominance relationships and suppose that these exert a decisive influence on the question who will obtain a territory (instead of the other way around). We assume that, because territories are limited in numbers, the higher-ranking individuals (say the top half of them) will acquire them, whereas individuals in the lower part of the hierarchy are unable to get one. Due to the bimodal distribution of dominance values among the bold birds, and the unimodal distribution of the cautious ones, the most extreme dominance positions in the colony will be occupied by bold ones, whereas the cautious ones are located in the middle of the hierarchy. Thus, among individuals in the top half of the hierarchy (the territory owners) the bolds will rank above the cautious, whereas in the bottom half of the hierarchy, namely among the individuals without a territory, the reverse is true (Fig. 24.13a). For this explanation to be proven correct, we must verify whether territory owners belong to the upper half of the dominance hierarchy or not. Thus, *DomWorld* produces new explanations for dominance relationships of these 'personality styles' in great tits.

An important question regarding personality is how different types, bold and cautious, may co-exist, and why one type does not take over. Although there are a number of explanations for various species, none of them applies to primates. Since in primates group survival and individual survival depend on competition within and between groups, Wantia and Hemelrijk have studied the two personality types in these contexts (Wantia 2007). They have found that risk-sensitive individuals out-competed obligate attackers in fights within groups, but that in fights between groups the obligate attackers did better: the higher the percentage of individuals who attacked obligatorily in fights between groups, the greater the chance of the group winning (Fig. 24.13b). The better performance within groups of risk-sensitive individuals was due to their more cautious and deliberate strategy: to attack when the chance of winning is high. Greater success by obligate attackers in fights between groups was a consequence of the higher dominance value of the highest-ranking individuals in groups with more obligate attackers. This is due to the steeper hierarchy as a consequence of the higher frequency of aggression in groups with more individuals that carry out obligatory attacks. Thus, whereas risk-sensitive individuals out-compete obligate attackers in conflicts within groups, the reverse happens in conflicts between groups. Since competition within and between groups is essential for primate societies (van Schaik and van Hooff 1983), and the success of both attack strategies depends on these contexts, we may imagine that a similar differential performance may contribute to the coexistence of bold and cautious primates.

24.3.8   Distribution of Tasks

It seems a miracle that a colony of social insects consisting of tens of thousands of individuals is able to cope with the huge socio-economic demands of foraging, building, cleaning nests and nursing the brood. It appears that group members are somehow able to divide the work efficiently among them. Such a division of labour is flexible, i.e. the ratio of workers performing different tasks varies according to changes in the needs and circumstances of the colony. There are several ways in which this task division may arise. Different mechanisms may operate in different species (for a review, see Beshers and Fewell 2001). Task division may be based on a genetic difference in predisposition (e.g. Moritz et al. 1996; Robinson 1998; Robinson and Page 1988) or a response threshold to perform certain tasks (Bonabeau et al. 1996b). Such a threshold may be combined with a self-reinforcing learning process (Gautrais et al. 2002; Theraulaz et al. 1998). Thus, after performing, the threshold is lessened (by learning), and after a long period of no involvement in the task, it increases by forgetting.

The execution of tasks may also be a consequence of dominance values, as is shown in a model of bumblebees (*Bombus terrestris*) developed by Hogeweg and Hesper (1983, 1985). This was based on an earlier experimental study (Honk and Hogeweg 1981) that showed that during the growth of the colony, workers develop into two types, the low-ranking, so-called 'common', and the high-ranking, so-called 'elite', workers. The activities carried out by the two types differ noticeably: Whereas the 'common' workers mainly forage and take rest, the 'elite' workers are more active, feed the brood, interact with each other and with the queen and sometimes lay eggs. In their study of the minimal conditions needed for the formation of the two types of workers (on the assumption that all workers are identical when hatching), Hogeweg and Hesper (1983, 1985) used an individual-based model based on empirical data concerning the time of the development of eggs, larvae, pupae, etc. Space in the model is divided into two parts, peripheral (where inactive common workers doze for part of the time) and central (where the brood is, and all interactions take place). The rules of the artificial, adult bumblebees operate 'locally' in so far as their behaviour is triggered by what they encounter. What they encounter in the model is chosen randomly from what is available in the space in which the bumblebee finds itself. For instance, if an adult bumblebee meets a larva, it feeds it; if it meets a pupa of the proper age, it starts building a cell in which a new egg can be laid, etc. All workers start with the same dominance value after hatching, with only the queen gifted with a much higher dominance rank. When an adult meets another, a dominance interaction takes place, the outcome of which (victory or defeat) is self-reinforcing. Dominance values of the artificial bumblebees influence almost all their behavioural activities (for instance, individuals of low rank are more likely to forage).

This model automatically, and unexpectedly, generates two stable classes, those of 'commons' (low-ranking) and 'elites' (high-ranking) with their typical conduct. This differentiation only occurs if the nest is divided into a centre and a periphery (as in real nests).

The flexibility of the distribution of tasks among individuals manifests when we take half the work force out of the model. In line with observations in similar experiments with real bumblebees, this reduction in work force causes the remaining ones to take over the work. In the model this arises because the decreased number of workers reduces the frequency of encounters among them and increases encounters between them and the brood (which has not been reduced in number). An increased rate of encounters with brood induces workers to collect food more frequently. Therefore, workers are absent from the next more often, and consequently, they meet other workers less frequently.

In real bumblebees the queen switches from producing sterile female offspring to fertile offspring (males and females) at the end of the season. Note that while females are produced from fertilised eggs, males are produced from unfertilised eggs. Usually females will develop into sterile workers, but if they are fed extra 'queen food' during larval development, they will develop into queens. The switch seems to take place at an optimal moment, because it occurs at the time when the colony is at its largest and can take care of the largest number of offspring. Oster and Wilson (1978) point out that it is difficult to think of a possible external signal that could trigger such a switch, because it takes three weeks to raise fertile offspring and during these three weeks there must still be enough food.

Hogeweg and Hesper (1983) discovered that no such external signal is needed in their bumblebee model, but that the switch originates automatically as if scheduled by a socially regulated 'clock'; it arises from the interaction between colony growth and stress development of the queen as follows. During the development of the colony, the queen produces a certain pheromone that inhibits extra feeding of larvae (which leads to queens) and worker ovipositions (i.e. unfertilised male eggs) by 'elite' workers. Just before she is killed, the queen can no longer suppress the 'elite' workers from feeding larvae to become queens and from laying drone eggs, because the colony has grown too large. Consequently, individual workers meet the queen less often and are less subjected to the dominance of the queen, so they start to lay unfertilised (drone) eggs. Furthermore, the stress on the queen increases whenever she has to perform dominance interactions with workers during her egg laying. When the stress on the queen has reached a certain threshold value she switches to producing male eggs (drones).

Because generative offspring are also sometimes found in small nests, Blom (1986) challenges the notion that dominance interactions induce stress in the queen and thus lead to this switch. However, in the model, it is not the number of workers that causes the switch: Hogeweg and Hesper (1985) have shown that in small nests the switch also appears to occur at the same time in the season as in large nests. For this they studied the bumblebee model for a reduced speed of growth. They found that the switch occurs at a similar moment due to the following complicated feedback process. If the colony grows faster, the heavy duties of caring for the brood leave the workers little time to interact with each other and the queen. Consequently, the dominance hierarchy among workers only develops weakly. Therefore, if workers interact with the queen, they do not pose much of a threat to her and as a result, the queen is not severely stressed and the colony can grow very large. In contrast, in a slowly growing colony, the small number of brood gives the workers little work and leaves them time to interact with each other. Consequently, their dominance relationships are clearly differentiated. Furthermore, by often interacting with the queen, they become on average higher-ranking themselves. In the end, the queen receives as much stress from frequent interactions with a few high-ranking workers in a slowly growing nest as from few interactions with very many low-ranking workers in a fast growing nest. Consequently, the switch to reproductive offspring takes place at about the same moment in both nests in the model.

24.4 Evaluation

The power of these kinds of models is the generation of phenomena that are emergent. These emergent phenomena lead to explanations that are more parsimonious than usual, because the patterns emerge from the interaction among individuals and their environment rather than from the cognition of an individual. These explanations can be tested empirically.

Considering the kind of questions posed in the models discussed above, it becomes clear that most of them can only be studied using certain kinds of agent-based models. Other kinds of models, such as partial differential equations based on density functions or even individual-based models of fluids and gases, cannot incorporate the complexity of the rules and/or emergent effects.

The behavioural rules of the agents in the agent-based models that were treated here were in all cases biologically inspired. In some cases, behavioural rules were based on precise experimental results specific to a certain species, such as in the case of group formation (Camazine et al. 2001; Jeanson et al. 2005). Usually, however, parameters were tuned only loosely to the real system (e.g. the angle of vision in fish is set at about 270° and that of mammals at 120°). Sometimes, mathematical equations were used that were developed for a different species, such as in the case of dominance interactions. Here, the equations were initially derived for the self-reinforcing effects in bumblebees (Honk and Hogeweg 1981) and subsequently extended and applied to explain social systems in primates (Hemelrijk 1999b). In all cases the behavioural rules are supposed to capture the essentials of those of real animals.

If, however, macro-patterns are obtained that resemble patterns observed in the real system, this is still no proof of the correctness of the rules in reflecting behavioural rules of real animals, because different sets of rules may lead to the same macro-pattern. The relevance of the rules and other aspects of the model could be studied further by investigating additional hypotheses and comparing further results of the model to natural systems.

Agent-based models appear to be particularly useful in showing the consequences of interactions among individuals for social structure and the reverse, i.e. from social structure to social interactions. In this way, they teach us about the integration of traits. This is particularly apparent when the models include a representation of space (time is always represented). The representation of time and space is of general importance, because all real individuals live in time and space and thus it is unrealistic to ignore these aspects. It appears that by representing the interaction of individuals with their social and physical environment in time and space, patterns result in such a way that explanations become unusually parsimonious in terms of their specific adaptive traits and the cognition needed (in line with findings of situated cognition Pfeifer and Scheier 1999).

These explanations give a new insight into the phenomenon of social complexity that may either be very general (e.g. conceptual model) or specific to the type of animal and the kind of behaviour under investigation. Usually, it leads to hypotheses that are easily tested in the real system. These hypotheses make unexpected connections between variables, such as between the production of sterile and reproductively active offspring with colony dynamics in bumblebees. Testing can be done by natural observation, for example, by studying the effects of different group sizes on group shape in case of fish. Furthermore, testing can be done by experimental interventions, such as by putting together individuals of both sexes in different group compositions to study effects on relative dominance of females to males.

While ignoring models of certain topic areas where no emergent effects appeared (so-called output-oriented models), we have here surveyed a large number of the most important models of social complexity in animals. Of course, it is impossible to discuss them all. Note that we have not treated models concerning biological evolution since only a few are related to social systems (such as Kunz et al. 2006; Oboshi et al. 2003; Ulbrich et al. 1996; Post et al. 2016). This is due to the fact that evolution usually happens on a far larger timescale than phenomena of social complexity. However, see Chap. 21 in this handbook (Chattoe-Brown and Edmonds 2017) on how concepts from biological evolution have influenced recent approaches in programming simulation models.

Furthermore, we have not treated a number of other complex phenomena, such as synchronisation of lighting behaviour of insects, temperature regulation of nests and the building of nests, because in these cases no agent-based models have been used, or the behaviour is insufficiently social. These topics are, however, covered by Camazine et al. (2001).

24.5 Future Work

With regard to areas that may be important for future work, I suggest (1) models of self-organisation that help us understand specific social systems better by being closely tuned to empirical systems, such as fish schools, insect colonies or primate groups, (2) models that present social behaviour and its ontogenetical development in greater detail and (3) evolutionary models that include also spatial effects.

Acknowledgements I would like to thank Andreas Flache and Rineke Verbrugge for comments on an earlier draft. I thank Daan Reid for correcting the English.

Further Reading

For further reading I recommend the book on self-organisation in biological systems by Camazine et al. (2001). This is an extensive introduction to almost all topics treated above and more (with the exception of task division and social systems of primates). Furthermore, new extensions of a number of the models are treated by Sumpter (2010). The above-mentioned book by Camazine and co-authors is also a good choice for teaching purposes, in addition to the well-written book by Resnick (1994). This latter book has been used in secondary schools and teaches to think in terms of complexity and self-organisation in general. For more advanced readers, I recommend "Self-organisation and Evolution of Social Systems" (Hemelrijk 2005). This is an edited book and contains recent articles on modelling of social systems, including those of humans.

References

Aureli, F., Cords, M., & Van Schaik, C. P. (2002). Conflict resolution following aggression in gregarious animals: a predictive framework. *Animal Behaviour, 64*, 325–343.

Aureli, F., et al. (2008). Fission-fusion dynamics: New research frameworks. *Current Anthropology, 49*, 627–654.

Axelsen, B. E., Anker-Nilssen, T., Fossum, P., Kvamme, C., & Nottestad, L. (2001). Pretty patterns but a simple strategy: Predator-prey interactions between juvenile herring and Atlantic puffins observed with multibeam sonar. *Canadian Journal of Zoology-Revue Canadienne De Zoologie, 79*, 1586–1596.

Bernstein, I. S., & Gordon, T. P. (1980). The social component of dominance relationships in rhesus monkeys (Macaca mulatta). *Animal Behaviour, 28*, 1033–1039.

Beshers, S. N., & Fewell, J. F. (2001). Models of division of labour in social insects. *Annual Review of Entomology, 46*, 413–440.

Blom, J. V. D. (1986). Reproductive dominance within the colony of Bombus terrestris. *Behaviour, 97*, 37–49.

Bonabeau, E., Theraulaz, G., & Deneubourg, J.-L. (1996a). Mathematical models of self-organizing hierarchies in animal societies. *Bulletin of Mathematical Biology, 58*, 661–717.

Bonabeau, E., Theraulaz, G., & Deneubourg, J.-L. (1996b). Quantitative study of the fixed threshold model for the regulation of division of labour in insect societies. *Proceedings of the Royal Society of London B, 263*, 1565–1569.

Buhl, J., et al. (2006). From disorder to order in marching locusts. *Science, 312*, 1402–1406.

Bumann, D., & Krause, J. (1993). Front individuals lead in shoals of 3-spined sticklebacks (Gasterosteus-Aculeatus) and juvenile roach (Rutilus-Rutilus). *Behaviour, 125*, 189–198.

Bumann, D., Krause, J., & Rubenstein, D. (1997). Mortality risk of spatial positions in animal groups: the danger of being in the front. *Behaviour, 134*, 1063–1076.

Caldecott, J. O. (1986). Mating patterns, societies and ecogeography of macaques. *Animal Behaviour, 34*, 208–220.

Camazine, S., Deneubourg, J.-L., Franks, N. R., Sneyd, J., Theraulaz, G., & Bonabeau, E. (2001). *Self-organization in biological systems. (Princeton studies in complexity)*. Princeton and Oxford: Princeton University Press.

Carere, C., Welink, D., Drent, P. J., Koolhaas, J. M., & Groothuis, T. G. G. (2001). Effect of social defeat in a territorial bird (Parus major) selected for different coping styles. *Physiology and Behavior, 73*, 427–433.

Chase, I. D., Bartelomeo, C., & Dugatkin, L. A. (1994). Aggressive interactions and inter-contest interval: how long do winners keep winning? *Animal Behaviour, 48*, 393–400.

Chattoe-Brown, E., & Edmonds, B. (2017). Evolutionary mechanisms. In *Simulating social complexity–A handbook*. Chapter 21 in this volume.

Cooper, M. A. (2004). Inter-group relationships. In B. Thierry, M. Singh, & W. Kaumanns (Eds.), *Macaque societies: A model for the study of social organization*. Cambridge: Cambridge University Press.

Couzin, I. D., Krause, J., Franks, N. R., & Levin, S. A. (2005). Effective leadership and decision-making in animal groups on the move. *Nature, 433*, 513–516.

Couzin, I. D., Krause, J., James, R., Ruxton, G. D., & Franks, N. R. (2002). Collective memory and spatial sorting in animal groups. *Journal of Theoretical Biology, 218*, 1–11.

Datta, S. B., & Beauchamp, G. (1991). Effects of Group Demography on Dominance Relationships among Female Primates .1. Mother-Daughter and Sister-Sister Relations. *American Naturalist., 138*, 201–226.

Deneubourg, J. L., & Goss, S. (1989). Collective patterns and decision-making. *Ethology Ecology and Evolution, 1*, 295–311.

Deneubourg, J. L., Goss, S., Franks, N., & Pasteels, J. M. (1998). The blind leading the blind: Modelling chemically mediated army ant raid patterns. *Journal of Insect behaviour, 2*, 719–725.

Dingemanse, N. J., & Goede, P. d. (2004). The relation between dominance and exploratory behavior is context-dependent in wild great tits. *Behavioral Ecology, 15*(6), 1023–1030.

Drews, C. (1993). The concept and definition of dominance in animal behaviour. *Behaviour, 125*, 283–313.

Ellis, L. (1991). A biosocial theory of social stratification derived from the concepts of pro/antisociality and r/K selection. *Politics and the Life Sciences, 10*, 5–44.

Enquist, B. J., & Niklas, K. J. (2001). Invariant scaling relations across tree-dominated communities. *Nature, 410*, 655–660.

Enquist, B. J., West, G. B., Charnov, E. L., & Brown, J. H. (1999). Allometric scaling of production and life-history variation in vascular plants. *Nature, 401*, 907–911.

Franks, N. R., Gomez, N., Goss, S., & Deneubourg, J.-L. (1991). The blind leading the blind in army ant raiding patterns: Testing a model of self-organization (Hymenopter: Formicidae). *Journal of Insect Behaviour, 4*, 583–607.

Gautrais, J., Theraulaz, G., Deneubourg, J. L., & Anderson, C. (2002). Emergent polyethism as a consequence of increased colony size in insect societies. *Journal of Theoretical Biology, 215*, 363–373.

Goldberg, T. L., & Wrangham, R. W. (1997). Genetic correlates of social behaviour in wild chimpanzees: Evidence from mitochondrial DNA. *Animal Behaviour, 54*, 559–570.

Goodall, J. (1986). *The chimpanzees of Gombe: Patterns of behaviour*. Cambridge, MA and London: Belknapp Press of Harvard University Press.

Gueron, S., Levin, S. A., & Rubenstein, D. I. (1996). The dynamics of herds: From individuals to aggregations. *Journal of Theoretical Biology, 182*, 85–98.

Guhl, A. M. (1968). Social inertia and social stability in chickens. *Animal Behaviour, 16*, 219–232.

Hamilton, W. D. (1971). Geometry for the selfish herd. *Journal of Theoretical Biology, 31*, 295–311.

Hemelrijk, C. K. (1996). Dominance interactions, spatial dynamics and emergent reciprocity in a virtual world. In P. Maes, M. J. Mataric, J.-A. Meyer, J. Pollack, & S. W. Wilson (Eds.), *Proceedings of the fourth international conference on simulation of adaptive behavior* (Vol. 4, pp. 545–552). Cambridge, MA: The MIT Press.

Hemelrijk, C.K. (1998, August 17–21). Risk sensitive and ambiguity reducing dominance interactions in a virtual laboratory. In R. Pfeifer, B. Blumberg, J.-A. Meyer, & S. W. Wilson (Eds.), *From animals to animals 5, proceedings of the fifth international conference on simulation on adaptive behavior* (pp. 255–262), Zurich, Switzerland. Cambridge, MA: MIT Press..

Hemelrijk, C. K. (1999a). Effects of cohesiveness on intersexual dominance relationships and spatial structure among group-living virtual entities. In D. Floreano, J.-D. Nicoud, & F. Mondada (Eds.), *Advances in artificial life. Fifth European conference on artificial life. (Lecture Notes in Artificial Intelligence, 1674)* (pp. 524–534). Berlin: Springer.

Hemelrijk, C. K. (1999b). An individual-oriented model on the emergence of despotic and egalitarian societies. *Proceedings of the Royal Society of London B: Biological Sciences, 266*, 361–369.

Hemelrijk, C. K. (2000). Towards the integration of social dominance and spatial structure. *Animal Behaviour, 59*, 1035–1048.

Hemelrijk, C. K. (2002a). Self-organising properties of primate social behaviour: A hypothesis on intersexual rank-overlap in chimpanzees and bonobos. In C. Soligo, G. Anzenberger, & R. D. Martin (Eds.), *Primatology and anthropology: Into the third millennium* (pp. 91–94). New York: Wiley.

Hemelrijk, C. K. (2002b). Understanding social behaviour with the help of complexity science. *Ethology, 108*, 655–671.

Hemelrijk, C. K., Wantia, J., & Daetwyler, M. (2003). Female Co-dominance in a virtual world: Ecological, cognitive, social and sexual causes. *Behaviour, 140*, 1247–1273.

Hemelrijk, C. K. (Ed.). (2005). *Self-organisation and evolution of social systems*. Cambridge: Cambridge University Press.

Hemelrijk, C. K., & Hildenbrandt, H. (2008). Self-organized shape and frontal density of fish schools. *Ethology, 114*, 245–254.

Hemelrijk, C. K., Hildenbrandt, H., Reinders, J., & Stamhuis, E. J. (2010). Emergence of oblong school shape: Models and empirical data of fish. *Ethology, 116*, 1099–1112.

Hemelrijk, C. K., & Hildenbrandt, H. (2012). Schools of fish and flocks of birds: Their shape and internal structure by self-organization. *Interface Focus, 2*, 726–737. https://doi.org/10.1098/rsfs.2012.0025

Hemelrijk, C. K., & Kunz, H. (2005). Density distribution and size sorting in fish schools: An individual-based model. *Behavioral Ecology, 16*, 178–187.

Hemelrijk, C. K., Meier, C. M., & Martin, R. D. (1999). 'Friendship' for fitness in chimpanzees? *Animal Behaviour, 58*, 1223–1229.

Hemelrijk, C. K., Meier, C. M., & Martin, R. D. (2001). Social positive behaviour for reproductive benefits in primates? A response to comments by Stopka et al. (2001). *Animal Behaviour, 61*, F22–F24.

Hemelrijk, C. K., van Laere, G. J., & van Hooff, J. (1992). Sexual exchange relationships in captive chimpanzees? *Behavioural Ecology and Sociobiology, 30*, 269–275.

Hemelrijk, C. K., van Zuidam, L., & Hildenbrandt, H. (2015). What underlies waves of agitation in starling flocks. *Behavioural Ecology and Sociobiology, 69*, 755–764. https://doi.org/10.1007/s00265-015-1891-3

Hemelrijk, C. K., & Wantia, J. (2005). Individual variation by self-organisation: a model. *Neuroscience and Biobehavioral Reviews, 29*, 125–136.

Hemelrijk, C. K., Wantia, J., & Isler, K. (2008). Female dominance over males in primates: Self-organisation and sexual dimorphism. *PloS One, 3*, e2678.

Hildenbrandt, H., Carere, C. C., & Hemelrijk, C. K. (2010). Self-organized aerial displays of thousands of starlings: A model. *Behavioural Ecology, 21*(6), 1349–1359.

Hogeweg, P., & Hesper, B. (1983). The ontogeny of interaction structure in bumble bee colonies: A MIRROR model. *Behavioral Ecology and Sociobiology, 12*, 271–283.

Hogeweg, P., & Hesper, B. (1985). Socioinformatic processes: MIRROR Modelling methodology. *Journal of Theoretical Biology, 113*, 311–330.

Honk, C. v., & Hogeweg, P. (1981). The ontogeny of the social structure in a captive Bombus terrestris colony. *Behavioral Ecology and Sociobiology, 9*, 111–119.

Hsu, Y., Earley, R. L., & Wolf, L. L. (2006). Modulation of aggressive behaviour by fighting experience: mechanisms and contest outcomes. *Biological Reviews, 81*, 33–74.

Huse, G., Railsback, S., & Ferno, A. (2002). Modelling changes in migration pattern in herring: collective behaviour and numerical domination. *Journal of Fish Biology, 60*, 571–582.

Huth, A., & Wissel, C. (1992). The simulation of the movement of fish schools. *Journal of Theoretical Biology, 156*, 365–385.

Huth, A., & Wissel, C. (1994). The simulation of fish schools in comparison with experimental data. *Ecological Modelling, 75/76*, 135–145.

Inada, Y., & Kawachi, K. (2002). Order and flexibility in the motion of fish schools. *Journal of Theoretical Biology, 214*, 371–387.

Itani, J. (1954). *The monkeys of Mt. Takasaki*. Tokyo: Kobunsha.

Jeanson, R., et al. (2005). Self-organized aggregation in cockroaches. *Animal Behaviour, 69*, 169–180.

Klimley, A. P. (1985). Schooling in Sphyrna lewini, a species with low risk of predation: a non-egalitarian state. *Zeitschrift fürTierpsychology, 70*, 279–319.

Koolhaas, J. M., de Boer, S. F., Buwalda, B., van der Vegt, B. J., Carere, C., & Groothuis, A. G. G. (2001). How and why coping systems vary among individuals. In D. M. Broom (Ed.), *Coping with challenge. Welfare including humans*. Berlin: Dahlem University Press.

Krause, J. (1993). The effect of 'Schreckstoff' on the shoaling behaviour of the minnow: a test of Hamilton's selfish herd theory. *Animal Behaviour, 45*, 1019–1024.

Krause, J., & Tegeder, R. W. (1994). The mechanism of aggregation behaviour in fish shoals: Individuals minimize approach time to neighbours. *Animal Behaviour, 48*, 353–359.

Kummer, H. (1974). *Rules of dyad and group formation among captive baboons (Theropithecus gelada)* (pp. 129–160). Basel: S. Karger.

Kunz, H., & Hemelrijk, C. K. (2003). Artificial fish schools: Collective effects of school size, body size, and body form. *Artificial Life, 9*, 237–253.

Kunz, H., Züblin, T., & Hemelrijk, C. K. (2006). On prey grouping and predator confusion in artificial fish schools. In L. M. Rocha et al. (Eds.), *Proceedings of the tenth international conference of artificial life* (pp. 365–371). Cambridge, MA: MIT Press.

Lee, S. H. (2006). Predator's attack-induced phase-like transition in prey flock. *Physics Letters A, 357*, 270–274.

Mazur, A. (1985). A biosocial model of status in face-to face primate groups. *Social Forces, 64*(2), 377–402.

Meier, C., Hemelrijk, C. K., & Martin, R. D. (2000). Paternity determination, genetic characterization and social correlates in a captive group of chimpanzees (Pan troglodytes). *Primates, 41*, 175–183.

Moritz, R. F. A., Kryger, P., & Allsopp, M. H. (1996). Competition for royalty in bees. *Nature, 384*(6604), 31.

Nottestad, L., & Axelsen, B. E. (1999). Herring schooling manoeuvres in response to killer whale attacks. *Canadian Journal of Zoology-Revue Canadienne de Zoologie, 77*, 1540–1546.

Oboshi, T., Kato, S., Mutoh, A., & Itoh, H. (2003). Collective or scattering: evolving schooling behaviors to escape from predator. In R. K. Standish, M. Bedau, & H. A. Abbass (Eds.), *Artificial life VIII* (pp. 386–389). Cambridge, MA: MIT Press.

Oster, G. F., & Wilson, E. O. (1978). *Caste and ecology in social insects*. Princeton: Princeton University Press.

Pagel, M., & Dawkins, M. S. (1997). Peck orders and group size in laying hens: 'future contracts' for non-aggression. *Behavioural Processes, 40*, 13–25.

Parrish, J. K. (1993). Comparison of the hunting behavior of 4 piscine predators attacking schooling prey. *Ethology, 95*, 233–246.

Parrish, J. K., & Viscido, S. V. (2005). Traffic rules of fish schools: A review of agent-based approaches. In C. K. Hemelrijk (Ed.), *Self-organisation and the evolution of social behaviour* (pp. 50–80). Cambridge: Cambridge University Press.

Pfeifer, R., & Scheier, C. (1999). *Understanding intelligence*. Cambridge, MA: MIT Press.

Pitcher, T. J., & Wyche, C. J. (1983). Predator avoidance behaviour of sand-eel schools: Why schools seldom split? In D. L. G. Noakes, B. G. Lindquist, G. S. Helfman, & J. A. Ward (Eds.), *Predators and prey in fishes* (pp. 193–204). Dordrecht: Dr. W. Junk Publications.

Post, D. J. v. d., Franz, M., & Laland, K. N. (2016). Skill learning and the evolution of social learning mechanisms. *BMC Evolutionary Biology, 16*, 166.

Preuschoft, S., Paul, A., & Kuester, J. (1998). Dominance styles of female and male Barbary macaques (Macaca sylvanus). *Behaviour, 135*, 731–755.

Procaccini, A., Orlandi, A., Cavagna, A., Giardiina, I., Zoratto, F., Santucci, D., et al. (2011). Propagating waves in starling, Sturnus vulgaris, flocks under predation. *Animal Behaviour, 82*, 759–765. https://doi.org/10.1016/j.anbehav.2011.07.006

Puga-Gonzalez, I., Butovskaya, M., Thierry, B., & Hemelrijk, C. K. (2014). Empathy versus parsimony in understanding post-conflict affiliation in monkeys: model and empirical data. *PloS One, 9*, e91262. https://doi.org/10.1371/journal.pone.0091262

Puga-Gonzalez, I., Hildenbrandt, H., & Hemelrijk, C. K. (2009). Emergent patterns of social affiliation in primates, a model. *PLoS Computational Biology, 5*, e1000630. https://doi.org/10.1371/journal.pcbi.1000630

Ramos-Fernández, G., Boyer, D., & Gómez, V. P. (2006). A complex social structure with fission-fusion properties can emerge from a simple foraging model. *Behavioral Ecology and Sociobiology, 60*(4), 536–549.

Resnick, M. (1994). *Turtles, termites and traffic jams: Explorations in massively parallel microworlds*. Cambridge, MA: MIT Press.

Reynolds, C. W. (1987). Flocks, herds and schools: A distributed behavioral model. *Computers and Graphics, 21*, 25–36.

Robinson, G. E. (1998). From society to genes with the honey bee. *American Scientist, 86*, 456–462.

Robinson, G. E., & Page, R. E. (1988). Genetic determination of guarding and undertaking in honey-bee colonies. *Nature, 333*, 356–358.

Romey, W. L. (1995). Position preferences within groups: do whirligigs select positions which balance feeding opportunities with predator avoidance? *Behavioral Ecology and Sociobiology, 37*, 195–200.

Rosenblum, L. A., & Nadler, R. D. (1971). The ontogeny of sexual behavior in male bonnet macaques. In D. H. Ford (Ed.), *Influence of hormones on the nervous system* (pp. 388–400). Basel: Karger.

Seyfarth, R. M. (1977). A model of social grooming among adult female monkeys. *Journal of Theoretical Biology, 65*, 671–698.

Smuts, B. B. (1987). Gender, aggression and influence. In B. B. Smuts, D. L. Cheney, R. M. Seyfarth, R. W. Wrangham, & T. T. Struhsaker (Eds.), *Primate societies* (pp. 400–412). Chicago: Chicago University Press.

Solé, R. V., Bonabeau, E., Delgado, J., Fernádez, P., & Marín, J. (2001). Pattern formation and optimization in army ant raids. *Artificial Life, 6*, 219–226.

Stanford, C. B. (1996). The hunting ecology of wild chimpanzees: Implications for the evolutionary ecology of Pliocene hominids. *American Anthropologist, 98*, 96–113.

Stanford, C. B. (1998). The social behaviour of chimpanzees and bonobos. *Current Anthropology, 39*, 399–420.

Stroebe, K., Nijstad, B. A., & Hemelrijk, C. K. (2016). Female Dominance in Human Groups: Effects of sex ratio and conflict level. *Social Psychological and Personality Science, 8*, 1–10. https://doi.org/10.1177/1948550616664956

Sumpter, D. J. T. (2010). *Collective animal behavior.* Princeton and Oxford: Princeton University Press.

te Boekhorst, I. J. A., & Hogeweg, P. (1994a). Effects of tree size on travelband formation in orang-utans: Data-analysis suggested by a model. In R. A. Brooks & P. Maes (Eds.), *Artificial life* (Vol. IV, pp. 119–129). Cambridge, MA: The MIT Press.

te Boekhorst, I. J. A., & Hogeweg, P. (1994b). Selfstructuring in artificial 'CHIMPS' offers new hypotheses for male grouping in chimpanzees. *Behaviour, 130*, 229–252.

Theraulaz, G., Bonabeau, E., & Deneubourg, J.-L. (1998). Threshold reinforcement and the regulation of division of labour in social insects. *Proceedings of the Royal Society of London B, 265*, 327–333.

Thierry, B. (1990a). Feedback loop between kinship and dominance: The macaque model. *Journal of Theoretical Biology, 145*, 511–521.

Thierry, B. (1990b). The state of equilibrium among agonistic behavior patterns in a group of Japanese macaques (Macaca fuscata). *Comptes rendus de l'Académie des sciences. Série III, Sciences de la vie, 310*, 35–40.

Thierry, B., Gauthier, C., & Peignot, P. (1990). Social grooming in Tonkean macaques (Macaca tonkeana). *International Journal of Primatology, 11*, 357–375.

Trivers, R. L. (1971). The evolution of reciprocal altruism. *Quarterly Review of Biology, 46*, 35–57.

Tutin, C. E. G. (1980). Reproductive behaviour of wild chimpanzees in the Gombe National Park. *Journal of Reproduction and Fertility. Supplement, 28*, 43–57.

Ulbrich, K., Henschel, J. R., Jeltsch, F., & Wissel, C. (1996). Modelling individual variability in a social spider colony (Stegodyphus dumicola: Eresidae) in relation to food abundance and its allocation. In: *Revue suisse de Zoologie (Proceedings of the XIIIth International Congress of Arachnology, Geneva, 3–8 September 1995), hors série, 1*, pp. 661–670.

van Schaik, C. P. (1989). The ecology of social relationships amongst female primates. In V. Standen & G. R. A. Foley (Eds.), *Comparative socioecology, the behavioural ecology of humans and other mammals* (pp. 195–218). Oxford: Blackwell.

van Schaik, C. P., & van Hooff, J. A. R. A. M. (1983). On the ultimate causes of primate social systems. *Behaviour, 85*, 91–117.

Verbeek, M. E. M., de Goede, P., Drent, P. J., & Wiepkema, P. R. (1999). Individual behavioural characteristics and dominance in aviary groups of great tits. *Behaviour, 136*, 23–48.

Vicsek, T., Czirók, A., Ben-Jacob, E., Cohen, I., & Shochet, O. (1995). Novel type of phase transition in a system of self-driven particles. *Physical Review Letters, 75*, 1226–1229.

Wantia, J. (2007). Self-organised dominance relationships: A model and data of primates. In *Wiskunde en Natuurwetenschappen* (p. 115). Groningen: Rijksuniversiteit Groningen.

Wilkinson, G. S. (1988). Reciprocal altruism in bats and other mammals. *Ethology and Sociobiology, 9*, 85–100.

Yerkes, R. M. (1940). Social behavior of chimpanzees: Dominance between mates in relation to sexual status. *Journal of Comparative Psychology, 30*, 147–186.

Chapter 25
Agent-Based Simulation as a Useful Tool for the Study of Markets

Juliette Rouchier

Abstract This chapter describes a number of agent-based market models. They can be seen as belonging to different trends in that different types of markets are presented (goods markets, with or without stocks, or financial markets with diverse price mechanisms or even markets with or without money), but they also represent different aims that can be achieved with the simulation tool. For example, it is possible to develop precise interaction processes to include loyalty among actors; try to mimic as well as possible the behaviour of real humans, which have been recorded in experiments; or try to integrate psychological data to show a diffusion process. All these market models share a deep interest in what is fundamental in agent-based simulation, such as the role of interaction, interindividual influence and learning, which induces a change in the representation that agents have of their environment.

Why Read This Chapter?
To understand the various elements that might be needed in a simulation of a market, including some options for implementing learning by agents in a market simulation, the role and kinds of market indicator and kinds of buyer-seller interaction (bargaining, loyalty based and reputation based). The chapter will give you an overview of the complexity of markets, including multi-goods markets and complicated/decentralized supply chains. It will help you understand financial markets (especially in contrast to markets for tangible goods) and the double-auction mechanism. Finally, it will give some indications of how such models either have informed or might inform the design of markets.

J. Rouchier (✉)
LAMSADE, Paris-Dauphine, 75775 Paris Cedex 16, France
e-mail: juliette.rouchier@dauphine.fr

© Springer International Publishing AG 2017   671
B. Edmonds, R. Meyer (eds.), *Simulating Social Complexity*,
Understanding Complex Systems, https://doi.org/10.1007/978-3-319-66948-9_25

25.1   Introduction

In recent years, there has been a growing recognition that an analysis of economic systems—which includes markets—as being complex could lead to a better under-standing of the participating individuals' actions. In particular, one element that could be incorporated in such analyses is the heterogeneity of agents and of their rationality. For example, the existence of multiple prices on a market for the same product, sold at the same moment and at the same place, cannot be captured in an equilibrium model, whereas it appears in real life and can be reproduced easily in an agent-based model (Axtell 2005).

This issue is at the centre of much debate among economists. In classical economy, agents are considered as rational, having a perfect knowledge of their environment, and hence are homogeneous. This view of agents stayed unchallenged for a while. Friedman (1953) argued, for example, that non-rational agents would be driven out of the market by rational agents, who would trade against them and simply earn higher profits. However, in the 1960s, the view on rationality evolved. Even Becker (1962) suspected that agents could be irrational and yet produce the same results as rational agents (i.e. the negative slope of market demand curves). However, the author who definitely changed the view on economic agents is Simon, who stated that any individual could be seen as a "bounded rational" agent, which means that it has an imperfect knowledge and has limited computing abilities (Simon 1955). In most markets, agents do not have perfect knowledge of the behaviour and preferences of other agents, which makes them unable to compute an optimal choice. If they do have perfect knowledge, they will require unlimited computational capacity in order to calculate their optimal choices.[1]

Indeed, for some contemporary authors, the understanding that one can get of real market dynamics is more accurate if the assumption of a representative agent or of the homogeneity of agents is dropped at the same time as the perfect knowledge assumption (Kirman 2001). The way bounded rationality is approached can be very formal and tentatively predictive (Kahneman and Tversky 1979), but some authors go even further by stating that the notion of rationality has been abandoned to be changed to the idea that agents possess rules of behaviours that they select thanks to diverse mechanisms, which are most of the time based on past evaluation of actions (Kirman 2001). Some authors also compare the different results one can get from a

---

[1]For example, an issue that anyone representing learning (not only on markets) has to face is the exploration-exploitation dilemma. When an action gives a reward that is considered as "good", the agent performing it has to decide either to continue with this action—and hence possibly miss other, more rewarding actions—or to search for alternative actions, which implies indeterminate results. Leloup (2002), using the multi-armed bandit (Rothschild 1974) to represent this dilemma, showed that a nonoptimal learning procedure could lead to a better outcome than an optimal—but non-computable—procedure.

representative agent approach and a bounded rationality approach for agents and try to integrate both simplicity and complexity of these two points of view (Hommes 2007).

To complete the view that agents have bounded rationality in a market and that they evolve through time, it is necessary to consider one more aspect of their belonging to a complex system: their interactions. The seminal works on markets with interacting heterogeneous agents date back to the beginning of the 1990s (e.g. Palmer et al. 1993; Arthur et al. 1997a); many of them were collected in the post-proceedings of a conference at the Santa Fe Institute (Arthur et al. 1997b). Since then, a large part of research in agent-based simulation concerns market situations, as can, for example, be seen in the cycles of WEHIA/ESHIA conferences and ESSA conferences. The former ones gather many physicists who apply dynamic systems techniques to representing heterogeneous interacting agents to deal with economics issues, and they tend to buy the "agent-based computational economics" (ACE) approach of simulated markets promoted by Leigh Tesfatsion[2] (Phan 2003). In the latter ones, not only economy but also sociology, renewable resources management and computer science participate and often try to generate subtle representations of cognition and institutions and a strong view of agents as computerized independent entities to deal with broad social issues and are thus being closer to multi-agent social simulation (MAS). We will refer to both terms (ACE or MAS) separately.

Not only the techniques can be different when studying markets with distributed agents, but also the aims can be. Some try to infer theoretical results about rationality and collective actions (Vriend 2000) or about market processes (Weisbuch et al. 2000). Others want to create algorithms to represent human rationality on markets and try to assess the value of these algorithms by comparing the simulated behaviour with actions of real humans in order to understand the latter (Hommes and Lux 2008; Duffy 2001; Arthur 1994). Eventually some explorations about the impact of diverse rationalities in a market context enable the identification of possible worlds using a sort of artificial society approach (Rouchier et al. 2001).

Being part of a handbook, this chapter should provide tools to be able to build and use agent-based simulation techniques to create artificial markets and analyse results. However, a "know-how" description of the building of artificial market is so dependent on the type of issue that is addressed, that it was decided instead to establish a classification of the type of markets and modelling techniques that can be found in agent-based simulation research. We are interested in representations of markets that focus on a micro-representation of decision processes, with limited information or limited computational abilities for agents and which take place in a precise communication framework. This form of representation forces authors to focus on new notions such as interaction or information gathering, and this induces new issues such as the bases for loyalty or the exploration-exploitation dilemma. The use of stylized facts is fundamental in this branch, where modellers try to mimic some elements of the world in their research, focusing either on representations

---

[2]http://www.econ.iastate.edu/tesfatsi/ace.htm.

of reasoning that are inferred from observed micro-behaviours or trying to mimic global behaviour through learning algorithms, thereby to stay closer to common view orthodox economics.

Contrary to this view, which does not distinguish among individual rationalities and assumes aggregate, centralized knowledge and decision-making, researchers involved in the use of multi-agent simulation usually try to understand the local point of view of agents and its influence on global indicators. Since the agent is then seen as unable to have complete knowledge, it has to accumulate data about its environment and treat those data according to its aims. The study of markets is interesting when it comes to this type of analysis because markets display a much more simple set of possible actions and motivations than many other social settings. Income and reproduction of activity are direct motivations, which imply choices in the short and the long term, prices and quantities are what have to be chosen (as well as sometimes acquaintances, in the case of bilateral bargaining), and information is limited to offers and demands, as well as responses to these two types of proposals.

This chapter presents the main fields of application for multi-agent simulation dealing with markets, to show how researchers have focused on different aspects of this institution and to conclude on the great interest of agent-based simulation when trying to understand the very dynamics of these social environments.

In the next section, we will describe the main notions that are covered by the term "market" in agent-based literature and also the main ways to represent rationality and learning that can be found. Subsequently, the main topics of market studies will be described in three parts. In Sect. 25.3, agents are on a market and actually meet others individually, having private interactions with each other. Choices that have to be modelled are about matching business partners, as well as all buying or selling decisions. In all other sections, agents are facing an aggregate market, and they have to make decisions based on global data, sometimes associated to networks. In part four, agents are either consumers or producers in a large market. In part five, we will deal with auctions, financial markets and market design.

25.2 Market and Agents' Reasoning

Although the study of market(s) has been predominantly carried out by economists, ethnographers and sociologists have also been active in this domain (Geertz et al. 1979; White 1995), and the field is now being developed through field studies and analysis. The main difference for those two approaches is that economists generally build models that are rather abstract and formal, whereas ethnologists and sociologists describe actual markets after observing them and generally produce models that are based on classifications of large amount of data. The notion of market itself has a double meaning, even more important with the increasing use of the Internet: it is at the same time the institution that enables individuals to coordinate their actions through the fixing of a price or, alternatively, a physical place where buyers meet sellers. There is no easy decision to choose the scale to study when dealing with market, neither is the limit of observation that is needed in the supply chain to understand a phenomenon easy to set. In agent-based simulation, markets (being open or closed to new entries) are represented as closed societies with specified agents that are possibly interconnected.

Simulations can be based on very specific case studies and in order to describe as accurately as possible the behaviour of real actors, but they can also be mainly theoretic and in an attempt to generate expected theoretical results. In all cases, a simulated market cannot be implemented as described in neoclassical economic theory since agents need to be independently specified and interact directly with one another during the simulation. For example, to develop a model where individual agents have to make a decision, a demand curve (that gives for any price of a product, the number of agents that are ready to buy or sell at this price) cannot be imposed on the model but has to be derived from the determinants of agent behaviour. One approach is to distribute reservation prices (the maximum price for buying or minimum for selling) among agents which can then be used to aggregate demand and offer curves.

The main elements to define a market model are given in the next section. We will then describe a few approaches to rationality and learning for agents in markets that depend on the type of market and goods that are represented. A similar analysis, more oriented towards consumers' behaviour, can be found in Jager (2007).

25.2.1 Main Elements to Build an Artificial Market

Several dimensions are important on a market, and each description of an element is easy to relate to a dimension in the building of an artificial system with agents (simply put, the market institution and the agents' rules of behaviour). Axtell (2005) proposes a very abstract view of decentralized exchange on an agent-based market, where he gives no explanation of the bargaining process that organizes the exchange but shows the existence of a computationally calculable equilibrium to increase all agents' utility. Here, the aim is to find out actual processes that can be used by modellers to represent markets.

The first distinction that can be made when developing a model is to know if one is building the representation of a *speculative market* or a *goods market*. What I call a speculative market, typically a financial one, is such that agents who have a commodity can keep it, sell it or buy it. They have to anticipate prices and wait to perform their actions so as to make the highest profit. Seminal works on agent-based simulation were related to speculative markets, which display interesting regularities in their stylized facts. A large body of literature has developed on this topic, which is also due to the fact that data to calibrate models are more easily available than for nonspeculative markets. On a goods market, agents have only one role, to sell or buy a certain number of units of one or more products, and they usually have a reservation price to limit the prices they can accept. The products can be perishable (with an intrinsic value that decreases over time) or durable so that stock management is an issue.

Then, both types of market can be organized either through auctions (with diverse protocols: double-auction, ascending, descending, with posted prices or continuous announcements) or via pairwise interactions which imply face-to-face negotiation (with many different protocols, such as "take-it-or-leave-it", one shot negotiation, a series of offers and counter-offers, the possibility for buyers to explore several sellers or not). In the design of the protocol, it can also be important to know if money exists in the system or if agents exchange one good for another directly.

Agents' cognition must entail some choice algorithm. Agent-based simulation is almost always used to design agents with bounded rationality because agents have limited computational abilities or limited access to information. They have tasks to perform, within a limited framework, and have to make decisions based on the context they can perceive. Most of the time, they are given a function, equivalent to a utility function in economics, that associates a value to each action, enabling the agents to classify the profit it gets and hence to compare actions. First, an agent must have constraints in its actions, in order to be able to make arbitration between all possible options:

- Each commodity is associated *a reserve price:* if a buyer (resp. seller) goes on a market, there is a maximum (resp. minimum) price it is willing to pay for the good.
- The importance of obtaining a commodity can be indicated by the price of entry to the market. Agents have a greater incentive to buy and get a 0 profit, rather than not buying. The constraint for selling can be the same.
- In some papers, the price is not represented in the system, and the acquisition of a product is limited by a utility function, where the agent acquires the product only if it makes enough profit.
- In the case of negotiation, time constraints are usually put on buyers who can visit a limited number of sellers, having hence a limit on their search for a commodity.
- There can be a discount factor: at each period, the risk of seeing the market close is constant, and hence agents never know if they will be able to trade at the next period.

The type of decisions that agents have to perform on a market:

- *For buyers*: how many units to buy, who to visit, how to decide to stay in a queue depending on its length, which price to propose/accept, which product to accept and more fundamentally to participate or not.
- *For sellers*: how to deal with the queue of buyers (first-come-first-served or with a preference for some buyers), which offer to make or accept and, in the case of repeated markets, how many units to buy for the next time step and, in the case of a market where quality is involved, which type of product to propose or build for the next time step

Due to the increasing complexity when adding another type of decision to a model, it is rare that all of these decisions will be made in one single model. For example, although interactions could potentially be represented in a continuous way, I know of no model where it is the case: all choices and meetings are made and messages sent at discrete time steps.

25.2.2 Agents' Learning

As said before, in most markets that are studied with agent-based models, the central element is that agents are heterogeneous in knowledge as well as in need. This situation can be decided from initialization or can emerge during the course of the simulation, while agents learn. Another element that is rarely given at initialization but is acquired by agents while learning is information about other agents' characteristics. In most cases, this learning takes place as a result of action, at the same time as the acquisition of an object or the acquisition of money.

The way learning is organized is generally linked to a performance of actions, with a selection of actions that "satisfice" or give the best performance. On a market, it is often assumed that agents are interested in getting the highest payoff for their individual actions: the performance is either the profit that agents get from their sells or the utility they get from consuming the product. In most models, learning agents have a set of predefined actions they can take, and they have to select the one they like the best, following a probabilistic choice.

One of the simplest learning models is reinforcement learning (Erev et al. 1999; Bendor et al. 2001; Macy and Flache 2002; see also Chap. 20 (Macy et al. 2017) in this volume), where agents attribute a probability of choice for each possible action that follows a logit function. The algorithm includes a forgetting parameter and a relative weight attributed to exploitation (going to actions known as having high value) and exploration (the random choice part in the system). The exploration parameter can be fixed during the simulation, where the level of randomness has to be chosen or can vary during the simulation (increase) so that there is a lot of exploration at the beginning of the simulation and, as time passes, agents focus on the "best" actions. This issue of which level of exploration and exploitation to put in a learning algorithm is of current concern in the literature about markets (Moulet and Rouchier 2007).

Another model of rationality for agents is based on a representation of strategies of other agents, fictitious play: a distribution of past actions is built by each agent, and they can then infer the most probable set of actions of others and hence choose their optimal behaviour (Boylan and El-Gamal 1993). The EWA model has been proposed by Camerer and Ho (1999) to gather both characteristics of these models: the agent not only learns what profit it got for each action but also computes notional gains for each possible action and attributes the resulting notional profit to each of those possible actions.

A slightly more complex representation of knowledge commonly found in the literature is the classifier system where each decision is made by considering a context, a choice and the past profit made in this precise context by making this special choice (Moulet and Rouchier 2007). This type of algorithm is very similar to what Izquierdo et al. (2004) call case-based learning, but it does not seem to be applied to market situations. In general the number of possible actions is fixed from the beginning, but the classifier system can be associated to a genetic algorithm that generates new rules over the time (Kopel and Dawid 1998). Genetic algorithm learning is also a quite usual way to represent learning, where the information that is used by agents to estimate the profit of each rule can be based on actual past actions (Vriend 2000) or also on the imaginary profit of all possible actions considering the actions of others (Hommes and Lux 2008). The presence of other agents can also be relevant information when agents use imitation or social learning.

Brenner (2006) undertook an extensive review of common learning processes.[3] In this paper, an interesting element arises: the way "satisficing" rationality can be developed by having agents not look for the best action but for one which enables them to get a "good enough" profit; the notion of "good enough", called "aspiration level", can then evolve during the simulation (Cyert and March 1963).

An alternative to learning algorithms that are only based on profit is to consider that agents have a social utility, which need not have the same dimensionality as profit. Indeed, it has been demonstrated that translating the social utility in costs or profits that can be added to monetary profit gives a dynamics of learning and behaviour that is radically different from a situation where agents reason in two dimensions, social and monetary, with a lexicographic ordering (Rouchier et al. 2001). A way to implement social utility without lexicographic ordering is to include in the utility the appreciation of similarity of members of the network, such as in consumer choice models (Delre et al. 2007).

In most models, action and information gathering are made in one action, and circulation of information as such is not really modelled. The reason is certainly because it would take modellers too far from the neoclassical economic approach to market, where the only information is the observation of transactions, sometimes of intermediate prices as in auctions or, sometimes, bargaining. One model of a market by Moss (2002) represents communication among agents before exchange takes place, and Rouchier and Hales (2003) (whose model evolved into the one of Rouchier (2004)) also allocate one period out of three every time step for agents to look for information.

---

[3]The main objection to Brenner's exposition is the lack of homogeneity of notation, which makes the algorithms difficult to compare and maybe to implement.

25.2.3   Indicators and Method

Several types of modelling can be found in papers about markets, just like in any other application domains of simulation. Some prefer to work at a purely abstract level, while others try to fit as well as possible data that they extract from observation and experience. Whatever the approach, indicators that are often observed in markets are prices, efficiency (the total profit that is extracted from agents compared to the maximum profit that could be extracted) and relative power of different agents. The notion of convergence is central to the modelling of markets, since most research refers to economics and has to compare results to economic static equilibrium. In some cases, what is observed is interaction patterns, which can be represented as the random part of the agents' choice when interacting, e.g. the number of different sellers that a buyer meets in a number of steps. In bargaining models and in general exchange models, the (non-)existence of an exchange is also something that is observed. Sometimes, the cognition of agents themselves is observed: their belief about others' preferences or even the distribution of propensities to choose sellers.

Among all the indicators that can be observed in the model, these last ones cannot be observed in real life and hence cannot be compared to human systems. In a lot of models, agents' behaviour is compared to that of humans in order to establish the validity of the cognitive model. The data are very rarely extracted from real-life situation (although it sometimes happens) but are mainly constructed via experiments. Experimental economists control all information circulation and record all actions of agents. It is thus possible to compare in a very precise and quantitative way the global behaviour of the group and individual behaviour, on one side with artificial agents and on the other side with human agents. Real-life situation can also be seen as the mix of human and of artificial agents, such as in financial online markets.

Other researchers do not want to match data too precisely. As Vriend (2006) says, agent-based models, like any other models, are abstract settings that have to be interpreted as such. The comparison between real and abstract data should go through a first step which is the building of stylized facts that are already a summary of human behaviours, where only the most striking elements are integrated. Vriend is much more interested in the reaction of his abstract model to changes in parameters and in its self-consistency. It could be said that by construction, a model can only capture a small part of human cognition, which is full of long-term experiments and memories and should not be compared to quantitative data without caution.

Eventually, some researchers want their model to influence real life and try to use the results they find to give advices on the way to build markets. Different ways to fit models with real life will be found in each example of a model—be it to fit precisely, to fit stylized facts or to be an abstract study of the effect of some rules in a social setting.

25.3   Buyer-Seller Interactions

In the literature about agent-based markets, a great attention has been given to the analysis of local interactions, hardly ever studied in classical economics, apart from rare exceptions (Rubinstein and Wolinsky 1985). An aim is to reproduce as well as possible the features of real markets. It is indeed to be noticed in real observation that buyers display regularity in the choice of sellers with whom to interact and that this regularity emerges in time with experience—this attempt to reproduce patterns of interaction is the only way to understand, rather than just describe, the way individuals coordinate (Weisbuch et al. 2000). Authors study local bargaining processes with great care, as well as repetition of interactions over several days, where choices are not only based on prices but also on the fact that buyers need to buy and sellers need to sell. The basic feature of these models is pairwise interactions on markets with several sellers and buyers where prices are not fixed and result from negotiation only. The number of visits buyers can undertake, the way sellers manage queues or the number of steps it takes to negotiate a price is different in all these systems that focus only on small parts of the whole set of stylized facts that are observable on such markets. Some aspects that are often studied are described in the following, and the subsequent choices for modelling the organization of pairwise interactions are given.

25.3.1   Bargaining Processes

Brenner (2002) studies agents' learning in a bilateral bargaining market, focusing on the convergence of prices and the dynamics of bargaining. There is one commodity in the market, and buyers and sellers meet at every time step to exchange it. Each buyer can choose one seller for each step, with the selection based on the price being acceptable. The sellers respond to buyers waiting in their queue in order of arrival by proposing a price. Buyers have to decide who to visit; sellers have to decide on the first price to propose and the number of subsequent proposals if the buyer rejects the offer, bargaining being costly for both agents. All decisions are made following *reinforcement learning* based on past experience. Hence, all choices are based on the satisfaction that is associated with each past action and on a rigidity variable. A buyer will continue to choose a seller as long as he is satisfied. His probability to change depends on his expectations with another agent. A seller also calculates the probability to change behaviour depending on his belief about what he would gain by performing another choice.

The rigidity parameter, which is the opposite of noise in the system, has a great impact on results. If rigidity is high, buyers keep visiting the same seller. The cost of bargaining also is important: if it is relatively high, sellers learn to offer the price they know to be acceptable to buyers, and they do not bargain after a few rounds. In this system, since the relations are so individual, the convergence of the price overall is not very fast and can be highly variable for a long time, although converging in the end. The model is extremely sensitive to all parameters that define aspiration levels for agents.

Brenner's paper is of the class that compares simulation to theoretical results. Here, sub-game equilibria are used to compare the possible outcomes and their efficiency to the generated prices and behaviours. There is no reference to any real-world data. However, it is interesting that both micro-behaviour (the number of bargaining steps) and macro-data (prices) are of importance, justifying an agent-based analysis.

Influenced by this paper, but referring to real data, Moulet and Rouchier (2007) reported a bargaining model based on two sets of data: qualitative, from a field study in the wholesale market of Marseilles (Rouchier and Mazaud 2004), and quantitative, giving all proposals, intermediate and final prices for a series of transactions in the same market (Kirman et al. 2005). Like the previous model, the market gathers buyers and sellers who meet at every time step. However, buyers can visit several sellers in one market opening. The choice for a buyer has several dimensions: to decide which seller to visit, to decide to accept an offer or to reject it and to propose a counter-offer or leave and which value to counter-offer. Sellers must choose the first price to offer and to accept buyer's counter-offers or not and the value of the second offer they can make. In this model, decisions evolve following classifier system learning, where each agent evaluates a list of possible options following his past experience. The results that are produced are compared with indicators derived from real-world data: values of offers and counter-offers of the agents that vary depending of the kind of product that is purchased and ex post bargaining power of sellers (which is the difference between the first offer and the price of transaction compared to the difference between the counter-offer and the price of transaction).

In the simulations, the values that are obtained fit the data quite well in that the observed bargaining sequences and agents' behaviours are reproduced. The two main parameters are the number of sellers that agents can meet (from one to four) and the speed of learning of sellers. The relative importance of learning for the agents can be seen as situating them in a negotiation for in-season goods and a negotiation for out-of-season goods. The model produces results similar to those of out-of-season goods when agents have to learn repeatedly, when there is no known market price but a lot of heterogeneity in the buyers' readiness to pay. In the case of in-season goods, the market price is more settled, and agents do not explore the possible values of goods as much, relying instead on their experience. Between the different in-season goods, the main difference could be the number of visits buyers make, but this number tends to reduce after a learning period, when buyers have selected their preferred seller. This aspect of the model—the growing loyalty of agents—is not the centre of the research and was represented mainly with the aim of matching the actual behaviours of the market actors. Other papers, described in the following section, are more focused on this issue.

Another direction for the study of bargaining processes is related to the creation of robots or artificially adaptive agents (AAA) to participate in electronic commerce

(Oliver 1997). Such models focus on complicated negotiations in that they integrate several dimensions of trade in the deal: price, quantity and delivery time. The main argument for the value of the algorithm that is proposed in the paper is that the agents learn to negotiate at least "as well as humans", which means that as many negotiations lead to an agreement as in human bargaining situations so that profit is extracted from both sides of the bargaining. The bargaining consists of several steps, where a customer reacts to the first offer by comparing its profit to a threshold and the offer is accepted if it is higher than the threshold and rejected with a counter-offer otherwise. Clearly, such models capture satisficing and bounded rationality rather than profit maximization. The bargaining can then carry on with several successive offers being made by customer and seller. Strategies for accepting and counter-offering evolve through a genetic algorithm. Five different games are used to test the learning, in a population of 20 agents with three rounds of bargaining at most, and each agent is given 20 chromosomes for decision-making. It is then proven that AAA perform better than random, that agents are able to learn general strategies that can be used against different bargaining partners and eventually that AAA perform as well as humans (depending on the game, sometimes better and sometimes worse, maybe depending on affective values for humans) in terms of number of agreements that are reached. This is an interesting result to consider when one wants to introduce artificial agents into electronic markets, since one wants to be able to reach as many agreements as possible.

25.3.2   Loyalty

Loyalty is present in quite a few models of markets where agents interact repeatedly. It is popular to deal with this topic with agents, mainly because it is related to two main advances of agent-based modelling: heterogeneity and interactions. There exist two representations of this loyalty in the literature: either fixed loyalties, assumed in order to understand its impact (Rouchier 2004), or emerging loyalties, as the result of endogenous interactions. Vriend refers to "endogenous interactions" when he uses individual learning to generate an evolution of interactions among agents (Vriend 2006). The idea is that agents learn to select which actions to perform as well as which agent to interact with; it is clear that this can lead to the apparition of loyalty and that it can take different regular patterns.

One main field where this loyalty issue has been important is the study of perishable goods markets (fruits and vegetables and fish). The participants of these markets are very dependent on the regularity—which implies predictability—of their interactions. The main reason is that buyers need to purchase goods almost every day: they have very little ability to stock, and they must provide their customers with all possible goods (a retailer can become unattractive to his customers just because of the lack of one commodity). In case of shortage, they need to have good relations with a seller to make sure the commodity will be available to them. Conversely, Rouchier (2004) shows in a model that the presence of loyal agents in a perishable goods market is necessary for the sellers to predict the right number of goods to provide every day. In this artificial market, two types of buyers interact with sellers: those that look for the cheapest prices ("opportunistic") and those that are faithful and try to get the product rather than to get it cheap ("loyal"). To be able to be opportunistic, agents first gather information about prices and then decide on the seller they want to meet to make the best transaction. The more opportunistic agents are present in the market, the more garbage is produced, and shortage occurs. Although there is some randomization of needs for the buyers, the presence of loyal agents makes the sellers estimate their stocks in the best way. This result holds for different learning algorithms (step-by-step learning, simple reinforcement learning and classifier systems). Considering that the field study on the fruits and vegetables market of Marseilles, France, showed that most of the agents are loyal (according to the definition of the model: first loyal and then try to find all the goods in a minimum of visits to sellers), this result can give a functional explanation of their action choices.

In a slightly different context, Rouchier et al. (2001) has represented the shape of emerging patterns of relations that could be created by two types of rationality with agents. The situation is a market-like situation, where offers are dependent on the situation in the preceding step, since the commodity is a renewable resource. Agents are herdsmen and farmers, with the latter selling access rights to their land. If none or if too many herdsmen are using the same land, it will get depleted, and hence the offer will be reduced. Two types of micro-behaviour are defined: either the herdsmen choose the cheapest farmers or they choose the ones that offered them access most often. In the first case, the simulations resulted in depletion of the resource, with congestion of demand for the cheapest farmers. The links that were created were highly stable (once an agent found the cheapest, it would not change), but on the other hand, agents could not readapt when there was a shock in the resource quantity (a drought) because everyone would converge to the same farms. With the second rationality, agents had a representation of a "good" farmer, which was only based on individual experience, and hence they would be heterogeneous. They would also have several "good" farmers to visit in case one was not available. This made them much more flexible in their choice, avoiding depletion of the resource, so everyone was better off. The macro-situation, although the process is different, also shows that a loyal micro-behaviour is a help to repartition of goods where there can be shortages. In this setting, the loyal micro-behaviour also enables a more equal repartition of gain among farmers as well as herdsmen.

Kirman and Vriend (2001) explored the emergence of loyalty in an artificial market based on a field study in the fish market of Marseille. The aim is to see loyalty emerge and in addition to see which emergent behaviour sellers display. They use classifier systems to represent learning, where their agents can have a lot of different actions, some of which are, a priori, not good for their profit. Through the exploration of good and bad possible actions, they select those that bring the highest profit in the past. Some buyers learn to be loyal, and those that learn this get higher profit than others in the long term (it is actually a co-evolution where sellers learn to offer lower prices to those that are loyal). The buyers are then differentiated:

their reservation price is heterogeneous (e.g. to represent that they do not sell their fish to the same population, some are in rich neighbourhood, some in poor ones). Sellers on the market learn to discriminate, and they offer higher prices to those that have higher reservation prices. Eventually, some of the sellers get themselves specialized since only low-price buyers can visit them. Using a very basic learning where agents are not rational but learn by doing, the results are very satisfying because they reproduce stylized facts of the fish market.

A third model represents the same market but refers more to quantitative data of this market (Weisbuch et al. 2000). The data represents sales that took place over more than 3 years and concern 237,162 individuals. From these, it is possible to observe that most buyers who are faithful to a seller buy a lot of quantities every month. The model was built in two parts: one which is simple enough to generate analytical results and a second that displays more realistic hypotheses. In the analytical model, agents use the logit function to select their action (basic reinforcement learning), which means that their choice depends on a β value, between 0 and ∞, which decrease gives a higher propensity to randomly test all sellers and which increase induces a higher propensity to look for the best past interaction. Agents can either imitate others or only base their choice on their own learning. The results can be found using the mean field approach, coming from physics. It is shown that there are radically different behaviours—either totally loyal or totally "shop around agents" depending non-linearly on β.

The model becomes more complex with sellers being able to sell at two different prices, high and low. What can happen in this system is that a buyer becomes loyal to a seller when the price is low and remains loyal even after the price has switched to high. The only important thing is that, as seen before, the product is actually provided. One indicator that is used to synthesize diverse information of the model is "order", which is defined as the number of loyal agents. The more regular the agents, the more ordered the society. Although the results, once interpreted, are very coherent with real data in terms of possible states of the market, it is a bit difficult to understand precisely some concepts of the paper because it refers mainly to physics indicators that are translated into social indicators, but this translation is not always straightforward.

25.3.3 Reputation of Sellers

Pinyol et al. (2007) have developed a market model as a benchmark for a reputation-based learning algorithm for agents in a social system. The model integrates quality and judgement of a relationship. Reputation is used in the group to enable agents to gather enough information in a context when it is scarce. The market that is used is a rather simple institution, where buyers have to select one seller at each time step to buy one unit of a commodity. The quality of the commodity is different for each seller. For a buyer, the acquisition of a commodity of lower quality will give less utility than the acquisition of a commodity of high quality. Sellers have a limited quantity of units, which they can sell at any period (the commodity is non-perishable) and they disappear from the system when they have sold everything. The most important information for buyers is the quality of the commodity that each seller offers. However, when the number of sellers is large, this information cannot be acquired efficiently if the buyer has to meet a seller to learn of the quality of his commodity. This is why information circulates among buyers, who communicate once every time step. A buyer who meets a seller forms an image of this seller; a buyer who gets information about a seller has access to a reputation of this seller. When giving information to another buyer, an agent can decide to give the direct knowledge it has (the image it formed of a seller) or the reputation it has already received (which is more neutral since it is not its own evaluation). Reputation can also circulate about the buyers and in that case concerns the validity of the knowledge they give about sellers. When a buyer is not satisfied with the information given by another buyer, it can also retaliate and cheat when this very agent asks him a question.

Pinyol et al. (2007) describe in detail the choices that agents make when asking a seller for a commodity, asking another buyer for information about a seller or a buyer, answering a question and the lying process.

The simulated market contains a large number of sellers (100 for 25 buyers). Simulation runs are defined by (a) the type of information that is used by agents (only image or image and reputation) and (b) the number of bad-quality sellers in the system (99, 95, 90 and 50). The addition of reputation to the system makes the difference between a normal learning mechanism where buyers select their favourite seller and a learning mechanism where agents aggregate information of different quality to (maybe) increase their performance. The results show that globally the agents indeed learn more efficiently when using reputation, in that the average quality that is bought is higher. The quantity of information that circulates is much higher, and this enables buyers to increase their utility. This social control mechanism is especially important when quality is really scarce (1% of good sellers). This result is all the more interesting since this is a very rare case of a simulated market where communication among sellers is represented, although this behaviour is commonly observed in real-life situations. For a more in-depth discussion of reputation, see Chap. 18 in this handbook (Giardini et al. 2017).

25.4   Consumers, Producers and Chains

Another way to look at the notion of goods markets is to consider large markets, where individual interactions are not important for the agents who do not record the characteristics of the ones they meet but only the fact that they can or cannot perform an exchange. A large market can indeed include numerous goods that are distributed among different other agents and not necessarily easy to access. Another interest in large market is to study endogenous preferences for goods and imagine their evolution depending on the type of good and some cognitive characteristics of agents. Eventually some authors are interested in the coordination process within the supply chain itself, where the issue is about the amount of information that each agent has to use to anticipate the needs of distant, end consumers.

25.4.1 Multi-goods Economy

As examples of a market with several goods, we will discuss two very abstract models where agents have to produce one commodity and consume others, which they can acquire only through exchanges with other agents. The first model was built to produce speculative behaviours in agents, which means acquiring a product that has no value for consumption but only a value for exchange (Duffy 2001); the second model's aim is to witness the emergence of commonly used equivalence value for the goods, which is interpreted as relative prices (Gintis 2006). Both models are interesting for their pure description of an abstract economy with minimalist but sufficient assumptions to induce economic exchanges. In the works cited here, the methodology used in realizing the models is slightly different: one is purely abstract, whereas the other tries to refer to experimental results and mimic human players behaviours.

In his paper, John Duffy (2001) designs a model that was originally proposed by Kiyotaki and Wright (1989) to induce some agents to store a good that is not their designated consumption good and is more costly to store than their own produced commodity, because they think it easier to exchange with others. There are three different goods in the economy; agents need to consume one unit of good to increase their utility and produce one unit of good each time they have consumed one. There are also three types of agents: agent type 1 needs good 1 and produces good 2, agent type 2 consumes good 2 and produces good 3 and agent type 3 consumes good 3 and produces good 1 (in short, agent type i consumes good i and produces good $i + 1$ modulo 3). Hence, agents have to exchange when they want to consume, and not all agents can be satisfied by just one exchange. Indeed, if two agents exchange their own production goods, one can be satisfied, but the other would get a useless good, which is neither its own production good nor its consumption good. In this economy, only bilateral trading exists, and it takes place after a random pairing of agents. This involves that some agents must keep a good for at least one time step after production before getting their consumption good.

In this economy, speculation is defined as the storage of the good $i + 2$, since the agent does exchange to get this good which it has not produced, only because of the chances to use it as an exchange good at the next time step. The economy is made nonsymmetric by having different costs for the storage of goods, here $0 < c_1 < c_2 < c_3$. The original Kiyotaki and Wright model is all about calculating, given the storage costs, the discount factor (the probability that the economy stops at the end of a time step) and the utility of consumption, based on which agents decide to get the most expensive good to store or keep their production good. In an economy with perfect information, the expected profit for each type of agent depends on the

proportion of agents of type i holding good $i + 1$ (their production good), which is
1—proportion of agents of type i holding good $i + 2$ (their "speculative" good).

With this model, John Duffy tries to investigate how agents could learn which
choice to make when they are able to acquire a good they do not consume. Especially
agents of type 1 are those that should hesitate, since good 3 is the most expensive.
A lot of models have been built on this topic (Basci 1999) already, but what
Duffy wants to produce is a setting that is close to laboratory experiments he has
been leading (Duffy and Ochs 1999) in order to be able to judge if his agents
are behaving in a way which is coherent with that of human actors. So, from a
theoretical setting, he builds experiments and simulations and compares all the
results this technique produces. In this paper, he therefore proposes an algorithm that
is close to his intuition of what individuals should do (he has also asked questions
to people involved in his experiments), and he then tries to mimic the results of
his experiments, at a global level and a local level. He also proposes some original
settings where he mixes human agents with artificial agents to test at the same time
his algorithm and how much he can make the human change behaviour depending
on the stimuli they get. He is satisfied with his results, where his model of learning
enables to reproduce human behaviour correctly. Unfortunately, the reproduction of
his model is not so straightforward (Rouchier 2003), but all in all, this description
of a very basic economy with few goods and where agents learn in an intuitively
plausible way is a very interesting example of market for agent-based modellers.

The paper by Gintis (2006) presents similarities, although the aim and the central
question are different. The economy that is presented can be seen in a very general
way but is only implemented in one setting, which is described here. In the economy,
there are three goods and 300 agents. Each agent can produce one good and needs
to consume both goods it cannot produce; hence, it is forced to exchange with other
agents. At the beginning of each period, an agent only holds the good it produces
(in a quantity that it can choose and which is costless) and can meet two agents,
each producing one of the good he needs to acquire. Each agent has a representation
of "prices", which is here defined as the equivalence quantity between two goods.
There is no common knowledge of prices, and each agent has its own representation.
When an agent meets another agent who can provide him with the needed good, he
offers to trade, by sending as a message its representation of relative "prices". The
exchange takes place at this exchange rate if it is acceptable to both agents, and
the exchanged quantities are the highest quantity that both can exchange. Agents
cannot choose who they meet; they just pick randomly from the other producers'
groups. After exchanging, they can consume, which gives them utility and defines a
performance for each individual. Learning in this system is an event that takes place
every 20 periods, where 5% of least performing agents (who get the lowest utility)
copy the price representation of the highest performing agents.

What is observed in the system is the efficiency of the represented market,
meaning the sum of all profits, compared to a setting where prices would be
public. When prices are public, all exchanges can take place since all agents
agree right away on the equivalence that is proposed and there is no refusal in
exchange. In the long term, the system converges to the highest possible efficiency,

so although the agents have private prices, these prices get to be close enough to have regular exchanges. This result in itself is not very surprising in terms of simulation (considering the process at stake) but is interesting in economics since it gives, at least, a process to attain to a common knowledge which is often presupposed to exist.

25.4.2 Adoption by Consumers

The study of the behaviour of large numbers of consumers facing the introduction of new products on a market is a topic that is very interesting to approach with agent-based simulation, since it allows, once more, looking for the influence of heterogeneity of individuals and of networks in the evolution of global results. Wander Jager is a prominent figure in this area of research, positioned between psychology and marketing. In a paper with Marco Janssen, he presents the basic model (Janssen and Jager 2003). The idea behind the study of the acquisition of a new product in a group is that agents have a preference that is based on two main parameters: the individual preference for the consumption of the product and the interest that the agent has to consume the same product as his acquaintances. Hence, a utility function depends on these two parameters, and this will influence an agent's decision to buy a new product. Agents are heterogeneous in such a system, and the representation of "early adopters" (in the real world, people who buy a product when it is just released) is modelled by a low need to conform to others' behaviour. On the opposite end of the spectrum, some agents buy a good only because a lot of their acquaintances have already acquired it.

In Janssen and Jager (2003), the influence of network size and topology is tested, as well as the influence of the utility brought by the product consumption and the way agents choose their action. One aspect that is studied is the type of cognitive process that can be used by the agent (repetition of the same action; deliberation to find a new action; imitation, where other agents' consumption is imitated; social comparison, where other agents are imitated based on their utility). This indicator is quite rare and shows the psychological grounding of the paper. It is interesting to observe that the cognitive process changes with the utility gained by the consumption of the considered product. Agents with a lot of links are very important for the spreading of product adoption: in a small-world network, many more products get adopted than in a scale-free network. A discussion is open here about the type of products, which certainly influence the way people copy others—it will be different for milk and for computers or clothes.

This last question is actually developed in a different paper: in Delre et al. (2007), the question that is at stake is to determine how to advertise efficiently depending on the type of good. Is it better to advertise a lot at the beginning of a campaign or after a moment; is it better to advertise to a large number of people or to disseminate information among only a few agents?

Two products are differentiated: a brown product, which is a high-tech and quite fancy product that can be compared with other agents of the network (e.g. CD, DVD player), and a white product which is related to basic need and is not really compared (fridge or washing machine). Agents are gathered in networks of different topologies. In this model, the heterogeneity in the utility formula is similar to the one in the preceding paper: each one is defined by a propensity to be influenced by others and to be an adopter of a new technology.

The first finding of this model is that the timing of promotion is important to the success of the campaign. For a first launch, the best strategy is to "throw gravel", i.e. to do a little advertising to many distant small and cohesive groups of consumers, who will then convince their network neighbours. Another element is that not too many people must be reached at first, since if they see that others have not adopted the good yet, they might not want it and become impossible to convince afterwards. This is mainly true for the white good, where it is better to advertise broadly when at least 10% of agents have already adopted the good, whereas with brown goods, adoption is much faster, and a campaign helps the takeoff.

The issue of the adoption of a practice within a social context has also been studied to understand the adoption of electronic commerce for consumers (Darmon and Torre 2004). The issue at stake is that it should be logical that everyone turns to electronic commerce, which radically reduces transaction and search costs, but we observe that a very small proportion of items are as yet traded via the Internet. This is mainly because consumers have not developed special abilities that are associated with this form of interaction and do not know how to reduce the risk of performing a bad transaction. To study the dynamics of adoption of electronic commerce and learning of agents in a risky setting, a simulation model has been built. The market is composed of agents who can produce a good and exchange it for the good they want to consume (hence, all agents are at the same time producers and end consumers). Agents and goods are located on a circle; the location of an agent defines the "quality" of the good it produces. For consumption, each agent is defined by an interval of quality: when consuming a good whose quality is within this interval (not including its own production good), it will get a strictly positive profit. The cost of production is heterogeneous and can be constant during the simulation or evolving.

When trading on the traditional market, an agent can identify the quality of a product offered by another agent, but it has to talk to many others before finding out who it can exchange with (depending on the number of agents and of its interval of choice). The authors also added a notion of friction, which is a probability of failing to trade when two agents meet. In the electronic market, an agent sees all other agents at once (no search cost) but cannot identify precisely the quality that is offered and evaluates it with an error interval. Hence, it potentially accepts goods with zero utility. Agents are heterogeneous in their ability to perceive the quality and their learning algorithm. If an agent learns via individual learning, then it eliminates an agent from its list of potential sellers whenever the previous trade brought no utility. If agents learn through collective learning, then a part of the whole society belongs to a community that shares information about the quality (location on the circle) of the seller they met at this time step; the agents not belonging to the community learn individually. In some simulations, for both types of learning, agents forget a randomly chosen part of what they learnt at each time step.

In the case of individual learning, the dynamics produced depends on the production cost, which can either change at all time steps (and hence all agents have the same average production cost over the simulation) or which can be constant and delineate populations with high or low production costs. When the production cost changes at each time step, the main result is that eventually all agents switch to the electronic market, with the move happening in stages. Those that have a good appreciation of quality go to the electronic market very fast because it is more profitable for them. Their departure from the traditional market reduces the probability of exchange for the remaining agents, who eventually move to the electronic market as well. When production costs are heterogeneous, some agents cannot switch from traditional to electronic because of their inadequate production cost. Hence, they never learn how to identify quality and stay in the traditional market. When agents forget part of what they have learnt, then the size of the electronic market does not get as large as with perfect learning, and a number of agents do not switch.

When agents participate in a community and exchange their information, the highest number of agents will switch to the electronic market, and overall the lowest number of agents is excluded from exchange. Three groups are created: agents belonging to the community, who get the highest payoff; agents with low production cost or high expertise, who can go on the electronic market and make a high profit; and the remaining agents, which sometimes cannot exchange. This result is rather coherent with what could be expected, but it is interesting to have it created with this location-based representation of quality that each individual wants to attain. It is especially clear that there is little risk that traditional markets should disappear if the main assumption of the model—that agents need an expertise that takes long to acquire before switching to the electronic market—is true.

25.4.3 Decentralized Supply Chain

Supply chains are an important aspect of economics, and they are often difficult to consider, mainly because their dynamics spread in two directions: (1) along the length of the chain, suppliers have to adapt to demand, and buyers have to adapt to the speed of production so as to be able to provide the end market with the right quantity of goods; and (2) another dimension is the fact that suppliers as well as buyers in the chain are substitutable and that each actor is itself in a market position and can choose between several offers or demands. In existing agent-based models, only the first issue is treated. The structure of these models is a series of agents (firms) that are each linked to two agents: a supplier and a client (except for first supplier and end consumer, of course, who are linked to only one firm). Each agent has to decide on its production level at each time step, knowing that it needs to use goods from the preceding firm in the production process. It must then anticipate the demand to order enough, before being able to transform the needed quantity. Of course, each firm takes some time (number of steps) to transform the product and be able to sell, and there is a cost in storing its own production when it is not completely sold.

One very important issue of these chains is at the centre of most research: how to avoid the so-called bullwhip effect. This effect is a mechanical dynamics that comes from the slow spreading of information and delay in answer because of the length of the production process in each firm. When there is variability in demand coming from end consumers, this variability increases a great deal when it goes up the chain, right up to the first producer who exhibits the highest variability. It can be very annoying for organizations to be trapped in such negative dynamics. Several authors propose algorithms for artificial agents that have to deal with the issue of anticipating demand at each stage of the chain. For example, Lin and Lin (2006) describe a system where artificial agents can interact with real agents (and hence be integrated in a real-life company to help decision-makers) in order to choose the right level of production and order to reduce costs. Several learning algorithms are tested and their efficiency attested, even in situations where the environment is dynamically evolving. The same issue is dealt with by others, for example, Kawagoe and Wada (2005), who propose another algorithm. They also propose a method to statistically evaluate the bullwhip effect. Their method is different from the usual frequency-based statistical measurement (like stochastic dominance) but is based on descriptive statistics.

25.5 Financial Markets and Auctions

Financial markets have been one of the first examples that were developed to prove the relevance of agent-based modelling. Arthur et al. (1997a) indeed reproduced some important stylized facts of asset exchanges on a market, and this paper is always cited as the first important use of this modelling technique for studying markets. Contrary to models that were presented before, there is no direct interaction among agents in these models, only observation of price patterns. One rare example presented here is an attempt to link a financial market to a consumer market such as the ones seen in previous sections. Another type of market that does not integrate any interaction in the economy is the representation of auctions.

25.5.1 Financial Markets

The literature on financial markets is very important in agent-based simulation and dates back to the 1990s (Arthur 1991, 1994; Arifovic 1996; Arthur et al. 1997b). This holds also true for the related branch of research, which is called *econophysics*:

the use of physics techniques to deal with economic issues in systems that are composed of a huge number of simple interacting actors (Levy et al. 2000). A comprehensive review of this topic (Lux 2009) describes the main stylized facts that can be found in financial markets (and hence are meant to be reproduced by simulation) and some models that are candidates for explaining these facts. Another review (Samanidou et al. 2007) describes several agent-based simulations models dealing with financial markets; unfortunately, these models do not achieve to reproduce the very general statistical regularities of these markets. As usual, in the following, I will describe only a selection of models and ways to represent agents' learning in the context of financial markets. The basic structure of the market, which defines the type of choice the agent has to make, can vary as well as the aim and methodology of the researcher building these models, and this is why discussing a few representative examples in detail seems a better idea than presenting very generic results.

One reason for using agent-based models is to be able to represent populations of heterogeneous agents. What is very often found is the representation of two types of agents with different reactions to information: chartists and fundamentalists. Fundamentalists base their investment decisions upon market fundamentals such as dividends, earnings, interest rates or growth indicators. In contrast, technical traders pay no attention to economic fundamentals but look for regular patterns in past prices and base their investment decision upon simple trend following trading rules. Computer simulations such as those of the Santa Fe Artificial Stock Market (LeBaron et al. 1999; but see also, e.g., Kirman 1991; Lux and Marchesi 1999, 2000) have shown that rational, fundamental traders do not necessarily drive out technical analysts, who may earn higher profits in certain periods. An evolutionary competition between these different trader types, where traders tend to follow strategies that have performed well in the recent past, may lead to irregular switching between the different strategies and result in complicated, irregular asset price fluctuations. Brock and Hommes (1998) have shown in simple, tractable evolutionary systems that rational agents and/or fundamental traders do not necessarily drive out all other trader types but that the market may be characterized by perpetual evolutionary switching between competing trading strategies. Non-rational traders may survive evolutionary competition in the market; see, for example, (Hommes 2001) for a survey.

In Hommes and Lux (2008), the chosen market model is the so-called cobweb model, which is a prediction model on a market, not an actual model of selling and buying for agents. The model offers, however, a rational expectation value, which serves as a benchmark. The methodology is to try to fit agents' behaviour in an artificial world to real behaviours of individuals in experiments. The game is such that participants of the experiments have no clear idea of the structure of the market but still have to predict the price of the next period. They neither know how many other agents are present nor the equation that calculates the future price based on the realized price and the expectations of all participants. The simulations are made based on rather simple models of agents including a genetic algorithm, simple learning that copies past prices and reinforcement learning. What interests the authors most is the GA learning, which is the only one to fit stylized facts in different treatments. What the GA learns about is a 40-bit string of 0 and 1 representing two values, α (the first 20 bits) and β (the remaining 20 bits), that predict the price at $t + 1$ depending on the price at t with $p(t + 1) = \alpha + \beta \, (p(t) - \alpha)$.

There are three runs both for experiments and for simulations, with one parameter defining the stability of the price (high, medium or low). The genetic algorithm being varied for different mutation rates is proven to be largely better than other learning procedures that have been implemented. "Better" means here that it fits the stylized facts that have been produced by humans in experiments: (1) the mean price is close to rational expectation, and the more stable the market, the closer the mean price is to this rational expectation value; and (2) there is no significant linear autocorrelation in realized market prices. The reason for the good fit of the GA given by the authors is really interesting because it is not obvious to imagine how GAs, which are random learning processes with selection, should be similar to human learning. The authors assume that the good fit is based on two facts: the fact that successes are selected positively and that there is heterogeneity in the strategies among the set that agents can use. Once the assessment of the model is done, it is used to question the stability of the results of the learning process. One question that arises is to wonder whether humans would adapt the same way when interacting in a very large group as they do in a small group of six. This opens many questions about the scalability of results concerning market dynamics.

In our second example, the interaction of agents is direct and not necessarily via the price system, as is usual in financial markets. Hoffmann et al. (2007) indeed consider that many agent-based simulations still take little interest in representing actual behaviours of decision-makers in financial markets. They argue that Takahashi and Terano (2003) is the first paper to integrate theories that come from behavioural finance and represent multiple types of agents, such as overconfident traders. In their own paper, Hoffmann et al. (2007) present their platform SimStockExchange™, with agents performing trades and making decision according to news they perceive and prices they anticipate. They argue that their model is based on several theories that are empirically sound and that they validated their model results against data over several years from the Dutch market. As usual, the platform allows many variations (increase the number of different shares of agents, change the characteristics of agents) but is tested only with some values of parameters.

Agents receive news that they forget after one time step and then can perform two types of action: either sell their stock (if they expect to lose at the next time step) or buy more shares (in the opposite case). To make sure that they are not making mistakes, agents can use risk-reducing strategies, which can be clarifying strategies (such as collecting more data) or simplifying strategies (i.e. imitating other agents), as well as purely individual (the first one) or social (the latter). In the presented simulation, strategies are always social, and hence agents' confidence, C, determines their use of risk-reducing strategies; the confidence values were deduced from empirical studies. Each agent is also defined by a tendency R to perform a simplifying strategy or a clarifying one. R and C are evaluated on the basis of surveys made with investors. Agents are imbedded in networks of two different topologies (torus, scale-free); agents may acquire information from their links or choose to imitate them. The market itself is designed as an order book, where proposals for sells and buys are written down with quantity and price and are erased as soon as an agent answers positively to a particular proposal. The market price is the average of all proposed bids and asks of the order book—hence it is not a realized price (average transactions' price) but an aggregation of desired prices for agents.

In the results, some statistical properties of the stock exchange have been reproduced. For example, with weekly data of Dutch stock exchange, linear autocorrelation can be observed, and this is better reproduced when a torus-shaped network is used rather than the scale-free one is used. With regard to volatility clustering, the torus network differs from both the scale-free network and the real data. This can be due to the high speed of information circulation reducing the shocks that it can cause. The main aspect of the SimStockExchange that needs improvement is the news arrival, which is a normal distribution around the present price. This might have a large impact since the use of different networks integrates the importance of information spreading.

25.5.2   Relation Between Two Markets

Sallans et al. (2003) report a model integrating two types of markets: a financial market and a goods market in the same system. Consumers, financial traders and production firms are interacting, and the aim is to understand how these two markets influence each other. The good is perishable and hence needs to be purchased regularly. Consumers make purchase decisions; firms get income from sales and update products and pricing policies based on performances; traders have shares, which they can hold, sell or buy. Firms decide upon the features of their products, which are represented as two binary strings of 10 bits. In choosing actions, the firm agent uses an action-value function that integrates expectations about future rewards (firms are not myopic agents) by taking into account the evolution of the price of its share in the financial market and the profit made by selling products on the goods market. Consumers have preferences for particular features of a product and its price and compare any available product to these preferences: they can choose not to buy if the product is too different from their preferences. In the financial market, agents build expectations and built representations of future values by projecting actual and past values into the future. They are divided into two groups: fundamentalists (use past dividend for projection) and chartists (use the history of stock prices); they are also heterogeneous regarding their time horizon. The market-clearing mechanism is a sealed bid auction, and the price is chosen to maximize the number of exchanges (and randomly among different prices if they produce the same trade volume).

Agents from the financial market and firm agents have different views on the future of a firm and evaluate future gains in a different way, which might impact the firm's performance negatively. The simulations' aim is to prove that the model can be used, in certain parameter settings, to reproduce stylized facts of markets.

Although the central issue is very interesting, the paper itself is not as helpful as it could be to understanding the dynamics of two markets. In particular, the stylized facts are not very explicit in the paper (appear only once at the end, when obtained results are given). They are classical in financial market analysis, but not clearly shown here: low autocorrelations in stock returns, high kurtosis in marginal return and volatility clustering. Hypotheses on behaviour are never explained; hence, there is no understanding of why the stylized facts can be achieved, apart from doing some random exploration of the parameter space. Thus, while the main issue of the paper is fascinating, the results are a bit frustrating, and the reciprocal influences of these two markets, so important in our real world, stay hidden.

25.5.3   Double Auctions

In economics double auction is a very fascinating topic, since it is an extremely stable market protocol in which predictions can be translated from theory to real life, which is not really the case for most economic systems. When putting real people in a double-auction setting, one can observe that the convergence to equilibrium price occurs. This does not mean that this protocol is efficient, since a lot of exchanges take place out of equilibrium price, but at least there is a tendency for the group to converge to a price where the highest number of exchange can be performed and hence the highest global profit can be extracted. Many authors have therefore wanted to reproduce a double-auction market in an artificial society in order to understand the source of this high efficiency.

The continuous double auction (CDA) is a two-sided progressive auction. At any moment in time, buyers can submit *bids* (offers to buy), and sellers can submit *asks* (offers to sell). Both buyers and sellers may also accept an offer made by others. If a bid or ask is accepted, a transaction occurs at the offer price. An improvement rule is imposed on new offers entering the market, requiring submitted bids (asks) at a higher (lower) price than the standing bid (ask). Each time an offer is satisfying for one of the participants, she announces the acceptance of the trade at the given price, and the transaction is completed. Once a transaction is completed, the market is cleared (meaning there is no standing bid or ask any more), and the agents who have traded leave the market. At that moment, similar to the opening of the market, the first offer can take any value, and this proposed price imposes a constraint on any following offer. When the market closes, after a time decided beforehand, agents who have not yet traded are not allowed to continue. In this market protocol, all market events are observed by all (bid, ask, acceptance and remaining time before market closing) and hence are said to be common knowledge.

Using this double-action setting, a seminal paper by Gode and Sunder (1993, 2004) shows the strength of institutional constraints on the actions of agents. In their model, agents are perfectly stupid from an economics point of view, since they have no understanding of their own interest and only follow a simple rule without any strategic planning. These so-called zero-intelligence agents are not allowed to sell (buy) lower (higher) than their reservation price, and they have to bid within the limits that have been put by others. With this rule, convergence of prices is obtained very fast. The approach in this paper is quite original in the behavioural economics literature in the sense that it is close to an "artificial life approach". The authors do not pretend to study human rationality but instead focus on the abstract reproduction of phenomena. It is interesting to note that is not so easy to design a double-auction market, especially in its continuity. Indeed, in a real situation, if two individuals have close reservation prices, they will often be able to buy or sell at the same moment. Who will be first is not obvious, since people have different aspirations for profit. Gode and Sunders randomly choose an agent between all buyers who can buy or make a bid and then randomly pick a seller among those who can sell or make an offer. After trying several methods, they decided on random selection, explaining that this is a good approximation to continuous double auctions.

Their work is widely criticized because (a) they are not interested in rationality but in a specific market protocol and (b) it cannot be generalized to other protocols (Brenner 2002). However, their result is important and led a lot of researchers to question it. For example, Brewer et al. (2002) show that humans are able to have markets converge when the context changes a lot, which Gode and Sunders' agents cannot do. They organize a double-auction market, in which agents participate in the public market but also receive offers from the experimenter privately. Only one offer is made at a time, and it is the same for all agents that are proposed the offer, since the equilibrium has to stay the same. The global equilibrium (which value is described in the paper) is thus constant, but individuals can have incentives not to participate in the public market if the offer is interesting. This does change the performance of zero intelligence a lot, since the prices do not converge anymore in simulations led with this new protocol. On the opposite, humans performing experiments attain convergence, which could mean that only very specific institutions constrain participants enough so that they have no choice but to converge, even while not understanding more than the rules (zero intelligence).

Inspired by Gode and Sunder, but also by the theoretical model of Easley and Ledyard (1993), Rouchier and Robin (2006) tried to establish the main elements a rational agent would need to be able to choose the right action in the context of a double auction. To differentiate among different possible learning procedures, a comparison with some experimental results was made. The learning procedure chosen is a simple algorithm that consists of making the agent revise its reservation price towards past average perceptible prices, depending on two variable elements. First, the duration after which an agent would change its reservation price (i.e. a buyer (seller) accepting higher (lower) prices), called the "stress time", could change—increasing after a successful transaction and decreasing after a day with no transaction. Second, the agent could either only perceive its own transactions or those of any successful transactions in the market. The paper demonstrated that agents learn faster to converge to the equilibrium price (making the highest global payoff) if they did not revise their stress time and had a global perception of prices. This quick learning would at the same time correspond best to the speed of convergence that could be found in experiments. What is a bit surprising in this result is that more "clever" agents (reacting to risk and failure from one day to another) would neither copy human behaviour well nor get to the equilibrium price very fast.

25.6 Market Design/Agent Design

In a chapter of the handbook for computational economics (Tesfatsion and Judd 2006), Marks (2006) reviews recent work in market design using agent-based simulation. *Market design* is the branch of economic research aiming to provide insights about which protocol, i.e. interaction structure and information circulation rules, is the best to obtain certain characteristics of a market. As said repeatedly in this chapter, this choice is crucial in having certain parts of a population gain more power than others or having efficiency attained in a short time. Hence, many scientists have been thinking about this issue, using the game theory (Roth et al. 1991), as well experimental economics, and more recently computational exploration. As seen before, sophisticated agents are not the ones who do best in market situations or copy human behaviour closest.

When designing a market protocol, it is important to see two challenges. First the "aim" of the protocol needs to be clear since not all positive aspects can be achieved in a single protocol (see, for example, Myerson and Satterthwaite 1983). For example, using Dutch auction has the advantage of being fast, whereas double auction is good because it extracts the highest global profit for all. On the other hand, one might wish to extract the highest profit for buyers only, for example. LeBaron (2001) explains that the fitness of a model is as important as all other elements (what is traded, the motivations of agents, how the interaction and information circulation is organized, etc.). To achieve this, trade-off between different characteristics is already a huge choice before starting the design.

Then one has to think on how to achieve this aim. It is indeed not easy to know how individuals will react to an interaction and information constraint. The basic use of agent-based simulation can then be to either test a certain agent behaviour and compare protocols to see what difference it makes in prices or other indicators (Moulet and Kirman 2008) or to test different learning algorithms in the same setting (Chan and Shelton 2001). Both approaches are uniquely developed using agent-based simulation and can indeed help understand the relation between participant behaviour and market protocol.

Many models, be it for computer scientists or economists, were designed to fit the context of the *electricity market*, which is crucial since problems can be very severe for society (when there are huge unpredicted shortages) and the variations in price can be very fast. The agents in those models are not designed to represent human rationality but to try to be as optimal as possible in the adaptation to the electricity market. Many market protocols can be used, although auctions (which are theoretically the most efficient of all market protocols) are most common. Bidding behaviours, but also the number of sellers and buyers, and the capacity to produce and sell (Nicolaisen et al. 2001) have an impact on the efficiency, and this can be explored. As said before, what is explored is the impact of the protocol on efficiency and market power. Two ways of learning are commonly used for the agents, and authors sometimes disagree over which one to choose: either social with a genetic algorithm or individual with reinforcement learning. While it is already well known that this has a huge impact on global results (Vriend 2000), in this chapter, we cannot decide on the best choice to make. However, to our view, most results cannot really be extended to real-life design since the representation of learning for agents can be badly adapted to the application context (necessity to have long learning in case of GA or even reinforcement learning).

One original approach that is cited by March (2007) is the "evolutionary mechanism design" (Phelps et al. 2002), where the strategies of three types of actors—sellers, buyers and auctioneers—are all submitted to evolution and selection (the fitness of the auctioneer's strategy being linked to the total profit of the participants). This approach is logically different since the protocol itself (via the auctioneer) is what evolves to get to a better result, with the characteristics of the participants being fixed (relative number of each and relative production and demand).

It is interesting to note that another branch of research deals with the representation of individual agents on large markets and is also quite close to an idea of design of markets, but from the opposite perspective: by introducing agents into real markets. Computer scientists interested in the analysis of cognition have the goal of making artificial agents as efficient as possible in a context of bidding in auctions, both from the point of view of the seller and the buyer (Kephart and Greenwald 2002). They are usually not interested in understanding human behaviour and decisions, but rather in explaining the properties that can emerge in markets in which many artificial learning agents interact (with each other or humans), differentiating their strategies, getting heterogeneous payoffs and creating interesting price dynamics. The focus lies mainly on information treatment. This applied approach is interesting in that many of its algorithms can also be used for economic analysis in the framework of models of the type that have been explored here. However, the aim is slightly different, since the indicator in the latter case is the individual success of a strategy, whereas the indicators for the previous works on markets are based on global properties of the system.

25.7 Concluding Remarks

This chapter is not a general review of market simulation in recent research; instead of giving many examples, we focused on a few to show the diversity of questions, models, rationality and eventual results that can be found in the literature, coming from different backgrounds (classical economy, experimental economy, computer science). The representation of a market is always linked to the purpose of the simulation study, and there is never just one way forward. The quantity and substitutability of goods; the possibility to interact with one or several sellers, with other buyers; and the memory of the agents themselves all depend on the type of issue, and this is why we have built the chapter in this manner: to give some ideas of the issues that have been addressed up until now with agent-based simulation. What is noticeable is the real difference between this approach and the classical approach in economics, where the dynamics are not regarded as a central question. However, the achievements with this new approach are now numerous enough to prove that agent-based simulation can participate in a better understanding of market protocols and behaviours of individuals on the market and enhance the institutional choices of politics. What can be noted in conclusion is that several issues are still at stake when it comes to the representation of markets.

First, like with most simulation models, the temporal issue is huge. Most models use discrete time to advance the simulation. This can lead to problems, for example, in an auction, where different agents might act precisely at the same time and have a different impact on prices than when they act sequentially. Some people are specifically working on this issue and build platforms that support a simulated continuous time[4] (Daniel 2006).

Another technical issue is the one of learning sequences of actions. In a situation where agents evaluate their actions with profit, if they have to perform several actions in a row (i.e. choosing a seller and then accepting a price or not), it is impossible to decide which of these actions is the reason for a success or a failure. Facing this issue, economic papers describe agents that associate the profit to all actions, as if they were separated. This is clearly not very satisfying in terms of logic, but no alternative modelling has been proposed yet.

Finally, there is a conceptual gap in all the cited models. As yet, another element has never been taken into account in the representation of agents' reasoning on markets, which would fit in models where agents try to maximize their profit by choosing the best strategies. In this case, they can scan past actions and the following profits or their past possible profit with all actions they could have undertaken and then select the best action in all contexts. While the latter strategy is a bit more general than the first one, neither lets the agents imagine that a change in their action will modify other agents' behaviour as well. This is strange enough, since a lot of people interested in game theory have been working on agents in markets, but none of them have produced models of anticipation of others' choices. In markets where bargaining is central, it could however be a central feature in the understanding of real human behaviour.

Acknowledgements I wish to thank Bruce Edmonds for his patience and Scott Moss and Sonia Moulet for their advice.

---

[4]Natlab, which can be found at http://www.complexity-research.org/natlab.

Further Reading

Arthur (1991) is one of the first models incorporating learning agents in a market. Lux (1998) describes a model of speculation on an asset market with interacting agents. Duffy (2001) was the first to attempt to link experimental data to simulation results in order to evaluate the kind of learning within a speculative environment. Jefferies and Johnson (2002) give a general overview of market models including their structures and learning by agents. Moulet and Rouchier (2007) use data on negotiation behaviours from a real market in order to fit the parameters of a two-sided learning model. Finally, Kirman (2010) summarizes many interesting dimensions that can be captured using agent-based models.

References

Arthur, W. B. (1991). On designing economic agents that behave like human agents: A behavioural approach to bounded rationality. *American Economic Review, 81*, 353–359.

Arthur, W. B. (1994). Inductive reasoning and bounded rationality. *American Economic Review, 84*, 406.

Arthur, W. B., Holland, J., LeBaron, B., Palmer, R., & Taylor, P. (1997a). Asset pricing under endogenous expectations in an artificial stock market. In *The economy as an evolving complex system II (Santa Fe Institute Studies in the Sciences of Complexity, XXVII)* (pp. 15–44). Reading, MA: Addison-Wesley.

Arthur, W. B., Durlauf, S., & Lane, D. (Eds.). (1997b). *The economy as an evolving complex system II (Santa Fe Institute Studies in the Sciences of Complexity, XXVII)*. Reading, MA: Addison-Wesley.

Arifovic, J. (1996). The behaviour of the exchange rate in the genetic algorithm and experimental economies. *Journal of Political Economy, 104*(3), 510–541.

Axtell, R. (2005). The complexity of exchange. *The Economic Journal, 115*(June), 193–210.

Basci, E. (1999). Learning by imitation. *Journal of Economic Dynamics and Control, 23*, 1569–1585.

Becker, G. (1962). Irrational behaviour and economic theory. *Journal of Political Economy, 70*, 1–13.

Bendor, J., Mookherjee, D., & Ray, D. (2001). Reinforcement learning in repeated interaction games. *Advances in Theoretical Economics, 1*(1). https://doi.org/10.2202/1534-5963.1008

Boylan, R., & El-Gamal, M. (1993). Fictitious play: A statistical study of multiple economic experiments. *Games and Economic Behaviour, 5*, 205–222.

Brenner, T. (2002). A behavioural learning approach to the dynamics of prices. *Computational Economics, 19*, 67–94.

Brenner, T. (2006). Agent learning representation: Advice on modelling economic learning. In *Handbook of computational economics* (Vol. 2, pp. 895–947).

Brewer, P. J., Huang, M., Nelson, B., & Plott, C. (2002). On the behavioural foundations of the law of supply and demand: Human convergence and robot randomness. *Experimental Economics, 5*, 179–208.

Brock, W. A., & Hommes, C. (1998). Heterogenous beliefs and routes to chaos in an asset pricing model. *Journal of Economic Dynamics and Control, 22*, 1235–1274.

Camerer, C., & Ho, T. H. (1999). Experience-weighted attraction learning in normal form games. *Econometrica, 67*, 827–874.

Chan, N.T., & Shelton, C.R. (2001) *An electronic market-maker* (Technical report, AI Memo 2001-005). Cambridge, MA: MIT AI Lab.

Cyert, R. M., & March, J. G. (1963). *A behavioural theory of the firm.* Upper Saddle River, NJ: Prentice-Hall.

Daniel, G. (2006). *Asynchronous simulations of a limit order book.* PhD thesis, University of Manchester, http://gillesdaniel.com/PhD.html.

Darmon, E., & Torre, D. (2004). Adoption and use of electronic markets: Individual and collective learning. *Journal of Articial Societies and Social Simulations, 7,* 2. http://jasss.soc.surrey.ac.uk/7/2/2.html.

Delre, S. A., Jager, W., Bijmolt, T. H. A., & Janssen, M. A. (2007). Targeting and timing promotional activities: An agent-based model for the takeoff of new products. *Journal of Business Research, 60,* 826–835.

Duffy, J. (2001). Learning to speculate: Experiments with artificial and real agents. *Journal of Economic Dynamics and Control, 25,* 295–319.

Duffy, J., & Ochs, J. (1999). Emergence of money as a medium of exchange: An experimental study. *American Economic Review, 89,* 847–877.

Easley, D., & Ledyard, J. (1993). Theories of price formation and exchange in double oral auction. In D. Friedman & J. Rust (Eds.), *The double-auction market: Institutions, theories, and evidence* (pp. 63–97). Reading, MA: Addison-Wesley.

Erev, I., & Roth, A. E. (1999). On the role of reinforcement learning in experimental games: the cognitive game-theoretic approach. In D. V. Budescu, I. Erev, & R. Zwick (Eds.), *Games and human behavior: Essays in honor of Amnon Rapoport* (pp. 53–77). Mahwah, NJ: Lawrence Erlbaum Associates.

Friedman, M. (1953). The methodology of positive economics. In M. Friedman (Ed.), *Essays in positive economics* (pp. 3–43). Chicago: The University of Chicago Press. [Republished in W. Breit & H. M. Hochman (Eds.) (1971), *Readings in microeconomics.* 2nd ed. New York: Holt, Rinehart and Winston, 23-47.]

Geertz, C., Geertz, H., & Rosen, L. (1979). *Meaning and order in Moroccan society: Three essays in cultural analysis.* Cambridge: Cambridge University Press.

Giardini, F., Conte, R., & Paolucci, M. (2017). Reputation. In *Simulating social complexity–A handbook.* Chapter 18 in this volume.

Gintis, H. (2006). The emergence of a price system from decentralized bilateral exchange. *The B. E. Journal of Theoretical Economics, 6*(1), http://works.bepress.com/hgintis/1/.

Gode, D. K., & Sunder, S. (1993). Allocative efficiency of markets with zero intelligence traders: markets as a partial substitute for individual rationality. *Journal of Political Economy, 101,* 119–137.

Gode, D. K., & Sunder, S. (2004). Double-auction dynamics: Structural effects of non binding price controls. *Journal of Economic Dynamics and Control, 28,* 1707–1731.

Hoffmann, A. O. I., Jager, W., & von Eije, J. H. (2007). Social simulation of stock markets: Taking it to the next level. *Journal of Artificial Societies and Social Simulation, 10*(2). http://jasss.soc.surrey.ac.uk/10/2/7.html.

Hommes, C. (2001). Financial markets as complex adaptive evolutionary systems. *Quantitative Finance, 1,* 149–167.

Hommes, C. (2007). *Bounded rationality and learning in complex markets* (Working paper, 07-01). Center for Nonlinear Dynamics in Economics and Finance (CeNDEF), University of Amsterdam. http://www1.fee.uva.nl/cendef/publications/papers/Handbook_Hommes.pdf.

Hommes, C., & Lux T. (2008). *Individual expectations and aggregate behavior in learning to forecast experiments* (Working paper, 1466). Kiel Institute for the World Economy. http://www.ifw-members.ifw-kiel.de/publications/individual-expectations-and-aggregate-behavior-in-learning-to-forecast-experiments/KWP_1466_Individual%20Expectations.pdf.

Izquierdo, L. R., Gotts, N. M., & Gary Polhill, J. (2004). Case-based reasoning, social dilemmas, and a new equilibrium concept. *Journal of Artificial Societies and Social Simulation, 7*(3). http://jasss.soc.surrey.ac.uk/7/3/1.html

Jager, W. (2007). The four P's in social simulation, a perspective on how marketing could benefit from the use of social simulation. *Journal of Business Research, 60*, 868–875.

Janssen, M. A., & Jager, W. (2003). Simulating market dynamics: Interactions between consumer psychology and social networks. *Artif Life, 9*(4), 343–356.

Jefferies, P., & Johnson, N. F. (2002). *Designing agent-based market models*. http://arxiv.org/abs/cond-mat/0207523

Kahneman, D., & Tversky, A. (1979). Prospect theory: An analysis of decision under risk. *Econometrica, 2*, 263–292.

Kawagoe, T., & Wada, S. (2005). A counterexample for the bullwhip effect in a supply chain. In P. Mathieu, B. Beaufils, & O. Brandouy (Eds.), *Artificial economics, agent-based methods in finance, game theory and their applications (Lecture Notes in Economics and Mathematical Systems, 564)* (pp. 103–111). Berlin: Springer.

Kephart, J., & Greenwald, A. (2002). Shopbot economics. *Autonomous Agents and Multi-Agent Systems, 5*(3), 255–287.

Kirman, A. (1991). Epidemics of opinion and speculative bubbles in financial markets. In M. Taylor (Ed.), *Money and financial markets* (pp. 354–356). London: Macmillan.

Kirman, A. (2001). Some problems and perspectives in economic theory. In G. Debreu, W. Neuefeind, & W. Trockel (Eds.), *Economic essays: A festschrift for werner hildenbrand* (pp. 231–252). Berlin: Springer.

Kirman, A. (2010). *Complex economics: Individual and collective rationality*. London: Routledge.

Kirman, A., & Vriend, N. J. (2001). Evolving market structure: An ACE model of price dispersion and loyalty. *Journal of Economic Dynamics and Control, 25*(3-4), 459–502.

Kirman, A., Schulz, R., Härdle, W., & Werwatz, A. (2005). Transactions that did not happen and their influence on prices. *Journal of Economic Behavior & Organization, 56*(4), 567–591.

Kiyotaki, N., & Wright, R. (1989). On money as a medium of exchange. *Journal of Political Economy, 97*, 924–954.

Kopel, M., & Dawid, H. (1998). On economic applications of the genetic algorithm: A model of the cobweb type. *Journal of Evolutionary Economics, 8*(3), 297–315.

LeBaron, B., Arthur, W. B., & Palmer, R. (1999). The time series properties of an artificial stock market. *Journal of Economic Dynamics and Control, 23*, 1487–1516.

LeBaron, B. (2001). Evolution and time horizons in an agent-based stock market. *Macroeconomic Dynamics, 5*(2), 225–254.

Leloup, B. (2002, June 24–27) Dynamic pricing with local interactions: Logistic priors and agent technology. In H. R. Arabnia & Y. Mun (Eds.), *Proceedings of the International Conference on Artificial Intelligence, IC-AI '02* (Vol. 1, pp. 17–23), Las Vegas, NV. CSREA Press.

Levy, M., Levy, H., & Solomon, S. (2000). *Microscopic simulation of financial markets: From investor behaviour to market phenomena*. San Diego, CA: Academic Press.

Lin, F.-R., & Lin, S.-M. (2006). Enhancing the supply chain performance by integrating simulated and physical agents into organizational information systems. *Journal of Artificial Societies and Social Simulation, 9*(4). http://jasss.soc.surrey.ac.uk/9/4/1.html.

Lux, T. (1998). The socio-economic dynamics of speculative markets: Interacting agents, chaos and the fat tails of return distribution. *Journal of Economic Behaviour and Organization, 33*(2), 143–165.

Lux, T. (2009). Stochastic behavioral asset-pricing models and the stylized facts. In T. Hens & K. R. Schenk-Hoppé (Eds.), *Handbook of financial markets: Dynamics and evolution* (pp. 161–215). Amsterdam: Elsevier.

Lux, T., Chen, S. H., & Marchesi, M. (2001). Testing for nonlinear structure in an 'artificial' financial. *Journal of Economic Behavior and Organization, 46*, 327–342.

Lux, T., & Marchesi, M. (1999). Scaling and criticality in a stochastic multi-agent model of a financial market. *Nature, 397*, 498–500.

Lux, T., & Marchesi, M. (2000). Volatility clustering in financial markets: A micro-simulation of interacting agents. *International Journal of Theoretical and Applied Finance, 3*, 675–702.

Macy, M. W., & Flache, A. (2002). Macy and Andreas Flache: Learning dynamics in social dilemmas. *PNAS, 99*(suppl 3), 7229–7236. https://doi.org/10.1073/pnas.092080099

Macy, M. W., Benard, S., & Flache, A. (2017). Learning. In *Simulating social complexity—A handbook*. Chapter 20 in this volume.

Marks, R. (2006). Market design using agent-based models. In: *Tesfatsion & Judd* (Vol. 2, pp. 1339–1380).

Moss, S. (2002). Policy analysis from first principles. *PNAS, 99*(3), 7267–7274.

Moulet, S., & Rouchier, J. (2007). The influence of sellers' beliefs and time constraint on a sequential bargaining in an artificial perishable goods market. *Journal of Economic Dynamics and Control, 32*(7), 2322–2348.

Moulet, S., & Kirman, A. (2008). *Impact de l'organisation du marché: Comparaison de la négociation de gré à gré et les enchères descendantes (Document de travail, 2008-56)*. Marseille: GREQAM.

Myerson, R. B., & Satterthwaite, M. A. (1983). Efficient mechanisms for bilateral trading. *Journal of Economic Theory, 29*, 265–281.

Nadal, J.-P., Phan, D., Gordon, M. B., & Vannimenus, J. (2005). Multiple equilibria in a monopoly market with heterogeneous agents and externalities. *Quantitative Finance, 5*(6), 557–568.

Nicolaisen, J., Petrov, V., & Tesfatsion, L. (2001). Market power and efficiency in a computational electricity market with discriminatory double-auction pricing. *IEEE Transactions on Evolutionary Computation, 5*, 504–523.

Oliver, J. (1997). Artificial agents learn policies for multi-issue negotiation. *International Journal of Electronic Commerce, 1*(4), 49–88.

Palmer, R., Arthur, W. B., Holland, J. H., LeBaron, B., & Taylor, P. (1993). Artificial economic life: A simple model for a stock market. *Physica D, 70*, 264–274.

Phan, D. (2003). From agent-based computational economics towards cognitive economics. In P. Bourgine & J.-P. Nadal (Eds.), *Cognitive economics* (pp. 371–398). Berlin: Springer.

Phelps, S., Parsons, S., McBurney, P., & Sklar, E. (2002). Co-evolution of auction mechanisms and trading strategies: Towards a novel approach to microeconomic design. In R. E. Smith, C. Bonacina, C. Hoile, & P. Marrow (Eds.), *Proceedings of ECOMAS-2002 workshop on evolutionary computation in multi-agent systems, at genetic and evolutionary computation conference (GECCO-2002)*.

Pinyol, I., Paolucci, M., Sabater-Mir J., & Conte, R. (2007, May 15) Beyond accuracy: Reputation for partner selection with lies and retaliation. In L. Antunes, M. Paolucci, & E. Norling (Eds.), *Multi-agent-based simulation VIII, International workshop, MABS*, revised and invited papers (Lecture Notes in Computer Science, 5003) (pp. 128–140), Honolulu, HI, USA. Berlin: Springer.

Roth, A., Prasnikar, V., Okuno-Fujizara, M., & Zamie, S. (1991). Bargaining and market behavior in Jerusalem, Ljubljana, Pittsburgh, and Tokyo: An experimental study. *The American Economic Review, 81*(5), 1068–1095.

Rothschild, M. (1974). A two-armed bandit theory of market pricing. *Journal of Economic Theory, 9*, 185–202.

Rouchier, J. (2003). Reimplementation of a multi-agent model aimed at sustaining experimental economic research: the case of simulations with emerging speculation. *Journal of Artificial Societies and Social Simulation, 6*(4). http://jasss.soc.surrey.ac.uk/6/4/7.html.

Rouchier, J. (2004). Interaction routines and selfish behaviours in an artificial market: transferring field observations of a wholesale fruits and vegetables market into a multi-agent model. *Ninth Annual Workshop on Economic Heterogeneous Interacting Agents (WEHIA 2004)*, Kyoto, Japan.

Rouchier, J., Bousquet, F., Requier-Desjardins, M., & Antona, M. (2001). A multi-agent model for describing transhumance in North Cameroon: Comparison of different rationality to develop a routine. *Journal of Economic Dynamics and Control, 25*, 527–559.

Rouchier, J., & Hales, D. (2003). *How to be loyal, rich and have fun too: The fun is yet to come* (Working Paper, 03B13). Groupement de Recherche en Économie Quantitative d'Aix-Marseille (GREQAM), France. http://www.greqam.fr/IMG/working_papers/2003/03b13.pdf.

Rouchier, J., & Mazaud, J.-P. (2004). Trade relation and moral link in transactions among retailers and wholesale sellers on the Arnavaux market. In *Proc. of 11th World congress for social economics*, 8–11 June 2004, Albertville, France.

Rouchier, J., & Robin, S. (2006). Information perception and price dynamics in a continuous double auction. *Simulation and Gaming, 37*, 195–208.

Rubinstein, A., & Wolinsky, A. (1985). Equilibrium in a market with sequential bargaining. *Econometrica, 53*(5), 1133–1150.

Sallans, B., Pfister, A., Karatzoglou, A., & Dorffner, G. (2003). Simulation and validation of an integrated markets model. *Journal of Artificial Societies and Social Simulation, 6*(4). http://jasss.soc.surrey.ac.uk/6/4/2.html.

Samanidou, E., Zschischang, E., Stauffer, D., & Lux, T. (2007). Agent-based models of financial markets. *Reports on Progress in Physics, 70*, 409–450.

Simon, H. (1955). A behavioural model of rational choice. *Quarterly Journal of Economics, 69*, 99–118.

Takahashi, H., & Terano, T. (2003). Agent-based approach to investors' behavior and asset price fluctuation in financial markets. *Journal of Artificial Societies and Social Simulation, 6*(3). http://jasss.soc.surrey.ac.uk/6/3/3.html.

Tesfatsion, L., & Judd, K. L. (2006). *Handbook of computational economics volume 2: Agent-based computational economics*. Amsterdam: Elsevier/North Holland.

Vriend, N. J. (2000). An illustration of the essential difference between individual and social learning and its consequences for computational analyses. *Journal of Economic Dynamics and Control, 24*, 1–19.

Vriend, N. J. (2006). ACE models of endogenous interactions. In L. Tesfatsion & K. J. Judd (Eds.), *Handbook of computational economics, Volume 2: Agent-based computational economics* (pp. 1047–1079). Amsterdam: Elsevier.

Weisbuch, G., Kirman, A., & Herreiner, D. (2000). Market organization and trading relationships. *The Economic Journal, 110*(463), 411–436.

White, H. (1995). Varieties of markets. In B. Wellman & S. D. Berkowitz (Eds.), *Social structures: A network approach*. New York: Cambridge University Press.

Chapter 26
Movement of People and Goods

Linda Ramstedt, Johanna Törnquist Krasemann, and Paul Davidsson

Abstract Due to the continuous growth of traffic and transportation and thus an increased urgency to analyze resource usage and system behavior, the use of computer simulation within this area has become more frequent and acceptable. This chapter presents an overview of modeling and simulation of traffic and transport systems and focuses in particular on the imitation of social behavior and individual decision-making in these systems. We distinguish between *transport* and *traffic*. *Transport* is an activity where goods or people are moved between points A and B, while *traffic* is referred to as the collection of several transports in a common network such as a road network. We investigate to what extent and how the social characteristics of the users of these different traffic and transport systems are reflected in the simulation models and software. Moreover, we highlight some trends and current issues within this field and provide further reading advice.

Why Read This Chapter?
To gain an overview of approaches to the simulation of traffic and transportation by way of representative examples and also to reflect on the characteristics and benefits of using social simulation as opposed to other methods within the domain. The chapter will inform both researchers and practitioners in the traffic and transportation domain of some of the applications and benefits of social simulation and relevant issues.

L. Ramstedt (✉)
Sweco, Stockholm, Sweden
e-mail: linda.ramstedt@sweco.se

J. Törnquist Krasemann
Blekinge Institute of Technology, Karlshamn, Sweden

P. Davidsson
Malmö University, Malmö, Sweden

© Springer International Publishing AG 2017
B. Edmonds, R. Meyer (eds.), *Simulating Social Complexity*,
Understanding Complex Systems, https://doi.org/10.1007/978-3-319-66948-9_26

26.1 Introduction

The continuous growth of traffic and transportation is increasing the interest in limiting their negative impact on society. Moreover, the different stakeholders involved in traffic and transportation are interested in utilizing the available resources in the best way possible. This has stimulated the development and use of both various policies and advanced transport infrastructure systems, as well as the deployment of information and communication technologies. In order to examine the effects of such developments prior to or during implementations, computer simulations have shown to be a useful approach. This chapter addresses modeling and simulation of traffic and transport systems and focuses in particular on the imitation of social behavior and individual decision-making.

Traffic and transport systems typically involve numerous different stakeholders and decision-makers that control or somehow affect the systems, and the prevalence of social influence is thus significant. In order to study the behavior of such systems and model them, it then becomes necessary to capture and include the significant social aspects to some extent, in addition to the flow of traffic or transport units that constitute the backbone of such systems. In this context, we consider a traffic or transport system as a *society*, consisting of *physical components* (e.g., cars, buses, airplanes, or parcels) and *social components* (e.g., drivers, passengers, traffic managers, transport chain coordinators, or even public authorities) where the interactions between them may play an important role. The social components determine the physical flow in their common *environment* (e.g., road or rail networks and terminals) in line with *external restrictions*, *internal intentions*, and so forth. Depending on the purpose of the simulation study and the sophistication and detail that is desired in the model, there are different approaches to incorporate social influence in such systems.

The purpose of this chapter is to present an overview of when and how social influence and individual behavior within the domain of traffic and transportation have been modeled in simulation studies. We also provide further reading advice to related approaches and highlight current issues. We divide the domain into transport and traffic. *Transport* is an activity where goods or people are moved between points in one or several traffic modes (road, rail, waterborne, or air). The types of vehicles we consider are train, truck, car, bus, ship, ferry, and airplane. While transport refers to the movement of something from one point to another, *traffic* refers to the flow of different transports within a network. One train set is thus a transport (and possibly also part of a transport chain) that takes part in the train traffic flow. Hence, a transport can be part of several traffic networks (air, waterborne, road, or rail), and a traffic network comprises several transports.

Typical points of interest related to transportation in this context are, for instance, to predict consequences of transport policies aimed at mediating governmental goals or objectives of companies and to design and synchronize bus timetables at certain stations to facilitate passenger flows. A typical issue that is interesting to study in the traffic context is the impact of driving assistance systems where the driving behavior in road networks is studied.

In the following sections, the types of simulation approaches that exist within the traffic and transportation domain, and the motivation behind them, are presented. The social aspects of the studied systems and models and how they have been accounted for are also described here. Finally, a concluding discussion is presented along with some further reading advice.

26.2 Traffic

This section addresses the modeling and simulation of individual and social behavior within traffic systems. There are several types of simulation models developed and applied which have different granularity, i.e., macro-, meso-, and microlevel models. The first two typically represent the traffic behavior by the use of equations, which are based on aggregated data. Thus, the traffic is modeled as a collection of rather homogenous entities in contrast to the microscopic models, which more in depth consider the *individual* characteristics of the traffic entities and how these influence each other and the traffic system. Since this handbook mainly addresses the modeling of social aspects, we will focus on microscopic models.

Within the domain of traffic system simulation, the dominating focus is on simulation of road traffic and car driver behavior to evaluate the quality of service in road traffic networks (Tapani 2008). The development and implementation of ADAS (advanced driver-assistance systems), ATIS (advanced traveler information systems), and road traffic control regimes, such as congestion taxation, stipulate an increasing interest in sophisticated simulation models. There are also approaches that study the behavior of traffic systems during extraordinary situations such as urban evacuations. Since users of road traffic systems normally do not communicate and interact with each other directly but rather indirectly due to the restrictions of the traffic system, the design of agent communication and protocols is mostly not considered in these simulation models. Focus is instead on individual behavior models.

The attention given to the simulation of social aspects in other modes of traffic such as air and railway traffic is, however, very limited, and we have not found any publications focusing on this specific subject. One reason may be that the traffic in these modes to a large extent is managed by central entities, such as traffic controllers. In this perspective, the behavior and interaction of the individual vehicles are less interesting to study. However, the interaction between traffic controllers in these traffic modes seems to be a domain that needs more attention.

Below we will provide a more in-depth presentation of social simulation in road traffic. Since the movement of people associated with a vehicle is the focus in this chapter, simulation of pedestrians and crowds will only be presented briefly in the last section on related research.

26.2.1 Road Traffic Simulation

In road traffic simulations, there are several distinctions made. First, there is a distinction made with respect to the problem area in focus and whether it concerns *urban*, *rural*, or *motorway* traffic. Furthermore, depending on the infrastructure modeled, there is a distinction between *intersection*, *road*, and *network* model. As an example, urban traffic is often modeled as a network (Pursula 1999).

One can see that there are three main categories of simulation models with increasing level of detail in driving behavior. First, there are the empirical macroscopic traffic models (e.g., software like TRANSYT provided by TRL Software) that focus on traffic flow analysis. Then we have the extended microscopic models with capabilities of representing individual driver behavior by use of different sub-models or rules for speed adaptation, car following, lane change, as well as intersection and roundabout movements if relevant (e.g., VISSIM for urban and freeway environments and TRARR, TWOPAS, and VTISim for rural road environments (Tapani 2008)). The third type of models has an even more complex representation of driver behavior by use of, e.g., neural networks (NN) (Lee et al. 2005; Dia and Panwai 2007) or discrete choice models (Dia 2002; Lee et al. 2005). In some cases, these behavior models can be dynamically configured to imitate driver behavior adaptation and the effect of learning from experience; see, e.g., the work by Rossetti et al. (2000).

These three types of models do complement each other, but with the growing need to evaluate the impact of investments in intelligent transport systems (ITS) such as ATIS and ADAS, and policies such as road pricing, the need to reflect the complexity of driver behavior in more detail becomes apparent (Tapani 2008). The current behavior sub-models for acceleration, lane changing, car following, and so forth are mainly equation based with threshold values. Toledo (2007) claims in a review of state of the art in road traffic driver behavior modeling that, in many cases, these sub-models are insufficient to adequately capture the sophistication of drivers and the impact of long-term driving goals and considerations. Henceforward, we will focus on the approaches that emphasize such advanced driving behavior modeling and refer to Pursula (1999), Mahmassani (2005), Toledo (2007), and Tapani (2008) for more information about related research.

In the third category of approaches mentioned, the individual drivers are usually modeled as individual autonomous vehicles represented by *intelligent agents*. The main differences between the equation- or rule-based behavior sub-models and the agent-based models are the increased reasoning capabilities and planning horizon considered. The traffic network flow in either case could be based on techniques like *cellular automata* (*CA*) (Nagel and Schreckenberg 1992; Esser and Schreckenberg 1997) and *queuing theory*, while the decision-making of the agent is based on a possibly more diverse set of objectives and influencing parameters. Sometimes, the agents use simple decision rules acting in a reactive manner rather than having sophisticated reasoning capabilities. We refer to Ehlert and Rothkrantz (2001), El Hadouaj et al. (2000), Wahle and Schreckenberg (2001), and Kumar and Mitra

(2006) for examples of such reactive behavior. However, these approaches are not considered part of the third category in this context, and this illustrates that agent technology does not necessarily imply, but only *enables*, a representation of more complex behavior.

The approaches and their models often focus on a certain traffic infrastructure depending on which aspect to study. Below, we address three main topics: simulation of road traffic in intersections, evaluation of ATIS and ADAS, and benchmarking of behavior modeling techniques.

26.2.1.1 Simulation of Road Traffic in Intersections

Urban traffic networks typically involve intersections with a complex coordination and interplay between vehicles. This is challenging to model, and it is difficult to prevent deadlocks from occurring during the simulation. One way to handle this is to use game theory and agents. Mandiau et al. (2008) propose a distributed coordination and interaction mechanism for the simulation of vehicles in T- and X-shaped intersections in urban road traffic networks. The vehicles are represented as agents who can choose between two actions: brake or accelerate. The coordination between the vehicles crossing the intersection is based on game theory, and a $2x2$ decision matrix for each pair of vehicles is used to compute the decisions. The approach is also extended to involve a larger traffic volume of n agents (vehicles), where the memory requirements then increase rapidly due to the $n(n-1)/2$ decision matrices. The mechanism does not prevent the occurrence of deadlocks but is able to resolve them.

Bazzan (2005) does also propose a distributed game theory-based approach for the coordination of road traffic signals at X-shaped intersections where vehicles are represented as autonomous intelligent agents. The agents have both local and global goals and are adaptive in the sense that they are equipped with a memory to register the outcome (i.e., payoff) of executed actions; via learning rules, they are able to incorporate their experience in the decision-making. Game theory has also been used in this context to model the route choice behavior (Schreckenberg and Selten 2004).

26.2.1.2 Evaluation of ATIS and ADAS

The benefits from and application of ATIS and ADAS have become more known and common in modern road traffic systems. An implementation is, however, often associated with large investments and limited insight in how the system and its users would respond to such an implementation in the short and long term. The use of simulation may then offer some information on possible consequences to guide the stakeholders. A typical question to address is how the individual drivers would change their choice of traveling with an increased access to traffic information. Dia (2002) investigates the implications of subjecting drivers to real-time information in a road traffic network. Each vehicle is represented as a BDI agent (Belief-Desire-Intention), which has a static route-choice multinomial logit model (Ben-Akiva and Lerman 1985). Static refers to the multinomial logit model not being updated during simulation and thus not making use of driver/vehicle experience. Panwai and Dia (2005) extend the approach using NN and then further improve it by giving each agent a memory and a dynamic behavior model based on NN (Dia and Panwai 2007). The behavior model is updated accordingly during the agent's journey allowing the agent to act on the information provided by an ATIS by reevaluating its strategy en route and possibly changing route if possible and beneficial.

Rossetti et al. (2000) focus on the departure time and route choice model of commuters in a road traffic network using the DRACULA (Dynamic Route Assignment Combining User Learning and microsimulAtion) simulation model. The drivers are represented by autonomous BDI agents, some of which have access to ATIS and thus more information about the current traffic situation than other drivers. The agent behavior model is dynamic in the sense that it incorporates the experience of the commuting driver on a daily basis, but does not allow the driver to change strategy en route. Before the commuter starts its journey and decides when to depart and which route to choose, it compares the predicated cost of choosing its usual daily route with the cost of the alternative route. If the cost of any of the alternatives is significantly lower than the cost of the daily route, the driver chooses that. Otherwise, it stays with its usual route. Once the agent has decided on departure time and route, it will follow that guided by car-following and lane-changing rules and cannot change its mind.

Chen and Zhan (2008) use the agent-based simulation system Paramics to simulate the urban evacuation process for a road traffic network and the drivers' route choice and driving behavior. The vehicles are modeled as individual agents that have local goals to minimize their travel time, i.e., to choose the fastest way. Based on traffic information, a car-following model, network queuing constraints, and a behavior profile (e.g., aggressive or conservative driving style), each agent dynamically, en route, reevaluates its driving strategy and route choice.

26.2.1.3   Benchmarking of Behavior Modeling Techniques

Since a number of alternative methods to model and simulate road traffic exist, a few studies have focused on comparisons to evaluate the strengths and weaknesses of some alternatives. Lee et al. (2005) compare and evaluate the use of three different route choice models for road traffic drivers with access to trip information: a traditional multinomial logit model, a traditional NN approach, and an NN combined with a genetic algorithm (GA). The initial attitude of the authors indicates a preference for the combined NN-GA solution proposing that it is better suited to consider the influence of nonlinearity and obscurity in drivers' decision-making. Based on their simulation experiments and the mean square error of the different route choice models, the NN-GA approach is said to be most appropriate.

Hensher and Ton (2000) make a similar comparison of the predictive potential of NN and nested logit models for commuter mode choice. They do not make any judgment about which method is most appropriate but conclude that both require a lot of historic data to train or construct the models. Due to the characteristics of NN, they are better at handling noise or lack of data than discrete choice models.

For more in-depth information about human behavior in traffic, we refer to Schreckenberg and Selten (2004), while Boxill and Yu (2000) present an overview of road traffic simulation models and Koorey (2002) an overview of software.

26.3   Transportation

In this section, approaches to simulating transportation systems are described, which include both transportation of freight and people. Transportation is often described as road, rail, waterborne, or air transportation, but transportation can also be *intermodal*. Intermodal transportation refers to a transport chain of two or more modes of transport where some modal shift activity takes place, for instance, at a terminal where loading and unloading of goods is done or at a train station where passengers transfer from train to bus. *Supply chains* are related to freight transportation, even if the focus mainly is on the product and its refinement processes, in contrast to freight transportation where the focus is on the vehicle and its operations.

Issues in the field of passenger transportation typically concern evaluation of policies for more efficient bus timetables and pricing policies. Another field within the domain is *emergency transportation*, which often concerns the planning of resources, such as ambulances or fire engines, in order to serve people in need efficiently with respect to costs, coverage equity, and labor equity. Other issues are the evaluation of different policies, such as dispatching policies. Goldberg (2004) has reviewed operations research approaches for emergency transportation and claims that mathematical programming currently is the dominating method used. He also states that simulation is a promising approach for future work related to emergency transportation, especially for vehicle relocation and dispatching, due to the complexity of the problem domain. However, we have not found any papers describing approaches of social simulation in the field of emergency transportation.

In papers describing simulation approaches to transportation with a focus on social aspects and individual behavior, we have only found models for road, rail, and waterborne transportation, i.e., we have not found any air transportation models.

26.3.1   Freight Transportation

Different approaches are used when modeling and simulating transportation. A common approach when analyzing transportation is the so-called four-step approach (production/attraction, distribution, modal split, and assignment), which primarily is developed for passenger transportation but also used for freight transportation (de Jong and Ben-Akiva 2007). This approach is on the macroscopic level, i.e., averaged characteristics of the population are in focus and aggregated data is used, in contrast to a microscopic level, which would include more details on the individuals of the population. A trend in the field of freight transport simulation for predicting probable consequences of transport policies is to include more details; see, e.g., de Jong and Ben-Akiva (2007) and Hunt and Gregor (2008). However, most of these approaches are still macroscopic approaches, but with some microscopic characteristics (Ramstedt 2008).

Traditional simulation approaches for simulating supply chains are discrete-event simulation and dynamic simulation (Terzi and Cavalieri 2004). In such models, the behavior of the individuals is often only represented as a set of actions related to a probability function of being executed, not capturing causal behavior of the simulated system.[1] Moreover, interactions between the individuals, such as negotiations, are not explicitly modeled in traditional supply chain models. Since modeling and simulating social aspects are the main concern here, we focus on simulation approaches that address interactions between individuals.

Most simulation approaches of freight transportation have a descriptive purpose, such as predicting the effects of different kinds of policies. For instance, Gambardella et al. (2002) who simulate intermodal transportation make use of multi-agent-based simulation to examine policies aimed at improving the operations at terminals, while simulation approaches of supply chains (Swaminathan et al. 1998; van der Zee and van der Vorst 2005) often focus on evaluating strategies such as VMI (vendor-managed inventories). Such studies are mainly of interest for private companies, even if they can be of interest for public authorities as well. Another example is provided by Davidsson et al. (2008), which studies the possible effects of transport policies in transport chains, which are of interest for public authorities. In this approach, the decision-making of actors in transport chains, for instance, regarding traffic mode choice, selection of supplier, etc., is simulated. New prerequisites as a consequence of transport policies have the potential to change these decisions so that a different system behavior occurs. There are also examples of models with a prescriptive purpose, such as to support the transport planning in order to improve the efficiency of the usage of transport resources (Fischer et al. 1999).

The simulated system in transport approaches typically consists of a network of links and nodes served by resources such as vehicles, which have a spatial explicitness and are time-dependent. In supply chains, the focus is more on the nodes and their processes, while the links are not explicitly considered.

In the domain of freight transportation and supply chains, the decision-making of stakeholders is typically modeled and simulated. To model the decision-making, agents representing real-world roles in transportation, such as customers, transport

---

[1]See Chap. 3 in this handbook (Davidsson and Verhagen 2017) for further general discussion.

planners, transport buyers, and producers, are typically implemented. Only in a few cases are physical entities, such as vehicles, also modeled as agents (Gambardella et al. 2002). Tasks, which are commonly performed by the agents, are selecting (1) *which* resources (e.g., terminals, transport, and production resources) to use and (2) *how* these resources should be used considering time, cost, availability, etc. These tasks are often performed as a consequence of a customer request with an aim of satisfying the demand based on cost, time, and availability. The behaviors of the agents are often implemented in terms of various types of algorithms or decision rules. Typically, the agents try to minimize their costs, e.g., labor or fuel costs, which occur as a consequence of performing a transport task between two nodes. Of course, restrictions and requirements from other agents – concerning, for instance, time of delivery – are also taken into account. The implemented algorithms in the agents can be rather complex, with optimization techniques and heuristics used to compute the decisions and actions (Holmgren et al. 2007; Fischer et al. 1999).

If some of the simulated agents represent physical entities, the locations of the entities and their characteristics are typically modeled. If the agents represent decision-makers, the responsibilities of the agents are also typically modeled, e.g., the responsibility of certain types of vehicles on certain infrastructure segments.

The interactions between the agents often take place as negotiations concerning, for instance, the cost and time of performing a task, such as transportation between two nodes. The negotiations are then carried out according to interaction protocols, and the corresponding information exchange between the agents is modeled explicitly. As an example, a customer agent requests information concerning possible transport solutions, or a transport planning agent requests information concerning available vehicles for the transport task (Davidsson et al. 2008).

The agent-based simulation models are implemented in different ways; in some cases multi-agent-based simulation platforms are used (e.g., Strader et al. 1998), while multi-agent system platforms are used in other cases (e.g., Davidsson et al. 2008; Fischer et al. 1999). It is also possible to implement the agent model without any platform; see, e.g., Swaminathan et al. (1998), Gambardella et al. (2002), and van der Zee and van der Vorst (2005). Using such platforms may often facilitate the implementation of the model. However, if the model is very complex, it can also cause problems due to the structure of the platform; see, for example, Davidsson et al. (2008) for further discussion.

26.3.2 Passenger Transportation

While more work has been done regarding modeling and simulation of freight transportation, there are some approaches concerning the transportation of passengers using rail and road. These are described here.

One example of the simulation of *bus transportation* is Meignan et al. (2007) where the main purpose of the simulations is to evaluate bus networks. The bus networks are assessed based on the interests of travelers, bus operators, and authorities with respect to, for instance, accessibility, travel time, passenger waiting time, costs, and profit. Therefore, a potential type of user of the simulation model is the manager of a bus network. The model includes a road network, a bus network, and a traveler network. Bus stops, bus stations, bus lines, and itineraries are part of the modeled system. There are two agent types: buses and passengers. A typical task of the bus agent is to perform a bus trip. The networks include spatial explicitness, and time is important for the agents since the bus routes are determined by timetables and the passengers have a need for travel at certain points in time. The model combines micro- and macroscopic approaches since the global traffic situation is taken into consideration in addition to the individual transports. Traffic assignment and modal choice of the overall demand are made on the macroscopic level with a discrete choice model. Interactions take place between the buses and passengers. For instance, the bus agents have to consider the travel demand of passenger agents, and the passenger agents have to consider the available bus agents. Moreover, the actual loading and unloading of passengers is one kind of interaction. Gruer et al. (2001) present a similar approach to evaluate the mean passenger waiting time at bus stops. Buses, stops, and road sections are modeled as agents, with the focus on the activities at the bus stops.

Work is also done in the *taxi domain* where, for instance, Jin et al. (2008) present a simulation model for planning the allocation of taxis to people requesting taxi transportation. The model can be used for evaluating different planning policies for how the allocation should be made, taking issues like vehicle usage, passenger waiting time, travel time, and travel distance into consideration. Four types of agents are included in the model: the user agent, the node-station agent, the taxi agent, and the transport administration agent. The taxi agent represents the physical taxis, while the other agents have different planning functions. The different agent types have different goals, and agreements are reached by negotiations between the agents through the user agent.

Li et al. (2006) present an example of a social simulation of *rail passenger transportation*. In this, an activity-based model is outlined for the evaluation of pricing policies and how they affect the traveler behavior. This approach is similar to an agent-based approach where passengers are modeled as agents. The focus in the model is on the traveler behavioral model, where the characteristics and preferences of travelers and their activities are modeled in terms of activity schedules. Typical tasks the traveler agent performs are scheduling and planning the journeys as well as executing these activities. The decisions are typically made based on generalized costs. In the presented model, the traveler agents can interact with a tool that provides information regarding available travel options by sending requests of possible travel options.

26.3.3 Related Research

Related to the simulation of passenger transportation is the simulation of pedestrians and crowds. Traditionally, pedestrian simulation has used techniques such as flow-speed-density equations, thus aggregating pedestrian movement into flows rather than a crowd of possibly heterogeneous individuals. Klügl and Rindsfüser (2007) propose an agent-based pedestrian simulation model of the movement and behavior of train passengers at the railway station in Bern, Switzerland. A similar approach is Qi et al. (2008), which models the alighting and boarding of passengers in the Beijing metro using a cellular automata approach. Pelechano et al. (2005) also discuss social aspects in the simulation of crowds and provide a review of different crowd simulation approaches.

One special case of crowd simulation is the simulation of emergency evacuation situations. Animal flocking behavior models are one type of models that have been applied here (see Chap. 24 in this handbook (Hemelrijk 2017) for an overview of modeling animal social behavior). For further information on simulation of emergency evacuation, we refer to Santos and Aguirre (2005), which provides a review of current simulation models.

Another type of related work is agent-based simulation of seating preferences of passengers in buses and airplanes (Alam and Werth 2008). The different agents are then representing different categories of people characterized by ethnicity, age, cultural background, and their respective seating preferences.

26.4 Discussion

In papers concerning simulation of traffic and transportation, different arguments are given to support the use of social simulation. A common argument for making use of multi-agent-based simulation in the transportation domain is that it enables capturing the complex interactions between individuals, such as coordination and cooperation (Fischer et al. 1999; Meignan et al. 2007; Liu et al. 2006), and consequently the emergent behavior of the system. Moreover, including autonomous and heterogeneous individuals and their behavior is also supported with the agent approach, as well as modeling and simulating distributed decision-making, which are important in the transportation domain (Meignan et al. 2007). Since multi-agent systems provide a modular structure of the system, the possibility to easily exchange or reuse different parts of the simulated system for different cases is facilitated (Swaminathan et al. 1998; van der Zee and van der Vorst 2005).

As pointed out earlier, there are new phenomena in the road traffic networks imposed by the increasing level of technologies facilitating driving as well as new motivations behind controlling and supervising networks and the effects of such support systems when the infrastructure capacity becomes a scarce resource. Toledo

(2007) argues that to capture the level of sophistication, the modeling capabilities need to improve where the use of agents can make a contribution.

In contrast to well-established software and methods for traffic and transport simulations, the data requirements are different, and data is not available to the same extent for the newer approaches mentioned in this chapter. In addition, the more novel approaches are all different and are often developed, used, and evaluated only by the researchers themselves. The level of maturity and acceptance reached by the traditional approaches and software by being used and evaluated by a large number of researchers during a long time is naturally difficult to compete with at this point. However, using social simulation, where individuals and their interactions are explicitly modeled, provides opportunities for validation due to the natural, structure-preserving representation of the system. For instance, the behavioral models of drivers or decision-makers can be validated by the actual drivers or real-world decision-makers.

It is possible to identify some general differences between traffic and transportation applications. In traffic approaches, the agents typically represent physical entities actually involved in the movement, i.e., vehicles or drivers are modeled as agents. In the simulation of freight transportation, on the other hand, the agents typically represent decision-makers such as customers or transport planners. Therefore, the physical representation of the agents is not of the same importance in these approaches. In freight transportation approaches, several agent types are typically necessary, as opposed to traffic approaches where typically only one agent type is modeled. Trends in the approaches to passenger transportation are not as obvious. However, passengers are typically represented as agents (one exception is the taxi transportation approach), and in the bus and taxi transportation approaches, vehicles are also represented as agents. Gruer et al. (2001) represent also bus stops and road sections as agents. Thus, physical entities may be represented as agents, like in the traffic approaches. The number of different agent types is smaller than in freight transportation, but there are more agent types than in traffic approaches.

The agent behavior models are typically of a lower level (more detailed) in the traffic approaches than in the transport approaches. One reason for this difference is that in traffic approaches, the personalities of the drivers often have a larger impact on the system and are therefore modeled with corresponding driving behaviors such as aggressive or calm driving style. The decision-making of the agents in traffic approaches is typically based on different rules where the choices or the planning in transport approaches is made based on the best performance metrics such as cost or time. The agent behaviors sometimes also include learning aspects, which provide the agents with a dynamic behavior by the use of, e.g., NN.

In transport approaches, the focus is not on modeling the different personalities of decision-makers but rather on modeling the different types of decision-making roles and their associated rational behavior. Moreover, (freight) transport approaches to model the negotiations and the interactions between the decision-makers are crucial in order to reach a solution. In traffic approaches, the interactions between the individuals are secondary, while the models of the individuals, their individual behavior, and consequently the system behavior are focused.

As far as we have seen, social simulation is not commonly applied to all modes in traffic and transportation. In the traffic domain, mainly road traffic is studied. Road traffic includes social aspects in terms of interactions between the vehicles. In air and rail traffic, the control typically takes place in a more centralized way with common objectives, which explain why these traffic modes include less social aspects in this context and therefore do not benefit as much from social simulation studies. For waterborne traffic, the infrastructure is not a scarce resource in the same sense as in the other modes; instead, the bottlenecks appear in the ports or other terminals. In the freight transportation domain, the social aspects mainly concern the interactions and decision-making of the actors in transport chains. The most common modes that are included here are road and rail, but also waterborne transportation is sometimes included. The types of decisions that are most often studied in the freight transport domain are related to planning and mode choice, which are a consequence of the interactions between actors in transportation. In passenger transportation, planning decisions are also sometimes simulated, but sometimes operational behavior, such as loading of passengers, is also simulated which is a consequence of the interactions between passengers and vehicles.

Further Reading

For further information about traffic simulation, we refer the interested reader to Chung and Dumont (2009), Tapani (2008), Toledo (2007), and Koorey (2002). Terzi and Cavalieri (2004) provide a review of supply chain simulation, while Williams and Raha (2004) present a review of freight modeling and simulation. For general information about transport modeling, we suggest to read Ortúzar and Willumsen (2001). For further information on how agent technologies can be used in the traffic and transport area, see Davidsson et al. (2005).

References

Alam, S. J., & Werth, B. (2008). Studying emergence of clusters in a bus passengers' seating preference model. *Transportation Research Part C: Emerging Technologies, 16*(5), 593–614.

Bazzan, A. L. (2005). A distributed approach for coordination of traffic signal agents. *Autonomous Agents and Multi-Agent Systems, 10*, 131–164.

Ben-Akiva, M., & Lerman, S. R. (1985). *Discrete choice analysis: Theory and application to travel demand.* Cambridge, MA: MIT Press.

Boxill, S., & Yu, L. (2000). *An evaluation of traffic simulation models for supporting ITS development* (Technical Report, SWUTC/00/167602-1). Houston, TX: Center for Transportation Training and Research, Texas Southern University.

Chen, X., & Zhan, F. B. (2008). Agent-based modeling and simulation of urban evacuation: Relative effectiveness of simultaneous and staged evacuation strategies. *Journal of the Operational Research Society, 59*(1), 25–33.

Chung, E., & Dumont, A.-G. (2009). *Transport simulation: Beyond traditional approaches.* Lausanne: EFPL Press.

Davidsson, P., Henesey, L., Ramstedt, L., Törnquist, J., & Wernstedt, F. (2005). An analysis of agent-based approaches to transport logistics. *Transportation Research Part C: Emerging Technologies, 13*(4), 255–271.

Davidsson, P., Holmgren, J., Persson, J. A., & Ramstedt, L. (2008, May 12–16). Multi agent based simulation of transport chains. In L. Padgham, D. Parkes, J. Müller, & S. Parsons (Eds.), *Proceedings of the 7th international conference on autonomous agents and multiagent systems (AAMAS 2008)* (pp. 1153–1160), Estoril, Portugal. Richland, SC: International Foundation for Autonomous Agents and Multiagent Systems.

Davidsson, P., & Verhagen, H. (2017). Types of simulation. In *Simulating social complexity—A handbook.* Chapter 3 in this volume.

de Jong, G., & Ben-Akiva, M. (2007). A micro-simulation model of shipment size and transport chain choice. *Transportation Research Part B, 41*(9), 950–965.

Dia, H. (2002). An agent-based approach to modelling driver route choice behaviour under the influence of real-time information. *Transportation Research Part C, 10*(5–6), 331–349.

Dia, H., & Panwai, S. (2007). Modelling drivers' compliance and route choice behavior in response to travel information. *Nonlinear Dynamics, 49*, 493–509.

Ehlert, P. A. M., & Rothkrantz, L. J. M. (2001). Microscopic traffic simulation with reactive driving agents. In *Proceedings of intelligent transportation systems 2001* (pp. 860–865). IEEE: Oakland, CA.

El Hadouaj, S., Drogoul, A., & Espié, S. (2000). How to combine reactivity and anticipation: The case of conflict resolution in a simulated road traffic. In S. Moss & P. Davidsson (Eds.), *Multi-agent-based simulation: Second international workshop, MABS; revised and additional papers (lecture notes in computer science, 1979)* (pp. 82–96), Boston, MA, Berlin: Springer.

Esser, J., & Schreckenberg, M. (1997). Microscopic simulation of urban traffic based on cellular automata. *International Journal of Modern Physics C, 8*(5), 1025–1036.

Fischer, K., Chaib-draa, B., Müller, J. P., Pischel, M., & Gerber, C. (1999). A simulation approach based on negotiation and cooperation between agents: A case study. *IEEE Transactions on Systems, Man, and Cybernetics–Part C: Applications and Reviews, 29*(4), 531–545.

Gambardella, L. M., Rizzoli, A., & Funk, P. (2002). Agent-based planning and simulation of combined rail/road transport. *Simulation, 78*(5), 293–303.

Goldberg, J. B. (2004). Operations research models for the deployment of emergency services vehicles. *EMS Management Journal, 1*(1), 20–39.

Gruer, P., Hilaire, V., & Koukam, A. (2001, October 7–10). Multi-agent approach to modeling and simulation of urban transportation systems. In *Proceedings of the 2001 IEEE international conference on systems, man, and cybernetics (SMC 2001)* (Vol. 4, pp. 2499–2504), Tucson, AZ.

Hemelrijk, C. (2017). Simulating complexity of animal social behaviour. In *Simulating social complexity–A handbook.* Chapter 24 in this volume.

Hensher, D. A., & Ton, T. T. (2000). A comparison of the predictive potential of artificial neural networks and nested logit models for commuter mode choice. *Transportation Research Part E, 36*(3), 155–172.

Holmgren, J., Davidsson, P., Persson, J.A., & Ramstedt, L. (2007, June 24–28). An agent based simulator for production and transportation of products. In *Proceedings of the 11th World Conference on Transport Research*, Berkeley, CA.

Hunt, J. D., & Gregor, B. J. (2008, September 25–27). Oregon generation 1 land use transport economic model treatment of commercial movements: Case example. In K. L. Hancock (Ed.), *Freight demand modeling: Tools for public-sector decision making; summary of a conference, (TRB conference proceedings, 40)* (pp. 56–60), Washington DC: Transportation Research Board.

Jin, X., Abdulrab, H., & Itmi, M. (2008, June 1–6). A multi-agent based model for urban demand-responsive passenger transport services. In *Proceedings of the international joint conference on neural networks (IJCNN 2008), part of the IEEE world congress on computational intelligence, WCCI,* IEEE. (pp. 3668–3675), Hong Kong, China.

Koorey, G. (2002). *Assessment of rural road simulation modelling tools* (Research Report, 245). Wellington: Transfund New Zealand. http://ir.canterbury.ac.nz/bitstream/10092/1561/1/12591251_LTNZ-245-RuralRdSimulatnTools.pdf

Klügl, F., & Rindsfüser, G. (2007). Large-scale agent-based pedestrian simulation, *MATES 2007, LNAI 4687*, (pp. 145–156), Springer-Verlag Berlin Heidelberg.

Kumar, S., & Mitra, S. (2006). Self-organizing traffic at a malfunctioning intersection. *Journal of Artificial Societies and Social Simulation, 9*(4). http://jasss.soc.surrey.ac.uk/9/4/3.html

Lee, S., Kim, Y., Namgung, M., & Kim, J. (2005). Development of route choice behavior model using linkage of neural network and genetic algorithm with trip information. *KSCE Journal of Civil Engineering, 9*(4), 321–327.

Li, T., van Heck, E., Vervest, P., Voskuilen, J., Hofker, F., & Jansma, F. (2006) Passenger travel behavior model in railway network simulation. In *Proceedings of the 2006 Winter Simulation Conference*. IEEE.

Mahmassani, H. S. (2005). Transportation and traffic theory: Flow, dynamics and human interaction. In *Proceedings of the 16th international symposium on transportation and traffic theory*. Oxford: Elsevier.

Mandiau, R., Champion, A., Auberlet, J.-M., Espié, S., & Kolski, C. (2008). Behaviour based on decision matrices for a coordination between agents in a urban traffic simulation. *Applied Intelligence, 28*(2), 121–138.

Meignan, D., Simonin, O., & Koukam, A. (2007). Simulation and evaluation of urban bus-networks using a multiagent approach. *Simulation Modelling Practice and Theory, 15*, 659–671.

Nagel, K., & Schreckenberg, M. (1992). A cellular automaton model for freeway traffic. *Journal de Physique I, 2*(12), 2221–2229.

Ortúzar, J. D., & Willumsen, L. G. (2001). *Modelling transport* (3rd ed.). Chichester: Wiley.

Panwai, S., & Dia, H. (2005, September 13–16). A reactive agent-based neural network car following model. In *Proceedings of the 8th international IEEE conference on intelligent transportation systems,* IEEE, (pp. 375–380), Vienna, Austria.

Pelechano, N., O'Brien, K., Silverman, B., & Badler, N. (2005). *Crowd simulation incorporating agent psychological models, roles and communication* (Technical Report). Pennsylvania: Center for Human Modeling and Simulation, University of Pennsylvania.

Pursula, M. (1999). Simulation of traffic systems–an overview. *Journal of Geographic Information and Decision Analysis, 3*(1), 1–8.

Qi, Z., Baomin, H., & Dewei, L. (2008). Modeling and simulation of passenger alighting and boarding movement in Beijing metro stations. *Transportation Research Part C, 16*, 635–649.

Ramstedt, L. (2008). *Transport policy analysis using multi-agent-based simulation*. Doctoral dissertation no 2008:09, School of Engineering, Blekinge Institute of Technology, Karlskrona.

Rossetti, R., Bampi, S., Liu, R., Van Vleit, D., & Cybis, H. (2000). An agent-based framework for the assessment of driver decision-making. In *Proceedings of the 2000 IEEE intelligent transportation systems* (pp. 387–392). Dearborn, MI: IEEE.

Santos, G., & Aguirre, B. E. (2005, June 10–11). A critical review of emergency evacuation simulation models. In R. D. Peacock & E. D. Kuligowski (Eds.), *Proceedings of the workshop on building occupant movement during fire emergencies* (pp. 27–52). Gaithersburg, ML: National Institute of Standards and Technology.

Schreckenberg, M., & Selten, S. (2004). *Human behaviour and traffic networks*. Berlin: Springer.

Strader, T. J., Lin, F. R., & Shaw, M. (1998). Simulation of order fulfillment in chains. *Journal of Artificial Societies and Social Simulation, 1*(2). http://jasss.soc.surrey.ac.uk/1/2/5.html

Swaminathan, J., Smith, S., & Sadeh, N. (1998). Modeling supply chain dynamics: A multiagent approach. *Decision Sciences Journal, 29*(3), 607–632.

Tapani, A. (2008). *Traffic simulation modeling of rural roads and driver assistance systems* (Doctoral thesis). Linköping: Department of Science and Technology, Linköping University.

Terzi, S., & Cavalieri, S. (2004). Simulation in the supply chain context: A survey. *Computers in Industry, 53*(1), 3–17.

Toledo, T. (2007). Driving behaviour: Models and challenges. *Transport Reviews, 27*(1), 65–84.

van der Zee, D. J., & van der Vorst, J. G. A. J. (2005). A modelling framework for supply chain simulation: Opportunities for improved decision making. *Decision Sciences, 36*(1), 65–95.

Wahle, J., & Schreckenberg, M. (2001). A multi-agent system for on-line simulations based on real-world traffic data. In *Proceedings of the 34th annual Hawaii international conference on system sciences (HICSS '01),* IEEE.

Williams, I., & Raha, N. (2004). *Review of freight modelling* (Final report). Cambridge: DfT Integrated Transport and Economics Appraisal.

Chapter 27
Modeling Power and Authority: An Emergentist View from Afghanistan

Armando Geller and Scott Moss

Abstract The aim of this chapter is to provide a critical overview of state-of-the-art models that deal with power and authority and to present an alternative research design. The chapter is motivated by the fact that research on power and authority is confined by a general lack of statistical data. However, the literal complexity of structures and mechanisms of power and authority requires a formalized and dynamic approach of analysis if more than a narrative understanding of the object of investigation is sought. It is demonstrated that evidence-driven and agent-based social simulation (EDABSS) can contend with the inclusion of qualitative data and the effects of social complexity at the same time. A model on Afghan power structures exemplifying this approach is introduced and discussed in detail from the data collection process and the creation of a higher order intuitive model to the derivation of the agent rules and the model's computational implementation. EDABSS not only deals in a very direct way with social reality but also produces complex artificial representations of this reality. Explicit sociocultural and epistemological couching of an EDABSS model is therefore essential and treated as well.

Why Read This Chapter?
To understand how an evidence-driven approach using agent-based social simulation can incorporate qualitative data, and the effects of social complexity, to capture some of the workings of power and authority, even in the absence of sufficient statistical data. This is illustrated with a model of Afghan power structures, which shows how a data collection process, intuitive behavioral models, and epistemological considerations can be usefully combined. It shows how, even with a situation as complex as that of Afghanistan, the object under investigation can shape the theoretical and methodological approach rather that the other way around.

A. Geller (✉)
Scencei–Analytics for All, 1420 Prince Street, Alexandria, VA, 22314, USA
e-mail: armando@scensei.ch

S. Moss
Scott and Linda Moss Associates, Brookcliffe House, Derbyshire and Koblenz University, Mainz, Germany
e-mail: scott@scott.moss.name

© Springer International Publishing AG 2017   721
B. Edmonds, R. Meyer (eds.), *Simulating Social Complexity*,
Understanding Complex Systems, https://doi.org/10.1007/978-3-319-66948-9_27

27.1 Introduction

Notions such as "power" and "authority" are redolent with meaning yet hard to define. As a result, the number of definitions of power and authority is overwhelming (Neumann 1950). Weber (1980, p. 53) defines power as the "probability that one actor within a social relationship will be in a position to carry out his own will in spite of resistance." Giddens (1976, p. 111) understands power in a relational sense as a "property of interaction," which "may be defined as the capability to secure outcomes where the realization of these outcomes depends upon the agency of others. He implies a major distinction between two types of resources in connection with power, (1) control over material resources and (2) authoritative resources. Parsons (1952, pp. 121–132) underlines the pragmatic character of the notion of power even more by stating that power is the capacity to achieve social and societal objectives, and as such can be seen as analogous to money. Power, consequently, is the basis of a generalized capacity to attain goals and to invoke consequences upon another actor (Moss 1981, p. 163).

Neither these nor any other definitions predominate, and the decision to apply a particular one is subjective and, if not based on normative grounds, most likely context-dependent. Hence, in this research it is not aimed at applying one particular theoretical approach to power and authority, but instead it is argued that such an approach can also be founded on available and contextual evidence. Evidence is understood as information that is derived from case studies, empirically tested theories, the high-quality media, and engagement with stakeholders, domain experts, and policy analysts and makers.

For heuristic reasons—and in awareness of the plethora of conceptual approaches to power—it is for the time being assumed that the social phenomena of power and authority occur in a two- (or more-) sided relationship. It is also assumed that power serves a purpose. What should be of interest to students of power has been identified by Lasswell (1936): "Who gets what, when, how." Who and what describe and explain structures; when and how describe and explain mechanisms[1] and processes. However, Lasswell ignores an important aspect: "why." Why does someone get something at a particular moment in time in a particular way? And more generally: Why did a particular condition of power form?

Castelfranchi (1999) already noted in the year 1990 that social power is a lacuna in social simulation and (distributed) artificial intelligence. This chapter shows that although power relations are ubiquitous in social systems, only a small number of relevant models have been developed. This is despite the fact that social simulation and in particular evidence-driven and agent-based social simulation (EDABBS) are valuable complementary techniques to orthodox approaches to the study of power and authority.

---

[1] Schelling (1998) understands a "social mechanism [. . .][as] a plausible hypotheses, or set of plausible hypotheses, that could be the explanation of some social phenomenon, the explanation being in terms of interactions between individuals and other individuals, or between individuals and some social aggregate." Alternatively, a social mechanism is an interpretation, in terms of individual behavior, of a model that abstractly reproduces the phenomenon that needs explaining.

A prime virtue of EDABSS is that it imposes on the modeler a requirement to devise an unambiguous formal meaning for such notions as power and authority and their corresponding concepts. The modeling process and thus formalization should begin by formulating questions that structure the rich body of narratives that are available to describe the *explanandum*:

- Under what conditions would you label someone as powerful or as being in a position of authority?
- Having labeled someone as being powerful or in a position of authority, how would you expect that person to behave?
- Having labeled someone as powerful or in a position of authority, how would you expect yourself/others to behave toward that person?

These questions are not abstract. They form part of a data collection strategy and aim at enriching general accounts of power and authority, such as that power is a form of relationship that is exercised between two or more individuals, by more specific and context-dependent forms of descriptions of power and authority. Often these descriptions concern actor behavior. Models based on such descriptions can be closer to the evidence, because this evidence can be straightforwardly translated into "if-then" rules that can be implemented in logic-like fashion in rule-based or standard procedural languages using their native if-then structures. The if part is determined by the first question and the then part by instances of the second or third questions. Our own preference is for an evidence-driven and declarative representation of power and authority in order to preserve as much as possible of the rich data drawn from case studies and evidence in general while maintaining conceptual clarity.

The computational modeling procedure described in this article is inspired by the idea to represent reality by means of modeling; it is driven by shortcomings and advantages of other methodological approaches; and it has matured out of research on cognitive decision-making as declarative and thus mnemonic implementations (Moss 1981; Moss 1998; Moss 2000; Moss and Edmonds 1997; Moss and Kuznetsova 1996; Moss et al. 1996) and on contemporary conflicts (Geller 2006b). Classical hermeneutic approaches, although they may be strong in argument, are often methodologically weak. However, they have an important "serendipity" function and creative role, very much like that of intuitive models (Outhwaite 1987). Traditional empirical approaches, such as statistical and econometric modeling, do not represent reality and have difficulties to produce insight into mechanisms, processes, and structures (cf. Shapiro 2005; Hedström and Swedberg 1998).[2] Moreover, regular incorporation of poor proxies and the use of inadequate data do not contribute to the plausibility of research results. However, rigorous formalization furnishes desirable clarity and comparability. Finally, qualitative and case-study-based analysis produces deep structural and processual insight as well as "thick description," although at a high—for some too high—price of idiography and thus lack of generalizability.

---

[2]See for a promising corrective Sambanis (2004).

None of these approaches is incompatible with EDABSS. But we argue here that a natural modeling procedure starts with an informed but intuitive and theoretical model that needs to be validated against reality. The intuitive model is to be validated at micro-level against largely qualitative evidence, so that a rich model develops. This qualitatively validated and enriched model is then formalized as a computational model. The social simulation's output should enhance our understanding of reality, which, in turn, necessitates adjustments in the model design. This process is the hermeneutic cycle of EDABSS (cf. Geller 2006b; Geller and Moss 2008b).

A selection of models dealing with power and authority is reviewed in Sect. 27.2. A critical appraisal of these models reveals the strengths and weaknesses of past modeling approaches and underlines the necessity of the research design presented here. The selection criteria are evidently influenced by our own modeling approach and are therefore subjective. However, models have also been chosen on a functional basis: How successful are they in describing and explaining their target system? We have also tried to choose exemplary models from a wide range of modeling philosophies, to elucidate conceptual differences among the discussed models. Section 27.3 comprises the dialectical result of 2 and discusses the analytical concepts and methodological tools applied here to analyze power and authority. The materialization of our approach is presented in Sect. 27.4, where the model implementation and simulated output is discussed. The model's context is conflict-torn Afghanistan. Section 27.5 concludes by embedding our modeling framework into a broader context of comparable social phenomena, such as conflict and organized crime, and promotes agent- and evidence-based social simulation as an efficient approach for the study of power and authority.

27.2   What Can We Learn from (a Selection of) Models on Power and Authority?

The development of a social simulation model can be informed by intuitive ideas, theory, or observation. For many simulations, the respective borderlines cannot be drawn unambiguously. Nevertheless, such a classification is superior to a more traditional one which only distinguishes between micro- and macro-models. Although agent-based models entail explicit micro-level foundations, for example, the micro foundations of econometric models are inherently and at best implicit. More importantly in relation to complexity, agent-based simulations often generate emergent phenomena at macro-level. From a modeling point of view, it is therefore more interesting to understand the level and nature of data that has guided the researcher to conceptualize a model in a particular way, to what model output this conceptualization has led, and how a design helped to better understand mechanisms, processes, and structures in a target system.

27.2.1 Modeling Ideas

A variety of models on power and authority are implemented not strictly based upon theory but rather on a mixture of intuition and existing theoretical research in a particular field. These models promise to lend insight into a usually only little defined social phenomenon in an explorative, and likely to be abstract, way and therefore operate as an *entrée* into the object of investigation's broader field. Robert Axelrod's emerging political actor model has been chosen because it epitomizes the prototype of an explorative model; Rouchier et al.'s model still exemplifies the want to explore, however, on a more evidence-oriented basis.

27.2.1.1 Emerging Political Actors

In a well-known agent-based model, Axelrod (1995) reasons about the emergence of new political actors from an aggregation of smaller political actors. His motivation is to explain the restructuring of the global political landscape after the end of the cold war and the fact that, although political scientists have a number of concepts and theories to analyze the emergence of new political actors, they lack models that account for this emergence endogenously.

In short, Axelrod's model of emerging actors is a well-structured and intelligible but empirically ungrounded, conceptual model. The core of his model is a simple dynamic of "pay or else" resulting in a "tribute system in which an actor can extract resources from others through tribute payments and use these resources to extract still more resources" (Axelrod 1995, p. 21). The model consists of ten agents distributed on a circle. Wealth is the only resource and is distributed to each agent randomly at the beginning of the simulation. In each simulation iteration, three agents are randomly chosen to become active. When active, agents can ask other agents for tribute. When asked, an agent has the choice between paying the demanded tribute or to fight, depending on his and the demander's resources. In case of paying the tribute, a specified amount of wealth is transferred to the demander; in case of fighting, each agent loses wealth equal to 25% of his opponents' wealth. After three iterations all agents exogenously receive again wealth.

The core of any agent-based model is the rules according to which the agents behave. In Axelrod's tribute model, the agents have to decide when to demand tribute and how to respond to such a demand. First, an active agent needs to decide whom it should address. "A suitable decision rule [. . .] is to choose among the potential targets the one that maximizes the product of the target's vulnerability multiplied by its possible payment" (Axelrod 1995, p. 24). If no agent is vulnerable enough, then no demand is made. The chosen agent responds by fighting only if t would cost less than paying the tribute.

Agents can form alliances. During the course of the simulation, agents develop commitments toward each other. Commitment between two agents increases if one agent pays tribute to the other agent and vice versa and if two agents fight another agent together. Commitment decreases if one agent fights at the opposite side of another agent. Alliance building indirectly increases an agent's resources and outreach, as it can only make a demand to an agent if it is either adjacent or indirectly connected via allied agents.

Axelrod gets six characteristic results: (1) The model does not converge to an equilibrium. (2) The model's history is fairly variable in terms of combinations of wealthy actors, fighting frequency, and overall wealth accumulation. (3) Agents do not only share the resources with their allies, they also share the costs and thus the risks. Major conflict can occur as agents can be dragged into fighting. (4) Because of asymmetric commitment, fighting can erupt among allies. (5) An alliance can host more than one powerful agent. (6) The initial wealth distribution is not an indicator for an agent's success. As an overall result, Axelrod reports that he was able to breed new political actors of a higher organizational level. These new political actors are represented in the model as so-called clusters of commitment.

Axelrod's model demonstrates the realist notion that states do not have friends but only interests. (Agents initially make commitments out of rational calculations—"pay or else"— not out of ideological considerations.) These interests, the model reveals, are attended most efficiently by joining coalitions. Thus, an effective powerful agent seeks cooperation of some form or the other.

Axelrod's model is convincing as long as he concentrates on his main task— letting new political agents emerge in the context of an explorative setup. But interpreting a positive feedback effect resulting from a total of ten agents as "imperial overstretch" or interpreting fighting between two agents of the same cluster of commitment as civil war rings a little hollow. As much as we can learn from Axelrod about how to setup and present a simple but innovative and explorative model, as little can we learn about how to discuss it, as Axelrod continually blurs the distinction between the model and reality which leads to over- and misinterpretation of his results. Hence, a number of open questions remain, the most important of which is whether the model would withstand even circumstantially founded cross-validation. This is naturally related to the question of how much explanatory power Axelrod's model holds. With agents so abstract from reality, agent behavior in Axelrod's model simply cannot provide any explanatory insight into real world actor behavior. And while it makes sense to claim that complex macro-level behavior can emerge from simple micro-level behavior—as is the case in Axelrod's model—this macro-outcome, again, is so abstract from reality that it does not allow for any insight into the real international system. Consequently, the emergent interplay between the micro- and the macro-level, so typical for complex systems, cannot have the effects it would have in reality. This accounts for another loss of explanatory power. Axelrod claims that his model's objective is to think in new ways about old problems. However, its lack of evidence-based foundations and its degree of abstraction from reality might well foster stereotyped perceptions of the international system instead of critical reflection about it.

An alternative is for agents and their environment to be derived either from an empirically well-tested theory or from qualitative real-world observations. Cederman (1997), discussed below, has advanced Axelrod's work into this direction.

27.2.1.2   An Artificial Gift-Giving Society

Rouchier et al. (2001) reported a model of an artificial gift-giving society that is loosely founded on ethnographic research. Gifts structure society and reproduce habits and values. The giving of gifts can also be a means of redistribution. The donation, reception, and reciprocation of a gift create relationships, which can become competitive. Thus, gifts can be a means to create authority, establish hierarchies, and uphold power structures.

The model's goal is to create an artificial society in which reputation emerges. The gift-giving society's agents are either occupied with working to accumulate resources, which enables them to give away gifts or by giving away gifts themselves. The motivation to give away gifts is twofold: On one hand gifts are given away because agents act according to their self-esteem, i.e., the desire to be able to make gifts that are acceptable to the group. On the other hand, agents give away gifts because they want to increase their reputation by swaggering. Agents are fully informed and share the same decision process to determine what actions to take. The artificial society's population consists of 50 agents.

Each agent has to decide at each time step what gift it wants to make to whom. This decision process stands at the core of the model. Agents can give away gifts, either for the sake of reputation or sharing. So-called "sharing gifts" are less costly than prestige gifts. The decision of whom to give what gift is socially embedded and the agent's rationality depends on its social position. Making a sharing gift to any agent adds to the social inclusion of an agent; making a prestige gift to an agent who is considered as being prestigious fosters a hierarchy among the agents. Therefore, receiving a sharing gift represents social acceptance; receiving a prestige gift represents the acceptance of social stratification. The better an agent is socially integrated, i.e., the higher its self-esteem, the higher is its motivation to give away gifts. At the same time, high social integration increases the likelihood that an agent is able to give away prestige gifts.

All agents exchange their gifts after each agent has decided to whom to give its gift. Subsequently the agents evaluate their ranks within the group and their reputation on the basis of the gifts they have received. Self-esteem and reputation are adapted according to the outcome of the gift-giving round. Then the next round starts.

The authors discuss their findings with regard to donation-reception dynamics evolving during the simulation, of which most are positive feedback loops (Rouchier et al. 2001, p. s.5). An agent's esteem increases with the number of gifts it has made and received, respectively. The higher this agent's esteem is, the higher the likelihood is that it receives even more gifts and that he can give gifts away. The same holds true with an agent's reputation, which increases with the reception of prestige gifts. The higher this agent's reputation is, the higher is the likelihood that it receives even more prestige gifts. Moreover, the authors find that esteem and prestige "go together." Both help the circulation of gifts, and then the creation of prestige reputation" (Rouchier et al. 2001: p. 5.3). The gift-giving model provides insight into the emergence of an elite group based on socially contextualized decision processes. Within the gift-giving model, the emergence of social power is explained by two factors: (1) the willingness and ability to become socially accepted and (2) the ambition to accumulate reputation. Gift-giving has been introduced and computationally implemented by (Rouchier et al. 2001) as a process of resource accumulation and redistribution, an important variant of strategic behavior in many other contexts.

The authors, while creating a naturalistic model, do not attempt to derive their agent rules directly from qualitative, in this case likely anthropological research. This would have allowed them to gain narrative insight into the dynamics of gift giving and receiving and would have enabled them to directly cross-validate their findings with accounts made by anthropologists. Micro-level explanation was not one of the modelers' main interests. Since the emergence of complex macro-level outcomes results from micro-level behavior and social interaction, the model is of limited usefulness in the analysis of real-world social complexity. Formulation of agent rules from anthropological evidence would also have enabled the authors to avoid the assumption that agents are totally informed, both, spatially and with regard to the internal state of other agents. The cognitive processing of information about other agents is an especially difficult task. For example, the model does not address the question of when, as well as to whom, agents give gifts. It was not a purpose of the Rouchier et al. (2001) model to capture strategic decision-making using incomplete information within a realistic context. To have done so would usefully have entailed reliance on more evidence-based information available in the anthropological literature and might then have provided an empirically better grounded account of social status and power.

27.2.2   Testing Theory

27.2.2.1   Power Games Are Not Games

Game theoretical applications comprise perhaps the most formalized, coherent, and theoretically unified approach to power, especially in international relations.[3] Regularly applied in military-political analysis and international-political economy, game theory has a straightforward conception of the nation state as interdependent, goal-seeking, and rational actor embedded in a context free of centralized, authoritative institutions (Snidal 1985). The application of game theoretical models in international relations raises questions such as "Who are the relevant actors?,"

---

[3]Models not discussed in this subsection but of further interest to the reader are Alam et al. (2005), Caldas and Coelho (1999), Guyot et al. (2006), Lustick (2000), Mosler (2006), Rouchier and Thoyer (2006), Saam and Harrer (1999), and Younger (2005). Particularly highlighted should be the work of Mailliard and Sibertin-Blanc (2010) who merge a multi-agent and social network approach to the complexity and transactional nature of power with approaches to power from the French school of sociology and develop against this background a formal logic system.

"What are the rules of the game?," "What are the choices available to each actor?," "What are the payoffs in the game?," and "Is the issue best characterized as single-play or repeated-play?" (Snidal 1985, p. 26). Abstract and simplified as they are, game theoretical models nevertheless intend replicating a particular social situation and aim at—if the actors' preferences, strategies, and payoffs are accurately modeled—generating testable predictions and understanding of fundamental processes governing the system of international relations.

If states are conceived as rational power maximizers in an anarchic system, then we talk about the realist paradigm of international politics. "Rationality in this Realist world centers on the struggle for power in an anarchic environment" (Snidal 1985, p. 39). However, compared with reality it would be misleading to conceive of states as self-defending entities of purely opposing interests. Rather, game theory teaches us that states exhibit a strategic rationality that incorporates the awareness that no state can choose its best strategy or attain its best outcome independently of choices made by others. This awareness is the birth of cooperation. Moreover, from so-called iterated and dynamic games, we learn that while states have incentives to defect in the short run, in the longer run, they can achieve benefits from cooperation through time (Snidal 1985).

Game theory has often been criticized on the grounds that it is unrealistic. Game theory can, of course, analyze very particular examples of world politics, such as the Cuban Missile Crisis, nuclear war in a bipolar world or the General Agreement on Trade and Tariffs (GATT). What can be expected, in more general terms, are explanations for meta-phenomena or for abstract representations of particular circumstances. We criticize game theoretical approaches in the social sciences from an evidence-based point of view. The difference between a game theoretic model and the target system is enormous. Everything that is a natural representation of the international system is a special and difficult case in game theory, for example, nonsymmetric n-person games involving intermediate numbers of states, and everything that is straightforwardly implementable in game theory is unrealistic. Yet analogies are quickly drawn between model results and the model's target system. Hence, findings rely on oversimplified model ontologies, which may lead to over-interpretation.

Game theory's simplistic ontologies also affect a model's simulation output, as Moss (2001) argues. An analysis of state-of-the-art game theory as represented by 14 game theoretic papers published in the *Journal of Economic Theory* in the year 1999 indicates that the game theoretic models' assumptions (perfect information) and implementations (e.g., Markov transition matrices and processes) preclude the emergence of self-organized criticality as reported by Bak (1997) and cannot capture the necessary interaction as a dynamic process. Game theoretic models on markets do not entail statistical signatures found in empirical data on markets, such as power-law distributed data, a characteristic of self-organized criticality. This critique applies to game theoretic models in international relations as well, as one of the observed regularities in the international system is the power-law distribution of wars (Richardson 1948). Whereas Cederman (2003) replicated Richardson's (1948) findings by means of agent-based modeling, to our knowledge there exists no game theoretical reproduction of this statistical signature.

Although game theory can make statements of the "when" and "why," it cannot say anything about the "how" and cannot give insight into the mechanisms and processes underlying the emergence of power structures. Therefore, game theoretical modeling is, like most statistical or econometric modeling, a type of black-box modeling. A rare exception to this is Hirshleifer's work (Hirshleifer 1991, 1995).

27.2.2.2 Exploring the Limits of Equation-Based Modeling

Hirshleifer (1991) published a seminal paper on the paradox of power. The paradox of power states that in case of conflict, a weaker contestant is able to ameliorate his position relative to the stronger actor because his inferiority makes him fight harder. In other words: "[N]on-conflictual or cooperative strategies tend to be relatively more rewarding for the better-endowed side" (Hirshleifer 1991, p. 178). Hirshleifer's modeling solution for this problem is based on the assumption that if there exists an equilibrium market economy, then there must also exist an equilibrium outcome if contestants in two-party interactions compete by struggle and organized violence. The model's assumptions are full information, a steady state, indifference toward geographical factors, and the nondestructive nature of fighting (Hirshleifer 1991, p. 198).

Hirshleifer's econometric model leads to a well-specified outcome from which a number of unequivocal results can be derived. When the paradox of power applies the "rich end up transferring income to the poor" and this "tends to bring about a more equal distribution of [. . .] income" (Hirshleifer 1991, p. 197). However, he also states that "the comparative advantage of the poor in conflictual processes can be overcome when the decisiveness of conflict is sufficiently great, that is, when a given ratio of fighting effort is very disproportionately effective in determining the outcome of conflict" (Hirshleifer 1991, p. 197). Hirshleifer validates his analytical results by providing circumstantial evidence for a number of examples from class-struggle within nation-states to firms (labor-management conflicts) and protracted low-level combat.

If one accepts economic theory's underlying assumptions, Hirshleifer's results are indeed compelling and apply to a wide range of social issues related to power. The paradox of power identifies, perhaps correctly, the expectations weak and strong actors can have when mutually entering power politics: when the stakes are high, the weak are likely to get crushed by the strong. This is why the paradox of power is likely to apply to more limited contests than to full-fledged conflict.

What can we learn from Hirshleifer with regard to modeling power and authority? He elegantly exemplifies state-of-the-art model conceptualization, presentation, and discussion of results, including the model's limitations. His presentation of the analytical tools and his disclosure of the model's underlying assumptions are exemplary and render his work amenable to critique. The same applies to the rigid and straightforward formalizations of the mechanisms underlying the paradox of power, which are, moreover, well-annotated and embedded in theory. Last, but not least, the model's scope and its delimitations are clearly marked by referring to a number of examples ranging from interstate war to the sociology of the family.

Hirshleifer informs us precisely of the outcomes from power struggles and of the factors that cause these outcomes; he fails to produce analytical insight into the structural arrangements and processes that lead to these outcomes as a consequence of the methodology itself as is demonstrated in another paper of his.

In "Anarchy and Its Breakdown" Hirshleifer (1995) models anarchy, a social state lacking *auctoritas*, as a fragile spontaneous order that may either dissolve into formless amorphy or a more organized system such as hierarchy. Interesting for our task is the fact that Hirshleifer produces results that indeed allow insight in terms of structure. He can, for example, demonstrate that the state of anarchy is stable if no actor is militarily dominant and if income is high enough to assure one's own and the group's survival. However, concrete processual insight can again not be delivered, and it is referred to circumstantial evidence to concretize particular aspects of the model. In fact circumstantial evidence is arguably the only validation that is feasible with these kinds of very abstract models as the statistical signature of the model's output is so ideal-typical that it is not comparable with empirical data. In short, orthodox theoretically informed models, such as econometric or statistical models, often do address those issues in which social scientists are really interested but cannot provide an explanation in involving complexity arising from social interaction.

One reason for this has already been identified above with relation to game theory, i.e., the preclusion of emergence of self-organized criticality. Another reason is highly unrealistic general assumptions, i.e., rational choice, and more specific unrealistic assumptions, e.g., the nondestructive nature of fighting (as such also identified by Hirshleifer (1991, pp. 196–199) himself) or monolithic actors lacking an internal state. Finally, methodological individualism completely ignores the micro-macro link as well as the heterogeneous nature of political actors. While the particular assumptions as well as the (homogeneous) agents could be chosen more realistically, methodological individualism is to most statistical and econometric modeling relying on homoskedasticity of data points and variances. Such models cannot be used to describe or explain the evolution of power structures. (For statistical models this holds true only, of course, if there is sufficient statistical data that describes power relations.) With this regard, Cederman's (1997) model is a paradigmatic shift.

27.2.2.3 When, Why, and How Nations Emerge and Dissolve

Cederman's (1997) model has been chosen, because it applies the analytical rigor of formalized approaches to agent-based modeling without relinquishing the latter's virtues. He raises fundamental questions regarding realist and methodologically orthodox approaches to international relations. He introduces an agent-based simulation of the emergence and dissolution of nations. His model is based on a 20×20 grid that is initially inhabited by "predator" and "status quo" states. Each period of time a state is randomly assigned to receive resources for attack and defense. Given they have an advantage in overall and local resources, respectively, predator states can attack and conquer neighboring states. Although Cederman applies neorealist principles to a dynamic and historical meta-context, his findings challenge the orthodox neorealist belief that applying alternative methods does make an epistemological difference. He reports that defense alliances and the dominance of defense in the international system are paradoxically destabilizing. While defensive systems of cooperation deter predator states from growing to predator empires, at the same time they make possible the buildup of a hegemonic predator actor, because once a predator has reached a critical mass it has, due to the defensive nature of the system, no rivals.

Cederman (1997, p. 136) asserts that "[s]tates have been mostly modeled as internally consolidated monoliths, albeit with emergent outer boundaries." He confronts this simplifying assumption by supplying his agents with an internal decision-making mechanism representing a nationalistic two-level-politics mechanism. State leaders cannot be concerned only with foreign affairs anymore but must also take into consideration domestic issues (cf. Putnam 1988). Lazer (2001) has stated that the insight gained from Cederman's nationalist implementation is not as striking as the one dating from his emergent polarity implementation. Perhaps this is true in terms of contents, but in an epistemological perspective, Cederman makes an important point: his nationalist model inspirits previously dead states. It is not enough to know that states do something, but from a social scientific point of view, it is essential to know why they do it and how. This affords realistic, i.e., evidence-based assumptions and implementations.

Although Cederman (2003) later on introduces technological change and power projection and Weidmann and Cederman (2005) introduce strategy and self-evaluation into the decision-making process, Cederman's models conform to traditional empirical perceptions of international relations. Consequentially, a number of issues relevant to the study of power and authority in contemporary conflicts remain untouched. The state, territorial conquest and consequentially the redrawing of borderlines have been important explanatory factors for conflicts since the emergence of territorial states, but they misconceive the nature of a great number of conflicts throughout history and consequentially can only partially explain the emergence of power structures in contexts where the nation state or any other type of centralized political power has only played a marginal role. And even where such well-defined territory existed, Cederman cannot explain the causes for conflict if they have been, for example, of ethnic, religious, or economic character. Other issues that should be taken into consideration are neo-patrimonialism, anomic states, genocide, transnational organized crime, and external intervention. Models analyzing such a reality must be multivariate and causal, allowing for an explorative framework, and—contrary to Cederman (1997)—be able to include atheoretical and evidence-based information, which is often of a qualitative type.

27.2.3   Toward Implementing Reality

Modeling reality is not just about modeling particular cases. Modeling reality is about the development of models that have explanatory power on both the micro- and macro-level and therefore give also insight into mechanisms, processes, and structures. A model can hardly claim to exhibit explanatory power when lacking pivotal aspects of a perceived reality and when abstracting too much from this reality. Every reasonable model of reality—i.e., a model that describes not only the who, what, and when but also the why and how—must entail construct validity.[4]

27.2.3.1   Explaining the Causes for Social Unrest and Organized Violence

Kuznar and Frederick (2007) propose a model in which they explore the impact of nepotism on social status and thus power. They rely on an innovative model architecture supported by relevant research results, which are not framed by dogmatic theory. An agent-based model is employed "to model the origins of the sort of wealth and status distributions that seem to engender political violence. In particular, we explore the minimum conditions required to evolve a wealth distribution from the mild inequalities seen in ancestral human societies of hunter-gatherers to the more extreme wealth inequalities typical of complex, stratified societies" (Kuznar and Frederick 2007, p. 31).

Wealth is implemented as a variable that takes an ideal cultural form over which actors would compete in a particular society. For hunter-gatherer societies, wealth is distributed in a sigmoid fashion, where agents in the convex parts of the distribution have more to gain than to lose when taking chances and therefore are risk prone. The model consists of three building blocks: the distribution of wealth, agent interaction, and nepotism. Wealth distributions in complex societies are, by contrast, exponential, with sigmoid oscillations around the exponential curve. Kuznar and Frederick (2007, p. 32) term this an expo-sigmoid curve. Agent behavior is modeled along the lines of a coordination game with two equilibria (either both players defect or both players cooperate) and a Nash optimum which is to play a mixed strategy of join and defect. Kinship is inherent to nepotism. Thus, agents with many kin and offspring perform better, i.e., have higher payoffs, in the coordination game and exhibit higher fertility due to the effect of cultural success.

The result Kuznar and Frederick are getting is that the effects of nepotism transform a sigmoid wealth distribution into an approximate expo-sigmoid distri-bution. The explanation for this is the emergence of a distinct three class society: the poor and their offspring get poorer, a small middle class gets richer without changing status, and the elites get richer and richer. Thus, the authors conclude the

---

[4]A model not discussed in this section but that is of excellent quality, both in terms of content and innovation is Guyot et al. (2006). The authors analyze and discuss the evolution of power relations on the basis of participatory simulations of negotiation for common pool resources.

positive feedback loop working between nepotism and cultural success increases the structural inequality between a powerful elite and the rest of the population and thus escalates social unrest and potential organized violence.

Whereas Axelrod (1995) over-interprets his model, Kuznar and Frederick explore the full potential of their research design: a simple, intuitive, and at the same time thoroughly grounded model is presented well, specified together with moderate but auspicious conclusions, which advance research and shed new light on a problem. The model, however, would be even more compelling if the authors would have had presented cross-validational results.

Model output should be, if possible, cross-validated against empirical data originating from the target system (Moss and Edmonds 2005). There are three strategies: (1) If the simulation leads to statistical output, this output is statistically analyzed and the resulting significant signatures are compared with the statistical signatures gained from data originating from the target system. Such signatures can, for example, be a leptokurtic data distribution, clustered volatility, power laws, or distinct (e.g., small world) patterns in a social network. If the model yields output of statistical nature but statistical data is not available for the target system, then validation must rely on qualitative data. In this case, validation must either (2) seek systematic structural and processual similarities between the model and the target system, e.g., cross-network analysis, or (3) find circumstantial evidence in the target system that can also be found in the simulation. In case of empirical data, scarcity (3) is often the last resort.

27.2.3.2 Power, Resources, and Violence

Geller (2006a, b) developed an agent-based model of contemporary conflict informed by evidence. The lack of a unified theory of contemporary conflict motivated an intuitive and explanatory model of contemporary conflict. This model is based on three types of interacting actors: a politician, businessman, and warrior. They engage in six interactions: (1) the politicization of the economy and (2) the military, (3) the economization of politics and (4) the military, (5) the militarization of politics, and (6) the economy. To ascertain if this intuitive and simple ontology can capture the main structural and processual characteristics of contemporary conflicts, ten cases, such as Afghanistan, Chechnya, and Sierra Leone, have been analyzed in a primary validation procedure against the backdrop of the intuitive model. The analytical results in the form of mini case studies based on secondary literature have been sent out to case experts for a critical review. None of the experts requested an essential revision of the analytical tool, the intuitive model.

As a next step, Geller enriched the theoretical background of the primarily validated intuitive model by consulting more relevant literature for further speci-fication of the model's structure and agency aspects in order to be able to model the computational model's agent rules. The basic idea is that politicians affiliate with businessmen and warriors to make good business and get protection, while businessmen affiliate with politicians to get access to the lucrative political arena, and warriors seek political representation by politicians. Businessmen affiliate with warriors for the same reason politicians do, to get protection, while warriors get money for their provided services. Warriors can kill warriors affiliated with other politicians or businessmen, whereas civilians are considered as being non-constitutive to the intuitive model, as they are introduced into the computational model in a reactive way, meaning that although they can affiliate themselves to politicians, they can be forcibly recruited and ultimately killed by warriors.

The model offers insight into the dynamics of power in contemporary conflicts. Contrary to the prevailing "greed and grievance" approach in current conflict studies, Geller's model demonstrates that a powerful agent is dependent on businessmen and warriors at the same time. Powerful is who is most socially embedded. He can also show that the main organizers of violence are the politicians and that the warriors need not exhibit enough organizational capacity for a fully fledged campaign of organized violence. Hence, the more fragmented the political landscape is, the greater is the magnitude of organized violence. Geller's results gain importance as they are cross-validated against statistical data describing the number of conflict-related victims on a daily basis in Northern Ireland, Iraq, and Afghanistan. Both the simulation output and the real-world data suggest that conflict-related victims are lognormally distributed over time (right-skewed), exhibiting outbreaks of violence unpredictable in magnitude and timing.

Modeling always involves a degree of arbitrariness. A modeler's task, then, should be to reduce arbitrariness by making the model's design as intersubjectively comprehensible as possible. Axelrod's emerging actors model is a good, but nevertheless simple, example of this. The more evidence oriented a model becomes, the more difficult it becomes to justify the various omissions, inclusions, and abstractions. Procedural programming is cumbersome in responding to idiographic challenges. As described in the next section, a declarative, rule-based approach is better suited to the translation of evidence-based information of actor behavior into agent rules.

27.2.4 Discussion

The synopsis presented above has revealed the many approaches through which the social phenomena of power and authority can be scrutinized: in models based on ideas interest-oriented states and gift-giving individuals have been implemented; in highly formalized theoretical models, agents are conceived as rationalist power maximizers or as neo-realist states internally and externally struggling for survival; detailed evidence collected from secondary literature is used for modeling processes and structures that entail a high construct validity. Power structures are complex as well as dynamic and emerge as a result of a multitude of structure generating inter-actions among self- and group-oriented actors encompassing manifold interests.

The discussed models allow for insight into dynamic model behavior, as well as drawing structural conclusions with regard to their object of investigation "whether it is theoretical or empirical by nature. Those models that feature an individual agent architecture also lend insight into aspects of agency. Only these cope with our stipulation that the *explanandi* of social simulations of power and authority must deal with structure and agency. Nevertheless, in most cases the agent rules have been implemented on a basis, which is underrepresenting evidence, bringing about the problem that structural emergence cannot be related clearly (i.e., intra-contextually) to agent behavior. Consequently, the analysis of agent behavior cannot be related to actor behavior in the target system. This lack of realism in model design renders validation attempts of simulation results against reality less plausible and informative. By contrast, homologue models of the type advocated for in this chapter enable the researcher to gain insight into structure and agency of the sort that is more directly linkable to actor behavior. As a result, validation results become more plausible and research can enter the hermeneutic cycle of EDABSS.

27.3   Evidence-Driven and Agent-Based Social Simulation[5]

EDABSS models seek homology. The modeled mechanisms, structures, and processes aim at resembling the mechanisms, processes, and structures identified in the target system. This has two reasons: (1) Construct validity renders validation more expressive. (2) An agent-based implementation of the type presented in the following sections is more than a mere input-output model. It is an "exhibitionist" model that allows to analytically focus on internal mechanisms, processes, and structures. From a socio-scientific standpoint, this can only be of interest, if the modeled mechanisms, processes, and structures exhibit construct validity in comparison with the target system—otherwise the model is just an arcade game (cf. Boudon 1998).

The key to homology lies in the agent design. It is the agents and their interactions, respectively, that trigger the evolution of emergent phenomena. Thus, at the bottom of EDABSS agent design lays an evidence-gathering process. Posing the right questions leads to a collection of data (evidence), which directly informs the modeling of agent behavior and cognition. We have presented such questions in Sect. 27.1.

27.3.1   Evidence-Based Modeling

The source of information for homologue models must be evidence-based. This refers to the fact that all information that is used during the process of model

---

[5]For a meta-theoretical discussion of what follows see Bhaskar (1979), Boudon (1998), Cruickshank (2003), Outhwaite (1987), Sawyer (2005), Sayer (1992, 2000).

design, whether derived from a single case or from a theory, must be empirically valid (see also Boero and Squazzoni 2005). The bulk of this data is of qualitative nature, stemming from one or a number of case studies. Case studies that give concrete information of actor behavior, in particular social circumstances, are of best use to EDABSS modelers. Such a presupposition excludes assumption-laden concepts such as rational choice or Belief, Desire, Intention (BDI). EDABSS's higher rational is to find models of social simulation on what is social reality and not what is methodologically convenient or theoretically desirable (cf. Shapiro 2005). It would be wrong to stipulate that all the details entailed in the dataset must also be recognized in an EDABSS model. Modeling is an intellectual condensation process, and it is the modeler who decides what particular aspects of a social phenomenon are crucial and need to be represented in a stylized way in the model.[6]

The extensive use of qualitative data in EDABSS can be a virtue in its own, when statistical data is scarce or not available at all. This applies to a variety of important topics in the social sciences, such as elites, power structures, conflict, or organized crime. Logically systematic statistical data collection in these areas of research is difficult. Although the same holds true for qualitative data collection as well, it is, nevertheless, better feasible. For example, researchers, journalists, or humanitarian aid workers very often have the opportunity to conduct an interview or to make an observation. Often this data becomes available to the public. EDABSS therefore fills an important lacuna that is set between abstract statistical modeling and idiographic case study research as it incorporates the advantages of both formalization and context sensitivity.

The integration of stakeholders in the modeling process plays an important role in EDABSS. Integrating stakeholders in the modeling process can be rewarding and delusive at the same time. Stakeholders are keepers of information that others do not have. For example, if a stakeholder is a powerful person, then s/he can be motivated in a semi-structured interview to reflect on why s/he thinks s/he is powerful, how s/he is acting as a powerful person, and how s/he expects others to behave toward her/him. On the other hand, stakeholder's accounts can be deliberately misleading. Consequentially, EDABSS modelers have to be familiar with qualitative data collection and analysis techniques.[7]

To presume that evidence-based modeling ignores theory is not justified. Evidence-based modeling is guided by theory in many respects. First, critical realism clearly defines a research project's *explanandi*: mechanisms, processes, and structures. Second, it highly depends on the researcher—and is not generic to evidence-based modeling—how much the research process is guided by theory.

---

[6]The term condensation is alternatively denoted by Stachowiak (1973) as reduction and by Casti (1997) as allegorization. Other important modeling principles are simplicity and pragmatism (Lave and March 1975).

[7]The literature on qualitative data research has grown considerably in the last years, and the interested reader is referred to, among many others, (Lazer 2001) and (Silverman 2004).

Third, evidence-based modeling seeks generalization by intensively studying a single unit for the purpose of understanding a larger class of (similar) units (Gerring 2004).[8]

27.3.2   Endorsements: Reasoning About Power and Authority

Whereas evidence-based models of social simulation incorporate a variety of structural and processual information of the target system, the actor's actual reasoning process cannot be derived from the data. Alternatively, the concept of endorsements is applied to couch an agent's reasoning process.

Power relations, as aforementioned, are interactions between at least two actors. The computational implementation of these interactions must be based on certain grounds. This can be knowledge an actor has about another actor; it can also be experiences an actor has made in the past with his environment. Endorsements are a "natural" way of implementing reasoning about this knowledge or experience.[9] They were introduced by Cohen (1985) as a device for resolving conflicts in rule-based expert systems. Endorsements can be used to describe cognitive trajectories aimed at achieving information and preferential clarity over an agent or object from the perspective of the endorsing agent himself. We use endorsements exactly in this sense, namely, to capture a process of reasoning about preferences and the establishment of a preferential ordering (Moss 1995, 1998, 2000; Moss and Edmonds 2005). Endorsements capture an agent's (the *endorser*'s) reasoning process about other agents (the *endorsees*). That process projects the endorser's internal preferences onto the endorsee. These preferences are represented by an endorsement scheme which is a collection of categories of possible characteristics of other agents. These categories of endorsements amount to a partial preference ordering of characteristics perceived in other agents. The ranking of collections of endorsements is an essentially arbitrary process. Cohen (1985) used a lexicographic ordering so that the preferred object (in this case, agent) would be that with the largest number of endorsements in the most valuable category. If several objects were tied at the top level, the second level of endorsements would be used to break the tie and then, if necessary, the third or fourth levels, etc. An alternative is to allow for a large number of less important endorsements to dominate over a small number of more important endorsements. One way of achieving this is to calculate endorsement values E for each endorsee as in Eq. (27.1) where b is the number base (the number of endorsements in one class that will be a matter of indifference with a single endorsement of the next higher class) and e_i is the endorsement class of the

---

[8]We are well aware of the ongoing discussion on induction with regard to case-study research and the interested reader may refer, among others, to Gomm et al. (1992), Eckstein (1992) and Stakes (1978).

[9]See for a more complete treatment of endorsements (Alam et al. 2010).

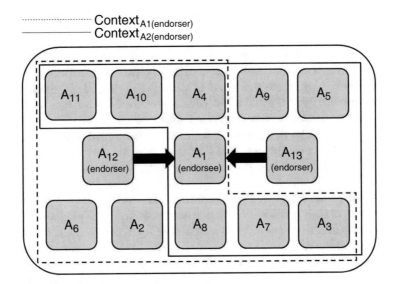

Fig. 27.1 Schematic representation of the embeddedness of the endorsement process

i_{th} endorsement token (Moss 1995). In choosing from among several other agents, an agent would choose the endorsee with the highest endorsement value E.

$$E = \sum_{e_i \geq 0} b^{e_i} - \sum_{e_i < 0} b^{|e_i|} \qquad (27.1)$$

The process of endorsing an agent must be thought of as being embedded in an agent's environmental context, i.e., his neighboring agents (see Fig. 27.1). The endorsement process allows an agent to find the agent most appropriate to him—it does not (and cannot) seek the best of all agents.

The main advantage in applying the idea of endorsements lies in the fact that they allow for combining the efficiency properties of numerical measures, with the richness and subtleties of non-numerical measures of interest or belief (Moss 1995).

The choice of endorsements and the conditions in which each endorsement will be attached is entirely context-dependent. Agents concerned with critical incidence management in water supply (Moss 1998) obviously have different criteria than agents embedded in the context of contemporary conflicts (Geller and Moss 2008a). While the former might be interested in actions and information models leading to the successful resolution of a complicated allocation problem, agents in models of contemporary conflict might be interested in with whom they should cooperate and whom they should shun or even fight. For example, it is of importance to an agent to know if its vis-à-vis is of the same ethnicity, religion, and kin; if it has lived a similar past; and if it is reliable, corrupt, or wealthy. Power and authority relations depend on knowledge about these kinds of questions. Section 27.4 addresses this as well as the question of how to translate the evidence into an adequate endorsement scheme in more detail.

27.3.3 Declarative Implementations of Agent-Based Models

A program is declarative if there is a set of statements on a database, rules have a set of conditions, which are statements with some values left open as variables, and consequents exist, which are another set of statements. When all of the statements in the conditions of a rule are matched by statements on the database, then the variables are given their specific values from the database statements and the consequent statements are added to the database. When a set of conditions are satisfied and a rule fires (i.e., puts its consequents on the database), then the state of the environment as represented by the database is changed, and perhaps other rules will now be able to fire and so on until all rules have fired and no further matches of conditions can be found on the database. The sequence of rules that will fire and the particular instantiations of their variable values are determined only as the program is running. The sequence of actions represents the process of agent behavior and leads in each case to a new state of the environment. If all agents are implemented declaratively, then they will be changing the state of the environment for one another and the pattern of rules, and therefore actions of all the agents taken together will be influenced by one another.

In these circumstances, the outcomes of such a model are usually impossible to predict with any exactitude.[10] Frequently, such models exhibit the sort of episodic volatility associated in the first section with complexity. The same effect can be achieved by other means, but declarative representations of agents have a number of virtues in terms of ease of development as new evidence becomes available and in terms of yielding comprehensible outputs stored as statements on the databases.

The assumption that the fulfillment of conditions triggers the execution of consequents marks, in a homologue model, a natural representation of actor behavior. Each actor behaves according to a defined set of rules. A rule fires only when the set of conditions attributed to this rule is satisfied. Accordingly, an agent's behavior is governed on the basis of the fulfillment of conditions. It is fairly straightforward to translate information on actor behavior obtained during the data collection process into conditions.

Recall the following question from the opening paragraph: "Under what conditions would you label someone as powerful?" A possible answer to this could be "If a person belongs to a well-known family." The condition for an agent providing the abovementioned answer to label another agent as powerful is fulfilled if this other agent belongs to a well-known family (whereas well known could translate into socially well connected). Similarly, an agent might provide the answer "If the agent has won five consecutive battles." In this case an agent has to follow another agent's battle history to be able to tell if it can label him powerful or not. The translation for a possible answer to the second question is analogous. "Having labeled someone as

---

[10]Hence, a declarative model architecture does not allow easily for exact simulation replication. Other disadvantages are that declarative models tend to be computationally expensive and ontologically complex.

being powerful, how would you expect that person to behave?" could be answered by saying "That the person is generous." However, the translation now includes two conditions. Firstly, an agent must have been already labeled powerful. Secondly, an agent must have experienced the powerful agent being generous to it.

If the conditions of a rule are satisfied, then its consequents are put into effect. Recall the third question stated in the introduction: "Having labeled someone as powerful, how would you expect yourself/others to behave towards that person?" A possible answer could be "Then I would subordinate myself to that person." The translation reads as follows: if an agent has labeled another agent as powerful, it then (as a consequence) subordinates itself to this agent.

Analogously to the examples given, all the collected information that bears relevance to the modeling process can be translated into declarative program code. This translation process is essentially an operationalization and formalization process of (sometimes vague) bits of qualitative information. Power and authority are not implemented as predefined entities but are "grown" artificially from a number of evidence-based rules that are proxies for dimensions of power and authority. Agents become powerful as a consequence of a variety of causally interconnected conditions and consequents.

27.4 Modeling Power and Authority: A Case from Afghanistan

This section presents an implementation of what has been discussed above theoretically. Reflections of power and authority in contemporary conflict are presented, from which an intuitive but evidence-informed model of power and authority in Afghanistan is derived. Against the background of this model and on the basis of qualitative data, answers in the form of evidence to the questions posed in the opening paragraph are presented. From these answers the agent rules are being developed and translated into program code.

27.4.1 Power and Authority in Contemporary Conflicts[11]

The anthropogenic nature of power structures (Popitz 1992) has been shown for a variety of conflict regions, including Afghanistan (Bayart et al. 1999; Reno 1998; Roy 1994, 1995). Sofsky (2002) argued that conflict societies are societies sui generis. They function according to their own social laws and are structurally and processually disjointed from societies lacking a comparable degree of organized violence. In conflict-torn societies virtually anything goes. This can be illustrated

---

[11]Parts of this and the next paragraph have been taken from Geller and Moss (2007).

by the concept of anomie. Anomie is the situation in which the upper and lower normative boundaries for the aspirations of members of a society are thrown awry (Marks 1974). An anomic situation emerges when the means to attain a specific goal, such as accumulation of wealth or power, run out of social control (Merton 1938). Accordingly, in a space emptied of restricting norms, i.e., an anomie, virtually everything goes along with the creation of power structures to one's own ideas and interests.

Anomic spaces are political spaces lacking strong modern institutions, such as the state's monopoly on organized violence, stability of the law, and protection of property rights. In these circumstances only highly adaptive stakeholders prevail. The socio-structural outcomes of this organizational process are manifold and so are the adopted means that serve one's interests.

In contemporary conflict societies this outcome is neo-patrimonialism (Geller 2006a; Medard 1990; Reno 1998). Weber (1980) understands patrimonial power as being based on authority, suppressed subjects, and paid military organizations, by virtue of which the extent of a ruler's arbitrary power as well as grace and mercy increases. Stakeholders interested in gaining power in contemporary conflict settings have to act neo-patrimonially to accumulate and redistribute material as well as social resources. The range of related activities is broad and includes corruption, clientelism, patronage, nepotism, praebendism, and so forth (cf. Medard 1990).

27.4.2 An Intuitive Model of Power and Authority in Afghanistan

Anthropogeneity, anomie, and neo-patrimonialism—or any other theoretical context relevant to a particular research project—have eminent ramifications for the perception of power and authority and henceforth for the development of the model at hand. The evidence presented below should therefore corroborate the implicit claim that anthropogeneity, anomie, and neo-patrimonialism amalgamate to a framework describing Afghan power structures and functioning as an evidence-informed theoretical framework that can be filled with the intricacies of the Afghan case. In the beginning of a research project, such a framework model provides a theoretically informed ontological entrée for the object of investigation.

The actual information the model at hand rests upon is derived either from data collected by ourselves or from relevant secondary data sources. The collected primary data stems from semi-structured interviews conducted with urban Afghan elites between May 2006 and October 2007. The secondary data stems from case studies, most of which are of anthropological type, reports published by nongovernmental organizations (NGOs) or non-state actors, such as the United Nations (UN) and the International Committee of the Red Cross (ICRC), or the print media.

Although 27 years of conflict accentuated two important factors in Afghan society, namely, ethnicity and religion, the traditional organizational principle of the *qawm* rested sound (Azoy 2003; Roy 1994, 1995; Shahrani 1998). Less mentioned, however, is a decline of norms and values in Afghan society leading to a Hobbesian form of society (Tarzi 1993). Today's Afghanistan can be characterized as an anomie (Geller 2010).

The causes for this development are complex but nevertheless directly linked to the Jihad of 1979 to 1989. Although trends of neo-patrimonial politics are already recognizable in the very beginning of the Jihad—and are indeed a characteristic of Afghan politics throughout history—the war's fundamental goals started to mutate with its increasing duration. Some of the adopted means of warfare have been traditional, such as organized violence, intrigue, alliance formation, and dissolution; others have been "imported," such as religious extremism and radicalization of ethnicity (cf. Geller 2010; Roy 1998).

The concept of *qawm* is context-dependent, defined by such social dimensions as family, kinship, ethnicity, and occupational groups and also more abstract but related concepts such as solidarity, rivalry, cooperation, and conflict. The notion of *qawm* also underlines the fluidity and contextual dependency of social relations in Afghanistan. Hence, *qawm* ontologies are devised to codify individual actors, their behavior and relations between actors, as well as social processes and structures arising as a result of social interaction (Dorronsoro 2005, pp. 10–11).[12] Each of those aspects is pertinent to the development of our models, which represent and clarify social processes associated with these overlapping identity spheres and the actors acting within them.

The notion of *qawm* varies not only in the literature but also among Afghans themselves. It can mean (extended) family, tribe, descent group, ethnicity, "people like us" (Tapper 2008, p. 88), an "occupational group" (Roy 1992, p. 75), and "persons who mutually assist each other" (Canfield 1973, p. 35), and it can connote a complex interpersonal "network" (Roy 1995, p. 22; Dorronsoro 2005, p. 10) of political, social, economic, military, and cultural relations (Mousavi 1997, pp. 46–48; Tapper 2008; Glatzer 1998, p. 174; Rasuly-Paleczek 1998. pp. 210–214; Roy 1995. pp. 21–25; Shahrani 1998, pp. 218–221).[13] In fact, our interview data suggests that these meanings are not mutually exclusive: *qawm* do not have clear boundaries nor do they divide Afghan society into mutually exclusive groups. "[A]n individual always belongs to more than one [*qawm*]" (Canfield 1988, p. 194).

qawm face competition with other *qawm* and internal competition among members of a *qawm* (Azoy 2003; Mousavi 1997, pp. 46–48; Roy 1994, p. 74; Roy 1995, pp. 21–22). *qawm* need to be sustained, and it is an Afghan leader's

---

[12]Monsutti (2004) "explores the basis of cooperation in a situation of war and migration" among the Hazara in Afghanistan through the concepts of solidarity and reciprocity. Nancy Tapper (1991) "reveals the structure of competition and conflict for the control of political and economic resources" through the concept of marriage.

[13]Whether a *qawm* denotes a group or a network is not clear from the evidence. Following Tapper's (2008) argument, a *qawm* can take the form of a group or a network, depending on the context.

ability to redistribute resources that makes him powerful and eventually successful (Roy 1994, p. 74). The ability to create a *qawm* for a particular aim is also perceived as a demonstration of power (Azoy 2003, p. 36). *qawm* still "have a powerful and pervasive effect on contemporary political discourse and the behavior of Afghans" (Shahrani 1998, p. 220) and have during the years of conflict often been misused by new elites for the pursuit of conflict and criminal aims (Canfield 1988; Rasuly-Paleczek 1998, pp. 210–214; Roy 1994; Rubin 1992; Shahrani 2002; Tapper 2008). Manifestations of such abusive behavior are, for example, corruption, drug production and smuggling, nepotism, massive organized violence, crime, and ethnic, political, and religious radicalization (Giustozzi 2006; Glatzer 2003; Rubin 1992, 2007; Schetter et al. 2007). *qawm* are not the cause for conflict in Afghanistan, as these causes are manifold (Shahrani 2002, p. 716; Dorronsoro 2005), but we will explore the usefulness of the notion of *qawm* in analyzing and understanding conflict in Afghanistan.

Figure 27.2 depicts an informed intuitive and ideal-typical representation of a *qawm*. It consists of ten actor types: politicians, religious leaders, commanders (meritocratic title for a militia leader), businessmen, warriors, civilians, farmers, drug farmers, organized criminals, and drug dealers. An important abstraction from reality is that in our model each actor has its distinct role, whereas in reality actors may incorporate a variety of roles. For example, a commander can be a (military) commander, a politician and a drug lord at the same time. We proxy individual role pluralism by mutual interdependence, i.e., each actor has virtues another actor may be in need of and vice versa, leading to mutual cooperation and interdependence. This, of course, is also a common pattern in reality, where there is no clear distinction between role incorporation and cooperation.

The following examples explain the *qawm* model in terms of agency. If a politician is in need of military protection, he approaches a commander. In return, a commander receives political appreciation by mere cooperation with a politician. If a businessman wants to be awarded an official construction contract by the government, he relies on a politician's political connections. In return, the politician receives a monetary provision, for example, bribes. If a politician wants beneficial publicity, he asks a religious leader for support. The religious leader, in return, becomes perceived as a religious authority. If a warrior seeks protection and subsistence for his family, he lends his services to a commander, who, in return, provides him with weapons, clothes, food, and/or money. If an organized criminal wants to carry drugs, he relies on the transport business of a businessman who, in return, receives a share of the drugs sold. If a drug farmer needs protection for his poppy fields, he affiliates with a commander, who, in return, receives a tithe of the drugs sold to a local drug dealer. According to Azoy (2003), such interactions are also guided by the following four social categories: kinship, residence, class, and religion. Our model represents this neo-patrimonial behavior. The links between the agents in Fig. 27.2 can also represent such categories.

The continuing existence of the *qawm* in times of severe social change as a means to organize and manage power cannot baffle the fact that the *qawm* itself has undergone configurational alteration. Protracted conflict deteriorated not only

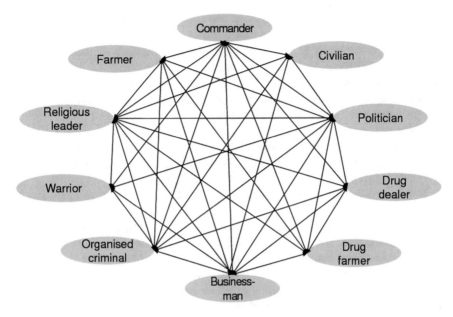

Fig. 27.2 A case study informed intuitive model of a *qawm*

social structure but also obliterated moral boundaries. Corruption, fraud, mistrust, crude materialism, and the like systematically found their way into Afghan society. The power of the *qawm* and of its members to constantly adapt to new states of anomie epitomizes the anthropogenic nature of power and authority (Geller 2010).

27.4.3 An Intuitive Model of Power and Authority in Afghanistan

27.4.3.1 Evidence

The collected interview data explicitly highlights three sources of power in current Afghanistan: ownership, reputation, and *qawm* (cf. Azoy 2003). Their meaning manifests when mirrored against the notions of *hisiyat* and *e'tibar*. *Hisiyat* and *e'tibar*, two Dari words that roughly translate into "character" and "credit." *Hisiyat* denotes qualities such as piety and wisdom; *e'tibar* is about meritocracy. A powerful actor must dispose of *hisiyat and e'tibar*.

Traditionally, ownership can be defined as land, access to water, livestock, and women. References to landownership were made often during interviews, whereas water and women have never been mentioned. Ownership of livestock was mainly mentioned to serve reputational means in order to increase one's own or someone else's reputation. The interview data and observations made during the field trips suggest that a modern comprehension of ownership has become more materialistic and less subsistence oriented; mundane symbols of power such as money, houses, and cars have increased in importance. This raises the issue of the sources for these goods. While some Afghans undoubtedly were able to build up prospering businesses or brought assets with them from exile, other sources remain dubious and are likely to include organized crime, corruption, and clandestinely working for foreign countries. Thus, a generalized answer to "When would you label someone as being powerful?" includes that, either being traditional or modern, or, more likely, a combination of both, a powerful individual must have access to ownership resources. "[. . .][P]ower without wealth is all but possible" (Azoy 2003, p. 30). But "[w]ealth is just a means to achieve prestige" (Roy 1994, p. 74).

In Afghanistan, authority is nourished by reputation. It is the "ultimate source of political authority" (Azoy 2003, p. 31). Reputation exhibits a static and dynamic component. The static component is closely related with ancestry. It is important what ethnicity, family, or tribe someone belongs to. Hazaras, one of the four major ethnicities in Afghanistan, are often regarded as working class, while Pashtun, the largest ethnicity, are often perceived as a warrior elite. A family may be regarded as politically powerful and/or religiously influential. To have roots in a political, religious, or scholarly family provides authority. This became obvious during interview sessions in a variety of ways. Interviewees have regularly been introduced or have introduced themselves by referring to their family either in a political or religious context. Two outspoken authoritarian interviewees, for example, claimed to be Sayyed, i.e., descendants from the Prophet, and scholars at the same time. The dynamic aspect of reputation relates to an individual's historicity. Individual politico-historical background is important. Individual histories link actors to different social groups and can provide them, depending on the social context, with esteem, such as in the case of having been a Mujahedin, a resistance fighter during the time of Soviet occupation. Thus a generic answer based on the interview data collected to "when would you label someone as being powerful?" would include that a powerful individual belongs to an important family and/or has played an important role in his past.

The *qawm* is the epitomization of network-based power structures in Afghanistan. A *qawm* is more than mere reference to an actor's ethnicity or tribe; it is more than an extended family. The *qawm* is an actively used instrument in the pursuit of power. An appropriate translation would therefore be "power basis": a number of people that can be mobilized to achieve a particular political aim. With changing political aims, this group of people may change as well. Thus a generalized answer to the question "when would you label someone as powerful?" includes that a powerful individual not only disposes of a *qawm* but also exerts control over it.

27.4.3.2 Model and Computational Implementation

Based on the evidence, how are a powerful actor and his entourage to be structurally modeled? First, it needs to be defined who is powerful and who is not by disposition. The evidence presented above indicates that the following agent types should be considered as being powerful in our model: politicians, businessmen, religious leaders, commanders, and organized criminals. These agents are computationally created as being powerful by definition, because they are politicians, religious leaders, commanders, etc.[14] Hence, social resources do not have to be distributed explicitly, agents are "born" possessing them—or not.

Secondly, agents must, at the initialization of the model, be equipped with material resources. These assets are given, pars pro toto, in the form of money, drugs, and land. The absolute amount of money distributed in the model is arbitrary as real numbers are not available. Money is distributed lognormally among agents of one particular agent type. A lognormal distribution of wealth also appears for the case of Afghanistan a plausible assumption (Limpert et al. 2001). Hence, a small number of agents are very rich, while most of the people are poor. The data record for land holdings, which is better than the one for wealth, suggests that holdings of land should be distributed lognormally as well (Wily 2004). Again, this means that a small number of agents possess a lot of land, while a large number of agents will only own little land. Last but not least, drug farmers receive some drugs in the beginning of the simulation and harvest drugs during the simulation in specified harvesting periods (UNODC 2006).

Thirdly, agents must be given an internal state, representing what has been identified above as *hisiyat* and *e'tibar*. This internal state is an agent's endorsement scheme, which is depicted in Table 27.1. Some of these labels (the endorsements) are static, others are dynamic. Static endorsements are attributed to each agent at the beginning of a simulation and cannot be changed during the course of the simulation: an agent is either an intellectual or he is not; he is either a Tajik, another important Afghan ethnicity, or he is not; he is either my brother or he is not, etc. For the time being it is not implemented that an agent's changing relationships are taken into account to form an individual political profile, hence the politico-military background is static. Some *hisiyat* endorsements are dynamic, such as loyalty, trustworthiness, neighborhood, or religiousness, and can change their respective values during a simulation run. All *e'tibar* endorsements are, by contrast, dynamic, as *e'tibar*, i.e., meritocracy, is inherently dynamic and must call for a dynamic conceptualization. Formerly reliable agents can become unreliable and previously successful agents can become unsuccessful, etc.

The endorsement scheme does not only depict an agent's internal state but also denotes the categories in which an agent reasons about other agents. For example,

---

[14]Note that we are not simulating the genesis of a powerful agent, but the emergence of power as a network-like structure in an evidence-based, artificial society. See for the qualitative description of such a genesis (Giustozzi 2006).

Table 27.1 Crosstabular presentation of an agent's endorsement scheme

	Static	Dynamic
hisiyat	Intellectual/non-scholarly	Loyal/disloyal
	Shared-ethnicity/different-ethnicity	Trustworthy/untrustworthy
	Shared-religion/different-religion	Is-neighbor/non-neighbor
	Is-kin/non-kin	Pious/sinful
	Politico-military-background	
e'tibar		Reliable/unreliable
		Successful/unsuccessful
		Capable/incapable

assume agent A is Tajik, a Mujahedin, successful, and neighbor of agent B. Further assume that agent B is Tajik as well, was also a Mujahedin, and evidently is also neighbor of A. It is then likely that agent B will look favorably at agent A and vice versa and that the two will establish an affiliation with each other. The endorsement scheme breaks down why an agent is more powerful than another agent: because it disposes over an internal state that is seen favorable by other agents. Or to put it differently: because it internalized a number of qualities that are perceived as symbols of power by other agents.

It is important to note that although an individual agent is defined at the moment of its creation as being powerful or not per se, nothing is said about how powerful it is going to be or spoken differently: how good of a neo-patrimonial agent it will be. The agent's performance as well as the social product of its performance, the *qawm*, are emerging out of the social simulation model and are not computational artifacts.

27.4.3.3 Behaving Powerfully

Having clarified what makes actors powerful in Afghanistan, it is now important to know how these powerful actors behave and how other actors behave toward powerful actors. The corresponding questions from the introduction are: "Having accepted someone being powerful and/or being an authority, how would you expect that person to behave?" And: "Having labeled someone as powerful/as being in a position of authority, how would you expect yourself/others to behave towards that person?" The freedom of choice an actor has to behave toward a powerful agent depends on a number of different factors (*hisiyat* and *e'tibar*) but is foremost based on the distinction whether an actor is powerful himself or not. The two constellations powerful actor/weak actor and powerful actor/powerful actor lead to different outcomes in the social organization. In our understanding these are patron-client relationships (powerful/weak) or affiliations (powerful/powerful). However, in both cases, accumulating and redistributing resources is in the center of actor behavior.

Weak actors only have little choice of how to behave toward a powerful actor and are forced into a patron-client relationship because of grievance. They are either locked in into economic dependency or are not even a member of the powerful actor's *qawm* and thus cosmos. In both cases weak actors can only submit themselves and must fully depend on the powerful actor's gratitude. Consequentially, in patron-client relationships, powerful behavior is more of a mundane sort, for example, supporting a client's family with food, clothes, and housing. Depending on what kind of client it is, the patron might ask for dog's body services in the case of a civilian, for protective services in the case of a warrior, or for *zakhat* (tithe) in the case of a farmer. In general, the patron demands loyalty for his support. The weak actor exerts power over the powerful actor indirectly insofar as it is harmful for a powerful actor's reputation to pay subsidies irregularly or not at all. The weak actor exerts power directly in case of grass-root opposition as a result of insensitive politicization by the powerful actor. This social relationship of dependence is standard in Afghanistan and every powerful man surrounds himself with such a "service force," be it small or big. Nevertheless, supporting the weak should not be considered as being unimportant, as they provide a basis of broad social support. Moreover, supporting the weak increases a powerful man's *e'tibar* as the following ideal-typical characterization of Pashtun men highlights: he is a man of honor, with prowess and pride, whose dignity does not forbid him to be attentive to authorities as well as the weak (Janata and Hassas 1975, p. 84) (translation ours).

Powerful behavior between two or more powerful agents is of a different nature. Powerful actors have the freedom to choose their behavior toward their vis-à-vis and can either act cooperatively, conflictously, or submissively. Which type of behavior is chosen, is based on a deliberate but nevertheless fragile assessment of who must be considered a supporter and who must be considered a spoiler or even a foe. Hence, whatever the project that is of concern to a powerful actor, it is intensively discussed with those people from his social circle who he thinks should be included in the decision-making process (Azoy 2003).

The evaluation of a project's—and ultimately of a powerful actor's behavior— supporters and opponents constitutes the initialization of a project-based *qawm*. The organization of a *qawm*, whether it is a generic *qawm* based on kinship or a temporal *qawm*, is a delicate operation. Exacerbating is the fact that potential supporters as well as potential enemies are competitors against whom the powerful actor must stand up—at all time. Hence, the creation and maintenance of *qawm* explains the volatile nature of cooperation, the often and sudden changes of alliances and the ubiquity of conflict (cf. Azoy 2003).[15]

In the case of cooperation, powerful actors establish affiliations between each other. Affiliations are relations between qualitatively equals. Consequentially, powerful agents do not give each other material support, but they provide each other with social resources. While powerful actors in patron-client relationships accumulate

---

[15] We do not consider the emergence of conflict in this chapter. See for a preliminary discussion (Gerring 2004).

social resources for redistributing material resources, powerful actors in affiliations mostly accumulate and redistribute social resources. Politicians guarantee that commanders are not denigrated as "warlords," commanders protect politicians, religious leaders openly designate politicians of being pious, politicians declare a religious leader their spiritual leader, businessmen financially support a politician's campaign, politicians provide businessmen with lucrative state contracts, etc. In short, powerful actors support each other in increasing their *hisiyat* and *e'tibar* record. The opposite, of course, exists as well. Powerful actors can actively engage in diminishing another actor's *hisiyat* and *e'tibar* record.

In summary, a powerful actor is able to control his *qawm* economically and socially. He is able to redistribute enough material and social resources to keep his *qawm* alive. If one would have to measure the power an actor in Afghanistan has, then it would not suffice to only count his material assets. A comprehensive measure of power would include this actor's reputation, measured in terms of his ability to "call on the services of supporters to help him in whatever enterprise" (Azoy 2003, p. 32).

27.4.3.4 Model and Computational Implementation

The computational implementation of the behavior of a powerful agent is straightforwardly informed by the evidence presented above. Powerful agents want to accumulate and redistribute material and social resources. For this reason their *hisiyat* and *e'tibar* needs to be relatively superior to their competitors'. Hence, a powerful agent's aim must be to establish as many favorable relationships as possible. He does this in two steps: first he reasons about which agents to endorse, and second he takes action on the basis of his decision he has taken in step one.

Step one consists of the endorsement process as explained in Sect. 27.3.3 and as contextualized in Sect. 27.4.3. Each agent continually checks all the agents visible to him—not all of them are—i.e., he projects his endorsement scheme upon them, rates the corresponding values, calculates E, compares all Es against each other, and chooses the one with the highest E to endorse. According to the evidence, the model implies that during the endorsement process, those agents are more likely to establish a relationship with each other if they are similar with regard to *hisiyat* endorsements and who exhibit higher values with regard to *e'tibar* endorsements. Two short examples shall clarify this, one for patron-client relationships and one for affiliations.

Powerful agents do not seek ordinary agents; rather they are sought out by the latter. But even though ordinary agents are in misery, they try to choose the powerful agent who is most suitable for them, given the choice. Assume a civilian, who is Tajik, a Mujahedin and in need of material support (see for what follows also Table 27.2). (In the simulation this is the case when the civilian's holdings of money are ≤ 0.) Before asking every politician for material support the agent can see, it checks which of the visible politicians are most suitable to him. Assume that there are two visible politicians. The civilian agent then projects its endorsement

Table 27.2 Cross-tabular presentation of a sample endorsement process between one civilian and two politicians

Endorsements	Civilian	Politician$_1$	Politician$_2$
Same/different ethnicity (± 1)	Tajik	Hazara/-2*	Pashtun/-2*
Same/different politico-military background (± 2)	Mujahedin	Mujahedin/4	Mujahedin/4
Reliable/unreliable (± 3)	–	Unreliable/-8*	Reliable/8
E	–	-6	10

The values in parentheses indicate the importance of an endorsement. Values on the left side of each dash represent "reality"; values on the right side of each dash represent the "weighed reality." Labels marked with an asterisk have been multiplied with -1, because in Eq. (27.1), they are part of the second Σ sign and are therefore subtracted

scheme upon each of the two politician agents. Recall that the endorsement scheme not only tells the civilian how important particular characteristics of these two politicians are to it but also what the categories of its own perceptions are. In Eq. (27.1) b denotes the number base for which every agent is randomly assigned a value $b > 0$. (For $b = 0$ the expression b^x is equal to 1, independently of the exponent x). In this example, the civilian is assigned $b = 2$. In Eq. (27.1) e_i denotes the value of the ith endorsement token. This value differs for each endorsement token and for each agent. In this example the endorsement tokens for the civilian are same-ethnicity/different-ethnicity, same-politico-military-background/different-politico-military-background, and reliable/unreliable. Each endorsement token is randomly assigned a value e_i. e_i differs for each agent. In the present example, the following values for e_i are assigned for the civilian: 1 for same-ethnicity, -1 for different-ethnicity, 2 for same-politico-military-background, -2 for different-politico-military-background, 3 for reliable, and -3 for unreliable. These values are on an ordinal scale and represent the civilian's endorsement scheme. The interpretation of this endorsement scheme is that the civilian perceives the labels same-ethnicity/different-ethnicity less important than the labels same-politico-military-background/different-politico-military-background and the labels same-politico-military-background/different-politico-military-background less important than the labels reliable/unreliable. Politician$_1$ is a Hazara, a Mujahedin and has never been endorsed reliable by the civilian. Politician$_2$ is an Uzbek, a Mujahedin and has been endorsed reliable in the past by the civilian. With regard to politician$_1$ the civilian reasons as follows: politician$_1$ is not of the same ethnicity, has the same politico-military-background, and is unreliable. Depending on this information and the values assigned above, (1) allows the calculation of E, which is: $2^2 - 2^{|-1|} - 2^{|-3|} = -6$. For politician$_2$ the civilian reasons as follows: politician$_2$ is not of the same ethnicity, has the same politico-military-background, and is reliable. Thus, E can be calculated as follows: $2^2 + 2^3 - 2^{|-1|} = 10$. Because politician$_2$ has a higher E than politician$_1$, the civilian decides to choose politician$_2$. This procedure can be extended to as many agents and to as large endorsement schemes as necessary.

Once the civilian has taken its decision, which politician to ask for support, the agent sends a message to this particular politician, requesting support. As long as the politician has enough money to support the civilian, it accepts the support request. The acceptance of the support request is tantamount to the establishment of a patron-client relationship. In each simulation iteration, the politician pays the civilian a defined amount of money, as long as it has enough money to do so. In return, the civilian is required to endorse the politician as being trustworthy. In the event that the politician cannot pay the civilian anymore, the latter endorses the politician being unreliable and untrustworthy, leading to a breakup of the patron-client relationship. The civilian then has to seek another politician, and the endorsement process starts again. All requests for support by ordinary agents follow this scheme.

Interactions between powerful agents processually do not differ substantially from the scheme described above. Consider the case where a commander wants a trustworthy assertion from a politician in order to not be denigrated as "warlord." Assume that two politicians are visible to the commander. Again, before sending off a message to the most suitable politician, the commander assesses which of the two politicians are most suitable to it, i.e., which one has the higher E. Although different and further endorsements might apply in this case, the endorsement process, as described in the civilian-politician case, does not change in essence. Once the commander has chosen a suitable politician—one, for example, who is of the same ethnicity, the same politico-military background, and who has a record of being trustworthy—it then sends a message to this politician requesting to be endorsed trustworthy. Because the commander has chosen a politician with a trustworthy record during its endorsement process, the politician is able to do so, i.e., to endorse the commander trustworthy. However, the politician demands a service in return. When dealing with a commander, this service is naturally protection. Hence, the politician agent sends a message to the commander agent, stating that it will endorse it trustworthy, but only if the commander agent can protect the politician agent. Naturally, the commander can provide protection only, if there is at least one warrior with whom he has a patron-client relationship, established according to the procedure described above. Given he can provide protection, he accordingly answers the politician's message. This mutual fulfillment of conditions triggers a number of mutual endorsements: the politician endorses the commander as not only being trustworthy, as requested, but also as being capable, as he is capable of providing protection; the commander endorses the politician as being trustworthy. Mutually endorsing each other is tantamount to establishing an affiliation. This affiliation holds as long as the conditions leading to it hold true: "Is the commander still able to provide protection?," "Is he still capable?," "Is it still trustworthy?," "Is the politician still trustworthy?" Both agents do this every simulation iteration. Moreover, all relationships, whether they are patron-client relationships or affiliations, break up when there is an agent (endorsee) found who fits an endorser better, because agents continually scan their neighborhood for better opportunities.

The example given above is representative for all the interactions between powerful agents: politicians request armed protection from commanders for a trustworthiness assertion; commanders approach businessmen to invest money; businessmen pay politicians for courtesy services; politicians request a pious assertion from a religious leader for a trustworthiness assertion; and religious leaders ask commanders for protection and, in return, assert a commander as being pious. All these affiliation interactions describe the neo-patrimonial usage of social capital as well as the accumulation and redistribution of material resources.

The quintessential mechanism is the endorsement scheme. It is the interface through which two agents communicate with each other, and it is the raster that filters those who are "endorsables" from those who are "unendorsables." If two agents match to establish a patron-client relationship or an affiliation, is ultimately decided via the mechanism of the endorsement scheme. Hence, the evidence that is enshrined in the endorsement scheme (cf. Sect. 27.4.3 and there in particular Table 27.1) finally leads to the emergence of social reality.

27.4.3.5   Emerging Structures: Simulation Results[16]

It would be beyond the scope of this chapter to discuss the simulation results and their validation in detail (see Chap. 9 (David et al. 2017), Chap. 10 (Evans et al. 2017), and Chap. 14 (Sawyer 2017) in this volume, also Moss 2007 for more on how to do this). Figure 27.3 depicts the output of a representative simulation run at time $t = 100$. There are ten different agent types and the total number of agents is 190: 6 politicians, 6 religious leaders, 6 businessmen, 6 organized criminals, 6 commanders, 10 drug dealers, 35 drug farmers, 35 farmers, 70 civilians, and 20 warriors. In summary, these effects represent the emergence of higher-order organizational structures or *qawm* out of the micro-processes and microstructures introduced above.

Agents affiliated with each other are linked via a line. Two distinct but nevertheless interconnected clusters of agents are apparent in the network. Each cluster consists of a variety of agent types. This means that the depicted clusters are not homogeneous organizations of power but rather heterogeneous concentrations of power generated by mutually dependent and interacting agents/actors. Agents assumed to be more powerful than others, i.e., politicians, commanders, religious leaders, and organized criminals are prevalent in each of the two dense clusters. The reasons for the evolution of this network of clustered affiliations are manifold: agents affiliate because they share the same ethnicity or religion, because they have established a business relationship, or because they seek protection with a commander. But in general, the clusters can be perceived as emergent properties of agent neo-patrimonial behavior as reified by our agent rules. The model generates data of the sort we expected and its output can therefore be considered as artificial representations of *qawm*.

A cogent argument for this claim is the fact that a cross-validational analysis between the network constructed from simulation output and a real network,

---

[16]This paragraph is a condensed version of Geller and Moss (2008a). See also Geller and Moss (2007, 2008b).

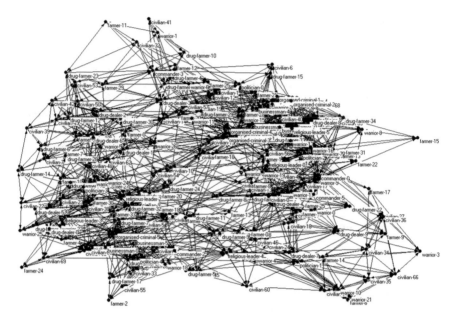

Fig. 27.3 Artificial Afghan power structures depicted as a relational network of Afghan agents

constructed by Fuchs (2005), was successfully conducted. Cross-validation provides a link between the model and its target system, i.e., reality (Moss and Edmonds 2005). Fuchs (2005) has collected open source data on Afghan power structures for the years 1992–2004/05 which can be compared with the AfghanModel network depicted in Fig. 27.3.[17]

The densities for the Fuchs and the AfghanModel networks are 0.0593 and 0.1943* (only strongmen, marked with *) and 0.0697 (208 agents, marked with [†]). The clustering coefficient for Fuchs is 0.428, while it is 0.542* and 0.432[†], respectively, for the AfghanModel. Both the Fuchs and the AfghanModel networks tend to be small world. They exhibit sub-networks that are characterized by the presence of connections between almost any two nodes within them. Most pairs of nodes in both networks are connected by at least one short path. There is an overabundance of hubs in both networks. The average geodesic distances for the Fuchs and the AfghanModel networks, 2.972, 3.331*, and 2.628[†], are relatively small compared to the number of nodes. Erdős-Rényi random networks of equal size and density as the Fuchs and the AfghanModel networks exhibit lower clustering

---

[17]Fuchs (2005) only collected data of elites. In order to compare her results with those generated from the simulation presented here, all ordinary agents, i.e., non-elites, had to be removed from the network to meaningfully calculate the desired network measures. Note that the simulation parameters remained unchanged. Note also that the two networks vary in size: 62 agents participate in the Fuchs network and 30 in the AfghanModel network. This can lead to boundary specification problems.

coefficients than the Fuchs and the AfghanModel networks, namely, 0.060, 0.212*, and 0.072[†]— an indication that neither the clustering of the Fuchs nor of the AfghanModel network is random. The two Erdős-Rényi random networks have geodesic distances (2.094*, 2.276[†]) which are of a comparable order as the geodesic distances that can be found in the AfghanModel network (3.331*, 2.628[†]). Thus, a structural and functional equivalence based on qualitative evidence between the model and the target system is observable.

27.5 Conclusions

The main task of this chapter was to provide a critical overview of state-of-the-art models that deal in alternative ways with power and authority and to present an alternative research design that overcomes the inefficiencies and shortcomings of these approaches. The work presented is motivated by the fact that research on power structures is confined on one hand by a general lack of statistical data. On the other hand, the literal complexity of power structures requires a formalized and dynamic approach of analysis if more than a narrative understanding of the object under investigation is sought. The case of Afghanistan has only been instrumentalized to exemplify such an approach, which would work without doubt for any other comparable contexts.

With regard to the case in hand the *explanandum* is power and authority in Afghanistan. The analysis focuses on power relations dominated by elites. The *qawm*, a fluid and goal-oriented solidarity network, has been identified in the data available to us as the pivotal structural and functional social institution to manage power and authority in Afghanistan. The totality of *qawm* in Afghanistan does not form a unified system of power but a cosmos of mutually interacting power systems. This qualitative analytical result has been reproduced by our simulation results and was subsequently cross-validated with independent out-of-sample network data. This cosmos is a root source for political volatility and unpredictability and ultimately an important explanatory factor for conflict in Afghanistan. For the time being, the latter only outlines an *a priori* statement that needs further corroboration.

The proposed approach starts with an evidence-informed but intuitive model of power and authority in Afghanistan. Such a model provides an intellectual entree, identifies and defines a target system, isolates relevant actors and generic actor behavior, and helps to address appropriate research questions. Based on this intuitive model, agent structures and rules are developed according to the evidence that is available. The aim is to develop a homologue, i.e., a construct valid model, which allows translation of the information describing the model with regard to structure and processes into program code. Weak agents have been implemented according to the rational that they seek affiliation with a powerful agent because of grievance and their will to survive; a powerful agent is implemented according to the rational that he affiliates himself with other powerful agents in terms of functional complementarity to subsist his solidarity network, i.e., his *qawm*, with the aim of consolidating or increasing his power. The general underlying notion of such behavior is neo-patrimonialism. The simulation results are not self-explanatory but need to be validated against reality. This not only provides new insight into the target system but also obliges to restart the research process as new evidence has become available. This is the hermeneutic circle of EDABSS, in which the manner of how to present simulation results meaningfully to stakeholders still needs to be considered as being in its experimental stage (though companion modeling has shown a very fruitful way forward). The approach introduced here does also constitute a consequent implementation of generative social science, for which, nevertheless, the research design still needs further formalization and clarification.

Special consideration was given to a cognitive process called "endorsements." The idea of endorsements serves two aims: *firstly*, to differentiate and define the relevant dimensions agents reason about and, *secondly*, to implement agent cognition in a natural way (being able to use the mnemonic tokens found in the evidence informing the model). Endorsements as they were introduced here lack two important features: that agent types should have randomized, type-specific endorsement schemes and that Eq. (27.1) to calculate E should allow for continuous data formalization. Alam et al. (2010) have proposed a solution for these problems.

What has been gained by this approach? First, tribute has been paid to reality by taking it seriously into account. Although evidence-driven modeling is about abstraction and formalization like any other modeling technique, it does it on the basis of evidence. Secondly, while agent-based modeling accounts in general for an epistemological shift from an intra-modeling view and disburdens modeling from serving only a mere input-output function, evidence-driven modeling facilitates cross-validation also between model and target system interagent mechanisms and on the systems' respective meso-levels. Thirdly, the inclusion and generation of narratives in evidence-driven modeling opens up new ways of engaging stakeholders and domain experts in the modeling process and in informing policy analysts and makers in their decision-making.

However, the heuristic usefulness and value of the applied approach can only be determined against the actual object under investigation, i.e., Afghanistan. What has been found by modeling Afghan power structures evidence-driven and agent-based that would have not been detected by applying only a hermeneutic analysis? Modeling requires not only abstraction but also formalization and thus disambiguation of the evidence describing the case at hand. In particular we were forced to dissect the notion of *qawm*, an inherently context-dependent and fuzzy concept, with regard to actor behavior and cognition, assigning clear meaning to all mechanisms deemed to be relevant to the model. This being only a beneficial side effect of evidence-driven social simulation, emerging social processes and structures are in the focus of description and analysis, social dynamics that could hardly be made graspable by pure narrative analysis. Here this is the evidence-driven generation of a model-based demonstration of an autopoietic system of power structures taking the form of a small world network.

From the point of view of complexity science, it is interesting to observe how the introduction of localized mechanisms of power generates order on a higher level, or to put it differently: how neo-patrimonial behavior and processes of accumulation and redistribution in a state of anomie create a social structure which can be found in Afghanistan, i.e., *qawm*. Further, EDABSS clarifies an important aspect of emergentism: it is not satisfactory to state that in agent-based modeling, agents constitute their own environment (cf. Cederman 2001). In lieu thereof, it should be replenished that agents and the emergent effects stemming from these agents' interactions as a whole constitute an agent's environment. Agent behavior on the micro-level and social structurization on the macro-level cannot be thought of as disjointed entities but emblematize the wholeness of what is in essence a complex system. EDABSS therefore is not only a methodological solution to the micro-macro gap problem but also an implementational one.

EDABSS implies more than inductive reasoning about social phenomena. It constitutes a consequent implementation of the generative social science paradigm. Evidence is the starting point of the research process and evidence denotes its end. It is the nature of the object under investigation that shapes the theoretical and methodological approach and not vice versa. This is perhaps EDABSS' most important virtue that it takes reality and its subjects seriously: during evidence collection, model development, and validation.

Acknowledgments We would like to thank Bruce Edmonds, Martin Neumann, and Flaminio Squazzoni for thoughtful and helpful comments. We also thank Zemaray Hakimi for translation and facilitator skills, Sayyed Askar Mousavi for advice in the data collection process, Shah Jamal Alam, Ruth Meyer, and Bogdan Werth for modeling support, and the Bibliotheca Afghanica for access to its library.

Further Reading

Whereas the literature on power and authority is overwhelming, published work on power and authority *and* modeling and simulation is, comparatively speaking, meager. For further reading we suggest Alam et al. (2005) and Rouchier et al. (2001) for models concerned with the emergence of structures and authority in gift exchange, Geller and Moss (2008a) and Alam et al. (2008) for empirical models relevant to power and authority, Axelrod (1995) and Cederman (1997) for applications of modeling power to conflict in international relations, Mailliard and Sibertin-Blanc (2010) for a good discussion of multiagent simulation and power from a sociological perspective, and finally Guyot et al. (2006) for a participatory modeling approach with relevance to power and authority.

References

Alam, S. J., Geller, A., Meyer, R., & Werth, B. (2010). Modelling contextualized reasoning in complex societies with 'endorsements'. *Journal of Artificial Societies and Social Simulation, 13*(4). http://jasss.soc.surrey.ac.uk/13/4/6.html.

Alam, S. J., Hillebrandt, F., & Schillo, M. (2005). Sociological implications of gift exchange in multiagent systems. *Journal of Artificial Societies and Social Simulation, 8*(3). http://jasss.soc.surrey.ac.uk/8/3/5.html.

Alam, S. J., Meyer, R., Ziervogel, G., & Moss, S. (2008). The impact of HIV/AIDS in the context of socioeconomic stressors: An evidence-driven approach. *Journal of Artificial Societies and Social Simulation, 10*(4). http://jasss.soc.surrey.ac.uk/10/4/7.html.

Axelrod, R. (1995). A model of the emergence of new political actors. In N. Gilbert & R. Conte (Eds.), *Artificial societies: The computer simulation of social life* (pp. 19–39). London/New York: Routledge.

Azoy, G. W. (2003). *Bukashi: Game and power in Afghanistan* (2nd ed.). Long Grove: Waveland Press.

Bak, P. (1997). *How nature works: The science of self organized criticality.* Oxford: Oxford University Press.

Bayart, J.-F., Ellis, S., & Hibou, B. (1999). *The criminalization of the state in Africa.* Bloomington/Indianapolis: Indiana University Press.

Bhaskar, R. (1979). *The possibility of naturalism: A philosophical critique of the contemporary human sciences.* Sussex: The Harvester Press.

Boero, R., & Squazzoni, F. (2005). Does empirical embeddedness matter? Methodological issues on agent-based models for analytical social science. *Journal of Artificial Societies and Social Simulation, 8*(4). http://jasss.soc.surrey.ac.uk/8/4/6.html.

Boudon, R. (1998). Social mechanisms without black boxes. In P. Hedström & R. Swedberg (Eds.), *Social mechanism: An analytical approach to social theory* (pp. 172–203). Cambridge: Cambridge University Press.

Caldas, J. C., & Coelho, H. (1999). The origin of institutions: socio-economic processes, choice, norms and conventions. *Journal of Artificial Societies and Social Simulation, 2*(2). http://jasss.soc.surrey.ac.uk/2/2/1.html.

Canfield, R. L. (1973). *Faction and conversion in a plural society: Religious alignments in the Hindu Kush.* Ann Arbor: University of Michigan.

Canfield, R. L. (1988). Afghanistan's social identities in crisis. In J.-P. Digard (Ed.), *Le fait ethnique en Iran et en Afghanistan* (pp. 185–199). Paris: Editions du Centre National de la Recherche Scientifique.

Castelfranchi, C. (1999). Social power: A point missed in multi-agent, DAI and HCI. In *Proceedings of the first european workshop on modelling autonomous agents in a multi-agent world* (pp. 49–62). Cambridge: Elsevier.

Casti, J. L. (1997). *Would-be worlds: How simulation is changing the frontiers of science.* New York: Wiley.

Cederman, L.-E. (1997). *Emergent actors in world politics: How states and nations develop and dissolve.* Princeton, NJ: Princeton University Press.

Cederman, L.-E. (2001). Agent-based modeling in political science. *The Political Methodologist, 10*(1), 16–22.

Cederman, L.-E. (2003). Modeling the size of wars: From billiard balls to sandpiles. *American Political Science Review, 97*(1), 135–150.

Cohen, P.R. (1985) Heuristic reasoning about uncertainty: An artificial intelligence approach. Boston: Pitman Advanced Publishing Program.

Cruickshank, J. (2003). Introduction. In J. Cruickshank (Ed.), *Critical realism: The difference that it makes* (pp. 1–14). London/New York: Routledge.

David, N., Fachada, N., & Rosa, A. C. (2017). Verifying and validating simulations. In *Simulating social complexity–A handbook.* Chapter 9 in this volume.

Dorronsoro, G. (2005). *Revolution unending*. London: Hurst.

Eckstein, H. (1992). Case study and theory in political science. In R. Gomm, M. Hammersley, & P. Foster (Eds.), *Case study method: Key issues, key texts* (pp. 119–164). London: Sage.

Evans, A., Heppenstall, A., & Birkin, M. (2017). Understanding simulation results. In *Simulating social complexity–A handbook*. Chapter 10 in this volume.

Fuchs, C. (2005). *Machtverhältnisse in Afghanistan: Netzwerkanalyse des Beziehungssystems regionaler Führer* (M.A. Thesis). University of Zurich, Zurich.

Geller, A. (2006a) *The Emergence of Individual Welfare in Afghanistan*. Paper presented at the 20th International Political Science Association World Congress, Fukuoka, July 9–13.

Geller, A. (2006b). *Macht, Ressourcen und Gewalt: Zur Komplexität zeitgenössischer Konflikte; Eine agenten-basierte Modellierung*. Zurich: vdf.

Geller, A. (2010). The political economy of normlessness in Afghanistan. In A. Schlenkhoff & C. Oeppen (Eds.), *Beyond the "Wild Tribes"—Understanding modern Afghanistan and its diaspora* (pp. 57–70). New York: Columbia University Press.

Geller, A. & Moss, S. (2007, 13–14 July). The Afghan nexus: Anomie, neo-patrimonialism and the emergence of small-world networks. *Proceedings of UK Social Network Conference*, (pp. 86–88). University of London.

Geller, A., & Moss, S. (2008a). Growing qawm: An evidence-driven declarative model of Afghan power structures. *Advances in Complex Systems, 11*(2), 321–335.

Geller, A. & Moss, S. (2008b, March 26–29). International contemporary conflict as small-world phenomenon and the hermeneutic net. *Paper presented at the annual international studies association conference*, San Francisco.

Gerring, J. (2004). What is a case study and what is it good for? *American Political Science Review, 98*(2), 341–354.

Giddens, A. (1976). *New rules of sociological method: A positive critique of interpretative sociologies*. London: Hutchinson.

Giustozzi, A. (2006). *Genesis of a "Prince": The Rise of Ismael Khan in Western Afghanistan, 1979-1992* (Crisis States Working Papers Series, 4). Crisis States Research Centre, London School of Economics, London.

Giustozzi, A. (2007). War and peace economies of Afghanistan's strongmen. *International Peacekeeping, 14*(1), 75–89.

Glatzer, B. (1998). Is Afghanistan on the brink of ethnic and tribal disintegration? In W. Maley (Ed.), *Fundamentalism reborn? Afghanistan and the Taliban* (pp. 167–181). New York: New York University Press.

Glatzer, B. (2003). *Afghanistan* (Studien zur Länderbezogenen Konfliktanalyse). Friedrich-Ebert Stiftung/Gesellschaft für Technische Zusammenarbeit.

Gomm, R., Hammersley, M., & Foster, P. (1992). Case study and theory. In R. Gomm, M. Hammersley, & P. Foster (Eds.), *Case study method: Key issues, key texts* (pp. 234–258). London: Sage.

Guyot, P., Drogoul, A., & Honiden, S. (2006). Power and negotiation: Lessons from agent-based participatory simulations. In P. Stone & G. Weiss (Eds.), *Proceedings of the fifth international joint conference on autonomous agents and multiagent systems (AAMAS)* (pp. 27–33). New York: ACM.

Hedström, P., & Swedberg, R. (1998). Social mechanisms: An introductory essay. In P. Hedström & R. Swedberg (Eds.), *Social mechanism: An analytical approach to social theory* (pp. 1–31). Cambridge: Cambridge University Press.

Hirshleifer, J. (1991). The paradox of power. *Economics and Politics, 3*(3), 177–200.

Hirshleifer, J. (1995). Anarchy and its breakdown. *Journal of Political Economy, 103*(1), 26–52.

Janata, A., & Hassas, R. (1975). Ghairatman–Der gute Pashtune: Exkurs über die Grundlagen des Pashtunwali. *Afghanistan Journal, 2*(2), 83–97.

Kuznar, L. A., & Frederick, W. (2007). Simulating the effect of nepotism on political risk taking and social unrest. *Computational and Mathematical Organization Theory, 13*(1), 29–37.

Lasswell, H. D. (1936). *Who gets what, when, how*. New York: Meridian Books.

Lave, C. A., & March, J. G. (1975). *An introduction to models in the social sciences*. New York: Harper & Row.

Lazer, D. (2001). Review of Cederman, L. E (1997) Emergent actors in world politics: How states and nations develop. *Journal of Artificial Societies and Social Simulation, 4*(2). http://jasss.soc.surrey.ac.uk/4/2/reviews/lazer.html.

Limpert, E., Stahel, W. A., & Abbt, M. (2001). Log-normal distributions across the sciences: Keys and clues. *Bioscience, 51*(5), 341–352.

Lustick, I. S. (2000). Agent-based modeling of collective identity: Testing constructivist theory. *Journal of Artificial Societies and Social Simulation, 3*(1). http://jasss.soc.surrey.ac.uk/3/1/1.html.

Marks, S. R. (1974). Durkheim's theory of anomie. *American Journal of Sociology, 80*(2), 329–363.

Mailliard, M., & Sibertin-Blanc, C. (2010, March 29). *What is power? Perspectives from sociology, multi-agent systems and social network analysis* (Paper presented at the second symposium on social networks and multiagent systems (SNAMAS 2010). Leicester, UK: De Montfort University). ftp://ftp.irit.fr/IRIT/SMAC/DOCUMENTS/PUBLIS/SocLab/SNAMAS-10.pdf.

Medard, J.-F. (1990). L'etat patrimonialise. *Politique Africaine, 39*, 25–36.

Merton, R. K. (1938). Social structure and anomie. *American Sociological Review, 3*(5), 672–682.

Monsutti, A. (2004). Cooperation, remittances, and kinship among the hazaras. *Iranian Studies, 37*(2), 219–240.

Mosler, H.-J. (2006). Better be convincing or better be stylish? A theory based multi-agent simulation to explain minority influence in groups via arguments or via peripheral cues. *Journal of Artificial Societies and Social Simulation, 9*(3). http://jasss.soc.surrey.ac.uk/9/3/4.html.

Moss, S. (1981). *An economic theory of business strategy: An essay in dynamics without equilibrium*. Oxford: Martin Robertson.

Moss, S. (1995). Control metaphors in the modelling of decision-making behaviour. *Computational Economics, 8*(3), 283–301.

Moss, S. (1998). Critical incident management: An empirically derived computational model. *Journal of Artificial Societies and Social Simulation, 2*(4). http://jasss.soc.surrey.ac.uk/1/4/1.html.

Moss, S. (2000). Canonical tasks, environments and models for social simulation. *Computational and Mathematical Organization Theory, 6*(3), 249–275.

Moss, S. (2001). Game theory: Limitations and an alternative. *Journal of Artificial Societies and Social Simulation, 4*(2). http://jasss.soc.surrey.ac.uk/4/2/2.html.

Moss, S. (2007). *Alternative approaches to the empirical validation of agent-based Models* (Technical report, CPM-07-178). Manchester: Centre for Policy Modelling, Manchester Metropolitan University.

Moss, S., & Edmonds, B. (1997). A knowledge-based model of context-dependent attribute preferences for fast moving consumer goods. *Omega–International Journal of Management Science, 25*(2), 155–169.

Moss, S., & Edmonds, B. (2005). Sociology and simulation: Statistical and qualitative cross-validation. *American Journal of Sociology, 110*(4), 1095–1131.

Moss, S., Gaylard, H., Wallis, S., & Edmonds, B. (1996). SDML: A multi-agent language for organizational modelling. *Computational and Mathematical Organization Theory, 4*(1), 43–69.

Moss, S., & Kuznetsova, O. (1996). Modelling the process of market emergence. In J. W. Owsinski & Z. Nahorski (Eds.), *Modelling and analysing economies in transition*. Warsaw: MODEST.

Mousavi, S. A. (1997). *The hazaras of Afghanistan: An historical, cultural, economic and political study*. New York: St. Martin's Press.

Neumann, F. L. (1950). Approaches to the study of power. *Political Science Quarterly, 65*(2), 161–180.

Outhwaite, W. (1987). *New philosophies of social science: Realism, hermeneutics and critical theory*. London: Macmillan Education.

Parsons, T. (1952). *The social system*. London: Tavistock.

Popitz, H. (1992). *Phänomene der Macht* (2nd ed.). Tübingen: Mohr Siebeck.

Putnam, R. D. (1988). Diplomacy and domestic politics: The logic of two-level games. *International Organization, 42*(3), 427–460.

Rasuly-Paleczek, G. (1998). Ethnic identity versus nationalism: The Uzbeks of north-eastern Afghanistan and the Afghan State. In T. Atabaki & J. O'Kane (Eds.), *Post-Soviet central asia* (pp. 204–230). London/New York: Tauris Academic Studies.

Reno, W. (1998). *Warlord politics and African states*. Boulder/London: Lynne Rienner Publishers.

Richardson, L. F. (1948). Variation of the frequency of fatal quarrels with magnitude. *Journal of the American Statistical Association, 43*(244), 523–546.

Rouchier, J., O'Connor, M., & Bousquet, F. (2001). The creation of a reputation in an artificial society organised by a gift system. *Journal of Artificial Societies and Social Simulation, 4*(2). http://jasss.soc.surrey.ac.uk/4/2/8.html.

Rouchier, J., & Thoyer, S. (2006). Votes and lobbying in the European decision-making process: application to the european regulation on GMO release. *Journal of Artificial Societies and Social Simulation, 9*(3). http://jasss.soc.surrey.ac.uk/9/3/1.html.

Roy, O. (1992). Ethnic identity and political expression in Northern Afghanistan. In J.-A. Gross (Ed.), *Muslims in Central Asia: Expressions of identity and change* (pp. 73–86). Durham/London: Duke University Press.

Roy, O. (1994). The new political elite of Afghanistan. In M. Weiner & A. Banuazizi (Eds.), *The politics of social transformation in Afghanistan, Iran and Pakistan* (pp. 72–100). Syracuse: Syracuse University Press.

Roy, O. (1995). *Afghanistan: From holy war to civil war*. Princeton, NJ: The Darwin Press.

Roy, O. (1998). Has Islamism a future in Afghanistan? In W. Maley (Ed.), *Fundamentalism reborn? Afghanistan and the Taliban* (pp. 199–211). New York: New York University Press.

Rubin, B. R. (1992). Political elites in Afghanistan: Rentier state building, Rentier state wrecking. *International Journal of Middle East Studies, 24*(1), 77–99.

Rubin, B. R. (2007). Saving Afghanistan. *Foreign Affairs, 86*(1), 57–78.

Saam, N. J., & Harrer, A. (1999). Simulating norms, social inequality, and functional change in artificial societies. *Journal of Artificial Societies and Social Simulation, 2*(1). http://jasss.soc.surrey.ac.uk/2/1/2.html.

Sambanis, N. (2004). Using case studies to expand economic models of civil war. *Perspectives of Politics, 2*(2), 259–279.

Sawyer, R. K. (2005). *Social emergence, societies as complex systems*. Cambridge: Cambridge University Press.

Sawyer, R. K. (2017). Interpreting and understanding simulations: The philosophy of social simulation. In *Simulating social complexity—A handbook*. Chapter in this volume.

Sayer, A. (1992). *Method in social science: A realist approach* (2nd ed.). London/New York: Routledge.

Sayer, A. (2000). *Realism and social science*. London/Thousand Oaks/New Delhi: Sage.

Schelling, T. C. (1998). Social mechanisms and social dynamics. In P. Hedström & R. Swedberg (Eds.), *Social mechanism: An analytical approach to social theory* (pp. 32–44). Cambridge: Cambridge University Press.

Schetter, C., Glassner, R., & Karokhail, M. (2007). Beyond warlordism: The local security architecture in Afghanistan. *Internationale Politik und Gesellschaft, 2*, 136–152.

Shahrani, M. N. (1998). The future of state and the structure of community governance in Afghanistan. In W. Maley (Ed.), *Fundamentalism reborn? Afghanistan and the Taliban* (pp. 212–242). New York: New York University Press.

Shahrani, M. N. (2002). War, factionalism, and the state in Afghanistan. *American Anthropologist, 104*(3), 715–722.

Shapiro, I. (2005). *The flight from reality in the human sciences*. Princeton, NJ: Princeton University Press.

Silverman, D. (Ed.). (2004). *Qualitative research: Theory, method and practice* (2nd ed.). London: Sage.

Snidal, D. (1985). The game theory of international politics. *World Politics, 38*(1), 25–57.

Sofsky, W. (2002). *Zeiten des Schreckens: Amok, Terror, Krieg*. Frankfurt A. M.: S. Fischer.

Stachowiak, M. (1973). *Allgemeine modelltheorie*. Wien/New York: Springer.

Stakes, R. E. (1978). The case study method in social inquiry. *The Educational Researcher, 7*(2), 5–8.

Tapper, N. (1991). *Bartered brides: Politics, gender and marriage in an Afghan Tribal Society*. Cambridge: Cambridge University Press.

Tapper, R. (2008). Who are the Kuchi? Nomad self-identities in Afghanistan. *Journal of the Royal Anthropological Institute, 14*, 97–116.

Tarzi, S. M. (1993). Afghanistan in 1992: A Hobbesian State of nature. *Asian Survey, 33*(2), 165–174.

UNODC (2006). *Afghanistan's drug industry: Structure, functioning, dynamics, and implications for counter-narcotics policy*. United Nations Office on Drugs and Crime/The World Bank.

Weber, M. (1980). *Wirtschaft und Gesellschaft: Grundriss der verstehenden Soziologie* (5th ed.). Tübingen: J.C.B. Mohr (Paul Siebeck).

Weidmann, N. & Cederman, L.-E. (2005). Geocontest: Modeling strategic competition in geopolitical systems. In K. Troitzsch (ed.) *Proceedings of the european social simulation association annual conference (ESSA)*, (pp. 179–185) Koblenz, Germany.

Wily, L. A. (2004). *Looking for peace on the pastures: Rural land relations in Afghanistan*. Kabul: Afghanistan Research and Evaluation Unit.

Younger, S. (2005). Violence and revenge in egalitarian societies. *Journal of Artificial Societies and Social Simulation, 8*(4). http://jasss.soc.surrey.ac.uk/8/4/11.html

Chapter 28
Human Societies: Understanding Observed Social Phenomena

Bruce Edmonds, Pablo Lucas, Juliette Rouchier, and Richard Taylor

Abstract The chapter begins by briefly describing two contrasting simulations: the iconic system dynamics model publicised under the *Limits to Growth* book and a detailed model of first millennium Native American societies in the southwest of the United States. These are used to bring out the issues of abstraction, replicability, model comprehensibility, understanding vs. prediction and the extent to which simulations go beyond what is observed. All of these issues are rooted in some fundamental difficulties in the project of simulating observed societies that are then briefly discussed. Both issues and difficulties result in three "dimensions" in which simulation approaches differ. The core of the chapter is a look at 15 different possible simulation goals, both abstract and concrete, giving some examples of each and discussing them. The different inputs and results from such simulations are briefly discussed as to their importance for simulating human societies.

Why Read This Chapter?

To get an overview of the different ways in which simulation can be used to gain understanding of human societies and to gain insight into some of the principle difficulties of these. The chapter will go through the various specific goals one might have in doing such simulation, giving examples of each. It will provide a critical view as to the success of reaching these various goals and hence inform about the current state of such simulation projects.

B. Edmonds (✉)
Centre for Policy Modelling, Manchester Metropolitan University, All Saints Campus, Oxford Road, Manchester, M15 6BH, UK
e-mail: bruce@edmonds.name

P. Lucas
School of Sociology, University College Dublin, Dublin 4, Ireland

J. Rouchier
LAMSADE, Paris-Dauphine, 75775 Paris Cedex 16, France

R. Taylor
Oxford Branch, Stockholm Environment Institute, Stockholm, Sweden

© Springer International Publishing AG 2017   763
B. Edmonds, R. Meyer (eds.), *Simulating Social Complexity*,
Understanding Complex Systems, https://doi.org/10.1007/978-3-319-66948-9_28

28.1 Introduction

Understanding social phenomena is hard. There is all the complexity found in other fields of enquiry but with additional difficulties due to our being embedded in what we are studying.[1] Despite these, understanding our own nature is naturally important to us, and our social aspects are a large part of that nature. Indeed, some would go as far as saying that our social abilities are *the* defining features of our species (e.g. Dunbar 1998). The project of understanding human societies is so intricate that we need to deploy all means at our disposal. Simulation is but one tool in this vast project, but it has the potential to play an important part.

This chapter considers how and to what extent computer simulation helps us to understand the social complexity we see all around us. It will start by discussing two simulations in order to raise the key issues that this project involves, before moving on to highlight the difficulties of understanding human society in more detail. The core of the chapter is a review of some of the different purposes that simulation can be used for, with examples of each. It then looks at a way of assessing the success and kind of purpose of simulations in terms of their inputs and outputs (Sect. 28.4).

28.1.1 Example 1: The Club of Rome's "Limits to Growth" (LTG)

In the early 1970s, on behalf of an international group under the name "The Club of Rome", a simulation study was published (Meadows et al. 1972) with the attempt to convince humankind that there were some serious issues facing it, in terms of a coming population, resource and pollution catastrophe. To do this, the group developed a system dynamics model of the world. They chose a system dynamics model because they felt they needed to capture some of the feedback cycles between the key factors—factors that would not come out in simple statistical projections of the available data. They developed this model and ran it, publishing the findings—a number of model-generated future scenarios—for a variety of settings and variations. The book (*Limits to Growth*) considered the world as a single system, and postulated some relationships between macro variables, such as population, available resources, pollution, etc. Based on the relationships it simulated what might happen if the feedbacks between the various variables were allowed to occur. The simulation outputs were the curves that resulted from this model as the simulation was continued for future dates. The results indicated that there was a coming critical point in time and that a lot of suffering would result, even if humankind managed to survive it.

---

[1]This embeddedness has advantages as well, such as prior knowledge.

The book had a considerable impact, firmly placing the possibility that humankind could not simply continue to grow indefinitely in the public mind. It also attracted considerable criticism (e.g. Cole et al. 1973), mainly based on the plausibility of the model's assumptions and the sensitivity of its results to those relationships. (For example, it assumed that growth will be exponential and that delay loops are extended.) The book presented the results of the simulations as predictions—a series of what-if scenarios. Whilst the authors did add caveats and explore various possible versions of their model, depending on what connections there turned out to be in the world system, the overall intent of the book was unmistakeable: that if we did not change what we were doing, by limiting our own economic consumption and population, disaster would strike. This was a work firmly in the tradition of Malthus (1798) who, 175 years earlier, had predicted a constant state of near starvation for much of the world based upon a consideration of the growth processes of population and agriculture.

The authors clearly hoped that by using a simulation (albeit a simplistic one by present standards), they would be able to make the potential feedback loops real to people. The simulation illustrated an understanding that the authors of LTG had. However, the model was not presented as such, but as something *more scientific* in some sense.[2] A science-driven study that predicted such suffering was a definite challenge to those who thought the problem was less severe.[3] By publishing their model and making it easy for others to replicate and analyse it, they offered critics a good opportunity for counter-argumentation.

The model was criticised on many different grounds, but the most effective was that the model was sensitive to the initial settings of some parameters (Vermeulen and de Jongh 1976). This raised the question whether the model had to be finely tuned in order to get the behaviour claimed and thus, since the parameters were highly abstract and did not directly correspond to anything measurable, questioned the applicability of the model to the world we live in. Its critics assumed that since this model did not produce reliable predictions, it could be safely ignored. It also engendered the general perception that predictive simulation models are not credible tools for understanding human socio-economic changes—especially for long-term analyses—and discouraged their use in supporting policy-making.

---

[2]The intentions of the authors themselves in terms of what they thought of the simulation itself are difficult to ascertain and varied between the individuals; however, this was certainly how the work was perceived.

[3]Or those whose vested interests may have led them to maintain the status quo concerning the desirability of continual economic growth.

28.1.2 Example 2: Modelling First Millennium Native American Society

A contrasting example to the Club of Rome model is the use of simulation models to assess and explore explanations of population shifts among the Native American nations, in the pre-Columbian era. This has been called "generative archaeology" (GA) by Kohler (2009). Here a spatial model of a population was developed, which was fitted to a wealth of archaeological and climatological data in order to find and assess possible explanations of the size, distribution and change in populations that existed in the first millennium AD in the Southwest US. This case offers a picture of settlement patterns in the context of relatively high-resolution reconstructions of changes in climate and resources relevant to the human use of these landscapes.

The available data in this case is relatively rich, allowing many questions to be answered directly. However, other interesting aspects are not directly answerable from a static analysis of the data, for example, those about the possible social processes that existed. The problem is that different archaeologists can inspect the same settlement pattern and generate different candidate processes (explanations) for its generation. Here agent-based modelling helps infer the social processes (which cannot be directly observed) from the detailed record over time. This is not a direct or certain inference, since there are still many unknowns involved in that process.

In (Kohler et al. 2005) and (Kohler et al. 2008),[4] agent-based modelling (ABM) has been mainly used to see what patterns one should expect if households were approximately minimising their caloric costs for access to adequate amounts of calories, protein, and water. The differences through time in how well this expectation fits the observed record and the changing directions of departure from those expectations provide a completely novel source of inference on the archaeological record. Simulations using the hypothesis of local food sharing during periods of mild food shortage may be compared to the fit in a simulation where food sharing does not occur. In this way we can get indirect evidence as to whether food sharing took place.

The ABM has hence allowed a comparison of a possible process with the recorded evidence. This comparison is relative to the assumptions that are built into the model, which tend to be plausible but questionable. However, despite the uncertainties involved, one is able to make *a useful* assessment of the possible explanation, and the assumptions are explicitly documented. This approach to processes that involve complex interaction would be impossible to do without a computer simulation. At the very least, such a process reveals new important questions to ask (and hence new evidence to search for) and the times when the

---

[4] For details of the wider project connected with these papers, see the Village Ecodynamics Project, http://village.anth.wsu.edu.

plausible explanations are demonstrably inadequate. However, for any real progress in explanation of such cases, a very large amount of data seems to have been required.

28.1.3   Some Issues that the Aforementioned Examples Illustrate

The previous examples raise a few issues, common with much social simulation modelling of human societies. These will now be briefly defined and discussed as an introduction to the problem of understanding social phenomena using simulation.

1. *Abstraction.* Abstraction is a crucial step in modelling observed social phenomena, as it involves choices about which aspects are salient in relation to the problem and what level of analysis is appropriate. The LTG example, being a macromodel, assumes that distributive aspects such as geography and local heterogeneity are less important with respect to feedbacks among global growth variables. In this model, the detail of the whole world is reduced to the interaction of a few numeric variables. The GA model was more specific and detailed, including an explicit 2D map of the area and the position of settlements at different times in the past. It is fair to say that the LTG model was driven by the goals of its modellers, i.e. showing that the coming crisis could be sharp due to slow feedback loops, whereas the GA model is rather driven by the available data, with the model being applied to a number of different questions afterwards.

2. *Replicability.* Replicability is the extent to which a published model has been described in a comprehensive and transparent way so that the simulation experiments can be independently reproduced by another modeller. Replicability may be considerably easier if care is taken to verify the initial model and if the original source code is effectively available and well commented. Here the LTG model was readily replicable and hence open to inspection and criticism. The GA models are available to download and inspect, but their very complexity makes them hard to replicate independently of the original implementation.

3. *Understanding the model.* A modeller's inferential ability is the extent to which one can understand one's own model. Evidence suggests that humans can fully track systems only for about two or three variables and five or six states (Klein 1998); for higher levels of complexity, additional tools are required. In many simulations, especially those towards the descriptive end of the spectrum, the agents can have many behavioural rules, which may interact in unpredictable ways. This makes simulations very difficult to fully understand and check. Even in the case of a simple model, such as the LTG model, there can be unexpected features (such as the fine-tuning that was required). Although the GA was rich in its depiction of the space and environment, the behavioural rules of subpopulations were fairly simple and easy to follow at the micro level. However, this does not rule out subtle errors and complexities that might result

from the interaction of the microelements. Indeed, this is the point of such a simulation that we cannot work out these complex outcomes ourselves, but require a computer program to do it.

4. *Prediction vs. understanding.* The main lesson to be drawn from the history of formal modelling is that, for most complex systems, it is impossible to model with accuracy their evolution beyond an immediate timeframe. Whilst the broad trends and properties may be forecast to some degree, the particulars, e.g. the timing and scale of changes in the aggregate variables, generally cannot (Moss 1999). The LTG model attempted to forecast the future, not in terms of the precise levels but in the presence of a severe crisis—a peak in population followed by a crash. The GA does not aim to predict any specific thing, but rather it seeks to establish plausible explanations for the data that is known. Most simulations of human society restrict themselves to establishing explanations, the simulations providing a chain of causation that shows that the explanation is possible.[5]

5. *Going beyond what is known.* In social science, there are many gaps in our knowledge, and social simulation methods may be well placed to address some of these gaps. Given some data, and some plausible assumptions, the simulations can be used to perform experiments that are consistent with the data and assumptions and then inspected to answer other questions. Clearly, this depends on the reliability of the assumptions chosen. In the GA case, this is very clear; a model with a food-sharing rule and one without can be compared to the data, seeing which one fits it better. The LTG model attempts something harder, making severe assumptions about how the aggregate variables relate; it "predicts" aspects of the future. In general: the more reliable the assumptions and data (hence the less ambitious the attempt at projection), the more credible the result.

A social scientist, who wants to capture key aspects of observed social phenomena in a simulation model, faces many difficulties. Indeed, the differences between formal systems and complex, multifaceted and meaning-laden social systems are so fundamental that some criticise *any* attempt to bridge this gap (e.g. Clifford 1986). Modellers have to face these difficulties, and these have an impact as to how social simulation is done and how useful (or otherwise) such models may be. We briefly consider six of these difficulties here.

- Firstly, there is the sheer difference in nature between formal models (i.e. computer programs) that modellers use as compared to the social world that we observe. The former are explicit, precise, with a formal grammar, predictable at the micro level, reproducible and work in (mostly) the same way regardless of the computational context. The latter is vague, fluid, uncertain, subjective, implicit and imprecise—which often seems to work completely different in similar situations and whose operation seems to rely on the rich interaction of

---

[5]Although in many cases this is dressed up to look like prediction, such as the fitting to out-of-sample data. Prediction has to be for data unknown to the modeller; otherwise the model will be implicitly fitted to it.

meaning in a way that is sometimes explicable but usually unpredictable. In particular, the gap between essentially formal symbols with precise but limited meaning and the rich semantic associations of the observed social world (e.g. as expressed in natural language) is particularly stark. This gap is so wide that some philosophers have declared it unbridgeable (e.g. Lincoln and Guba 1985, Guba and Lincoln 1994).

- Secondly, there are the sheer variability, complication and complexity of the social world. Social phenomena seem to be at least as complex as biological phenomena but without the central organising principle of evolution as specified in the neo-Darwinian synthesis. If there are any general organising principles (and it is not obvious that this is the case), then there are many of these, each with differing (and sometimes overlapping) domains of application. In that sense, it is clear that a model will always capture only a small part of the phenomenon among many other related aspects, hence reducing drastically the possibility to predict with any degree of certainty.
- Then there is the sheer lack of adequate multifaceted data about social phenomena. Social simulators always seem to have to choose between longitudinal studies, narrative data, cross-sectional surveys or time-series data. Having all of these datasets about a single social process or event is to date very unlikely. There does not seem to be the emphasis on data collection and measurement in the social sciences that there is in some other sciences and certainly not the corresponding prestige for those who collect it or invent ways of doing so.
- There is the more mundane difficulty of building, checking, maintaining, and analysing simulations (Galán et al. 2009). Even the simplest simulations are beyond our complete understanding, indeed that is often why we need them, because there is no other practical way to find out the complex ramifications of a set of interacting agents. This presence of emergent outcomes in the simulations makes them very difficult to check. Ways to improve confidence that our simulations in fact correspond to our intentions for them[6] include: unit testing, debugging, and the facility for querying the database of a simulation (see Chap. 9 (David et al. 2017) in this handbook). Perhaps the strictest test is the independent replication of simulations—working from the specifications and checking their results at a high degree of accuracy (Axtell et al. 1996). However, such replication is usually very difficult and time-consuming, even in relatively simple cases (Edmonds and Hales 2003).
- Another difficulty is that of the inevitability of background assumptions in all we do. There is always a wealth of facts, processes and affordances giving meaning to, and providing the framework for, the foreground actions and causal chains that we observe. Many of these are not immediately apparent to us since they are part of the contexts we inhabit and so are not perceptually apparent. This is the same as in other fields, as it has been argued elsewhere, the concept of causation

---

[6]In terms of design and implementation, if one has a good reference case in terms of observed data then one can also check one's simulation against this.

only makes sense within a context (Edmonds 2007). However, it does seem that context is more critical in the social world than elsewhere, since it can change not only the outcomes of events but also their very meaning (and hence kind of social outcome). Whilst in other fields it might be acceptable to represent extra-contextual interferences as some kind of random distribution or process, this is often manifestly inadequate with social phenomena (Edmonds and Hales 2005).

- The uncertainty behind the foreground assumptions in social simulation is also problematic. Even when we are aware of all of the assumptions, they are often either too numerous to include in a single model, or we simply lack any evidence as to what they should be. For example, many social simulation models include some version of inference, learning or decision-making within the agents of the model, even when there is no evidence as to whether this actually corresponds to the one used by the observed actors. It seems that often it is simply hoped that these details will not matter much in the end—thus becoming a rarely checked, and sometimes wrong, aspect of simulations (Edmonds 2001; Rouchier 2001).

- Finally, there is a difficulty from the nature of simulation itself. Simulation will demonstrate *possible* processes that might follow from a given situation (relative to the assumptions on which the simulation is built). It does not show all the possibilities, since it could happen that a future simulation will produce the same outcomes from the same set-up in a different way (e.g. using a different cognitive model). Thus, simulation differs in terms of its inferential power from analytic models (e.g. equation-based ones), where the simplicity of the model can allow formal proofs of a general formulation of outcomes that may establish the necessity of conditions as well as their adequacy. This difficulty is the same for many mathematical formulations since, in their raw form, they are often unsolvable. Hence, either one has to use numerical simulation of results (in which case one is back to a simulation) or one has to make simplifying assumptions (in which case, depending on the strength of these assumptions, one does not know if the results still apply to the original case).

These difficulties bring up the question of whether some aspects of societies can be at all understood by means of modelling. The hypothesis asserting that simulation is a credible method to better explore, understand or explain social processes is implicitly tested in the current volume and is discussed in some detail below. We are not going to take any strong position but will restrict ourselves to considering examples within the context of their use.[7] Agent-based social simulation is not a magic bullet and is not yet a mature technique. It is common sense in the social simulation community that best results will be achieved by combining social simulation with other research methods.

---

[7]Obviously, we *suspect* it can be a useful tool; otherwise we would not be bothering with it.

28.2 Styles of Modelling and Their Impact on Simulation Issues

28.2.1 Models of Evidence vs. Models of Ideas

One response to the above difficulties is not to model social phenomena directly, but rather to restrict ourselves to modelling *ideas* about social phenomena. This is a lot easier, since our ideas are necessarily a lot simpler and more abstract than the phenomenon itself (and can be formalised with the notion of pattern modelling (Grimm et al. 2005) rather than strict adequacy to data). Some ideas need modelling, in the sense that the ramifications of the ideas are themselves complex. These kinds of models can be used to improve our understanding of the ideas, and later this understanding can be applied in a rich, flexible and context-sensitive way. This distinction is made clear in (Edmonds 2001).

Of course, to some extent, any model is a compact abstraction of the final target of modelling. There will, presumably, be *some* reason why one conceptualises what one is modelling in terms of evidence or experience by someone, and there will always be *some* level of theory/assumption that motivates the decision as to what can be safely left out of a model. Thus, all models are somewhat about ideas, and, presumably, all models have *some* relation to the evidence. However, there is still a clear difference between those models that take their primary structure from an idea and those whose primary considerations come from the available evidence. For example, the former tend to be a lot simpler than the latter. The latter will tend to have specific motivations for each feature, whilst the former will tend to be motivated in general terms. These two kinds of simulation have a close parallel with the theoretical and phenomenological models identified by Cartwright (1993).

Unfortunately, these kinds of model are often conflated in academic papers. This seems frequently not deliberate, but rather due to the strong theoretical spectacles (Kuhn 1962) that simulation models seem to provide. There is nothing like developing and playing with a simulation model to make one *see* the world in terms of that model. It is not only that the model is your creation and your best effort in formulating an aspect of the social world but also that you have interacted with it and changed it to include the features that you, the modeller, think it should have. Nevertheless, whatever the source, it can take some careful "reading between the lines" to determine the exact nature of any model and what it purports to represent.

28.2.2 Modelling as Representation of Social Phenomena vs. as an Intervention in a Social Process

It must be said that some simulation models are not intended to *represent* anything but are rather created for another purpose, such as a tool for demonstrating an approach or an intervention in a decision-making process. This may be deliberate and explicit, or not, for various different reasons. Of course if a computer model does not represent anything at all, it is not really a simulation but simply a computer program, which may be presented in the style of a simulation. Also for a simulation to be an effective tool for intervention, it has to have some credibility with the participants.

However, in some research, representation is either not the primary goal or the object of representation is deliberately subjective in character. Thus, in some participatory approaches (see Chap. 12, Barreteau et al. 2017), it may be the primary goal to raise awareness of an issue, to intervene in or facilitate a social process like a negotiation or consensus process within a group of people. The modeller may not focus as much on whether the model captures an objective reality but rather on how stakeholders[8] understand the issues and processes of concern and how this might influence the outcomes. This does not mean that there will be *no* elements that are objective and/or representative in character—for example, such models might have a well-validated hydrological component to them—but that the parts of the model that *are* the focus are checked against the opinions of those being modelled or those with an interest in the outcomes rather than any independent evidence.

Of course, this is a matter of degree—in a sense most social simulations are a mixture of objective aspects linked to observations and other aspects derived from theories, opinions, hypotheses and assumptions. In participatory approaches, the modellers seek not to put their own ideas forward but rather take the, possibly more democratic, approach of being expert facilitators expressing the stakeholders' opinions and knowledge. Whilst some researchers might reject such ideas as too "anecdotal" to be included in a formal model, it is not obvious that the stakeholders' ideas about the nature of the processes involved (e.g. how the key players make decisions) are less reliable than the grander theories of academics. However, researchers do have a professional obligation to be transparent and honest about their opinions, documenting assumptions to make them explicit and, *at least*, not state things that they think are false. Thus, although social simulations are not a world away from more traditional models of using simulation, they do have some different biases and characteristics.

28.2.3   Context and Social Simulation

Human knowledge, but particularly human *social* knowledge, is usually not context-free. There is a set of background assumptions, facts, relationships and meanings that are necessary (to understand the situation) and generally known but not made explicit. These background features can all be associated with the context of the knowledge (Edmonds 1999). In a similar way, most social simulation happens *within* a particular context as given, for example, the environment in

---

[8]I.e. those who are part of or can influence the social phenomenon in question.

which racial segregation occurs might be obvious to all concerned. This context is sometimes indicated in papers but is often left implicit since it is associated with the many background assumptions that can be safely ignored, either because they are irrelevant or they do not change in that context. Social simulation would probably be impossible if one was not able to assume a context whose associated assumptions need not be questioned for a given model (Edmonds 2010). Without an effective restriction of scope, every social simulation model would have to include *all* potential aspects of human behaviour and social interaction. Whilst such assumptions concerning the context are common to almost all fields of knowledge, they are particularly powerful in the social sciences because we unavoidably use our folk knowledge[9] of social situations to make sense of the studied social phenomena. This process of (social) context identification is often automatic, so that we correctly identify the appropriate context without expending much conscious thought. For this reason, the context is often left implicit, despite the fact that it is can be crucial to the construction and interpretation of a simulation. This leaves the decisions as to what to implement as foreground, deliberate decisions.

Choosing a social context that is relatively identifiable and self-contained is important if one is seeking to represent some evidence in a simulation. Being able to include all the important factors of some social process and obtain some evidence for their nature allows the building of simulations that are not misleading in the sense of not missing out factors that might critically change the outcomes. Clearly the more restricted the context, the easier the representational task. However, in this case one does not know whether what one learns from the simulation is applicable in other contexts. Using a simulation developed for one context and purpose for a different context and/or purpose might well lead to misleading conclusions (Edmonds and Hales 2005; Lucas 2010; Edmonds 2017).

Those simulations that are more focused on exploring an idea will often seek to transcend context, in the hope that the models will have some degree of generality—these often deliberately ignore any particular context. Although these may seem general, their weakness can become apparent when its applicability is tested. Here the *ideas* they represent might give some useful insights but may be misleading if taken as *the* defining feature of a specific case study. Clearly, a simulation that is claimed to have general applicability needs to have been validated across the claimed scope before being relied upon by others. To date, *no* social simulation has been found to be generally applicable beyond theoretical and illustrational purposes (Lucas 2011).

---

[9]Folk knowledge is the set of widely held beliefs about popular psychological and social theories; this is sometimes used in a rather derogatory way even when the reliability of the academic alternatives is unclear.

28.3   A Plethora of Modelling Purposes with Examples

Given the different purposes for which simulation models are used (see Epstein 2012; Edmonds 2017), they will be considered in groups of those with similar goals. It is only relative to their goals that simulation efforts can be judged. Nowadays it is widely acknowledged that authors should clearly state the purpose of their models before describing how the model is constituted (Grimm et al. 2006, 2017). Firstly, however, it is worth reviewing two goals that are widely pursued in many other fields but have not been convincingly attained with respect to the simulation of human society.

The first of these goals is that of *predicting* what will definitely happen in unknown circumstances. In other words, social simulation cannot yet make accurate and precise predictions. The nearest social simulations come (to our knowledge) is predicting some outcomes in situations where the choices are very constricted, and the data available is comprehensive. The clearest case of this is the use of microsimulation models to predict the outcome of elections once about 30% of the results are known (Curtis and Frith 2008). This constitutes hardly new or unknown circumstances, and is still not immune from surprises, since such predictions can be wrong. The microsimulation model relies on the balance between parties in each constituency and then translates the general switches between parties (and non-voters) to the undeclared results. Thus, although it is a prediction, its very nature rules out counter-intuitive or surprising predictions and comes more into the category of extending known data rather than prediction. The gold standard for prediction is that of making predictions of outcomes that are unexpected but true.[10]

The second goal that simulations do not achieve is to decisively *test* sociological hypotheses—in other words, they do not convincingly show that any particular idea about what we observe occurring in human societies can be relied upon or comprehensively ruled out. Here the distinction between modelling what we observe and modelling our ideas is important. A simulation that attempts to model what we observe is a contingent hypothesis that may always be wrong. However, social simulations of evidence are always dependent on a raft of supportive assumptions— that the simulation fails to reproduce the desired outcomes may be due to a failure of any of its assumptions. Of course, if such a model is repeatedly tested against evidence and fails to be proved wrong, we may come to rely upon it more (Popper 1963), but this success may be for other reasons (e.g. we simply have not tested it in sufficiently diverse conditions). Hypothesis testing using simulations is always relative to the totality of assumptions in the simulations, and thus the gain in

---

[10]This is when prediction is actually useful, for if it only gives expected values one would not need the simulation.

certainty is, *at best*, incremental and relative.[11] Thus, the core assumptions of a field may be preserved by adjusting "auxiliary" aspects (Lakatos and Musgrave 1970).

If a simulation is about ideas, then a very restricted kind of test is possible: a counterexample. If it has been assumed that factor *A* will lead to result *B*, then one might be able to show that this might not be the case in a plausible simulation. Indeed, the simulation may show that to obtain result *B* from factor *A* an additional and implausible assumption *C* is necessary. This does *prove* that "it is not necessarily the case that *A* leads to *B*", but it may shift the burden of proof back onto those who have assumed *A* will lead to *B*. This very restricted test is only useful if the context of causation between *A* and *B* is appropriately identifiable. This case of using a simulation to establish counterexamples is considered below.

A particular case of seeking for counterexamples is that of testing for the "existence of a sufficient condition" for some particular results. For example, it may be possible to show that there is no need to add some particular hypothesis to see a phenomenon take place, as in economics where it can be shown that in many cases the assumption of perfect rationality for agents does not need to be made.[12]

One might be disappointed that simulation provides neither predictions nor proofs (in the stronger senses of those terms), but that does not stop them being useful in other ways, which the sections below illustrate.

In the following, we look at how simulations might contribute to the understanding of human societies in a number of different ways, with examples from the literature. Unfortunately many articles describing social simulation research do not make their goals explicit (as advocated by ODD, see Polhill et al. 2008 and Chap. 15 Grimm et al. 2017); therefore, the categorisation below is that of the chapter's authors and not necessarily the category that the authors of the papers discussed would choose. In addition, it appears that some researchers have multiple purposes for their simulations or simply have not thought about their goals clearly.

28.3.1 Abstract Goals

First, we consider simulations that have more abstract goals, i.e. these tend to be more about *ideas* and *theories* than observed evidence (as discussed above in Sect. 28.2.1).

---

[11]If a simulation is not directly related to evidence but is more a model of some ideas, then it might be simple enough to be able to test hypotheses, but these hypotheses will then be about the abstract model and not about the target phenomena.

[12]This fact has led some to argue that such assumptions of perfect rationality should be dropped and that it might be better to adopt a more naturalistic representation of human's cognition (Gode and Sunder 1993; Kirman 2011).

28.3.1.1 Illustration of Ideas

Simulations can be good ways of making processes clear, because simulations are what simulations *do*. Thus, if one can unpack a simulation to show *how* the outcomes result from a set-up or mechanism, then this can demonstrate an idea clearly and dramatically. Of course, if how the outcomes emerge from the set-up in a simulation is opaque and/or difficult to understand, then this is not an effective technique. For this reason, relatively simple simulations prevail that are specifically designed to bring out the focus idea.

An example is (Rouchier and Thoyer 2006) which models voting and lobbying in the EU decision-making process. It does make strong assumptions about how the voting strategies might operate, but it does not pretend to be a descriptive model. Instead, it makes clear how the links between public opinion, lobbying groups and elected representatives might operate at the national scale as well as the EU one.

Another example is (Gode and Sunder 1993), a fairly simple demonstration that in some cases market institutions are so constraining that agents do not even need to be clever to achieve excellent results in this setting. They take the example of continuous double auction (CDA), a two-sided progressive auction, which is the protocol that is most used in financial markets. At any moment, buyers can submit *bids* (offers to buy). Similarly, sellers can submit *asks* (offers to sell). Both buyers and sellers may propose an offer or accept the offer made by others. The main constraint is an improvement rule, imposed on new offers entering the market, which requires submitted bids/asks at a price higher/lower than the standing bid/asks. Each time an offer is satisfying for one of the participants, he or she announces the acceptance of the trade at the given price, and the transaction is completed. Once a transaction is completed, the agents who have traded leave the market, and the bid-offer process starts again following the same rule starting from any price. The result of Gode and Sunder's simulation is that even with completely stupid agents, who know nothing of the market and only follow two constraints, the bid-offer rule described above and not selling below or buying above their reservation price, the market converges and enables agents to get excellent profits. This paper shows how institutional constraints might act to ensure a reasonable allocation of goods when agents are very clear about the value of things they want to sell or buy, and that this does not *require* any other substantive rationality by the agents. This result cannot necessarily be extended to any observed markets, which are most of the time complex, where the agents do have intelligence, where the value of items might be unclear and where there might be many other social and institutional mechanisms, but at least this result clarifies an idea about why protocols of this kind might be important.

The OpenABM project[13] has made significant progress in the development of a community of people using illustrative models to facilitate the communication of ideas. Working with others, this group in particular promotes the educational value of agent-based models.

---

[13]http://www.openabm.org.

A particular case of using a simulation to illustrate an idea is that of using a simulation in teaching. Whilst demonstrating an idea to one's peers might lead one to choose a simulation that emphasises the idea's generality and power, in teaching one may well choose to simplify and highlight certain features of the idea that will be important later on. This is a matter of degree but tends to result in simulations of a slightly different kind.

For example, researchers at Oxford University Department of Computer Science have developed a web application to assist students (particularly nonprogrammers) in understanding the behaviour of systems of interactive agents (Kahn and Noble 2009). They model, for example, the dynamics of epidemics in schools and workplaces and effect of vaccination or school closing/quarantine periods upon spread of disease in the population (Scherer and McLean 2002). The students can quickly and easily test different policies and other parameter combinations or in intensive sessions can work through a series of guided steps to build models from pre-existing modular components or "micro-behaviours"—a process called "composing". The models can also be run, saved and shared through a web browser in order to facilitate discussion and collaboration as well as ownership of the ideas and creative thinking.

28.3.1.2   Establishing The Possibility of a Process

A simulation can be used to show *how* a mechanism might result in certain outcomes, and thus establish that a proposed process is possible, demonstrated by enfolding the process in the simulation. This established plausibility of the process is relative to the plausibility of the assumptions behind the simulation—clearly, if the simulation is one that could not convincingly be related to any observed system, then one would not have established that the process is possible in any encountered system, but only be a *theoretical* possibility. This does not require that the simulation is an accurate representation of any observed system since all that is required is that one could imagine that a version of the target process in the simulation *could* occur in a real system.

A classic example of this is Axelrod's (1984, 1997) work on the evolution of cooperation. Previous models in evolutionary biology had suggested that cooperative behaviour would not be selected within an evolutionary setting, as any group of co-operators would be vulnerable to a single non-cooperative invader or mutant. Axelrod's books describe simulations in which a population of competing individuals evolved, playing repeated games against others. Some cooperative strategies, in particular "tit-for-tat" (cooperate unless your partner did not last time), were shown to survive and flourish in many game set-ups. Although the simulations described were highly speculative and abstract, they did firmly establish that it was possible that cooperative strategies might evolve within an evolutionary setting, where selfish strategies had a short-term advantage.

One use for establishing the possibility of a process is as a *counterexample* to an existing assumption or accepted theory, if the process demonstrated contradicts the assumption. Thus, the simulations of Axelrod above can also be seen as a counterexample to the assumption that cooperative behaviour cannot survive in an evolutionary setting.

The particular case of the Schelling (1969, 1971) model can be classified in this trend. Through very simple simulations, which Schelling ran by hand at the time, he discovered that segregation could be attained at a group level even though each individual agent had no strong preference for segregation. This paper was important, because it was one of the first examples of emergent phenomena applied to social issues. However, the most important element was the positive result obtained with the model. Schelling used a very intuitive (though not necessarily realistic) way of describing the change of location of agents in a city where they are surrounded by neighbours, which can be of two distinct types: identical to themselves or different. Each agent decides if it is satisfied with its location by judging if the proportion of neighbours that are different is acceptable to it. If this is not the case, it moves to a new location. Even when each agent accepts up to 65% of agents different to itself in its neighbourhood, high levels of segregation in the global society of agents result. This is a counterexample to the assumption that segregation results from a high level of intolerance to those of different ethnic origins, since one can see from the simulation that high levels of segregation in cities could be due to the movement of people at the edges of segregated areas who are in regions dominated by those of different ethnicities. Of course, what this does not show is that this is what actually causes segregation in cities; it merely undermines the assumption that it *must* be due to high levels of intolerance.

28.3.1.3 Understanding the Properties of an Abstract Model

With some analytic mathematical models and very few, very simple simulation models, one might seek to *prove* some properties of that model, for example, the parameter values under which a given outcome is reached. If this is not possible (the usual case), then one has two basic options: to simplify the original to obtain a model that is analytically tractable or to simulate it. If the simplifications that are necessary to obtain a tractable model are well understood and plausible, then the simplified model might be trusted to approximate the original model (although it is always wise to check). If it is the case that to obtain an analytically tractable model one has to simplify *so much* that the relationship between the simplified and the original model is suspect (e.g. by adding implausibly strong assumptions), then one cannot say that the simplified model is *about* the same things as the original model. At best, the simplified model might be used as an analogy for what was being modelled—it cannot be relied upon to give correct answers about the original target of modelling. In this case, if one wants to actually model the original target of modelling, then simulation models are the only option. In this case, one might wish to understand the simulation itself by systematically exploring its properties, such as doing parameter sweeps. In a sense, this is a kind of pseudo-maths, trying to get a grasp of the general model properties when analytic proof is not feasible.

An example of such an exploration is (Biggs et al. 2009). The authors examined regime shifts using a fisheries food web model, in particular looking at the existence of turning points in a system with two attractors (piscivore- and planktivore-dominated regimes). Anthropogenic drivers were modelled as gradual changes in the amount of angling and shoreline development. Simulations were carried out to investigate the onset of regime shifts in fish populations, the possibilities to detect these changes and the effectiveness of management responses to avert the shift. In relation to angling, it was found that shifts could be averted by reducing harvesting to zero at a relatively late stage (and well into the transition to alternate regime), whereas with shore development, it required action to be taken substantially earlier, i.e. the lag time was substantially longer between taking action and the resultant shift. The behaviour of different indicators to anticipate regime shifts was examined. This is an example of a mathematical model with stochastic elements that is solved numerically by means of a simulation.

Such stylised models, although based on well-understood processes, are carica-tures of real systems and have a number of simplifying assumptions. Nevertheless, they may provide an insight that would be applicable to many types of real-world issues. In contrast to this, some seek to understand the properties of some very abstract models, aiming to uncover some structures and results that might be quite generally applicable. This is directly analogous to mathematics that seeks to establish some general structures, theorems and properties that might later be usefully applied as part of the extensive menu of tools that mathematics applies. In this case, the usefulness of the exercise depends ultimately on the applicability of the results *in practice*. The criteria by which pure mathematics is judged can be seen as distinguishing those that are likely to be useful in the future: soundness, generality and importance.

An example of where the study of an abstract class of mechanisms has been explored thoroughly to establish the general properties is the area of social influence, in particular the sub-case of opinion dynamics. It can be found in works that use physics methodologies (Galam 1997) or those adapted from artificial life (Axelrod 1997). The topic in itself is extremely abstract and cannot be validated against data in any direct manner. In particular, the notion of culture or opinion that is studied in these models is so abstract that sociologists find this hard to accept (von Randow 2003). In this area, the most studied mechanism is the creation of consensus or convergence of culture represented by a single real number or a binary string (Galam 1997; Deffuant et al. 2000; Axelrod 1997). Many variations and special cases of these classes of model exist, for a survey see (Lorenz 2007). Some of these studies have indeed used a combination of parameter sweeps of simulations and analytic approximations to give a comprehensive picture of the model behaviour (Deffuant and Weisbuch 2007). Other merely seems to point out possible variations of the model.

Sometimes the exploration of abstract properties of models can result in sur-prises, showing behaviour that was contrary to expectations, so this category can overlap with the one discussed in the next section.

28.3.1.4 Exploration of the Safety of Assumptions in Existing Models

This is similar to the previous goal, but instead of trying to establish the behaviour of the model *as it is*, one might seek to explore what happens if any of the assumptions in the model is changed or weakened. Thus, here one is seeking to explore a space of possibilities *around* the original model. The idea behind this is often that one has a hypothesis about a particular assumption the model is based upon. For example, one might suspect that one would get very different outcomes if one varied some mechanism in the model in (what might seem) a trivial manner. Or one suspects that a certain assumption is unnecessary to the outcomes and can be safely dropped. Hence, for this goal, one is essentially comparing the behaviour of the original model to that of an altered or extended model.

For example, Izquierdo and Izquierdo (2006) carried out a systematic analysis of the effect of making slight modifications to structural assumptions in the prisoner's dilemma game: in the population size, the mutation rate, the way that pairings were made, etc., all of which produced large changes in the emergent outcome—the frequency of strategies employed. The authors conclude that "the type of strategies that are likely to emerge and be sustained in evolutionary contexts is strongly dependent on assumptions that traditionally have been thought to be unimportant or secondary" (Izquierdo and Izquierdo 2006, 181).

How cooperation emerges in a social setting was first fashioned into a game-theoretical problem by Axelrod (1984). The outcome was long thought to be dependent upon the defining questions such as which strategies are available, what are the pay-off values for each strategy, the number of repetitions in a match, etc., whereas other structural assumptions, supposed to be unimportant, were ignored. On further investigation, however, conclusions based on early work were shown to be rather less general than would be desired and sometimes actually contradicted by later work.

A different case is explorations of the robustness of the simulation described in (Riolo et al. 2001). This showed the emergence of a cooperative group in an evolutionary setting similar to the Axelrod one mentioned above. Here each individual had a characteristic (modelled as a number between 0 and 1) and a tolerance in a similar range. Individuals were randomly paired, and if the difference between their partner's and their own characteristic was less than or equal to their tolerance, they cooperated; otherwise they did not. As a result, a group of individuals with similar characteristics formed that effectively shared with each other. However, later studies (Roberts and Sherratt 2002 and Edmonds and Hales 2003) probed the robustness of the model in a number of ways—crucially by altering the rule for cooperation from "cooperate if the difference between my partner's and my own characteristic is *less than or equal* to my tolerance" to "if the difference between my partner's and my own characteristic is *strictly less than* my tolerance", i.e. from "≤" to "<". With this change, the crucial result—the emergence of a cooperative group—disappeared. It turned out that the (Riolo et al. 2001) effect relied on the existence of a group of individuals with *exactly* the same characteristic having to cooperate, since the smallest tolerance possible was zero. When the existence of completely selfish individuals was made possible by this change, the cooperation disappeared.

28.3.1.5 Exploring Counterfactual Possibilities

We only observe a few of the possible configurations of the social phenomena around us. Thus, it is natural to wonder what might happen if events or processes were other than what is observed or known to be the case. This is the world of artificial societies, where possible worlds loosely related to the one observed are explored. Sometimes an analogy with artificial life is made, where alternative algorithmic versions of life in the broadest sense are specified and experimented with—not life-as-it-is but life-as-it-might-have-been.

An extreme example of this is Jim Doran's model of a society with knowledge of the future (Doran 1997)—this can be thought of as what a society might be like whose members' predictions of the future happen to be correct. Clearly, this does not hold in any observed human society.

Such explorations might not contribute much to the understanding of our society, but it may inform the design of distributed computational systems where the components have a need to flexibly organise themselves in a way analogous, *but not identical to*, how humans organise (see Chap. 23 Hales 2017).

28.3.2 Concrete Goals

Here we consider some of the goals that are more at the concrete and descriptive end of the simulation spectrum. These tend to be *more* concerned to relate to the available evidence and tend to be more specific. In the subsections below, the "plausibility" of assumptions, results and simulations is a frequent issue. The simulation of human societies has not yet reached the situation where there is enough evidence to obtain much more than simple plausibility. At this current stage of social simulation, getting close enough to be deemed a "plausible" model is difficult enough, and there is almost never data enough to justify a stronger claim. Thus, claims of anything stronger should be treated with appropriate scepticism.

28.3.2.1 Building Towards Realism

One common approach is to start with a fairly simple model that is easier to understand and then to add aspects and mechanisms that are thought to be significant features of the observed system. That is, to add an additional level of realism to make the model more plausible or useful in some way, e.g. as a thought experiment. This is sometimes known as the TAPAS approach ("Take A Previous model and

Add Something"). It is consistent with the engineering principle of "KISS"—keep it simple, stupid. Here, one starts simply and adds more features/aspects one at a time and only if the simple approach turns out to be inadequate for some purpose.

Accordingly, Izquierdo (2008) starts with some standard models of the iterated prisoner dilemma games and adds some more "realistic" features, such as case-based learning and reasoning. A key idea in this is to maintain rigorous understanding of the extended model but take a step towards models that might eventually be validated against observed data from human interactions.

Whether one would, in fact, reach useable and valid models by this means is contested, with the alternative approach being to start with a complex model that reflects the evidence as well as possible and then seek for understanding and simplifications of this (Edmonds and Moss 2005).

To investigate the social aspects of socio-environmental systems, often some highly complicated models have to be used that include the relevant biophysical dynamics, coupled with social simulation. Rather than developing all components of the simulation model "from scratch" (and because the biophysical parts are relatively universal), such models may have a modular architecture designed to be reusable. It may therefore be more accurate to refer to the software as a "toolkit" from which various sub-models can be configured depending on the desired purpose of a particular study. In the area of land use simulation, PALM (Matthews 2006) is one such integrative model, and FEARLUS (Polhill et al. 2001, 2008) is part of another longstanding approach to socioecological modelling. With each iteration, the toolkit obtains further refinement and new features—whilst the level of understanding of its user(s) increases. The social simulator is interested in what additional complexity the human interaction part brings and to what extent it adds realism to the model's behaviour when compared with observed evidence.

28.3.2.2 Extending Evidence to Extrapolate to Unobserved Cases

Data about social systems is often limited to measurements from a limited number of observed cases. Thus, there may be many cases where one would like to estimate the outcome. Of course, one could use simple statistical techniques such as linear interpolation or similar to do this, but such techniques depend upon assumptions concerning the regularity of the results with respect to small changes in the set-up which may be implausible for some social systems. In this case, one might simulate the system using plausible assumptions, validate it against the known observations and then find the outcomes for set-ups that are different to those observed. For the results of this to be reliable, the simulation needs to be well validated, to correctly indicate the observed cases, to not differ very much from the observed cases (in contrast to the case described in Sect. 28.3.2.1), and for any unvalidated assumptions to be of a mild and uncontroversial nature.

The plausibility of the results from such experiments depends upon the validity of the original measurements as well as the generality of the assumptions (which must be plausible for the unobserved as well as observed cases).

For example, Brown and Harding (2002) use a microsimulation model to extend regional socio-demographic (census) data to cases that are not directly observed (synthetic householder-level records for each spatial district). The extension is attempted with deliberately cautious assumptions.

The "Sienna" programme (Snijders et al. 2010) fits a particular class of dynamic network model to "waves" of panel data. Simplifying a little, what happens is that the modeller specifies some basic assumptions (e.g. symmetry of network links) along with more than one set of panel data concerning the properties of the nodes at certain points in time. The algorithm then finds the dynamic network model that is consistent with the given specified constraints and that most closely fits the data. This is directly analogous to the process of fitting a line to a set of values using minimum total squared errors (or similar). What one gets out of this are some "surprise free" projections to network and node properties for times other than those given in the waves of panel data. This is not simulation in the same sense as other simulations mentioned here, since what is simulated is not a kind of process (that is given in the base specification of the family of models this technique uses) but rather a set of structures and values that fit given data in a statistical sense. When this technique is reliable and what its particular biases are have not yet been established.

28.3.2.3 Establishing the Consistency of a Process/Assumption with Evidence

Oftentimes a social process is not included in a study because it is not considered valid in the same way as a physical or biological principle might be. This is particularly true in historical examples where social processes are less in evidence. Going back to our second example of generative archaeology (Sect. 28.1.2), there are few archaeological findings that suggest a particular social structure and set of social processes, hence the need often for guesswork and the resulting coexistence of many competing theories. This is an area where social simulation can make an important contribution.

Perhaps the most well-known example is the Artificial Anasazi simulation model (Axtell et al. 2002). The objective was to see if a model could be constructed broadly consistent with available evidence—the number of households settled in part of the US southwest region over the period 800 to 1350. The performance of the model was impressive in its convergence upon the actual historical time series after a calibration of several parameters (a "fitting" process), which suggested new social explanations regarding the apparent land abandonment after 1350 might be possible. Interestingly a later paper (Janssen 2009) demonstrates that the model fit is mainly explained by two parameters related only to the model's carrying capacity. The author argues that a more insightful basis might be to generalise the target domain, working initially from less concrete goals rather than fitting a particular case and focusing on one evident and quantifiable trend (such as population). If the evidence base is broadened to include more ethnographic knowledge, this approach would resemble the pursuit of abstract goals as discussed in Sect. 28.3.1.

Data about real-world social networks introduced at the design or validation stages can be a valuable way of checking the consistency of a model. For example, Guimera et al. (2005) reconstruct the history of team collaborations in five different scientific and artistic fields and the development of corresponding collaboration networks. The authors develop and parameterise a probabilistic model of team selection. Using real data on team sizes, along with estimation of probabilistic parameters, to control the team assembly mechanism, the characteristics of the resulting networks (the degree distribution and the largest component) are compared with the real ones (independently for each of the five cases). The interest is in the transition of the collaboration network from "isolated schools" to an "invisible college"—the point at which the largest component of the network contains 50% or more of the nodes (which is the case for all representative fields). All simulated network measurements are shown to be in close agreement with the real networks, which establishes the plausibility of the proposed team selection mechanism. However, being a probabilistic model, it does not attribute any particular decision process to this mechanism that might be able to reveal new questions.

Another example is in White (1999), which attempts to evaluate some statistical assumptions against data about marriage systems in different cultures using a "controlled simulation".

28.3.2.4 Analysis of Influence Factors

In any complex system, it is very difficult to estimate the importance of different factors on particular outcome measures or results. This is due to the "nonlinearity" in many social systems where a normally insignificant factor can trigger a system-wide change in behaviour. However, given a trusted simulation model of the system, one can perform experiments to determine the importance of each factor in the class of simulation set-ups that are run. Thus, one does not have to determine the relative importance of factors on an a priori basis; one can simply run the experiments and measure the outcomes. Clearly, this approach depends on having a reliable simulation model.

(Saqalli et al. 2010) investigate a simulation model of the development over several generations of a rural agrarian society to weigh the importance of several different model parameters on simulation results. In simulation experiments reported, four parameters were assessed in relation to six state variables—with measurements taken at the end of the run. The model was based on a case study of the Nigerian Sahel, typified as a low-data situation where, in particular, little has been published on the social factors governing access to economic activities (including off-farm activities so often neglected as an important revenue generating source) or on intra-household dynamics (which the authors recognise as having a complex structure). The objective was to assess the robustness of results against variation in socio-economic and biophysical parameters to show that it is "constrained by the different parameters of its structure" (Saqalli et al. 2010: para. 3.6). This step provides the researcher with an improved understanding of the range of possible model outcomes and what might constitute a significant or meaningful difference when comparing outcomes. It is worth noting, however, that the single-parameter approach neglects any possible parametric interaction that could be identified from a pairwise analysis of influence factors.

A very different example of this is Yang et al. (2009), which studies the factors that influenced success in the system of Chinese civil service exams that existed in the Imperial era in Mainland China. The simulation model used historical data from civil service records and some assumptions to assess the importance of factors such as class, wealth and family connections in terms of success at passing this exam (and hence obtaining a coveted civil service post). It is difficult to see how such indications about events that are otherwise lost in the past could be obtained, although this is open to the criticism of being unfalsifiable.

The disadvantages of this approach are that the assessment of influence is only as good as the simulation model, and it only samples particular sets of initial conditions—it does not rule out the case where very special values of parameters cause totally different outcomes (unless one happens to be lucky and sample these).

28.3.2.5   Assessment of Policy Options

Recently more and more articles have appeared in the literature featuring ABMs that address policy-making in contemporary issues such as developmental sustainability and climate change adaptation. For example, Berman et al. (2004) consider eight employment scenarios defined by different policies for tourism and government spending, as well as different climate futures, for an ABM case study of sustainability in the small arctic community of Old Crow, Yukon. Scenarios were developed with the input of local residents: tourism being a policy option largely influenced by the autonomous community of Old Crow (stemming from their land rights) and attracting great local interest. In ABM, policy options are often addressed as a certain type of scenario (scenarios are discussed in Sect. 28.3.2.9), embedding the behaviour of actors within a few possible future contexts. The attraction of this approach is that the model could potentially be used as a decision support tool, in a form that is familiar to many analysts, to provide answers to very specific policy questions. The merit is that it can improve the reckoning of human and social factors and information into the issues at stake; the drawback is the multiplication of uncertainties, not least of which is that we do not convincingly know how social actors might adapt (even if the possible policy options are more concrete).

For example, Alam et al. (2007) investigate the outcomes indicated by a complex and detailed model of a village in the Sekhukhune district of South Africa. This model in particular looks at many aspects of the situation, including social network, family structure, sexual network, HIV/AIDS spread, death, birth, savings clubs, government grants and local employment prospects. It concludes with hypotheses about this particular case. This does not mean that these outcomes will actually occur, but this does provide a focus for future field research and may provide thought for policy-makers.[14]

28.3.2.6 Social Engineering: "Designing" Better Systems

Market design is the branch of economic research aiming to provide insights into which market protocol, i.e. interaction structure and information circulation rules, is the best to obtain certain characteristics of a market. Agent-based simulation seems to be a good method to test several such protocols and see their influence on economic performances, e.g. efficiency, fairness and power repartition (Marks 2007). Each protocol is already known for its advantages and disadvantages (e.g. whilst Dutch auction is fast, double auction tends to extract the highest global profit). Since not every desirable aspect can be achieved with a single protocol, one has to choose the aim to attain (LeBaron 2006). Assuming agents act rationally, it is then possible to compare protocols to see what difference they make in prices or other indicators (e.g. Kirman and Moulet 2008). Many studies have been designed within the context of electricity markets, which are very crucial since unpredicted shortages are a problem and prices vary very quickly, and involve a comparison of protocol (e.g. Nicolaisen et al. 2001). One can also note the use of "evolutionary mechanism design" (Phelps et al. 2002; March 2007) where strategies of three types of actors— sellers, buyers and auctioneers—are all submitted to evolution and selection, so the actual organisation of the market evolves, while the context of production and demand is fixed. In today's economy, more and more artificial agents really interact—either in bidding on consumers' sites or even in financial markets (Kephart and Greenwald 2002)—so there is some convergence between real markets and artificial systems, which utilise market mechanisms. For a more detailed discussion of modelling and designing markets, see Chap. 25 in this handbook (Rouchier 2017).

28.3.2.7 Data Integration

A mundane and sometimes overlooked aspect of the scientific process is simple description. That is, recording what has been observed in a suitable form. Traditionally these forms have included the likes of narratives, logs, videos, measurements and pictures. However, simulations can also be used as a sort of description, where the aim is not to express a *theory* about a mechanism, but rather to integrate as much of the relevant evidence about what is observed as possible about a particular target. Simulation has some advantages in such a process, since it can allow the

---

[14]Although in this particular case, it did not as the model indicated outcomes that the policy-makers preferred to ignore, being not compatible with the actions they had already decided to take.

integration of several different kinds and levels of evidence in one framework. For example, aspects of narrative texts can be incorporated within the behavioural rules of an agent; the social network of subcommunities can be compared to those that result from the simulation; time-series data can be compared to the corresponding time series derived from measurements on the simulation outcomes and survey data compared to the equivalent answers at instances of the simulation runs. Such integration is far from easy, since some aspects are programmed directly (e.g. agent behaviour), whilst others have to be achieved in terms of the results (e.g. aggregate statistics about the outcomes). Achieving any particular set of outcomes in a social simulation is difficult due to the prevalence of unpredictable interactions and effects (i.e. emergence), so the achievement of a data-integration model is not an easy one. Such models are not entirely (or solely) a description since the structure of a simulation sometimes brings into question the consistency of the various parts of the evidence. Thus, if it is difficult to square an account of how individuals behave with some of the outcomes, one may be forced to make some choices, including possibly adding in aspects that are not directly observed. This is all right as long as these are clearly documented—they can provide fertile issues for future data collection. However, such data-integration models do not aim for a level of generality beyond the particular case study (or studies) focused on. In this way, they can avoid "high" theory to motivate simulation features where this is not supported by the evidence with respect to the target case. It is not that there is *no* theory in such simulations— any description or abstraction, however mild, will rely on some theory, but the point is that in a descriptive simulation such theory is either well established or relatively mundane.

Examples of simulations that intend to be descriptive in this sense include (Christensen and Sasaki 2008) which aims at producing a simulation of the evacuation from a particular building, with a view to a future evaluation of evacuation plans and facilities, in particular with regard to disabled people. It uses many particulars of the building structure but makes assumptions (albeit of a plausible variety) about how people behave when evacuating. Likewise, Terán et al. (2007) aim to simulate land use and users within a forest reserve with a view to producing a computational representation of this. As in similar simulations, there is a mixture of assumptions— some that are backed by evidence, and some that are just plausible guesses. This simulation is loosely validated against some data and broadly confirms the results found from some other models. The ultimate use of this (and similar models) is not specified.

Such simulations can take a long time to construct, involving many iterations of model development as well as being complicated and slow to run. The advantage of such models is that they are a precise and coherent representation of a set of evidence—in a sense an encapsulation of a particular case study.[15] This can be the basis for further experiments/inspection, which, in turn, can lead to more abstraction. This kind of process can result in modelling theories of the processes

---

[15]To be precise: a possible encapsulation of a particular set of evidence on the case study.

observed within the data-integration model with simpler models whose properties are easier to establish but whose outcomes can be checked against targeted experiments on the data-integration model.

28.3.2.8 Finding New Questions and Areas of Ignorance, Hypothesis Suggestion

Another use of a simulation is as an aid to good observation. That is, suggesting issues and questions that should be sought in order to gain an adequate observational coverage. The simulation is developed as in the data-integration case above, including different aspects of the observational evidence that are available. It is often the case that only when one tries to simulate a process, the gaps in our knowledge become clear. Thus, building a simulation *as one is observing* can help direct the data-gathering research in order to complete an adequate computational description. In this sense, it forms a similar role to simulation in some cognitive science (Newell 1990; Sun 2005).

For example, Moss (1998) exhibits a simulation built on a mixture of bases: (a) an assumed but plausible cognitive architecture that captures how one might divide a problem into subproblems until they are doable, (b) some suggestions elicited from an expert from the domain and (c) plausible guesses for the remainder. This model attempted to examine behaviour in the face of crises (defined as when one unwanted event causes another in an out-of-control chain), in particular how the rotating of crisis management teams and the information they pass on to the next team might impact their effectiveness at fighting the crisis. The results were not independently validated, but this is not the point of this simulation. As the author says:

". . . results obtained with the North West Water model indicate a clear need for an investigation of appropriate organizational structures and procedures to deal with full-blown crises".

In contrast, Younger (2005) is a very much more abstract model, which is only loosely built upon evidence, but with the same broad aim of suggesting hypotheses—in this case, hypotheses concerning the occurrence of violence and revenge within egalitarian societies. Clearly, the plausibility of the hypotheses or questions suggested by a simulation will be greater when the simulation is more firmly rooted in evidence. However, hypotheses and questions can be worthwhile investigating whatever their source, and at least having a simulation grounds and defines the question in a precise way, making clear what it might explain and the sort of other issues and questions that might accompany it.

28.3.2.9 Creation/Critique of Scenarios

Berman et al. (2004) present an example of scenarios being used to constrain models to produce simulations of the wider consequences of those scenarios (as measured by relevant socio-economic or environmental indicators or by their possible influence on human institutions) that can then be used to inform discussions with stakeholders and may ultimately produce a better understanding of such changes. Bharwani et al. (2005) use climate change scenarios to investigate adaptive decision-making among villagers in the Limpopo province of South Africa, focusing on the use of seasonal forecast information in farming strategies. Data from the Hadley Centre climate model—HadAM3—showing a 100-year drying trend with increasing potential evapotranspiration (PET) were used as model input (providing PET and precipitation values). Results show that a degree of resilience to these changes is afforded when the forecast is correct 85% of the time so that farmers establish increased trust in, and use of, seasonal forecasts. They are able to choose cropping strategies that are suited to climate change, though this behavioural shift may only occur over a very long timeframe.

Bharwani et al. (2005) introduce the use of scenarios into the methodology in a further and very interesting way: by postulating them as "drivers" of actors' decision-making processes. In this ethnographic approach, the authors combine simplified scenarios across different domains (irrigation, forecast and market) asking respondents what they would do under each scenario, in a given context. This information was then used to produce the model rules for the agents' decision-making.

In either case, where conventional scenarios used in future planning can seem rather terse and lacking in specifics—which may be a limitation to their subsequent use in policy discussion—simulation outputs that explore scenarios offer a great deal of detailed information "that would be difficult to imagine otherwise" (Berman et al. 2004, p. 410). Moreover, this can apply at different levels of analysis from trends in macro variables down to the impacts on different sectors and regions, as well as differentiated impacts for agents fitting any given "profile" in which the analyst is interested. Perhaps greater care has to be taken, however, in the use of model-generated scenarios, to ensure that these are not taken as "more accurate predictions" by virtue of being "computed" stories rather than conventional "imagined" stories.

Scenarios are often used in policy discussions, e.g. climate change. However, they are usually somewhat vague and/or only described in qualitative terms. Simulations can be used to produce consistent scenarios or to produce models that instantiate aspects of given scenarios.

28.3.2.10   Intervention with Stakeholders

Instead of developing a simulation to represent some aspect of society, one can also try to use a simulation to intervene *in* society. That is, use a simulation to change some interaction between stakeholders, for example, to facilitate collective decision-making or mutual understanding. One well-known approach is the Companion Modelling approach, which has been developed since a decade (see Chap. 12, Barreteau et al. 2017).

An example is demonstrated by Etienne (2003). Here, a model is used in conjunction with a role-playing game to show chosen participants the issues that can arise when several users compete on a pastoral resource. The model building process was an integration of multidisciplinary knowledge acquired on French Mediterranean silvopastoral systems into a model capable of representing the interactions between ecological dynamics and social behaviours. In order to help foresters and livestock farmers to better integrate these interactions into their planning work, a multi-agent system was designed to simulate different management strategies and to compare their impact on forest quality. This model was coupled with a role-playing game (RPG) initially developed as a didactic support to silvopastoral training programmes, and very soon, it proved useful in the negotiations and interactions between livestock farmers and foresters involved in the management of the same forest. The tool revealed itself flexible enough to make it possible to play with actively involved stakeholders such as the current users of the resource (local farmers and foresters), with potential regulators of the system (managers or administrators), technical experts (extensionists, technicians) or learners concerned with the topic (students, scientists).

This model is effectively an intervention between the livestock farmers and foresters by being a subtle mediating tool, allowing the stakeholders to play at decision-making, to educate them in the possible effects of their choices and to thus encourage debate and introspection. This model has also been used for didactic purposes (Sect. 28.3.1.1) by getting agronomy students to play it.

28.4   Inputs and Results of Simulation Models

One method of assessing the use, and ultimately the success, of a simulation for understanding aspects of society is to tease out what has gone into making a simulation model, the *input*, and how the results from the simulation are interpreted and used, the *output*. These, the input and the output, together form the mapping from the computer program and its calculation from and to the target of study. They are crucial parts of what characterises a simulation, even if they tend to be described in a less formal manner than the simulation code and behaviour.

28.4.1   Inputs

What is put "in" to the design of a model tends to be more explicitly distinguished in papers than what comes "out". This might be because the "job" of a modeller is seen as a process of deciding what processes and structures will go into a model and because the inputs are under the control of the modeller in a way that the results certainly are not and hence can be displayed and talked about with greater confidence. However, all social simulations are based on a raft of different assumptions, settings and processes. These are somewhat separated out for analysis here.

28.4.1.1   Evidence-Based Assumptions

If there is some evidence about the nature or extent of the processes that are being observed, then this can be used to inform the set-up or structure of a simulation. For example, evidence from social psychology might be used to inform the specification of the behavioural rules of a set of agents in a simulation, or the narrative account of a participant used as the basis for programming a particular agent.[16] Of course, it is rare that such evidence constrains the possible settings and algorithms completely but rather that it partially constrains these or constrains them in conjunction with additional assumptions from another source. Clearly, the more assumptions can be constrained by evidence (either directly or as the result of previous research) the better. The presence of other assumptions and inputs does not make a simulation useless, especially if documented, but any results are then relative to the assumptions. If assumptions that were included are completely misguided and critically affect the results, then this would seriously limit the usefulness of the model with respect to the observed world.

28.4.1.2   Indirectly Inferred Settings

In situations where some parameters are unknown and where there is a relative abundance of time-series data, one can attempt to infer the values of these parameters by seeing which parameter values result in the model giving the best fit to a segment of the time-series data. This is a sort of evidence-based setting, but it often seems to be used when the parameters concerned do not have any discernable meaning in terms of the target of modelling. A tradition of using a certain kind of decision or learning algorithm in an agent might lead to this algorithm being "fitted" to an initial segment of the data (so-called in-sample data) *even when it is unlikely*[17] *that the algorithm corresponds to how the target agents think*. Thus, the credibility of this technique is dependent on the reliability of the other assumptions in the model and the meaning of the parameters being fitted. If the parameter was a scaling parameter, then this might be a sensible way to proceed.

---

[16]This can either be done directly as a translation of an interview text into programmed rules or used to check that such programming is correct by comparing the resulting behaviour of an agent against what happens when the simulation is run. Thus, there is not a clear distinction between verification and validation from evidence. In a sense, this second method is verification since the programming is rejected until correct, but, on the other hand, this is part of the production of a simulation, which may only be completed later for its validation as a whole.

[17]Unlikely with regard to the psychological or sociological evidence about the target subjects.

28.4.1.3 Documented Theoretical Assumptions

Clearly, researchers do not invent all the details and algorithms of their model from the ground up, but are doing their research with knowledge of certain approaches and algorithms and within a community of other research, with established techniques and traditions. Thus, many parts of a simulation model will be based on (parts of) other models or algorithms from other fields. Many models in economics will use a decision algorithm based on constrained comparisons of predicted utility, and other models might import techniques from the fields of artificial intelligence or evolutionary computation. It seems impossible to avoid all such theoretical assumptions; however, there are distinctions to be made in terms of the *strength* of the assumptions, the likely *biases* behind such assumptions and the degree to which they are *evidence based*.

"Strong" assumptions are those that are surprising or seem to specify conditions that are rarely observed. Thus, an assumption that an agent has *in effect* a perfect model of the economy in its head is a very strong assumption, since even experts find it difficult to understand the economy as a whole. Strong assumptions are often introduced to allow analytically solvable models to be specified and used, for example, the assumption of perfect information in game theory. Whilst analytically tractable models were necessary when there was no other avenue for the precise modelling of many kinds of phenomena, the advent of cheap computing power and accessible simulation platforms means that often more appropriate methods are now available, with analytic models possibly being used to check or understand the reference simulation model rather than being the focus. Clearly, all other things being equal, weaker assumptions are preferable to strong ones—the stronger an assumption, the more evidence is needed to justify its use. In any case, all such assumptions should be as fully documented as possible.

28.4.1.4 Explored Conditions

In much simulation work, there will be a focus hypothesis or set of hypotheses that are being investigated. In these cases, it is usual to try the simulation using that hypothesis and then compare the results to those coming from a version of the simulation with a different hypothesis implemented. This provides evidence about the possible effects of that hypothesis on the outcomes, allowing comparison with evidence and possible subsequent inference as to which is more likely to be the case. The clearest case of this is testing the significance of the inclusion of a hypothesis against that of a *null model*[18] to see if the properties of the results that are deemed significant indeed result from the hypothesis or from other aspects of

---

[18]A "null" model is a model version where the claimed causal mechanism is eliminated to see if the resultant "effect" would have arisen as the result of background (e.g. random) mechanisms anyway.

the model. Thus, a simulation of a stock market might compare the results obtained with intelligent agents (that notice patterns in pricing and try and exploit these) to the results obtained with agents that buy and sell at random. Unfortunately, it is sometimes the case that a simulation is presented purporting to show the significance of a hypothesis without indicating what the comparison case is.

28.4.1.5   Randomness and Other Essentially Arbitrary Assumptions

A simulation modeller is often faced with deciding how to design a part of a simulation model for which there is neither evidence nor any tradition of modelling to guide them. In such a case, one might simply make that aspect random. For example, where it is unknown how a kind of choice is made in the modelled situation, it might be implemented as a random choice in a simulation model of that situation.[19] This is usually done in conjunction with a "Monte Carlo" approach that runs the simulation a number of times and averages the resulting different sets of outcomes. Presumably, this is done under the assumption that the introduced randomness will be averaged out, leaving only the effects of the other design settings. However, this assumption is rarely proven but often simply remains a hope. Of course, if it can be shown that the value of the particular input does not influence those aspects of the results that are deemed significant by a series of simulation experiments (or otherwise), then a random input or process might well be acceptable. However, in this case, a constant value might be simpler and have the same effect.[20]

We suspect that many uses of randomness in simulations are in the nature of a programming "stub"—that is, a stand-in that the programmer intends (or intended) to expand to a more plausible algorithm at a later date. Whilst this is perfectly acceptable during model development and to some extent inevitable given that researchers always have time constraints, such stubs are likely targets for model criticism by other researchers. At the very least, some exploration of them to assess the extent to which they affect those aspects of the results deemed significant is advisable.

Randomness can be considered as a special case of a broader class of assumptions: those that are added into the model simply to get it to run, and for no theoretical or evidence-related reason. We hope that these are honestly declared

---

[19] Another option is to try all the possibilities exhaustively in a series of simulations or by using techniques such as constraint logic programming, but these are technically difficult and require a lot of computational power.

[20] There are possible reasons why a constant value might not work, for example, when the input provides some mechanism of symmetry breaking.

rather than "dressed up" under some other categories, although often these are excused under the broad umbrella of "simplicity".[21]

28.4.1.6 Undocumented Assumptions

It is not feasible to document all of the assumptions in a model. Firstly, this might take too much space in a single paper,[22] and secondly, many might be previously established and well known to those in a particular field of work. However, it is also likely that researchers are simply not aware of all the assumptions inherent in their simulation models, due to the limitation of human cognition.[23] Clearly, it is part of the job of other researchers to point out undocumented assumptions where these can be shown to be significant.[24]

28.4.2 Outputs

A similar set of distinctions can be made about what comes out of a simulation, the results. There is not an obligation to describe all the outputs from a simulation, but rather one tends to get a sample of results, which typically is composed of: sample results, sensitivity analyses, evidence of validation and the outcomes from experiments designed to test a hypothesis. However, not all the details of the results are considered as equally significant—we now consider each of these in order of increasing significance.

- *Firstly*, there are those aspects of the results that are considered as artefacts of the model, for example, the randomness that might have been input into the model.
- *Secondly*, there are those features that might be considered to reflect some of the model structure and the processes that result from them. These features may not be judged as reflecting those parts of the simulation that reflect what is being modelled, but may be caused by theoretical or arbitrary assumptions that were put in. These features of the results may well not be so much of a surprise to the modeller.

---

[21] There is nothing wrong with assumptions that had to be made due to constraints on resources, such as time, expertise or computing power, but it is simply disingenuous to pretend that this is sanctioned by a higher "virtue".

[22] However, this is a poor excuse given the ease with which a relatively complete technical paper can be archived and then cited by a journal article or report discussing the model.

[23] Alternatively it may be because the simulation designers had not thought about what they were doing.

[24] It is trivial to point out that a simulation has missed out some assumption or other, but this is not very useful. It is far more useful to point out *how* and *why* an assumption might be important and for *which* purposes.

- *Thirdly*, there are those features of the results that are interpreted as indicating something about what is being modelled. For example, they may suggest a hypothesis about those phenomena. They indicate a possibility that may be inherent in what is being modelled or that is inherent in the target of modelling. This may well go beyond what can be directly validated in the model but, for example, track counterfactual possibilities concerning what *might* have occurred.
- *Lastly*, there are those features that would be positively expected of the phenomena being modelled. That is, if they were *not* present, this would be taken as evidence that there was something amiss with the model. In other words, they are a necessity of the phenomena. It is against this category of results that models are validated.

It is not easy to distinguish these different categories of significance in terms of the results, since the causation within a model can be very complicated, being a result of many model aspects interacting together. It is also usually the case that the modeller has hypotheses (or assumptions) about what aspects of the results are significant in which ways, and this is crucially useful information to impart to a reader interested in the results. However, this is often left implicit.

One might justifiably criticise many social simulations in terms of the lack of empirical grounding of both inputs and outputs. Many social simulations have only the weakest connection with anything observed—the inputs are largely assumption based, and indeed often highly artificial—the outputs only relating in the broadest way to any data and then only in terms of a few aspects of the possible outputs (i.e. only a few selected aspects are deemed significant to what is observed and then in the loosest, "hand-waving", manner). It may well be that simulating human society is just very, *very* difficult, but one suspects that it is simply easier to stick to considering abstract ideas.

28.5   Conclusion

Simulation has undoubtedly helped to improve our understanding of human society, although in a number of different and usually indirect ways. It is fair to say that, so far at least, this has served to improve our understanding of some societal processes and our ideas *about* society rather than directly in terms of being able to strongly predict aspects of society or conclusively test hypotheses about society.

Simulation is not a replacement for other ways of understanding society;[25] it is simply a flexible way of precisely modelling it in a way that can represent some of the dynamic and complex aspects of it. It can be especially productive in conjunction with other approaches. For example, analytic models can be used to check the outputs and properties of a simulation model and help us understand

---

[25] At least, not in any of the cases we have as yet come across.

the model, and, conversely, a simulation can be used to probe and check some of the simplifications and assumptions employed in an analytic model. In participatory models, social science techniques of engagement and elicitation can be applied to inform the construction of agent-based social simulations as well as the simulations suggesting what might be usefully investigated in terms of the collection of new data.

Clearly social simulation has some way to go in terms of the maturity of its method and the reporting and use of simulation models. There are still areas in which the methodology needs substantial improvement and standardising. There are also significant unresolved issues, such as how to decide what level of detail to include and to what extent one should rely on prior theory.

We predict that simulation will be even more significant in helping us understand human society in the future, *in particular where it is used in close conjunction with other relevant approaches.*

Further Reading

A more general and simpler introduction to varying modelling purposes can be found in Chap. 4 (Edmonds 2017). The best general introduction to social simulation is (Gilbert and Troitzsch 2005) which covers general issues and gives code examples. For a wider range of views on social simulation, the published papers from the US National Academy of Sciences colloquium on "Adaptive Agents, Intelligence, and Emergent Human Organization: Capturing Complexity through Agent-Based Modeling" (PNAS 2002) give a good cross-section of the different approaches people take to this area. It is difficult to point to further good sources as this topic is so diverse, but the Journal of Artificial Societies and Social Simulation has many accessible papers.

References

Alam, S. J., Meyer, R., Ziervogel, G., & Moss, S. (2007). The impact of HIV/AIDS in the context of socioeconomic stressors: An evidence-driven approach. *Journal of Artificial Societies and Social Simulation, 10*(4).http://jasss.soc.surrey.ac.uk/10/4/7.html.

Axelrod, R. (1984). *The evolution of cooperation.* New York, NY: Basic Books.

Axelrod, R. (1997). *The complexity of cooperation.* Princeton, NJ: Princeton University Press.

Axtell, R., Axelrod, R., Epstein, J. M., & Cohen, M. D. (1996). Aligning simulation models: A case study and results. *Computational and Mathematical Organization Theory, 1*(2), 123–141.

Axtell, R. L., Epstein, J. M., Dean, J. S., Gumerman, G. J., Swedlund, A. C., Harburger, J., et al. (2002). Population growth and collapse in a multi-agent model of the Kayenta Anasazi in Long House Valley. *Proceedings of the National Academy of Sciences, 99*(3), 7275–7279.

Barreteau, O., Bots, P., Daniell, K., Etienne, M., Perez, P., Barnaud, C., et al. (2017). Participatory approaches. doi:https://doi.org/10.1007/978-3-319-66948-9_12.

Berman, M., Nicolson, C., Kofinas, G., Tetlichi, J., & Martin, S. (2004). Adaptation and sustainability in a small arctic community: Results of an agent-based simulation model. *Arctic, 57*(4), 401–414.

Bharwani, S., Bithell, M., Downing, T. E., New, M., Washington, R., & Ziervogel, G. (2005). Multi-agent modelling of climate outlooks and food security on a community garden scheme in Limpopo, South Africa. *Philosophical Transactions of the Royal Society B, 360*(1463), 2183–2194.

Biggs, R., Carpenter, S. R., & Brock, W. A. (2009). Turning back from the brink: Detecting an impending regime shift in time to avert it. *Proceedings of the National Academy of Sciences (PNAS), 106*, 826–831.

Brown, L., & Harding, A. (2002). Social modelling and public policy: Application of microsimulation modelling in Australia. *Journal of Artificial Societies and Social Simulation, 5*(4). http://jasss.soc.surrey.ac.uk/5/4/6.html.

Cartwright, N. (1993). *How the Laws of Physics Lie*. Oxford: Oxford University Press.

Christensen, K., & Sasaki, Y. (2008). Agent-based emergency evacuation simulation with individuals with disabilities in the population. *Journal of Artificial Societies and Social Simulation, 11*(3). http://jasss.soc.surrey.ac.uk/11/3/9.html.

Clifford, J. (1986). *Writing culture: The poetics and politics of ethnography*. Berkeley, CA: University of California Press.

Cole, H. S. D., Freeman, C., Jahoda, M., & Pavitt, K. L. (Eds.). (1973). *Models of doom: A critique of the limits to growth*. New York: Universe Books.

Curtis, J., & Frith, D. (2008). Exit polling in a cold climate: The BBC–ITV experience in Britain in 2005. *Journal of the Royal Statistical Society: Series A (Statistics in Society), 171*(3), 509–539.

David, N., Fachada, N., & Rosa, A. C. (2017). Verifying and validating simulations. doi:https://doi.org/10.1007/978-3-319-66948-9_9.

Deffuant, G., & Weisbuch, G. (2007). Probability distribution dynamics explaining agent model convergence to extremism. In B. Edmonds, C. Hernandez, & K. G. Troitzsch (Eds.), *Social simulation: Technologies, advances and new discoveries* (pp. 43–60). Hershey, PA: IGI Publishing.

Deffuant, G., Neau, D., Amblard, F., & Weisbuch, G. (2000). Mixing beliefs among interacting agents. *Advances in Complex Systems, 3*(1), 87–98.

Doran, J. E. (1997). Foreknowledge in artificial societies. In R. Conte, R. Hegselmann, & P. Tierna (Eds.), *Simulating social phenomena (Lecture notes in economics and mathematical systems, 456)* (pp. 457–469). Berlin: Springer.

Dunbar, R. I. M. (1998). The social brain hypothesis. *Evolutionary Anthropology, 6*(5), 178–190.

Edmonds, B. (1999, September 9–11). The pragmatic roots of context. In P. Bouquet, L. Serafini, P. Brezillon, M. Benerecetti, & F. Castellani (Eds.), *Modelling and using context, second international and interdisciplinary conference, CONTEXT'99, proceedings (Lecture notes in artificial intelligence, 1688)* (pp. 119–132), Trento, Italy. Berlin: Springer. http://cfpm.org/cpmrep52.html.

Edmonds, B. (2001). The use of models–making MABS actually work. In S. Moss & P. Davidsson (Eds.), *Multi agent based simulation (Lecture notes in artificial intelligence, 1979)* (pp. 15–32). Berlin: Springer.

Edmonds, B. (2007). The practical modelling of context-dependent causal processes—A recasting of Robert Rosen's thought. *Chemistry and Biodiversity, 4*(1), 2386–2395.

Edmonds, B. (2010, June 23–25). *Context and social simulation* (Paper presented at the IV edition of Epistemological Perspectives on Simulation (EPOS2010)Hamburg, Germany). http://cfpm.org/cpmrep210.html.

Edmonds, B. (2017). Five different modelling purposes. doi:https://doi.org/10.1007/978-3-319-66948-9_4.

Edmonds, B., & Hales, D. (2003). Replication, replication and replication-some hard lessons from model alignment. *Journal of Artificial Societies and Social Simulation, 6*(4). http://jasss.soc.surrey.ac.uk/6/4/11.html.

Edmonds, B., & Hales, D. (2005). Computational simulation as theoretical experiment. *Journal of Mathematical Sociology, 29*(3), 209–232.

Edmonds, B., & Moss, S. (2005). From KISS to KIDS—An 'anti-simplistic' modelling approach. In P. Davidsson et al. (Eds.), *Multi agent based simulation 2004 (Lecture Notes in Artificial Intelligence, 3415)* (pp. 130–144). Berlin: Springer.

Epstein, J. (2012). Why model? *Journal of Artificial Social Societies Simulation, 11*(4). http://jasss.soc.surrey.ac.uk/11/4/12.html.

Etienne, M. (2003). SYLVOPAST: A multiple target role-playing game to assess negotiation processes in sylvopastoral management planning. *Journal of Artificial Societies and Social Simulation, 6*(2). http://jasss.soc.surrey.ac.uk/6/2/5.html.

Galam, S. (1997). Rational group decision making: A random field Ising model at $T = 0$. *Physica A, 238*, 66–80.

Galán, J. M., Izquierdo, L. R., Izquierdo, S. S., Santos, J. I., Del Olmo, R., López-Paredes, A., & Edmonds, B. (2009). Errors and artefacts in agent-based modelling. *Journal of Artificial Societies and Social Simulation, 12*(1). http://jasss.soc.surrey.ac.uk/12/1/1.html.

Gilbert, N., & Troitzsch, K. (2005). *Simulation for the Social Scientist* (2nd ed.). Open University Press.

Gode, D. K., & Sunder, S. (1993). Allocative efficiency of markets with zero intelligence traders: Markets as a partial substitute for individual rationality. *Journal of Political Economy, 110*, 119–137.

Grimm, V., Revilla, E., Berger, U., Jeltsch, F., Mooij, W. M., Railsback, S. F., et al. (2005). Pattern-oriented modeling of agent-based complex systems: Lessons from ecology. *Science, 310*(5750), 987–991.

Grimm, V., Berger, U., Bastiansen, F., Eliassen, S., Ginot, V., Giske, J., et al. (2006). A standard protocol for describing individual-based and agent-based models. *Ecological Modelling, 198*(1–2), 115–126.

Grimm, V., Polhill, J. G., & Touza, J. (2017). Documenting social simulation models: The ODD protocol as a standard. doi:https://doi.org/10.1007/978-3-319-66948-9_15.

Guba, E. G., & Lincoln, Y. S. (1994). Competing paradigms in qualitative research. In N. K. Denzin & Y. S. Lincoln (Eds.), *Handbook of qualitative research* (pp. 105–117). London: Sage.

Guimera, R., Uzzi, B., Spiro, J., & Amaral, L. A. (2005). Team assembly mechanisms determine collaboration network structure and team performance. *Science, 308*(5722), 697–702.

Hales, D. (2017). Distributed computer systems. doi:https://doi.org/10.1007/978-3-319-66948-9_23.

Izquierdo, S. S., & Izquierdo, L. R. (2006, September 20–23). On the Structural Robustness of Evolutionary Models of Cooperation. In E. Corchado, H. Yin, V. J. Botti, & C. Fyfe (Eds.), *Intelligent data engineering and automated learning-IDEAL 2006, 7th international conference, proceedings (Lecture notes in computer science 4224)* (pp. 172–182), Burgos, Spain. Berlin: Springer.

Izquierdo, L.R. (2008). *Advancing learning and evolutionary game theory with an application to social dilemmas* (PhD Thesis), Manchester Metropolitan University, http://cfpm.org/theses/luisizquierdo/.

Janssen, M. A. (2009). Understanding artificial anasazi. *Journal of Artificial Societies and Social Simulation, 12*(4). http://jasss.soc.surrey.ac.uk/12/4/13.html.

Kahn, K., & Noble, H. (2009, March 02–06). The modelling4all project—A web-based modelling tool embedded in Web 2.0. In O. Dalle et al. (Eds.), *Proceedings of SIMUTools '09, 2nd international conference on simulation tools and techniques* (p. 50), Rome, Italy. Brussels: ICST, Article.

Kephart, J. O., & Greenwald, A. R. (2002). Shopbot economics. *Autonomous Agents and Multi-Agent Systems, 5*(3), 255–287.

Kirman, A. (2011). *Complex economics: Individual and collective rationality*. London: Routledge.

Kirman, A. & Moulet, S. (2008). *Impact de l'organisation du marché: Comparaison de la négociation de gré à gré et des enchères descendants* (Working Papers, halshs-00349034). HAL, Centre pour la communication scientifique directe. http://halshs.archives-ouvertes.fr/docs/00/34/90/34/PDF/DT2008-56.pdf.

Klein, G. (1998). *Sources of power: How people make decisions.* Cambridge, MA: MIT Press.

Kohler, T. (2009) Generative Archaeology: How even really simple models can help in understanding the past. *Invited talk at 6th conference of the European social simulation association.* Guildford, United Kingdom: University of Surrey.

Kohler, T. A., Gumerman, G. J., & Reynolds, R. G. (2005). Simulating ancient societies. *Scientific American, 293*(1), 76–82.

Kohler, T. A., Varien, M. D., Wright, A., & Kuckelman, K. A. (2008). Mesa Verde Migrations: New archaeological research and computer simulation suggest why ancestral Puebloans deserted the northern Southwest United States. *American Scientist, 96*, 146–153.

Kuhn, T. (1962). *The structure of scientific revolutions.* Chicago: University of Chicago Press.

Lakatos, I., & Musgrave, A. (Eds.). (1970). *Criticism and the growth of knowledge.* Cambridge: Cambridge University Press.

LeBaron, B. (2006). Agent-based computational finance. In Tesfatsion & Judd (Eds.), *Handbook of computational economics* (Vol. 2. North-Holland, pp. 1187–1232).

Lincoln, Y. S., & Guba, E. G. (1985). *Naturalistic inquiry.* Beverly Hills, CA: Sage Publications.

Lorenz, J. (2007). Continuous opinion dynamics under bounded confidence: A survey. *International Journal of Modern Physics C, 18*(12), 1819–1838.

Lucas, P. (2010). *Conventional social behaviour amongst microfinance clients—a behavioural and financial case study* (PhD thesis). Centre for Policy Modelling, Manchester Metropolitan University.

Lucas, P. (2011). Usefulness of simulating social phenomena: Evidence. *AI & SOCIETY, 26*(4), 355–362.

Malthus, T. (1798). *An essay on the principle of population.* London: Johnson. Transcript available online at http://socserv2.mcmaster.ca/~econ/ugcm/3ll3/malthus/popu.txt.

Marks, R. E. (2007). Validating simulation models: A general framework and four applied examples. *Computational Economics, 30*(3), 265–290.

Matthews, R. B. (2006). The people and landscape model (PALM): Towards full integration of human decision-making and biophysical simulation models. *Ecological Modelling, 194*, 329–343.

Meadows, D. H., Meadows, D., Randers, J., & Behrens, W. W., III. (1972). *The limits to growth: A report for the Club of Rome's project on the predicament of mankind.* New York: Universe Books.

Moss, S. (1998). Critical incident management: An empirically derived computational model. *Journal of Artificial Societies and Social Simulation, 1*(4). http://jasss.soc.surrey.ac.uk/1/4/1.html.

Moss, S. (1999). Relevance, realism and rigour: A third way for social and economic research (Report no. CPM-99-56). Manchester, UK: Centre for Policy Modelling, Manchester Metropolitan University. http://cfpm.org/cpmrep56.html.

Newell, A. (1990). *Unified theories of cognition.* Cambridge, MA: Harvard University Press.

Nicolaisen, J., Petrov, V., & Tesfatsion, L. (2001). Market power and efficiency in a computational electricity market with discriminatory double-auction pricing. *IEEE Transactions on Evolutionary Computation, 5*(5), 504–523.

Phelps, S., McBurney, P., Parsons, S., & Sklar, E. (2002). Co-evolutionary auction mechanism design: A preliminary report. In J. Padget, O. Shehory, D. Parkes, N. Sadeh, & W. E. Walsh (Eds.), *Agent-mediated electronic commerce IV, designing mechanisms and systems (Lecture Notes in Computer Science, 2531)* (pp. 193–213). Berlin: Springer.

Polhill, J. G., Gotts, N. M., & Law, A. N. R. (2001). Imitative versus non-imitative strategies in a land use simulation. *Cybernetics and Systems, 32*(1–2), 285–307.

Polhill, J. G., Parker, D., Brown, D., & Grimm, V. (2008). Using the ODD protocol for describing three agent-based social simulation models of land-use change. *Journal of Artificial Societies and Social Simulation, 11*(2). http://jasss.soc.surrey.ac.uk/11/2/3.html.

Popper, K. (1963). *Conjectures and refutations: the growth of scientific knowledge*. London: Routledge.

PNAS. (2002). Colloquium papers. *Proceedings of the National Academy of Sciences, 99*(s3). http://www.pnas.org/content/99/suppl. 3.toc#ColloquiumPaper.

Riolo, R. L., Cohen, M. D., & Axelrod, R. (2001). Evolution of cooperation without reciprocity. *Nature, 411*, 441–443.

Roberts, G., & Sherratt, T. N. (2002). Does similarity breed cooperation? *Nature, 418*, 499–500.

Rouchier, J. (2001). *Est-il possible d'utiliser une définition positive de la confiance dans les interactions entre agents?* Paper presented at Colloque Interactions, Toulouse, May 2001.

Rouchier, J. (2017). Agent-based simulation as a useful tool for the study of markets. doi:https://doi.org/10.1007/978-3-319-66948-9_25.

Rouchier, J., & Thoyer, S. (2006). Votes and lobbying in the European decision-making process: Application to the European regulation on GMO release. *Journal of Artificial Societies and Social Simulation, 9*(3). http://jasss.soc.surrey.ac.uk/9/3/1.html.

Saqalli, M., Bielders, C. L., Gerard, B., & Defourny, P. (2010). Simulating rural environmentally and socio-economically constrained multi-activity and multi-decision societies in a low-data context: A challenge through empirical agent-based modeling. *Journal of Artificial Societies and Social Simulation, 13*(2). http://jasss.soc.surrey.ac.uk/13/2/1.html.

Schelling, T. (1969). Models of segregation. *American Economic Review, 59*(2), 488–493.

Schelling, T. (1971). Dynamic Models of Segregation. *Journal of Mathematical Sociology, 1*, 143–186.

Scherer, A., & McLean, A. (2002). Mathematical models of vaccination. *British Medical Bulletin, 62*(1), 187–199.

Snijders, T. A. B., Steglich, C. E. G., & van de Bunt, G. G. (2010). Introduction to actor-based models for network dynamics. *Social Networks, 32*, 44–60.

Sun, R. (2005). *Theoretical status of computational cognitive modelling (Technical report)*. Troy, NY: Cognitive Science Department, Rensselaer Polytechnic Institute.

Terán, O., Alvarez, J., Ablan, M., & Jaimes, M. (2007). Characterising emergence of landowners in a forest reserve. *Journal of Artificial Societies and Social Simulation, 10*(3). http://jasss.soc.surrey.ac.uk/10/3/6.html.

Vermeulen, P. J., & de Jongh, D. C. J. (1976). Parameter sensitivity of the 'Limits to Growth' world model. *Applied Mathematical Modelling, 1*(1), 29–32.

von Randow, G. (2003). When the centre becomes radical. *Journal of Artificial Societies and Social Simulation, 6*(1). http://jasss.soc.surrey.ac.uk/6/1/5.html.

White, D. R. (1999). Controlled simulation of marriage systems. *Journal of Artificial Societies and Social Simulation, 2*(3). http://jasss.soc.surrey.ac.uk/2/3/5.html.

Yang, C., Kurahashi, S., Kurahashi, K., Ono, I., & Terano, T. (2009). Agent-based simulation on women's role in a family line on civil service examination in Chinese history. *Journal of Artificial Societies and Social Simulation, 12*(2), 5. http://jasss.soc.surrey.ac.uk/12/2/5.html.

Younger, S. (2005). Violence and revenge in egalitarian societies. *Journal of Artificial Societies and Social Simulation, 8*(4). http://jasss.soc.surrey.ac.uk/8/4/11.html.

Chapter 29
Some Pitfalls to Beware When Applying Models to Issues of Policy Relevance

Lia ní Aodha and Bruce Edmonds

Abstract This chapter looks at some of the ways things can go wrong when mathematical or computational models are applied to inform policy on important issues. It looks at some of the pitfalls in the model construction and development phase, including choosing assumptions, the effect of 'theoretical spectacles', over-simplified models, not understanding model limitations, and not testing a model enough. It then goes on to discuss the pitfalls that can occur when a model is applied to inform policy, including entrenched policies based on models with little or no evidential support and how models can narrow the evidential base considered. It also looks at confusions concerning model purpose and kinds of question they may answer, when models are used out of context, asking unreasonable things of models, when the uncertainties are too great, when models give a false sense of security, and when the focus should be on values rather than facts. This discussion is then illustrated with two examples, one economic and one from fisheries. It concludes that most of these problems stem from the interface between the modelling and policy worlds. It ends with some simple recommendations to reduce these mistakes.

Why Read This Chapter?
We have compiled this chapter so that the reader may become aware of a number of ways in which complex models can do more harm than good or mislead those who use them. Such awareness may help you avoid these pitfalls as well as help others to avoid them. In particular, you will learn about some of the dangers that may arise when models escape from their modelling enclosure to be used in the policy or public arenas to inform decision making.

L.n. Aodha (✉) • B. Edmonds
Centre for Policy Modelling, Manchester Metropolitan University, All Saints, Manchester, M15 6BH, UK
e-mail: niaodhal@tcd.ie

© Springer International Publishing AG 2017
B. Edmonds, R. Meyer (eds.), *Simulating Social Complexity*,
Understanding Complex Systems, https://doi.org/10.1007/978-3-319-66948-9_29

29.1 Introduction

We all use models all the time, albeit usually informal mental models but sometimes mathematical or computational models. These models help us think about situations we encounter—both familiar and unfamiliar. While such models, from the very informal to the most formal, can be helpful, they can also work to limit our understanding—biasing and even constraining how we think about things or how we *might* think about things. An important characteristic of models is that they are simplified descriptions—representations that are designed by humans. As such, it is worth remembering that, in much the same way that certain arguments or concepts can work to obscure more than they illuminate (Moore 2017), so too can the most sophisticated models. In this chapter, we are considering the impact (and hence pitfalls) of relying on formal models—that is, mathematical or computational models. This is what we will mean when we talk about 'models' here; if we mean informal models, we will explicitly say so as in 'mental models' or 'informal models'.

On balance, it may be argued that people are relatively good at reflecting and negotiating how we collectively think about things, via social and political processes—although it should be clarified here that we do not all necessarily possess the same negotiating capacities or opportunities to affect these processes in a similar manner. What is relatively new in these social and political negotiating arena(s), and what we need to carefully consider, is that formal models are increasingly being introduced into (and their legitimacy questioned within) these processes, in terms of their results or their underlying ideas. This chapter looks at some of the dangers that this might introduce.

A complex simulation model can be a powerful tool—capturing and integrating knowledge that would be almost impossible to do in other ways and then facilitating calculations from that knowledge. However, these complex tools can be difficult to construct (adequately) and even more difficult to use (appropriately). Consequently, there is always an underlying chance of fooling yourself or others when building or using them and hence the possibility of prompting bad decisions. Furthermore, complex models can act as a mistake amplifier, making small mistakes have big consequences. As the saying goes

> To err is human, but to really screw things up you need a computer.

This chapter looks at some of these pitfalls in the hope that we might raise awareness of them and their consequences. We have structured our account according to two phases—firstly, the construction phase (and all that goes or should go with that), and secondly, the application phase when the model has been released to be used in the wider world.

In the event that models are to be used to inform policy, both of these stages merit close consideration, not only by those involved in developing the models but also by those that will be affected by them. As such, we lay out some of the core issues that may arise within these stages and highlight how these might present pitfalls for the modeller, for the policy maker, and for wider society. This rough categorisation is simply to aid the reader by giving them some structure and is not to be taken as definitive—any of the pitfalls might affect anyone.

Not everyone involved and implicated by this process is a modeller or policy maker. The shortcomings of the interaction between modellers, the modelling process, policy makers, and the entire policy process are likely to be felt most by many facets of society that may have had very little (or no) bearing on or input into this process. A wider awareness by stakeholders and the public of the pitfalls may encourage them to be more critical of model-informed outcomes and to direct debate more towards the options being considered and the decisions being made.

29.2 Constructing a Model: Pitfalls for Modellers to Avoid and Policy Makers to Ask About

There are a number of pitfalls that can occur in the construction phase of the model. Although there is no shortage of evidence of poor modelling practices and the negative consequences these can entail for society, many aspects of models are not usually subject to close examination by people outside the original modelling team[1] (Saltelli and Funtowicz 2014). As such, the points made here are aimed at modellers. However, they hint at the type of questions that those who are considering using a model (i.e. policy makers) should ask when presented with a model, regardless of how impressive it may look. Doing so may go some way to avoid some of the potential pitfalls outlined for the following phase of model application.

It is worth stressing, at this stage, what a model is. Models are abstractions, formal constructions that represent aspects of the world. They are created by someone, somewhere, for a particular purpose, most likely to answer a particular question and most certainly drawing on various assumptions. Indeed, there's no getting away from the fact that in engaging in the activity of modelling, assumptions about reality have to be made. While these assumptions might be more or less reliable (given the context and purpose of the model), there are potential traps that even the most experienced, competent, and respected modellers may fall into. The overarching point we wish to make is that a combination of reflexivity and transparency is key to avoiding some of these pitfalls.

---

[1] At best, examination is by a few in the same domain as themselves—people who likely have the same assumptions and worldview. Thus, many models are not *effectively* critiqued in an independent manner.

29.2.1   Modelling Assumptions

How we choose to see the world is itself a complex and subtle process that is not well understood. However, how we view the world, and how we choose to represent the world impacts both the questions we ask about the world, and how we try to answer them (Benessia et al. 2016). In this sense, how we conceive of a problem matters a great deal because it frames how we try and solve that problem (Moore 2017). The institutionalisation of seeing the world through numerical abstraction as the most authoritative way of seeing has a long history and one that is deeply embroiled in facets of power (Bavington 2009, 2010, 2015; Moore 2017; Scott 1998). It is especially interesting that post 2007 crash we continue to view words largely as 'interesting points of view', while numbers 'never lie' (or at least are more convincing), and difficult calculations or simulations often retain an unwavering authority. What is perhaps even more worrying in this, however, is that this institutionalisation runs so deep, and is so ingrained in our thought, that often we fail to recognise that numerical abstractions and complex models can be as laden with 'points of view' as other forms of knowledge. Failing to recognise this is the first pitfall that both modellers and policy makers are likely to fall into.

As per our aforementioned statement, all models are built on assumptions about reality, and this includes complex models. These assumptions are both implicit and explicit. Some may be based on theory, others on empirical evidence, maybe a mix of both or, perhaps, something altogether more ad hoc, such as those derived from tradition. We may not even be aware of the implicit assumptions. These assumptions will determine what goes into a model and, perhaps more importantly, what we leave out of it (Sterman 2002). Together these will have a bearing on our entire conceptualisation of the problem at hand.

Some of the assumptions we make will be somewhat reasonable, while others may be downright unreasonable. A good example of this comes from the models that are employed in fisheries management, some of which assume that nature is a stable system (although most current thinking acknowledges that it is anything but stable). The corollary of this is that goals or policies designed according to this assumption (e.g. maximum sustainable yield (MSY), which has become the de facto goal of most fisheries management regimes today) may turn out to be fairly unreasonable themselves (Bavington 2015). Models populated with homogenous rational economic agents are another good example. Stiglitz (2011) has highlighted that many of the standard economic models so deeply implicated in the last financial crash had critical omissions, along with a raft of incorrect assumptions, oversimplifications, or the 'wrong' simplifications. In turn, policies that have been designed according to these models were, and continue to be, worryingly dysfunctional.

29.2.2 Theoretical Spectacles

As indicated, many things can colour our perception of reality and thus the assumptions that we make—for a good discussion on this, see Glynn (2015). Among the factors he highlights are experiential and environmental biases (including disciplinary biases) that—regardless of how 'objective' they view themselves—scientists are subject to. Their disciplinary orientation will most certainly entail some commitment to a worldview that leans towards some value systems over others or to depicting aspects of that worldview over others in their models. Thomas Kuhn described this effect as wearing 'theoretical spectacles' (Kuhn 1962)—the theories one believes lead one to only notice the aspects of the world that fit the theories, and not those that do not.

This is often inadequately considered by those engaged in building such models. In fact, as Sterman (2002) has observed, narrow modelling assumptions are a common occurrence, even in work that has been published in highly respected journals. Although some complex models (e.g. agent-based simulations) allow the avoidance or widening of some of these assumptions, it would be wrong to think that such models are free from unconsidered or oversimplistic assumptions that may critically affect the results they give. Thus, a potential pitfall for modellers is failing to sufficiently consider and/or critique assumptions that underlie a model's construction and how useful, or dangerous, these may become down the line.

Modellers tend to spend a significant amount of time with their models—deeply engaged in constructing them, thinking about them, and tuning them. Thus, the danger of 'theoretical spectacles' is particularly acute for modellers as they often learn to see the world 'through' their models and begin fitting it to adhere to what they perceive to be true, developing a strong confirmation bias (Sterman 2002). This process can result in modellers making models that fit 'their version of reality' quite well, but it may not necessarily reflect observed 'reality' very well.[2] What we perceive to be true is based on our assumptions, and if these are not subjected to sufficient independent or reflexive examination and critique, then dangerous, brittle, or simply wrong assumptions may be included in the models we use. The upshot of this is that we get models built on bad foundations that may perform in a completely inadequate, indeed sometimes catastrophically mistaken, way.

29.2.3 Oversimplified Models

For models to be understandable, they need to maintain a certain level of simplicity. However, it is always worth keeping in mind that while simplifying reality it is likely to become more removed from that which we are trying to represent. As such, it is

---

[2]This is shorthand for saying the model's assumptions make a model useless in terms of its purpose.

helpful to be aware that a model might not necessarily prove encompassing enough to incorporate alternative understandings, experiences, values, or needs of those whose reality is being abstracted. It merits consideration that in our abstractions, that which we exclude is likely to matter for something, or someone, somewhere. In this sense, it is worth remembering that highly stylised interpretations can work to colour our vision from the beginning (Moore 2017)—the more stylised they are, the stronger the 'colouring' might be (since it makes for a more attractive and portable story).

Agent-based models are generally more straightforward in how they represent the world—allowing one computational entity for each actor, for example. This means that they do not need such simplifying assumptions as models, which represent populations as abstractions. However, they are still subject to the same pressures as other kinds of model, and there is a strong academic and publication bias towards simpler models.

Notwithstanding this, many models—even the simplest—can be useful, and one way or another we are all working off *some* kind of model (even if only a mental or informal model). Given that we cannot observe or measure everything, everywhere all of the time, simplifications can and do help us understand some complex processes that we may otherwise not understand (Glynn 2015). The pitfall here for the modeller to avoid, however, is one of *over*simplification, and any type of model can fall guilty to this charge. Complex computational models can become oversimplified, if the modeller may be constrained by time limitations, information limitations, their aforementioned worldviews, computational capacities, and so on. The danger here arises when simplification leads to a level of abstraction that misses key mechanisms and aspects that really *are* important in the process we are trying to understand. For example, this might happen through choosing to restrict what goes into a model to available numerical data—because it is easier than dealing with non-numerical data (Sterman 2002). Oversimplification, subsequently, can lead to many kinds of error, including human errors, computer errors, incorrect/misleading results, biassed or limited interpretations, and, ultimately, bad decisions (Glynn 2015). Whilst it is very difficult to be sure as to what aspects are crucial to include or to include every tiny nuance that might be relevant or important (even ethnographers struggle with this one), the task here is to be upfront about and reflect on these simplifications and try to catch them out if they are oversimplifications. At the very least, simplifying assumptions should be clearly acknowledged and documented.

29.2.4   Underestimating Model Limitations

Given the two previous points that have been made—that models are built on assumptions about reality and even the most complex entails a fair amount of simplification—it is not difficult to make the point that all models are going to have some limitations. The usefulness of a model is going to be constrained, so that an overoptimistic selling of your model for any and all purposes is not likely to end well—for anyone. Building a representation of the human or natural world—or both—is hard, and it would be a mistake to think otherwise. Building an oversimplified/over-prescriptive model and putting too much faith into what it can tell us can have many negative repercussions.

Some idea of a model's limitations can be gleaned through the assumptions that are built into it—it is unlikely to work well in situations where the assumptions do not hold. Other clues to a model's limitations can be found by running the model under many different considerations, e.g. its sensitivity analysis. However, the final arbiters of a model's limitations are usually only apparent when the model is used in practice. Thus, models need to be continually reviewed as to their continuing suitability and usefulness.

Particular care is needed when the model is being applied in a context that is very different to the one it was designed for or tested within. In a way, each time a model is applied in a different context, its utility there should be separately established, and not taken for granted. The more different the situation, the more it needs retesting, but this is often not done due to the cost of this. It is much easier just to reuse the model and hope for the best—easier in the short-run, that is.

One subtle way that a model can be used beyond its limitations is when it is subsumed as a sub-model of a bigger, more complex model. In such cases, the failures of the model can be masked by all the other things going on and not noticed. However, since models can 'amplify' error and bias, it might have an even bigger impact on the results.

Thus, it is important to remember that even the best models have limitations. Models are not (or almost never) general-purpose tools but more specific encapsulations of knowledge that have a quite specific scope of use. In many cases, if one does not know whether a model is being used beyond its scope, then it might be better to simply not use it at all—sometimes it is better to know the limitations of one's knowledge than to think one has some idea (or baseline) of what is happening.

29.2.5   Not Checking and Testing a Model Thoroughly

Clearly if there is any danger that a model might be used to inform real-world decisions, then the modellers and/or model users have a duty to check and test the model for its intended purpose as carefully as possible. However, these issues are dealt with extensively in other chapters in the Methodology section of this handbook.

29.3 Unleashing the Model: Pitfalls for Modellers, Policy Actors, and Society

Saltelli and Funtowicz (2014) have made the case that models should never make it onto the policy arena without undergoing rigorous and independent sensitivity auditing. This is a fairly reasonable suggestion, given that many of the pitfalls that can occur during the development phase (which should include testing) of the model can have fairly serious implications in the event that it is applied, and its results are taken seriously, with little question. There have been some very public examples of this over the past number of years that have had implications for those who are in the business of developing models, those in the business of designing policy, and, in turn, those who have to live with the consequences of these policies (e.g. Cavero and Poinasamy 2013; Cassidy 2013; Pierce 2008). Two such examples will be discussed in the final section. However, even in the event that best practice has been strictly adhered to prior to the model's application, there are still a number of pitfalls to be avoided at this stage of the process. Thus, some of the big traps to be navigated at this stage relate to the state of the evidence base and confusion over what a model can deliver on and what it realistically cannot. Points which merit consideration here include the purpose for which the model was built, the conditions it was built to satisfy, high levels of uncertainty, and the inability to answer the less 'scientific' questions that are being asked (and that, arguably, call for less 'scientific' answers altogether).

29.3.1 From Lack of Evidential Support, Mistaken or Misleading Models, to Entrenched But Ill-Informed Policies

As we have indicated, models are constructed using little snippets of information, about somebody, something, or some event or process that has been observed. It is worth reiterating here that even for the most complex models, the 'real world' is going to be more complex than the pieces of information we have on it. This also holds true for 'good' models. However, if things go wrong in the construction phase, you may well be dealing with a fairly inaccurate model. It is also worth reiterating here that sometimes models are not grounded in empirical evidence but rather in tradition, that is, disciplinary theories that may well never have been proved beyond theory. Some of these models may be relying not only on tacit but wholly unverified assumptions (Saltelli and Funtowicz 2014). It seems reasonable to suggest that models based loosely on real evidence, or perhaps none at all, are going to throw up problems at some stage if used in guiding the policy formation process. An easy target here are the aforementioned but often used, well-versed models on the policy scene coming from economics. While models such as these may make policy formation somewhat easier, in that they give a representation of something in a way that gives it an appearance of manageability, it is worth considering and questioning in detail their assumptions and the actual wisdom that they encapsulate.

Unfortunately, history has taught us that mistaken or misleading models can quickly gain traction. Particularly, if they appear to offer a workable solution that is amenable to policy making (see the example in Sect. 29.5.1). Furthermore, once these models (or the policies they have justified) become institutionalised—even when our knowledge has progressed so that we can see that the models (and thus the policies they inform) are underpinned by incorrect assumptions—it may be very tempting to continue to use them because the alternative is too messy or appears too hard. Essentially the pitfall here is that models can become so embedded within the policy making process that they are difficult to change. This may be for a variety of reasons, including that they reinforce particular interests or simply out of sheer habit. 'We've been doing it this way for thirty years, so it must be right' is most certainly a pitfall policy makers can (and do) fall into (see Rosewell 2017 p. 163). While this may make the game of policy making easier, it may not make for the best societal outcomes, and these may range from minor to fairly catastrophic consequences.

29.3.2   Model Spread

One of the big advantages of formal models is that they can be copied and used extensively with little effort. This can have big advantages in terms of allowing others to inspect, critique, and improve these models, but it also has downsides. One of these downsides is that models, once made and accepted in some way, tend to proliferate. That is, they tend to spread as if on their own accord. Of course, the ease of their reuse means that it is tempting to reuse them with little care or attention, in particular, care to retest or otherwise evaluate the applicability of a model for each area of application. In addition, once a model becomes widespread, then others take this as a mark of its suitability, so that it spreads even more.

An example of this is when the 'Black-Scholes' formula (Black and Scholes 1973) and its extensions became a common basis on which to price many kinds of financial derivatives. However, it later turned out that in other than the circumstances, it was originally conceived for it gives misleading prices, e.g. in the presence of extreme price changes, long-term price variation, or when dynamic hedging is not possible. The prevalence of this model has even been blamed for the bank crash of 2007/2008 (Stewart 2012). As the famous investor, Warren Buffet, put it in a letter to shareholders 'The Black–Scholes formula has approached the status of holy writ in finance … If the formula is applied to extended time periods, however, it can produce absurd results. In fairness, Black and Scholes almost certainly understood this point well. But their devoted followers may be ignoring whatever caveats the two men attached when they first unveiled the formula'.[3]

---

[3]http://www.berkshirehathaway.com/letters/2008ltr.pdf (accessed 1 June 2017)

29.3.2.1   Narrowing the Base Even Further

Another way that models, once unleashed into the policy making process, can affect the evidence base is through narrowing it. The case of the Newfoundland cod, mentioned above, indicates how models can work to constrain the evidence base, therefore limiting decision making. In this sense, a policy maker pitfall would be narrowing the evidence base to the part, which is seen as authoritative, and all other evidence is sidelined. This raises further questions in relation to what a model may and may not be able to adequately capture, and whether these may be other sources of evidence better suited to that task. It further raises questions in relation to the institutionalisation of what we deem to be authoritative evidence.

This point is very much related to our earlier point regarding the 'theoretical spectacles'—we all wear some type of spectacles that have been coloured by our environment and our need to navigate it. A model might be built from one viewpoint using a particular set of scientific spectacles and used in accordance with the different spectacles of a policy maker. These spectacles might bias or limit out vision in innumerable ways. While these limited viewpoints might be ok within their original context of development and use, they may not adequately capture things outside of it. For example, this view might not be compatible with the spectacles those operating in the context, with which the policy is concerned or will be employed in, are wearing.

This, of course, remains a challenge in policy making today, despite the widespread rhetoric in favour of stakeholder engagement, participatory governance, and human dimensions. Pearce et al. (2014, p. 163) have made the case that the tendency to prioritise technical data (numerical data and the output of formal models) over all else is still a firm feature of the policy making process. They highlight that studies indicate that the 'prevailing order' of the evidence-based policy process remains firmly rooted in traditional power hierarchies that are buttressed by a technocracy. In contrast, qualitative research and local knowledge are marginalised, so that a belief in the superiority of scientific methods from the natural sciences remains entrenched. Further, Saltelli and Giampietro (2017) have argued that modelling, when unleashed onto this space, can actually exacerbate this.

This becomes a pitfall in the sense that, even though a model can help us to understand something in a way we previously were unable to, it might effectively limit consideration of other understandings 'out there' that are likely to require consideration or perhaps might even trump the model itself. The danger then becomes that models may work to propagate established forms of thinking to the detriment of all others.

Arguably, given the special status we bestow on models (perhaps arising from their impressive appearance or the authority they gain from their scientific status), it is worth considering that these processes are imbricated in power in a number of ways. In this sense, the representations we present and use can work to cement this. The authoritative role of models may help justify the centralisation of decision making or perpetuate a top-down hierarchical mode of regulation – precisely the mode of management that many have increasingly recognised as suffering a crisis of legitimacy and from which purportedly we are moving away from. Given the special status this type of model can command, it can be used to justify decisions—which may not necessarily have the best outcomes.

Modellers, given the authoritative position of science, or at worst the appearance of science, can trip into the pitfall of further contributing and perpetuating these hierarchies, which may result in poorer answers than may have been available elsewhere. In this way, models may work to further exclude or obscure other ways of knowing, other ways that might prove to be a better answer to the current complex global challenges that policy makers and society have to deal with. In this sense, they may perpetuate the failure to integrate or deal seriously with other forms of knowledge.

29.4 Some Other Things to be Aware Of

29.4.1 Confusion Over Model Purpose

Good models will have, or should have, a clearly stated purpose—at least those that are applied to issues of real importance. Such a model will have been designed with that purpose in mind and tested with respect to this. If it is used for another purpose, then it is likely to fail at this. Therefore, that model will only be able to help when used for its particular purpose, e.g. for scenarios where that kind of role is required. These kinds of confusion are dealt with in Chap. 4 (Edmonds 2017).

29.4.2 Confusion Over the Kind of Question a Model Can Answer

A related confusion is when a model is designed to answer a question or inform thinking about one kind of issue is assumed helpful for a different question or issue. Take, for example, the bioeconomic models of fisheries management. These models are built using biological and economic parameters and largely ignore social parameters. They are designed with these objectives in mind. Proponents of these types of model are sometimes explicit about this and may indicate that although other social objectives like employment, equality, or biodiversity conservation are important, they are not explicitly modelled (e.g. see Costello et al. 2016). As a policy maker, it is worth considering whether these models may be the most appropriate tool for suitable policy formulation or to which the extent they should be relied upon. Interestingly, after years of managing fisheries based on these models, the poor social outcomes with respect to fisheries management are often lamented, though, arguably, are not at all unexpected.

As a policy maker, one should be aware that a lack of clear purpose for a model is far too common (Chap. 4). It is therefore sensible to inquire carefully into this and consider whether the purpose of the model is compatible with the kind of policy one is trying to design and whether it meets one's objectives. Indeed, Sterman (2002) argues that, along with incorrect or missing assumptions, models often fail because more basic questions about the suitability of the model for the intended purpose were not asked. As such, the pertinent questions to ask become: Are the assumptions being made in line with the purpose of the model? What would this mean in terms of policy output? What kind of contradictions might this lead to?[4]

A pitfall for modellers here would be failing to declare whether the model is suitable for the particular purpose or whether it might hold up under different conditions, falling into the trap of being too policy prescriptive on questions that are inherently political, rather than scientific. Sometimes answering these questions will require much more than a model that has been built with specific, perhaps narrow, objectives in mind, using specific assumptions as to how society is, rather than how it could or ought to be.

29.4.3 When Models Are Used Out of the Context they were Designed For

Context matters! While a set of assumptions may accurately hold in one context, they might not in another—other factors could come into play in a new context that change the outcomes or may even negate them entirely. For example, with bioeconomic models in fisheries, scientists often acknowledge that the effects of their policy prescriptions, according to their models, assumptions, and goals, are likely to be context specific and depend on the social, economic, and ecological objectives within any given context. The danger here is that the policy maker is not adequately aware or fails to consider this declared context sensitivity but rather goes off the tagline whereby the solution is posited without the necessary caveats. So although scientists may make certain caveats about their model explicitly clear, this does not necessarily mean that they are heard. Furthermore, these may be lost as the model moves up the chain to where policy will actually be implemented.

29.4.4 What Models Cannot Reasonably Do

It is worth highlighting that there are some things that models just cannot do. In these cases, the policy maker should not attempt to ask such questions of a model, nor should a modeller present (spurious) answers if asked.

---

[4]Giampietro and Saltelli (2014) provide a discussion on these questions in relation to the ecological footprint.

Some of the biggest questions we are trying to answer today simply cannot be answered by science, certainly not alone anyhow, no matter how much we dress them up with science (Weinberg 1972). Three conditions give rise to such questions. Firstly, there are questions that science may not be capable of answering due to limited resources. Secondly, there are questions whereby the subject matter is just too variable to measure according to narrow positivistic frames (Weinberg explicitly places the social sciences as such a case). Thirdly, there are the types of questions or issues that involve moral and aesthetic judgements—they are not about 'facts' but values, although some questions may have elements of both (Weinberg 1972).

For example, if a policy maker asks a model to predict the consequences of a particular policy and this is simply not predictable, then it is wrong to provide that prediction, even with caveats (because the modeller knows that the caveats will be ignored). If a policy maker tries to off-load the responsibility of a decision to the outcomes of a model, then this too should be resisted—it is the place of modellers to advise but policy makers to decide.

29.4.5   Uncertainty Is Too Great

All models entail a level of uncertainty. This uncertainty usually increases exponentially with the complexity of the system we are trying to understand. A number of authors have highlighted this in relation to climate and hence climate change. In this area, reasonable predictions are simply not feasible given the huge uncertainties this kind of modelling entails (Saltelli et al. 2015; Saltelli and Giampietro 2017).[5] Others have highlighted the total inaccuracy of these for comparing the possible damage (i.e. climate change costing) (Stern 2016). There are simply too many processes involved here that we do not have an adequate understanding of, and as such, models of this kind ought not to be used for justifying policy decisions (Saltelli et al. 2015), and this is likely to stand regardless of how super our 'super computers' get. The modeller pitfall that arises here is ignoring or hiding the uncertainties in their models, while for the policy maker, it is allowing yourself to believe that we can quantify everything, including the uncertainty—which often we can't (Saltelli et al. 2015)—or failing to check that levels of uncertainty have been over- or underestimated. Good science should be very cautious about giving the impression that its outcomes are more accurate than they merit, unfortunately, when scientists get involved in the policy process, there is a temptation to capitalise on their scientific status and use numbers or numerical representation to make their conclusions seem more certain and dramatic.

Saltelli and Funtowicz (2014) point out that a good indication that something may be suspicious about a model is if the information or numbers it offers up are

---

[5]However, model-based explanations of why climate change has been happening are well founded.

too precise—something that provides the accuracy this implies just is not in line with what is usually possible with science.

Furthermore, engaging in this type of speculation has societal implications and throws up pitfalls for society, who may have little bearing on the actual policy process, by providing fuel for sceptics. Saltelli et al. (2015) show that introducing models into debates in relation to climate change may have done more harm than good, with the authors stating that society is potentially in danger of endless debates over uncertainties and competing arguments. The authors further highlight another, just as serious, issue—with excessive confidence in our ability to model the future, we may well commit to policies that reduce, rather than expand, available options and thus our ability to cope with what comes in the future.

29.4.6   A False Sense of Security

This point is interrelated with many of the previous points. As we have pointed out, just because models look very impressive or authoritative or provide us with a graspable number, they may not always actually be that impressive or authoritative, and the number may well be just as useful as one that was written down randomly on a sheet of paper. While at face value they might be quite enticing, they can lure us into a false sense of security and actually prevent us from doing anything useful, safe in the illusion that we can predict and hence manage the changes that are predicted. History has taught us that such an approach does not necessarily end well. See examples.

Having a tool that can provide us with some kind of forecasts, there is the risk of relying on this, as a mechanism through which to avoid responsibility for perhaps the worst-case scenario or the unknown scenario. In this sense, we tend to focus on the best-case/most tolerable scenario and use a model to justify this restricted focus, rather than consider the full range of possibilities. For example, is it reasonable to assume that we can continue on our current growth trajectory while solving the ecological crises we are facing and the increasingly social facets of each of these and stay within the best-case scenario limit of, say, climate change?

29.4.7   Not More Facts but Values!

As much of our discussion has indicated, there are just some things that a model cannot do or compensate for. Models cannot provide us with or replace a moral or an ethical vision. They are unlikely, on their own, to provide us with a clear answer to the some of the hardest questions that the policy making process busies itself with, or with avoiding.

For example, a model may (somewhat) adequately capture some economic numbers, but it may not be able to capture what is important to real people. These personal values may be hard to capture or measure and so are not easily quantified. Examples of this may be cultural attachments to a place, which we might only garner through qualitative judgements. It is also unclear as to whether a model can really give a voice to or include those that they seek to represent adequately—regardless of how participatory the approach employed has been. In this sense, there are always going to be qualitative judgements to be made.

There is much scholarship based on looking at 'what could be', rather than drawing on models that look at 'what might be' based on assumptions about 'what is'—and this might be a more useful consultative tool for some issues. While these models may present us with some alternative course of action in relation to a specific question, they may not present us with any real alternatives for the future. If we are interested in articulating what 'could be' in a meaningful sense, models may not prove to be very useful. So although a model may give us some sense of how things are, from a particular perspective, they might not be so good at telling us about how things should be or how things could be (for an anthropological discussion related to this, see Holbraad et al. 2014).

29.5 Two Examples

29.5.1 An Economic Example

The 2008 crash and the recent financial crisis give ample evidence as to how models can go wrong—the pitfalls modellers can fall into, the pitfalls policy makers can fall (or jump) into when consulting models, and the severe societal consequences that this can entail. However, one really worth citing here, even though it has been widely cited elsewhere (e.g. Saltelli et al. 2015; Saltelli and Giampietro 2017; Saltelli and Funtowicz 2014), is the Reinhart and Rogoff case. This case is particularly illustrative of the far-reaching and long-term consequences for the day-to-day lives of people potentially entailed when what turns out to be a seriously flawed model is used to justify particular policies.

This case exemplifies what can go wrong—from dodgy assumptions and basic coding errors in the construction phase to the uptake (of flawed results) and implementation (along with institutionalisation) of very prescriptive policies in the use phase, that once unleashed resulted in really devastating societal outcomes. This example is particularly pertinent as it shows both the short-term and longer-term implications, many of which continue to reverberate in the lives of ordinary people today. It is at least partly imbricated in the current political climate, not only in Europe but on the other side of the Atlantic as well.

In 2010 Reinhart and Rogoff, two Harvard economists, published a study based on a model, that would provide the impetus for the implementation of severe austerity measures in many countries during the economic crisis. Their paper 'Growth in a Time of Debt' was widely publicised and actively drawn on by policy makers. It argued that high debt had a negative impact on growth, and once this passed a threshold of 90% this would, potentially, become dangerous and actually impede growth (Cassidy 2013; Rogoff and Reinhart 2010; Saltelli and Funtowicz 2014; Saltelli et al. 2015; Saltelli and Giampietro 2017).

This was taken up by debt-facing policy makers on both sides of the Atlantic, and subsequently used to justify huge cuts in government spending and the implementation of austerity measures and packages in some countries, particularly across the EU (Cassidy 2013). A 2013 report by Oxfam highlights the shift towards austerity that occurred in 2010, which marked a turn from earlier more interventionist approaches to the crisis. In the UK, for example, prior to 2010, the government had taken the track of implementing a stimulus package, which included increased spending on social housing and education. This contrasted with post 2010 spending cuts aimed at reducing the deficit. Cassidy (2013) details references to the Reinhart and Rogoff paper being made by George Osborne in the House of Commons. Similar changes to public spending were implemented across the Eurozone and elsewhere.

Three years later, the work of Reinhart and Rogoff was replicated, and it turned out to contain some basic errors (Cassidy 2013). The authors of this replication, Herndon et al. (2013) found: '*that selective exclusion of available data, coding errors and inappropriate weighting of summary statistics*' had led to serious miscalculations and inaccurate representations with respect to the relationship between public debt and growth. Unfortunately, this came too late for the people who were subjected (and continue to be subjected) to the policies the original paper had justified (Cassidy 2013; Saltelli and Funtowicz 2014; Saltelli et al. 2015; Saltelli and Giampietro 2017). Indeed, many countries within the EU, under a great deal of 'encouragement' from the EU, have now institutionalised austerity via changes to legal mechanisms that mandate a balanced budget (Bruff 2016).

Austerity policies have had some wide-ranging effects. A 2013 Oxfam report (Cavero and Poinasamy 2013) documented the implications of the austerity programmes that have been implemented across Europe, arguing that with inequality and poverty on the rise, Europe is facing a lost decade, with an additional 15–25 million people facing the prospect of living in poverty by 2025 if austerity measures continue. Indeed, decreased provision of public services, regressive taxation policies, rising inequality, persisting unemployment (in particular youth unemployment in some countries), increased food insecurity (as seen the widespread popping up of food banks), health implications, lower income, debt burdens, and widespread discontent is evident in many countries (Bruff 2016). Perhaps some of the worst of these effects have been felt within the 'bail out' countries like Greece, but they have certainly not been restricted to these countries. These effects have been felt and continue to be felt at the individual, household, societal, and wider political level, even within those countries now drawn as 'good examples' of austerity, such as Ireland (Bruff 2016). While there are obviously other things at play here (and we cannot just blame the model), the model certainly is in some way culpable for how this has played out. In this sense, this story is not intended to prove that the opposite policies would have turned out differently. We also do not know that the politicians involved would not have pursued the same policies anyway. What it does show is how sloppy modelling can be used to justify the policies that politicians choose, giving these more credibility than they might otherwise have, and help to insulate them against criticism and debate and thus to institutionalise the choices.

29.5.2   A Socioecological Example

Bad models are, of course, not only confined to the world of economics and finance (although at times it may seem that way). The second example we draw on is one from fisheries management, with respect to the collapse of the Newfoundland cod. This example, again, gives us a sense of many of the issues and how they overlap. It also serves to illuminate how models can serve to override other sources of knowledge.

The story of fisheries and their often lamented status is of course a straight-forward one—overfishing or 'too many fishermen catching too much fish'. This narrative, while often cited, is problematic, in that it gives us little insight into the reality of modern fishing and all that goes with it, including the way modern fisheries are managed. Certainly, it gives us little indication of the historical, political, economic, and ecological contexts of these endeavours and the relations that structure them. Within this world of fish, fishermen, fisherwomen, scientists, and managers (state and increasingly non-state actors), models feature highly—from population models of fish stocks and bioeconomic models of efficiency to increasingly complex models of a variety of aspects of fisheries, including agent-based models.

On the 2 July 1992, Canada's fisheries minister, John Crosbie, placed a moratorium on all cod fishing off the northeast coast of Newfoundland and Labrador. That day 30,000 people lost their jobs and hundreds of years fishing ended. The cod were declared commercially extinct (Bavington 2010). What happened?

Much work has been done in this area, and many predictable answers have been put forth. A lot of these tell a simple story of overfishing, environmental conditions, and poor management. However, a number of authors (Bavington 2010; Finlayson 1994) have looked at the role of fisheries science in this story, arguing it and its models played a pivotal role in the collapse of the Newfoundland cod stocks. As the fishery and the fisheries management surrounding it developed, the management game became one of counting how many fish there were in the sea and predicting how many fish could be caught (Bavington 2010; Finlayson 1994), which in turn fed back onto the scientists engaged in making those predictions, leading to the development of increasingly intricate mathematical models. Partly due to the traditions in the field and the increasingly complex data they were trying to fit, these became more and more divorced from reality during this development (Finley 2008, in Bavington 2015).

Finlayson (1994) details the series of scientific blunders (based on models) that were made in the years leading up to the moratorium, in spite of repeated concerns being raised by inshore fishermen, with respect to the status of the cod. Despite a number of Commissions and corresponding reports investigating the status of the cod in the years leading up to the collapse, despite protestations from the inshore sector they were seeing declining catches, the science depicted an increasing resource base. Successive failures were made in making adequate inferences in relation to the overall stock health—for example, the Kirby Report indicated that any reported decrease in profitability was merely down to cost-price squeeze. This report led to more development and investment in the fishery driven by both the state and individuals. Scientists and the fisheries department throughout much of the 1980s estimated a 15% annual rate of growth in the stock—figures that were consistently slated by inshore fishermen. Similarly, the subsequent Alverson Commission was formed to investigate the declines being reported by inshore fishermen but cited environmental influences on the annual inshore migrations of the stock. Again such findings were contested by the inshore sector.

It was not until 1989 that this erroneous forecast for fish stocks was corrected. The fisheries department issued its annual assessment based upon revised mathematical models to generate stock estimates from research and catch data, which indicated that abundance had been overestimated by as much as a factor of two. The subsequent Harris Commission found that the fisheries department's estimates of stock strength were based upon data, methodologies, and models of such poor or uncertain quality as to be essentially useless as a rational basis for management or commercial planning.

The executive summary of the Harris Report (1990, p. 2) states that:

> During the next seven years the euphoria that had been engendered by the declaration of the exclusive economic zone was reinforced by the steady growth of the stock, by continually improving catches, and by the belief that the FO.I objective was, indeed, being met. In those circumstances, scientists, lulled by false data signals and, to some extent, overconfident of the validity of their predictions, failed to recognize the statistical inadequacies in their bulk biomass model and failed to properly acknowledge and recognize the high risk involved with state-of-stock advice based on relatively short and unreliable data series. Furthermore, the Panel is concerned that weaknesses in scientific management and the peer review process permitted this to happen.

Finlayson (1994, pp. 12–15) argues that social dynamics were certainly at play in generating some of the stock assessments in this case. In this instance, scientists and policy makers had become so committed to their description of reality (despite its wild inaccuracy) and: *'the idea of a strongly rebuilding Northern cod stock was so powerful that it can be shown to have been read back into ambiguous data through analytical models built upon necessary but hypothetical assumptions about population and ecosystem dynamics. Further, those models required considerable subjective judgement as to the choice of weighting of the input variables'* (Finlayson 1994, p. 13).

29.6 Conclusion

There are many pitfalls in both the modelling and policy arenas, and many of these feedback upon one another. However, each of these arenas has its own experts and professionals who will (hopefully) be aware of their own kinds of pitfall. It is perhaps when the policy and modelling world interact that many of the worst mistakes are made: when the policy actors do not understand the models or when the modellers do not understand, or assume adequate responsibility for, the consequences of their modelling. Thus, particular care needs to be taken when describing the capabilities or reliability of models to non-modellers, and policy actors need not to delegate their decision making to a complex model that they do not understand but retain their critical faculties.

It is also worth highlighting that the efficacy of a model is likely to depend on the question under investigation. Technical questions may not pose such a problem; however, more complex problems will likely increase the urgency of the points that have been raised above and are likely to require more information and wider consideration than simply drawing on *a* model.

The demarcation line between these two worlds is blurry in more ways than most of us like to admit, and this is not something new. As Weinberg (1972) pointed out:

> The politician, or some other representative of society, is then expected to say whether the society ought to proceed in one direction or another. The scientist and science provide the means; the politician and politics decide the ends—this view of science is of course oversimplified. Ends and means are hardly separable no matter how straight forward the question. (Weinberg 1972).

However, the interface between scientists/modellers and the policy world is one that has increasingly come under scrutiny. Given the points that have been made in this chapter, we suggest the following for:

- Stop using the word predict and stop expecting the word predict. Be very sceptical about any models that claim to be able to predict more than anything else.
- Use models to increase the number of alternative futures that might occur, rather than to reduce the apparent uncertainty.
- Ensure that models are re-evaluated frequently, especially when being used in a new context.
- Make effort to ensure that the models, the assumptions they are made from, and the whole policy process are open to scrutiny from all those affected.
- Even when a model is helpful by informing the formulation of a good policy, it cannot *decide* the policy. Deciding a policy is, and should remain, a political and not a technical process.
- Try and ensure that research and models that focus on what is happening now do not distract from the question of what 'could be'—the choices we have for the future.

The worst-case scenario is: What if these models or the ways they have been taken up are just *plain wrong*[6]? Society has to live with the consequences. Thus, it is also important to remember we are not using these models in a vacuum; we are using them in a social, economic, political, and environmental context that involves complex relations and power hierarchies. Ignoring the context and just focussing on the technical aspects of modelling may lead to bad outcomes for everyone.

Further Reading

For further information, we suggest you read the following four reports. Some of these are books, others are articles.

Bavington, D. (2010). *Managed annihilation: an unnatural history of the Newfound-747 land cod collapse*. Vancouver: UBC press.

Cavero, T., & Poinasamy, K. (2013). A cautionary tale: The true cost of austerity and inequality in Europe. Oxfam International.

Cassidy, J. (2013). The Reinhart and Rogoff controversy: A summing up, available at http://www.newyorker.com/news/john-cassidy/the-reinhart-and-rogoff-controversy-a-summing-up

Harris, L. (1990). *Independent review of the northern cod stock: Executive summary, and recommendations*, available at http://www.dfo-mpo.gc.ca/Library/114277.pdf

Other useful reading includes the following.

European Commission. (2015). *Workshop 'significant digits responsible use of quantitative information'*, at https://ec.europa.eu/jrc/en/event/conference/use-quantitative-information

Glynn, P. D. (2015). Integrated Environmental Modelling: human decisions, human challenges. *Geological Society, London, Special Publications, 408*(1), 161–182.

Pierce, A. (2008). *The Queen asks why no one saw the credit crunch coming*, available at http://www.telegraph.co.uk/news/uknews/theroyalfamily/3386353/The-Queen-asks-why-no-one-saw-the-credit-crunch-coming.html

Pilkey, O. H., & Pilkey-Jarvis, L. (2007). *Useless arithmetic: why environmental scientists can't predict the future*. Columbia University Press.

Saltelli, A., & Funtowicz, S. (2014). When all models are wrong. *Issues in Science and Technology, 30*(2), 79–85.

---

[6]Whilst models are a tool rather than a picture (see Chap. 4), some are so useless at what they are supposed to do, that it makes sense to call them wrong.

References

Bavington, D. (2009). Managing to endanger: Creating manageable cod fisheries in Newfoundland & Labrador, Canada. *Maritime Studies (MAST), 7*, 99–119.

Bavington, D. (2010). *Managed annihilation: An unnatural history of the Newfoundland cod collapse*. UBC press.

Bavington, D. (2015). Marine and freshwater fisheries in Canada: Uncertainties, conflicts, and hope on the water. In B. Mitchell (Ed.), *Resource and environmental management in Canada*. Canada: OUP.

Benessia, A., Funtowicz, S., Giampietro, M., Guimarães Pereira, A., Ravetz, J., Saltelli, A., et al. (2016). *Science on the Verge*. Tempe, AZ: The Consortium for Science, Policy and Outcomes at Arizona State University.

Black, F., & Scholes, M. (1973). The pricing of options and corporate liabilities. *Journal of Political Economy, 81*(3), 637–654.

Bruff, I. (2016). Scandalous obfuscations of crisis in Europe. In: *Progress in political economy (PPE)*. Available from: http://ppesydney.net/scandalous-obfuscations-of-crisis-in-europe/.

Cavero, T., & Poinasamy, K. (2013). *A cautionary tale: The true cost of austerity and inequality in Europe*. Oxford: Oxfam International.

Costello, C., Ovando, D., Clavelle, T., Strauss, C. K., Hilborn, R., Melnychuk, M. C., et al. (2016). Global fishery prospects under contrasting management regimes. *Proceedings of the National Academy of Sciences, 113*(18), 5125–5129.

Edmonds, B. (2017). Five different modelling purposes. doi:https://doi.org/10.1007/978-3-319-66948-9_4.

Finlayson, A. C. (1994). *Fishing for truth: A sociological analysis of northern cod stock assessments from 1977 to 1990*. St. John's, Newfoundland: Institute of social and economic research, Memorial University of Newfoundland.

Finley, C. (2008). A political history of maximum sustainable yield, 1945–1955. In D. Starkey, P. Holm, & M. Barnard (Eds.), *Ocean's past: Management insights from the history of marine animal populations* (pp. 1989–1206). London: Earthscan.

Giampietro, M., & Saltelli, A. (2014). Footprints to nowhere. *Ecological Indicators, 46*, 610–621.

Glynn, P. D. (2015). Integrated environmental modelling: Human decisions, human challenges. *Geological Society, London, Special Publications, 408*(1), 161–182.

Harris, L. (1990). *Independent review of the northern cod stock: Executive summary, and recommendations*, Available at: http://www.dfo-mpo.gc.ca/Library/114277.pdf.

Herndon, T., Ash, M., & Pollin, R. (2013). Does high public debt consistently stifle economic growth? A critique of Reinhart and Rogoff. *Cambridge Journal of Economics, 38*(2), 257–279.

Holbraad, M., Pedersen, M. A., & de Castro, E. V. (2014). The politics of ontology: Anthropological positions. *Cultural Anthropology, 13*. Available at: https://culanth.org/fieldsights/462-the-politics-of-ontology-anthropological-positions.

Kuhn, T. (1962). *The Structure of Scientific Revolutions*. Chicago: University of Chicago Press.

Moore, J. W. (2017). The Capitalocene, Part I: On the nature and origins of our ecological crisis. *The Journal of Peasant Studies, 44*(3), 594–630.

Pearce, W., Wesselink, A., & Colebatch, H. (2014). Evidence and meaning in policy making. *Evidence & Policy, 10*(2), 161–165.

Rogoff, K., & Reinhart, C. (2010). Growth in a time of debt. *American Economic Review, 100*(2), 573–578.

Rosewell, B. (2017). Complexity science and the art of policy making. In J. Johnson, P. Ormerod, B. Rosewell, A. Nowak, & Y. C. Zhang (Eds.), *Non-equilibrium social science and policy* (pp. 159–178). Switzerland: Springer.

Saltelli, A., & Funtowicz, S. (2014). When all models are wrong. *Issues in Science and Technology, 30*(2), 79–85.

Saltelli, A., Stark, P. B., Becker, W., & Stano, P. (2015). Climate models as economic guides scientific challenge or quixotic quest? *Issues in Science and Technology, 31*(3), 79–84.

Saltelli, A., & Giampietro, M. (2017). What is wrong with evidence based policy, and how can it be improved? *Futures, 91*, 62–71, Available at: https://doi.org/10.1016/j.futures.2016.11.012.

Scott, J. C. (1998). *Seeing like a state: How certain schemes to improve the human condition have failed*. Yale University Press.

Sterman, J. D. (2002). All models are wrong: Reflections on becoming a systems scientist. *System Dynamics Review, 18*(4), 501–531.

Stern, N. (2016). Economics: Current climate models are grossly misleading. *Nature, 530*(7591), 407–409.

Stewart, I. (2012). *The mathematical equation that caused the banks to crash*. The Observer, 12 February 2012, https://www.theguardian.com/science/2012/feb/12/black-scholes-equation-credit-crunch. Accessed 01 June 2017.

Stiglitz, J. E. (2011). Rethinking macroeconomics: What failed, and how to repair it. *Journal of the European Economic Association, 9*(4), 591–645.

Weinberg, A. M. (1972). Science and trans-science. *Minerva, 10*(2), 209–222.

Erratum to: Verifying and Validating Simulations

Nuno David, Nuno Fachada, and Agostinho C. Rosa

Erratum to:
Chapter 9 in: B. Edmonds, R. Meyer (eds.),
Simulating Social Complexity,
Understanding Complex Systems,
https://doi.org/10.1007/978-3-319-66948-9_9

The original version of this chapter was inadvertently published with an entry omitted in the reference list. The missing reference reads as given below:

Fachada, N., Lopes, V. V., Martins, R. C., & Rosa, A. C. (2017b). Model-independent comparison of simulation output. *Simulation Modelling Practice and Theory, 72*, 131–149.
doi:10.1016/j.simpat.2016.12.013,
http://www.sciencedirect.com/science/article/pii/S1569190X16302854

The updated online version of this chapter can be found at
https://doi.org/10.1007/978-3-319-66948-9_9

Index

A

Aalaadin metamodel, 578
Abelson's and Bernstein's model, 15–17
ABMS, *see* Agent-based modelling and
    simulation (ABMS)
ABMs, *see* Agent-based modelling (ABM)
Absorbing class, 314, 316–318, 321
ABSS model, *see* Agent-based social
    simulation (ABSS) model
Accuracy in parameter estimation (AIPE), 245,
    246
Actors, resources, dynamics and interactions
    (ARDI), 280
ACT-R model, 29, 30, 94
Advanced driver-assistance systems (ADAS),
    709–710
Advanced traveler information systems (ATIS),
    709–710
Afghanistan power and authority models
    actor's behavior, 748–750
    endorsements, 750–753, 756
    intuitive model, 742, 745, 755
    ownership, 745
    powerful agents, 747–748
    power sources, 745
    power structures, 753–755
    reputation, 746
    weak actor/powerful actor, 748–750
AfghanModel network, 754–755
Agent-based computational economics (ACE),
    673
Agent-based modelling (ABM), 7, 20, 149,
    182, 198, 282, 418–419, 766, 767,
    785
    abstraction, 125

computational simulation, determination of
    number of runs, 229–230
    AIPE, 245, 246
    configurations of parameters, 231–232
    emergent properties, study of, 230
    hypothesis generation, 230–231
    KISS and KIDS model, 232
    measurement, 231
    realistic situation, analysis of, 231
    statistical power analysis, 236–248
    testing theory (*see* Testing theory)
design, implementation and use
    computer scientist's role, 129
    modeller's task, 127–129
    programmer's job, 129–130
    stages, 128
    thematician role, 127
vs. neural networks, 142, 147, 153
ODD protocol (*see* Overview, design
    concepts and details (ODD)
    protocol)
representation, 126
symbolic systems
    computer modelling, 124–125
    formal languages, 122–123
    natural language, 122
toolkits, 193
Agent-based modelling and simulation
    (ABMS)
    agriculture
        adaptive watershed management, 592
        Agripolis model, 591
        economic aspects, 590–591
        pest and disease incursions, 591
    coordination, 577–578

Printed in the United States
By Bookmasters